Introduction to Energy, Renewable Energy, and Electrical Engineering

Introduction to Energy, Renewable Energy, and Electrical Engineering

Essentials for Engineering Science (STEM) Professionals and Students

Ewald F. Fuchs
University of Colorado
Boulder, CO, USA

Heidi A. Fuchs
Lawrence Berkeley National Laboratory
Berkeley, CA, USA

WILEY

This edition first published 2021

Registered Office(s)
John Wiley & Sons, Inc., 111 River Street, Hoboken, NJ 07030, USA

Editorial Office
John Wiley & Sons, Inc., 111 River Street, Hoboken, NJ 07030, USA

For details of our global editorial offices, customer services, and more information about Wiley products visit us at www.wiley.com.

Wiley also publishes its books in a variety of electronic formats and by print-on-demand. Some content that appears in standard print versions of this book may not be available in other formats.

Library of Congress Cataloging-in-Publication Data is applied for

Hardback: 9781119448334

Cover Design: Wiley
Cover Image: © Joe Therasakdhi/Shutterstock

Set in 9.5/12.5pt STIXTwoText by SPi Global, Pondicherry, India

10 9 8 7 6 5 4 3 2 1

Contents

Acknowledgments

This book is based on the author's experience gained in teaching Electrical and Computer Engineering (ECEN) 3030, Circuits for Non-Majors [1] at the University of Colorado for eight years, as well as the input provided by undergraduate and graduate students, and the second author's research experience at Lawrence Berkeley National Laboratory, whose mission is to foster groundbreaking fundamental science that enables transformational solutions for energy and environment challenges, using interdisciplinary teams and by creating advanced new tools for scientific discovery. The encouragement and support of Dipl.-Ing. Dietrich J. Roesler, formerly with the US Department of Energy (DOE), Washington, DC, one of the first professionals initiating the research of photovoltaic power plants [2] as part of the mission, is greatly appreciated. The authors wish to express gratitude to wife/mother Wendy L. Fuchs and son/brother Franz S. Fuchs for their help in shaping and proofreading the manuscript. Lastly, the work on numerical field calculation [3] initiated by the late Professor Edward A. Erdelyi is reflected in part of this book by assisting in the visualization of magnetic fields as they occur in electric apparatus.

References

1 Fuchs E.F. Undergraduate course, ECEN 3030: circuits for non-majors, Fall 2011. Prerequisites: APPM 2350 Calculus III. University of Colorado at Boulder, Department of Electrical, Computer, and Energy Engineering. http://thiscourse.com/colorado/ecen3030/fa10 (accessed 11 May 2020).

2 Dugan, R.C., Jewell, W.T., and Roesler, D.J. (1983). Harmonics and reactive power from line-commutated inverters in proposed photovoltaic subdivision. ID: 6637722, DOE Report Contract Number W-7405-ENG-2 (1 January 1983).

3 Trutt, F.C., Erdelyi, E.A., and Jackson, R.F. (1963). The non-linear potential equation and its numerical solution for highly saturated electrical machines. *IEEE Trans. Aerosp.* 1 (2): 430–440.

2018

Ewald F. Fuchs, Professor
Department of Electrical,
Computer, and Energy Engineering
University of Colorado at Boulder
Boulder, CO, USA
ewald.fuchs@colorado.edu
ewald.fuchs@gmail.com

Heidi A. Fuchs, Senior Research Associate
Energy Efficiency Standards Group
Energy Technologies Area
Lawrence Berkeley National Laboratory
Berkeley, CA, USA
hfuchs@lbl.gov
heidi.fuchs@gmail.com

Summary

This textbook breaks new ground in introducing students and professionals in science, technology, engineering, and mathematics (STEM) disciplines to electric energy associated with renewable resources, energy conservation, and laws of thermodynamics converting heat to mechanical power such as Carnot, Rankine, Brayton, and Ericsson cycles as they apply to internal combustion engines as well as to steam, gas, and oil turbines. It compares the energy content of common fuels, which is important for deciding how energy conservation and CO_2 reduction might take place. Next, it discusses various heat pump principles such as water–water, air–water, and air-conditioning systems; one of these principles is illustrated with yearlong measured data of a single-family (zero net energy) house where a photovoltaic (PV) plant delivers more energy than heating and cooling the house requires. The coefficient of performance based on combined heating and cooling performance (CHCP) is applied to this zero-emission house.

For storing the surplus energy of PV and wind power plants, this text shows how pumped-storage hydroelectricity and either battery or fuel cell storage plants are indispensable in conjunction with reformers changing methane to hydrogen for electric cars, as well as electrolysis for hydrogen generation from water and electric surplus energy. Wind power plants are analyzed based on the Lanchester–Betz–Joukowsky limit. In addition, design data of the abovementioned single-family house PV plant are explained and evaluated for performance. Further, this book assesses the advantages and disadvantages of AC versus DC transmission/distribution systems, including understanding the nature of electricity and its manufacturing, which are essential for the development of smart power systems. Associated with this manufacturing process are the costs of electricity in renewable energy systems.

In this book, circuit analyses are based on laws of electrical and mechanical engineering properties of electric charges, forces, conductors, insulators, semiconductors, instantaneous voltage, current, power, and work. For the analysis of electric networks, this volume describes Ohm's and Kirchhoff's laws as well as Thévenin's and Norton's theorems, nodal analysis, loop and mesh analysis, source exchange or transformation, and Wheatstone and Thomson bridges. The principle of superposition lends itself to complicated linear circuits. Transient DC and steady-state phasor AC analytic circuit investigations based on hand calculations are compared to linear and nonlinear PSPICE and Mathematica solutions. Also explained are instantaneous and steady-state power analyses, coupled magnetic circuits including design of permanent magnets, and single- and three-phase transformers, AC frequency characteristics of filters, and trigonometric and complex Fourier analyses. Operational amplifiers are important for control applications, consisting of semiconductor diodes and switches for voltage and current controllers, rectifiers, and inverters. Circuit models of semiconductor diodes including limiters and various switches (e.g. BJT MOSFET, SCR, triac, IGBT, and GTO) are introduced so that they can be readily used for PSPICE solutions.

In addition, this book provides novel applications for uncontrolled and controlled AC–DC converters (rectifiers), AC voltage and current regulators, and DC–AC converters (inverters) to PV and wind power plants and considers the efficiency increase and the resulting power quality problems resulting from these nonlinear components. DC machines and their commutator actions serve as role models for electronic commutation employed for rectifiers and inverters, which are indispensable for the formation of smart power systems. Permanent-magnet, induction, and synchronous machine variable-speed and torque drives with high switching frequencies are analyzed as used in electric/hybrid automobiles, wind power plants, and trains, by dealing with flux weakening (FW) and introducing compensation of flux weakening (CFW) at high angular velocities, reducing the need for high battery voltages (e.g. 800 V) in automobiles; in addition, various permanent-magnet characteristics are relied on for automobile-drive designs. For the determination of machine parameters, such as linear and nonlinear (operating-point-dependent) leakage, mutual, and self-inductances, magnetic field computation based on finite-difference (FD) and finite-element (FE) techniques is reviewed and employed for permanent-magnet, induction, and synchronous machines.

This text teaches by example. The book's innovative approach applies to 125 application examples and 157 problems highlight the teaching approach so that STEM students and professionals can better understand the electrical engineering and sciences associated with renewable energy issues by working through them. 449 reference citations are listed, which permit the reader to further delve into renewable energy projects. PSPICE and Mathematica software programs are included, for which personal computers suffice. Finally, the ~309-page Instructor Manual provides solutions to the abovementioned problems.

Key Features

- Real-world energy measurements of a single-family house with zero CO_2 emissions including PV plant, groundwater–water heat pump, and groundwater cooling.
- Design and performance of permanent-magnet, induction, and synchronous machine drives for electric/hybrid automobiles and rail drives for starting, rated operation, flux weakening (FW), and compensation of flux weakening (CFW), with torque and speed control within wide speed range.
- Design and performance of wind power and PV plants.
- Introduction to magnetic field analyses based on finite-element (FE) and finite-difference (FD) methods with respect to permanent, induction, and synchronous machine drives.
- Explanation of the nature of electricity and its manufacturing.
- The book's novel approach applies to 125 practical application (example) problems with solutions.
- 157 problems at the end of chapters, with solutions in the ~309-page Instructor Manual.
- 449 references – mostly journal articles and conference papers – as well as national and international standards and guidelines.

Preface

This book is intended for undergraduate students in non-electrical engineering disciplines such as architectural, civil, and mechanical engineering, physics, and chemistry, as well as professionals in related fields who want to acquaint themselves with electrical engineering as applied to energy or specifically to renewable energy. It is assumed that the reader has good high school and/or applied mathematics educational college background such as Calculus III – covering vector analysis, complex numbers, differentiation, ordinary differential equations, and partial derivatives as well as integration. This text integrates energy and electrical engineering technologies, starting from basic principles and covering detailed analyses of renewable and electric energy issues and their solutions. Application examples highlight conventional and renewable energy problems. Software programs such as Mathematica and PSPICE (Personal Computer Simulation Program with Integrated Circuit Emphasis) are applied to solve component and system problems.

This book has evolved chiefly from the content of undergraduate electrical and energy engineering courses taught by the first author at the University of Colorado at Boulder and also draws on the second author's research experience at the University of California Berkeley's Energy and Resources Group and at the Lawrence Berkeley National Laboratory. It is suitable for students who do not have educational background in energy and electrical engineering, as well as for practicing engineers who want to refresh and enhance their knowledge in these disciplines through 125 application example solutions, 157 problems and their solutions in the 309-page Instructor Manual, and more than 449 reference citations (mostly journal and conference papers, as well as national and international standards and guidelines). The extensive and detailed inclusion of practical applications stresses learning by example. Up-to-date references are given through the inclusion of Internet addresses. The "Système International" (SI) of units has been used throughout with some reference to the American/English system of units. Note that mass has the units (e.g. kg or lb) and weight has the units (e.g. kg-force or lb-force), where lb stands for "librum" as used in the English technical literature.

P.1 Key Features

- Provides theoretical and practical insight into renewable energy problems.
- 125 practical application (example) problems with solutions, some implemented in PSPICE and Mathematica.
- A total of 157 problems at the end of the chapters dealing with practical applications to electric, electronic, and renewable/energy engineering. Solutions to these problems are provided in a 309-page Solution Manual.

P.2 The Climate Dilemma

Figure P.1 illustrates the present warming distribution throughout the earth as published by *National Aeronautics and Space Administration*'s (NASA) Goddard Space Flight Center [1] where the measured global temperature during 2014 – indexed to the average values during the twentieth century – has increased by 0.68 °C. This average temperature increase, driven chiefly by human influence on the global carbon cycle as illustrated in Figure P.2 through deforestation and unrelenting emissions of greenhouse gases, captures our rationale for writing this book.

A 2008 paper by NASA scientist James Hansen [3] shows that the true gravity of the situation due to human, or anthropogenic, effects on the environment includes impacts on biophysical environments, biodiversity, and other natural resources. Hansen set out to determine what level of atmospheric carbon dioxide (CO_2) society should aim for "if humanity wishes to preserve a planet similar to that on which civilization developed and to which life on Earth is adapted." His climate models show that exceeding 350 parts per million (ppm) CO_2 in the atmosphere would likely have catastrophic effects. We have already surpassed that limit because environmental monitoring showed concentrations of around 400 ppm in 2014. This is particularly problematic because CO_2 and other greenhouse [4] gases (see Table P.1) remain in the atmosphere for a long time. Even if we shut down every fossil-fueled power plant today, existing greenhouse gases will continue to warm the planet.

The climate conundrum [5] is best illustrated in Figure P.3, which shows two sharply divergent pathways for CO_2 emissions: "business as usual" and the "best-case scenario" of CO_2 emissions in billions of metric tons. Figure P.4 depicts the CO_2 accumulations in the atmosphere for "business as usual," "best-case

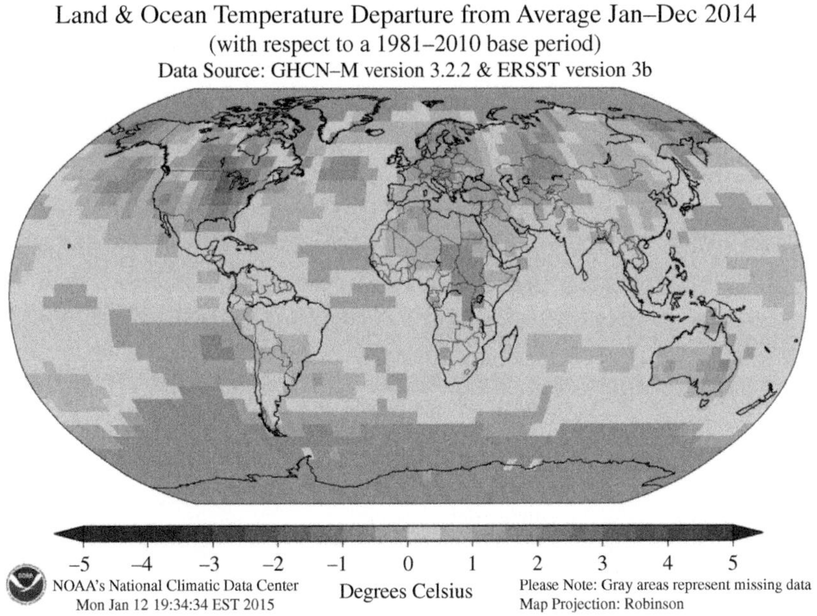

Figure P.1 Measured temperature anomalies during 2014 referring to the average values during the twentieth century, where the global average temperature in 2014 has increased [1] by 0.68 °C.

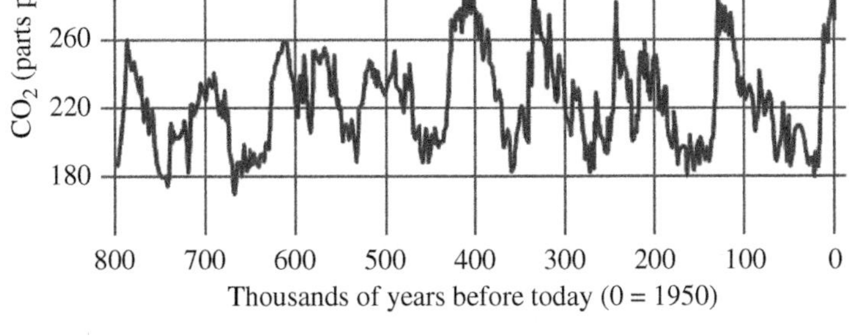

Figure P.2 CO_2 variation and increase [2] during the past 800 000 years.

Table P.1 Greenhouse gases and their chemical formula, anthropogenic sources, atmospheric lifetime, and global warming potential (GWP) [4].

Greenhouse gas	Chemical formula	Anthropogenic sources	Atmospheric lifetime [1](years)	GWP [2] (100-year time horizon)
Carbondioxide	CO_2	Fossil fuel combustion, land-use conversion, cement production	~100 [1]	1
Methane	CH_4	Fossil fuels, rice paddies, waste dumps	12 [1]	25
Nitrousoxide	N_2O	Fertilizer, industrial processes, combustion	114 [1]	298
Tropospheric ozone	O_3	Fossil fuel combustion, industrial emissions, chemical solvents	Hours-days	NA
CFC-12	CCL_2F_2	Liquid coolants, foams	100	10 900
HCFC-22	CCl_2F_2	Refrigerants	12	1 810
Sulfur hexafluoride	SF_6	Dielectric fluid	3 200	22 800

scenario," "Hansen model," and "safety threshold" scenarios. Achieving the deep cuts in carbon emissions required to step down from the business-as-usual trajectory to one with increased probability of climate stability is a monumental task. Fossil fuels and extractive industries are currently deeply interconnected with Western lifestyles, infrastructure, and typical economic development pathways. Making the leap will require political commitment to pricing carbon, continued innovation in low-carbon energy and storage, and the imagination and flexibility to envision – and bring about – a more resilient future.

P.2.1 Is the Renewable Energy Approach Too Optimistic?

Figures P.1–P.4 and Table P.1 require that the continued accumulation of CO_2 and other greenhouse gases in the atmosphere must be avoided by relying on renewable energy based on solar, wind, hydro, and biomass power plants in Germany as indicated in Figure P.5. While during summertime these renewable energy sources

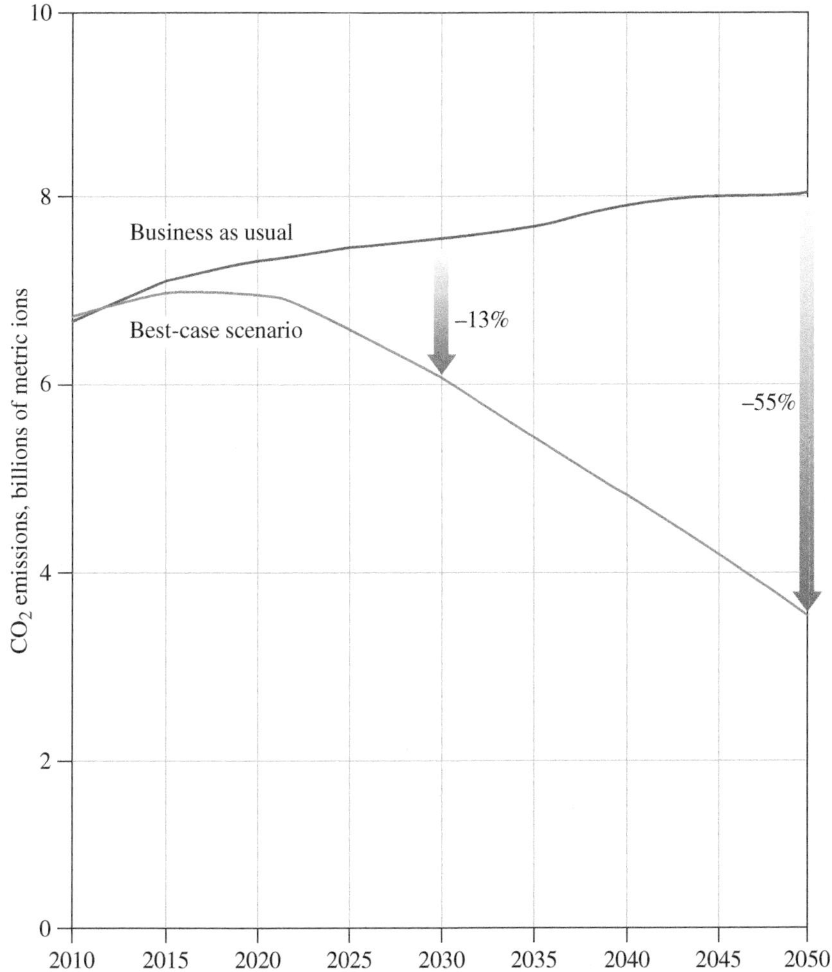

Figure P.3 Possible emissions pathways [5], billions of metric tons of CO_2.

generate to date more than 30% of electric energy usage, this is not so during winter when clouds/fog and calm weather conditions prevail with wind speeds too low to operate wind turbines either onshore or offshore as indicated in Figure P.5, where 26 000 wind turbines and 1.2 million solar plants within the German grid failed for almost two weeks to generate expected sufficient energy. This was due to regional atmospheric high pressure resulting in fog and still air conditions as frequently occur in Central Europe during cold weather. Such a "foggy calm [6, 7]" – called

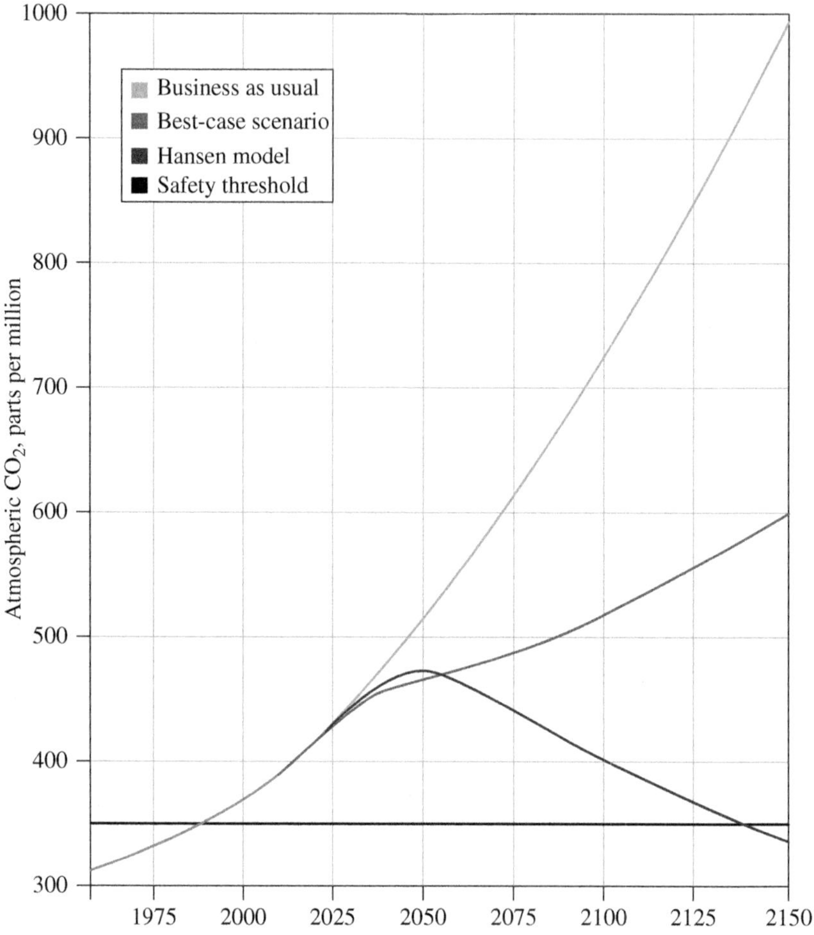

Figure P.4 Divergent scenarios [5] for atmospheric CO_2 in parts per million (ppm).

in German "Dunkelflaute" – brought the grid near its limits, where 80 GW must be available to serve peak-load conditions as indicated in Figure P.5.

During the period as indicated in Figure P.5, about 90% of the electric energy had to be provided by natural gas, coal, and nuclear power plants, the latter two of which are scheduled to be discontinued or decommissioned during the coming decade. Such a shutdown appears to be not prudent because short-term (e.g. battery) and long-term (e.g. pumped hydro) storage plants [8, 9] are unable to have an output power of 70 GW even only during a few days. Note that 70 GW is the power provided, for example, by either 35 nuclear and 70 coal or natural gas-fired plants

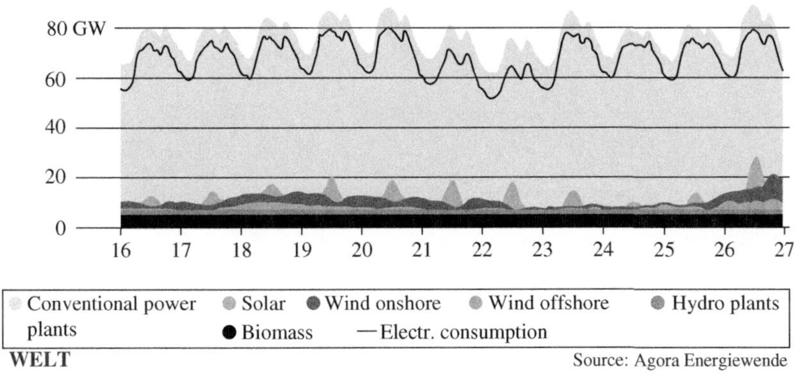

Figure P.5 Electric power (generation and consumption) requirements in Germany during 16–27 January 2017. The lightly-shaded area, which exceeds the full line indicating consumption, represents generation, while the full line indicates consumption: generation = (consumption +losses), that is, generation > consumption.

each having a rated output of 1 GW and 500 MW, respectively, or 300 hydro storage plants with each 233 MW output power rating. Clearly such a switch from conventional (e.g. coal, natural gas, nuclear) plants to storage plants appears to be not feasible. The transition from combustion engines (e.g. gasoline, diesel) to electric drives in automobiles, trucks, and trains will lead to significantly increased additional power consumption that cannot be provided by renewable energy during the times as indicated in Figure P.5. As a result, the strategy is to keep natural gas power plants each having an output of 200 MW operating together with some of the nuclear power plants that serve as a frequency leader. The supply of electric energy of Guangdong in southern China via high-voltage DC (HVDC) lines led to a breakup [10] of the alternating current (AC) grid due to the lack of a frequency-leading AC power plant and through the lack of sufficient reactive power to control voltage and frequency (Figure P.6). It is well known that inverters converting direct current (DC) to AC at the point of common coupling (PCC) can supply reactive power only if the DC voltage at the DC input voltage of the inverters is significantly increased, as discussed in Chapters 10 and 12. The renewable energy approach, where 100% of the energy is generated and transmitted via DC transmission lines, appears to be clearly too optimistic: there is indeed a climate dilemma.

P.2.2 Prerequisites for the Understanding of the Material as Offered in This Book

As a prerequisite to this effort, students must understand the following core competencies based on Applied Mathematics (APPM) 2350 Calculus III:

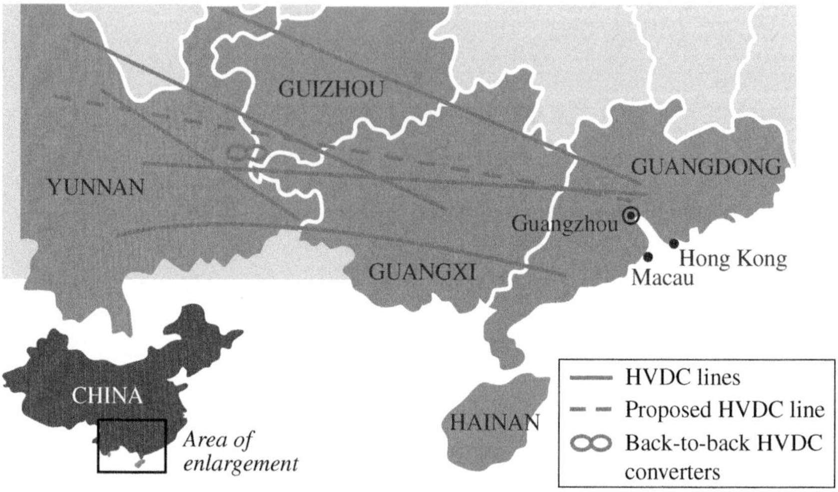

Figure P.6 The breakup of China's southern power grid due to the concentration of high-voltage DC transmission lines in Guangdong, making the AC grid unstable.

1) Single-variable calculus (e.g. differentiation, integration),
2) Functions of several variables,
3) Vectors and vector operations,
4) Curves in space,
5) Multivariable functions,
6) Partial differentiation, multiple integrals,
7) Line integrals, e.g. Stokes' theorem and Gauss' theorem,
8) Complex numbers, complex calculus in Gaussian complex plane.

In summary, the intent of this text is to educate undergraduate students in non-electrical engineering disciplines, physics, and chemistry as well as professionals in related fields who want to acquaint themselves in electrical engineering as applied to conventional and renewable energy.

References

1 NOAA (2014). Global climate report – January 2014. NOAA National Centers for Environmental Information, State of the Climate (published online February 2014). https://www.ncdc.noaa.gov/sotc/global/201401 (accessed 31 May 2020).

2 Proxy (Indirect) Measurements. Data source: Reconstruction from ice cores. Credit: NOAA. https://climate.nasa.gov/vital-signs/carbon-dioxide (accessed 31 May 2020).

3 Hansen, J., Sato, M., Kharecha, P. et al. (2008). Target atmospheric CO_2: where should humanity aim? *Open Atmospheric Science Journal* 2: 217–231. https://doi.org/10.2174/1874282300802010217.
4 Tans, P. and Keeling, R. http://www.c2es.org/content/main-greenhouse-gases (accessed 31 May 2020).
5 Koningstein, R. and Fork, D. (2014). Energy's creative destruction. *IEEE Spectrum* (December 2014): 26–31.
6 ntv.de, kst (2017). Kohlekraft muss einspringen Dunkelflaute zeigt Makel des Ökostroms. Foto: picture alliance/dpa. https://www.n-tv.de/wirtschaft/Dunkelflaute-zeigt-Makel-des-Okostroms-article19691796.html (accessed 31 May 2020).
7 Wetzel, D. (2017). Die "Dunkelflaute" bringt Deutschlands Stromversorgung ans Limit. https://www.welt.de/wirtschaft/article161831272/Die-Dunkelflaute-bringt-Deutschlands-Stromversorgung-ans-Limit.html (accessed 31 May 2020).
8 Fuchs, E.F. and Masoum, M.A.S. (2011). *Power Conversion of Renewable Energy Systems*, 2nd Printing. New York: Springer 692 pp.
9 Masoum, M.A.S. and Fuchs, E.F. (2008). *Power Quality in Power Systems and Electrical Machines*, 2e. Amsterdam: Academic Press/Elsevier 1123 pp.
10 Fairley, P. (2016). Why Southern China broke up its power grid. *IEEE Spectrum* (December 2016): 13. 14.

Glossary of Symbols, Abbreviations, and Acronyms

A

a	armature
a_0	average term of Fourier analysis
$a_{\ell+1/2}$	area of quadrilateral (one irregular third) of triangular mesh
$a_{\ell+1/2}^{\text{total}} = 3a_{\ell+1/2}$	area of triangular mesh
A	appliances, forcing term, unit of ampere, cross section of magnet, ampere
A_ℓ	vector potential at node ℓ
A_{o}	vector potential at node o
A_3	constant
A_{c}	cross section of iron core
A_{g}	cross section of air gap of an iron core
A_{v}	open-loop voltage gain
A, B, C	phase sequence A, B, C of line A, line B, and line C
ab	(voltage) from a to b
AB	(voltage) from line A to line B
a, b, c	phase sequence lines a, b, c
AC	alternating current
a_n	coefficient of Fourier term $a_n \cos n\omega_0 t$
an	phase a (a) to neutral (n)
anrms	phase a (a) to neutral (n) root-mean-square (rms) value
APPM	applied mathematics
ASIC	application-specific circuit
AWG	American wire gauge
alpha	firing angle α of thyristor bridge
α	damping factor/coefficient, angle

$\angle -90^\circ = e^{-j90^\circ}$ angular displacement/shift of time function
$\angle\theta_v$ indicates phase shift voltage $v(t)$ by either θ_v degrees (°) or θ_v radians (rad)

B

B boron, base, flux density, susceptance, viscous or fluid friction coefficient
$B(t)$ magnetic flux density as a function of time
$\vec{B}$ magnetic flux density vector
$\left|\vec{B}\right|$ magnitude of magnetic flux density vector
B_m susceptance of main (m) winding of transformer
B_1 constant
B_2 constant
B_g magnetic flux density within air gap
B_ϕ magnetic field density component in Φ direction of (r, Φ, z) coordinate system
$B_\phi(r)$ magnetic field density component in Φ direction as a function of r coordinate of (r, Φ, z) coordinate system
BW = 1/RC bandwidth
bat battery
B_{earth} magnetic DC field density of Earth caused by north and south poles of Earth
b_n coefficient of Fourier term $b_n \sin(n\omega_0 t)$
bn phase b (b) to neutral (n)
B_{ϕ_in} magnetic field density component in Φ direction inside of round wire
B_{ϕ_out} magnetic field density component in Φ direction outside of round wire
B-H *B-H* characteristic
BMW Bayerische Motoren Werke AG
bpm beats per minute
BJT bipolar junction transistor
BW bandwidth of resonant circuit
β angle

C

C	combined, cooling, collector, capacitance, conductance, contour length of integration path
c	subscript for capacitance (c), constant (c) for windage mechanical torque
c_1	coefficient linking Φ_{11} and i_1
c_2	coefficient linking Φ_{12} and i_2
ccw	counterclockwise
CAGR	compound annual growth rate
CBA	phase sequence C, B, A
cd	candela
CFL	compact fluorescent light (bulb)
CFW	compensation of flux weakening
CHCP	combined heat, cooling, and (electric) power
CHP	combined (thermal) heat and (electric) power
CMRR	common-mode rejection ratio
cn	phase c (c) to neutral (n)
CMOS	complementary metal–oxide semiconductor
$\cos\theta$	cosine of fundamental power factor angle θ between transformer load voltage and load current, fundamental power factor
$\cos(\Phi_{\mathrm{Vp}} - \Phi_{\mathrm{Ip}}) = \cos\Phi_{\mathrm{Vp\text{-}Ip}}$	power factor due to angles of $\widetilde{V}_{\mathrm{p}}$ and $\widetilde{I}_{\mathrm{p}}$
COP	coefficient of performance
CO_2	carbon dioxide
$\widetilde{C}_n$	coefficient of exponential Fourier analysis term
cum	cumulative

D

d	diode, developed, direct, differentiation
D	drain terminal, diode, subscript for determinant D, symbol for denominator, iron stack depth
$D(j\omega)$	denominator of transfer function
dot (•)	dot convention/marking
dB	decibel: dB (decibel)=20log $M(\omega)$
DC	direct current
DFIG	doubly fed induction generator

diff	differential
$\vec{dl}$	incremental (differential) length vector
$\overrightarrow{dS}$	incremental (differential) surface vector
dt	time increment/differential
dV	differential volume
DOE	US Department of Energy
dryer	clothes dryer
$	dollar symbol
Δ	three-phase winding connected in delta/Δ/triangle, delta/deviation/ incremental change
ΔT	torque difference
Δ-Δ or Δ/Δ	primary and secondary windings of three-phase transformer are connected in delta
Δ-Y or Δ/Y	primary winding of three-phase transformer is connected in delta and secondary winding in wye

E

e	electric
e_a	induced voltage of DC machine
e(t)	induced voltage as a function of time
E	energy, emitter
$\tilde{E}_1$	phasor-induced voltage of winding 1
$\tilde{E}_2$	phasor-induced voltage of winding 2
$\tilde{E}_p$	induced phasor voltage in primary winding
$\vec{E}$	electric field intensity/strength vector
$energy_c$	energy/work stored in iron core
$energy_g$	energy/work stored in air gap
ECEN	electrical, computer engineering
ECEEN	electrical, computer, energy engineering
ECPM	electronically commutated permanent-magnet machine
ECPMM	electronically commutated permanent-magnet machine
eff	effective
EG_i	energy (E) fed to utility grid (G)
EGG	Erneuerbare-Energien-Gesetz (German Renewable Energy Sources Act)
EHC_i	energy (E) for heating (H) and cooling (C)
EI_i	energy (E) by PV system feeding two inverters
EKG	electrocardiogram

ELA_i	energy (E) for lighting (L) and appliances (A)
EnEV	Energieeinsparverordnung (German Energy Saving Ordinance)
ERCOT	Electric Reliability Council of Texas
ERG	Energy (E) and Resources (R) Group (G)
ETN	European Type Number for batteries
EU	European Union
Eq	equation
eq	equivalent
E_x	*x*-component of electric field intensity/strength
E_y	*y*-component of electric field intensity/strength
E_z	*z*-component of electric field intensity/strength
$E(t)$	electric field intensity/strength as a function of time
€	euro symbol
ε	permittivity of dielectric, steady-state error/deviation
ε_0	permittivity of free space
ε_r	relative permittivity

F

f	field, frequency
$\vec{F}$	(mechanical) force vector
F	farad, unit of capacitance, energy functional for finite element method, feedback
$\mathcal{F}$	magnetomotive force
$\vec{F}_{\text{attraction}}$	attraction force due to magnetic fields of two magnets
$\vec{F}_{\text{repulsion}}$	repulsion force due to magnetic fields of two magnets
fe	ferrum (Latin for iron)
FET	field-effect transistor
FD	finite difference
FE	finite element
FM	figure of merit
FW	flux weakening
$f(t)$	periodic function of time t
ft	feet
ϕ	phase angle, power factor angle
Φ	magnetic field flux, useful flux, iron-core flux
ϕ_s	phase angle of source voltage $\widetilde{V}_s(\omega)$

ϕ_{Is}	phase angle of source current $\widetilde{I}_s(\omega)$ as a function of angular velocity or frequency ω
$\Phi(t)$	magnetic flux as a function of time
$\phi(\omega)$	phase angle of phasor transfer function $G(j\omega)$ as a function of angular velocity or frequency ω
$\Phi_{\mathrm{m}}(t)$	mutual/main flux as a function of time
Φ_{fringing}	fringing magnetic flux
Φ_{useful}	useful magnetic flux
Φ_{12}	mutual flux originating in winding 2 linking winding 1
Φ_{21}	mutual flux originating in winding 1 linking winding 2
Φ_{11}	total flux generated by coil 1
Φ_{22}	total flux generated by coil 2
Φ_{leak1}	leakage flux linked with coil 1
Φ_{leak2}	leakage flux linked with coil 2
Φ_1	total flux linked with coil 1
Φ_2	total flux linked with coil 2
Φ_{A}	flux in transformer A
Φ_{B}	flux in transformer B
Φ_{C}	flux in transformer C
$\Phi_{\mathrm{total}} = \Phi_{\mathrm{A}} + \Phi_{\mathrm{B}} + \Phi_{\mathrm{C}}$	total flux in common limb of bank of three-phase transformer consisting of three-single phase transformers A, B, C
$\phi_{\widetilde{I}_{s'}}$	phase angle of secondary current $\widetilde{I}_{s'}$ referred to primary side

G

g	ground, air-gap length
G	grid, gate, conductance $G = 1/R$, giga, closed-loop gain
$G(j\omega)$	phasor transfer function
G_{c}	core-loss conductance
G_{D}	differential (D) mode gain
G_{v}	closed-loop voltage gain
G_{CmRR}	common-mode rejection ratio gain
GLC	conductance–inductance–capacitance network
gen	generator
giga	10^9
G_{i}	current gain
$G_{\omega\mathrm{m}}$	angular velocity gain
G_{rect}	thyristor rectifier gain
$G_{\mathrm{V}}^{\mathrm{NI}}$	closed-loop voltage gain of noninverting operational amplifier

G_V^{UG}	closed-loop voltage gain of unity gain operational amplifier
G_V^{IN}	closed-loop voltage gain of inverting operational amplifier
G_V^{HP}	closed-loop voltage gain of high-pass filter
$\|G_V^{LP}(\omega)\|$	magnitude of closed-loop voltage gain of low-pass filter
$\|G_V^{HP}(\omega)\|$	magnitude of closed-loop voltage gain of high-pass filter
GmbH	Gesellschaft mit beschränkter Haftung (German for company with limited liability)
GSP	gas and steam (electric) power
GTO	gate turn-off thyristor
GW	gigawatt
GWP	global warming potential

H

H	heating, magnetic field intensity/strength, henry unit symbol for inductance
HV	high voltage
HP	high pass
hp	horsepower: 1 hp = 745.7 watts ≈ 746 W, or 1.3392 horsepower = 1000 watts.
$\vec{H}$	magnetic field intensity vector
$\|\vec{H}\|$	magnitude of magnetic field intensity vector
H_c	magnetic field intensity in iron core
$\vec{H}_g$	vector of magnetic field intensity/strength in air gap
$\vec{H}_{core}$	magnetic field intensity/strength vector of iron core
$H_{core}=\|\vec{H}_{core}\|$	magnitude of magnetic field intensity vector within iron core
$H_g=\|\vec{H}_g\|$	magnitude of magnetic field intensity/strength vector within air gap (g)
H_r	magnetic field intensity/strength r-component of (r, Φ, z) polar system
H_Φ	magnetic field intensity/strength Φ-component of (r, Φ, z) polar system
H_{Φ_in}	magnetic field intensity component in Φ direction inside of round wire
$\vec{H}_{\Phi\_in}$	magnetic field intensity vector in Φ direction inside of round wire

H_{Φ_out}	magnetic field intensity component in Φ direction outside of round wire
H_z	magnetic field intensity/strength z-component of (r, Φ, z) polar system
H_x	x-component of magnetic field intensity vector $\vec{H}_{core}$
hom	homogeneous
HVDC	high-voltage DC line
HTE	high-temperature electrolysis
H_2O	chemical symbol for water
Hz	hertz, cycles per 1 second

I

i	unit directions of (i, j, k) rectangular system, current, integration
i_+	zero input current at positive terminal of operational amplifier
i_-	zero input current at negative terminal of operational amplifier
I, II, III, IV	identification of step function
IC	integrated circuit
IN	inverting configuration of operational amplifier
$i_a(t)$	armature current as a function of time
i_a^*	reference/nominal/rated (∗) value of armature (a) current
$i_p(t)$	primary current as a function of time
$\tilde{I}_s(\omega)$	phasor source current as a function of angular velocity or frequency
i_A^p	primary current through winding/phase A
i_B^p	primary current through winding/phase B
i_C^p	primary current through winding/phase C
i_a^p	secondary current through winding/phase a
i_b^p	secondary current through winding/phase b
i_c^p	secondary current through winding/phase c
$\tilde{i}_p$	primary phasor current
$\lvert\tilde{I}_p\rvert$	magnitude of phasor current in primary winding
$\tilde{i}_c$	core-loss phasor current
$\tilde{i}_\phi$	excitation phasor current
$\tilde{I}_m$	magnetizing phasor current
$\tilde{I}'_s$	secondary phasor current referred to primary side of transformer

$\widetilde{I}_{load} = \widetilde{I}_s$	phasor load current of single-phase transformer referred to secondary side
$\widetilde{I}'_s = \widetilde{I}'_{load}$	phasor load current of single-phase transformer referred to primary side
$i_s(t)$	secondary current as a function of time
I	inverter, integral, inductance, current, DC current
$\widetilde{I}_2$	phasor current of secondary winding 2
$\widetilde{I}^*_2$	phasor conjugate (∗) current of secondary circuit
$\widetilde{I}_s$	phasor source current of primary circuit
$\widetilde{I}'_s$	secondary phasor current of transformer referred to primary side
$\left\vert\widetilde{I}'_s\right\vert$	magnitude of secondary phasor current of transformer referred to primary side
$\widetilde{I}'_C$	phasor capacitor current of transformer referred to primary side
$\widetilde{I}_c$	phasor current through core-loss resistance R_c of transformer
$\widetilde{I}_1$	phasor input current of primary circuit
$\widetilde{I}_2$	phasor output current of secondary circuit
$i(t)$	current as a function of time
$i_1(t)$	primary current as a function of time function
$i_2(t)$	secondary current as a function of time function
i_2	current in secondary winding 2
i_1	current in primary winding 1
$\widetilde{I}_1$	phasor current in primary winding 1
I_{DC}	DC current
I_{in}	DC current inside of round wire up to radius r_{in}
I_o	current $i(t)$ at $t = 0$
$\widetilde{I}'_o$	component current of $\widetilde{I}_o = \widetilde{I}'_o + \widetilde{I}''_o$
$\widetilde{I}''_o$	component current of $\widetilde{I}_o = \widetilde{I}'_o + \widetilde{I}''_o$
∞	symbol for infinity
ICE	internal combustion engine
in	input
IM	induction or asynchronous machine
Im	indicates imaginary part of a complex number
$\left\vert\widetilde{I}_M\right\vert$	absolute value or magnitude of phasor $\widetilde{I}_M$
I_M	amplitude of time function $i(t)$
IM	induction motor/machine
IGBT	insulated-gate bipolar transistor

incandescent	Edison-type light bulb
$\widetilde{I}^*_{\mathrm{rms}}$	complex conjugate of $\widetilde{I}_{\mathrm{rms}}$

J

j	complex number operator indicating imaginary part of phasor, unit component of (i, j, k) rectangular system
$j\omega L$	reactive impedance of inductance L
$j\omega L_1$	reactive impedance of self-inductance L_1
$j\omega L_2$	reactive impedance of self-inductance L_2
$j\omega M$	reactive impedance of mutual inductance M
JAZ	Jahresarbeitszahl (German for seasonal performance factor of heat pump)
J	unit of energy joules
$\vec{J}$	current density

K

k	unit component of (i, j, k) rectangular system, coupling coefficient of two windings/coils, kilo, Boltzmann constant
krad/s	kilo radians per second
$(k_e\Phi)$	field flux of DC machine
K_1	constant
K_2	constant
KCL	Kirchhoff's current law
KVL	Kirchhoff's voltage law
kVA	kilovolt-ampere
kg	kilogram is unit of mass
kg-force	kilogram-force is unit of weight
K	kelvin

L

l	leakage
lx	lux
$\mathrm{length}_{\mathrm{core}}$	iron-core length

limit	symbol for limiter or limit operation
length	total length of turn
L	lighting, lower, line, symbol for inductance
L_a	armature inductance of DC machine
L_s	self-inductance
L_m	mutual inductance
LP	low pass
LV	low voltage
L_{11}	self-inductance of primary winding 1
L_{12}	mutual inductance between winding 1 and winding 2
L_{21}	mutual inductance between winding 2 and winding 1
L_{22}	self-inductance of secondary winding 2
L_1	self-inductance of primary winding 1 when $L_1 = L_{11}$
L_2	self-inductance of secondary winding 1 when $L_2 = L_{22}$
l_c	iron-core length
LED	light-emitting diode
ℓ	leakage, general subscript for vector potential A_ℓ
ℓ_{core}	core length
ℓ_g	air-gap length
lagging	lagging power factor (e.g. current lags voltage)
leading	leading power factor (e.g. current leads voltage)
lb	librum (Latin for pound)
L-L	line-to-line
L-Lsecondary_rms	secondary line-to-line root-mean-square value
lighting	lighting parameter
lighting_rms	rms value of lighting parameter (e.g. voltage)
lm	lumen
loadold	indicates load (load) before (old) power factor correction
loadnew	indicates load (load) after (new) power factor correction
LHS	left-hand side
L_m	mutual/magnetizing inductance of T-equivalent circuit
$L_{\ell p}$	leakage inductance of primary winding
$L'_{\ell s}$	leakage inductance of secondary winding referred to primary side of transformer
LNG	liquid natural gas
LPG	liquid petroleum gas, propane
λ	flux linkages

M

m	milli, mechanical, magnet, unit of meter, magnetizing, motor, slope of a linear function
M	mutual inductance if $L_{12} = L_{21} = M$, stands for main power system, mega
mmf	magnetomotive force
$M(\omega)$	magnitude of phasor transfer function $G(j\omega)$ as a function of angular velocity or angular frequency ω
MATLAB	software of the US American company MathWorks for the solution of mathematic problems
max	maximum
micro	10^{-6}
milli	10^{-3}
ms	milliseconds
mH	millihenries
mV	millivolts
mA	milliampere
MHD	magneto-hydro dynamic
MKSA	meter-kilogram-second-ampere system of units
MKSAK	meter-kilogram-second-ampere-kelvin system of units
mmf	magnetomotive force
MOS	metal-oxide semiconductor
MOSFET	metal–oxide–semiconductor field-effect transistor
MPP	maximum power point
MW	megawatt
MVA	megavolt-ampere
μ	permeability of (magnetic) material, 10^{-6}
$\mu_0 = 4\pi 10^{-7}$ H/m	permeability of free space
μ_r	relative permeability of (magnetic) material

N

n	neutral, running index, $\pm 1,\ \pm 2,\ \pm 3, \ldots$
n_s	neural point of secondary winding connected in wye
n_p	neural point of primary winding connected in wye
n_m	mechanical speed
n_{load}	neural point of wye-connected load
nano	10^{-9}

N	north (pole), neutral, Norton, running index, subscript for numerator, number of turns
$N(j\omega)$	numerator of transfer function $G(j\omega)$
N_1	number of turns of primary winding (1) of transformer
N_2	number of turns of secondary winding (2) of transformer
N_p	primary number of turns of transformer
N_s	secondary number of turns of transformer
NI	noninverting configuration of operational amplifier
NASA	US National Aeronautics and Space Administration
NdFeB	neodymium–iron–boron magnet material
NMOS	n-type metal–oxide–semiconductor
NOCT	nominal operating cell temperature
Norton theorem	impressed current source theorem
NREL	National Renewable Energy Laboratory

O

oc	open circuit
OP	operational
ORC	organic Rankine cycle
out	output
ω	angular frequency or angular velocity
ω_0	angular frequency or angular velocity at resonance, undamped natural angular frequency, center frequency
ω_m	mechanical angular velocity
ω_m^*	reference/nominal/rated (∗) value of mechanical (m) angular velocity
ω_{HI}	upper (HI) break or cutoff frequency, half-power angular frequency
ω_{LO}	lower (LO) break or cutoff frequency
ω_{LO}, ω_{HI}	half-power frequencies
Ω	unit for ohms
ωL_{load}	reactance due to load inductance
ωL_{load1}	reactance due to load 1 inductance
ωL_{load2}	reactance due to load 2 inductance
ωM	mutual reactance due to coupling of two windings
ωL_1	self-reactance of winding 1
ωL_2	self-reactance of winding 2
$\omega L'_{load}$	reactance due to load impedance referred to primary side

$\omega L^{\Delta}_{\text{load1}}$	per phase reactance of delta-connected first load
$\omega L^{\Delta}_{\text{load2}}$	per phase reactance of delta-connected second load
$\omega L^{\text{Y}}_{\text{load1}}$	per phase reactance of Y-connected first load
$\omega L^{\text{Y}}_{\text{load2}}$	per phase reactance of Y-connected second load
ωLls	leakage reactance of secondary winding
$\omega Lls'$	leakage reactance of secondary winding referred to primary side of transformer
$\omega L^{\Delta}_{\text{load}}$	per phase reactance of delta-connected load
ω_{Llp}	leakage reactance of primary winding
ωL_{m}	reactance of magnetizing/cross-branch of transformer
ωL_{line}	per phase reactance of distribution/transmission line
$\omega_{\text{m}}(t)$	time-dependent mechanical angular velocity

P

p	pole, peak, parallel, primary
P	performance, phosphorous, proportional control, average or real power, proportional
$\widetilde{p}_1, \widetilde{p}_2, \ldots, \widetilde{p}_n$	poles of transfer function $G(j\omega)$ from 1, ..., n
PD	proportional–differential
PI	proportional–integral
PID	proportional–integral–differential
P_1	input real power
P_2	output real power
PF	power factor
$p(t)$	instantaneous power
P_{rat}	rated real output power
P_{actual}	actual real output power
$p_{\text{in}}(t)$	input power as a function of time
$p_{\text{out}}(t)$	output power as a function of time
P_{in}	input real power
P_{out}	output real power
P^{p}_{cu}	copper loss of primary winding
P^{s}_{cu}	copper loss of secondary winding
P_{fe}	iron-core loss
PCC	point of common coupling
ph	phase
PLL	phase-lock loop

per	period
PF	displacement or fundamental power factor
PEM	proton exchange membrane
PEMFC	proton exchange membrane fuel cell
pico	10^{-12}
phaserms	phase (phase) root-mean-square (rms) value
PM	permanent magnet
PMG	permanent-magnet generator
PMM	permanent-magnet machine
pn	positive and negative types
ppm	parts per million
PSPICE	Personal Simulation Program with Integrated Circuit Emphasis
pu	per unit
PV	photovoltaics
pw	pulse width
pw/per	pulse width per period
PWM	pulse-width-modulated
$p_{\text{oscillation}}(t)$	oscillating power
π	pi, or 3.141 59...

Q

Q	electric charge, imaginary or reactive power
$Q = 1/(\omega_0 \text{RC}) = 1/\text{BW} = \text{RC}/\sqrt{\text{LC}} = R\sqrt{L/C} = \sqrt{\text{LC}}/\text{RC}$	quality factor
q	quadrature
$q(t)$	electric charge as a function of time
quadrant	one-quarter area of complex/ Gaussian plane

R

r	rotor
R	radius, resistance
R_a	armature resistance of DC machine
RC	resistor–capacitor network
$\mathfrak{R}$	reluctance, magnetic resistance
R_c	core-loss resistance, resistance of iron core
R_g	resistance of air gap of iron core

$\mathfrak{R}_g$	reluctance or magnetic resistance of air gap of iron core
$\mathfrak{R}_c$	reluctance or magnetic resistance of iron core
R_p	resistance of primary winding
R_s	resistance of secondary winding
R'_s	secondary resistance referred to primary side of transformer
R_c	loss (series) resistance of capacitor
R_1	resistance of primary winding 1
R_2	resistance of secondary winding 2
R_{line}	per phase resistance of distribution/transmission line
RLC	resistor–inductor–capacitor network
rad	radians
rat	rated
Re	indicates real part of a complex number
R_{load}	load resistance
R_{load1}	resistance of load 1
R_{load2}	resistance of load 2
R^{Δ}_{load}	per phase resistance of delta-connected load
R^{Δ}_{load1}	per phase resistance of delta-connected first load
R^{Δ}_{load2}	per phase resistance of delta second connected load
R^{Y}_{load1}	per phase resistance of Y-connected first load
R^{Y}_{load2}	per phase resistance of Y-connected second load
RHS	right-hand side
r_{in}	radius inside of round wire
r_{out}	radius outside of round wire
rms	root-mean-square
rmsload	subscript for root-mean-square value of a load quantity (e.g. voltage)
rmssource	indicates root-mean-square value of a source quantity (e.g. voltage)
rpm	revolutions per minute
rps	revolutions per second
$(R - jX)$	complex conjugate of $(R + jX)$
(r, Φ, z)	r coordinate, Φ coordinate, z coordinate

S

s	source, stator, synchronous, slip, unit of second, abbreviation symbol for first derivative $\frac{d\ldots}{dt}$, secondary, abbreviation for $j\omega$ =s, second
s^2	abbreviation symbol for second derivative

S	south (pole), surface, source terminal
$S = \lvert\tilde{S}\rvert$	apparent power, magnitude of complex power
$\tilde{S}$	complex power
s_1, s_2	two roots of characteristic equation
sc	short circuit
$\tilde{S}_{out}$	complex output power
S_{rat}	rated apparent power
S_{actual}	actual apparent power
SEER	seasonal energy efficiency ratio
SCR	silicon-controlled rectifier
SI	Système International d'Unités (French for International System of Units)
Si	silicon
SiC	silicon carbide
single	single-phase value
SM	synchronous machine
SOFC	solid oxide fuel cell
SS	steady state
start	starting
SM	synchronous motor/machine
STC	Standard Test Condition
STEM	science, technology, engineering, and mathematics
sub	substation
SUM	summation
sys	system
SWM	Stadtwerke München
switch	mechanical or electronic switch

T

t	turn, time variable
T	unit of Tesla, period length/duration, time constant, transformer, absolute temperature
$\vec{T}$	torque in vector notation
$T(t)$	time-dependent torque
t_d	delay time
T_e	electric torque
T_c	mechanical torque due to Coulomb (c) friction

T_s	mechanical torque due to standstill (s)
T_L	mechanical load torque
T_m	mechanical torque performing useful work
TGV	Train à Grande Vitesse (French high-speed trains)
th	theoretical
TH	thyristor, Thévenin (TH)
trans	transresistance
Thévenin theorem	impressed voltage source theorem
THD_i	total harmonic distortion of current
THD_v	total harmonic distortion of voltage
three	three-phase value
τ	time constant, time constant RC
τ_p	time constant for peak rectification
τ_{RC}	time constant for RC circuit
τ_{RL}	time constant for RL circuit
θ	phase shift of a time function, power factor angle
θ_v	phase shift of a voltage $v(t)$
θ_i	phase shift of a current $i(t)$
θ_m	mechanical displacement angle
θ_n	phase shift of Fourier component
θ_p	phase shift angle between v_A^p and i_A^p

U

U	stands for upper
USA	United States of America
UG	unity gain

V

V	voltage, unit of voltage, volt
$\widetilde{V}_o$	phasor output voltage
V_{DD}	positive DC supply voltage of operational amplifier
V_{SS}	negative DC supply voltage of operational amplifier
$\widetilde{V}_{s1}$	phasor source 1 input voltage
$\widetilde{V}_{s2}$	phasor source 2 input voltage

$V_{harmonic}$	harmonic voltage magnitude
V^{p}	primary phase voltage of transformer
V^{s}	secondary phase voltage of transformer
V_1	primary phase voltage of transformer
V_2	secondary phase voltage of transformer
V_{CA}	magnitude of primary voltage between winding terminals A and A
V_{Anp}	magnitude of primary voltage between terminal A and neutral point n_p of primary winding A
V_{ans}	magnitude of secondary voltage between terminal a and neutral point n_s of secondary winding a
$\widetilde{V}_1$	phasor voltage across primary winding 1
$\widetilde{V}_2$	phasor voltage across secondary winding 2
$\widetilde{V}_C$	phasor voltage across capacitor C
$\widetilde{V}_p$	primary phasor voltage
v_+	positive input voltage of operational amplifier
v_-	negative input voltage of operational amplifier
$\lvert\widetilde{V}_p\rvert$	magnitude of phasor voltage at primary side of transformer
$\widetilde{V}_{out}$	phasor output voltage of secondary circuit
$\widetilde{V}'_{load} = \widetilde{V}'_s$	phasor load secondary voltage of single-phase transformer referred to primary side
$\widetilde{V}'_s$	phasor secondary voltage of transformer referred to primary side
$\lvert\widetilde{V}'_{load}\rvert$	magnitude of the load phasor voltage of transformer referred to primary side
$\widetilde{V}_{Rp}$	phasor voltage across the primary resistance R_p
$\widetilde{V}_{Xlp}$	phasor voltage across the primary leakage reactance X_{lp}
$\widetilde{V}_{Xls'}$	phasor voltage drop across leakage reactance of secondary side referred to primary side of transformer
$\widetilde{V}_{Rs'}$	phasor voltage drop across winding resistance of secondary side referred to primary side of transformer
$\widetilde{V}_s$	phasor voltage of source
$\widetilde{V}_s(\omega)$	phasor source voltage as a function of angular velocity or frequency ω
$\widetilde{V}_o(\omega)$	phasor output voltage as a function of angular velocity or frequency ω
$\widetilde{V}_s^{(1)}$	phasor load/secondary voltage of single-phase transformer where superscript (1) indicates the first step of drawing the phasor diagram

$\tilde{V}_p^{(15)}$	phasor primary voltage of single-phase transformer where superscript (15) indicates the fifteenth step of drawing the phasor diagram
$v(t)$	voltage as a function of time function
$v_1(t)$	primary voltage as a function of time function
$v_p(t)$	primary voltage as a function of time function
$\tilde{v}(t)$	phasor of voltage time function $v(t)$
VA	unit for apparent power S called volt-ampere
V_A^p	voltage across primary winding of phase A
V_B^p	voltage across primary winding of phase B
V_C^p	voltage across primary winding of phase C
V_a^p	voltage across primary winding of phase a
V_b^p	voltage across primary winding of phase b
V_c^p	voltage across primary winding of phase c
$\|\tilde{V}_{Anp}\|$	magnitude of primary phase voltage of transformer with winding A
$\|\tilde{V}_{Bnp}\|$	magnitude of primary phase voltage of transformer with winding B
$\|\tilde{V}_{Cnp}\|$	magnitude of primary phase voltage of transformer with winding C
$\|\tilde{V}_{ans}\|$	magnitude of secondary phase voltage of transformer with winding a
$\|\tilde{V}_{bns}\|$	magnitude of secondary phase voltage of transformer with winding b
$\|\tilde{V}_{cns}\|$	magnitude of secondary phase voltage of transformer with winding c
$\tilde{V}_{Anp}$	phasor primary phase voltage of transformer with winding A
$\tilde{V}_{Bnp}$	phasor primary phase voltage of transformer with winding B
$\tilde{V}_{Cnp}$	phasor primary phase voltage of transformer with winding C
$\tilde{V}_{ans}$	phasor secondary phase voltage of transformer with winding a
$\tilde{V}_{bns}$	phasor secondary phase voltage of transformer with winding b
$\tilde{V}_{cns}$	phasor secondary phase voltage of transformer with winding c
$\tilde{V}_{CA}$	phasor primary line-to-line voltage of transformer between windings C and A
$\tilde{V}_{AB}$	phasor primary line-to-line voltage of transformer between windings A and B
$\tilde{V}_{BC}$	phasor primary line-to-line voltage of transformer between windings B and C
$\tilde{V}_{aa'}$	phasor secondary phase voltage of transformer between terminals a and a′
$\tilde{V}_{bb'}$	phasor secondary phase voltage of transformer between terminals b and b′

$\widetilde{V}_{cc'}$ phasor secondary phase voltage of transformer between terminals c and c′

V_{Anp} magnitude of primary phase voltage of transformer with winding A

$V_{aa'}$ magnitude of secondary phase voltage of transformer between terminals a and a′

$|\widetilde{V}_{AA'}|$ magnitude of primary phase voltage of transformer between terminals A and A′

$|\widetilde{V}_{BB'}|$ magnitude of primary phase voltage of transformer between terminals B and B′

$|\widetilde{V}_{CC'}|$ magnitude of primary phase voltage of transformer between terminals C and C′

$|\widetilde{V}_{aa'}|$ magnitude of secondary phase voltage of transformer between terminals a and a′

$|\widetilde{V}_{bb'}|$ magnitude of secondary phase voltage of transformer between terminals b and b′

$|\widetilde{V}_{cc'}|$ magnitude of secondary phase voltage of transformer between terminals c and c′

$V_{AA'}$ magnitude of primary phase voltage of transformer between terminals A and A′

$V_{aa'}$ magnitude of secondary phase voltage of transformer between terminals a and a′

V_{BA} magnitude of primary line-to-line voltage of transformers between terminals B and A

V_{ba} magnitude of secondary line-to-line voltage of transformers between terminals b and a

$|\widetilde{V}_{BA}|$ magnitude of primary line-to-line voltage of transformers between terminals B and A

$|\widetilde{V}_{CB}|$ magnitude of primary line-to-line voltage of transformers between terminals C and B

$|\widetilde{V}_{AC}|$ magnitude of primary line-to-line voltage of transformers between terminals A and C

$|\widetilde{V}_{ba}|$ magnitude of secondary line-to-line voltage of transformers between terminals b and a

$|\widetilde{V}_{cb}|$ magnitude of secondary line-to-line voltage of transformers between terminals c and b

$|\widetilde{V}_{ac}|$ magnitude of secondary line-to-line voltage of transformers between terminals a and c

VAr unit for imaginary/reactive power Q called volt-ampere reactive

$v_c(t = -0)$ $v_c(t)$ voltage before switching occurs at $t = 0$

$v_L(t = +0)$ $v_L(t)$ voltage after switching occurs at $t = 0$

$v_C(t = \infty)$ $v_C(t)$ at an infinite amount of time

V_M	amplitude of time function $v(t)$
VSI	voltage source inverter
V_C	capacitor voltage

W

w	waveshaping
W	work/energy, iron-core width
W_C	work stored in capacitor
W_{core}	energy stored in iron core
W_{gap}	energy stored in air gap
$W_L(t)$	energy/work stored in inductor as a function of time
Ws	watt-seconds

X

x	x-coordinate of x-y spatial rectangular system, variable of integration
(x, y, z)	Cartesian coordinate system with x, y, z components
$x(t)$	x-variable as a function of time
X	reactance, variable for voltage and current
X_M	amplitude of time function $x(\omega t)$
X_m	magnetizing/cross-branch reactance of transformer
X_{ls}	leakage reactance of secondary winding
X_{line}	per phase reactance of distribution/transmission line
X'_{ls}	secondary leakage reactance referred to primary side of transformer
X_{lp}	leakage reactance of primary winding
$X_{rms} = X_{eff}$	root-mean-square (rms) or effective (eff) value of a quantity
X_{load}	load reactance

Y

y	y-coordinate of x-y spatial rectangular system, subscript refers to admittance $\widetilde{Y}$
Y	winding connected in wye/Y/star

$\widetilde{Y}(\omega)$	phasor admittance as a function of angular velocity or frequency ω
Y-Y or Y/Y	primary and secondary windings of three-phase transformer are connected in wye or star
Y-Δ or Y/Δ	primary winding of three-phase transformer is connected in wye and secondary winding in delta
$\widetilde{Y}_{\phi}$	cross-branch phasor admittance of transformer
$\widetilde{Y}$	admittance phasor
Y_{M}	amplitude of sinusoidal/cosinusoidal time function $y(\omega t)$

Z

z	z-coordinate of a spatial x, y, z or r, φ, z system
$\widetilde{Z}$	impedance phasor
$\widetilde{z}_1, \widetilde{z}_2, \ldots, \widetilde{z}_m$	zeros of transfer function $G(j\omega)$ from 1, ..., m
$\widetilde{Z}(\omega)$	phasor impedance as a function of angular velocity or frequency ω
$\widetilde{Z}_{\mathrm{C}}$	capacitive phasor impedance
$\widetilde{Z}_{\mathrm{L}}$	inductive phasor impedance
Z	Zener diode
$\vec{z} = x + jy$	complex number
$\widetilde{Z}_{\mathrm{load}}$	phasor load impedance
$\widetilde{Z}^{\Delta}_{\mathrm{load}}$	phasor load impedance when in Δ connection
$\widetilde{Z}^{\mathrm{Y}}_{\mathrm{load\_}i}$	phasor impedance of ith Y-connected transformer load
$\widetilde{Z}^{\Delta}_{\mathrm{load\_}i}$	phasor impedance of ith Δ-connected transformer load
$\widetilde{Z}^{\Delta}_{\mathrm{load1}}$	phasor impedance of first Δ-connected transformer load
$\widetilde{Z}^{\Delta}_{\mathrm{load2}}$	phasor impedance of second Δ-connected transformer load
$\widetilde{Z}^{\mathrm{Y}}_{\mathrm{load1}}$	phasor impedance of first Y-connected transformer load
$\widetilde{Z}^{\mathrm{Y}}_{\mathrm{load2}}$	phasor impedance of second Y-connected transformer load
$\widetilde{Z}^{\mathrm{Y}}_{\mathrm{load}}$	phasor impedance of Y-connected transformer load
$\widetilde{Z}_{\mathrm{load1}}$	phasor impedance of first connected transformer load
$\widetilde{Z}_{\mathrm{load2}}$	phasor impedance of second connected transformer load
$\widetilde{Z}_{\mathrm{line}}$	phasor transmission/distribution line impedance
$\widetilde{Z}_{\mathrm{in}}$	phasor input impedance of primary circuit
$\widetilde{Z}_{\mathrm{reflect}}$	phasor reflected impedance of secondary circuit as viewed from primary circuit

$\widetilde{Z}_{\mathrm{out}}$	phasor output impedance of secondary circuit
$\widetilde{Z}_{\mathrm{s}}$	impedance of secondary winding of transformer
$\widetilde{Z}'_{\mathrm{s}}$	impedance of secondary winding referred to the primary side of the transformer
$\widetilde{Z}_{\phi}$	cross-branch/magnetizing impedance of transformer
$\widetilde{Z}_{\mathrm{p}}$	impedance of primary winding of transformer
$\widetilde{Z}'_{\mathrm{load}}$	phasor load impedance of secondary circuit referred to primary circuit of transformer
$\widetilde{Z}_{\mathrm{total}}$	phasor total impedance
Z_{out}	output impedance
Z_{in}	input impedance
ZCS	electronic zero-current switch
zigzag	zigzag transformer (a special type of three-phase transformer)
$\widetilde{Z}_{\mathrm{load}}$	phasor value of load impedance
ZnO	zinc oxide

About the Companion Website

Do not forget to visit the companion website for this book:

www.wiley.com/go/fuchs/renewableenergyandelectricalengineering

There you will find valuable material designed to enhance your learning, including:

- Learning outcomes for all chapters
- Color version of figures
- Instructor Manual (309 pages) contains solutions to Problems (157) listed at the end of chapters

Scan this QR code to visit the companion website

1

Basic Concepts

To make reading this chapter easier, each abbreviation will be defined when used for the first time, for example, "internal combustion engines (ICE)." In addition, there is a list of abbreviations and symbols listed in the *Glossary of Symbols, Abbreviations, and Acronyms* section.

This first chapter provides an overview of basic principles related to electrical engineering as applied to renewable energy. We intend it to be useful for reference throughout your perusal of later chapters. First we introduce basic energy conversion processes, such as converting heat or chemical energy to either mechanical or electrical energy, covering Carnot efficiency, Rankine, Brayton, and Ericsson cycles, conventional turbines, and internal combustion engines (ICE). Next we discuss renewable generation technologies such as hydro turbines, heat pumps, wind turbines, thermal solar, and photovoltaics (PV) and consider the associated capacity factors. We then turn to electrical engineering topics using Coulomb's law as an entry point to electric charge, one of the fundamental dimensions of physics. Here we present the properties of conductors, insulators, and semiconductors; the relationships between current, voltage, power, and work; advantages and disadvantages of alternating current (AC) and direct current (DC) distribution and transmission systems; ideal passive electric circuits; and galvanic elements, batteries, electrolysis, and fuel cells. Finally, we demonstrate the application of these principles through the solution of examples and refer to helpful software available in the public domain. At the end we close with a set of problems, which can serve as homework assignments. Appendix 1.A lists the design information of a PV power plant for a single-family house, Appendix 1.B closely describes the nature of electricity and its manufacturing, and Appendix 1.C explains the cost of electricity in a renewable energy system.

Introduction to Energy, Renewable Energy, and Electrical Engineering: Essentials for Engineering Science (STEM) Professionals and Students, First Edition. Ewald F. Fuchs and Heidi A. Fuchs.

Companion website: www.wiley.com/go/fuchs/renewableenergyandelectricalengineering

1.1 Energy Conservation: Laws of Thermodynamics

Of primary importance for the survival of society as we know it is the energy propelling our appliances and machinery in residential, commercial, and industrial settings, as well as drive/propulsion systems for vehicles on farmlands, roads, rails, sea lanes, and in the air. The Industrial Revolution was the transition to new manufacturing processes from about 1760 to 1830, which still continues today. Energy conversion is the engineering technology that deals with the production of electrical, mechanical, and/or thermal power for industrial processes and for heating from other available energy sources. The first law of thermodynamics states that energy can neither be created nor destroyed: only its form (e.g. heat, mechanical, electric) can change. The ability to convert energy to its different forms is limited, as described by the second law of thermodynamics.

1.2 Converting Heat to Mechanical Power

The efficiency of converting heat to either mechanical (e.g. turbine) or electrical (e.g. thermocouple) energy – or work, as it is called by physicists – is strongly influenced by the temperature of the heat involved in the conversion process. Figure 1.1 illustrates the block diagram and conversion efficiencies (Eq. 1.1a–e) of a combined (thermal) heat and power (CHP) plant that generates electric power and

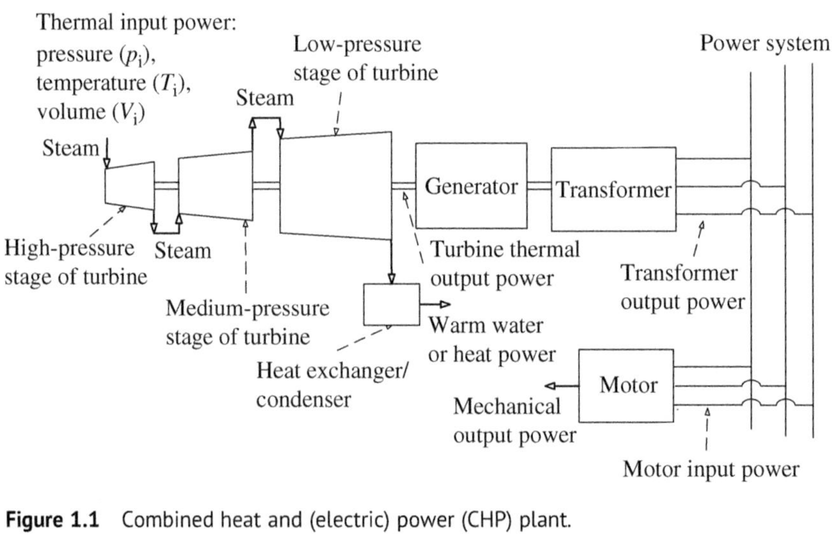

Figure 1.1 Combined heat and (electric) power (CHP) plant.

warm water for district heating, as well as supplies – through a generator and transformer – a power system with a connected motor, resulting in mechanical output power.

In so doing, it is important to minimize the waste heat (e.g. heat loss of turbines, or losses in generators, transformers, and motors) during the conversion process. Note the efficiencies of such power generation are defined by the power efficiencies η^{i}_{power} (Eq. 1.1a–e), where i stands for the different type of application, such as turbine, generator, transformer, system, and motor:

$$\eta^{\text{turbine}}_{\text{power}} = \frac{\text{turbine output power}}{\text{steam input power}} \approx (25-30)\%, \tag{1.1a}$$

$$\eta^{\text{generator}}_{\text{power}} = \frac{\text{generator output power}}{\text{turbine output power}} \approx (96-97)\%, \tag{1.1b}$$

$$\eta^{\text{transformer}}_{\text{power}} = \frac{\text{transformer output power}}{\text{generator output power}} \approx (98-99)\%, \tag{1.1c}$$

$$\eta^{\text{system}}_{\text{power}} = \frac{\text{motor input power}}{\text{transformer output power}} \approx (98-99)\%, \tag{1.1d}$$

$$\eta^{\text{motor}}_{\text{power}} = \frac{\text{mechanical output power}}{\text{motor input power}} \approx (95-97)\%. \tag{1.1e}$$

Steam turbines, in which the blade's velocity scales with the velocity of the steam passing over it, can have more than one stage that extracts energy from the steam; such compounding allows higher efficiencies. If three-stage steam turbines are operated in series with a one-stage gas turbine, one obtains a gas and steam (electric) power (GSP) plant with an overall power efficiency of about 60%. In recently constructed power plants, the waste heat is utilized in so-called combined heat, cooling, and (electric) power (CHCP) plants; see Figure 1.1 for CHP. The combination of heat, cooling, and electric generation increases total overall power efficiency up to 90%. Gas turbines have greater power efficiency than steam turbines, as will be shown later.

Energy efficiencies, η_{energy}, are mostly smaller than the rated power efficiencies, η_{power}, due to low-power operation periods where the power efficiency is relatively small. In most cases, the nameplate of an apparatus gives the power efficiency and sometimes, when the operating time is important, also lists energy efficiency.

The generation of electricity via solar PV cells is governed by the solar cell energy conversion efficiency, $\eta_{\text{solar\_cell}}$, the ratio of the maximum electric output power of a solar cell to the incident power of sunlight at a maximum insolation of $E = 1\ \text{kW/m}^2$ as it occurs on Earth. This conversion efficiency is the percentage of the insolation power to which the cell is exposed that is converted into electric energy [1]. This is calculated by dividing a cell's power output (in watts) at its maximum power point (P_m) by the input light power per unit area (E, in W/m^2) and the surface area of the solar cell (A_c, in m^2):

$$\eta_{solar\_cell} = \frac{P_m}{E \cdot A_c} \approx 18 - 44\%. \tag{1.2}$$

Under laboratory conditions, researchers have achieved solar cell efficiencies of more than $\eta_{solar\_cell} = 44\%$. For commercially available photovoltaic plants, the overall efficiency [2] (including PV array, maximum power tracker, DC–DC boost converter, and DC–AC inverter, plus transformer) is

$$\eta_{PV\_plant} \approx 12 - 16\%. \tag{1.3}$$

The efficiency of average commercial wafer-based silicon modules decreases in real-world application about 0.50% per year [3, 4] through material aging and also depends upon temperature: increasing temperature decreases the efficiency of most conducting materials [5].

1.2.1 Carnot Cycle, Carnot Machines, and Carnot Efficiency

Nicolas Léonard Sadi Carnot (1796–1832), a French military engineer and physicist, established theoretical principles of thermodynamics as applied to heat engines, systems that convert heat or thermal energy to mechanical energy, such as gas, steam, and ICE. In his 1824 published monograph, *Reflections on the Motive Power of Fire and on Machines Fitted to Develop that Power*, Carnot gave the first successful theory of heat engine maximum efficiency [6]. Rudolf Clausius and Lord Kelvin then formalized the second law of thermodynamics and defined the concept of entropy using Carnot's theory.

Figure 1.1 gives an example of a heat engine, representing a steam turbine power plant with high-, medium-, and low-pressure turbines. The Carnot cycle converts heat energy into useful mechanical work. Figure 1.2 depicts the cycle of a periodic machine undergoing repetitive states. A Carnot machine is an ideal heat engine consisting of a cylinder filled with ideal gas where a piston can be displaced, as illustrated in Figure 1.3. Note that "ideal" means there are no losses. The ideal gas has a certain pressure p, temperature T, and volume V. p is measured in (kg-force)/m^2, T is measured in kelvins, and V is measured in m^3. During one cycle the cylinder must have opposing (contradicting) properties: in Figure 1.3 the thermal conductivity, б, between cylinder and piston is infinite during expansion and zero during compression. Of course, such contradicting conductivity properties cannot be maintained in real-world engines. Therefore, the efficiency of a Carnot

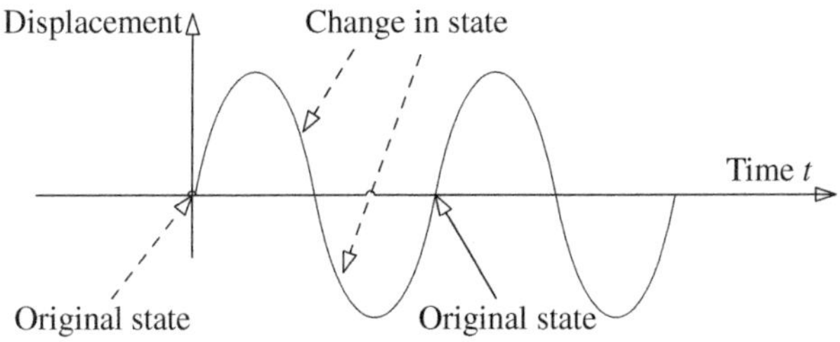

Figure 1.2 Definition of a (Carnot) cycle. The vertical y-axis represents the periodic displacement of the piston, and the horizontal x-axis denotes the time t during which the periodic displacement takes place.

Figure 1.3 Carnot (piston) machine.

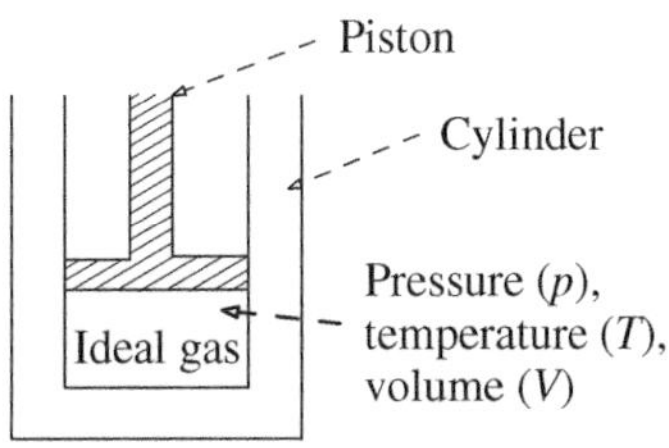

cycle is much less than 100%: in the range of 30% for combustion and steam engines. Figure 1.4a illustrates the (pTV) diagram of Carnot cycle [6, 7].

Due to these idealized assumptions and Figure 1.4a, the maximum theoretical efficiency, or Carnot efficiency [8],

$$\eta_{\text{Carnot}} = \frac{T_{\text{i}} - T_{\text{f}}}{T_{\text{i}}} = \frac{T_1 - T_2}{T_1} \approx 64\% \tag{1.4}$$

can never be reached – and neglects any compressor action required to increase the pressure of the medium used, such as steam or gas. Here, $T_1 = T_{\text{i}}$ and $T_2 = T_{\text{f}}$ are the input (i) and final (f) output temperatures, respectively. For the adiabatic process defined in Figure 1.4a, available software programs can determine the Carnot [9] efficiency. For all practical applications, the Rankine, Brayton, and Ericsson cycles take into account the losses associated with compression of the steam or gas. Usually the (Ts) diagram is used instead of the (pTV) diagram. The (Ts) diagram is a function of the entropy $s \equiv \left(\frac{p}{p_{\text{t1}}}\right)^{(\gamma-1)/\gamma}$, a pressure parameter, where the specific heat ratio $\gamma = C_{\text{p}}/C_{\text{v}}$ and p, p_{t1} are pressure values. Subscript "t" stands for "tube" representing a nozzle. Figure 1.4b illustrates the work produced by a heat engine consisting of a compressor, combustor, and turbine.

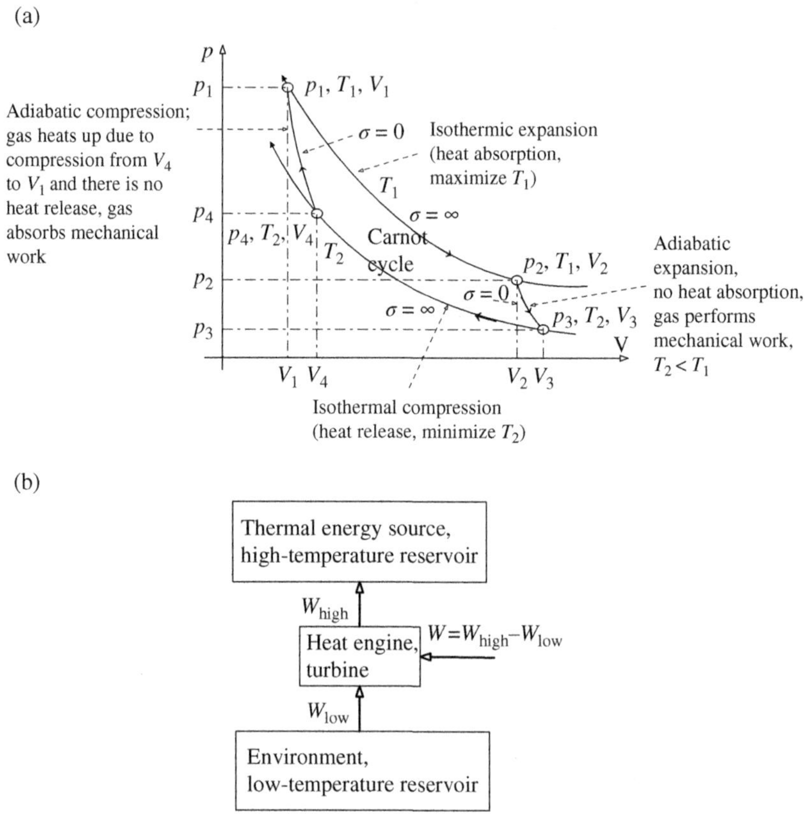

Figure 1.4 (a) Summary of (pTV) diagram of Carnot cycle [7]. (b) Heat flow in a heat engine or turbine producing work $W = W_{high} - W_{low}$. The term environment denotes the low states of T, p, and V at the output of the heat engine or turbine.

Application Example 1.1 Adiabatic process and Carnot efficiency

Calculate the performance of a steam power plant with a generator output of P_{gen} = 1.5 MW at a generator efficiency of $\eta_{gen} = 95\%$. The overall efficiency of the steam turbine, including that of the generator and the pump to apply input pressure, is $\eta_{electric} = 18\%$ at an input temperature of $T_i = 246\,°C = 520\,K$. The efficiency of the pump is $\eta_{pump} = 70\%$. Determine input (p_i) and output (p_f) pressures of the steam turbine and the associated steam output temperature of the turbine.

Solution

The rated output power of the steam turbine is $P_{turbine} = P_{gen}/(\eta_{gen} \cdot \eta_{pump}) = 1.5\,MW/(0.95 \cdot 0.7) = 2.26\,MW$ and the Carnot efficiency is $\eta_{Carnot} = \frac{T_i - T_f}{T_i} = \frac{246 - T_f}{246} = \frac{\eta_{electric}}{\eta_{gen} \cdot \eta_{pump}} = \frac{18\%}{0.95 \cdot 0.7} = 27.1\%$, resulting in $T_f = 179.33\,°C = 452.48\,K$.

Using the software [9], one obtains the results in Figure E1.1.1. For the adiabatic process of the steam turbine, the calculation yields $p_\mathrm{i} = 1.68 \cdot 10^9$ Pa and $p_\mathrm{f} = 1.20 \cdot 10^9$ Pa, resulting in the work of $W = 0.5685 \cdot 10^{10}$ J, which corresponds to the output power $P = W/3600\ \mathrm{s/h} = 1.579$ MW.

Adiabatic process

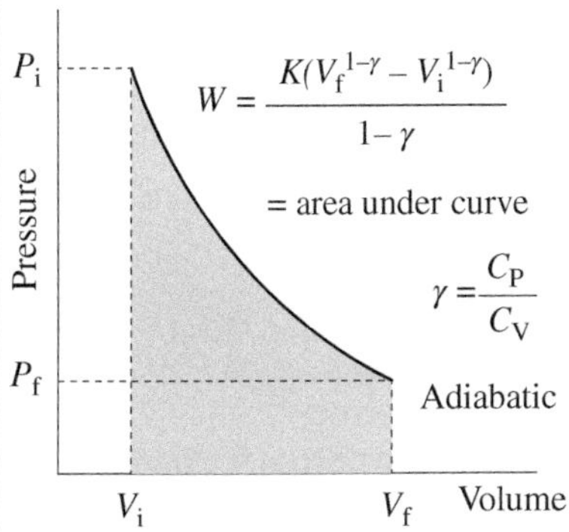

For an ideal gas consisting of $n = 6\,990\,000$ moles of gas, consider an adiabatic process which involves expansion from

$V_\mathrm{i} = 18\ \mathbf{m^3}$

to = $V_\mathrm{f} = 22\ \mathbf{m^3}$

at initial temperature $T_\mathrm{i} = 520.735\,7043$ K

Note: If initial volume or temperature are changed, a new value of pressure will be calculated to correspond to the number of moles of gas above.

With the initial volume and temperature specified, the initial pressure is determined by the ideal gas law:

Using $P = \dfrac{nRT}{V}$, $P_\mathrm{i} = \ldots$ kPa $= 1.681\,352\,162 \times 10^9$ Pa

(Note: If the initial pressure value is changed, then a new value of moles will be calculated to be consistent with the initial volume and temperature specified above. If you wish to restrict the number of moles to your supplied value, then enter that value plus values for initial volume and temperature - the required pressure will be calculated.)

The detailed behavior of the pressure and volume depends upon the specific heats of the gas:

Constant pressure $C_\mathrm{P} = 20.814\,51$ J/mol K

$$R = C_\mathrm{P} - C_\mathrm{V}$$

Constant volume $C_\mathrm{V} = 12.5$ J/mol K

The specific heats determine the ratio, $\gamma = \dfrac{C_\mathrm{P}}{C_\mathrm{V}} = 1.665\,160\,799$

Figure E1.1.1 Adiabatic calculations for the determination of input and output pressures of steam turbine p_i and p_f, respectively.

Note that the value of γ is determined by the gases, the most common values being $\gamma_{air} = 1.4$ and $\gamma_{monoatomic\ gas} = 1.66$.

The adiabatic condition can be applied to determine the constant K.

$PV^{\gamma} = K = 2\,069\,613\,533$

The work done by the gas can then be determined. If the final volume is less than the initial, then work is done on the gas and the work will be negative.

$$W = \frac{K(V_f^{1-\gamma} - V_i^{1-\gamma})}{1-\gamma} = \cdots J = 0.568\,529\,57' \times 10^{10}\,\text{J}$$

The final pressure for the process can be determined from the adiabatic condition:

$P_f = \cdots\ kPa = 1.203\,759\,230 \times 10^9$ Pa

and the final temperature can be obtained from the ideal gas law.

$T_f = 455.667\,942\,2$ K

The specified initial temperature for the work calculation must be in Kelvins, so it must be coverted from the other temperature scales.

$T_i = 520.735\,704\,3\ \text{K} = 247.585\,704\,3\ °\text{C} = 477.654\,267\,8\ °\text{F}$

Figure E1.1.1 (Continued)

1.2.2 Rankine Cycle

The Rankine cycle [10], the namesake of Scottish engineer William Rankine, governs steam and oil turbines. Its efficiency (Figure 1.5a) is less than that of the ideal Carnot cycle, which generates an ideal unobtainable high-efficiency value. Figure 1.5b shows the corresponding (Ts) diagram of the Rankine cycle. The expander work $(h_4 - h_5)$ is reduced by the pump work $(h_2 - h_0)$, while the heat input is $(h_3 - h_2)$. The efficiency of the cycle is (see Figure 1.5b)

$$\eta_{\text{Rankine}} = \frac{(h_4 - h_5) - (h_2 - h_0)}{(h_3 - h_2)} = \phi \cdot \eta_{\text{Carnot}} = \phi \frac{(T_i - T_f)}{T_i}. \tag{1.5}$$

The efficiency of the Rankine cycle is given by the fraction ϕ as compared with the Carnot cycle. When the working fluid is water steam, $\phi \approx 0.5$; thus, when Carnot efficiency is 64%, the Rankine efficiency is about 30%.

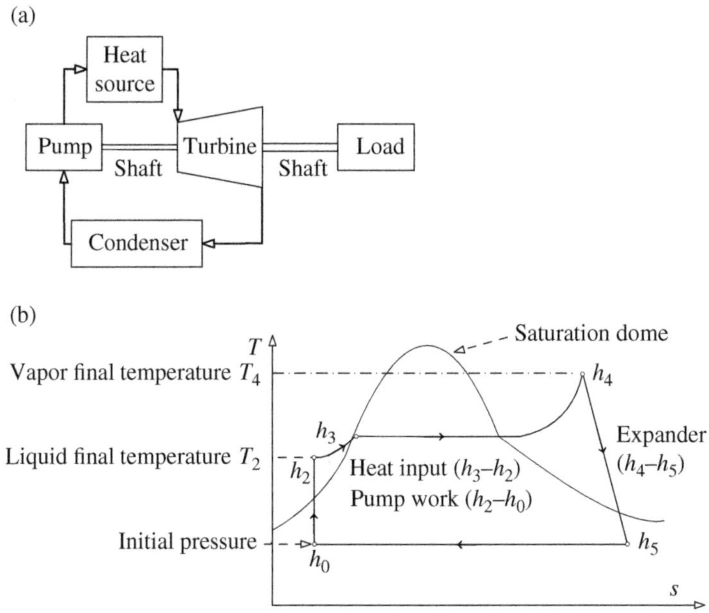

Figure 1.5 (a) Rankine cycle. (b) (*Ts*) diagram of Rankine cycle. Below the saturation dome, which represents the continuous (*Ts*) function, a mixture of liquid and vapor exists. On the left-hand side of the dome, the liquid phase prevails, and on the right-hand side, the vapor phase does.

1.2.3 Brayton Cycle

Turning from steam turbines, Figure 1.6a shows a block diagram of the Brayton cycle [10, 11], which applies mainly to gas turbines and was named after American inventor George Brayton. The (*pV*) diagram of the Brayton cycle is depicted in Figure 1.6b; alternatively, the (*Ts*) diagram of the Brayton cycle is shown in Figure 1.6c.

The theoretical (th) efficiency of the Brayton cycle is [11]

$$\eta_{\text{th,Brayton}} = 1 - \left(\frac{T_1}{T_2}\right) \cdot \left(\frac{\frac{T_4}{T_1} - 1}{\frac{T_3}{T_2} - 1}\right) \tag{1.6}$$

1.2.4 Ericsson Cycle

An alternative approach to the Brayton cycle is the Ericsson cycle (after Swedish-American engineer John Ericsson) [12], which makes use of isothermal compression and expansion for applications such as magnetohydrodynamic (MHD) power

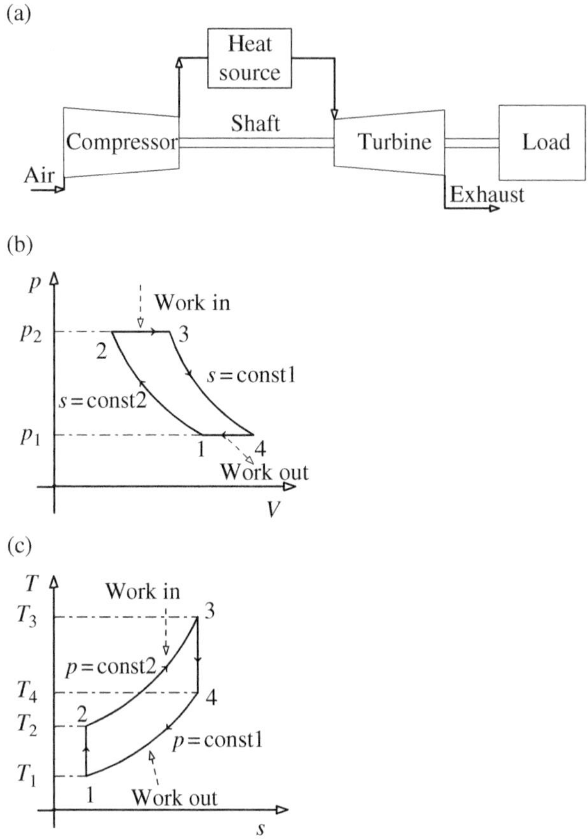

Figure 1.6 (a) Brayton cycle. (b) (pV) diagram of Brayton cycle, where s is parameter. (c) (Ts) diagram of Brayton cycle, where p is parameter.

plants. MHD plants operate without moving parts at high temperatures using conductive plasma for the conductor that is moved through a magnetic field to create electric current. Figure 1.7a illustrates the (pV) diagram and Figure 1.7b the (Ts) diagram.

The theoretical (th) efficiency of the Ericsson cycle is

$$\eta_{\text{th,Ericsson}} = 1 - (T_{\text{L}}/T_{\text{H}}). \tag{1.7}$$

1.2.5 Internal Combustion Engines

ICE can be analyzed based on the Diesel and Otto cycles, which have characteristics similar to those mentioned earlier. Figure 1.8a compares [10] diesel, gas, and steam turbine efficiencies. Diesel engines are very efficient and represent a

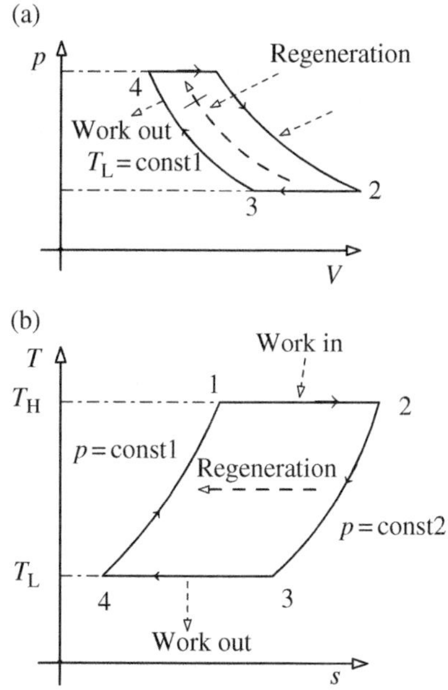

Figure 1.7 (a) (*pV*) diagram of Ericsson cycle, where *T* is parameter. (b) (*Ts*) diagram of Ericsson cycle where, *p* is parameter.

formidable competitor for electric car drives, though both have differing advantages (e.g. high efficiency of electric car and of diesel engines) and drawbacks (e.g. particulate matter, CO_2, NO_x for diesel engines) when considered on the systems level.

Breathing particulate matter can damage the lungs: Figure 1.8b illustrates the clean particulate matter filter before (top) and after (bottom) it is used for 336 hours = (7 days/week)·(6 hours/day)·(8 weeks) by a human in a large city of about 1.5 million people. This shows that the particulate matter represents a serious health problem and ICE and other sources (e.g. heating) should be replaced by electric car drives and heat pumps, respectively, whose fuel stems from renewable energy sources.

Figure 1.8c illustrates the NO_2 pollution in several larger German cities as of 2017. For example in München (Munich), the NO_2 content of air in the inner city was 78 μg/m^3 in 2017, reduced by 2 μg/m^3 since 2016. The maximum value as permitted by the European Union (EU) is 40 μg/m^3. The Umweltbundesamt (Federal Environmental Agency) has recently achieved a German court ruling where such high values are deemed illegal, and cities are entitled to ban diesel automobiles with high NO_x in the exhaust fumes. One remedy is NO_x pollution filters employing AdBlue, which is a NO_x reduction agent consisting of 32.5% urea (CH_4N_2O) and 67.5% deionized water. It is standardized as AUS 32 in ISO 22241.

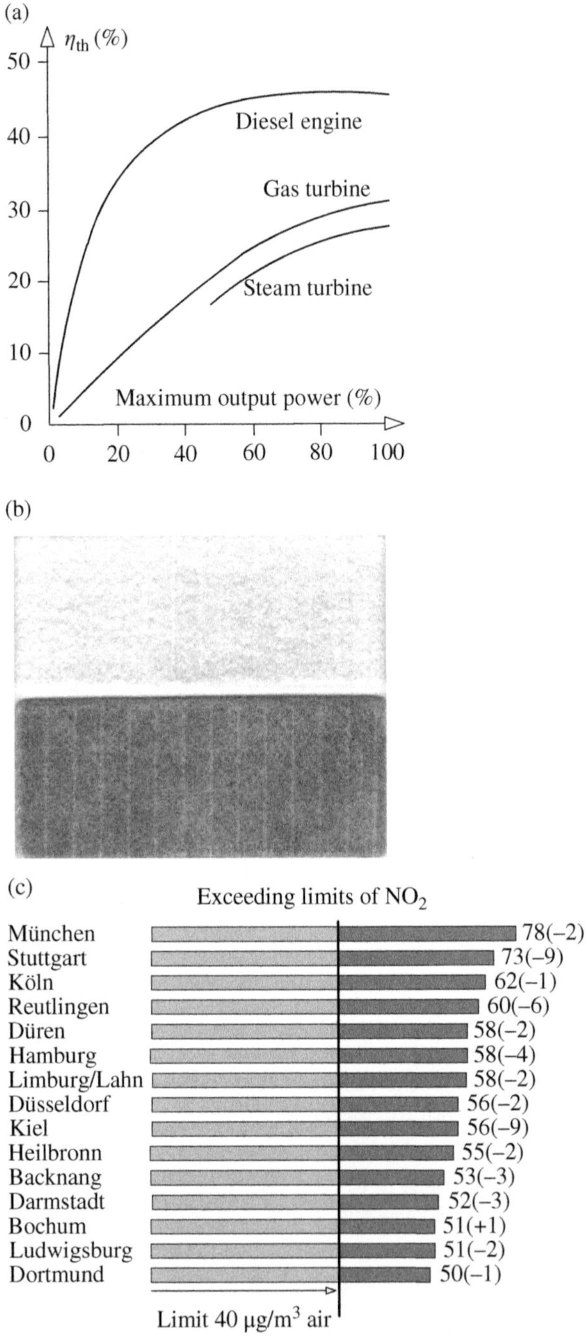

Figure 1.8 (a) Theoretical (th) efficiency characteristics η_{th} of diesel engine, gas, and steam turbines as a function of output power [10]. (b) Particulate matter filter of a compressor used for assisted breathing before (top) and after (bottom). (c) NO_2 generation from diesel automobiles in large German cities in 2017; the shown values have the unit $\mu g/m^3$ air. *Source:* https://www.umweltbundesamt.de/presse/pressemitteilungen/luftqualitaet-2017-rueckgang-der.

1.2.6 Steam, Gas, and Oil Turbines

The Rankine and Brayton cycles govern steam and gas turbines, respectively. Organic Rankine cycle (ORC) technology enables solutions with well-proven components for heat sources with low temperatures (e.g. 800 °C). The ORC module can generate power output from 300 kW to 2 MW [13]. The working medium is a chlorine-free, nontoxic substance with zero ozone depletion potential, for example, thermo-oil as used in Application Example 1.2.

Waste heat with a temperature of 300 °C is just enough to drive the ORC module and can be used as demonstrated in Application Example 1.2 [14, 15]. Heat recovery will increase the efficiency and thus decrease the costs of fuel and energy consumption required for this process.

Application Example 1.2 Combined heat and electric power plant (CHP) [14, 15]

In *1995* a cooperative CHP fueled by timber and wood chips arose in Dobbiaco–San Candido in the South Tyrol region of Italy. Today it has 850 members. The plant supplies 1200 customers with 55 000 MWh district heat with a total pipe length of 90 km and 10 000 MWh of electric energy to 3500 residences. There are three furnaces and boilers with a rated thermal power of $P_{thermal} = 18$ MW and an electrical output power $P_{electric} = 1.5$ MWe. A turbine based on thermo-oil is illustrated in Figure E1.2.1.

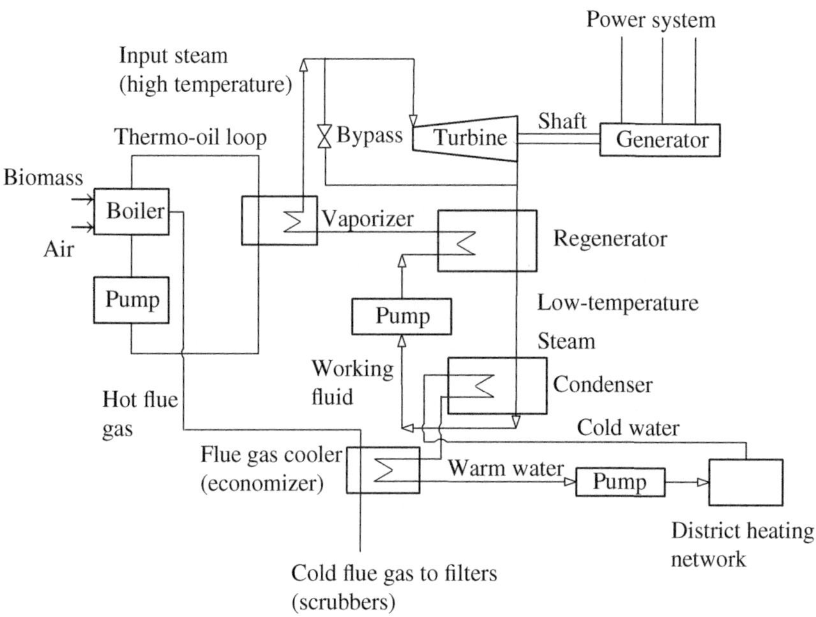

Figure E1.2.1 Components of the CHP Dobbiaco–San Candido, Italy.

Technical data: The residents of Dobbiaco and San Candido save €4 million annually compared to heating oil consumption. Forty-two percent of South Tyrol is covered with forest, an important economic factor that is a source of clean, sustainable energy – assuming sustainable harvesting methods – as the raw material for the CHP is wood. Nine hundred degree celsius is the ideal temperature for burning wood chips. Multi-cyclones, electrical filters, and flue gas condensation processes remove dust/particles from the flue gases.

Description of the heat and power production system: The heat generated by a biomass boiler is transmitted to an ORC process via a thermo-oil cycle. Thermo-oil is used as a heat transport medium for the ORC process, which enables simultaneous operation of a boiler and an associated thermo-oil cycle. The principle of the electricity generation corresponds to that of a conventional water-steam process, except that instead of water, it employs an organic working fluid. For the ORC, the organic working fluid is vaporized through a heat exchanger with the thermo-oil in an evaporator (hot side of the ORC) and thereafter expanded in a turbine. The turbine drives a generator from which electricity is produced.

The temperature of the steam of the organic working fluid leaving the turbine is decreased in a so-called regenerator – through internal heat exchange – so that the temperature of the liquid working fluid is decreased to the point where it has not yet condensed. A subsequently connected condenser further cools the organic working fluid, and it completely condenses via heat exchange with the water of the process and heat system (cold side of the ORC). The total heat from the condenser is available for the residential, commercial, and industrial heat supply systems of the communities of Dobbiaco and San Candido.

The condensed working fluid is pumped to obtain enough pressure necessary for steam turbine operation after the working fluid passes through the regenerator in which – through heat exchange with the thermo-oil – the working fluid will be preheated (internal heat recovery). In an economizer, the preheated working fluid temperature is further increased through heat exchange with the condenser's thermo-oil (hot side of ORC) before it again enters the evaporator and the cycle is thus completed. The electrical efficiency is about 18%, which means the remaining energy is utilized in the district heating network.

Summary: Not much energy is lost and no cooling towers are needed, since the flue gas leaves the plant at about (30–40) °C. The boiler power is 10 MW at a furnace temperature of 950 °C, the thermo-oil temperature is (250–300) °C, the overall efficiency is near 100%, and the flue gas temperature before cooling is (150–200) °C.

Table 1.1 Energy content of various fuels usable for transportation per gallon (gal), pound (lb-force), kilowatt hours (kWh), and British thermal unit (Btu).

	Gasoline/ E10	Low-sulfur diesel	Biodiesel	Propane (LPG)	Liquefied natural gas (LNG)	Methanol	Hydrogen	Electricity
Energy content	120 388–124 340 Btu/gal	138 490 Btu/gal	127 960 Btu/gal for B100	91 420 Btu/gal	23 726 Btu/(lb-force)	65 200 Btu/gal	61 013 Btu/(lb-force)	3412 Btu/kWh

1.2.7 Energy Content of Common Fuels (e.g. Gasoline, Diesel, Methanol, Hydrogen)

Table 1.1 illustrates the energy content of various fuels [16] used now or in future transportation applications. As expected, diesel has the largest energy content per unit of weight, while electricity as stored in batteries and capacitors has the lowest. Note that 1 kWh corresponds to 3412.142 Btu.

1.3 Heat Pumps and Air-Conditioning Units

Both heat pumps and air conditioners are based on the same concept: in the case of a heat pump, the conversion of heat from a low-temperature to a high-temperature reservoir occurs, and in the case of an air-conditioning unit or refrigerator, the inverse conversion applies. Figure 1.9a illustrates the schematic of a heat pump, where the work $W = W_{\text{high}} - W_{\text{low}}$ is the thermal input to the high-temperature reservoir and W_{high} and W_{low} are the heat values of the high- and low-temperature reservoirs, respectively.

Figure 1.9b shows the detailed operational principle of the heat pump [17] consisting of electrical drive energy; a heat source of air, ground, or groundwater; and a heat pump consisting of compressor and pressure reducer (expansion component) applied to an environmentally benign coolant such as R410A, a greenhouse gas approved by the Kyoto Protocol. However, with a global warming potential of GWP = 1925,[1] this coolant must not be released into the atmosphere and must remain in the evaporator and condenser. Either air, ground, or groundwater heat

1 GWP compares the amount of heat the greenhouse gas in question traps in the atmosphere relative to an equivalent mass of carbon dioxide.

(a)

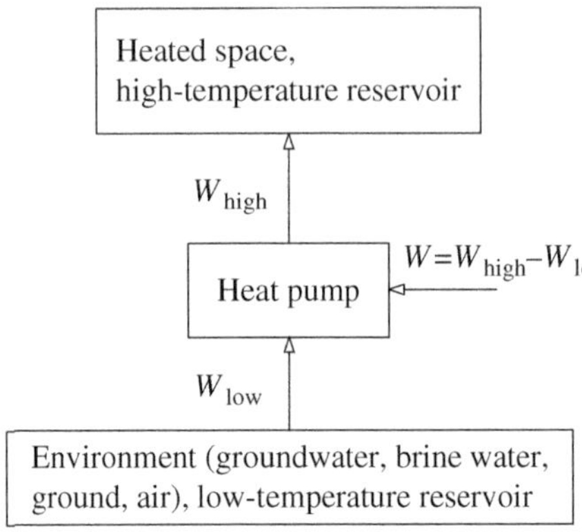

(b)

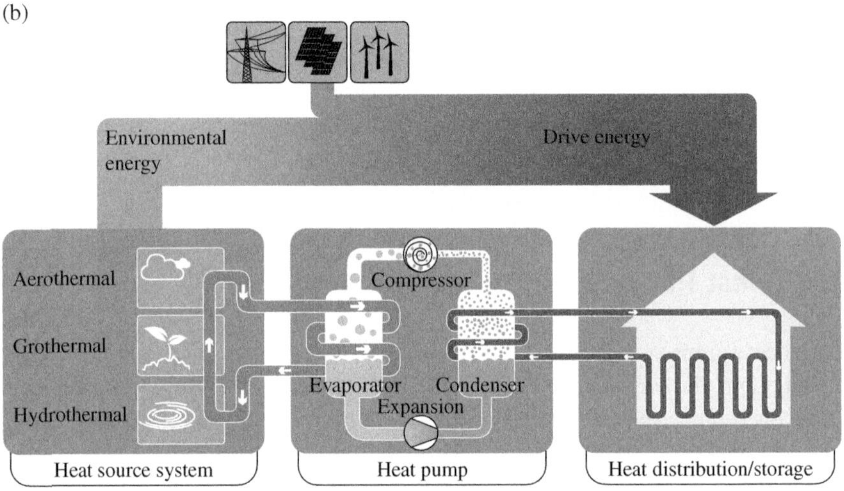

Figure 1.9 (a) Schematic of heat pump. (b) Operating principle of a heat pump [17]: the circuit of a heat pump works like a refrigerator, but inversely. (c) Air–water heat pump [17]. (d) Brine–water heat pump [17]. (e) Vertical geothermal collector heat pump [17]. (f) Groundwater–water heat pump [17] with suction well on the left (warm water) and injection well on the right (cold water). (g) Photograph of installed groundwater–water heat pump for 220 m^2 residence: heat pump (1), water reservoir for floor heating (2) at $T_{heating} = 45\,°C$, water reservoir for warmwater applications (3) at $T_{warm\text{-}water} = 55\,°C$ where the water temperature is increased from 45 to 55 °C via electric rheostats, one of the inverters of PV system (4), and (h) its hydraulic schematic [18]. The *y*-axes represent (i) heating output power (kW), (j) electric power input (kW), (k) coefficient of performance (COP) in per unit (–). The parameter identifications "1," "2," "3," and "4" correspond to the warmwater temperatures of 35, 45, 55, and 60 °C, respectively [18]. The *x*-axes represent the temperature of groundwater in °C entering the heat pump, and (ℓ) Efficiency improvement of fireplace [19] efficiency through heat exchanger/storage and heat pump. *Source*: Bundesverband Wärmepumpe e.V., Hauptstraße 3, 10827 Berlin.

(c)

(d)

Figure 1.9 (Continued)

(e)

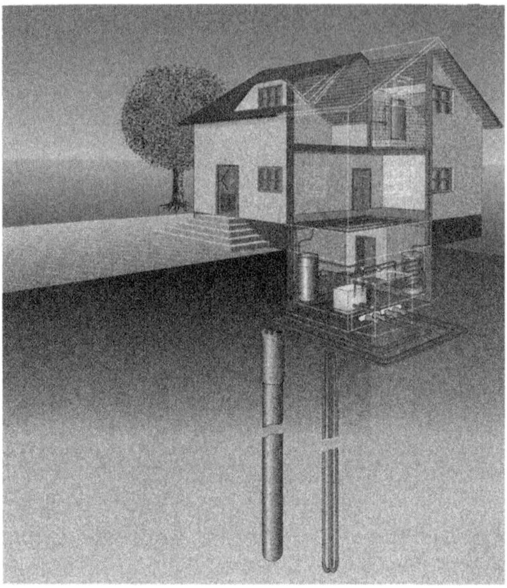

(f)

A water–water heat pump utilizes the heat of the groundwater: a delivery well (source of groundwater) and a swallow well (sink of groundwater) are required, where groundwater is available in sufficient amounts and quality

Figure 1.9 (Continued)

(g)

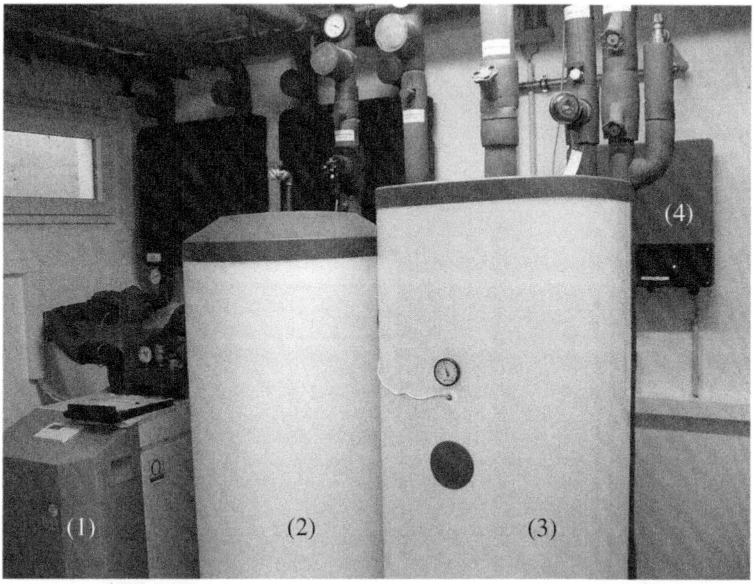

(h)

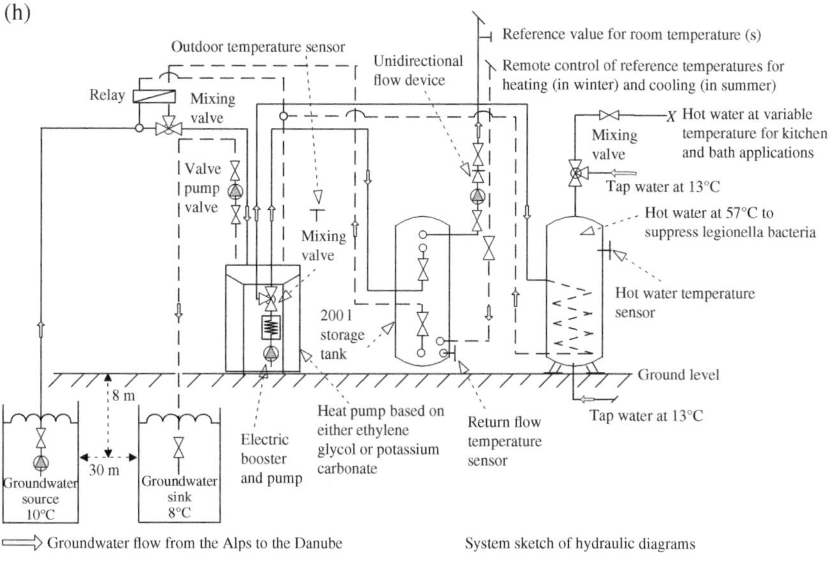

Figure 1.9 (Continued)

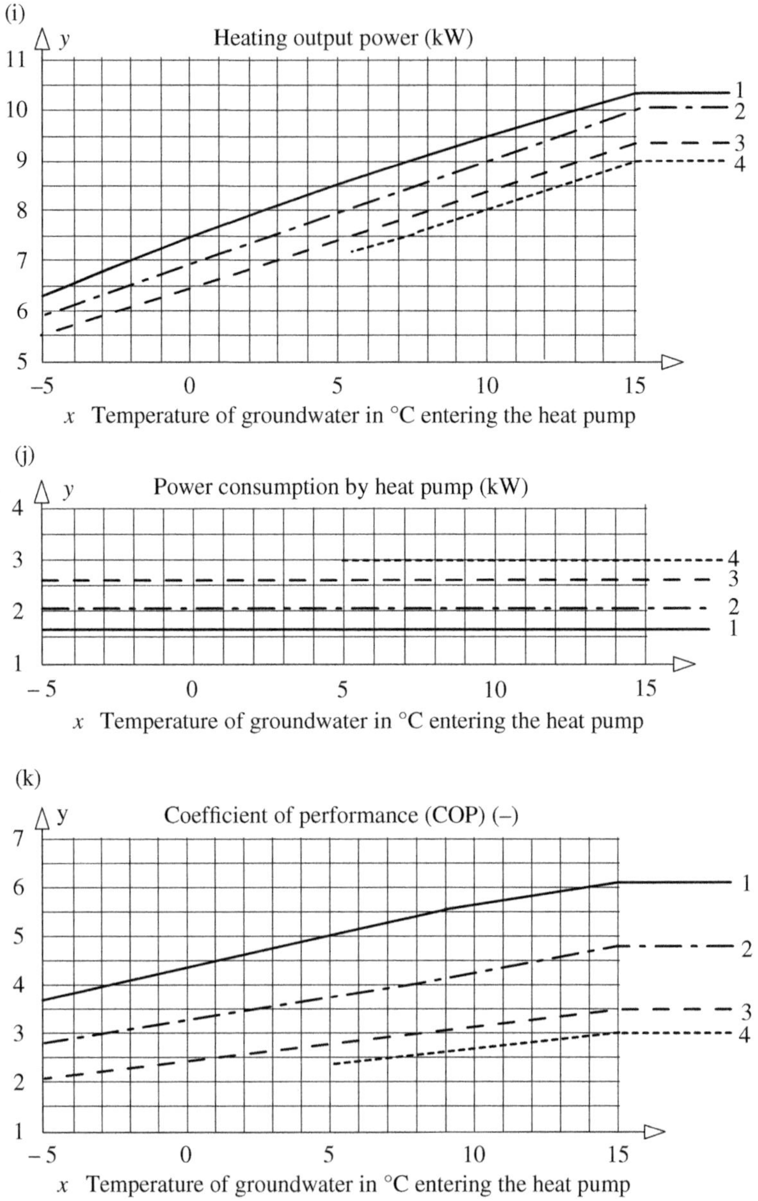

Figure 1.9 (Continued)

(1)

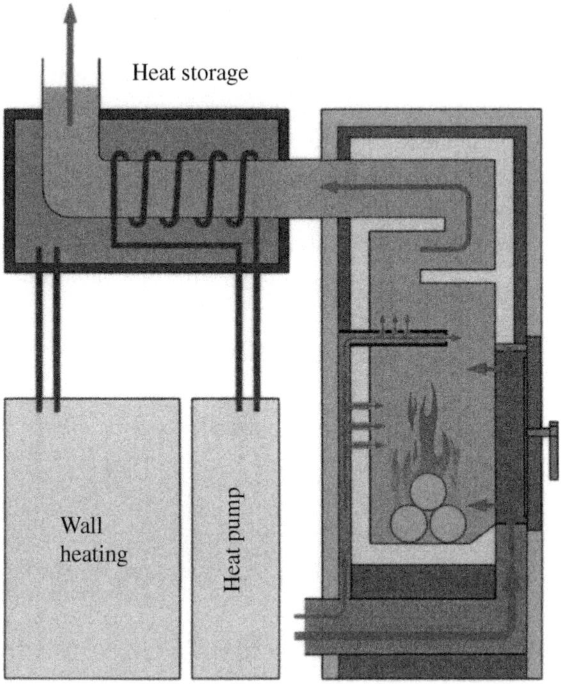

Figure 1.9 (Continued)

pumps can be employed to heat residences. While air-heat pumps are useful down to outside temperatures of a few negative degrees (e.g. −5 °C), the groundwater heat pump works down to an outside temperature of −20 °C due to the groundwater temperature of (8–10) °C during the entire year. The coolant R410A has a low boiling temperature (e.g. below 8 °C) and becomes gaseous in the evaporator.

1.3.1 Heating Cycle Of Heat Pump

By compressing the evaporated coolant, its temperature increases (e.g. from $T_{groundwater} = 8$ °C to $T_{heating} = 45$ °C). Thereafter, in the condenser, the hot evaporated coolant becomes fluid, which transfers the electrical drive energy and the absorbed environmental energy – which includes the energy gained due to the heat pump process, as specified by the coefficient of performance (COP) – to the central heat distribution system. Through expansion via the pressure reducer, the coolant reduces its temperature again and changes from a fluid to vapor due to its low boiling point. This cycle is repeated when the fluid coolant's temperature increases during the groundwater loop and becomes vaporized again. Figure 1.9c–f illustrate the application of air–water, brine–water, vertical geothermal heat collector, and groundwater–water (such that the operation mode can be switched between summer and winter operation mode) heat pumps [17] installed in residences, respectively. The assistance of A. Vogel (private communication, September 2017) for modifying these illustrations is appreciated.

Figure 1.9g depicts the installation components of an 8 kW groundwater–water heat pump, and Figure 1.9h illustrates its hydraulic schematic [18]. For passive cooling, the 8 °C groundwater bypasses the heat pump and flows through a heat exchanger supplying the piping system embedded in floors with cold water, limiting the residence's temperature to <24 °C.

The above heat pump has the characteristics for heating power output with an electric power input; the coefficient of performance is defined as COP = (power output)/(power input) and is illustrated in Figure 1.9i–k. From these diagrams we can conclude that for a groundwater temperature of 8 °C, at a warmwater temperature $T_{\text{heating}} = 45\,°\text{C}$, the heating output power is $P_{\text{out-heating}} = 8.5\,\text{kW}$. The input electric power is $P_{\text{in-electric}} = 2.2\,\text{kW}$, and $\text{COP} = (8.5\,\text{kW}/2.2\,\text{kW}) \approx 4.0$, resulting in a seasonal energy efficiency ratio (SEER) [20] of

$$\text{SEER} = 3.413 \cdot \text{COP} = 13.65. \tag{1.8a}$$

This is an acceptable value, but it could be increased – at higher cost – to SEER = 19. Figure 1.9i–k depict the heating output power (kW), electric input power (kW), and the coefficient of performance, which is the ratio of the heat delivered (output) divided by the heat equivalent of the energy input measured in per unit or (pu), also denoted by no unit (–). The parameter identifications "1," "2," "3," and "4" correspond to the warmwater temperatures of 35, 45, 55, and 60 °C, respectively.

1.3.2 Combined Heating and Cooling Performance (CHCP) Coefficient of a Residence

The groundwater heat pump of Figure 1.9h–k is installed in a single-family residence (see Application Example 1.6) where the groundwater has a temperature range of 8–10 °C and the heat pump increases the temperature for radiant floor heating to about 30 °C. During the winter season, the heat pump boosts the floor heating to about maximum of 35 °C. Additional electricity boosts the temperature (via rheostats) of the hot water tank to about 56 °C during the entire year. During the summer season, the groundwater is fed via a heat exchanger through the floor piping in passive cooling mode, bypassing the heat pump where electricity is relied on for pumping action, to maintain the room temperature at or below 24 °C maximum. Clearly this combined heating and cooling cannot be adequately described by the well-known coefficient of performance (COP) of a heat pump. The CHCP is defined as the "Jahresarbeitszahl (JAZ)" [21], translated as the annual COP, which does not seem to be defined or used in English technical literature:

$$\text{CHCP} = \text{JAZ} = \frac{\text{heat}/(\text{kWh} \cdot \text{a})}{\text{electricity}/(\text{kWh} \cdot \text{a})}$$

$$= \frac{(\text{heat for heating warm water})/(\text{kWh} \cdot \text{a})}{\frac{\text{electr.energy}_{\text{heat pump}}}{\text{kWh} \cdot \text{a}} + \frac{\text{electr.energy}_{\text{resistive heating}}}{\text{kWh} \cdot \text{a}} + \frac{\text{electr.energy}_{\text{cooling}}}{\text{kWh} \cdot \text{a}}} \tag{1.8b}$$

In the numerator of Eq. (1.8b), the cooling energy is not included because it was not measured, while in the denominator the electric energy feeding the water pump for the cooling water is included.

Clearly CHCP = JAZ < COP as defined in Eq. (1.8a), where the coefficient of performance is

$$\text{COP} = \frac{\text{power output}}{\text{power input}}.$$

Representative single-family residence values for the terms in Eq. (1.8b) (see Application Example 1.6 and Figure E1.6.5) were measured during a roughly four-year period (22 January 2013 to 13 December 2016) as follows:

$$\text{Heat generated by heat pump for floor heating up to about } 30\,^\circ\text{C} = 36.89\ \text{MWh}, \tag{1.8c}$$

$$\text{Heat generated by heat pump for hot water tank up to about } 30\,^\circ\text{C} = 10.81\ \text{MWh}, \tag{1.8d}$$

$$\text{Heat generated by resistive heating for floor heating up to about } 35\,^\circ\text{C} = 0.6051\ \text{MWh}, \tag{1.8e}$$

$$\text{Heat generated by resistive heating for hot water tank up to about } 56\,^\circ\text{C} = 2.334\ \text{MWh}. \tag{1.8f}$$

Adding Eq. (1.8c–f) yields total generated heat energy during the above time frame of roughly 4 years = 50.639 MWh. (1.8g)

To provide this total heat during winter and the cooling during summer, 15.488 MWh of electricity was required during the time frame from 22 January 2013 to 13 December 2016. This results in

$$\text{CHCP} = \text{JAZ} = 50.639\ \text{MWh}/15.488\ \text{MWh} = 3.27, \tag{1.8h}$$

which is clearly less than the well-known coefficient of performance of a heat pump COP = 4.00 as specified above when the heat pump is used for heating only and no additional resistive heating during winter season and cooling during summer season are employed. The above CHCP = 3.27 value is an average value during the roughly four years.

Because the cooling energy is not included in the numerator of Eq. (1.8b) while the electricity for pumping the cooling water is included in the denominator, the actual CHCP = JAZ is somewhat larger than 3.27. However, the electric energy used for cooling (pumping of water) to arrive at a maximum of 24 °C in the entire residence is small compared with the electric energy employed for heating based on heat pump and additional resistive heating to arrive at 35 and 56 °C, respectively. For water-to-water heat pumps, a JAZ of at least 3.8 should be obtained [22]. Equation (1.8h) is on the low side because in the numerator of Eq. (1.8b) the cooling energy is not included (it could not be measured).

Application Example 1.3 Operating point of groundwater to water heat pump for 220 m^2 residence

Determine for a groundwater temperature of $T_{groundwater} = 8$ °C and a warmwater flow temperature of 45 °C the coefficient of performance, the electrical input power of $P_{in\text{-}electric}$, and the heating output power $P_{heating}$.

Solution

For a groundwater temperature of $T_{groundwater} = 8$ °C, one obtains from Figure 1.9k for a warmwater flow temperature of 45 °C the COP = 4.0, resulting with Figure 1.9j in an electrical power consumption of $P_{in\text{-}electric} = 2.1$ kW and with Figure 1.9i in a heating power output of $P_{heating} = 8.5$ kW.

Heat pumps are employed in many other applications aside from whole-home space conditioning. For example, Figure 1.9ℓ shows an inexpensive improvement to the heating efficiency of a fireplace [19].

1.4 Hydro Turbines

Turning to energy generation sources on the more renewable end of the spectrum, we see that different common types of water turbines exist, such as Pelton wheels, Francis turbines, mixed-flow turbines, Kaplan turbines, and bulb-type turbines; they all convert the potential and kinetic energy of water into mechanical work

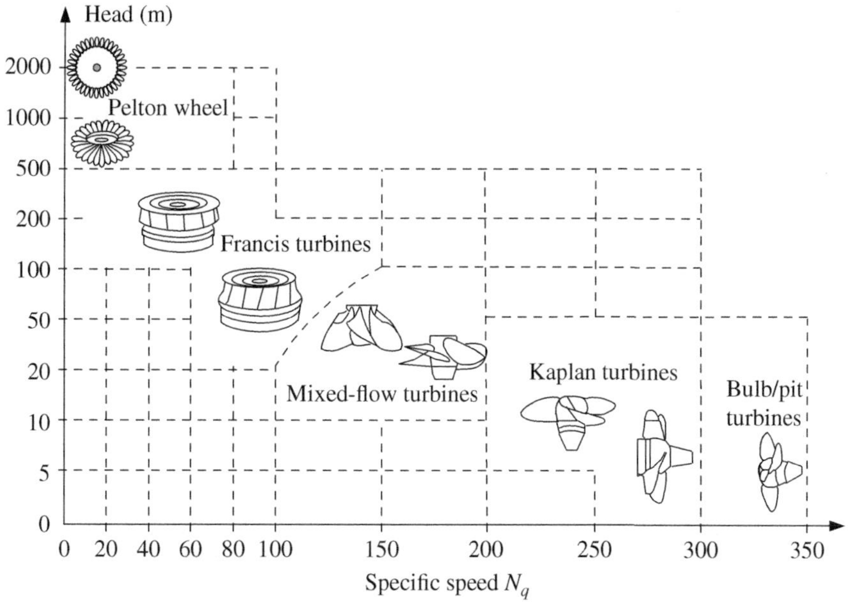

Figure 1.10 Representative types of water turbines [20, 23–25].

[20, 23–25]. Figure 1.10 shows the water head required as a function of the specific speed N_q:

$$N_q = n\frac{Q^{0.5}}{H^{0.75}}, \tag{1.9}$$

where n is the mechanical speed in rpm, Q is the flow of water in (m^3/s), and H is the head of water measured in m.

Most existing hydropower plants cannot be used for frequency and load control due to conventional induction and synchronous generators feeding into the power system. The electric frequency of these generators f is rigidly related to the angular velocity of the water turbine $\omega_{turbine}$, as will be discussed in Chapter 12 for conventional polyphase electric machines. Due to the inherently slow control of water inflow to the turbine, the nominal frequency of $f = 60$ Hz corresponding to 16.667 milliseconds (ms) cannot be changed quickly. However, if doubly fed induction generators (DFIGs [20]) are employed, quick control responses are possible (within a few 60 Hz cycles). Hydro-storage plants are important for the increased reliance on intermittently operating renewable energy sources such as photovoltaic and wind power [20], which require energy compensation if either too little or too much energy is generated.

Application Example 1.4 Hydropower plant with Francis turbine
A $P_{hydro} = 5$ MW hydropower plant consists of a Francis turbine, a conventional induction generator with a synchronous speed $n_s = 90$ rpm and a rated speed $n = 92.7$ rpm resulting in a slip of $s = (n_s - n)/n_s = -3\%$ connected to the utility system.

a) If the power efficiencies of the water turbine and the induction generator are $\eta_{turbine} = 0.75$ and $\eta_{gen} = 0.90$, respectively, compute the required input power of the water turbine $P_{turbine}^{required}$.
b) Provided the water head is $H = 49.213$ ft $\equiv 15$ m and the water flow is $Q = 2118.88$ cfs (cubic feet per second) $\equiv 60$ m^3/s, compute the mechanical power available at the turbine input [20, 23–25]:

$$P_{kW} = \frac{H \cdot Q \cdot W \cdot 0.746}{550},$$

where H is the water head in feet (ft), Q is the flow of water in cubic feet per second (cfs), W is the weight of a cubic foot of water $W = 62.4$ (lbs-force)/ft^3, 0.746 is a constant used to convert horsepower (hp) to kilowatts (kW), and 550 is a constant used to convert {ft-(lbs-force)}/s into horsepower.
c) How does $P_{turbine}^{required}$ compare with P_{kW}?
d) Compute the specific speed $N_q = n\frac{Q^{0.5}}{H^{0.75}}$, where n is the speed measured in rpm, Q is the flow of water measured in (m^3/s), and H is the head of water measured in m.
e) Is the selection of the Francis-type turbine according to Figure 1.10 justified?

Solution

a) $P_{turbine}^{required} = P_{hydro}/(\eta_{turbine} \cdot \eta_{gen}) = 5\text{ MW}/(0.75 \cdot 0.9) = 7.41\text{ MW}$.
b) $P_{kW} = \dfrac{H \cdot Q \cdot W \cdot 0.746}{550} = 8.83\text{ MW}$.
c) $P_{turbine}^{required} < P_{kW}$. Turbine output power must be increased.
d) $N_q = n\dfrac{Q^{0.5}}{H^{0.75}} = n_s(1+s)\dfrac{Q^{0.5}}{H^{0.75}} = \dfrac{90(1.03)7.746}{7.622} = 94.208$.
e) The Francis turbine is acceptable because a Francis turbine can be operated at a specified speed range of $N_q = 70$–120 as is shown in Figure 1.10.

1.5 Wind Power and the Lanchester–Betz–Joukowsky Limit

Cumulative global wind power capacity rose from almost 60 GW in 2005 to just under 370 GW by the end of 2014 (an increase of 517% in 10 years) [26]. Wind power farms are installed either on land as depicted in Figure 1.11 or offshore

Figure 1.11 Wind power farm.

on shallow continental shelves. While installation and maintenance costs are much higher for offshore plants, they experience more constant wind speeds, and often strongest wind during the day and late afternoons, when electricity demand generally peaks. In either case, wind power plants must be turned off if the wind velocity is either too low or too high, and they should preferably be variable-speed type to accommodate changing wind conditions. It is also desirable to avoid mechanical gears in order to minimize maintenance.

Since winds crossing a wind turbine are not robbed of all their velocity, the share of kinetic energy in wind that can be converted to electricity is limited. The maximum power that can be extracted from any wind turbine design is given by the Lanchester–Betz–Joukowsky wind turbine limit [27]:

$$P_{\max}\big|_{c_\mathrm{p} = 0.592\,59} = \frac{8}{27} \cdot \rho \cdot A \cdot v^3. \tag{1.10}$$

The power coefficient c_p has a maximum value of $16/27 = 0.592\,59$; in other words, this represents the upper limit of recoverable wind energy. For a given c_p, one obtains the maximum output power $P_{\max}$ of the wind turbine at the air density ρ, the area swept by wind turbine blades A, and the wind velocity v. The disadvantage of wind energy plants is their discontinuous generation. One noteworthy example occurred during the March 2015 Niklas storm in Germany: hurricane-force wind conditions with unprecedented speed between 100 and 160 km/h forced the shutdown of wind power plants totaling 2.3 GW during high-wind periods, and additional conventional generation of 20.3 GW had to be employed during low-wind periods. This was necessary to stabilize generation, as reported by the German transmission grid operators TenneT, 50Hertz, Amprion, and TransnetBW [28]. Because of protests of the people in Bavaria and other German states, the wind energy from the offshore plants north of Germany will be transmitted via underground DC cables to the south in order to mitigate environmental concerns [29, 30]. Wind farm developers have noticed that winds may be weakened after turbines are installed: dense arrays of turbines produce drag, which depletes local or even regional wind resources [31]. Large onshore wind farms can extract to about 1.6 MW/km^2, while large wind farms offshore on the North Atlantic can extract about 6 MW/km^2.

Application Example 1.5 Analysis of a 2.3 MW wind power plant at an altitude of 728 m

A wind turbine has an output power of $P_{\text{out}} = 2.3$ MW without taking into account the efficiencies of gear, generator, and converter(s). The coefficient of power is $c_{\text{p}} = 0.5$. Find the required area A swept by the turbine blades and the blade length R for the rated wind speed of $v = 12$ m/s [27]. The air density at an altitude of 728 m is $\rho \approx 1.14$ kg/m^3.

Solution

From Eq. (1.10) one obtains

$P_{\text{out}}|_{c_{\text{p}}=0.5} = \frac{0.5}{0.592\,59} \cdot \frac{8}{27} \cdot \rho \cdot A \cdot v^3 = \frac{0.5}{0.592\,59} \cdot \frac{8}{27} \cdot 1.14 \cdot A \cdot 10^3 = 2.3\,\text{MW}$ or $A = 8.07 \cdot 10^3 = \pi R^2$, resulting in a blade length of $R = 50.68$ m.

1.6 Thermal Solar and PV Plants

Thermal solar sources have enormous potential to save 20% of the natural gas and oil consumed by residences, industry, and commercial buildings. Solar thermal is especially useful for low and midrange temperatures. Figure 1.12a illustrates the rapid growth of such assisting energy sources in Germany [32]. Figure 1.12b presents an example of how thermal solar heat can augment the heating system within a residence.

The global PV market has exhibited rapid expansion in the last two decades; the compound annual growth rate (CAGR) for PV installations [2] between 2000 and 2014 was 44%. In 2014, Germany accounted for more than 20% (38 GW) of cumulative installed capacity worldwide (183 GW), and solar PV met close to 7% of Germany's electricity demand – despite Germany's middling daily insolation energy

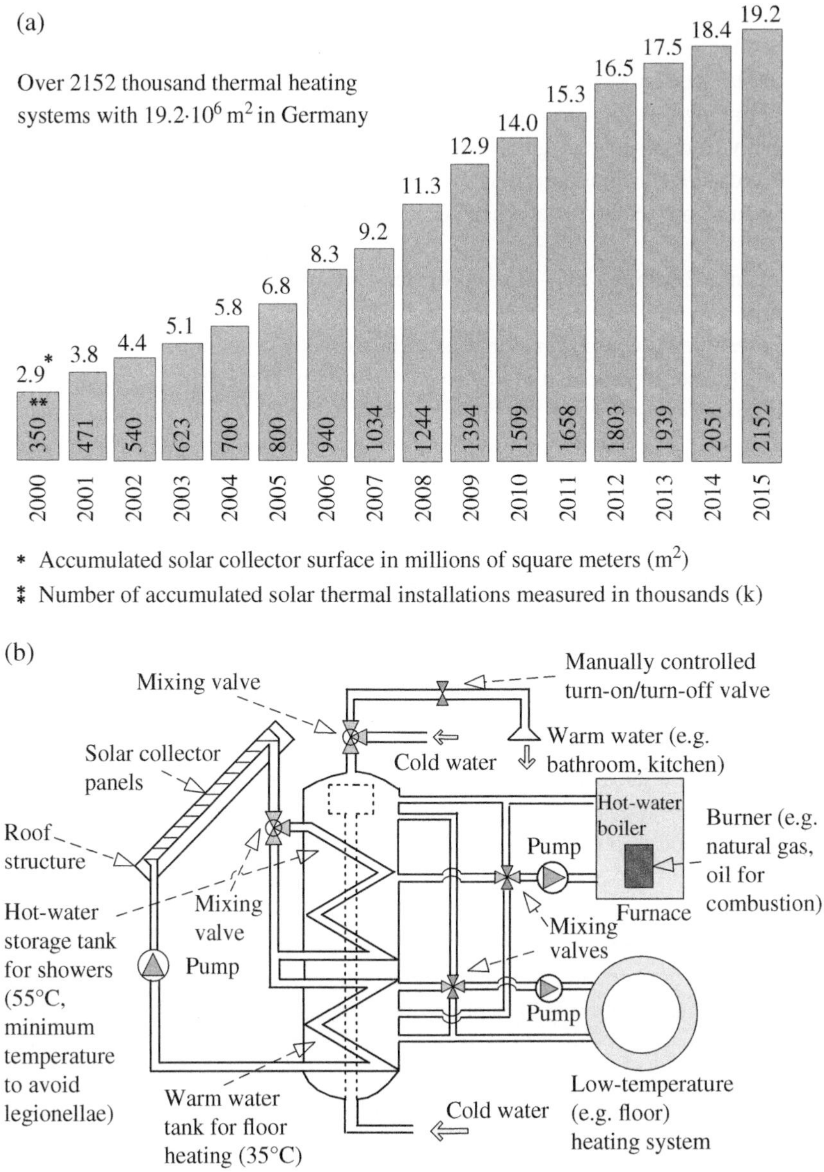

Figure 1.12 (a) Growth of solar-heating power plants in Germany [32]. (b) Utilization of thermal solar heat within a residence [33].

(kWh/m^2/day). The share of EU installed capacity fell to 48% in 2014 (from 58% in 2013), while those in China and Taiwan rose to 17% from 13% the year before [2]. As generous feed-in tariff schemes like those in Germany and Hawaii taper off, novel financing mechanisms, technology innovation, and national climate commitments will drive further growth worldwide. Of course, one should note that PV plants do not continuously generate power; the above 7% relates to the energy provided, and not to the installed rated peak power capacity, which is larger.

Application Example 1.6 Energy-independent residence [34], Munich, Germany

A good example for energy consumption reduction of a residential dwelling, a single-family house, is presented in Figure E1.6.1a and b. The 220 m^2 house with traditional three levels (basement, first and second floors) was built in 2013 with insulating clay bricks combined with humidity-absorbing plaster [35], a rated peak power PV plant of 6.48 kWp at €2163/kW without tax or $2552/kW including tax, and triple-pane windows [36]. Note that the inverter output power is limited to 70% based on STC (see Appendix 1.A), resulting in (4.53 kWh)/(6.48 kWh) = 70%. This 70% limit is imposed by the utilities/government in Germany in order to mitigate the variability of the PV plant output [37–39].

The "energy disclosure" of this building [34] is depicted in Figure E1.6.2, where the annual in-house energy consumption (*Endenergiebedarf*) per square meter is 18 kWh/(m^2·year). The annual primary energy consumption (*Primärenergiebedarf*), which represents the in-house energy consumption for heating and cooling as well as the energy required to process and provide these services (e.g. energy required for natural gas drilling and pipeline operation), is 46 kWh/(m^2·year).

These two values represent predicted values – based on calculations – before the house was built. As of 2014 the energy savings regulations (*Energieeinsparverordnung, EnEV*) in Germany require that the total energy consumption of newly constructed residences be reduced by 25% relative to 2012 where 124.3 kWh/(m^2·year) was required to heat one flat of a multifamily house [40] and that furnaces/boilers more than 30 years old must be replaced with more efficient ones.

Solution

To check the energy (E) generation and consumption of the house, daily electricity meter readings of the following energy components were recorded and averaged for each week during one year from 14 July 2014 to 14 July 2015:

a) Generated DC energy (EI_i) by PV system feeding two inverters (I) during week i,
b) Supplied energy (EG_i) to the utility grid (G) by PV system during week i,
c) Consumed energy (EHC_i) for heating (H) and cooling (C) during week i,
d) Consumed energy (ELA_i) by lighting (L) and appliances (A) during week i.

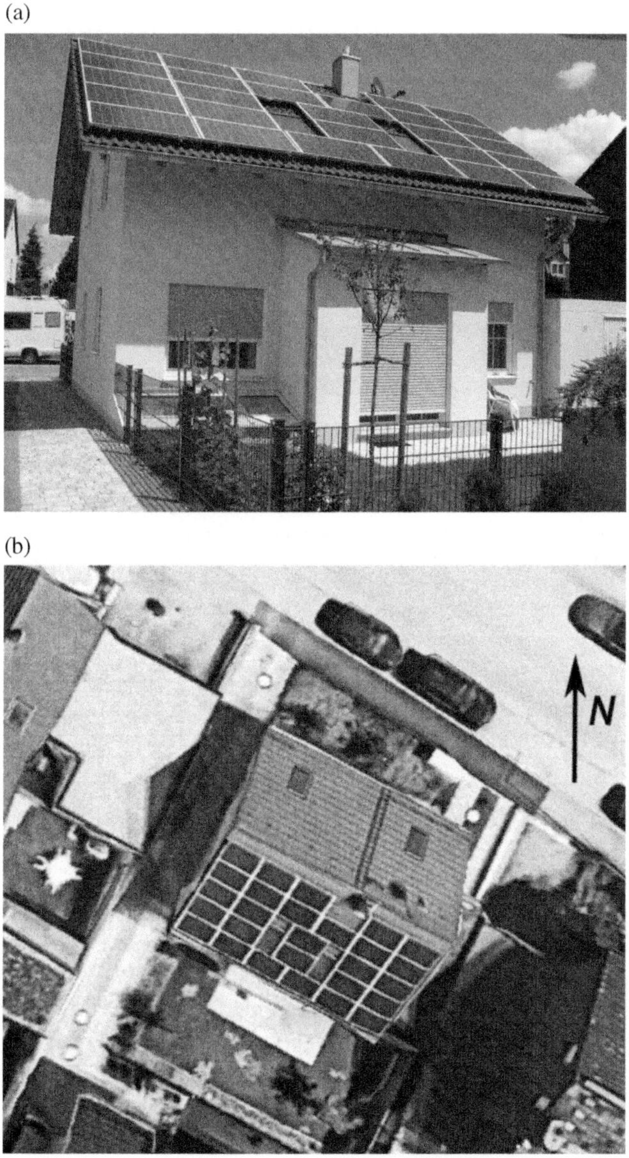

Figure E1.6.1 (a) High-efficiency, single-family dwelling with PV system and groundwater–water heat pump (courtesy of E.F. Fuchs). (b) aerial view and orientation of residence shown in (a) in Munich with latitude of 48.1351 °N and longitude of 11.5820 °E. The roof pitch is 35° and the geographic orientation is (180° + 36°) = 216°, where north corresponds to 0°.

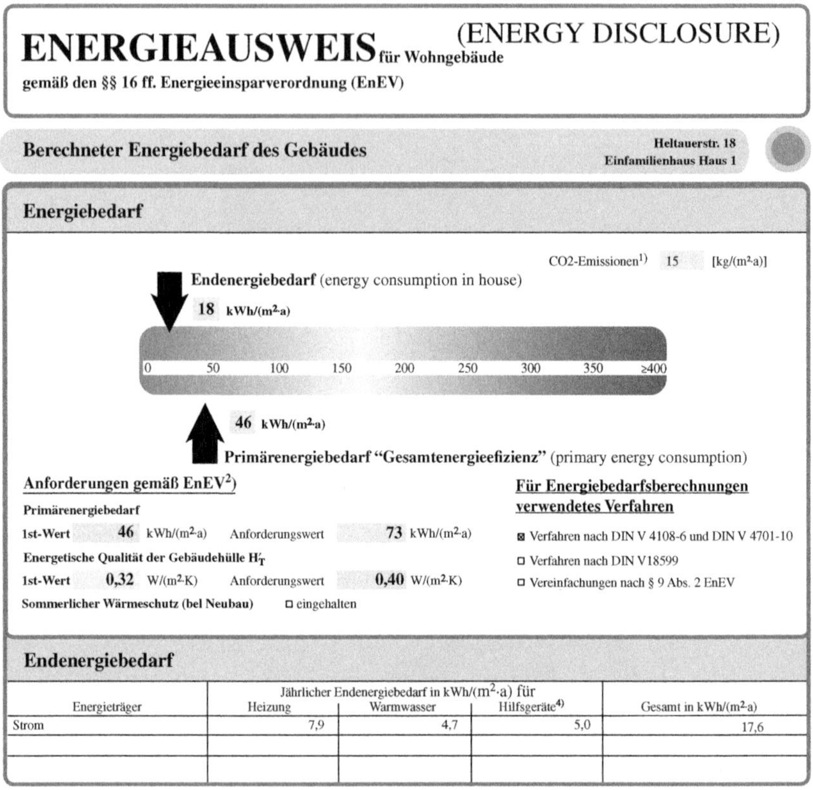

ENERGIEAUSWEIS für Wohngebäude (ENERGY DISCLOSURE)

gemäß den §§ 16 ff. Energieeinsparverordnung (EnEV)

Berechneter Energiebedarf des Gebäudes

Heltauerstr. 18
Einfamilienhaus Haus 1

Energiebedarf

CO2-Emissionen[1) 15 [kg/(m²·a)]

Endenergiebedarf (energy consumption in house)

18 kWh/(m²·a)

0 50 100 150 200 250 300 350 ≥400

46 kWh/(m²·a)

Primärenergiebedarf "Gesamtenergieeffizienz" (primary energy consumption)

Anforderungen gemäß EnEV[2)

Primärenergiebedarf

Ist-Wert 46 kWh/(m²·a) Anforderungswert 73 kWh/(m²·a)

Energetische Qualität der Gebäudehülle H'_T

Ist-Wert 0,32 W/(m²·K) Anforderungswert 0,40 W/(m²·K)

Sommerlicher Wärmeschutz (bei Neubau) ☐ eingehalten

Für Energiebedarfsberechnungen verwendetes Verfahren

☒ Verfahren nach DIN V 4108-6 und DIN V 4701-10

☐ Verfahren nach DIN V18599

☐ Vereinfachungen nach § 9 Abs. 2 EnEV

Endenergiebedarf

Energieträger	Jährlicher Endenergiebedarf in kWh/(m²·a) für Heizung	Warmwasser	Hilfsgeräte[4)	Gesamt in kWh/(m²·a)
Strom	7,9	4,7	5,0	17,6

Figure E1.6.2 Energy disclosure for single-family dwelling of Figure E1.6.1a without PV system [34].

The floor heating of the house maintained the bedroom temperatures at 17–20 ° C and all other rooms at 20–22 °C. The residence is equipped with a 7.64 kW (rated power) groundwater–water heat pump [18]. During summer the passive cooling of the house is via floor-piped cooling using groundwater (at 8 °C), maintaining room temperatures at <24 °C. The above-described generation (EI_i), energy fed into grid (EG_i), energy requirements for both heating (including hot water heating during entire year for kitchen and baths) and cooling modes (EHC_i), and the energy supplied by the utility (ELA_i) are reflected in Figure E1.6.3 and Table E1.6.1.

The abovementioned costs and reimbursements per kWh are not based on net metering as done in some states in the United States, where private solar customers may sell excess energy back to the grid at retail price. Some of the states, such as Hawaii, Nevada, Arizona, Maine, and Indiana, have decided to phase out net metering [41], because programs have spurred explosive growth in the rooftop

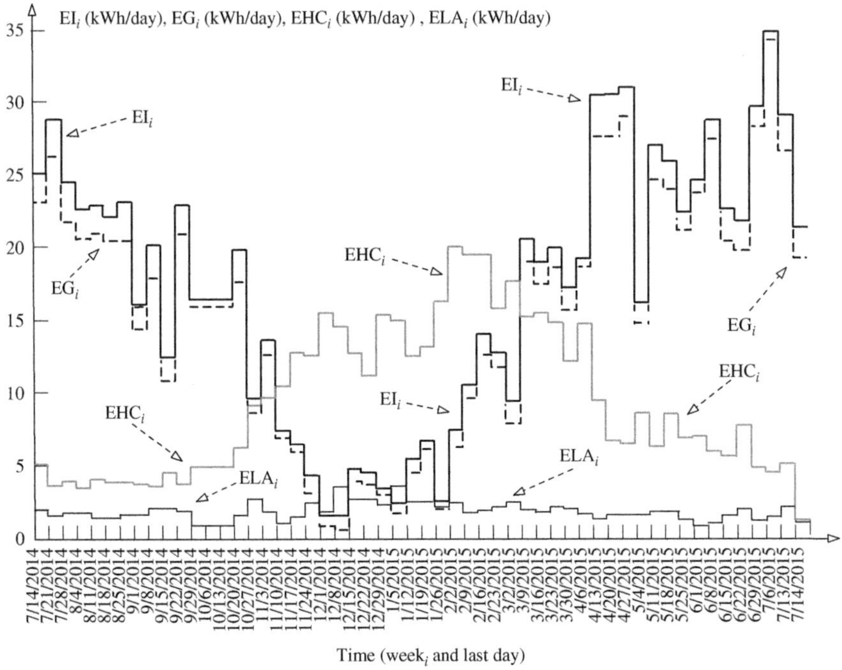

Figure E1.6.3 Weekly average energy readings per day for EI_i, EG_i, EHC_i, and ELA_i taken over one year from 14 July 2014 to 14 July 2015, where cost of high tariff for EHC_i was €0.16/kWh = US$0.19/kWh, cost of low tariff for EHC_i was €0.15/kWh = US$0.18/kWh, and cost for ELA_i was €0.21/kWh = US$0.25 plus a flat rate/base price of €125/year = US$151/year, without value-added tax of 19%. For the supplied energy to the grid EG_i, the utility pays €0.1854/kWh = US$0.22/kWh, which is less than ELA_i due to the charge to be paid for energy storage in the grid.

solar market. Utilities argue that net metering can be unfair to homeowners who do not want or cannot afford their own solar installations: in short, net metering does not pay any charge to the utility for storing electricity, and the utilities say it means that fewer ratepayers cover the costs of traditional power generation.

The predicted value of 18 kWh/(m^2·year) can be checked using some of the values (EHC_i) related to either Figure E1.6.3 or Table E1.6.1 as follows:

With $7\cdot\sum_{i=1}^{52}\text{EHC}_i + \text{EHC}_{7/14/2015} + 7\cdot\sum_{i=1}^{52}\text{ELA}_i + \text{ELA}_{7/14/2015} = (3438 + 610)$ kWh/year = 4048 kWh/year, energy consumption in house without PV system = {energy input or energy consumption in house on 14 July 2015 (EHC + ELA) – energy input or energy consumption in house on 14 July 2014 (EHC + ELA)}/(220 m^2·year) = {10 030 – 5982 kWh}/(220 m^2·year) = 4043 kWh/(220 m^2 · year) = 18.4 kWh/(m^2·year), neglecting the additional energy gain through the

Table E1.6.1 Daily average energy readings per week from 14 July 2014 to 14 July 2015.

End date of week	PV-generated DC energy EI_i (kWh/d)	Energy supplied to utility grid EG_i (kWh/d)	Energy for heating and cooling EHC_i (kWh/d)	Energy for lighting and appliances ELA_i (kWh/d)	Water use (m^3/d)
07/21/2014	28.71	26.14	3.57	1.57	0.29
07/28/2014	24.29	21.71	3.93	1.64	0.29
08/04/2014	22.43	20.43	3.36	1.64	0.29
08/11/2014	22.71	20.86	4.00	1.29	0.29
08/18/2014	22.00	20.29	3.86	1.29	0.29
08/25/2014	23.00	20.29	3.86	1.57	0.43
09/01/2014	16.00	14.14	3.71	1.57	0.29
09/08/2014	20.00	17.43	3.57	2.00	0.29
09/15/2014	12.29	10.71	4.43	2.00	0.29
09/22/2014	22.89	20.83	3.68	1.96	0.29
09/29/2014	16.23	15.81	4.77	0.73	0.00
10/06/2014	16.23	15.81	4.77	0.73	0.00
10/13/2014	16.23	15.81	4.77	0.73	0.00
10/20/2014	18.70	17.60	6.01	1.42	0.29
10/27/2014	9.57	8.43	9.00	2.57	0.29
11/03/2014	13.43	12.29	9.57	1.71	0.14
11/10/2014	7.09	6.73	10.27	0.82	0.00
11/17/2014	6.34	5.84	12.73	1.32	0.14
11/24/2014	4.14	3.00	12.43	2.29	0.14
12/01/2014	1.14	0.71	15.14	1.86	0.14
12/08/2014	1.14	0.43	14.29	3.29	0.29
12/15/2014	4.71	3.86	12.57	2.57	0.29
12/22/2014	4.57	3.57	11.00	2.57	0.14
12/29/2014	3.14	2.86	15.14	2.14	0.14
01/05/2015	2.14	1.57	14.86	3.29	0.29
01/12/2015	5.14	4.14	12.29	2.29	0.14
01/19/2015	6.71	6.00	13.00	2.29	0.29
01/26/2015	2.41	1.96	16.04	1.96	0.00
02/02/2015	7.16	6.04	19.82	2.18	0.29
02/09/2015	10.29	9.43	19.29	1.57	0.14

Table E1.6.1 (Continued)

End date of week	PV-generated DC energy EI$_i$ (kWh/d)	Energy supplied to utility grid EG$_i$ (kWh/d)	Energy for heating and cooling EHC$_i$ (kWh/d)	Energy for lighting and appliances ELA$_i$ (kWh/d)	Water use (m^3/d)
02/16/2015	13.86	12.29	19.29	1.71	0.29
02/23/2015	12.71	11.57	15.71	2.00	0.14
03/02/2015	9.14	7.86	17.43	2.29	0.29
03/09/2015	20.43	18.86	15.00	1.86	0.14
03/16/2015	18.86	17.43	15.14	1.71	0.29
03/23/2015	19.71	18.43	14.71	2.00	0.14
03/30/2015	17.00	15.57	12.00	1.86	0.29
04/06/2015	19.00	18.52	14.52	1.43	0.14
04/13/2015	30.14	27.62	9.19	1.14	0.29
04/20/2015	30.14	27.43	6.43	1.43	0.14
04/27/2015	30.71	28.71	6.29	1.43	0.43
05/04/2015	16.00	14.43	8.43	1.43	0.14
05/11/2015	26.86	24.43	6.14	1.57	0.43
05/18/2015	25.71	23.71	8.43	1.57	0.29
05/25/2015	22.18	21.00	6.70	1.07	0.00
06/01/2015	24.38	23.50	6.81	0.75	0.00
06/08/2015	28.45	27.07	5.78	0.89	0.43
06/15/2015	22.29	20.14	5.57	1.29	0.29
06/22/2015	21.43	19.43	7.57	1.71	0.14
06/29/2015	29.29	28.00	4.71	1.00	0.29
07/06/2015	34.57	34.00	4.43	1.14	0.43
07/14/2015	27.75	25.25	4.50	0.88	0.13
Year (cumulative)	6128	5625	3438	610	79

Total energy produced and consumed for data year 2014–2015 (difference cumulative energy as of 07/14/2014 and as of 07/14/2015).

Readings were taken at the end of each day for 52 weeks; the final week included 1 extra day to total 365 days for the year.

Note that $EI = 7 \cdot \sum_{i=1}^{52} EI_i + EI_{7/14/2015} = 6128$ kWh/year, $EG = 7 \cdot \sum_{i=1}^{52} EG_i + EG_{7/14/2015} = 5625$ kWh/year, $EHC = 7 \cdot \sum_{i=1}^{52} EHC_i + EHC_{7/14/2015} = 3438$ kWh/year, and $ELA = 7 \cdot \sum_{i=1}^{52} ELA_i + ELA_{7/14/2015} = 610$ kWh/year.

coefficient of performance [30] COP_H. Note that the above values 10 030 and 5982 kWh were obtained without rounding errors from actual electricity meter readings EHC and ELA that occurred during 365 days of meter readings on which Figure E1.6.3 and Table E1.6.1 are based. Taking into account the coefficient of performance $COP_H = 3.98$, the energy consumption in the house (heating including warm water for kitchen and bathrooms and cooling) is analyzed in Problem 1.17, which leads to a much higher value than 18 kWh/($m^2 \cdot$ a).

Comparison with prior (2013–2014) year's energy consumption [30]: Energy consumption in house without PV system = {energy input or energy consumption in house on 14 July 2014 (EHC + ELA) – energy input or energy consumption in house on 14 July 2013 (EHC + ELA)/(220 m^2·year) = {5982 – 2013 kWh}/(220 m^2·year) = 3974 kWh/220 m^2 =18.04 kWh/(m^2·year).

Note the above values 5987 and 2013 kWh were obtained without rounding errors from actual electricity meter readings EHC and ELA, which occurred during 365 days of meter readings.

The output inverter energy graph EI_i – from the middle of July 2014 to the middle of July 2015 – is a function of insolation, temperature, and dirt and snow on panels, including cloudiness. A Fourier analysis reveals that the fundamental corresponds to a cosine function modulated by the changing insolation. Figure E1.6.4 supports this cosine function [42] when the maximum possible hours of sunshine are considered for the same time frame. Figure E1.6.5 depicts the measured decrease of the PV module output power EI_{year} (kWh) for four years: 2014 [30], 2015, 2016, and 2017.

Figure E1.6.5 illustrates measured energy generation by the two inverters EI (kWh/year) during a four-year time frame. The measured DC energy of the two inverters (Figure E1.6.1) are:

EI_{2014} = (8960 – 2383) kWh/year = 6577 kWh/year during 7/14/2013 to 7/14/2014,

EI_{2015} = (15 088 – 8960) kWh/year = 6128 kWh/year during 7/14/2014 to 7/14/2015,

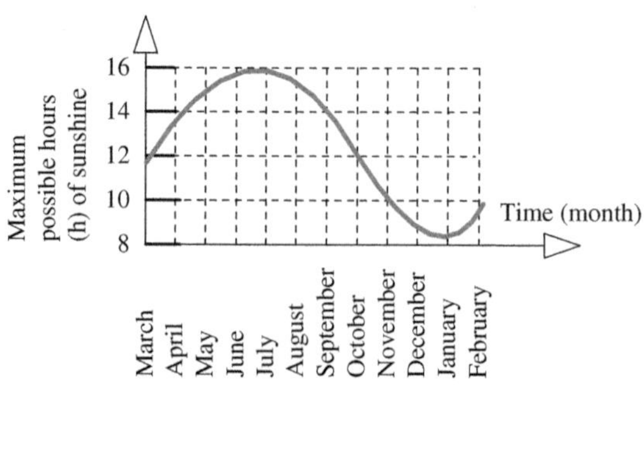

Figure E1.6.4 Maximum possible hours of sunshine in Munich in particular on 26 August 2015 at 1:45 p.m.

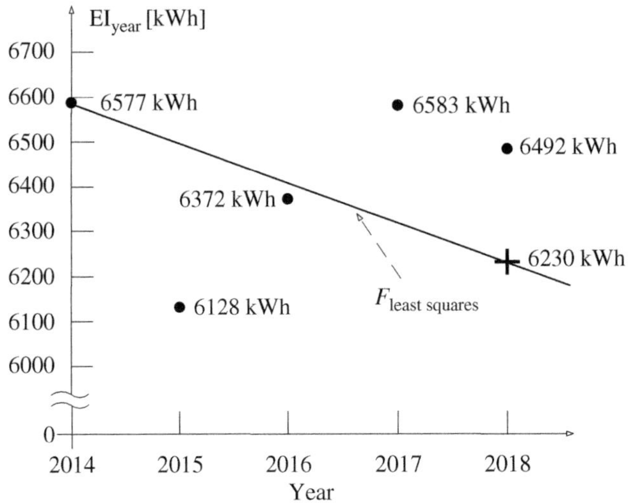

Figure E1.6.5 Reduction of generation of PV modules from 2014 to 2018, where the function $F_{\text{least square}}$ represents the least squares approximation.

EI_{2016} = (21 460 − 15 088) kWh/year = 6372 kWh/year during 7/14/2015 to 7/14/2016, leap year (+1 day),

EI_{2017} = (28 043 − 21 460) kWh/year = 6583 kWh/year during 7/14/2016 to 7/14/2017,

EI_{2018} = (34 535 − 28 043) kWh/year = 6492 kWh/year during 7/14/2017 to 7/14/2018.

Note that the year 2015 was to date the hottest on record [43], and therefore the generation is relatively low because the warmer the summer is, the greater is the humidity of the air causing clouding. In addition, as PV panels warm, the (electric) efficiency of the solar cells decreases. A least squares analysis yields the decreasing function $F_{\text{least squares}}$ of Figure E1.6.5; therefore the reduction of generation is reduced due to aging and dirt on panels [44]; it varies as a function of insolation as well. According to Figure E1.6.5, the measured reduction of lifetime is obtained from $[(F_{\text{least squares}}(2014) - F_{\text{least squares}}(2018)]/[4 \cdot F_{\text{least squares}}(2014)] = [6577 - 6230]/[4 \cdot 6577] = 0.0132$ or 1.32% per year.

Water consumption per year: It is notable that the yearly freshwater consumption of the residence, including a 200 m^2 lawn area as depicted in Figure E1.6.1, <80 m^3 or 108 liters per person and per day (see Table E1.6.1). The average water consumption in Germany is 120 liters per person per day, while the corresponding amount in Las Vegas [45] is 446 liters.

Application Example 1.7 Comparison of the yearly energy consumption of a household-type refrigerator and a smartphone

According to Stadtwerke München GmbH, the municipal utility in Munich, a small refrigerator (with a volume of 100 liters or 3.53 ft^3) uses 180 kWh/year, and a smartphone with its infrastructure such as the transmission towers that wirelessly transmit communication signals, as well as recharging the battery, requires 360 kWh/year [46]. Note that there are many transmission towers between sender and receiver, where each pair of towers must be located within viewing distance to be able to transmit high-frequency signals within the bandwidth of about 850–2100 MHz. Figure E1.7.1 illustrates this surprising difference in energy consumption. Such high frequencies might be also causing environmental concerns [30] such as radio and television interference and the hypothesis that high frequencies may cause leukemia in children through "electro-smog" emanating from cell towers at very high frequencies in the mega- to gigahertz range. Figure E1.7.2 shows the presently permitted general electric field intensity E measured in volts/meter or V/m – called "electro-smog" for various European countries: the maximum permitted values are 61 V/m in Switzerland, Luxembourg, Liechtenstein, and Germany. Figure E1.7.3 presents recommended precautionary electric field intensity limits as advocated mainly by Switzerland, San Marino, Russia, Poland, Monaco, Luxembourg, Lithuania, Liechtenstein, Italy, Bulgaria, Belgium, Salzburg, and Electric Norm with a maximum value of equal or less than 7 V/m. Note that Germany and China appear not to have precautionary limits on "electro-smog." Salzburg in Austria appears to have great concerns about "electro-smog" limiting the electric field intensity to 0.02 V/m.

Figure E1.7.1 One smartphone (including its use of transmission towers) can have greater energy consumption (360 kWh) than a 100-liter compressor-type refrigerator (180 kWh) with a high COP [46]. *Source*: Published in MDirekt-Magazin 20/01/2015, @ bsa Marketing GmbH, Munich

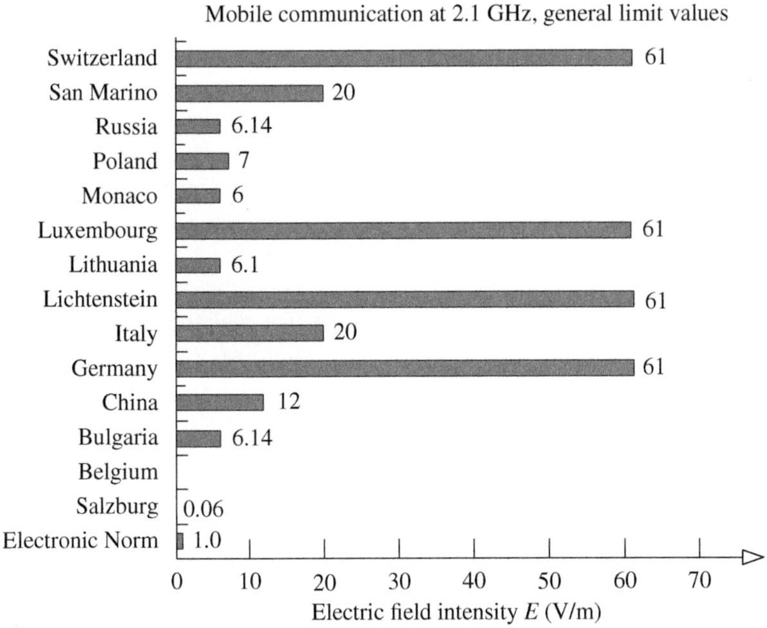

Figure E1.7.2 Presently permitted general "electro-smog" limit values for 2.100 GHz. *Source:* Bundesamt für Strahlenschutz, Salzgitter, 2016.

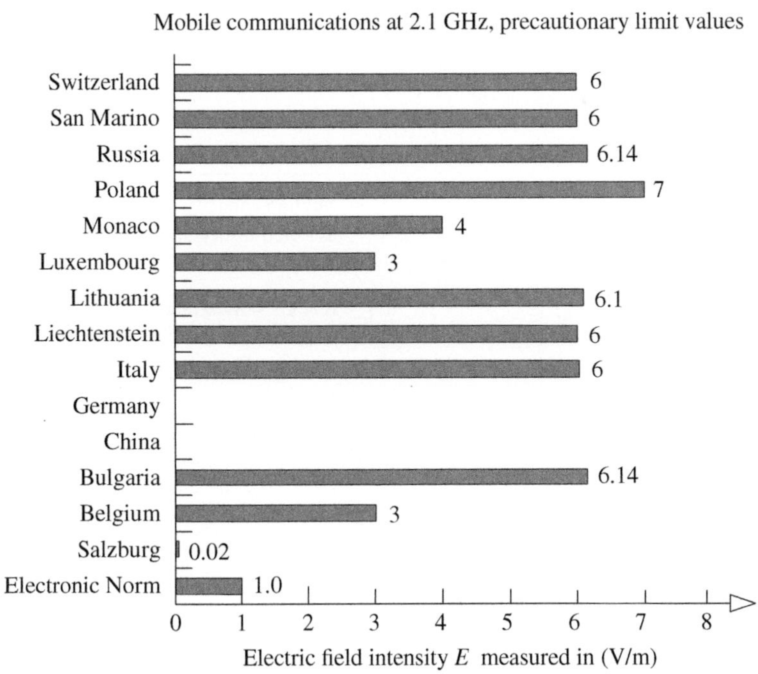

Figure E1.7.3 Recommended precautionary "electro-smog" limit values for 2100 MHz. *Source:* Bundesamt für Strahlenschutz, Salzgitter, 2016.

1.7 Capacity Factors

Not all power plants operate all the time. The *capacity factor* refers to a measure of actual output over a period of time [47, 48]. Capacity factors for various types of energy systems, as recorded by US Department of Energy (DOE)'s National Renewable Energy Laboratory (NREL), are given in Figure 1.13. For example, the capacity factor of geothermal power facilities is very efficient, usually ≈90%. This is on par with other base-load power sources like coal and nuclear and is much greater than intermittent sources such as wind and PV plants. PV and ocean-wave technologies have low capacity factors, but the lowest is that for natural gas combustion turbines, as these are typically used as peak power plants that generally run only under peak electricity demand.

1.8 Force Calculations Based on Coulomb's Law

We now pivot to the world of electricity, electromagnetism, and electronics. The fundamental unit of charge on the atomic level is a single proton or electron. However, electrical engineers deal primarily with electric charge q measured in coulombs (C) or ampere-seconds (As). One coulomb represents the charge of $6.25 \cdot 10^{18}$ protons. The change of the charge with respect to time is called current,

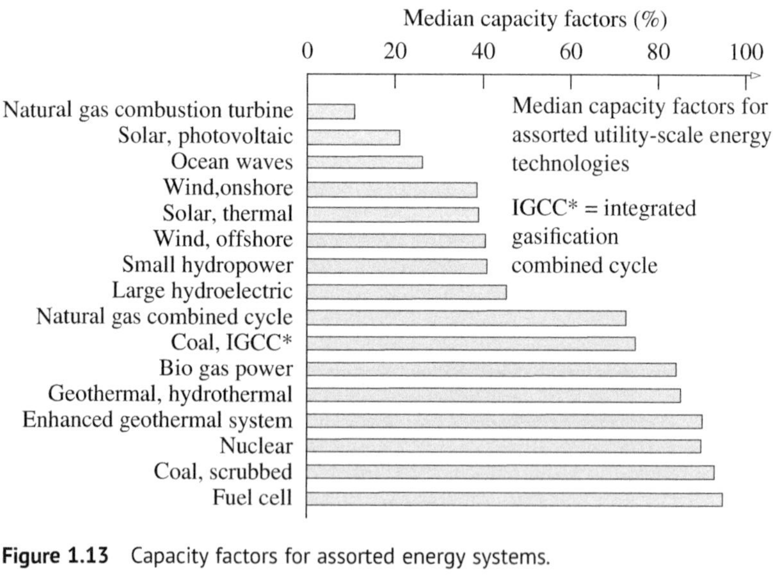

Figure 1.13 Capacity factors for assorted energy systems.

$$i = \frac{dq}{dt}, \tag{1.11}$$

measured in amperes (A); current can be thought of as a change of charge with respect to time, so one ampere equals one coulomb per second. This electric-charge change with time, representing current, $i = (dq/dt)$, induces the voltage v in conductors measured in volts (V), generating instantaneous work or energy measured in watt-seconds (Ws) or volt-ampere-seconds (VAs). Instantaneous work divided by time t is defined as instantaneous power p measured in volt-amperes (VA) or watts.

1.8.1 Electric Charge

If an electron leaves its parent atom, it becomes a free electron and leaves behind a "hole"; that is, the parent atom becomes an ion or a positive charge carrier. An electron has the charge $q_o = 1.6020 \cdot 10^{-19}$ As. A capacitor can temporarily store electrons that are endowed [49] with voltage v as shown in Figure 1.14a. For example, rubbing a plastic ruler on fabric can charge a capacitor with endowed electrons, thus converting free electrons into endowed electrons that are capable of useful work. The endowed electrons can then be discharged by connecting a resistor across the capacitor as illustrated in Figure 1.14b. Thus, the endowed electrons heat up the resistor by discharging, and endowed electrons become free electrons again. This concept is applied to the generation of electric energy within power plants and transmission and distribution systems in Appendix 1.B. Note that a resistor R measured in ohms (Ω) consists of a conducting wire of length ℓ, cross section A, and resistivity ρ such that

$$R = \rho \frac{\ell}{A}. \tag{1.12}$$

Resistivity is a property inherent to a particular material, related to atomic interactions that hinder charge carriers' movement due to movements of electrons into and from the valence and conduction bands of ions. Resistance of a material results from friction-like interactions between electrons or ions in motion and other material particles. A substance's molecular makeup and form determine its resistance. For example, in a wire, resistance declines with cross-sectional area and rises with the length of the wire; however, it increases with temperature.

The symbol of a capacitor should not be confused with that of a normally open and closed relay switch as is depicted in Figure 1.4c and d. Unfortunately many textbooks use the wrong symbol (see Figure 1.14c) for a capacitor.

The correct way to depict a capacitor is shown in Figure 1.14e.

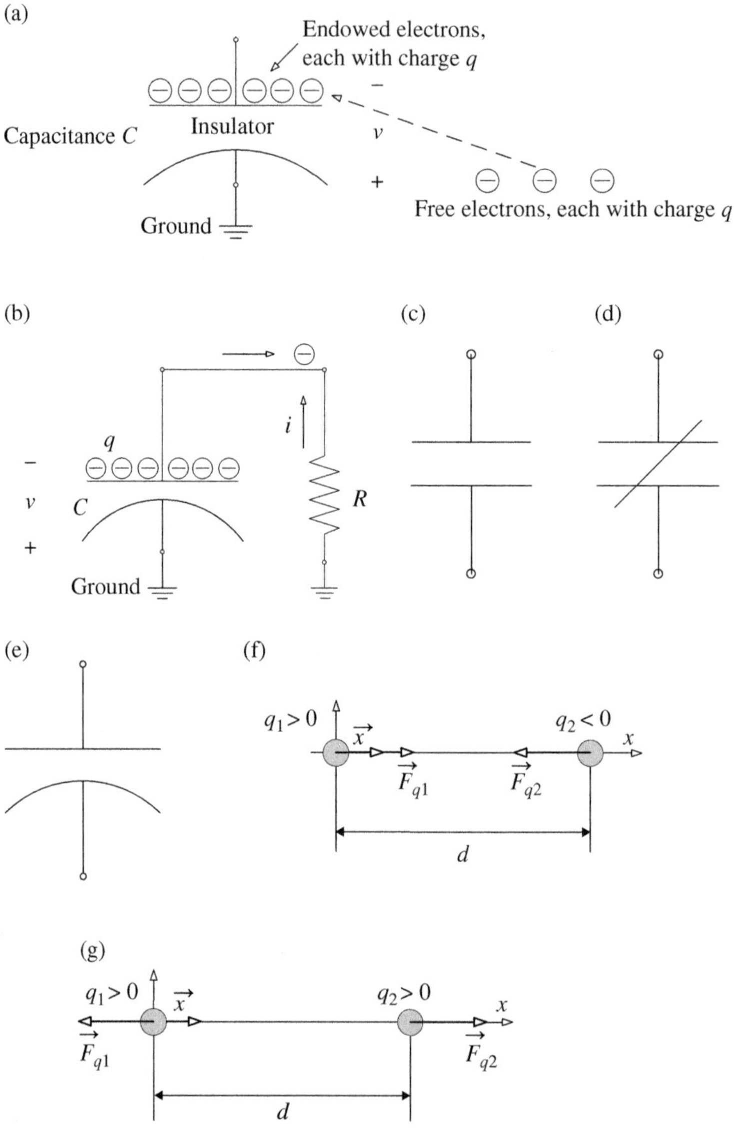

Figure 1.14 (a) Charged capacitor with electrons endowed [49] with voltage *v*. (b) Discharging capacitor with resistor *R* connected across voltage *v*. Symbols of a normally (c) open relay switch, and normally closed (d) relay switch. (e) Symbol for a capacitor. (f) Attraction of negative and positive charges. (g) Repellence of either two positive or two negative charges.

1.8.2 Electrostatic Force

Coulomb's law gives the electrostatic force between two charges q_1 and q_2 that are a distance d apart:

$$\left|\vec{F}\right| = F = k\frac{q_1 \cdot q_2}{d^2}, \tag{1.13}$$

where the constant $k = 8.99 \cdot 10^9\,\text{Nm}^2/\text{C}^2$ and C represents the unit coulomb. Positive ($q_1 > 0$) and negative ($q_2 < 0$) charges attract one another as illustrated in Figure 1.14f by virtue of the forces $\vec{F}_{q1} = k\frac{q_1 \cdot q_2}{d^2}\left(\vec{x}\right)$ and $\vec{F}_{q2} = k\frac{q_1 \cdot q_2}{d^2}\left(-\vec{x}\right)$ with (q_1, $q_2 > 0$) follows $\left|\vec{F}_{q1}\right| = \left|\vec{F}_{q2}\right|$, while either two positive or two negative charges repel one another as shown in Figure 1.14g, as defined by $\vec{F}_{q1} = k\frac{q_1 \cdot q_2}{d^2}\left(-\vec{x}\right)$ and $\vec{F}_{q2} = k\frac{q_1 \cdot q_2}{d^2}\left(\vec{x}\right)$ with (q_1, $q_2 > 0$) follows $\left|\vec{F}_{q1}\right| = \left|\vec{F}_{q2}\right|$.

Application Example 1.8 Force calculation

The three point charges $q_1 = -2$ milliC, $q_2 = 4$ milliC, and $q_3 = 10$ milliC, where milliC stands for millicoulombs, are placed in a straight line in the x-direction as indicated in Figure E1.8.1. The distances are $d_{12} = 0.05$ m, $d_{13} = 0.15$ m, and $d_{23} = 0.10$ m. Calculate the magnitudes of the resultant forces acting on q_1, q_2, and q_3 in newtons (1 newton = 1 N ≡ 1 kg·m/s^2) and the directions of the resultant forces acting on the three q_1, q_2, and q_3, employing the unit vector $\vec{x}$. If $q_1 = -6$ milliC, calculate the magnitude and direction of the charge q_3 so that the resultant force on q_2 is zero.

Solution

Note that positive and negative charges attract one another, positive charges repel each other, and negative charges repel one another as well. It is best to deal with absolute charge values $|q_i|$ and to include the directional information in the unit vector $\vec{x}$.

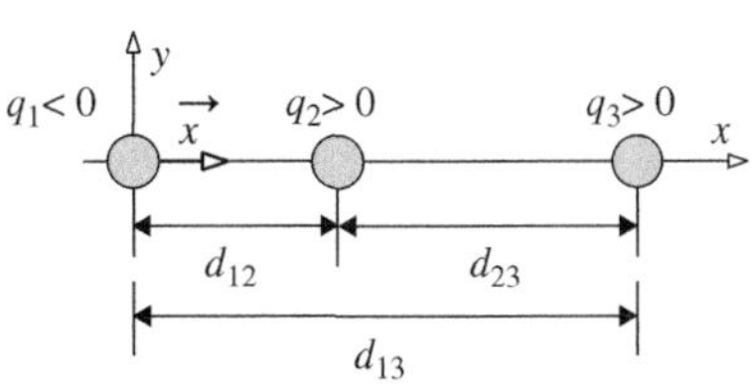

Figure E1.8.1 Arrangement of three charges along the x-axis.

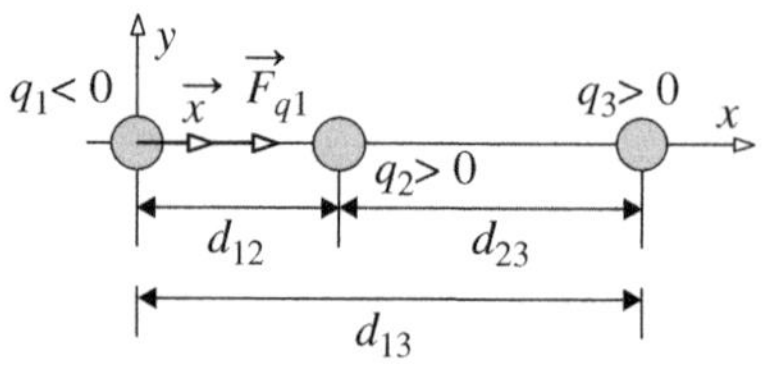

Figure E1.8.2 Calculation of force applied to q_1.

Calculation of force applied to q_1: The forces applied to q_1 by q_2 and by q_3 are to the right in $\vec{x}$-direction. Figure E1.8.2 illustrates the direction of the force $\vec{F}_{q1}$, which becomes

$\vec{F}_{q1} = k\frac{|q_1||q_2|}{d_{12}^2}\left(\vec{x}\right) + k\frac{|q_1||q_3|}{d_{13}^2}\left(\vec{x}\right)$, that is, both q_2 and q_3 attract q_1 to the right in $\left(\vec{x}\right)$-direction or

$$\vec{F}_{q1} = 8.99 \cdot 10^9 \left[\frac{2 \cdot 10^{-3} \cdot 4 \cdot 10^{-3}}{(0.05)^2} + \frac{2 \cdot 10^{-3} \cdot 10 \cdot 10^{-3}}{(0.15)^2}\right]\left(\vec{x}\right)$$
$$= 3.675 \cdot 10^7 \left(\vec{x}\right) \text{ N.}$$

Calculation of force applied to q_2: The forces applied to q_2 by q_1 and by q_3 are to the left in $\left(-\vec{x}\right)$-direction. Figure E1.8.3 illustrates the direction of force $\vec{F}_{q2}$.

$\vec{F}_{q2} = k\frac{|q_1||q_2|}{d_{12}^2}\left(-\vec{x}\right) + k\frac{|q_2||q_3|}{d_{23}^2}\left(-\vec{x}\right)$, that is, both q_1 and q_3 apply forces to q_2 in $\left(-\vec{x}\right)$ - direction or $\vec{F}_{q2} = 8.99 \cdot 10^9 \left[\frac{4 \cdot 10^{-3} \cdot 2 \cdot 10^{-3}}{(0.05)^2} + \frac{4 \cdot 10^{-3} \cdot 10 \cdot 10^{-3}}{(0.10)^2}\right]$ $\left(-\vec{x}\right) = 6.47 \cdot 10^7 \left(-\vec{x}\right)$ N.

Calculation of force applied to q_3: The force applied to q_3 by q_1 is to the left in $\left(-\vec{x}\right)$ direction and the force applied to q_3 by q_2 is to the right in $\left(\vec{x}\right)$ direction. Figure E1.8.4 illustrates the direction of the forces applied to q_3.

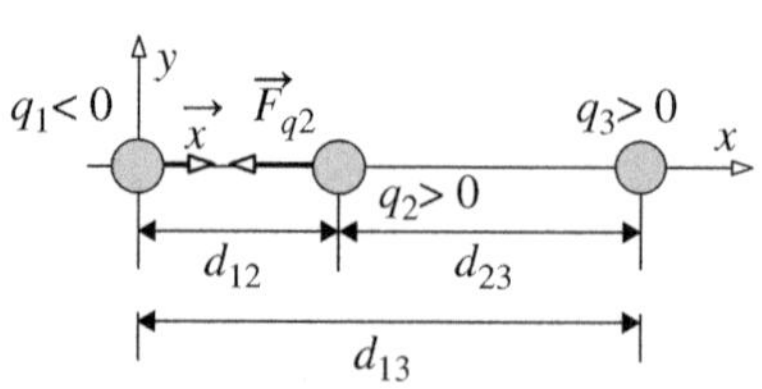

Figure E1.8.3 Calculation of force applied to q_2.

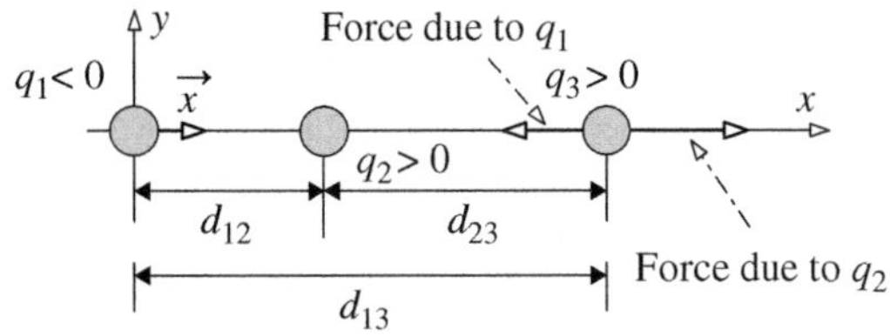

Figure E1.8.4 Calculation of force applied to q_3.

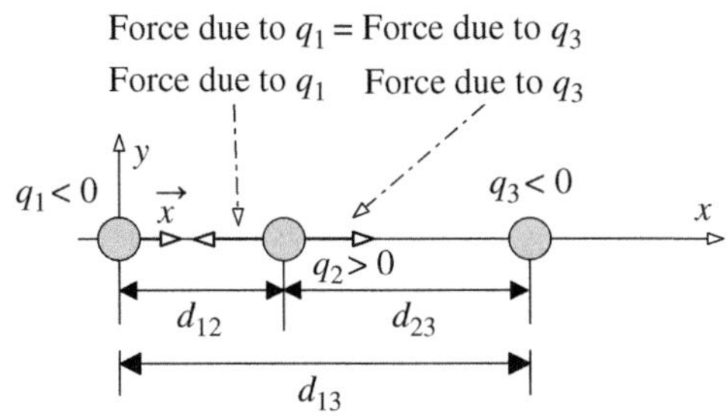

Figure E1.8.5 Force distribution so that the resultant force acting on q_2 is zero.

$$\vec{F}_{q3}=k\frac{|q_1||q_3|}{d_{13}^2}\left(-\vec{x}\right)+k\frac{|q_3||q_2|}{d_{23}^2}\left(\vec{x}\right),\ \vec{F}_{q3}=8.99\cdot10^9\left[-\frac{2\cdot10^{-3}\cdot10\cdot10^{-3}}{(0.15)^2}+\frac{10\cdot10^{-3}\cdot4\cdot10^{-3}}{(0.10)^2}\right]\left(\vec{x}\right)=2.7977\cdot10^7\left(\vec{x}\right)\text{N}.$$

If $q_1 = -6$ milliC, $q_2 = 4$ milliC, one obtains the magnitude and direction of the charge q_3 so that the resultant force on q_2 is zero as illustrated in Figure E1.8.5. One concludes that the charge q_3 must be negative. $\vec{F}_{q2} = k\frac{|q_2||q_1|}{d_{12}^2}\left(-\vec{x}\right) + k\frac{|q_2||q_3|}{d_{23}^2}\left(\vec{x}\right)$ or $\vec{F}_{q2}=8.99\cdot10^9\left[-\frac{4\cdot10^{-3}\cdot6\cdot10^{-3}}{(0.05)^2}+\frac{4\cdot10^{-3}q_3}{(0.10)^2}\right]\left(\vec{x}\right)=0$, resulting in $q_3 = -24\cdot10^{-3}\,\text{C} = -24$ milliC.

1.9 Conductors, Insulators, and Semiconductors

A *conductor* has low resistance to the flow of charge q, that is, $R \to 0$. All metals conduct, as do fluids with dissolved ions like rainwater and tap water (but not pure distilled water). An *insulator* does not permit the motion of charge q, that is, $R \to \infty$; typical insulators are glass, ceramics, or plastics. A *semiconductor*'s conductivity falls somewhere between that of a conductor and an insulator; the motion of charge increases the temperature of a conductor and increases its resistance and thus the losses of the conductor, while doping with impurity atoms increases the rate of change of charge, which is current [50]. Typical semiconducting materials such as silicon and germanium are crucial to electronics manufacturing.

1.10 Instantaneous Current *i* and Voltage *v*

Current i is defined through the rate of change of charge with time:

$$i = \frac{\Delta q}{\Delta t} = \frac{\mathrm{d}q}{\mathrm{d}t} = \frac{\mathrm{d}q^{+} + \mathrm{d}q^{-}}{\mathrm{d}t}, \tag{1.14}$$

where $\mathrm{d}q^{+} \equiv$ incremental positive charge transferred, that is, holes caused by stripped electrons, and $\mathrm{d}q^{-} \equiv$ incremental negative charge transferred, that is, stripped electrons, as shown in Figure 1.15. Note that the positive charge carriers – the atoms stripped of electrons – are not mobile and are embedded in the atomic structure, the so-called lattice. The current is a scalar quantity and requires a sign convention. The movement of electrons through a conductor (wire) segment is illustrated in Figure 1.16. Currents range from 10^{-14} A through nerve cells to 10^{-4} A for integrated circuits, to 10^{-1} A causing ventricular fibrillation that is fatal to humans [51], to 10 A for typical household appliances, and to 10^{6} A for lightning bolts.

1.10.1 Instantaneous Voltage *v*, Work/Energy *work*, and Power *p*

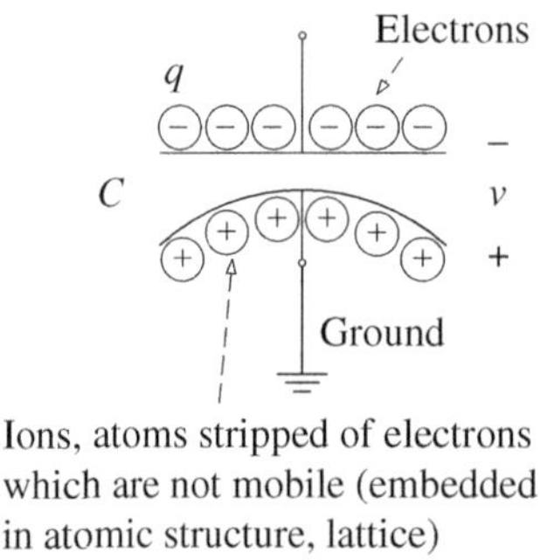

Figure 1.15 Negative (e.g. electrons) and positive (e.g. atoms stripped of electrons) charge carriers residing on capacitor plates.

The quantity that causes charges to move through a load R (e.g. light bulb) is the voltage (V) or the potential (Φ) difference

$$V = V_{\mathrm{A}} - V_{\mathrm{B}} = \Phi_1 - \Phi_2 \tag{1.15}$$

as illustrated in Figure 1.16. The conventional direction of the current i is in the opposite direction of the electron flow, as indicated in Figure 1.14b. The range of voltages is 10^{-10} V for communication applications (e.g. radio) to 10^{-4} V for electrocardiograms (Elektrokardiogram or EKG), to 10 V for integrated circuits, to 120–220 V for house wiring, and to 10^{8} V for lightning bolts.

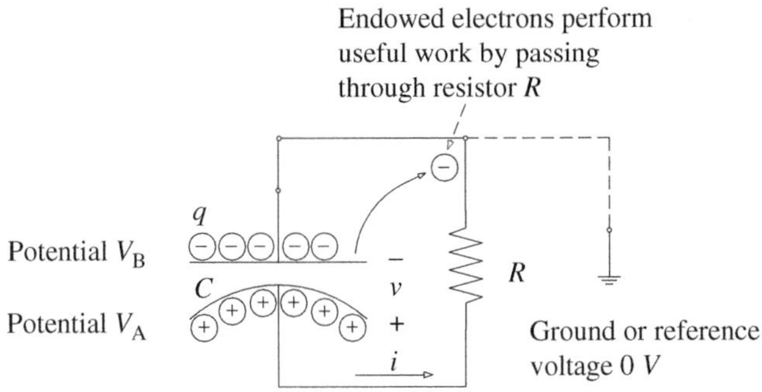

Figure 1.16 Voltage *v* causes electrons to move through load resistance *R* including current direction in opposition to electron motion.

For $v = V_A - V_B$, $V_A > V_B$, and $V_B = 0$, one obtains as a function of the differential energy *work* the instantaneous voltage

$$v = V_A = \frac{\mathrm{d}work}{\mathrm{d}q} \tag{1.16}$$

or d*work* = *v*d*q*. With d*q* = *i*d*t* one gets

$$\mathrm{d}work = v\mathrm{d}q = vi\mathrm{d}t = p\mathrm{d}t, \tag{1.17}$$

where *p* is the instantaneous power measured in volt-ampere VA, watt or W. The range of powers is 10^{-6} W for communication (antenna) applications to 10^{-3} W for EKGs, to 10 W for integrated circuits, to $5 \cdot 10^3$ W for residences, to 10^6 W for lightning bolts, and to 10^{12} W = 1000 GW for the entire US power system, encompassing the Eastern, Western, and Electric Reliability Council of Texas (ERCOT) grids. Unfortunately, lightning bolts last only a few microseconds (μs), so the energy that could be utilized by collecting lightning bolts is of negligible importance.

1.11 The Question of Frequency: AC Versus DC Distribution and Transmission Systems

While Thomas Edison [52] was a proponent of unipolar DC electricity as indicated in Figure 1.17, Nikola Tesla [53] favored AC, as shown in Figure 1.18. Although Tesla's approach was adopted more than 100 years ago by the electrical engineering community for distribution and transmission systems, it appears that high-voltage DC technology may be experiencing a resurgence in transmission [54] and possibly distribution systems [55] either on power lines or in buried earth

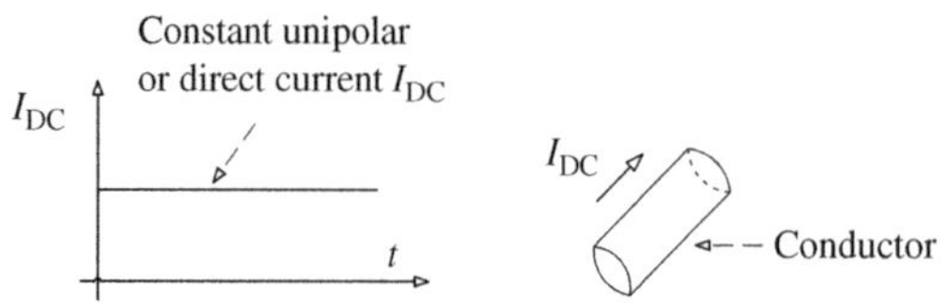

Figure 1.17 Constant unipolar direct current (DC).

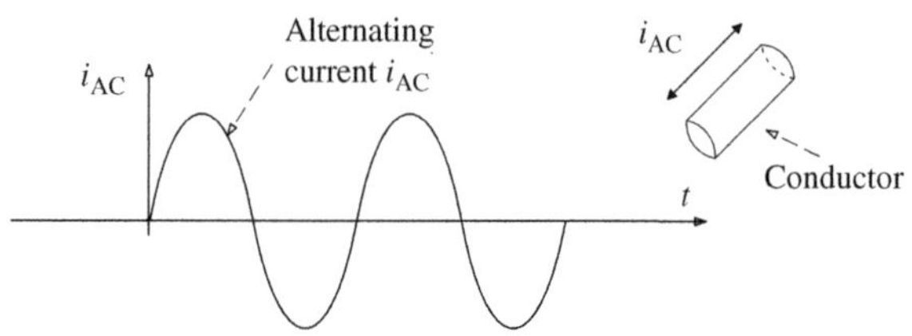

Figure 1.18 Alternating current (AC) with frequency *f*.

cables. The major advantage of DC transmission systems is the lack of stability problem associated with the power transfer along very long AC transmission lines (e.g. 1000 km and longer). However, one aspect of DC distribution systems may threaten the safety of building residents. It is well known that both AC voltages and currents are subjected to the skin effect [56]: AC current mainly flows on the surface of the human body, while DC signals may flow through the entire body – including vital organs. One mA of DC current through the heart may be fatal.

In nature AC signals are prevalent, such as the heart pulsation as shown in Figure E1.9.1 for an adult person from 60 to 80 beats per minute (bpm) and Figure E1.9.2 for a newborn where the rate is about twice that of an adult. Also, the heart pulse rate is dependent upon the stress experienced by the human body, as illustrated in Figures E1.10.1 and E1.10.2 as a function of hiking up a mountain range, starting out at 59 bpm and reaching 163 bpm. Hearing relies also on alternating audio signals, as shown in Figure E1.11.1 for a young adult for the right ear. The hearing area of Figure E1.11.2 is obtained by averaging the perceived signals as a function of the frequency. Most healthy persons hear sound frequencies from about 100 Hz to 20 kHz, while animals hear audio signals within a much larger frequency range. Breathing is another example where nature relies on repetitive patterns. These examples demonstrate that alternating signals commonly occur in nature, and frequently it is wise to copy biological phenomena in technological developments.

A few examples where engineers have copied the repetitive nature of living bodily functions relate to variable-speed drives (see Chapters 10 and 12) where the inverter fundamental frequency varies about from 10 to 180 Hz, while the switching frequency varies within 5–20 kHz. In general power system frequencies range nominally from 16.6667 Hz = $16^{2/3}$ Hz (German railway) to 50–60 Hz. Satellite TVs operate with 10^8 to 10^{10} Hz, microwaves from 10^{10} to 10^{12} Hz, X-rays from 10^{18} to 10^{20} Hz, and gamma rays about 10^{22} Hz.

Application Example 1.9 Measurement of pulse in beats per minute (bpm)

Figures E1.9.1 and E1.10.2 show the heartbeat of a 12-year old and that of a newborn child, respectively: the pulse (bpm) of the newborn is about twice as high as that of the 12-year old, which helps to detect pregnancies.

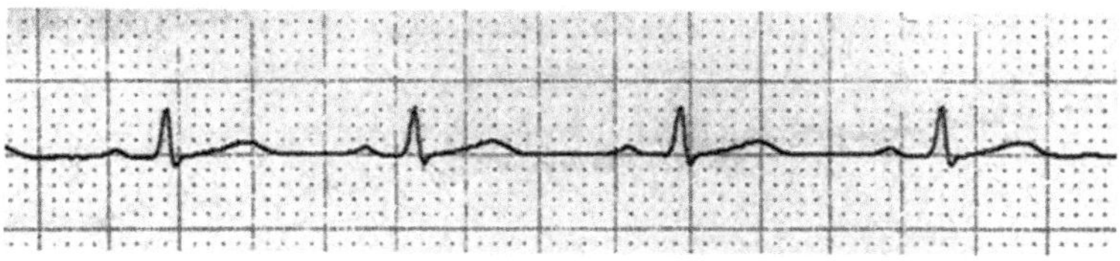

Figure E1.9.1 Heart pulsations of 12-year-old person with 82 bpm.

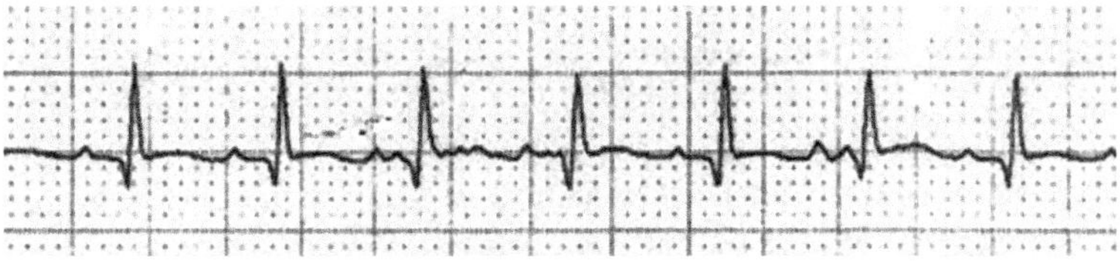

Figure E1.9.2 Heart pulsations of a newborn with 150 bpm.

Application Example 1.10 Measurement of pulse in beats per minute (bpm) hiking up an alpine mountain

Figure E1.10.1 shows the location of the mountain range Drei Türme at the Swiss-Austrian border, climbed by an adult whose bpm was recorded during the climb with appropriate resting periods during which the beats per minute increased. This example demonstrates that the bpm can increase by a factor of about two (Figure E1.10.2) for a middle-aged adult without adverse side effects.

Figure E1.10.1 View of the mountain range Drei Türme at the Swiss-Austrian border with Tschagguns-Lindauer Hütte.

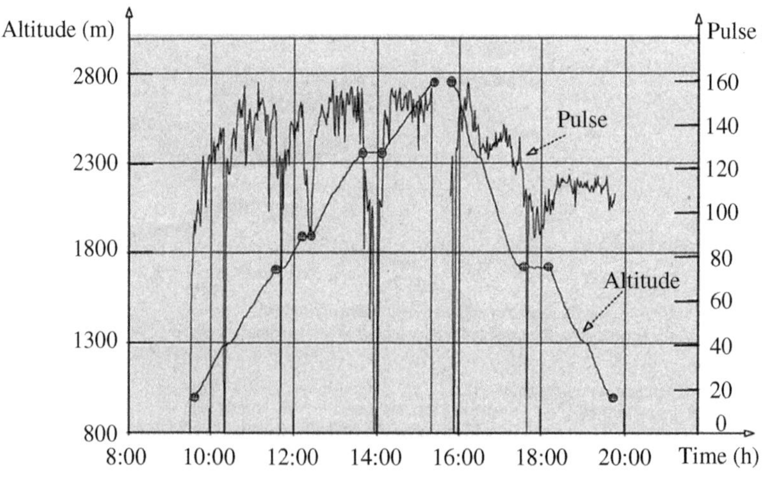

Figure E1.10.2 Changing bpm of an adult hiking up Drei Türme from minimum altitude of 985 m to maximum altitude of 2767 m. During resting periods the bpm sensor was turned off, and that is the reason why the bpm drops during these times to 0. *Source:* Courtesy E.F. Fuchs.

Application Example 1.11 Measurement of hearing sound intensity and hearing area

The human ear can detect frequencies from about 100 Hz to 20 kHz. Due to some environmental noise (e.g. explosions), hearing of some people is impaired, and the sound frequency curve is not constant. In some cases there exist hearing holes (K. Walliser, private communication, September 2017) where certain frequencies cannot be heard, as shown in Figure E1.11.1, where at about 1 kHz and from 5 to 10 kHz higher sound intensity must exist in order to perceived by the young adult. It is well known that some animals have a wider hearing spectrum, such as birds.

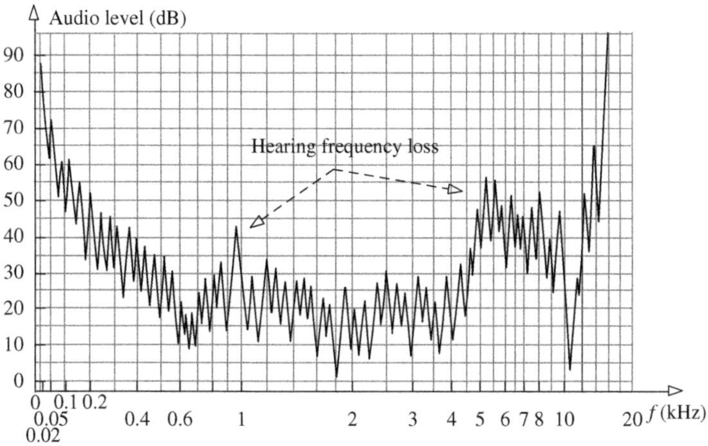

Figure E1.11.1 Hearing sound intensity as a function of frequency for a young adult (right ear). Note a 3 dB (decibel) increase the required sound intensity by a factor of two.

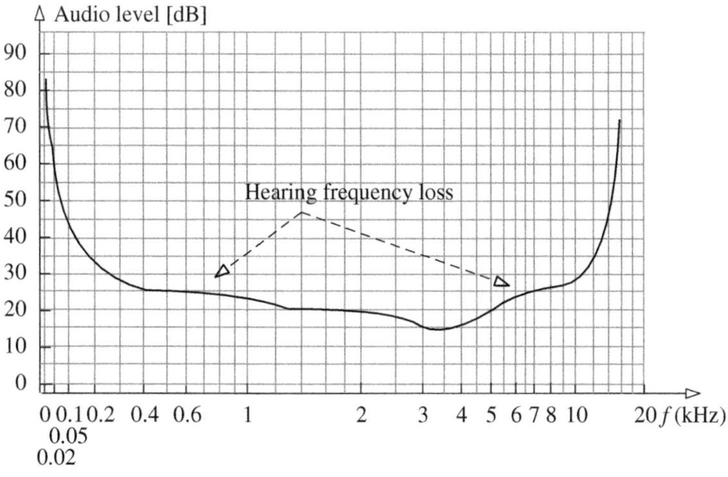

Figure E1.11.2 Hearing area intensity as a function of frequency for a young adult (right ear). Note a 3 dB (decibel) increase the required sound intensity by a factor of two.

1.12 Reference Directions and Polarities of Voltages and Currents

Voltage is the difference of two potentials Φ_A and Φ_B. In Figure 1.19a $V_1 = \Phi_A - \Phi_B = V_A - V_B = +2\,V - 0\,V = +2\,V$, where the reference potential is that of the ground potential $V_B = \Phi_B = 0\,V$. If we go from the +ve sign to the −ve sign, we encounter a positive voltage or potential difference.

Figure 1.19b and c depict two identical circuits and potentials, but with different reference direction of voltage. If we go in Figure 1.19b from the +ve sign to the −ve sign, we encounter a negative voltage or potential difference: $V_1 = V_A - V_B = \Phi_A - \Phi_B = -2\,V - 0\,V = -2\,V$. In Figure 1.19c, going from the +ve sign to the −ve sign, one encounters a positive voltage or potential difference: $V_1 = V_B - V_A = \Phi_B - \Phi_A = 0\,V - (-2\,V) = +2\,V$.

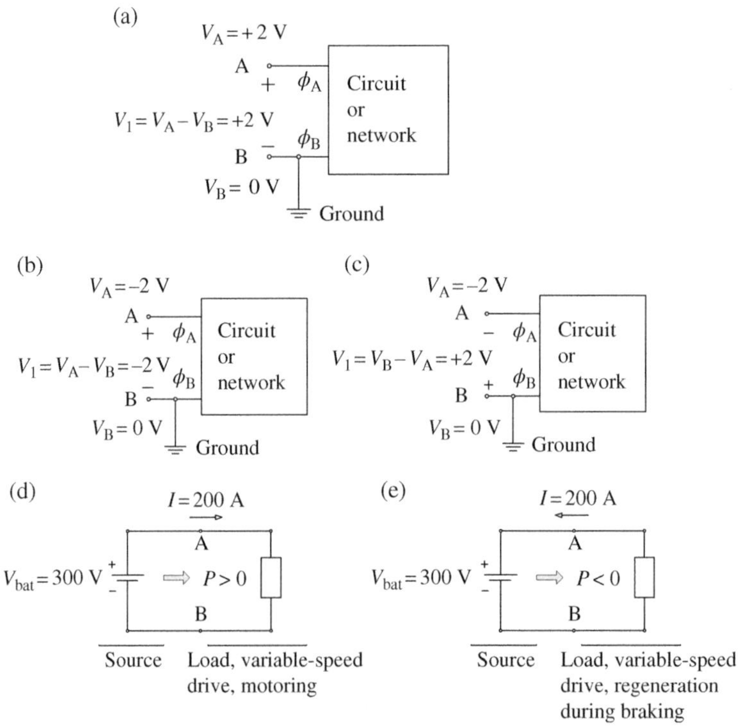

Figure 1.19 (a) Definition of voltage $V_1 = V_A - V_B$ as a function of voltages ≡ potentials $V_A \equiv \Phi_A$ and $V_B \equiv \Phi_B$. Note that voltages are defined as potential differences. (b and c) Two identical circuits and voltages/potentials ($V_A \equiv \Phi_A$ and $V_B \equiv \Phi_B$), but with different reference direction. (d) *Consumer system* of voltage–current notation, and (e) *generator system* of voltage-current notation.

Voltage and current relations are explained in Figure 1.19d, in which the battery is the source supplying current to the load, which might be, for example, a variable-speed drive of an electric car during motoring operation. In this so-called consumer system of voltage–current notation, the current leaves the positive terminal of the source and enters the load at its positive terminal. The power ($p > 0$) or energy (*work*) is supplied by the source (e.g. battery) and consumed in the load.

Figure 1.19e illustrates that the battery is now the load, absorbing current from the source, which might be a variable-speed drive of an electric car during braking/regeneration operation. In this so-called generator system of voltage–current notation, the current leaves the positive terminal of the source (variable-speed drive during regeneration) and enters the load (e.g. battery) at its positive terminal. The power ($p < 0$) or energy (*work*) is supplied by the source consumed in the load (e.g. battery).

1.13 Power *p*

We define the instantaneous power p as a function of the energy *work* and time t by Eq. (1.18a and b)

$$\mathrm{d}work = v\mathrm{d}q = vi\mathrm{d}t = p\mathrm{d}t, \tag{1.18a}$$

or the instantaneous power p

$$p = \mathrm{d}work/\mathrm{d}t = vi. \tag{1.18b}$$

Integration over time results in the energy *work*

$$work = \int_{t_1}^{t_2} p\mathrm{d}t = \int_{t_1}^{t_2} vi\mathrm{d}t, \tag{1.18c}$$

where the time instants t_1 and t_2 are positive, as negative time instants do not exist. In general (e.g. time-dependent instantaneous voltages and currents), one can draw an equivalent circuit of Figure 1.20, and Figure 1.21 shows two equivalent circuits for DC voltages and currents.

1.14 Ideal Passive Electric Circuit Elements

Figure 1.22a depicts an ideal passive electric circuit with a load consisting of either a resistor R, a capacitor C, or an inductor L, as defined in Figure 1.22b.

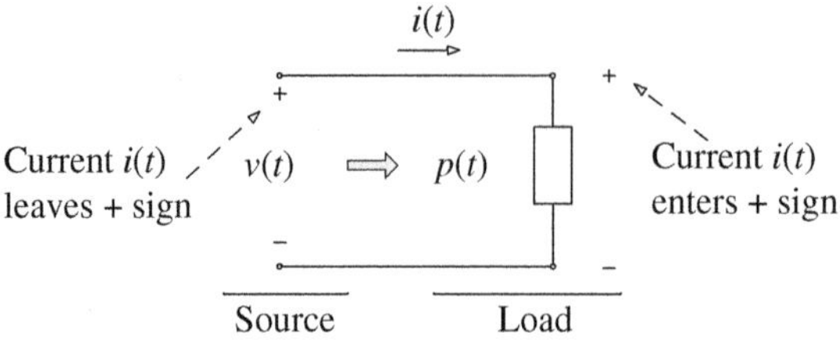

Figure 1.20 Equivalent circuit for time-dependent instantaneous voltages $v(t)$ and currents $i(t)$.

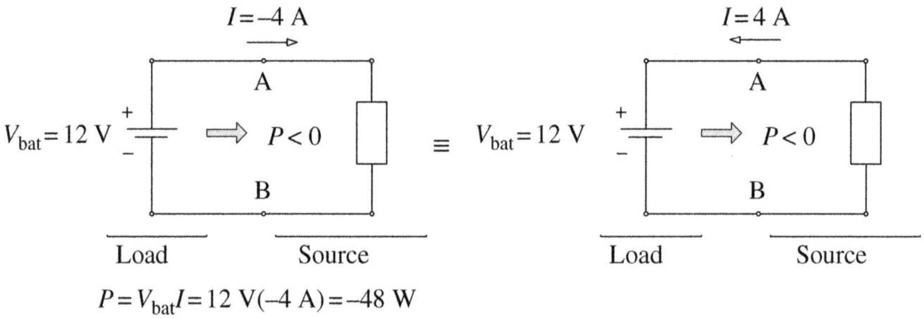

Figure 1.21 Two equivalent circuits for DC voltage V_{bat} and current I.

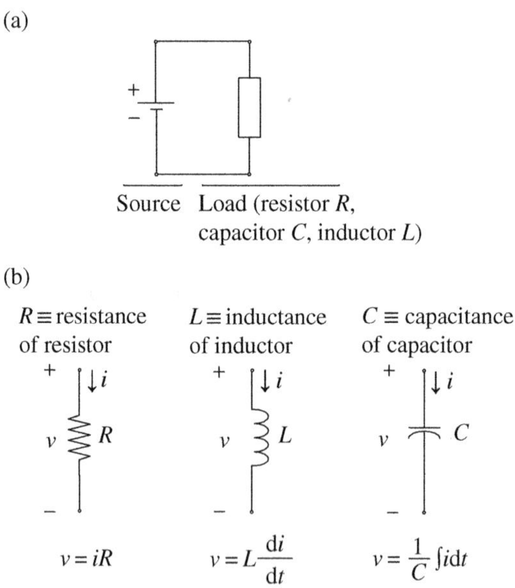

Figure 1.22 (a) Ideal passive DC circuit with either resistor R, capacitor C or inductor L as load. (b) Definition of resistance R, inductance L, and capacitance C in terms of algebraic, differential, and integral equations, respectively.

1) The relation $v = iR$ is an algebraic relation and can be expressed in terms of i as the algebraic relation $i = v/R$.
2) Correspondingly, the differential equation $v = L(di/dt)$ can be expressed in terms of i as the integral equation $i = (1/L)\int v dt$.
3) The integral equation $v = (1/C)\int i dt$ can be expressed in terms of v as the differential equation $i = C(dv/dt)$.

The unit of R is ohms or Ω, that of L is henries or H, and that of C is farads or F.

1.15 Independent and Dependent Voltage and Current Sources

Voltage and current sources can be either independent or dependent. A dependent source is controlled by an independent source, and a source can either be of a DC source or of a time-dependent (e.g. AC) type. Figure 1.23a defines a time-independent DC and a time-dependent voltage source. Similarly, Figure 1.23b depicts a time-independent DC current source and a time-dependent current source. Figure 1.23c and d present dependent voltage and current sources, where a diamond- or rhombus-type symbol indicates a dependent source. Applications of Figure 1.23c and d are amplifier circuits, which increase the power of a signal.

1.16 Galvanic Elements, Voltaic Series, and Lead–Acid Batteries

A galvanic element harnesses electrical energy from spontaneous reduction–oxidation (redox) reactions between chemical reagents; one type consists of a zinc (Zn) electrode, an electrolyte (e.g. $CuSO_4$), and a copper (Cu) electrode. If one connects the Zn and Cu electrodes by a conducting wire, the Zn^{++} ions migrate from the Zn electrode into the electrolyte, and electrons wander from the Zn electrode to the Cu electrode; that is, the Zn electrode becomes smaller, losing Zn atoms (Zn → Zn^{++} ions + electrons). Correspondingly, the electrons flowing from the Zn electrode combine with Cu^{++} ions residing within the electrolyte and form copper atoms, which increase the volume of the copper electrode ($Cu^{++} + 2e \rightarrow Cu$).

The direction the electrons travel (here from the Zn electrode to the Cu electrode) depends upon the potential difference $\Delta\varphi_{\text{metal-electrolyte}}$ that exists between the metal and electrolyte. The electron flow is such that the metal (here Zn) separates into ions and electrons having a higher (than Cu) potential difference

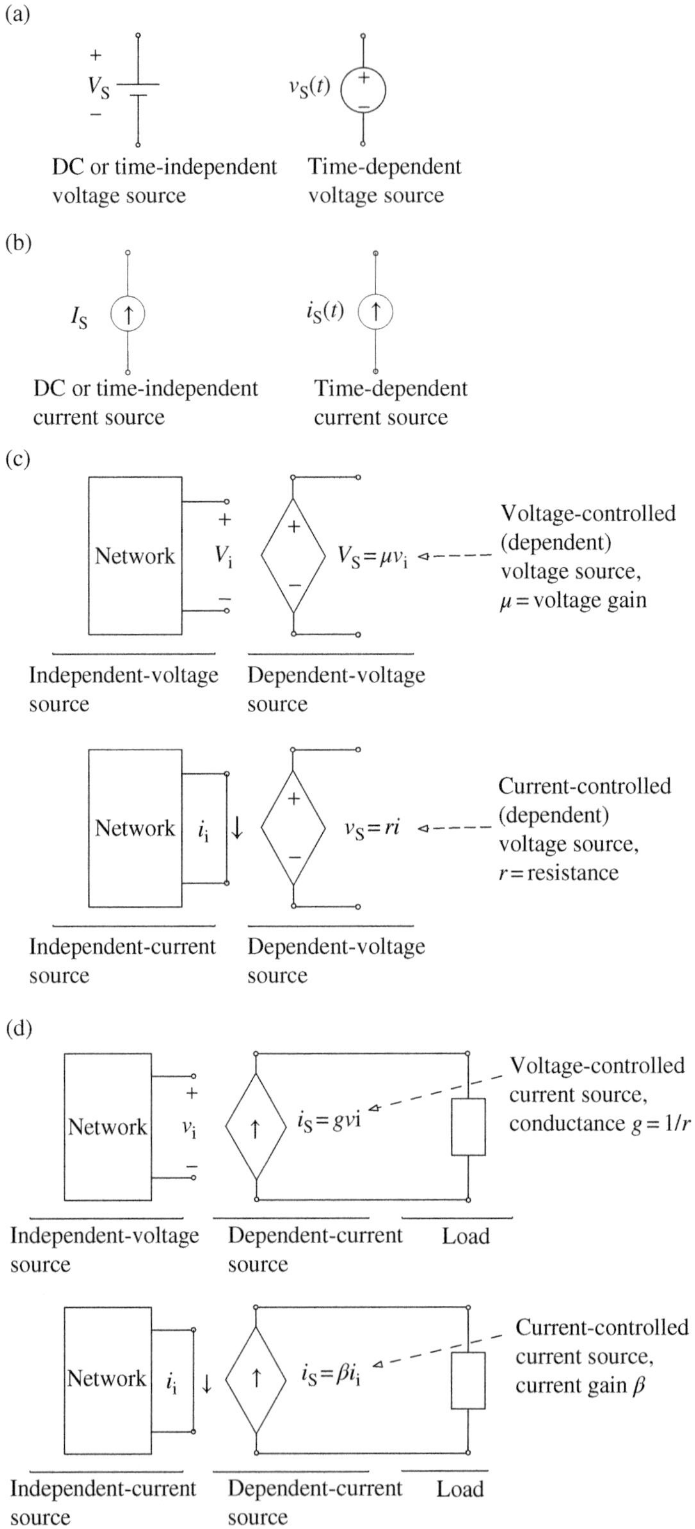

Figure 1.23 (a) Time-independent DC and time-dependent voltage sources. (b) Time-independent DC and time-dependent current sources. (c) Dependent voltage sources. (d) Dependent current sources.

Table 1.2 Voltaic series.

Row 1	Metal electrode	Li	K	Na	Mg	Zn
	$\Delta\varphi_{\text{metal-electrolyte}}$ (V)	−3.02	−2.92	−2.71	−2.35	−0.762
		1				
Row 2	Metal electrode	Fe	Cd	Ni	Pb	H_2
	$\Delta\varphi_{\text{metal-electrolyte}}$ (V)	−0.44	−0.402	−0.25	−0.126	0
		2				
Row 3	Metal electrode	Cu	Ag	Hg	Au	
	$\Delta\varphi_{\text{metal-electrolyte}}$ (V)	+0.345	+0.80	+0.86	+1.5	

$\Delta\varphi_{\text{metal-electrolyte}} = \Delta\varphi_{\text{Zn-CuSO4}}$ between metal (e.g. Zn) and electrolyte. The metal with the higher potential serves as the positive pole of the galvanic element.

When choosing a combination of electrodes from the voltaic series of Table 1.2, one can arrange the metals of the voltaic sequence or series such that all metals to the right (e.g. Fe, Cd, Ni, Pb, H_2, Cu, etc.) of a certain metal (e.g. Zn) form the positive pole. Note that Table 1.2 is displayed in three separate sets of rows (labeled Row 1 to Row 3) for ease of display, but that Row 2 can be considered to be appended to the right of Row 1 (i.e. Fe is directly to the right of Zn) and Row 3 to the right of Row 2. For example, Zn is the negative pole with respect to Fe, Cd, ..., Au electrodes, but a Li electrode forms the negative pole with respect to K, Na, Mg, Zn, Fe, ..., Au electrodes.

The voltage measured between the two electrodes of Figure 1.24b is the difference in voltage between the Zn electrode and the electrolyte ($CuSO_4$) and the Cu electrode and the electrolyte. It is not possible to directly measure the potential difference between an electrode and the electrolyte ($CuSO_4$); that is, only $v = \Delta\varphi_{\text{metal1-metal2}} = \Delta\varphi_{\text{Cu-Zn}} = -\Delta\varphi_{\text{Zn-Cu}}$ can be measured.

Vanadium redox-flow batteries [57] might be useful for storing excess power due to excess PV and wind power: rated power of 200 MW and energy storage of 800 MWh appear to be feasible.

Application Example 1.12 Lithium-gold battery

A battery consists of lithium (Li) cathode and gold (Au) anode. Determine the voltage developed across one Li–Au cell.

Solution

A Li electrode and an Au electrode generate the largest potential difference: $\Delta\varphi_{\text{Li-Au}} = -3.02\text{ V} - 1.5\text{ V} = -4.52\text{ V}$.

A lead–acid battery [58–60] works somewhat differently than batteries governed by voltaic series and is based on the polarization of positive and negative charges. Two lead (Pb) plates residing within a sulfuric acid (H_2SO_4) solution are covered with a $PbSO_4$ layer. During the charging, lead forms on the cathode, and PbO_2

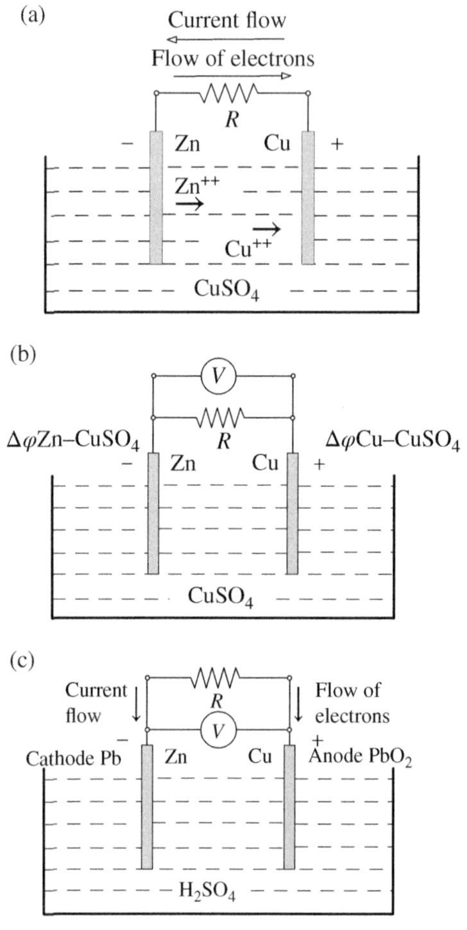

Figure 1.24 (a) Current flow of a galvanic element consisting of Zn and Cu electrodes and $CuSO_4$ electrolyte. (b) Potential distribution across a galvanic element consisting of Zn, Cu electrodes and $CuSO_4$ electrolyte. (c) Configuration of a (fully charged) lead–acid battery with current flow during discharge.

forms a deposit on the anode. In doing so the following chemical processes take place:

On cathode: $PbSO_4 + H_2 \rightarrow Pb + H_2SO_4$,

On anode: $PbSO_4 + SO_4 + 2H_2O \rightarrow PbO_2 + 2H_2SO_4$.

During discharging, the electrons flow from cathode to anode (or the current flows from anode to cathode) until the above chemical processes have been reversed and $PbSO_4$ forms on the cathode and anode. The voltage across the two lead plates is $e_0 = 2.02$ V for one set of plates. If six sets of lead plates are connected in series, the total terminal voltage is $E_0 = 6e_0 = 12.12$ V; the positive pole (anode) is covered with PbO_2, which gets reduced as the discharging progresses. The energy efficiency of a lead–acid battery (η_B) is such that one can recover about 83–90% of the energy with which the battery has been charged. Depending upon the temperature, the discharging capabilities of a battery decline with negative (e.g. −10 °C) temperatures. This represents a problem for electric/hybrid cars.

Application Example 1.13 Electric automobile

The large-scale introduction of electrical automobiles hinges on two important issues: mileage range and fast charging of (heavy) batteries. If the weight of cars is reduced, then their energy consumption can be either reduced or their mileage can be increased based on the figure of merit (FM) relation [20]:

$$\text{FM} = \frac{(\text{weight in tons} - \text{force}) \cdot (\text{distance in miles})}{(\text{energy used at wheel from battery in kWh})} = 4 - 10.$$

In particular the use of carbon fiber [61] and aluminum will additionally decrease the weight of automobiles, as does replacing mechanical gears with electric gears, so that the distance range of an electric car is 233 km = 145 miles. An automobile with a carbon-fiber passenger cell and an aluminum-drive module at a curb weight of 1.0 ton-force has a mileage of 145 miles per battery charge of 30 kWh, as used at the wheel. Determine the figure of merit FM and the weight of the car without batteries.

Solution

The figure of merit FM = (1.0 ton-force)·(145 miles)/(30 kWh) = 4.8. Note that the battery cannot be completely discharged and the battery must have an energy content of about 30 kWh·1.3 = 39 kWh. The weight of the 39 kWh battery is 39 kWh/(0.2 kWh/kg-force) = 195 kg-force. That is, the car without the battery weighs 805 kg-force. The auxiliary energy needs of a car, such as heating and control applications, require additional energy. In addition, the battery performance during the cold season is decreased so that the range for one charge is significantly lowered to about 160 km [62, 63]. For this reason, energy contents of up to 100 kWh are considered for electric cars and 250 kWh for small trucks.

Application Example 1.14 Selection of a battery for an automobile

Batteries age faster if a few large-deep discharges have occurred. Lead–acid batteries can have about 400 charge/discharge cycles, while lithium-ion batteries can have up to 2000. Most internal combustion engine cars have lead–acid batteries. One differentiates between the battery used only for starting and that for start-stop automatic and regeneration applications as they occur in electric/hybrid automobiles. The European Type Number (ETN code) provides information about the construction and performance of batteries [64–66].

1.17 Electrolysis

The technology of hydrogen production through conventional water electrolysis is well established [67]. Conventional electrolysis splits water into its components – hydrogen (H) and oxygen (O) – by supplying a water tank (Figure 1.25) with an electrical current. The electric charge breaks the chemical bond between the hydrogen and oxygen and splits apart the atomic components. The resulting ions form at two poles: the positively charged anode and the negatively charged cathode. Hydrogen ions gather at the cathode and react with it to form hydrogen gas, which is then collected. Oxygen goes through a similar process at the anode. The main drawbacks of conventional electrolysis for large-scale hydrogen production are the amount of electricity required and the high cost of membrane (diaphragm) production. The efficiency of conventional electrolysis, where the losses heat up the electrolyte, is about 80%.

At high-temperature electrolysis (HTE), process heat increases some of the energy required to split the water into hydrogen and oxygen instead of electricity, thus reducing the overall energy required. HTE uses a device very similar to a solid oxide fuel cell (SOFC). Essentially, the electrolytic cell consists of a solid oxide electrolyte with conducting electrodes deposited on either side of the electrolyte. A mixture of steam and hydrogen at (750–950) °C is supplied to the cathode side of the electrolyte. The electrical potential draws oxygen ions O^+ through the electrolyte; the ions then combine to O_2 on the cathode side. The steam–hydrogen mixture exits, and the water and hydrogen gas mixture is passed through a separator to separate hydrogen. Because using process heat directly is more efficient than first converting heat into electricity, the overall efficiency of the high-temperature system is much higher than that of conventional electrolysis. That assumes, of course, that one has readily available a non-fossil-fuel-based source of high process heat (e.g. either an advanced high-temperature nuclear reactor or an adapted solar thermal energy system).

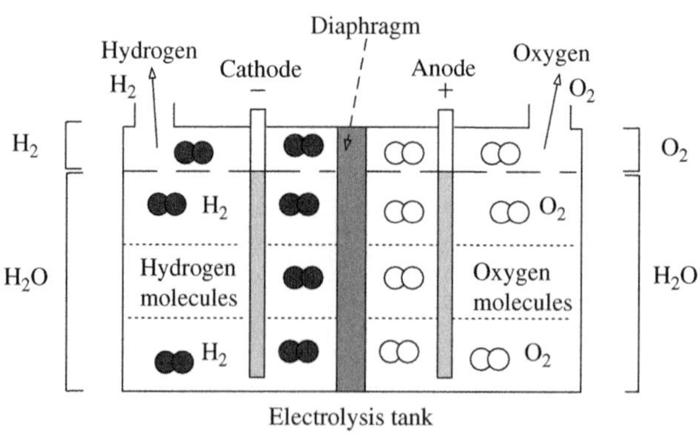

Figure 1.25 Conventional electrolysis.

1.18 Flow Batteries and Fuel Cells

Fuel cells and batteries are similar because they use a chemical reaction to provide electricity. Most batteries rely on chemical reactants, usually metal compounds like lithium, zinc, or manganese. Once used up, one must recharge the battery, which has a limited lifetime of about 4–10 years. Flow batteries, also known as redox (reduction–oxidation) flow [57] batteries, are similar to fuel cells in that the reactants are externally stored and supplied to the battery. Until now, flow batteries have not been used for electric cars. A fuel cell generates electricity through reactants, e.g. hydrogen and oxygen, stored in a tank [68]. A fuel cell and a flow battery will produce electricity as long as they have a fuel supply; that is, a fuel cell vehicle is refueled instead of recharged. Unfortunately, hydrogen is dangerously explosive, and for this reason fuel reformers are employed to reform methane – stored in an external tank in an electric car – into hydrogen. Presently the most common battery for electric car applications is the lithium-ion battery. Proton exchange membrane fuel cells (PEMFC) are a type of fuel cell being considered for transport, stationary, and portable applications. In order to use regeneration in fuel cell-powered cars, batteries are required for accepting the regenerated/braking energy.

1.19 Reformer

Fuel cells employ an attractive power generation technology that converts chemical energy directly and with high efficiency into electricity – all while causing little pollution. Most fuel cells require hydrogen as the fuel, but viable near-term applications will likely need to use the more readily available hydrocarbons, such as methane. Present-day fuel cell electric vehicles therefore include a reformer that

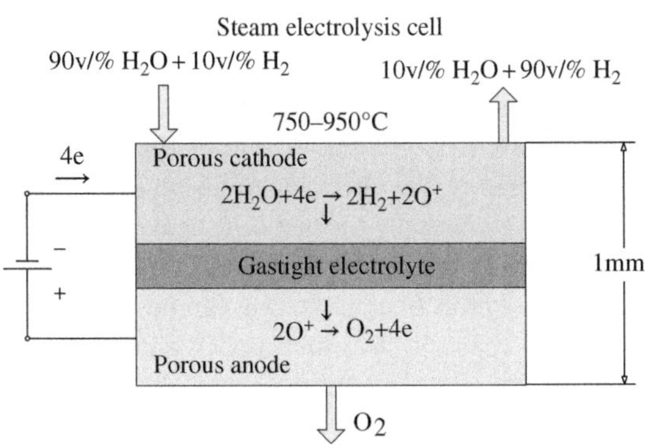

Figure 1.26 High-temperature [67] electrolysis, where "v/%" stands for "volume percent."

converts hydrocarbon fuel at high temperature (e.g. 600 °C) into hydrogen [69]. The disadvantages of a reformer are its long time constant and high operating temperature; additionally, a battery might be needed to speed up the dynamics of an electric drive and to exploit braking energy through regeneration.

1.20 Energy Storage Plants

Most renewable energy sources operate intermittently (e.g. wind, PV), and therefore energy storage plants [20, 30] are required to absorb energy when there is too much generation, as well as release energy during low power generation periods. One can classify storage plants into short-duration (e.g. battery, flywheel, hydro plants with DFIGs, and gas turbines fed by biogas) and long-duration (e.g. conventional hydro plants) sources. Integrating sufficient responsive energy storage into the electrical grid is one of the major obstacles to successfully addressing the climate challenge and is consequently increasing in importance as a research area. Any endowed electron – which contains energy – generated in the power system whose energy must be momentarily consumed in the served load of the power system result in a free electron, which does not have the potential to do useful work. If this energy balance is not maintained, the nominal power system frequency $f = 60$ Hz either decreases or increases, leading to instability or a collapse of the power system, if the rate change of frequency [70] is too large. Usually frequency variations within 59–61 Hz are permitted. Short-duration energy storage plants generally do not face environmental opposition in contrast to long-duration plants such as pump hydropower plants, because they do not take up valuable land – they can be buried [71, 72].

1.21 Current Projects and Issues with Potential Solutions

The world community faces problems with respect to energy and water supplies as well as pollution (e.g. NO_x, particles, CO_2, methane). A few of them will be addressed, such as provision of heating energy, the operation of fusion as it occurs on the sun, and the non-saline groundwater supply throughout the earth. Application Example 1.15 demonstrates – albeit on a small scale – the utilization of geothermal heat. Provided earthquakes due to this heat extraction can be avoided, it represents minimum generation of heat-trapping gases such as CO_2 and methane.

Application Example 1.15 Geothermal district heating in Munich [73–76]
Munich's municipal utility, Stadtwerke München GmbH (SWM), has one of the largest district heating networks in Europe. The network currently encompasses

800 km in length throughout the city and mainly harnesses the waste heat from electricity generation via combined heat and power (CHP) plants, which are thus around 90% efficient. SWM recently made public its goal for district heating to be powered entirely by renewable sources by 2040 [76].

This will primarily be accomplished by further exploiting Munich's excellent geothermal resources. Directly underneath the city, at a depth of around 3000 m, is a massive thermal aquifer of porous limestone, the Malm, containing water at the temperature of the surrounding earth, around 100 °C. Further south toward the Alps, the depth of this limestone layer increases, as does the temperature, to 140 °C under the village of Sauerlach [76], as seen in Figure E1.15.1.

Hot water brought up from underground runs through a heat exchanger, which transfers geothermal heat to the water in the district heating network, such that the underground water returns to the aquifer at a much cooler temperature, leaving the water table unaffected (see Figure E1.15.2). For times of peak demand, waste heat from CHP – or other future thermal generation methods – remains as a supplementary heat source.

As of 2013 two geothermal district heating plants were already in operation in the east and south of Munich, in Messestadt Riem and Sauerlach, respectively, and two additional plants are to be completed by 2019. However, to achieve its ambitious renewable energy goals for the heating network, SWM must build up to five additional geothermal plants by 2025 [75]. Thus, from November 2015 through

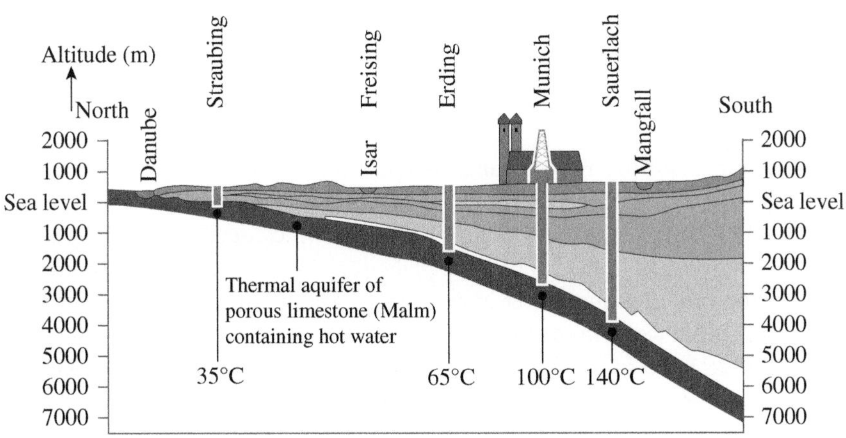

Figure E1.15.1 North–south cross section of Alpine foothills near Munich [73–76].

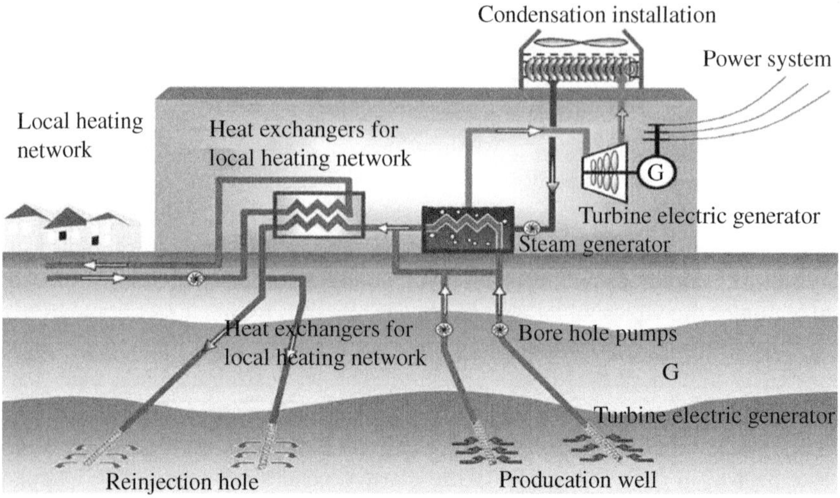

Figure E1.15.2 Schematic of geothermal district heating plant in Messestadt Riem, Munich [73–76].

March 2016, SWM conducted "vibro-seismic" research over 170 km^2; these precise measurements of thermal water depth will allow SWM to find the most appropriate sites for these plants [74–76]. This approach may hold promise in other areas with favorable hydrothermal resources and geology.

Application Example 1.16 demonstrates the availability of freshwater accumulated during the past 50 years.

Application Example 1.16 Global distribution of modern groundwater depth

Figure E1.16.1 maps, for the first time, "modern" (recharged less than 50 years ago) groundwater volumes around the globe [77]. While there are an estimated 22.6 million km^3 of groundwater in the upper 2 km of Earth's crust, dominating other volumes of water in the hydrologic cycle, less than 6% of this water is modern, meaning humans are overdrawing many shallow aquifers, with important implications for water, energy, food, and ecological systems. Although freshwater near coastal communities can be produced from saltwater through desalination [78], this approach is limited due to the reliance on electrical energy for booster and high-pressure pumps providing, for example, 1200 psi of pressure to achieve seawater reverse osmosis.

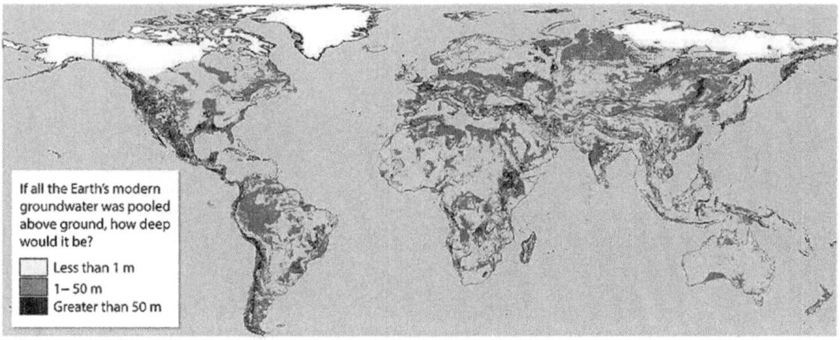

Tom Gleeson, Kevin M. Befus, Scott Jasechko, Elco Luijendijk and M. Bayani Cardenas (2015). *Nature Geoscience*, DOI: 10.1038/NGEO2590.

Figure E1.16.1 Global distribution of modern groundwater depth.

Application Example 1.17 attempts to explain the difficulties scientists have in copying on Earth the controlled fusion of hydrogen to helium.

Application Example 1.17 Controlled fusion

Calculate the number of fusions per second required for the controlled fusion of hydrogen to form helium on the sun, which occurs at a solar-core temperature of 14 million kelvins.

1) How much energy is released when four hydrogen atoms fuse to form one helium atom?
2) If the power output from the sun is approximately 4×10^{26} J/s, estimate the number of fusion reactions per second that must occur to generate the above-mentioned power.

Solution

Use $E = mc^2$. In this case the mass difference $(m_{4H} - m_{1He})$ of the four hydrogen atoms fused to one helium atom and converted into energy is 0.048×10^{-27} kg. With $c = 3 \times 10^8$ m/s, then one fusion reaction to form one helium atom generates the energy $E = (0.048 \times 10^{-27}\ \text{kg})(3 \times 10^8\ \text{m/s})^2 = 4.3 \times 10^{-12}$ J. This energy released in the fusion of four hydrogen atoms may not seem to be very much, but there are many hydrogen atoms in the sun. The amount of energy we receive from the sun is $E_{\text{total}} = 4 \times 10^{26}$ J/s; therefore, 4×10^{26} J/s/$(4.3 \times 10^{-12}$ J$) = 10^{38}$ fusions per second occur. Figure E1.17.1 depicts the controlled fusion of hydrogen to helium atoms on the surface of the sun [80]. Reproducing the sun's controlled fusion process on Earth is attempted with the Wendelstein [79] fusion reactor (see Figure E1.17.2). Note that a nuclear fusion bomb is based on a not-controlled

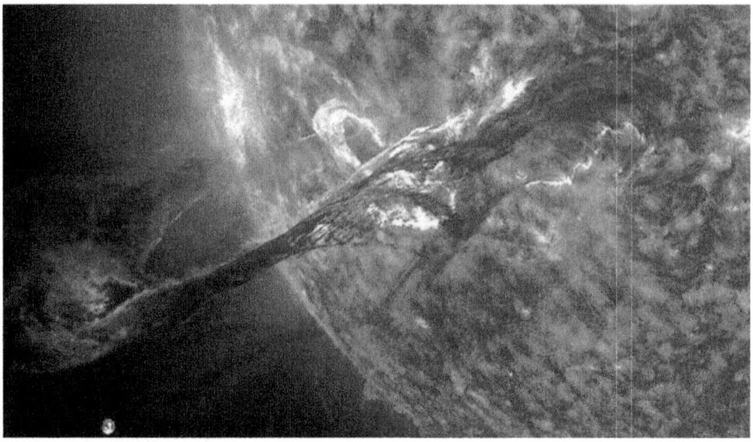

Figure E1.17.1 Depicts the controlled fusion [80] of hydrogen to helium atoms as it occurs on the surface of the sun.

Figure E1.17.2 Wendelstein [79] fusion reactor, not yet operational.

phenomenon. The difficulty associated with the operation of fusion reactors is to maintain a very high temperature – as it occurs on the sun – and at the same time to prevent an explosion in order to extract the fusion energy in form of heat in a controlled manner by containing the hydrogen-ion plasma by means of magnetic fields, because no material is able to withstand a temperature of about 14 million kelvins.

Light pollution of the sky cannot be easily prevented since human settlements must employ light on the ground. It appears very difficult to achieve light on the ground to increase safety for population and to prevent the illumination of the sky as shown in Figure E1.18.1 for Berlin, Germany.

Application Example 1.18 Light pollution due to street and building lighting in Berlin

Figure E1.18.1 shows the illumination of Berlin as viewed from outer space. Due to the great impact of lights within and above human settlements – so-called light smog – caused by illumination of streets and buildings of modern civilization (e.g. cities), the stars of the sky above settlements become less visible for bird migration, which relies on stars for orientation. Thus the yearly bird migration is impaired, and the birds traveling during fall and spring, e.g. between Europe and Africa, becomes impossible for some species. It is well known that insects such as butterflies and others – worldwide there are about 120 000 species of

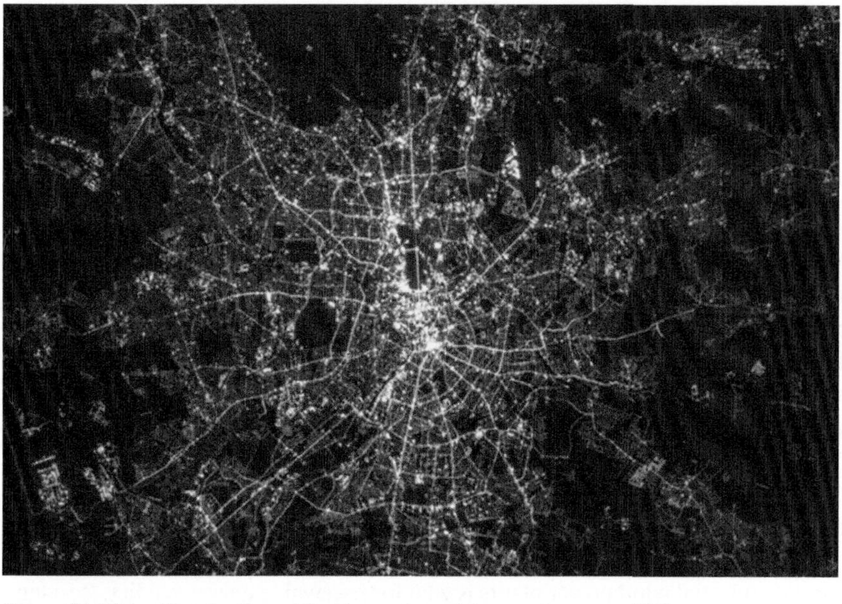

Figure E1.18.1 Illumination of Berlin as viewed from outer space [81].

flies – congregate around light sources and eventually die because of exhaustion, thus reducing the food supply for birds and marine species (e.g. fish, frogs) and limiting pollination of food crops and other plants. Baby turtles hatched at the beach orient their travel to the sea based on moon and stars. Lights of nearby cities disorient them, and instead of diving into the sea, they end up on city streets where they are killed by cars. Human sleep is affected as well, if the street light cannot be blocked from entering sleeping quarters: it is well known that human liver function is diminished by lights between 1:00 and 3:00 in the morning, and the flicker of fluorescent street lighting can induce seizures. The color of light plays an important role in the well-being of humans and animals, and the best are those colors between 2700 and 3000 K, and not used in outer building lights (Figure E1.18.1). There are more reasons (e.g. humans cannot discern stars such as the North Star and other star patterns, formerly relied on for navigation on land and sea) why such "light smog" should not exist above cities and other populated areas.

1.22 Software in Public Domain (e.g. PSPICE, Mathematica, MATLAB/Simulink)

Commonly available engineering and science software programs are PSPICE [82], Mathematica [83], and MATLAB/Simulink [84]. Some of them will be applied to the solution of application examples within chapters and problems listed at the end of chapters.

1.23 Summary

We describe basic energy conversion processes converting heat and chemical to mechanical energy based on the adiabatic process and Carnot efficiency; the related Rankine cycle applies to steam turbines, and the Brayton Cycle is used to analyze natural gas turbines. Both types of turbines will be important for renewable energy generation, such as woodchip and natural gas-fired power plants. Various (e.g. short- and long-term pump storage) hydropower plants will gain in importance as well as energy conservation techniques by employing groundwater–water heat pumps (such that the operation mode can switched between summer and winter operation mode) and geothermal power plants combined with district heating of large cities (see Application Example 1.15). The analysis of horizontal and axial wind power plants is vital for renewable energy studies. Residential, commercial, and industrial buildings with rooftop PV power plants are essential to conserve and generate heating and air-conditioning power.

Coulomb's law provides an entry into electrical engineering topics, starting with the discussion of electrostatic forces due to distributed electric charges and advancing to the definition of conductors, insulators, semiconductors, voltages, currents, power, and energy. Lithium-ion and flow batteries as well as the application of fuel cells are indispensable for the design of electric automobiles. According to automobile manufacturers [85], the most promising single-family car design has an electric drive consisting of pressurized hydrogen tank, fuel cell, lithium-ion battery, and onboard charger and electric motors, where a single tank permits a travel range of 500 km ≈ 312 miles. The weight of the entire car is about 1.9 ton-force: 50% of this weight is required for the electric drive. To reduce the overall weight, aluminum or carbon fiber is relied on for the car construction, and the mechanical gear is replaced by an electric gear. The 1.9 ton-force car accelerates at high speed only very slowly [86] due to the low torque produced during flux weakening (FW) operation. To increase torque at high speed, compensation of flux weakening (CFW) is proposed [87–90]. The efficiencies of electrolysis, fuel cell, electric machine, onboard electric charger, and battery are 66, 60, 90, 90, and 80%, so the overall efficiency of the entire drive is about 26% neglecting regeneration. This is not much better than the efficiency of an internal combustion with about 30%; however, the electric drive neither generates particulate matters nor nitric oxides (NO_x), which is a very important advantage.

Appendices 1.A and 1.B present an example for the design of a PV power plant on the roof of a single-family house and explain the nature of electricity and its manufacturing, respectively. Appendix 1.C details the incurring costs of electricity per kWh in a renewable energy system; as can be noted, these costs appear to be very large. It appears that the transition from a conventional grid consisting mainly of coal, nuclear, and natural gas-type power plants to renewable ones with solar and wind power sources will not be easy, although E. L. Doub (private communication) is of contrary opinion based on the public speeches of T. Seba (private communication). At the same time, it is critical for a reasonably stable climate system to move rapidly away from carbon-based fuels.

Problems

1.1 Adiabatic process and Carnot efficiency

The Carnot efficiency of Eq. (1.4) $\eta_{\text{Carnot}} = \dfrac{T_{\text{i}} - T_{\text{f}}}{T_{\text{i}}}$ is an ideal (e.g. 60%) value neglecting the energy required for the compressor and the losses of the turbine through expanding the medium (e.g. hot steam, gas or air). It is an adiabatic process where idealized conditions are assumed as defined by Sadi Carnot [6]. For a steam turbine with $P_{\text{rated}} = 583\ \text{MW}$ output power, an input temperature of $T_{\text{i}} = 1565.3\ {}^\circ\text{F} = 815.85\ {}^\circ\text{C} = 1089\ \text{K}$ at

$p_i = 4.6432 \cdot 10^6$ psi $\equiv 32.014 \cdot 10^9$ Pa, and an output (final) temperature of $T_f = 873.25\,°F = 740.51$ K, compute the output pressure p_f of the steam turbine using the software [9].

1.2 Rankine cycle applied to steam turbines

For the initial temperature $T_i = 500\,°C$ and final (output) temperature $T_f = 250\,°C$ of a water steam turbine of 2.5 MW with $\phi = 0.5$, determine the Rankine efficiency $\eta_{Rankine}$.

1.3 Rankine cycle applied to steam turbines [20]

Large steam power (nuclear) plants can have ratings up 1200 MW. The prime mover consists of compressor and steam turbine and is governed by the Rankine cycle, taking into account the losses of the compressor and the turbine. Provided the combustor efficiency is $\eta_{combustor} = 0.84$, the compressor efficiency is $\frac{P_{out}^{compressor}}{P_{in}^{compressor}} = \eta_{compressor} = 70\%$, the steam turbine efficiency is $\frac{P_{out}^{steam\ turbine}}{P_{in}^{steam\ turbine}} = \eta_{steam\ turbine} = 82\%$, and the steam turbine output power is $P_{out}^{turbine} = W/3600$ s/h $= 598$ MW as indicated in Figure P1.3.1. Calculate based on the software [9] the required input power to the combustor $P_{in}^{combustor}$ for the temperature and pressure conditions of Problem 1.1.

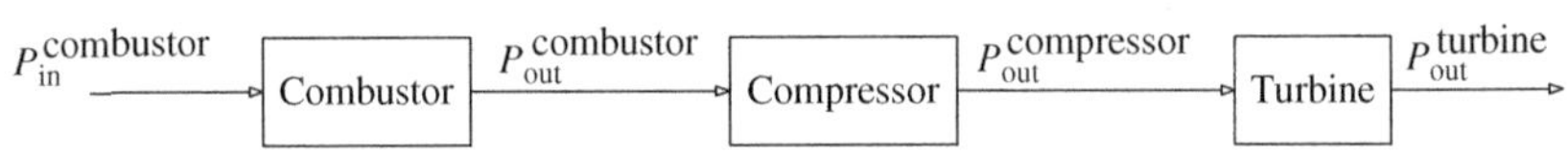

Figure P1.3.1 Block diagram of combustor, compressor, and steam turbine with efficiency and power values $P_{in}^{combustor}$, $P_{out}^{combustor}$, $P_{out}^{compressor}$, and $P_{out}^{turbine}$.

1.4 Brayton cycle applied to gas turbines

For the temperatures $T_1 = 120\,°C$, $T_2 = 220\,°C$, $T_3 = 370\,°C$, $T_4 = 220\,°C$ of a gas turbine. Determine the Brayton efficiency $\eta_{th,\ Brayton}$.

1.5 Brayton cycle applied to gas turbines

Large gas turbine power plants can have ratings up 400 MW. The prime mover consists of compressor and natural or biogas turbine and is governed by the Brayton cycle, taking into account the losses of the combustor, compressor, and the gas turbine.

Provided the combustor efficiency is $\eta_{\text{combustor}} = 0.90$, the compressor efficiency is $\frac{P_{\text{out}}^{\text{compressor}}}{P_{\text{in}}^{\text{compressor}}} = \eta_{\text{compressor}} = 80\%$, the gas turbine efficiency is $\frac{P_{\text{out}}^{\text{steam turbine}}}{P_{\text{in}}^{\text{steam turbine}}} = \eta_{\text{steam turbine}} = 87\%$, and the gas turbine output power is $P_{\text{out}}^{\text{turbine}} = W/3600\ \text{s/h} = 188\ \text{MW}$ as indicated in Figure P1.3.1. Calculate based on the software [9] the required input power to the combustor $P_{\text{in}}^{\text{combustor}}$ for the temperature and pressure conditions of Problem 1.1.

1.6 Ericsson cycle applied to gas turbines

For the temperatures $T_{\text{H}} = 1000\ °\text{F}$, $T_{\text{L}} = 700\ °\text{C}$ of a gas turbine, determine the Ericsson efficiency $\eta_{\text{th, Ericsson}}$.

1.7 Hydropower plant with Pelton turbine

A $P_{\text{hydro}} = 50$ MW hydropower plant consists of a Pelton turbine, a 20-pole induction generator with a synchronous speed of $n_{\text{s}} = 180$ rpm, and a slip of $s = -3\%$ connected to the utility system.

If the power efficiencies of the water turbine and the induction generator are $\eta_{\text{turbine}} = 0.80$, $\eta_{\text{gen}} = 0.95$, respectively,

a Compute the required input power of the water turbine $P_{\text{turbine}}^{\text{required}}$.

b Provided the head of the water is $H = 300\ \text{m} = 984.6\ \text{ft}$ and the water flow is $Q = 850$ cfs (cubic feet per second) $\equiv 24\ \text{m}^3/\text{s}$, compute the mechanical power available at the turbine input [20, 23–25]

$$P_{\text{kW}} = \frac{H \cdot Q \cdot W \cdot 0.746}{550},$$

where H is the head of the water in feet (ft), Q is the flow of water in cubic feet per second (cfs), W is the weight of a cubic foot of water $W = 62.4$ (lbs-force)/ft^3, 0.746 is a constant used to convert horsepower (hp) to kilowatts (kW), and 550 is a constant used to convert {ft-(lbs-force)}/s into horsepower.

c How does $P_{\text{turbine}}^{\text{required}}$ compare with P_{kW}?

d Compute the specific speed $N_{\text{q}} = n\frac{Q^{0.5}}{H^{0.75}}$, where n is the turbine speed measured in rpm, Q is the flow of water measured in (m^3/s), and H is the head of water measured in m.

e Is the selection of the Pelton-type turbine according to Figure 1.10 justified?

1.8 Operating point of groundwater–water heat pump for a 400 m^2 residence

Determine for a feed groundwater temperature of $T_{\text{groundwater}} = 6\,°\text{C}$ and a warmwater flow temperature of 55 °C the coefficient of performance COP, the electrical power consumption of $P_{\text{in-electric}}$, and the heating output power P_{heating}.

1.9 Analysis of a 5 MW wind power plant at an altitude of 1500 m

A wind turbine has an output power of $P_{\text{out}} = 5$ MW, without taking into account the efficiencies of gear, generator, and converter(s). The coefficient of power is $c_{\text{p}} = 0.45$, and the plant is at an altitude of 1500 m. Find the required area A swept by the turbine blades and the length of a blade for the rated wind speed [27] of $v = 12$ m/s. The air density [20] at an altitude of 1500 m is $\rho = 1.0583$ kg/m^3.

1.10 Energy consumption of house (kWh) per square meter (m^2) and per year

During one year the energy used for heating and cooling is EHC = 10 128 kWh/year and for lighting and appliances ELA = 1779 kWh/year. Determine the energy consumption (kWh) of house per square meter (m^2) and per year, that is, energy consumption kWh/(m^2·year) of 300 m^2 house.

1.11 Calculation of electrostatic forces

The three point charges $q_1 = -2$ milliC, $q_2 = 4$ milliC, and $q_3 = 10$ milliC, where milliC stands for millicoulombs, are placed in a triangular pattern in the $y - x$ plane as indicated in Figure P1.11.1. The distances are d_{12}, d_{13}, and

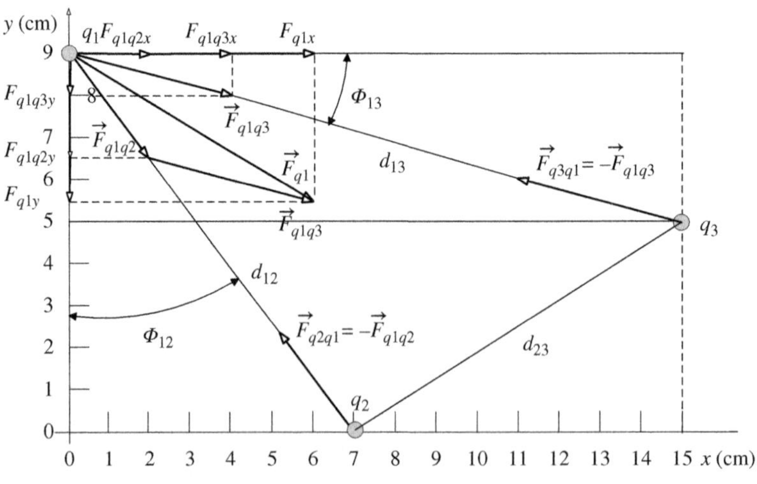

Figure P1.11.1 Arrangement of three charges in the $x - y$ plane.

d_{23}. Calculate the magnitude of the resultant force $\left|\vec{F}_{q1}\right|$ acting on q_1 due to q_2 and q_3 in newtons (N) and its direction.

1.12 Lithium–copper battery

A battery consists of a lithium (Li) cathode and a copper (Cu) anode. Determine the voltage developed across one Li–Cu cell.

1.13 Design of family electric car

An electric car has a total (including five persons) weight of 1430 kg-force, where the lithium-ion battery weighs 430 kg-force and it can travel with one battery charge 340 km. The stored battery energy is 60 kWh.

a Determine the stored specific energy kWh/(kg-force) of the battery.

b Calculate the figure of merit [20] FM = (tons-force)(miles)/(energy used at the wheels E_{wheels}), and assume that the entire energy stored in the battery (60 kWh) is applied at the wheels.

1.14 Comparison of two electric passenger automobiles

The mileage range of a passenger car must be 300 miles. Determine for a figure of merit FM = 7.0 the total weight if the automobile weighs without batteries either 1 ton-force (small car) or 2.8 tons-force (large car). Assume for a lithium–nickel–cobalt–aluminum oxide battery an energy density of 0.2 kWh/kg-force.

1.15 Electric automobile with fuel cell and variable-speed electric drive

If a proton exchange membrane (PEM) fuel cell [20, 30] is powered with pure hydrogen, it has the potential to convert $\eta_{\text{PEM}} = 80\%$ of the energy content of the hydrogen into electrical energy. The in-series connected motor-inverter drive has an efficiency of $\eta_{\text{motor-inverter}} = 80\%$. That gives an overall efficiency of $\eta_{\text{overall}} = 64\%$.

a Compute the weight of hydrogen (W_{hydrogen}) required for a $d = 400$ mile trip based on the above efficiencies and a figure of merit of the electric car FM = 5.0. The total weight of the car is $W_{\text{car}} = 2.8$ (tons-force), which includes the weight of the drive unit (PEM, reformer, small 20 kWh backup lithium–nickel–cobalt–aluminum oxide battery with an energy density = 0.2 kWh/(kg-force) with rated power $P_{\text{rated}} = 100$ kW of inverter with permanent-magnet motor). The efficiency of the fuel cell is $\eta_{\text{fuel cell}} = 64\%$.

b If the fuel source is not hydrogen, then the vehicle will also need a reformer converting hydrocarbon (e.g. methane) or alcohol fuels into hydrogen. This reforming process generates heat and other gases besides hydrogen, which must be eliminated. The hydrogen that exits the reformer is not pure, and this lowers the efficiency of the fuel cell to about $\eta_{\text{fuel cell+reformer}} = 50\%$. Compute the weight of methane (W_{methane}) required.

c For the hydrogen fuel case, what is the payload of the car consisting of weight of persons and luggage, provided the car weighs without drive components $W_{\text{frame+wheels}} = 1$ (ton-force)?

1.16 Evaluation of energy measurements for a single-family house

a For Table P1.16.1 plot in a similar manner as for Application Example 1.6 the energy parameters EI and EHC.

Table P1.16.1 Daily average energy readings per week: 07/15/2015 through 07/14/2016.

End date of week	PV-generated DC energy EI_i (kWh/d)	Energy supplied to utility grid EG_i (kWh/d)	Energy for heating and cooling EHC_i (kWh/d)	Energy for lighting and appliances ELA_i (kWh/d)	Water use (m^3/d)
07/21/2015	33.71	32.43	5.00	0.71	0.29
07/28/2015	26.71	24.86	4.00	1.00	0.43
08/04/2015	28.29	26.29	4.14	1.00	0.29
08/11/2015	36.29	33.86	4.43	1.00	0.43
08/18/2015	22.57	21.00	5.00	1.14	0.43
08/25/2015	25.86	23.43	6.00	1.43	0.29
09/01/2015	32.71	30.57	5.29	1.57	0.43
09/08/2015	15.71	13.86	6.86	1.86	0.29
09/15/2015	22.29	20.43	7.43	1.57	0.29
09/22/2015	19.57	18.00	8.43	1.57	0.29
09/29/2015	14.14	12.00	9.29	2.00	0.29
10/06/2015	23.57	22.14	8.86	1.71	0.29
10/13/2015	10.86	9.14	11.29	2.43	0.29
10/20/2015	6.86	5.86	12.14	2.00	0.29
10/27/2015	10.57	9.29	9.43	2.00	0.14
11/03/2015	14.14	13.57	10.43	1.43	0.14
11/10/2015	16.43	14.29	9.71	2.57	0.29

Table P1.16.1 (Continued)

End date of week	PV-generated DC energy EI_i (kWh/d)	Energy supplied to utility grid EG_i (kWh/d)	Energy for heating and cooling EHC_i (kWh/d)	Energy for lighting and appliances ELA_i (kWh/d)	Water use (m^3/d)
11/17/2015	9.29	8.57	9.14	2.00	0.29
11/24/2015	8.57	7.29	12.43	3.29	0.29
12/01/2015	2.43	1.57	14.57	3.14	0.29
12/08/2015	8.71	7.71	12.29	2.29	0.29
12/15/2015	7.43	6.00	13.71	3.00	0.29
12/22/2015	4.43	3.43	14.86	2.71	0.14
12/29/2015	8.71	8.14	15.71	1.43	0.14
01/05/2016	3.57	2.57	19.00	3.57	0.29
01/12/2016	3.86	3.29	17.29	2.43	0.14
01/19/2016	2.57	2.00	20.71	3.14	0.29
01/26/2016	7.12	6.10	20.40	2.71	0.14
02/02/2016	9.45	8.62	13.88	1.29	0.14
02/09/2016	7.71	6.43	16.71	2.29	0.29
02/16/2016	10.43	9.14	16.71	2.43	0.14
02/23/2016	8.14	7.00	16.29	2.57	0.29
03/01/2016	11.90	10.48	16.90	2.05	0.14
03/08/2016	12.67	11.10	17.38	1.95	0.29
03/15/2016	13.29	11.57	18.14	2.57	0.29
03/22/2016	25.00	23.43	14.00	2.00	0.29
03/29/2016	16.71	16.00	14.14	1.00	0.00
04/05/2016	23.14	21.29	10.14	1.29	0.29
04/12/2016	18.57	16.86	12.57	1.57	0.29
04/19/2016	15.57	13.71	12.00	1.71	0.14
04/26/2016	24.57	22.43	12.00	1.43	0.29
05/03/2016	27.40	25.02	11.86	1.65	0.29
05/10/2016	31.78	31.11	9.00	0.56	0.00
05/17/2016	18.97	17.73	10.43	0.94	0.14
05/24/2016	22.00	19.86	10.71	1.57	0.29

(Continued)

Table P1.16.1 (Continued)

End date of week	PV-generated DC energy EI_i (kWh/d)	Energy supplied to utility grid EG_i (kWh/d)	Energy for heating and cooling EHC_i (kWh/d)	Energy for lighting and appliances ELA_i (kWh/d)	Water use (m^3/d)
05/31/2016	27.86	26.57	14.00	0.86	0.00
06/07/2016	24.00	21.71	10.71	1.00	0.29
06/14/2016	22.14	19.43	12.29	1.71	0.29
06/21/2016	24.57	22.43	10.57	1.14	0.14
06/28/2016	30.71	28.00	11.29	1.29	0.43
07/05/2016	26.00	24.29	8.00	1.14	0.14
07/14/2016	27.22	24.89	14.00	1.33	0.33
Total energy produced and consumed for data year 2015–2016 (difference of cumulative energy as of 07/14/2015 and as of 07/14/2016)					
Year (cumulative)	6402.00	5837.00	4309.00	661.00	91.00

Note: Average for the last week includes 2 extra days to total 366 days for (leap) year.

b Determine the size of a battery (assume for a lithium–nickel–cobalt–aluminum oxide battery an energy density of 0.2 kWh/kg-force) to store the energy generated by the PV plant (EI) as listed in Table P1.16.1. Is such a storage battery feasible?

c Is the energy stored on the battery sufficient to operate the heat pump during winter season?

d Check whether the predicted yearly energy consumption of the 220 m^2 house (see Figure E1.6.2) given by 18 kWh/($m^2 \cdot a$) is correct, and neglect the influence of coefficient of performance [30] $COP_H = 3.98$.

1.17 Evaluation of energy measurements for a single-family house with latitude 48.1351 °N and longitude 11.5820 °E.

Check whether the predicted yearly energy consumption of the 220 m^2 house (see Figure E1.6.2) given by 18 kWh/($m^2 \cdot a$) is correct by taking into account the coefficient of performance [30] $COP_H = 3.98$ for the values as listed in Table P1.17.1.

Table P1.17.1 Daily average energy readings per week: 07/15/2016 through 07/14/2017.

End date of week	PV-generated DC energy EI_i (kWh/d)	Energy supplied to utility grid EG_i (kWh/d)	Energy for heating and cooling EHC_i (kWh/d)	Energy for lighting and appliances ELA_i (kWh/d)	Water use (m^3/d)
07/21/2016	28.57	25.86	10.14	1.14	0.29
07/28/2016	24.71	22.57	8.43	1.14	0.29
08/04/2016	26.43	24.14	11.43	1.43	0.29
08/11/2016	20.00	18.00	11.43	1.29	0.29
08/18/2016	27.14	25.71	11.43	1.29	0.29
08/25/2016	29.43	27.00	12.86	1.86	0.43
09/01/2016	28.00	26.43	8.43	1.86	0.29
09/08/2016	26.14	24.86	8.71	1.29	0.14
09/15/2016	27.19	25.50	10.83	1.76	0.29
09/22/2016	13.38	11.93	10.17	1.38	0.14
09/29/2016	27.14	24.71	11.00	1.57	0.43
10/06/2016	13.86	11.71	12.86	1.86	0.29
10/13/2016	13.57	11.71	17.57	2.14	0.14
10/20/2016	12.86	11.00	15.00	2.14	0.29
10/27/2016	10.57	8.71	13.57	2.14	0.29
11/03/2016	13.00	11.86	14.86	1.43	0.14
11/10/2016	8.86	7.71	16.57	2.43	0.14
11/17/2016	5.43	3.86	17.29	3.29	0.29
11/24/2016	6.14	4.86	16.00	3.14	0.29
12/01/2016	5.86	4.43	17.57	3.71	0.14
12/08/2016	7.29	5.86	20.00	2.71	0.29
12/15/2016	5.43	4.14	15.57	2.86	0.29
12/22/2016	3.71	2.71	15.71	3.29	0.14
12/29/2016	4.29	3.43	14.14	2.14	0.14
01/05/2017	4.43	3.57	19.29	3.43	0.29
01/12/2017	1.00	0.71	16.43	4.00	0.29
01/19/2017	1.43	0.71	19.43	4.14	0.14
01/26/2017	7.14	5.86	22.29	2.29	0.29
02/02/2017	10.14	8.86	18.57	2.71	0.14
02/09/2017	7.71	6.29	16.00	2.57	0.29

(Continued)

Table P1.17.1 (Continued)

End date of week	PV-generated DC energy EI_i (kWh/d)	Energy supplied to utility grid EG_i (kWh/d)	Energy for heating and cooling EHC_i (kWh/d)	Energy for lighting and appliances ELA_i (kWh/d)	Water use (m^3/d)
02/16/2017	15.57	14.14	17.14	2.29	0.29
02/23/2017	9.86	8.57	14.43	2.14	0.14
03/02/2017	15.86	14.00	14.00	2.57	0.29
03/09/2017	13.43	12.43	13.43	1.86	0.14
03/16/2017	23.43	21.14	13.29	1.71	0.29
03/23/2017	15.86	14.29	10.71	2.00	0.29
03/30/2017	28.57	26.14	10.71	1.71	0.14
04/06/2017	20.43	18.14	9.29	1.57	0.29
04/13/2017	25.86	23.86	9.45	1.21	0.29
04/20/2017	15.29	13.71	11.98	1.50	0.14
04/27/2017	18.71	16.71	12.57	1.71	0.29
05/04/2017	21.29	19.00	12.14	1.43	0.14
05/11/2017	22.29	20.14	10.00	1.71	0.29
05/18/2017	29.00	26.57	7.14	1.71	0.29
05/25/2017	28.71	27.00	6.86	1.14	0.29
06/01/2017	34.86	32.00	10.43	1.14	0.29
06/08/2017	24.43	22.26	10.72	1.37	0.29
06/15/2017	34.14	32.46	6.56	1.06	0.14
06/22/2017	34.00	32.29	13.43	1.00	0.29
06/29/2017	27.29	24.71	9.71	1.43	0.43
07/06/2017	27.50	25.39	11.86	1.46	0.14
07/14/2017	29.50	28.58	11.02	0.75	0.13
Total energy produced and consumed for data year 2016–2017 (difference of cumulative energy as of 07/14/2016 and as of 07/14/2017)					
Year (cumulative)	6586.50	5966.42	4774.17	721.25	89.00

Note: Average for the last week includes 1 extra day to total 365 days for year.

1.18 Determination of energy generation of PV plant on single-family residence with area = 70 m^2 panel surface (see Application Example 1.6) with latitude 48.1351 °N and longitude 11.5820 °E.

1.19 The objective of this problem is to investigate the influences of geographic location (latitude, longitude), geographic orientation, and roof pitch on the energy generation of two PV power plants.

A single-family residence has PV panels on the residence and on the adjacent garage (D.A. von Boyen and G.H. von Boyen, private communication, April 2018) with latitude = 48.0695 °N and longitude = 9.9516 °E. Details with respect to the two sets of PV panels are represented in Figure P1.19.1.

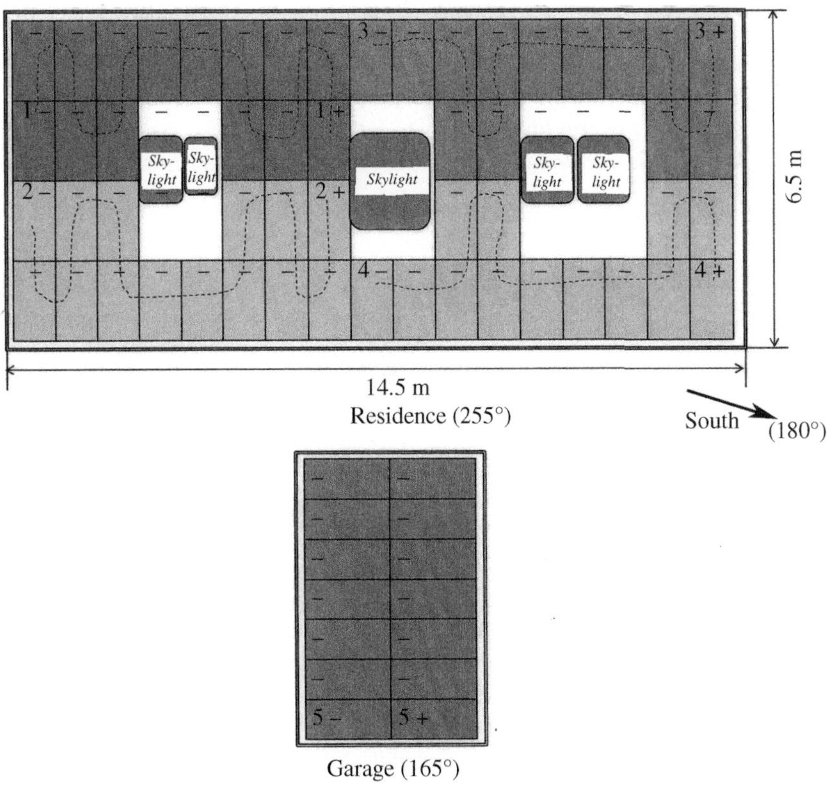

Figure P1.19.1 Configuration of PV panels on residence and on garage. The roof's pitch of the residence is 30° and that of the garage 35°. The geographic orientation of the PV panels on the roof of the residence is 255° and that of the garage is 165°, where north corresponds to 0°.

On the residence's roof, reside 54 PV panels with an area of $54 \cdot (1.581 \cdot 0.810)\ m^2 = 69.153\ m^2$ and on the roof of the garage 14 PV panels with an area of $14 \cdot (1.581 \cdot 0.810)\ m^2 = 17.929\ m^2$. The total energy generation (DC energy at the three inverter inputs) of the 68 PV panels is from 4/13/2017 to 4/13/2018 $E_{total} = 11\,765$ kWh.

In Application Example 1.6 the PV system installed on the roof of a single-family dwelling is discussed, whereby the location is defined by latitude = 48.1351 °N and longitude = 11.5820 °E. The roof's pitch is 35° and the orientation of the panels is 216°, producing within the time frame 7/17/2017 to 7/14/2018 (see Figure E1.6.5), the aging-adjusted energy value of $E_{Figure\ 1.6.5\_2018} = 6230$ kWh.

a Compute the energy density (energy generated per 1 m^2) for the PV system with latitude = 48.0695 °N and longitude = 9.9516 °E and for the PV system of Application Example 1.6 with latitude = 48.1351 °N and longitude = 11.5820 °E.

b Explain why the two single-family dwellings result in similar energy density values energy/m^2.

c Would you recommend installing also PV plants on northern parts of a roof?

References

1 U.S. Department of Energy (2014). *Photovoltaic Cell Conversion Efficiency Basics* (6 September 2014).

2 Fraunhofer Institute for Solar Energy Systems ISE. (2015). *Photovoltaics Report* (26 August 2015). https://www.ise.fraunhofer.de/de/downloads/pdf-files/aktuelles/photovoltaics-report-in-englischer-sprache.pdf (accessed 11 May 2020).

3 ub.de Fachwissen GmbH. http://www.photovoltaik.org/wissen/degradation-von-solarmodulen (accessed 11 May 2020).

4 E. Kaplani. Design and performance considerations in stand-alone PV powered telecommunication systems. *Journal of Engineering Science and Technology Review* 5(1) (2012) 1–6. http://jestr.org/downloads/volume5issue1/fulltext01050112.pdf (accessed 11 May 2020).

5 Ginley, D.S. and Cahen, D. (2011). *Fundamentals of Materials for Energy and Environmental Sustainability*. Cambridge University Press.

6 Dixit, U.S., Hazarika, M., and Paulo Davim, J. (2017). *A Brief History of Mechanical Engineering*. Springer.

7 Fuchs, E.F. (2010). Conventional and renewable energy issues. University of Colorado at Boulder, Lecture ECEN 5017, Fall 2010.

8 Pisupati, S. EGEE 102: Energy conservation and environmental protection. The Pennsylvania State University. https://www.e-education.psu.edu/egee102/node/1942 (accessed 11 May 2020).

9 Nave, C.R. http://hyperphysics.phy-astr.gsu.edu/hbase/thermo/adiab.html#c3 (accessed 11 May 2020).

10 Decher, R. (1994). *Energy Conversion*, 676. Oxford: Oxford University Press.

11 Brayton cycle. http://web.mit.edu/16.unified/www/SPRING/propulsion/notes/node27.html (accessed 11 May 2020).

12 Chegg, Inc. The Stirling and Ericsson cycles. https://www.chegg.com/homework-help/definitions/the-stirling-and-ericsson-cycles-5 (accessed 11 May 2020).

13 http://www.siemens.com/press/pool/de/feature/2014/energy/2014-04-orc/brochure-siemens-orc-module-e.pdf (accessed 11 May 2020).

14 Fernheizkraftwerk Toblach-Innichen Genossenschaft. (2005). *Vom Fernheizkraftwerk*, Toblach, Italy (June 2005), 35 p.

15 Turboden ORC – Modul ORC, Brecia, Italy. www.turboden.it (accessed 11 May 2020).

16 Alternative Fuels Data Center. Fuel properties comparison. http://www.afdc.energy.gov/fuels/fuel_comparison_chart.pdf (accessed 11 May 2020).

17 Bundesverband Wärmepumpe e.V. (2014). Heizen mit der Wärme aus Wasser, Erde und Luft, Funktionsprinzip einer Wärmepumpe. *Siedlung und Eigenheim, Zeitschrift für Haus, Wohnen und Garten*, Organ des Eigenheimerverbandes Bayern e.V. (October 2014), pp. 390–391.

18 Stiebel Eltron GmbH & Co. KG (2012). *7.64 kW Wasser Wärmepumpe (Groundwater Heat Pump) WPF 7 Basic*. Holzminden, Germany.

19 SANA CASE-Werner Müller, Drestedt.

20 Fuchs, E.F. and Masoum, M.A.S. (2011). *Power Conversion of Renewable Energy Systems*, 692. New York, 2nd Printing: Springer.

21 http://www.jahresarbeitszahlen.info/index.php/jahresarbeitszahl/cop-und-jaz (accessed 11 May 2020).

22 Bundesverband Wärmepumpe e.V. Heizung und Kühlung zugleich. *Siedlung und Eigenheim, Zeitschrift für Haus, Wohnen und Garten*, Organ des Eigenheimerverbandes Bayern e.V. (January 2018), pp. 20–22.

23 Cowdrey, J. (2004). *Boulder's Municipal Hydroelectric System*, 7 July 2004. Colorado: City of Boulder, 16 pp.

24 Zipparro, V.J. and Hasen, H. (1993). *Davis' Handbook of Applied Hydraulics*, 4e. New York: McGraw-Hill.

25 American Society of Mechanical Engineers (1996). *The Guide to Hydropower Mechanical Design*. HIC Publications.

26 Global Wind Energy Council (2015). *Global Wind Statistics 2014*. http://www.gwec.net/wp-content/uploads/2015/02/GWEC_GlobalWindStats2014_FINAL_10.2.2015.pdf (accessed 11 May 2020)

27 Bergey, K.H. (1979). The Lanchester-Betz limit. *Journal of Energy* 3 (6): 382.
28 http://www.zeit.de/wirtschaft/2015-04/sturm-netzbetreiber-strom-kosten-niklas (accessed 11 May 2020).
29 http://www.pro-erdkabel-neuss.de/bundesnetzagentur.html (accessed 11 May 2020).
30 Masoum, M.A.S. and Fuchs, E.F. (2015). *Power Quality in Power Systems and Electrical Machines*, 2e. Amsterdam: Academic Press/Elsevier, 1123 p.
31 Fairley, P. (2017). *For More Power, Float Wind Turbines Far Out at Sea*, 9–10. *IEEE Spectrum International.*
32 BSW-Solar, BDH (2015). Zwei Millionen Solarheizungen in Betrieb, Stand 1/2015. https://deinenergieportal.de/?p=4253 (accessed 11 May 2020).
33 Hilz, W. Warmth for the house from the sun. Sonnenhaus-Institut e.V. https://www.sonnenhaus-institut.de/?s=W%C3%A4rme+f%C3%BC+Haus+von+der+Sonne (accessed 11 May 2020).
34 König, B. (2013). Energy disclosure for house in Munich. König-Tragwerksplanung, Regen, Germany.
35 http://www.clay-works.com/blog/why-use-clay-plaster (accessed 11 May 2020).
36 http://www.livingwoodwindows.co.uk/whats-new/why-choose-triple-glazed-windows (accessed 11 May 2020).
37 https://www.solarwirtschaft.de/presse/pressemeldungen/pressemeldungen-im-detail/news/einspeisemanagement-kleiner-solarstrom-anlagen-uebergangsfrist-endet-zum-jahreswechsel.html (accessed 11 May 2020).
38 https://www.photovoltaik-web.de/photovoltaik/wechselrichter/wechselrichter-70-prozent-regelung-eeg-2012 (accessed 11 May 2020).
39 https://www.geoplex.de/fileadmin/user_upload/PDF_PV/Gesetzgebung_Photovoltaik_seit_2012.pdf (accessed 11 May 2020).
40 http://www.energieheld.de/energiesparen/heizkosten-sparen/durchschnittliche-heizkosten-eines-wohnhauses (accessed 11 May 2020).
41 Tabuchi, H. (2017). After rapid growth, rooftop solar programs dim under pressure from utility lobbyists. *The New York Times* CLXVI (57,653): 13.
42 Ippen, D. and Döserm, A. (2015). Das Wetter am Mittwoch, 26,8,2015: Sonnenstunden in Oberbayern. *Münchener Merkur* 195 (26 August 2015): 14.
43 Fritz, A. November, autumn and year-to-date were all hottest on record for Earth, NOAA says. https://www.washingtonpost.com/news/capital-weather-gang/wp/2015/12/17/november-autumn-and-year-to-date-were-all-hottest-on-record-for-earth-noaa-says (accessed 11 May 2020).
44 Höhentinger, S. (2017). Photovoltaicanlagen richtig warten. *Eigenheimer actuell, Zeitschrift für Haus, Wohneigentum und Garten* (May 2017): 23.
45 ntv.de. Wasser-Cops drehen Las Vegas den Hahn zu. http://www.n-tv.de/panorama/Wasser-Cops-drehen-Las-Vegas-den-Hahn-zu-article15939596.html (accessed 11 May 2020).

46 Stadtwerke München GmbH (2015). Macht die Grösse den Unterschied? *MDIREKT, Das Service-Magazin der Stadtwerke München* (January 2015): 22.

47 NREL. Utility-scale energy technology capacity factors. https://www.nrel.gov/analysis/tech-cap-factor.html (accessed 11 May 2020).

48 Geoenergist. Leading news: Geothermal industry to meet next week. https://geoenergist.wordpress.com/2015/02/19/leading-news-geothermal-industry-to-meet-next-week/#more-4116 (accessed 11 May 2020).

49 Franck, J. (1925). Transformation of kinetic energy of free electrons into excitation energy of atoms by impacts. Nobel Price Lecture, pp. 79–108. https://www.nobelprize.org/prizes/physics/1925/franck/biographical (accessed 11 May 2020).

50 Sedra, A.S. and Smith, K.C. (1998). *Microelectronic Circuits*, 4e, 1237. Oxford: Oxford University Press.

51 Giovinazzo, P. The fatal current. https://www.physics.ohio-state.edu/~p616/safety/fatal_current.html (accessed 11 May 2020).

52 http://inventors.about.com/library/inventors/bledison.htm (accessed 11 May 2020).

53 History.com. Nikola Tesla. http://www.history.com/topics/inventions/nikola-tesla (accessed 11 May 2020).

54 Fairley, P. (2013). Germany takes the lead in HVDC. *IEEE Spectrum* (29 April 2013), 18 pp. https://spectrum.ieee.org/energy/renewables/germany-takes-the-lead-in-hvdc (accessed 11 May 2020).

55 Starke, M., Tolbert, L. M., and Ozpineci, B. AC vs. DC distribution: a loss comparison. http://web.eecs.utk.edu/~tolbert/publications/pestd2008_starke.pdf (accessed 11 May 2020).

56 Chemandy Electronics. Skin effect depth calculator. http://chemandy.com/calculators/skin-effect-calculator.htm (accessed 11 May 2020).

57 Yang, Z.G. (2017). *Is This the Ultimate Grid Battery?* 36–41. *IEEE Spectrum International.*

58 Bode, H. (1977). *Lead-Acid Batteries.* New York: Wiley.

59 Appelbaum, J. and Weiss, R. (1982). An electrical model of the lead-acid battery. In: *International Telecommunications Energy Conference, 1982*, INTELEC 1982 (3–6 October 1982), pp. 304–307.

60 Zhan, C.-J., Wu, X.G., Kromlidis, S. et al. (2003). Two electrical models of the lead-acid battery used in dynamic voltage restorers. *IEE Proceedings of the Generation, Transmission and Distribution* 150 (2): 175–182.

61 n-tv.de Anleger feiern SGL carbon. http://www.n-tv.de/wirtschaft/Anleger-feiern-SGL-Carbon-article12290021.html (accessed 11 May 2020).

62 ADAC Motorenwelt (2015). Heft 4 (April 2015), p. 24. www.adac.de.

63 Voelker, J. Top 10 tech cars of 2009. *IEEE Spectrum* (1 April 2009), 23 pp. https://spectrum.ieee.org/energy/fossil-fuels/top-10-tech-cars-of-2009 (accessed 11 May 2020).

64 ETN (European Type Number). http://www.auto-einbau.de/auto-elektrikelektronik/autobatterie/etn-nummer.html (accessed 11 May 2020).

65 Code-Knacker. Starterbatterien. http://www.code-knacker.de/starterbatterie.htm (accessed 11 May 2020).

66 Akkumulatorenfabrik MOLL GmbH + Co. KG. https://www.moll-batterien.de/de/batterien/pkw (accessed 11 May 2020).

67 Herring, S. (2004). Milestone for H_2 production by high-temperature electrolysis. *Green Car Congress, Energy, Technologies, Issues and Policies for Sustainable Mobility* (19 November 2004).

68 Basu, S. (2007). *Recent Trends in Fuel Cell Science and Technology*. New York: Springer https://link.springer.com/book/10.1007/978-0-387-68815-2 (accessed 11 May 2020).

69 PowerCell. The reformer explained. http://www.powercell.se/technology_head/the-reformer (accessed 11 May 2020).

70 Fuller, J.F., Fuchs, E.F., and Roesler, D.J. (1988). Influence of harmonics on power system distribution protection. *IEEE Transaction on Power Delivery* TPWRD-3 (2): 546–554. (PES/IEEE Prize Paper Award 1989).

71 Steffen, B. Prospects for pumped-hydro storage in Germany. http://citeseerx.ist.psu.edu/viewdoc/download?doi=10.1.1.472.9368&rep=rep1&type=pdf (accessed 11 May 2020).

72 Rehman, S., Al-Hadhrami, L.M., and Alam, M. (2015). Pumped hydro energy storage system: a technological review. *Renewable and Sustainable Energy Reviews* 44: 586–598.

73 Stadtwerke München. Geothermie. https://www.swm.de/privatkunden/unternehmen/energieerzeugung/geothermie.html (accessed 11 May 2020).

74 Stadtwerke Muenchen. M/Fernwärme. https://www.swm.de/privatkunden/m-fernwaerme.html (accessed 11 May 2020).

75 Stadtwerke München. Geothermie. https://www.swm.de/privatkunden/unternehmen/energieerzeugung/erzeugungsanlagen/geothermie.html (accessed 11 May 2020).

76 Stadtwerke München. https://www.swm.de/dam/jcr:e54a313a-e1f4-4545-9b60-58cba5f3d5a0/Animation%20Geothermie.swf (accessed 11 May 2020).

77 Gleeson, T.; Befus, K. M.; Jasechko, S.; Luijendijk, E.; and Cardenas, M. B. (2016). "The global volume and distribution of modern groundwater," *Nature Geoscience*, 9:161–167 (published 16 November 2015). https://www.nature.com/articles/ngeo2590 (accessed 11 May 2020).

78 Youtube. How seawater desalination works. https://www.youtube.com/watch?v=mZ7bgkFgqJQ (accessed 11 May 2020).

79 Ingenieur.de. http://www.ingenieur.de/Fachbereiche/Kernenergie/So-Fusionsreaktor-Wendelstein-7-X-besser (accessed 11 May 2020).

80 n-tv.de. Sonnen-Doppelsturm trifft auf die Erde. http://www.n-tv.de/wissen/Sonnen-Doppelsturm-trifft-auf-die-Erde-article13594966.html (accessed 11 May 2020).

81 https://eol.jsc.nasa.gov/DatabaseImages/ESC/large/ISS035/ISS035-E-17210.JPG (accessed 11 May 2020).

82 OrCAD. OrCAD downloads. http://www.orcad.com/resources/orcad-downloads (accessed 11 May 2020).

83 Wolfram. Latest features in Mathematica 10. http://www.wolfram.com/mathematica/new-in-10 (accessed 11 May 2020).

84 MathWorks. de.mathworks.com (accessed 11 May 2020).

85 Rudschies, W. (2017). Mobilitaet von morgen, Teil III: Brennstoffzelle/Wasserstoff. *ADAC Motorenwelt* (4): 20–24.

86 Haeussler, J. (2017). Gas geben, Dampf ablassen. *ADAC Motorenwelt* (3): 56–57.

87 Fuchs, E.F. and Myat, M.H. (2010). Speed and torque range increases of electric drives through compensation of flux weakening. In: *Proceedings of the 20th International Symposium on Power Electronics, Electrical Drives, Automation, and Motion*, Pisa, Italy (14–16 June 2010), pp. 1569–1574. IEEE PES (Power Engineering Society).

88 Fuchs, E.F. (2012). Alternating current machine with increased torque above and below rated speed for hybrid/electric propulsion systems. U.S. Patent No. 8,183,814 B2 (22 May 2012).

89 Fuchs, E.F. (2013). Alternating current machine with increased torque above and below rated speed for hybrid/electric propulsion systems. U.S. Patent No. 8,436,569 B2 (7 May 2013).

90 Hosad, D.M., Chandana, M., Bhaskar, R., and Ranjit, S. (2017). An integrated starter-generator and winding configuration for hybrid vehicles. *International Journal of Innovative Research in Electrical, Electronics, Instrumentation and Control Engineering* 5 (2): 62–67. Report of Lawrence Berkeley National Laboratory, LBNL-2001074, October 2017, 48 p.

Appendix 1.A Design Data of Photovoltaic Power Plant of Figure E1.6.1

A single-family residence was constructed in 2012–2013 in southeast Germany, with a PV power plant on the roof (see Figure E1.6.1). The 27 solar panels are arranged on the roof in two parallel DC circuits as depicted in Figure 1.A.1 feeding two (single-phase) inverters. The first DC circuit consists of 14 panels in series and the second one of 13 in series. The total peak (p as used in kWp) power is $P_{\text{peak}} =$

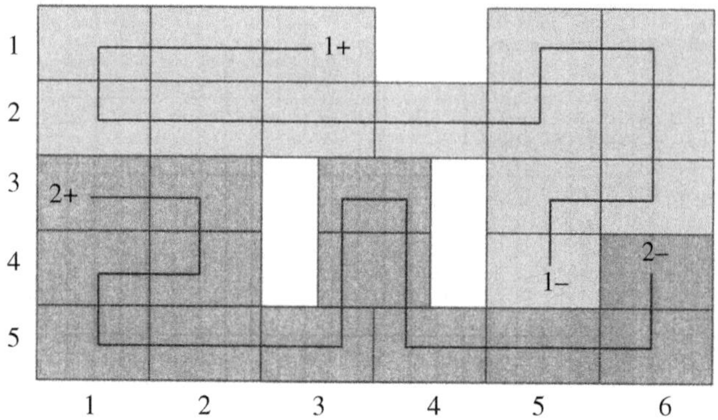

Figure 1.A.1 27 Rooftop 27 solar panels consisting of two independent DC circuits, view from south, type of panels: 27 × Aleo A18/240, power 6.48 kWp. Panels identified by string 1 (from 1+ to 1−) with 14 panels feeding inverter 1 SB 3000TLST-21. Panels identified by string 2 (from 2+ to 2−) with 13 panels feeding inverter 2 SB 3000TLST-21.

6.48 kWp. The rated peak output power of one panel is 6480/27 W = 240 W. The product warranty is 10 years and the power warranty is 25 years [1].

1.A.1 Electric Data at STC

At Standard Test Conditions (STC), the nominal DC voltage and current of one panel are 29.5 V and 8.13 A, respectively. The no-load voltage is 37.0 V, short circuit current is 8.65 A, and the efficiency is 14.6%.

1.A.2 Electric Data at NOCT

At Nominal Operating Cell Temperature (NOCT), the nominal DC output power is 173 W, resulting in a reduced peak output power of 4.67 kWp whereby voltage and current are 28.3 V and 6.11 A, respectively. The corresponding no-load voltage is 34 V and a short circuit current of 6.72 A with an efficiency of 13.1%.

The above two sets of data indicate maximum possible DC output power of 4.67 kWp. It was observed on a very sunny day in June of 2016 that inverter 1 had a maximum AC power output of $P_{\text{inverter1peak}} = 2.35$ kWp and inverter 2 the corresponding value of $P_{\text{inverter2peak}} = 2.18$ kWp resulting in $P_{\text{totalpeak}} = P_{\text{inverter1peak}} + P_{\text{inverter2peak}} = 4.53$ kWp. This means the inverter output power is limited to 70% based on STC, resulting in (4.53 kWh)/(6.48 kWh)=70% or limited based on NOCT giving (4.53 kWh)/(4.67 kWh) = 97%. The 70% limit is imposed by the utilities in Germany in order to mitigate the variability of the PV plant output [2–4]. The total maximum generated AC output energy of the two inverters on 24 June 2016 was

40 kWh. Note the 70% reduction is applied only if the maximum output exceeds 4.53 kW, which is reached only during summer with high insolation. The overall energy efficiency loss is therefore not 30% but in the range of a few percent only. The 70% limit went into effect at the beginning of 2013 in order to protect the grid from too much power during high insolation. The inverter output power in this case is thus limited to 4.53 kWp.

The question remains whether either STC or NOCT should be the reference efficiency: if NOCT is assumed, then the cost per kW installed capacity is 30% larger than for STC. It is an unsolved question as to which one of the efficiency values is valid. The cost (in euro) of the PV plant was without tax €14 016.81, that is, for STC one obtains €14 016.81/(6.48 kWp) = €2163/kWp, and for NOCT, one gets €14 016.81/(4.67 kWp) = €3001.5/kWp. The latter cost (€3001.5/kWp) is more realistic, while the first one (€2163/kWp) is more appealing to the buyer, however, overly optimistic and indeed not correct.

The efficiency is reduced as the insolation is reduced; for example, for 1000 W/m^2 insolation, the efficiency is 13.1%, and at 200 W/m^2 the efficiency is less than 6%. The manufacturer guarantees a measuring accuracy of ±3% for $P_{\text{peak at STC}}$ and ±10% for all other values. The mechanical pressure on the panels can be 5400 Pa and the maximum DC voltage is 1000 V_{DC}. The reverse current I_R can be 15 A. The temperature coefficient $\alpha(I_{sc})$ is +0.04 (%/K), the temperature coefficient β (V_{oc}) is −0.34 (%/K), and the temperature coefficient $\gamma(P_{MPP})$ is −0.46 (%/K).

The material consists of polycrystalline silicone with 60 cells per panel and 3 bypass diodes. The weight is 21 kg-force and the cell dimensions are 156 × 156 mm^2. One panel has the dimensions 1660 × 990 mm^2. As measured during a four-year time frame (see Figure E1.6.5), the yearly energy efficiency decrease is about 1.32%.

1.A.3 Conclusions with Respect to the Rating of PV Power Plants

The power rating relying on laboratory STC does not apply (6.48 kWp) in practice; thus it appears that the buyer/consumer is misled. It would be more appropriate to tell the consumer/buyer the actual power rating based on the NOCT, which is substantially less, 4.67 kWp as compared with 6.48 kWp, where p stands for peak. In addition, the buyer of the PV plant should be informed of the 70% limit, leading to 4.53 kWp instead of 6.48 kWp, resulting in €14 016.81/(4.53 kWp) = €3094/kWp.

1.A.4 Inverter Sunny Boy 3000TL Data

The maximal DC input voltage is 750 V and the maximum DC power at $\cos\theta = 1$ is $P_{max} = 3200$ W. There is no service factor (e.g. 1.15). Maximum power point (MPP) voltage range is 213–500 V; the rated input voltage is 400 V, the minimum input voltage is 125 V, and the input voltage, where start-up of inverter occurs, is 150 V. Maximum input current to the inverter is 15 A, and maximum input current per string is 15 A. The number of independent MPP inputs is 1, and the number of strings per MPP input is 2.

The AC output power is 3000 W at a maximum apparent power of 3000 VA. Available AC rated voltages are 220/230/240 V. The AC voltage range is 180–280 V, and the available frequencies are 50 Hz, 60 Hz $\pm$ 5 Hz. The rated frequency and rated voltage for this application are 50 Hz and 230 V, respectively. The maximum AC output current is 13.1 A at a total power factor of 1. The displacement power factor can be adjusted from 0.8 overexcited to 0.8 underexcited. The number of phases is 1 (single phase). Efficiency is in the range of 96.1–97%. The AC current has at rated power a third harmonic of about 0.85%. All other harmonic current components are smaller.

1.A.5 Conclusions with Respect to the Rating of Inverters

The total maximum inverter output power is 6.0 kW, which is less than 6.48 kWp. In addition, the service factor of the inverters is 1.0, and for this reason the inverters support the rating of 4.53 kW, resulting in an actual service factor of 4.53/6.00 = 0.756, which means the two inverters are over-dimensioned, which is standard practice.

References

1 Bauer, M. (July 2015). Elektro Kaiser GmbH, Dokumentation PV-Anlage für Heltauer Str. 18, Auftrags-Nr.: AU001937, 10, p. 4.

2 DZ-4. 70-Prozent Regelung bei Photovoltaik-Anlagen. https://www.dz-4.de/ratgeber/70-prozent-regelung (accessed 11 May 2020).

3 Märtel, C. Ist eine begrenzte Einspeiseleistung von 70% zu empfehlen? https://www.photovoltaik-web.de/photovoltaik/wechselrichter/wechselrichter-70-prozent-regelung-eeg-2012 (accessed 11 May 2020).

4 Geoplex. Neue Vorschriften für Photovoltaikanlagen ab 1. Januar 2012. https://www.geoplex.de/fileadmin/user_upload/PDF_PV/Gesetzgebung_Photovoltaik_seit_2012.pdf (accessed 11 May 2020).

Appendix 1.B The Nature of Electricity and Its Manufacturing

1.B.1 Tangible Personal Property

Electrons are particles with a finite mass m_{electron} and charge q that can be measured and touched and therefore possess tangible personal properties [1, 2]. Since electric current consists of an organized flow of endowed electrons [3] that can perform (useful) work, electricity possesses tangible personal properties as well. Electricity is the layman's term for electric power and electric energy. We refer to the composition of electricity as power = (voltage · current) or electric energy = (voltage · current · time). There is no single scientific definition of "electricity," as it is also used to refer to electric voltage, current, lightning in the sky, charge, and electromagnetic waves, just to name a few terms.

1.B.2 Manufacturing of Electric Energy

An electric utility company manufactures, produces (in plant), and processes (in transmission and distribution systems) electric energy. The production of electric energy begins in a generation plant and ends at the customer's meter. The electric utilities throughout the world generate alternating voltage–current, referred to as alternating current (AC). With AC, during the positive current half cycle, the electrons move forward a certain distance and then backward a certain distance.

1.B.3 Free and Endowed Electrons

In the course of the manufacturing/production of electricity, a utility company changes the character of the electrons: free electrons become endowed electrons that are able to do useful work, and endowed electrons become free electrons by passing through a load (e.g. light bulb) where they give up the potential to do useful work. Voltage, current, and time are instrumental to the delivery of energy. Without changing voltage, a utility company could not deliver the electricity efficiently, and voltage is instrumental in producing current.

1.B.4 Processing of Electric Energy

Electricity processing occurs throughout the electric power manufacturing system, and this processing is necessary to ensure that the product ultimately delivered to the customer is at suitable voltage and current levels at the customer's meter including acceptable frequency and power quality [4]. The processing of electricity occurring in the transmission and distribution systems is the same as it occurs in

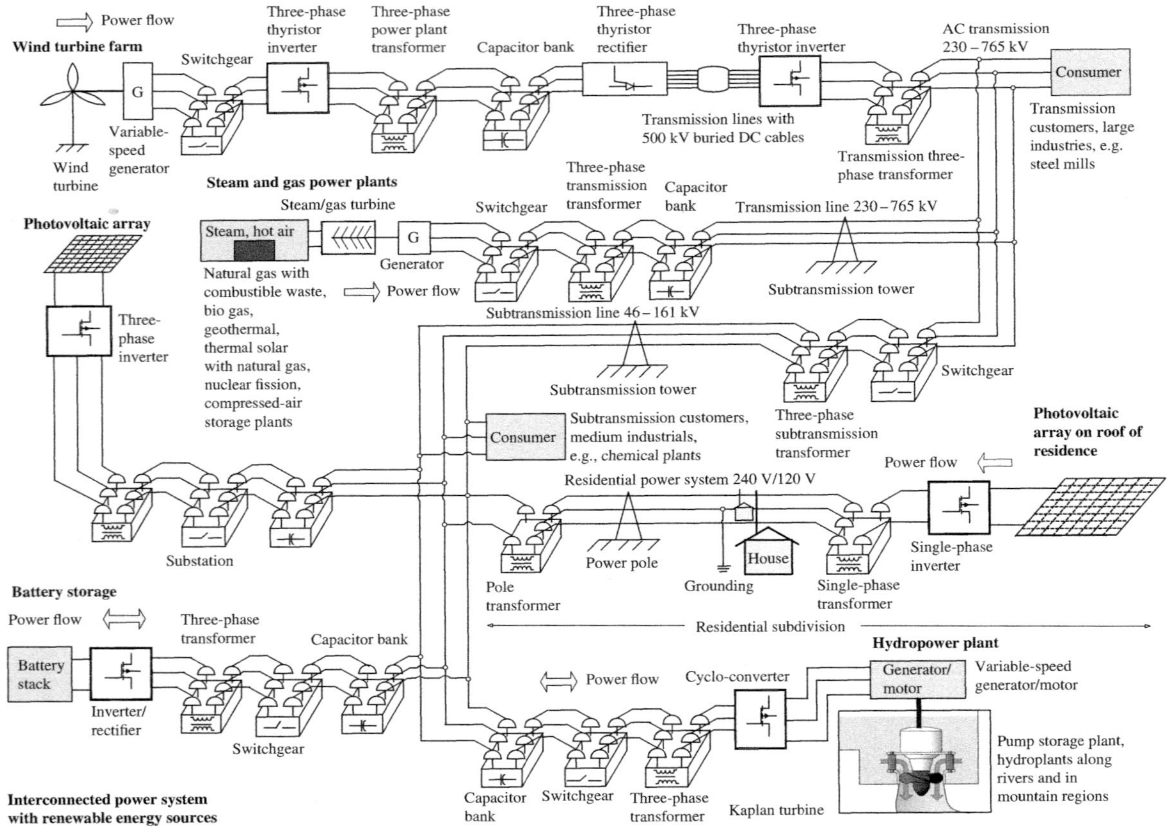

Figure 1.B.1 Basic elements of a power generation, transmission, and distribution system representing a large machine from the furnace to the consumer point of delivery, which is the electricity meter owned by the utility [4, 5].

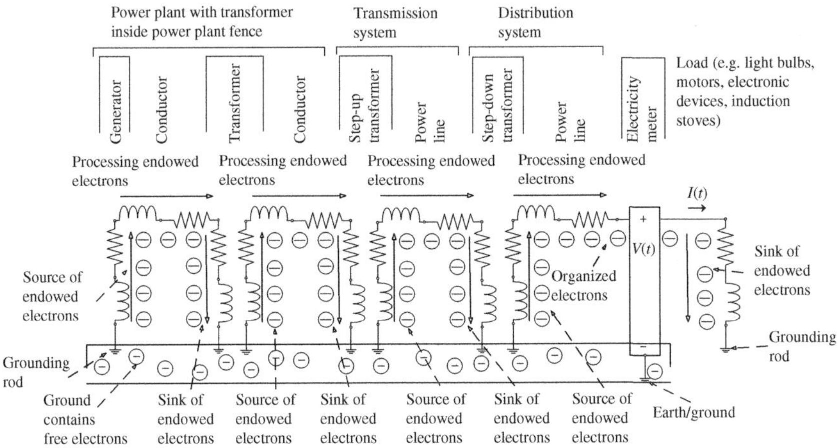

Figure 1.B.2 Generator with transformer inside the power plant fence, step-up transformer connected to transmission line, and step-down transformer connected to distribution line, meter, and load (single-phase diagram).

the generators at the generation plant [5]. Figure 1.B.1 illustrates schematically a utility system consisting of power plant, transmission system and distribution system, including loads [4, 5].

In Figure 1.B.2 the processing can be best understood by examining the migration of endowed electrons that compose the electricity. The same physical phenomena that take place in a generator also occur in transformers and transmission/distribution lines throughout the electric system: the generator equips free electrons with voltage, and they become endowed or imbued electrons (possessing tangible personal property) within an electric circuit consisting of a generator, transmission and distribution line conductors, step-up and step-down transformers, and loads. The transmission and distribution conductors and various transformers process electricity like the generator (e.g. source/sink of endowed electrons) and can therefore be considered to be an extension of the generator and its step-up transformer, which are located within the power plant fence.

1.B.5 Sources and Sinks of Endowed Electrons

In Figure 1.B.2 the term "source" is used to show the source of the endowed electrons: free electrons are endowed with voltage by either the generator or the secondary windings of the various transformers. These endowed electrons lose their potential to do useful work while applying their energy to the primary windings of the transformers and to the loads and "sink" back into ground to become free electrons again. The electrons must be continually processed by the equipment in the entire electric system as they migrate from the generator to reach the customer's meter in an endowed state. Without a connection to the transformers and the ancillary equipment within the electric system, these endowed electrons cannot flow from the generation plant to the customer's meter. Through the sources and sinks, the endowed electrons that began at the generation plant make their way to the customer's meter. These endowed electrons cannot travel to the customer's meters on their own. They must be replenished with the use of transformers and auxiliary equipment located in the electric system.

1.B.6 Valence and Conduction Bands of Copper Atoms [6]

A small section of a transmission and distribution line wire is shown in Figure 1.B.3 that takes part in the process of conducting electric energy. Similarly to semiconductor materials [7], there are valence and conduction bands [8] associated with generator, transformer, transmission, and distribution conductors. A hole is left behind within the atom structure (shell) of a copper atom if an electron leaves in a random manner its parent atom due to temperature; thus copper atoms freely provide electrons for the conduction process, leaving behind a copper ion.

1.B.7 Electron Migration

As indicated in Figure 1.B.4, electron migration within a copper conductor is caused by the electric field $\vec{E}$ due to the voltage across the generator, transmission, and distribution systems including the loads. In this case the electron flow is from left to right due to the potential being negative on the left-hand side. In a statistical manner the atoms of the conductor provide electrons for the electric energy flow: it is an interactive process between organized (endowed/imbued with potential/voltage) electrons and the copper (or aluminum) atoms. The individual electrons move with drift velocity (few mm/s), while the entire group of electrons within a

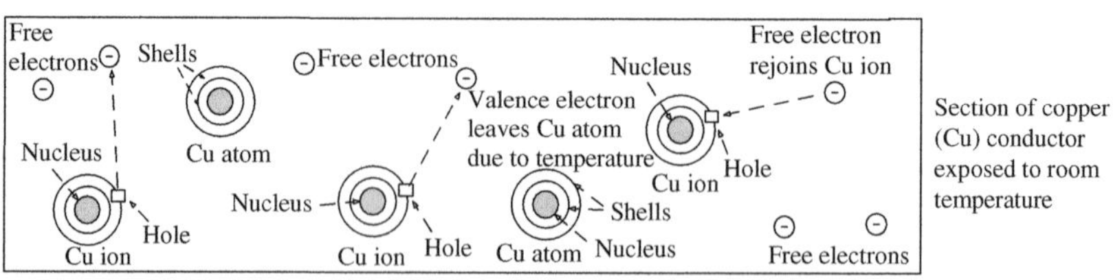

Figure 1.B.3 Generation of free electrons within a copper (Cu) conductor due to temperature.

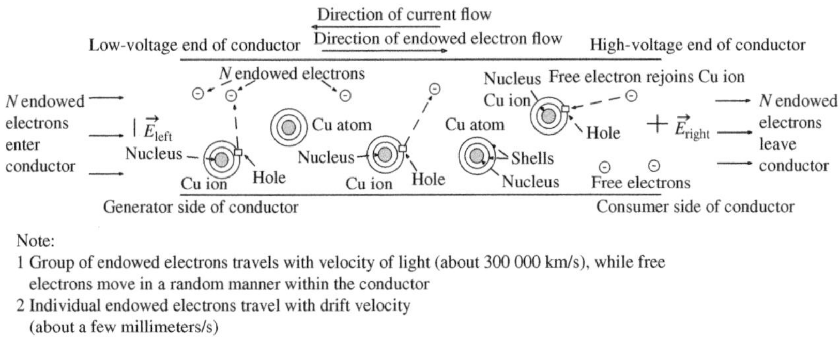

Figure 1.B.4 Process of endowed electron migration within a copper/aluminum conductor section during a positive half-wave of alternating voltage. Note the statistical interactive process of endowed electron flow together with copper/aluminum ions. $\vec{E}_{\text{left}}$ and $\vec{E}_{\text{right}}$ are the electric fields imposed by the generated voltage in the generator.

conductor moves at the speed of light (299 792.458 km/s). The valence electrons stemming from the copper atoms repel each other.

1.B.8 Conductor Atoms/Ions Continually Interact with Free and Endowed Electrons

Figures 1.B.2–1.B.4 illustrate that the atoms/ions of the transmission and distribution portions continually interact with the endowed electrons. The transmission and distribution portions of the electric system not only provide a means of transport for the electricity but, more importantly, process the electrons in order to allow them to maintain their endowed nature.

1.B.9 Deployment of Step-Up and Step-Down Transformers

The electric circuit of Figure 1.B.5 consists of generator, transmission/distribution line conductors, step-up and step-down transformers, and customer load. Load or energy loss refers to any use of the endowed electrons, such as lighting equipment, motor drives, or computers. Loss can occur in the coils of a transformer or in other equipment within the electric system. The circuit of Figure 1.B.5 illustrates that with the use of transformers a large part of the generated voltage (120 V at the generator) arrives at the load (119.9 V at the residence). This underscores the need for the deployment of transformers if the efficient processing of energy is required. Through the use of transformers the voltage–current magnitudes are changed and processed; that is, the composition of electricity is altered.

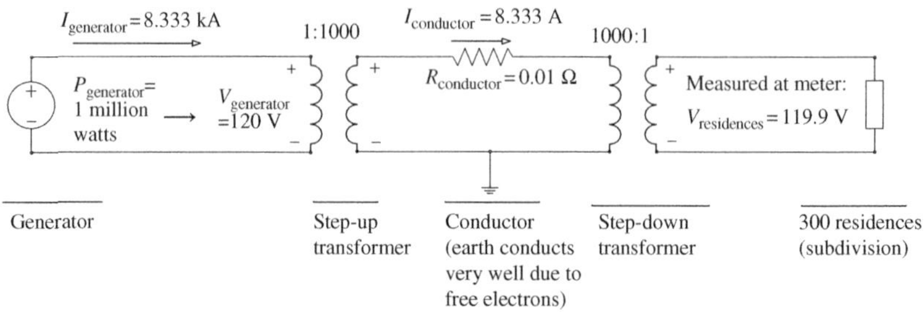

Figure 1.B.5 Electricity processing along a conductor and employing transformers (single-phase diagram).

1.B.10 Voltage and Current Changes

Electricity exits the generation plant generator at a voltage that is too high for use by utility customers. However, this voltage is not high enough for the transmission of the electricity over the long distance it needs to reach utility customers.

When the electricity leaves the generator at about 20 kV, it enters the power plant step-up transformer within the power plant fence, which transforms the voltage to about 138 kV.

1.B.11 Regulatory Accounting Procedure

Most state governments in the United States have regulatory accounting procedures in place that confirm that manufacturing of electric energy within the power plant fence occurs and benefits from capital tax exemption.

1.B.12 Efficient Transmission of Electric Energy and Transformer Action

After leaving the power plant step-up transformer, the voltage is increased from about 138 to 345 kV in the step-up transformer of the transmission system so that it can be effectively and efficiently transmitted over a large distance to reach utility customers. Inside a transformer, the entering current produces a magnetic field in the core of the transformer. The magnetic field induces, depending upon the number of turns of the windings, an output voltage. In other words, the output voltage of the electricity can be changed and an appropriate output current results – based on Ampere's law [5] – depending upon the load of the transformer. When transmitting electricity, it is more efficient to transmit the electricity at a high voltage and low current rather than a low voltage and high current because higher currents produce heat, resulting in power loss. When a manufacturing electric utility company increases the voltage, it decreases current and thus maintains the same level of electric power. Increasing the voltage allows the utility company to operate effectively and efficiently to move the electric power over a long distance. When current is lower, less energy is lost by dissipation as the electricity flows over a given distance within the electric power manufacturing system.

1.B.13 Voltage Levels Usable by Customers

Electric energy at 345 kV is too high for a utility customer to use. Electricity must be provided to the customer at a low voltage (e.g. 120–240 V). If it is not, the electric appliances and equipment in homes and businesses will not operate properly and may be damaged. Through the use of transformers and other (e.g. conductors) as well as ancillary (e.g. filters, controllers) equipment in the electric system, the voltage and current (consisting of organized electrons) are changed and processed. Figure 1.B.5 further illustrates the change in the composition of the electricity. Moving from left to right, we see that the composition of the electricity has changed from 120 to 120 000 V (on the secondary winding of the step-up transformer) as electricity moves from the generator to a conductor via a step-up transformer.

1.B.14 Energy Becomes a Finished Good with Specified Composition (*e.g.* Mandated Voltage, Current, Frequency, Quality, and Safety)

As electricity approaches a utility customer's site, it must go through a step-down substation transformer, in which the voltage is reduced to a range between 4000 and 46 000 V. Electricity then flows through a final or pole transformer near or at the utility customer's point of use. Here, the voltage is further reduced to approximately 120–240 V. Only after this point at the customer's meter does the electric power first become usable to a typical residential customer. This is where the electric power has become a finished good. Until this point, processing of the electricity has continuously occurred all the way from the plant generator to the meter at the customer's point of use. Throughout the utility company's electric power manufacturing system, electricity continues to be processed and transformed as its voltage, current, and frequency are changed and affected. This is a change to the composition, quality, and character of the electricity until it is at the specified parameters that are usable by its customers at the meter.

1.B.15 Water Pipeline and Electric Conductor Analogy Is Erroneous

In everyday life complicated physical phenomena are frequently explained based on well-known systems, which can be easily understood by a layperson. One example is the analogy of a conductor processing endowed electrons, which represent electric current and a water pipeline. A pipeline serves as a conduit for water, and there is no interaction between pipe and water. In a conductor the endowed electrons interact with the atoms/ions of the conductor where the electrons move in a random/statistical manner. Conductor atoms/ions are very willing to participate in the conduction process of endowed electrons because copper–aluminum atoms easily give up electrons residing in the outer shells due to temperature providing electrons from the valence band to the conduction band where electrons become free. The analogy of a conductor with a water pipeline is not correct as is illustrated in Figure 1.B.6 where a hole in the pipe permits the water molecules to leave the pipe.

If a hole is drilled into a conductor as is illustrated in Figure 1.B.7, then neither free nor endowed electrons will leave the conductor because the electrons continually interact with the conductor atoms/ions.

In addition, the increase of the pressure in a water pipe increases the velocity of the water molecules, while the increase of the voltage does not increase the speed of light of electricity. All water molecules entering the pipe leave it at the other end provided there is no hole in the pipe. This is not so in case of endowed electrons because of the sinks and sources associated with transformers: the endowed electrons injected by the generator into the transmission and distribution network do not necessarily end up at the meter.

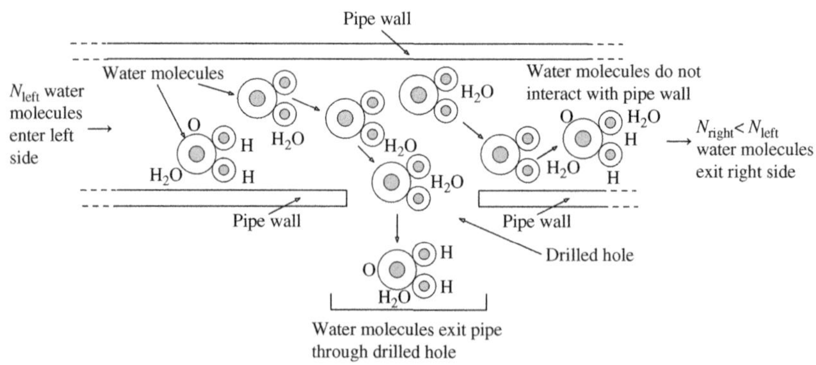

Figure 1.B.6 Some water (H_2O) molecules exit pipe through drilled hole.

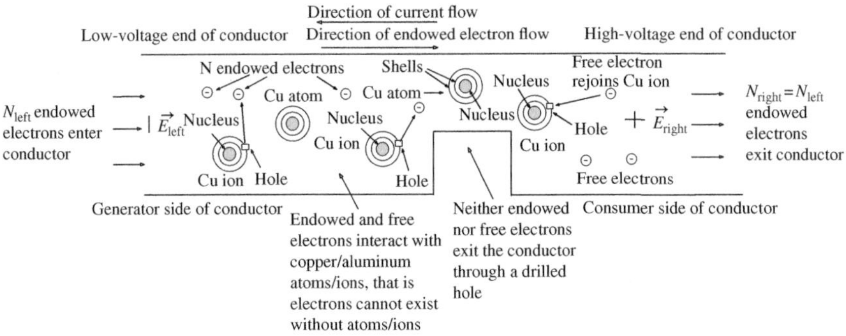

Figure 1.B.7 Neither free nor endowed electrons leave the conductor through a hole.

1.B.16 An Analogy Between Information Transmission via Cellular Phone and Electric Energy Supply via Transmission Lines Is Erroneous

Cellular phones rely on radio towers that transmit information signals with very low energy content (e.g. milliwatts) via microwaves in the frequency range from 850 to 2100 MHz to the customer (see Application Example 1.7). In contrast, electric energy transmission occurs with very large energy content (e.g. 1000 GW in the United States) in the frequency range from 0 Hz for DC transmission, 16.667 = 16⅔ Hz for the German railroad system, and 50–60 Hz for AC electric power systems throughout the world. High-frequency electromagnetic fields can be transmitted through the air, while low-frequency electromagnetic fields cannot exist without a conductor, e.g. either AC or DC transmission lines. It is well known that electromagnetic fields along transmission lines can be either cancelled or reduced by appropriate transmission line configurations (Figure 1. B.8) as used by all utility companies. This means the cellphone analogy is erroneous, because microwaves are instrumental for transmitting information signals. This reliance on electromagnetic fields is not a prerequisite for transmitting electric energy. To minimize electromagnetic field exposure to the surrounding environment, utilities employ reverse-phased, double-circuit three-phase transmission lines, as illustrated in Figure 1.B.8. Two three-phase systems are suspended on transmission towers so that the electromagnetic fields cancel or are greatly reduced. In Figure 1.B.8 the electromagnetic fields of the three-phase system ABC on the left-hand side is counteracted/cancelled (reduced by about a factor of two) by the electromagnetic fields of the three-phase system CBA on the right-hand side.

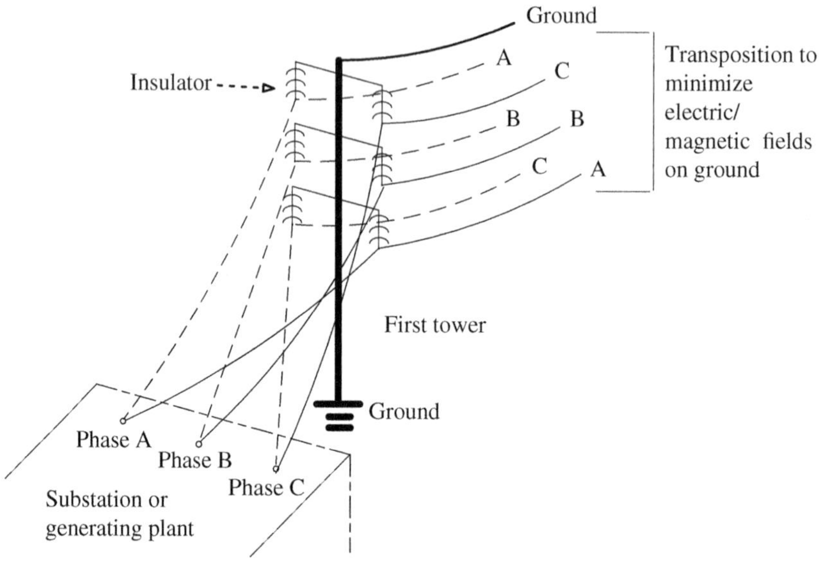

Figure 1.B.8 Reduction of magnetic fields by employing a reverse-phased, double-circuit three phase line [9].

1.B.17 Vehicle of Energy Transmission

The vehicle for energy transmission is the endowed electrons and not the electromagnetic fields, because they are cancelled in reverse-phased, double-circuit three-phase transmission lines as applied by all utilities.

1.B.18 Court Decisions in Which Electric Energy Has Been Determined to Be Manufactured

In 1926 the South Carolina Supreme Court held that a corporation engaged in the business of generating electricity was "engaged in the business of manufacturing" and thereby subject to a tax imposed on manufacturers [10]. The South Carolina Supreme Court reaffirmed that conclusion in 1930, holding that Duke Power, a direct predecessor of Duke Energy, was entitled to claim an exemption for manufactories defined as a physical plant, or building, where manufacturing occurs [11]. In 2014 the South Carolina Supreme Court held up the decision that Duke Energy is in the business of manufacturing electricity [12].

References

1 Nolo. Tangible personal property. http://www.nolo.com/dictionary/tangible-personal-property-term.html (accessed 11 May 2020).
2 http://taxes.about.com/od/statetaxes/a/tangible-personal-property.htm (accessed 11 May 2020).
3 Franck, J. (1926). Transformation of kinetic energy of free electrons into excitation energy of atoms by impacts. Nobel Lecture (11 December). https://www.nobelprize.org/prizes/physics/1925/franck/lecture (accessed 11 May 2020).
4 Masoum, M.A.S. and Fuchs, E.F. (2015). *Power Quality in Power Systems and Electrical Machines*, 2e, 1123. Amsterdam: Academic Press/Elsevier.
5 Fuchs, E.F. and Masoum, M.A.S. (2011). *Power Conversion of Renewable Energy Systems*, 692. New York, 2nd Printing: Springer.
6 Nave, R. Conductors and insulators. http://hyperphysics.phy-astr.gsu.edu/hbase/electric/conins.html (accessed 11 May 2020).
7 Sedra, A.S. and Smith, K.C. (1998). *Microelectronic Circuits*, 4e, 1237. Oxford: Oxford University Press.
8 Nave, R. Band theory of solids. http://hyperphysics.phy-astr.gsu.edu/hbase/solids/band.html (accessed 11 May 2020).
9 Leeper, E.A. (2001). *Silencing the Fields: A Practical Guide to Reducing AC Magnetic Fields*. Symmetry Books, 340 pp.
10 Columbia Ry., Gas & Elec. Co. v. Query, 134, S.C. 319, 132, S.E. 611 (1926).
11 Duke Power Co. v. Bell, 156 S.C. 299, 152 S.E. 865 (1930).
12 FindLaw. Duke Energy Corporation v. 2012 213180. http://caselaw.findlaw.com/sc-court-of-appeals/1680169.html (accessed 11 May 2020).

Appendix 1.C The Cost of Electricity in a Renewable Energy System

The switch from a conventional electric power system to that with renewable sources is very expensive. Indeed, Denmark [1], which has the most advanced renewable system based on wind power sources, charges more than 30 ct/kWh, while in Germany the cost for 1 kWh is approaching 30₵ expressed in euro (€) (Figure 1.C.1).

The breakdown [2] of the electric energy cost per kWh during 2016 is for Germany as follows:

1) Tax: 23%.
2) Reallocation charge based on the renewable energy law (Umlage gemäss des Erneuerbaren Energie Gesetzes, EEG): 22.1%.

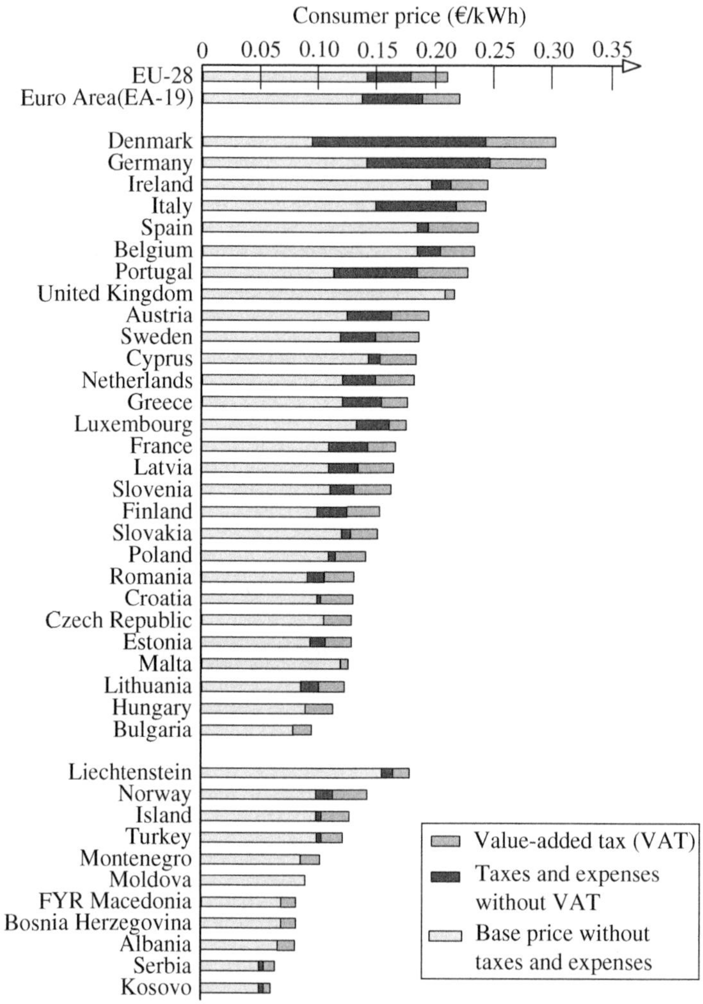

Figure 1.C.1 Electricity charges in cents/kWh expressed in euro (€) for various European countries.

3) Charge paid to the transmission line operator: 22%.
4) Costs for generation and operation of power plants including distribution: 21.8%.
5) Franchise tax paid to communities for using public roads for distribution systems: 5.8%.
6) Operation of measuring equipment (e.g. electricity meters): 2.1%.

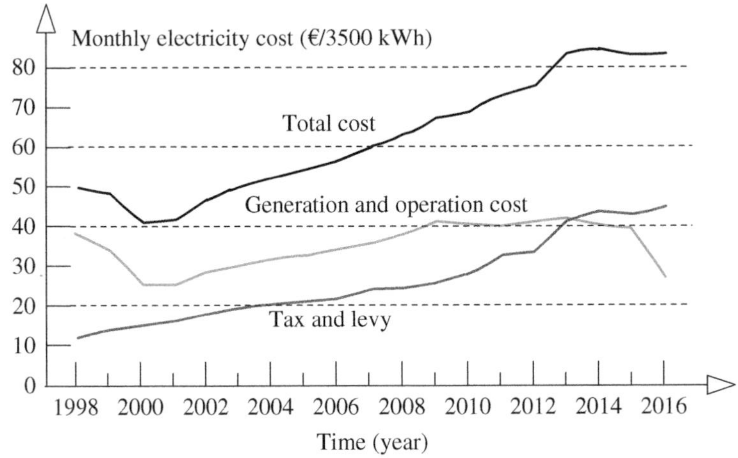

Figure 1.C.2 Increase of monthly electricity bill in euro (€) for a residential house with a yearly consumption of 3500 kWh from 1998 to 2016.

7) Levy for combined heat and power generation law: 1.5%.
8) Large electricity consumers do not pay under certain circumstances charges to the transmission line operator and these revenue losses must be paid by the remaining customers: 1.3%.
9) Offshore liability levy: 0.1%.
10) Voluntary disconnect charges for periods during short-power conditions: 0.006 ct/kWh.

From this list, one concludes that tax and levy represent more than 50% of the entire electricity cost measured in ct/kWh. The increase of the monthly electricity cost of a residence consuming 3500 kWh/year from 1998 to 2016 is illustrated [2] in Figure 1.C.2.

1.C.1 Summary

The replacement of conventional power system sources such as coal, natural gas, and nuclear plants by renewable plants such as solar, wind, biomass, and hydro plants will result in increased costs due to the intermittency of such plants and their control requiring storage elements such as batteries and pump-storage hydro plants.

References

1 Eurostat. Electricity prices for household consumers, second half 2015 (1) (EUR per kWh) YB16.png. http://ec.europa.eu/eurostat/statistics-explained/index.php/File:Electricity_prices_for_household_consumers,_second_half_2015_(%C2%B9)_(EUR_per_kWh)_YB16.png (accessed 11 May 2020).

2 Keller, S. (2017). Strompreise auf Rekordniveau. *Eigenheimer actuell, Zeitschrift für Haus, Wohneigentum und Garten* (April 2017), pp. 6–9.

2

Electric Circuit Laws

Electric circuit laws are valid for direct current (DC) and alternating current (AC) circuits, as well as for time-dependent or time-varying DC voltages and currents $v_{DC}(t)$ and $i_{DC}(t)$. While in power and energy applications AC circuits promoted by Tesla [1] were dominant in the past, the DC approach originally favored by Edison [2] appears to have new and important applications in the area of DC transmission lines and simple PV power plants, where DC is supplied to appliances and no DC–AC conversion is used. Figure 2.1a–c depict time-dependent AC voltage and current $v(t) = v_{AC}(t)$, and $i(t) = i_{AC}(t)$, respectively; time-independent DC voltage and current $v(t) = V_{DC}$ = constant1 and $i(t) = I_{DC}$ = constant2, respectively; and time-dependent DC voltage and current $v(t) = v_{DC}(t)$ and $i(t) = i_{DC}(t)$, respectively.

2.1 Ohm's Law and Instantaneous Electric Power *p*(*t*)

For any instantaneous voltage $v(t)$ (measured in volts or V) and instantaneous current $i(t)$ (measured in amperes or A), in the time domain of Figure 2.2, Ohm discovered that the quotient of both quantities is the resistance R of an electric circuit [3]:

$$R = v(t)/i(t), \tag{2.1}$$

where the resistance $R > 0$ is measured in ohms.

The instantaneous power $p(t)$ (which is the most general form of electric power) is defined as the product of $v(t)$ and $i(t)$:

$$p(t) = v(t)i(t) = R[i(t)]^2 = [v(t)]^2/R = [v(t)]^2G = [i(t)]^2/G, \tag{2.2}$$

where the conductance $G = 1/R$ is measured in $1/\Omega$ or in siemens or S; a short circuit exists if $R = 0$ or $G = \infty$, and for an open circuit $R = \infty$ or $G = 0$.

Introduction to Energy, Renewable Energy, and Electrical Engineering: Essentials for Engineering Science (STEM) Professionals and Students, First Edition. Ewald F. Fuchs and Heidi A. Fuchs.

Companion website: www.wiley.com/go/fuchs/renewableenergyandelectricalengineering

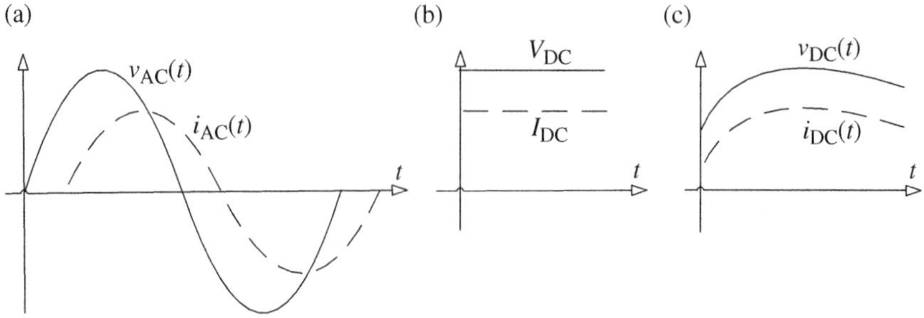

Figure 2.1 (a) Time-dependent AC voltage $v_{AC}(t)$ and current $i_{AC}(t)$, (b) time-independent DC voltage V_{DC} = constant1 and current I_{DC} = constant2, (c) time-dependent or time-varying DC voltage and current $v_{DC}(t)$ and $i_{DC}(t)$, respectively.

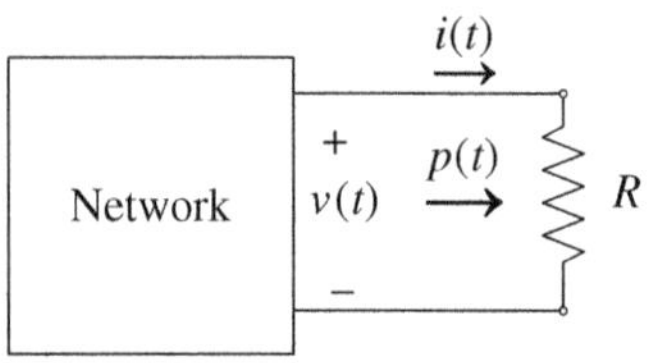

Figure 2.2 Definition of ohmic resistance R measured in Ω.

2.2 Kirchhoff's Current and Voltage Laws (KCL) and (KVL), Respectively

Kirchhoff's current law (KCL) [4] is called the first Kirchhoff's law and is defined as the summation of currents

$$\sum_{j=1}^{N} i_j(t) = 0 \tag{2.3}$$

at a node with N entering currents, as depicted in Figure 2.3. Note that (independent) current sources are defined as positive if the arrow points to the summation node and negative if the arrow points away from the node.

Kirchhoff's voltage law (KVL) [4] is called the second Kirchhoff's law and is defined as the summation of voltages

$$\sum_{j=1}^{N} v_j(t) = 0 \tag{2.4}$$

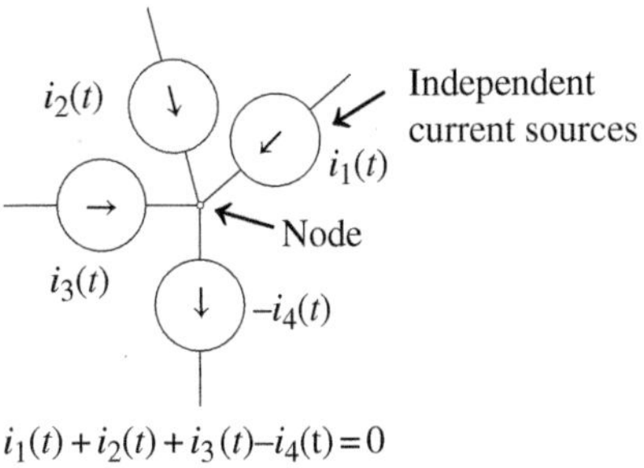

Figure 2.3 Kirchhoff's current law at a node, where $N = 4$.

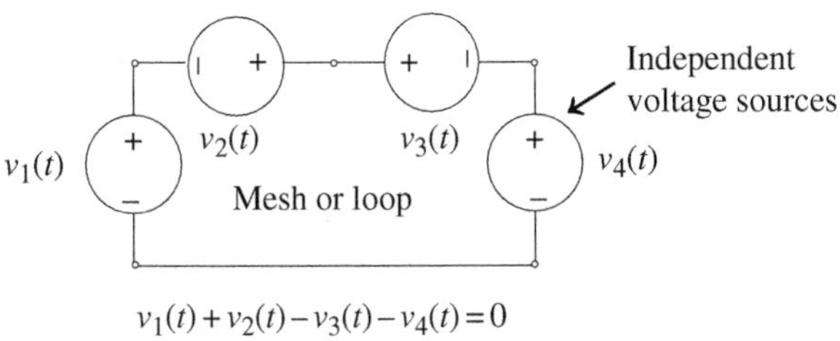

Figure 2.4 Kirchhoff's voltage law within a mesh or loop, where for $N = 4$ voltages $\sum_j^N v_j = 0$, that is, the sum of four series-connected voltages is zero.

within a mesh with N branch voltages as depicted in Figure 2.4. In general a mesh represents a closed-loop connection of circuit elements such as series-connected voltage and current sources as well as circuit components, such as resistors. In the mesh of Figure 2.4, voltage sources are connected in series only.

Note that as a running index, the j has been chosen in Eqs. (2.3) and (2.4) and not the commonly used i, which we rely on throughout the book as the symbol for current i. Voltages with − to + signs are counted as positive, and voltages with + to − are assumed to be negative.

There are independent voltage and independent current sources, as well as dependent voltage and dependent current sources. Independent voltage and current sources are shown in Figure 2.5a and b, respectively. In these sources the voltages and currents are independent of any other parameter, except possibly the time t for AC or general time-dependent voltages.

Dependent voltage sources are a function of other parameters such as voltage gain μ and resistance r, respectively, as indicated in Figure 2.5c and d.

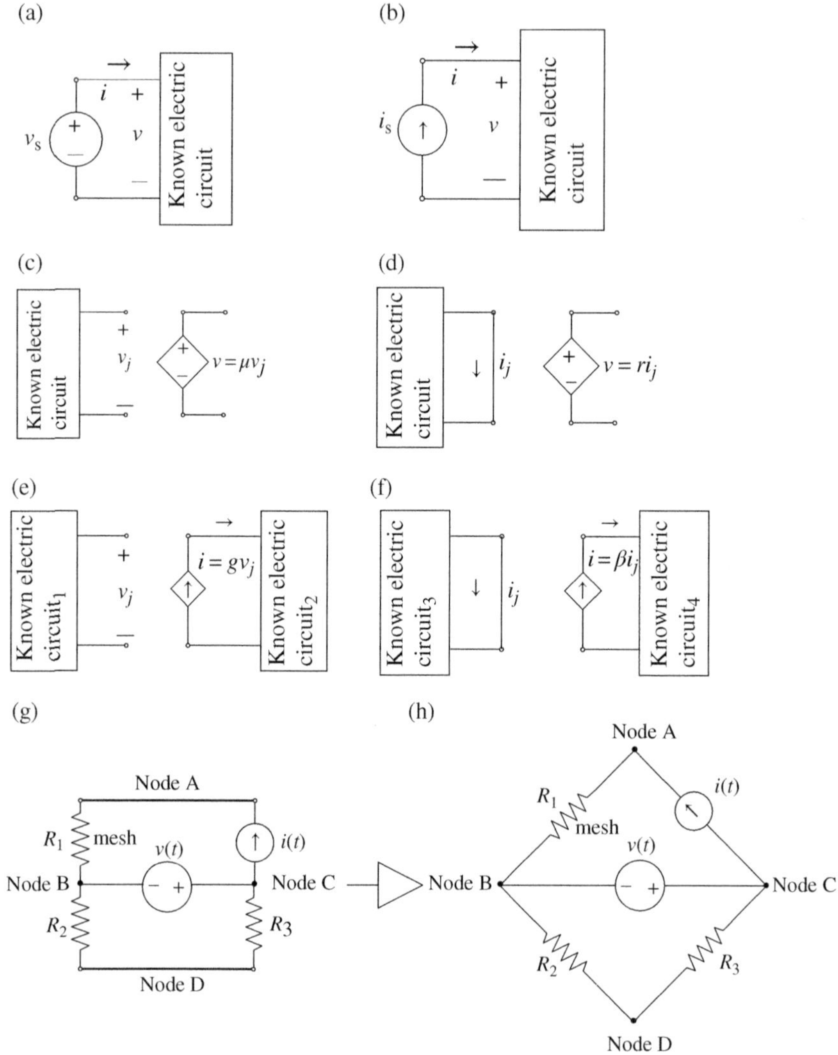

Figure 2.5 (a) Independent voltage source characterized by a circle with + and − signs inside, indicating polarities. The independent voltage source is applied to a circuit with known elements. (b) Independent current source characterized by a circle with an arrow inside indicating the flow direction of the current. The independent current source is applied to a circuit with known elements. (c) Dependent voltage source indicated by a rhombus/diamond-shaped symbol with + and − signs inside indicating polarities. The independent voltage v_j stems from the known electric circuit, and the dependent voltage is $v = \mu \cdot v_j$, where μ represents a voltage gain. (d) Dependent voltage source indicated by a rhombus/diamond-shaped symbol with + and − signs inside indicating polarities. The independent current i_j stems from the known electric circuit, and the dependent voltage is $v = r \cdot i_j$, where r represents a resistance. (e) Dependent current source indicated by a rhombus/diamond-shaped symbol, with an arrow inside indicating current direction. The independent voltage v_j stems from the known electric circuit$_1$, and the dependent current is $i = g \cdot v_j$, where g represents a conductance supplying current to known electric circuit$_2$. (f) Dependent current source indicated by a rhombus/diamond-shaped symbol, with an arrow inside indicating current direction. The independent current i_j stems from the known electric circuit$_3$, and the dependent current is $i = \beta \cdot i_j$, where β represents a current gain supplying a known electric circuit$_4$. (g) and (h) Definition of mesh, node, and independent voltage and current sources within a circuit, including its equivalent circuit.

Dependent current sources are a function of other parameters such as conductance g and current gain β, respectively, as indicated in Figure 2.5e and f.

The bridge-type electric circuit of Figure 2.5g defines meshes, nodes, independent voltage, and current sources $v(t)$ and $i(t)$, respectively. By redrawing Figure 2.5g one obtains the equivalent circuit of Figure 2.5h.

2.3 Application of KVL to Single-Loop Circuits

Figure 2.6 shows a single-loop circuit with two voltage sources connected in series via four resistors.

The application of $\sum_{j=1}^{N} v_j(t) = 0$ to the circuit of Figure 2.6 is as follows:

1) A voltage of a source {e.g. $v_1(t)$} is counted positive when we go from the negative terminal to the positive terminal.
2) A voltage is counted negative when we go from the positive to the negative terminals of $i_1(t)R_1$, $i_1(t)R_2$, $i_1(t)R_3$, $i_1(t)R_4$, and $v_2(t)$.

For the application of KVL to the loop or mesh as indicated in Figure 2.6, one obtains

$$+v_1(t) - i_1(t)R_1 - i_1(t)R_2 - i_1(t)R_3 - v_2(t) - i_1(t)R_4 = 0, \quad (2.5)$$

or

$$[v_1(t) - v_2(t)] = i_1(t)[R_1 + R_2 + R_3 + R_4]. \quad (2.6)$$

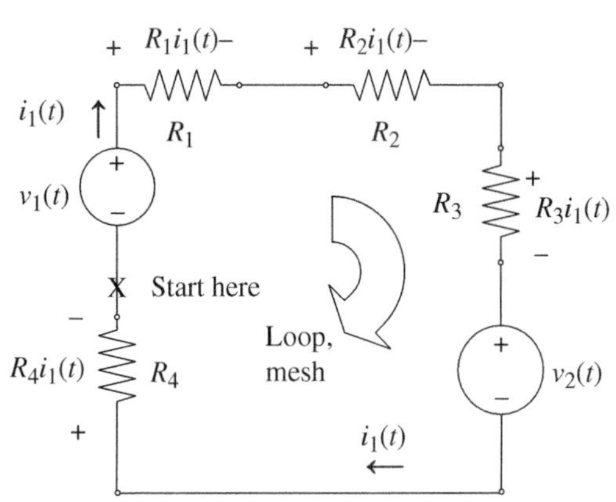

Figure 2.6 Application of KVL to single-loop circuit.

2.3.1 Voltage Division or Voltage Divider

Frequently, the applied voltage $v(t)$ divides into voltage components across connected resistors. As illustrated in Figure 2.7, the application of the procedure explained above yields for the KVL:

$$v(t) - i(t)R_1 - i(t)R_2 = 0 \tag{2.7}$$

or

$$v(t) - i(t)[R_1 + R_2] = 0. \tag{2.8}$$

Solving Eq. (2.8) for $i(t)$ is synonymous with the application of Ohm's law, leading to

$$i(t) = \frac{V(t)}{(R_1 + R_2)}; \tag{2.9}$$

the voltage drop across R_1 is then

$$v_{R1}(t) = i(t)R_1 = \frac{v(t)}{(R_1 + R_2)}R_1; \tag{2.10}$$

correspondingly

$$v_{R2}(t) = i(t)R_2 = \frac{v(t)}{(R_1 + R_2)}R_2. \tag{2.11}$$

The results of Eqs. (2.10) and (2.11) can be checked by summing the voltages in the mesh of Figure 2.7:

$$v(t) - v_{R1}(t) - v_{R2}(t) = 0, \tag{2.12}$$

or

$$v(t) = \frac{v(t)R_1}{(R_1 + R_2)} + \frac{v(t)R_2}{(R_1 + R_2)} = v(t)\frac{(R_1 + R_2)}{(R_1 + R_2)} = v(t). \tag{2.13}$$

Conclusions

The voltage across series-connected resistances divides in proportion the resistances.

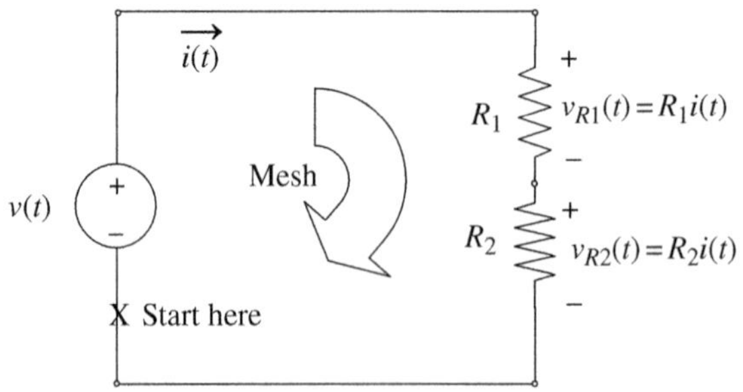

Figure 2.7 Application of voltage divider rule.

2.4 Single-Node Pair Circuits

An important circuit for the analysis of electronic amplifiers is the single-node pair. This represents an example for nodal analysis discussed in Section 2.6.

The detailed equivalent circuit to be analyzed is shown in Figure 2.8 with the current sources $i_A(t)$, $i_B(t)$, and $i_C(t)$ and the load resistances R_1, R_2, R_3, and R_4 with the output voltage $v(t)$.

KCL is applied to the upper node, where $i_A(t)$ and $i_B(t)$ are entering the upper node and are defined as positive, while the currents $i_C(t)$, $i_1(t)$, $i_2(t)$, $i_3(t)$, and $i_4(t)$ are leaving the upper node and are counted as negative:

$$i_A(t) - i_1(t) + i_B(t) - i_2(t) - i_C(t) - i_3(t) - i_4(t) = 0, \tag{2.14}$$

or defining the summation of the source currents as

$$i_o(t) = i_A(t) + i_B(t) - i_C(t) = i_1(t) + i_2(t) + i_3(t) + i_4(t). \tag{2.15}$$

Applying Ohm's law yields with Eq. (2.15):

$$\begin{aligned} i_o(t) = \frac{v_A(t)}{R_1} + \frac{v_A(t)}{R_2} + \frac{v_A(t)}{R_3} + \frac{v_A(t)}{R_4} &= v_A(t)\left[\frac{1}{R_1} + \frac{1}{R_2} + \frac{1}{R_3} + \frac{1}{R_4}\right] \\ &= v_A(t)[G_1 + G_2 + G_3 + G_4], \end{aligned} \tag{2.16}$$

or

$$i_o(t) = v_A(t)\frac{1}{R_p} = v_A(t)G_p, \tag{2.17}$$

where p stands for parallel.

The parallel-connected resistors are function of R_p and the conductance G_p:

$$\frac{1}{R_p} = \frac{1}{R_1} + \frac{1}{R_2} + \frac{1}{R_3} + \frac{1}{R_4} = G_1 + G_2 + G_3 + G_4 = G_p. \tag{2.18}$$

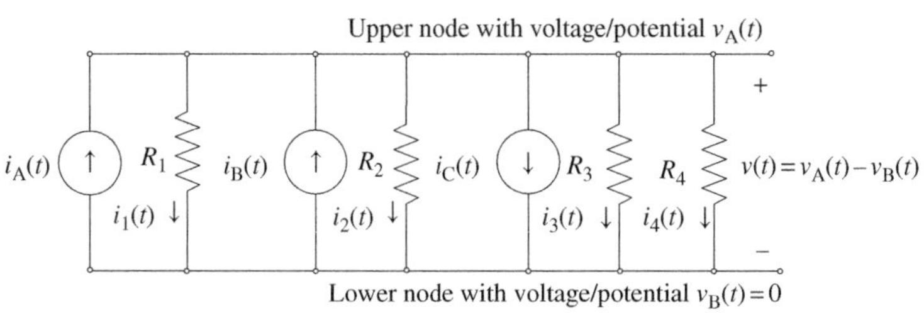

Figure 2.8 Detailed equivalent circuit for single-node pair.

Note that

$$\frac{1}{R_p} = \frac{1}{R_1} + \frac{1}{R_2} + \frac{1}{R_3} + \frac{1}{R_4} = \frac{R_2R_3R_4 + R_1R_3R_4 + R_1R_2R_4 + R_1R_2R_3}{R_1R_2R_3R_4}, \tag{2.19}$$

or

$$R_p = \frac{R_1R_2R_3R_4}{R_2R_3R_4 + R_1R_3R_4 + R_1R_2R_4 + R_1R_2R_3}. \tag{2.20}$$

The simplified or reduced equivalent circuit is given in Figure 2.9.

Conclusions

1) Several independent current sources (e.g. $i_A(t)$, $i_B(t)$, $i_C(t)$) connected in parallel can be replaced by one equivalent source $i_o(t)$ whose value is the algebraic sum of the individual sources.
2) The equivalent conductance G_p of any number of resistors in parallel is the sum of the individual conductances.

2.4.1 Current Division

In the detailed parallel equivalent circuit of Figure 2.10, where the current source $i(t)$ is in parallel to the two resistors R_1 and R_2, the parallel connection of the resistors can be expressed by one resistor R_p:

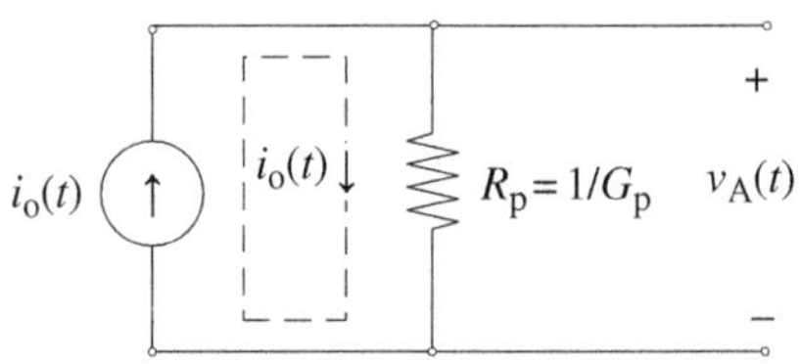

Figure 2.9 Reduced equivalent circuit for single-node pair where $v_A(t) = R_p i_o(t)$.

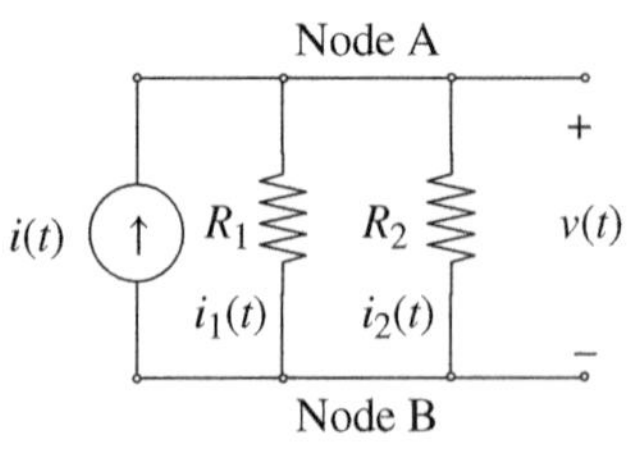

Figure 2.10 Detailed parallel equivalent circuit illustrating current division.

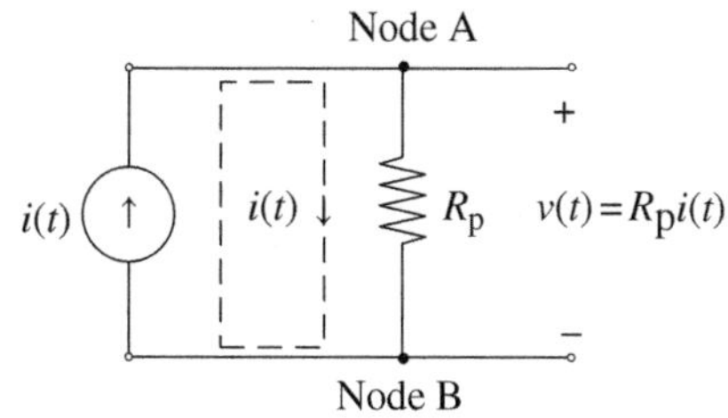

Figure 2.11 Reduced parallel equivalent circuit illustrating current division.

$$\frac{1}{R_p} = \frac{1}{R_1} + \frac{1}{R_2} = \frac{R_1 + R_2}{R_1 R_2}, \tag{2.21}$$

or

$$R_p = \frac{R_1 R_2}{R_1 + R_2}. \tag{2.22}$$

The detailed equivalent circuit of Figure 2.10 can be reduced to that of Figure 2.11.

The output voltage $v(t)$ as a function of R_p and $i(t)$ is

$$v(t) = R_p i(t) = \frac{R_1 R_2}{R_1 + R_2} i(t), \tag{2.23}$$

and the voltage drop in Figure 2.10 across R_1 with $i_1(t)$ is

$$v(t) = R_1 i_1(t). \tag{2.24}$$

Equation (2.23) must be equal to Eq. (2.24); therefore

$$R_1 i_1(t) = \frac{R_1 R_2}{R_1 + R_2} i(t) \tag{2.25}$$

or

$$i_1(t) = \frac{R_2}{R_1 + R_2} i(t) \tag{2.26}$$

and correspondingly,

$$i_2(t) = \frac{R_1}{R_1 + R_2} i(t). \tag{2.27}$$

Conclusions

The current divides in inverse proportion to the parallel-connected resistances.

2.5 Resistor Combinations

Resistors can be either connected in series or in parallel to one another. Figure 2.12 depicts the series connection and Figure 2.13 the parallel connection.

For N resistors connected in series, one gets, where s stands for series,

$$R_s = R_1 + R_2 + \cdots + R_N. \tag{2.28}$$

For N resistors connected in parallel, one gets, where p stands for parallel,

$$\frac{1}{R_p} = \frac{1}{R_1} + \frac{1}{R_2} + \cdots + \frac{1}{R_N} \tag{2.29}$$

or

$$R_p = \frac{1}{\frac{1}{R_1} + \frac{1}{R_2} + \cdots + \frac{1}{R_N}} = R_1 // R_2 // \ldots R_N, \tag{2.30}$$

where the symbol // is a shorthand expression for parallel connection.

As an example for the manipulation of resistors, the so-called Δ-Y transformation is used to simplify a network, as demonstrated in Application Example 2.1.

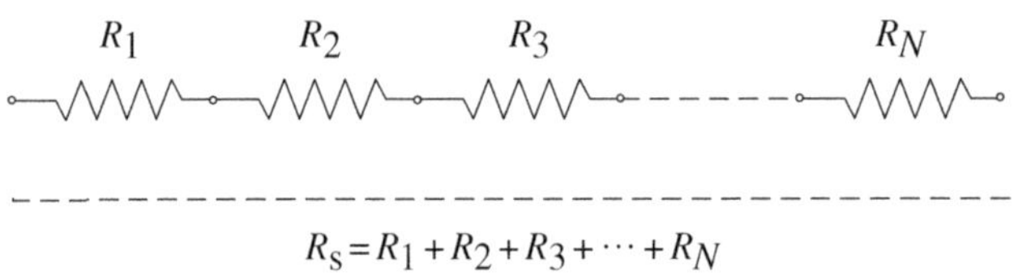

Figure 2.12 Series connection of N resistors.

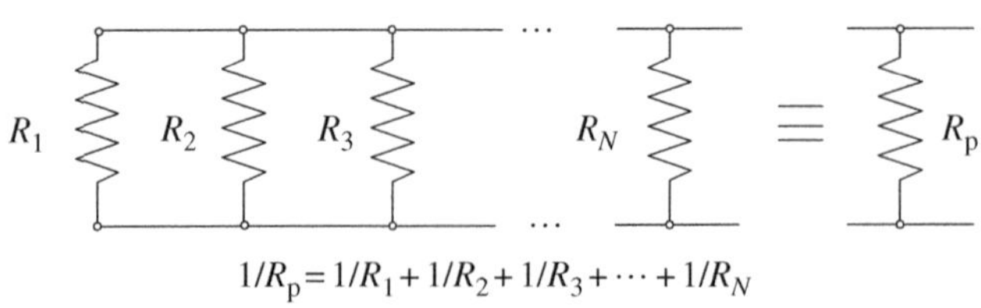

Figure 2.13 Parallel connection of N resistors.

Application Example 2.1 Simplification/transformation/reduction of a bridge circuit

Figure E2.1.1 shows a bridge-type circuit, which is frequently relied on to measure quantities such as temperature deviations. A (Δ-Y) transformation is used to simplify this bridge circuit so that it becomes amenable to simpler analytical expressions.

$$R_{ab} = R_a + R_b = \frac{R_2(R_1 + R_3)}{R_1 + R_2 + R_3} = \frac{R_1 R_2}{R_1 + R_2 + R_3} + \frac{R_2 R_3}{R_1 + R_2 + R_3}. \tag{E2.1.1a}$$

With cyclic (ab → bc → ca) replacement of indices, one obtains

$$R_{bc} = R_b + R_c = \frac{R_2 R_3}{R_1 + R_2 + R_3} + \frac{R_3 R_1}{R_1 + R_2 + R_3}, \tag{E2.1.1b}$$

$$R_{ca} = R_c + R_a = \frac{R_3 R_1}{R_1 + R_2 + R_3} + \frac{R_1 R_2}{R_1 + R_2 + R_3}; \tag{E2.1.1c}$$

by identification

$$R_a = \frac{R_1 R_2}{R_1 + R_2 + R_3} = \frac{R_1 R_2}{\sum_{j=1}^{3} R_j}, \tag{E2.1.2a}$$

with cyclic (a → b → c) replacement follows

$$R_b = \frac{R_2 R_3}{R_1 + R_2 + R_3} = \frac{R_2 R_3}{\sum_{j=1}^{3} R_j}, \tag{E2.1.2b}$$

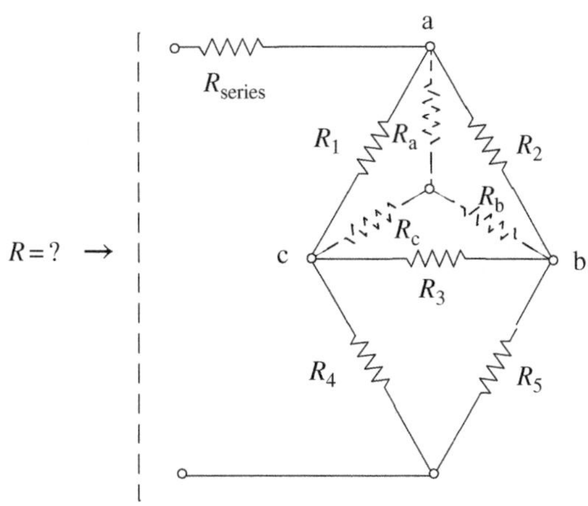

Figure E2.1.1 Bridge-type circuit with nodes a, b, and c, which are instrumental for the replacement of the Δ defined by R_1, R_2, and R_3 by an equivalent Y circuit defined by R_a, R_b, R_c in Figure E2.1.2.

$$R_c = \frac{R_3 R_1}{R_1 + R_2 + R_3} = \frac{R_3 R_1}{\sum_{j=1}^{3} R_j}. \tag{E2.1.2c}$$

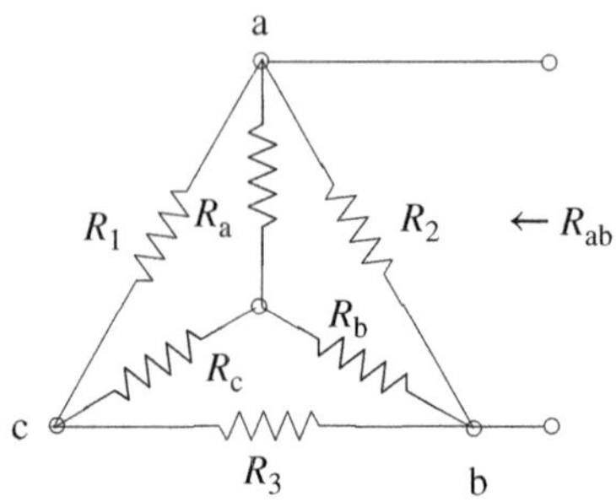

Figure E2.1.2 Replacement of Δ by equivalent Y circuit.

Expressing R_1 as a function of R_a, R_b, and R_c yields

$$R_1 = \frac{R_a R_b + R_b R_c + R_c R_a}{R_b}; \tag{E2.1.3a}$$

through cyclic $(1 \to 2 \to 3)$ replacement in the denominator, only one obtains

$$R_2 = \frac{R_a R_b + R_b R_c + R_c R_a}{R_c}, \tag{E2.1.3b}$$

$$R_3 = \frac{R_a R_b + R_b R_c + R_c R_a}{R_a}. \tag{E2.1.3c}$$

For $R_a = R_b = R_c$ $(=R_Y)$ and $R_1 = R_2 = R_3$ $(=R_\Delta)$, one gets from Eq. (E2.1.2a)

$$R_Y = \frac{1}{3} R_\Delta, \tag{E2.1.4a}$$

$$R_\Delta = 3R_Y. \tag{E2.1.4b}$$

Application Example 2.2 Bridge circuit in series with resistor R_{series}

Figure E2.2.1 shows a bridge circuit in series with a resistor R_{series}. Find the input resistor R_{AB} between terminals A and B.

Solution

Replacing the Δ-circuit a, b, c with the resistors R_1, R_2, R_3 by a Y circuit with the resistors R_a, R_b, R_c leads to Figure E2.2.2.

The equivalent Y-resistors R_a, R_b, R_c as a function of the resistors Δ-connected resistors R_1, R_2, R_3 are

$$R_a = R_1 R_2/(R_1 + R_2 + R_3) = 18 \cdot 12/(18 + 12 + 6) = 6\,\Omega, \tag{E2.2.1a}$$

$$R_b = R_2 R_3/(R_1 + R_2 + R_3) = 12 \cdot 6/(18 + 12 + 6) = 2\,\Omega, \tag{E2.2.1b}$$

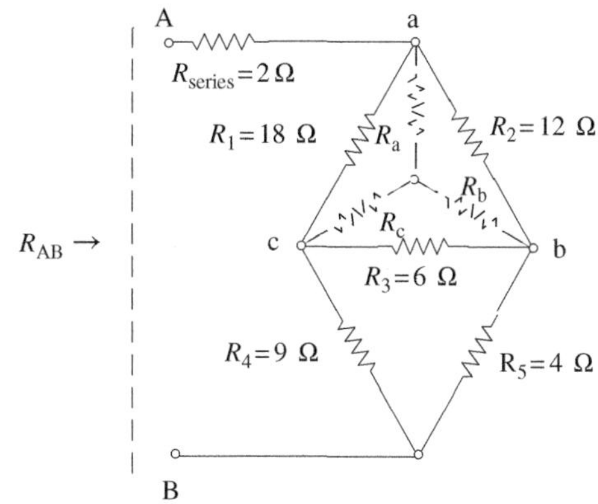

Figure E2.2.1 Bridge circuit in series with the resistor R_{series}.

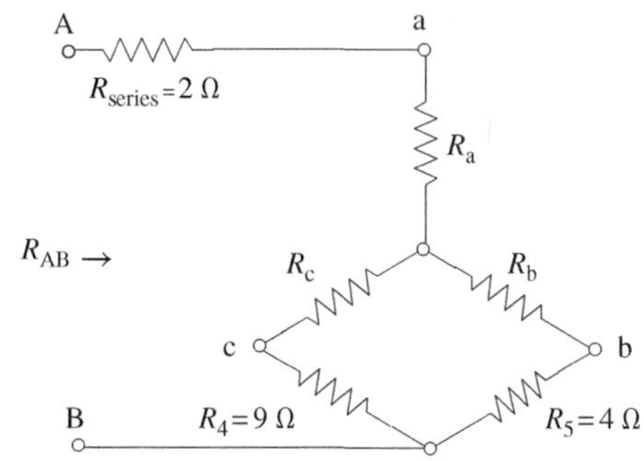

Figure E2.2.2 Reduced circuit, equivalent to circuit of Figure E2.2.1.

$$R_c = R_3 R_1/(R_1 + R_2 + R_3) = 6 \cdot 18/(18 + 12 + 6) = 12\,\Omega. \tag{E2.2.1c}$$

The input resistance R_{AB} is now

$$\begin{aligned} R_{AB} &= R_{series} + R_a + (R_b + 4\,\Omega) \cdot (R_c + 9\,\Omega)/(R_b + 4\,\Omega + R_c + 9\,\Omega) \\ &= 2\,\Omega + 6\,\Omega + (2\,\Omega + 4\,\Omega)(12\,\Omega + 9\,\Omega)/(2\,\Omega + 4\,\Omega + 12\,\Omega + 9\,\Omega) \\ &= 12\,\Omega. \end{aligned} \tag{E2.2.2}$$

2.6 Nodal Analysis

Nodal analysis consists of the application of KCL and Ohm's law.

In this procedure one selects one node as reference or ground node with zero voltage (0 V) and defines the voltage at every other node with respect to this reference node.

Application Example 2.3 A three-node network using KCL and Ohm's law

Figure E2.3.1 illustrates the three-node network where node #3 is assumed as ground having 0 V potential and the independent currents $i_A(t)$ and $i_B(t)$ are assumed to be known.

Solution

1) Application of KCL at node #1 leads to

$$i_A(t) - i_1(t) - i_2(t) = 0. \tag{E2.3.1}$$

2) Use of Ohm's law yields by replacing $i_1(t)$ and $i_2(t)$ in the above equation to

$$i_A(t) - \frac{[v_1(t) - v_2(t)]}{R_1} - \frac{v_1(t)}{R_2} = 0, \tag{E2.3.2}$$

or

$$i_A(t) - v_1(t)\left(\frac{1}{R_1} + \frac{1}{R_2}\right) + \frac{v_2(t)}{R_1} = 0, \tag{E2.3.3}$$

resulting with $G_1 = 1/R_1$ and $G_2 = 1/R_2$ in

$$i_A(t) = v_1(t)(G_1 + G_2) - v_2(t)G_1. \tag{E2.3.4}$$

3) Application of KCL at node #2 gives

$$i_1(t) - i_3(t) - i_B(t) = 0. \tag{E2.3.5}$$

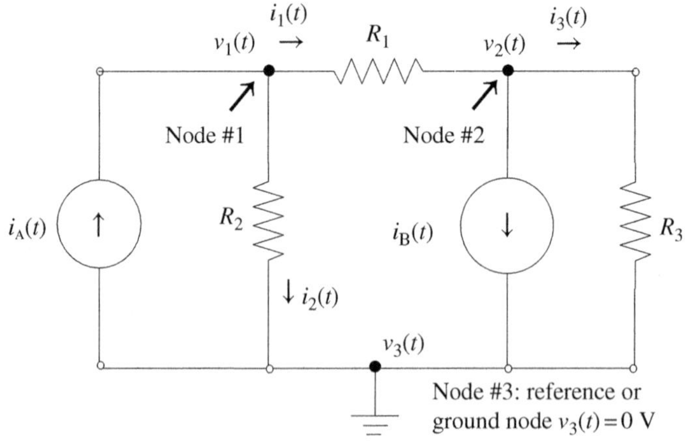

Figure E2.3.1 Three-node network with known currents $i_A(t)$ and $i_B(t)$.

4) Use of Ohm's law and replacing $i_1(t)$ and $i_3(t)$ generates

$$[v_1(t) - v_2(t)]\frac{1}{R_1} - v_2(t)\frac{1}{R_3} - i_B(t) = 0, \tag{E2.3.6}$$

or by introducing conductances

$$i_B(t) = v_1(t)G_1 - v_2(t)(G_1 + G_3). \tag{E2.3.7}$$

Equations (E2.3.4) and (E2.3.7) represent two equations with the two unknowns $v_1(t)$ and $v_2(t)$.

Conclusions

In the nodal analysis the unknown quantities are the node voltages based on KCL and Ohm's law. The two Eqs. (E2.3.4) and (E2.3.7) can be solved via:

1) Gaussian elimination
2) Cramer's rule
3) MATLAB

2.7 Loop or Mesh Analysis

KVL and Ohm's law are employed in this method to determine the currents in circuits. Once the currents are known, Ohm's law can be used to calculate the desired voltages.

Application Example 2.4 Two-mesh circuit with mesh currents $i_1(t)$ and $i_2(t)$

Figure E2.4.1 depicts a two-mesh circuit with two independent voltage sources. The mesh currents $i_1(t)$ and $i_2(t)$ are to be calculated.

Solution

KVL applied to mesh #1:

$$v_1(t) - i_1(t)R_1 - [i_1(t) - i_2(t)] \ R_2 = 0. \tag{E2.4.1}$$

KVL applied to mesh #2:

$$-v_2(t) - R_3 i_2(t) - [i_2(t) - i_1(t)] \ R_2 = 0, \tag{E2.4.2}$$

or

$$i_1(t)(R_1 + R_2) - i_2(t)R_2 = v_1(t), \tag{E2.4.3}$$

$$i_1(t)R_2 - i_2(t)(R_2 + R_3) = v_2(t). \tag{E2.4.4}$$

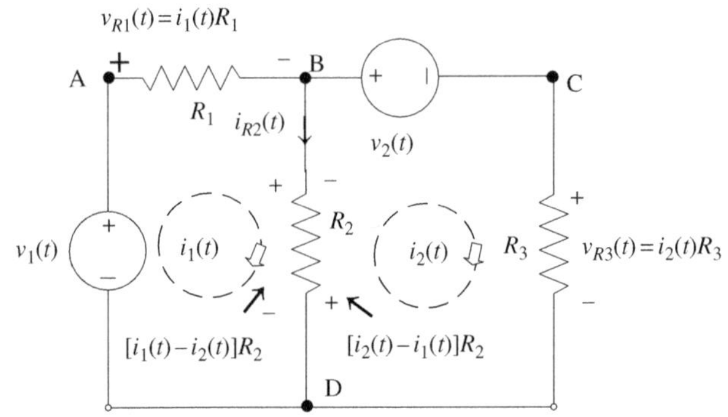

Figure E2.4.1 Two-mesh circuit.

Equations (E2.4.3) and (E2.4.4) can be solved for the currents $i_1(t)$ and $i_2(t)$ by the same methods as indicated in Application Example 2.3, and the node voltages can be found by applying Ohm's law.

2.8 Superposition

Superposition can be applied to linear circuits (see Figure 2.14a), but cannot be applied to nonlinear circuits (see Figure 2.14b).

An example for a linear characteristic is given by Ohm's law $v = \text{const.} \cdot i$ where const. is identical with the resistance R, and examples for nonlinear characteristics are diode characteristic and transformer $v = f(i)$ saturation curve as shown in Figure 2.14b.

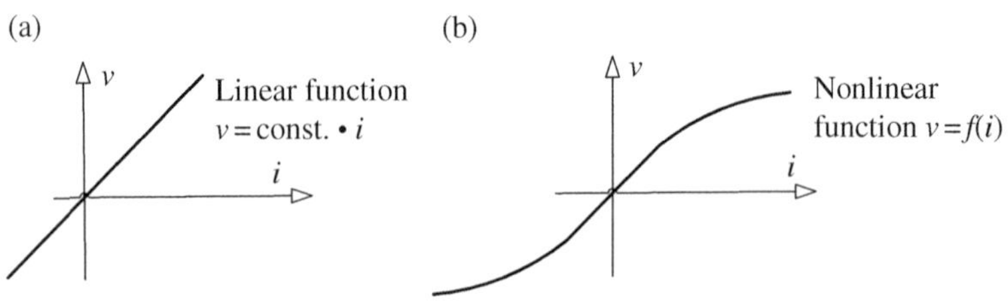

Figure 2.14 (a) Linear and (b) nonlinear voltage–current characteristics $v = (\text{const.} \cdot i)$ and $v = f(i)$, respectively.

2.8.1 Principle of Superposition

In a linear circuit, if two or more independent sources (either voltage or current) are present, the net or total voltage or current at some location of the network is the algebraic sum of the results of each independent source operation alone.

Application Example 2.5 Circuit with two independent sources $v_A(t)$ and $i_A(t)$

Figure E2.5.1 shows the circuit with the independent sources $v_A(t)$ and $i_A(t)$. The task is to find $v_o(t) = v_o'(t) + v_o''(t)$, where the voltage $v_o'(t)$ is due to the current source $i_A(t)$ and voltage $v_o''(t)$ is due to the voltage source $v_A(t)$.

Solution

Step #1: Set $v_A(t) = 0$, representing a short circuit, to obtain reduced circuit of Figure E2.5.2.

Redrawing the circuit of Figure E2.5.2, one obtains the circuit of Figure E2.5.3 for which Ohm's law is applied.

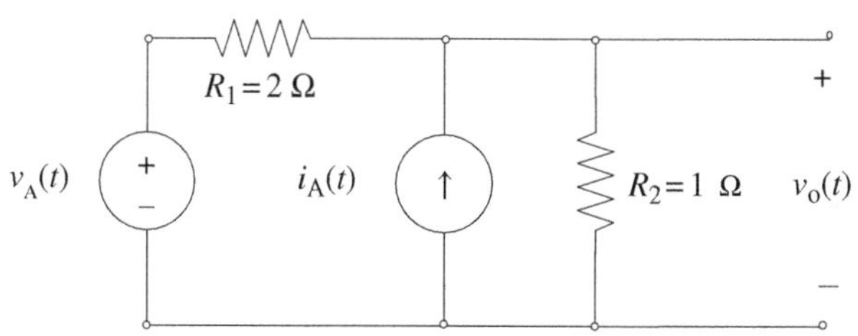

Figure E2.5.1 Two independent sources $v_A(t)$ and $i_A(t)$ within an electric circuit.

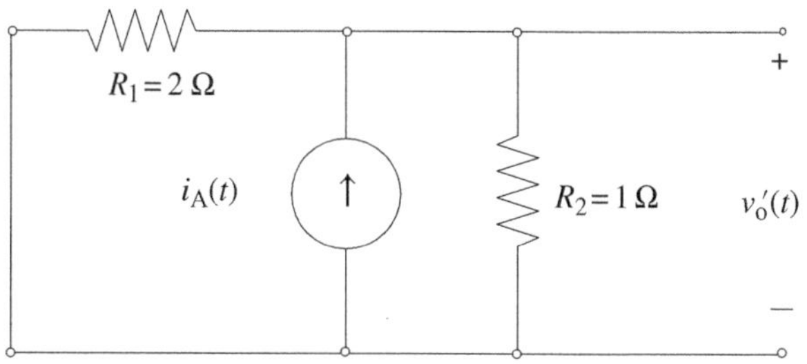

Figure E2.5.2 Reduced electric circuit with $v_A(t) = 0$.

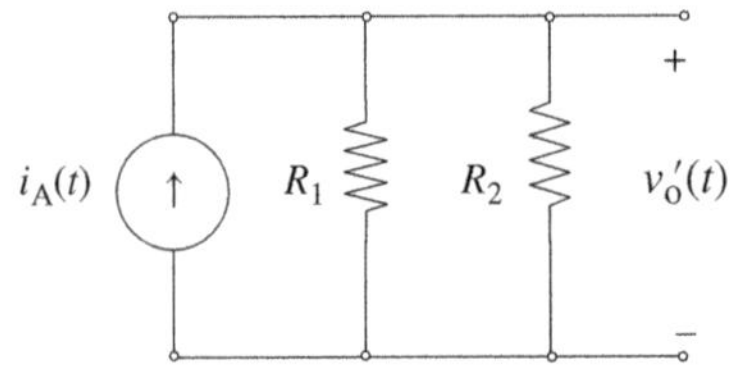

Figure E2.5.3 Redrawing of electric circuit of Figure E2.5.2.

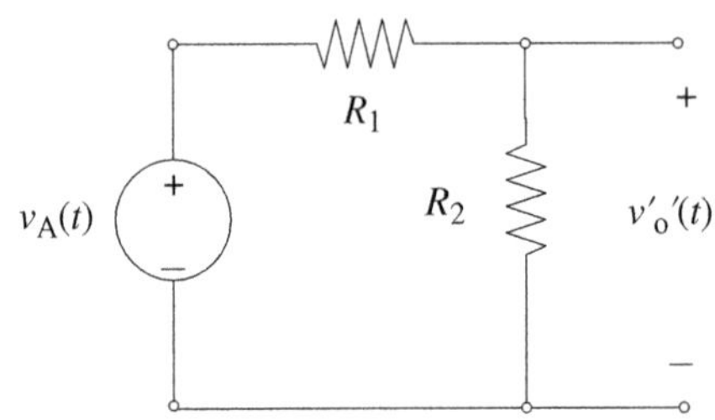

Figure E2.5.4 Reduced electric circuit with $i_A(t) = 0$.

The resistor combination R_1, R_2, which is in parallel, can be indicated in a symbolic manner as $R_1//R_2$, where "//" indicates that there is a parallel connection. Therefore, $R_1//R_2 = \dfrac{R_1R_2}{R_1 + R_2}$ or the application of Ohm's law yields

$$v_o'(t) = i_A(t)\frac{R_1R_2}{R_1 + R_2} = \left(\frac{2 \cdot 1}{3}\right)i_A(t) = \left(\frac{2}{3}\right)i_A(t). \tag{E2.5.1}$$

Step #2: Set $i_A(t) = 0$, representing an open circuit, to obtain reduced circuit of Figure E2.5.4.

The application of the voltage divider rule yields

$$v_o''(t) = \frac{R_2}{R_1 + R_2}v_A(t) = \left(\frac{1}{3}\right)v_A(t). \tag{E2.5.2}$$

Step #3: Superposition means that we add the two components $v_o'(t)$ and $v_o''(t)$; therefore

$$v_o(t) = v_o'(t) + v_o''(t) = \frac{2}{3}i_A(t) + \frac{1}{3}v_A(t). \tag{E2.5.3}$$

Conclusions

The principle of superposition states in general that in any linear circuit containing independent sources, the current or voltage at any point in the network may be calculated as the algebraic sum of the individual contributions of each source acting alone.

2.9 Source Exchange/Transformation

The source exchange or transformation of sources is aimed at simplifying the network by applying KCL, KVL, and Ohm's law.

Step #1: Figure 2.15 depicts an independent current source $i(t)$ with ohmic internal resistance R_i and load R_L.

The application of KCL leads to

$$i(t) = i_L(t) + v_L(t)/R_i \tag{2.31}$$

or

$$i(t)R_i = i_L(t)R_i + v_L(t). \tag{2.32}$$

Step #2: Figure 2.16 depicts an independent voltage source $v(t)$ with ohmic internal resistance R_v and load R_L.

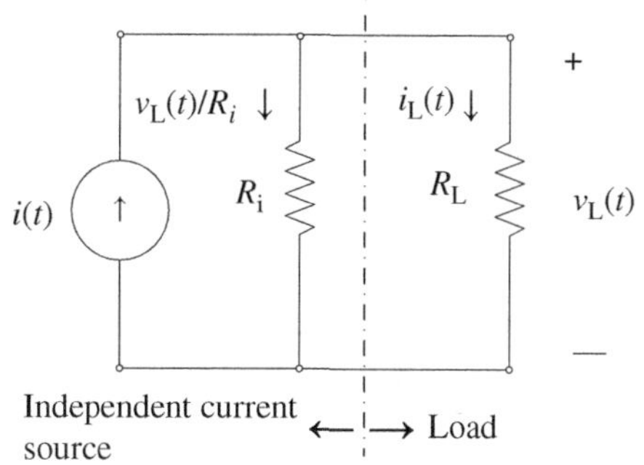

Figure 2.15 Independent current source $i(t)$ with load resistor R_L.

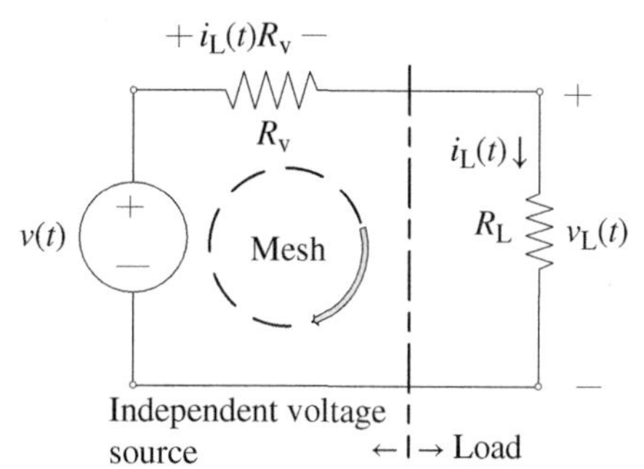

Figure 2.16 Independent voltage source $v(t)$ with load resistor R_L.

The application of KVL leads to

$$v(t) = i_L(t)R_v + v_L(t). \tag{2.33}$$

Step #3: Comparison of coefficients of Eqs. (2.32) and (2.33) leads to

$$v(t) = i(t)R_i \tag{2.34a}$$

and

$$R_i = R_v. \tag{2.34b}$$

2.10 Thévenin's and Norton's Theorems

Thévenin's and Norton's theorems represent equivalent voltage and equivalent current sources, respectively. They are used to simplify electric networks.

Thévenin's (abbreviated by TH) theorem tells us that we can replace the entire electric network, exclusive of the load resistor R_L, by an equivalent circuit that contains only an independent open-circuit (represented by oc) voltage source V_{oc} in series with a resistor R_{TH}.

Step #1: Figure 2.17 represents a given circuit consisting of a network containing independent voltage–current sources where the load resistor R_L is identified.

Step #2: Remove load resistor R_L and find open-circuit (oc) voltage V_{oc} as shown in Figure 2.18.

Step #3: With load resistor removed, find the short-circuit (sc) current I_{sc} between terminals A and B as illustrated in Figure 2.19.

Step #4: Connect load resistor R_L to Thévenin-adjusted equivalent and simplified network as depicted in Figure 2.20. The Thévenin-adjusted load resistor has the resistance value $R_{TH} = V_{oc}/I_{sc}$.

Norton's theorem tells us that we can replace the entire electric network of Figure 2.17, exclusive of the load resistor R_L, by an equivalent circuit that contains

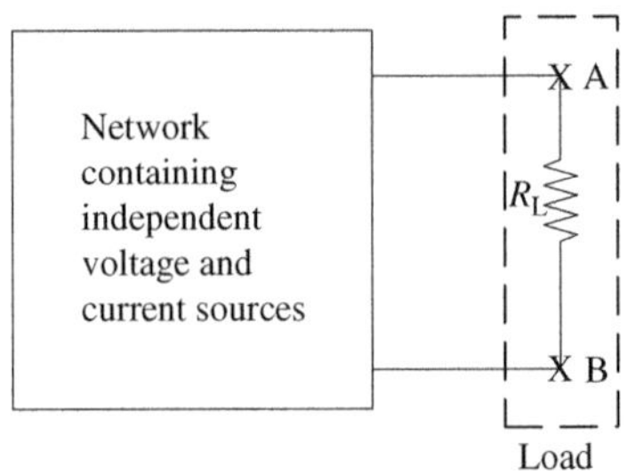

Figure 2.17 Given electric circuit.

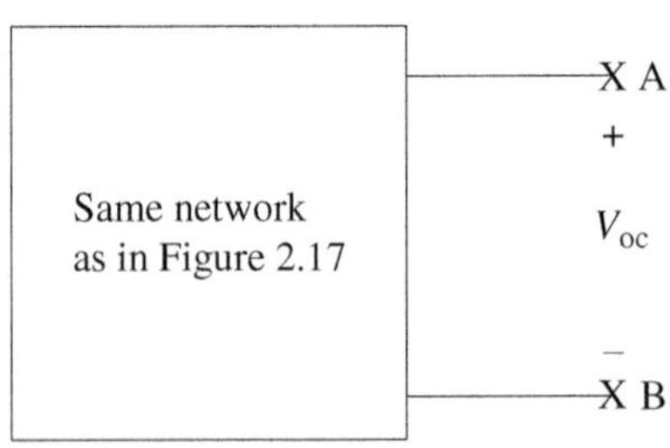

Figure 2.18 Network of Figure 2.17 but with removed load resistor R_L; definition of open-circuit (oc) voltage V_{oc}.

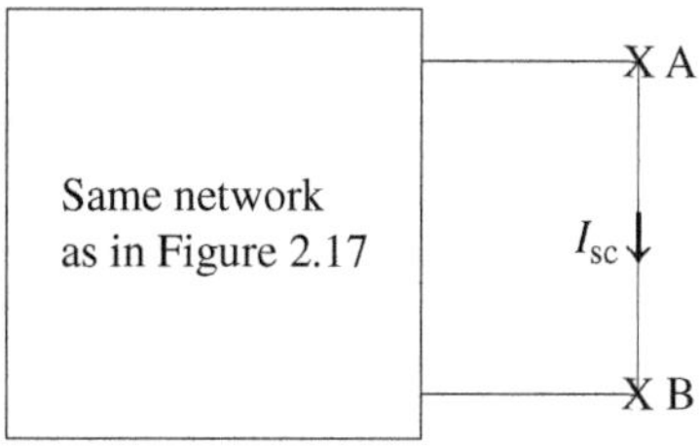

Figure 2.19 Calculation of short-circuit (sc) current I_{sc} flowing from terminals A to B.

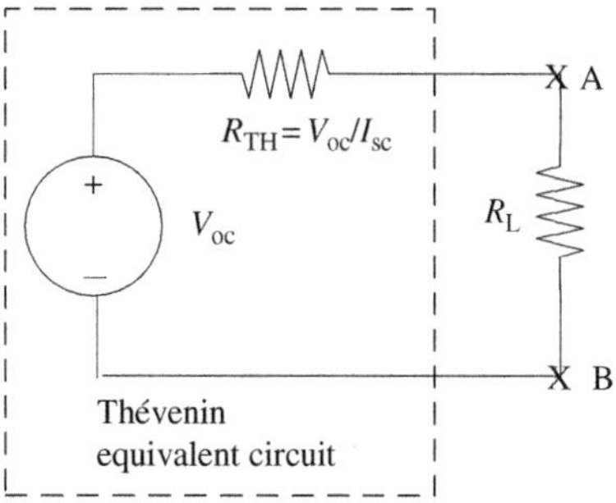

Figure 2.20 Thévenin (TH)-adjusted equivalent network with load resistor R_L connected.

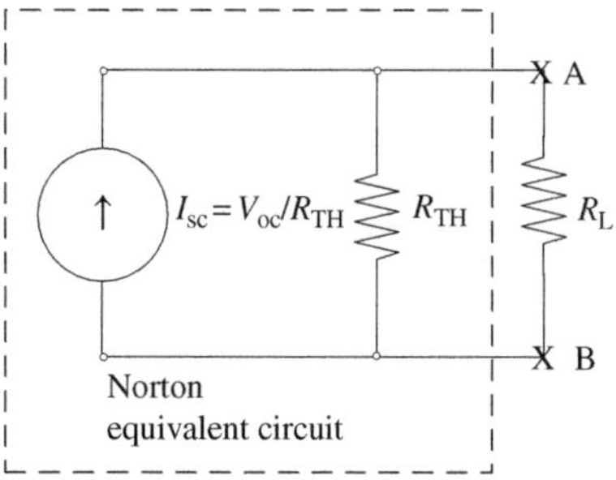

Figure 2.21 Norton equivalent electric circuit with load resistor R_L connected.

only an independent short-circuit (represented by sc) current source I_{sc} in parallel with resistors R_{TH} and R_L.

Step #1: Figure 2.21 represents a given circuit consisting of an equivalent network containing independent current source I_{sc} and Thévenin resistance R_{TH}, where the load resistor R_L is identified. The short-circuit current is defined as $I_{sc} = V_{oc}/R_{TH}$.

Application Example 2.6 Derivation and use of Thévenin and Norton equivalent circuits for a given network

Solution

Step #1: Given network of Figure E2.6.1, calculate V_{out} across resistor R_3 of Figure E2.6.1 by using Thévenin's theorem.

Step #2: Determine the open-circuit (oc) voltage V_{oc} across terminals A and B by using the voltage divider rule:

$$V_{AB} = V_{oc} = \frac{R_4}{R_1 + R_4} V_1 = \frac{6}{3+6} \cdot 36\ \text{V} = 24\ \text{V}. \tag{E2.6.1}$$

Step #3: Determine the Thévenin resistance R_{TH} by setting $V_1 = 0$ as depicted in Figure E2.6.2 resulting in Figure E2.6.3, where

$$R_{TH} = (R_1 R_4)/(R_1 + R_4) = 3 \cdot 6/(3+6) = 2\,\Omega. \tag{E2.6.2}$$

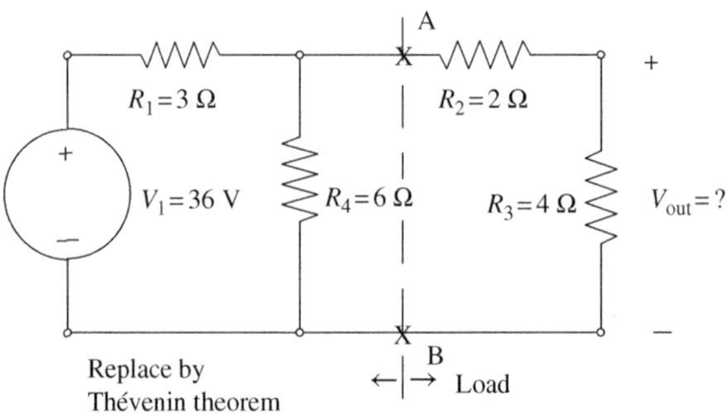

Figure E2.6.1 Given network.

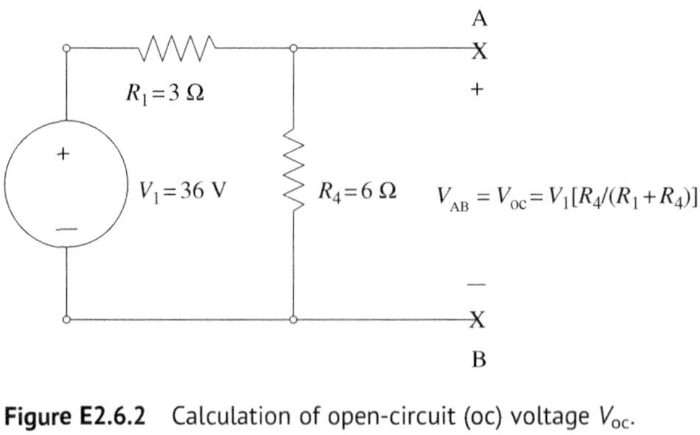

Figure E2.6.2 Calculation of open-circuit (oc) voltage V_{oc}.

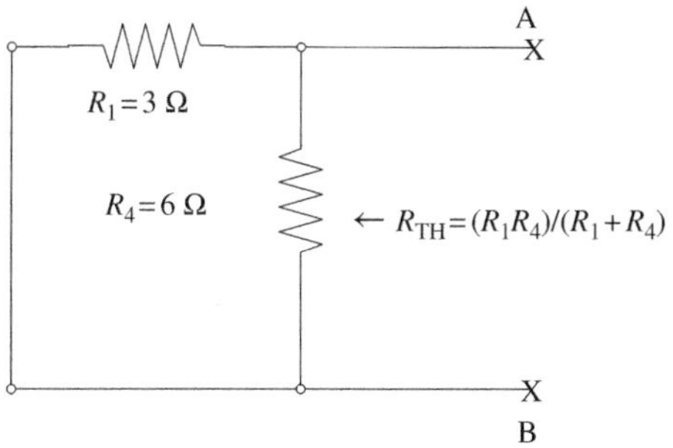

Figure E2.6.3 Calculation of Thévenin resistance R_{TH}.

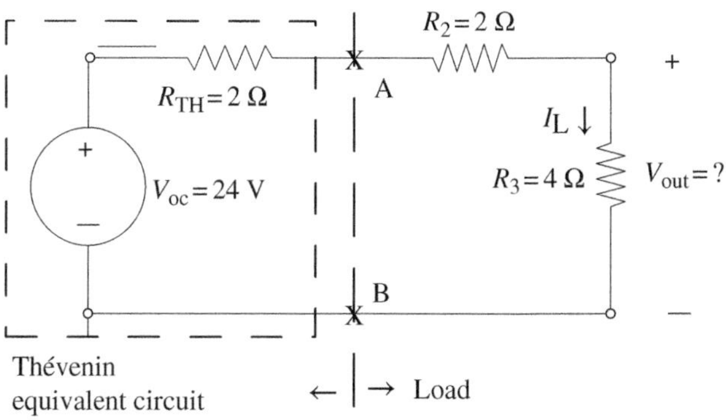

Figure E2.6.4 Thévenin equivalent circuit with load consisting of R_2 and R_3.

Step #4: Thévenin equivalent circuit resistor with R_{TH} and R_2 including load resistor $R_L = R_3$ as illustrated in Figure E2.6.4. The output voltage V_{out} is obtained by applying the voltage divider rule:

$$V_{out} = \frac{R_3}{R_3 + R_2 + R_{TH}} \cdot V_{oc} = \frac{4\,\Omega}{(4 + 2 + 2)\Omega} \cdot 24\,\text{V} = 12\,\text{V}. \tag{E2.6.3}$$

Alternatively the output voltage can be determined from the calculation of the load current I_L and by applying Ohm's law across R_3 to calculate V_{out}:

$$I_L = \frac{V_{oc}}{(R_{TH} + R_2 + R_3)} = \frac{24\,\text{V}}{(2 + 2 + 4)\,\Omega} = 3\,\text{A} \tag{E2.6.4}$$

and

$$V_{out} = R_3 I_L = 4\,\Omega \cdot 3\,\text{A} = 12\,\text{V}. \tag{E2.6.5}$$

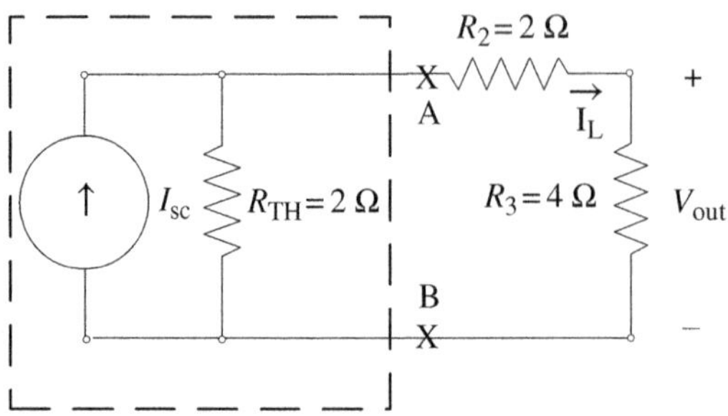

Figure E2.6.5 Norton equivalent circuit and load.

Step #5: Norton equivalent circuit and load

The short-circuit current for the Norton equivalent circuit (see Figure E2.6.5) is obtained from Figure E2.6.4 as

$$I_{sc} = V_{oc}/R_{TH} = 24\,\text{V}/2\,\Omega = 12\,\text{A}. \tag{E2.6.6}$$

Use the current divider rule (Figure E2.6.5) to find the load current I_L

$$I_L = \frac{R_{TH}}{R_{TH} + R_2 + R_3} I_{sc} = \frac{2\,\Omega}{(2 + 2 + 4)\,\Omega} 12\,\text{A} = 3\,\text{A} \tag{E2.6.7}$$

and Ohm's law to determine the output voltage V_{out}

$$V_{out} = I_L R_3 = 3\,\text{A} \cdot 4\,\Omega = 12\,\text{V}. \tag{E2.6.8}$$

2.10.1 Equivalency of Thévenin and Norton Circuits

Figure 2.22a shows the Thévenin (see Figure E2.6.4) and Figure 2.22b the Norton (see Figure E2.6.5) equivalent circuits, both connected to the load resistor R_L. In the following it will be shown that both equivalent circuits satisfy Ohm's law across the load, that is, $v_L = R_L i_L$.

Step #1: One can write for Figure 2.22a with the voltage divider rule for the load voltage v_L as a function of the open-circuit voltage v_{oc}

$$v_L = \left(\frac{R_L}{R_{TH} + R_L}\right) v_{oc}, \tag{2.35}$$

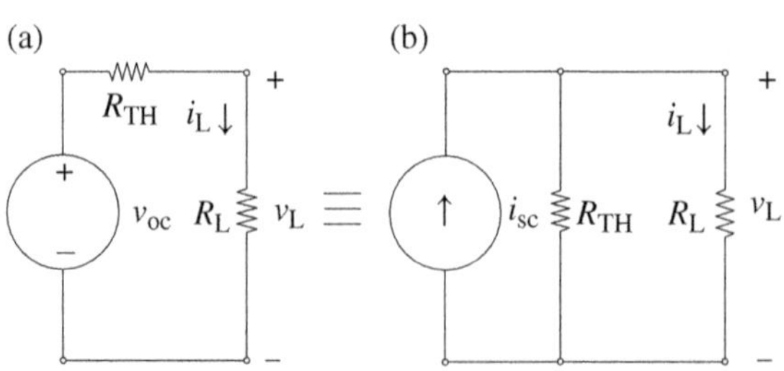

Figure 2.22 (a) Thévenin and (b) Norton equivalent circuits.

or solve for the open-circuit voltage

$$v_{oc} = \left(\frac{R_{TH} + R_L}{R_L}\right) v_L. \tag{2.36}$$

The application of Ohm's law to the circuit of Figure 2.22a leads to the load current as a function of v_{oc}:

$$i_L = \left(\frac{1}{R_{TH} + R_L}\right) v_{oc}. \tag{2.37}$$

Introducing Eq. (2.36) into Eq. (2.37) leads to

$$i_L = \left(\frac{R_{TH} + R_L}{R_{TH} + R_L}\right) \frac{1}{R_L} v_L = \frac{v_L}{R_L}, \tag{2.38}$$

which is Ohm's relation applied to the load resistor R_L.

Step #2: One can write for Figure 2.22b with the current divider rule for the load voltage as a function of the short-circuit current i_{sc}

$$v_L = \left(\frac{R_{TH} R_L}{R_{TH} + R_L}\right) i_{sc}, \tag{2.39}$$

or solve for the short-circuit current

$$i_{sc} = \left(\frac{R_{TH} + R_L}{R_{TH} R_L}\right) v_L. \tag{2.40}$$

The application of the current divider rule to the circuit of Figure 2.22b leads to the load current as a function of i_{sc}:

$$i_L = \left(\frac{R_{TH}}{R_{TH} + R_L}\right) i_{sc}. \tag{2.41}$$

Introducing Eq. (2.40) into Eq. (2.41) leads to

$$i_L = \left(\frac{R_{TH}}{R_{TH} + R_L}\right) \left(\frac{R_{TH} + R_L}{R_{TH} R_L}\right) v_L = \frac{v_L}{R_L}, \tag{2.42}$$

which represents Ohm's law applied to the load resistor R_L as a function of the load voltage v_L.

Conclusions

Both the Thévenin equivalent circuit (impressed equivalent voltage source) and the Norton equivalent circuit (impressed equivalent current source) lead to the same answer for the load voltage and current; that is, $v_L = i_L R_L$.

2.11 Wheatstone and Thomson Bridges

Wheatstone [5] and Thomson [6] bridges are well-known circuits for measuring resistance values: while the first one is suitable for larger (e.g. $0.1–10^6\ \Omega$) resistances, the latter is suitable for very small (e.g. micro ohm) resistances. Figure 2.23 depicts the circuit of a Wheatstone bridge consisting of three known resistors R_2, R_3, R_4 and the unknown resistor $R_1 = R_x$. A galvanometer or high-sensitivity current meter G is relied on to make the current I_5 across the bridge zero. The resistance of G is R_5 ($\approx 1\ \text{m}\Omega$).

If $R_1 = R_x$ is to be measured, then the known resistance values R_2, R_3, and R_4 are changed until voltage V_5 is zero. Then according to Kirchhoff

$$V_1 = V_3, \tag{2.43}$$

and

$$V_2 = V_4, \tag{2.44}$$

$$I_1 = I_2, \tag{2.45}$$

and

$$I_3 = I_4. \tag{2.46}$$

From Eqs. (2.43)–(2.46) follow with

$$I_1 R_x = I_3 R_3, \tag{2.47}$$

and

$$I_2 R_2 = I_4 R_4 \tag{2.48}$$

the relation for the balanced bridge

$$R_x / R_2 = R_3 / R_4 \tag{2.49}$$

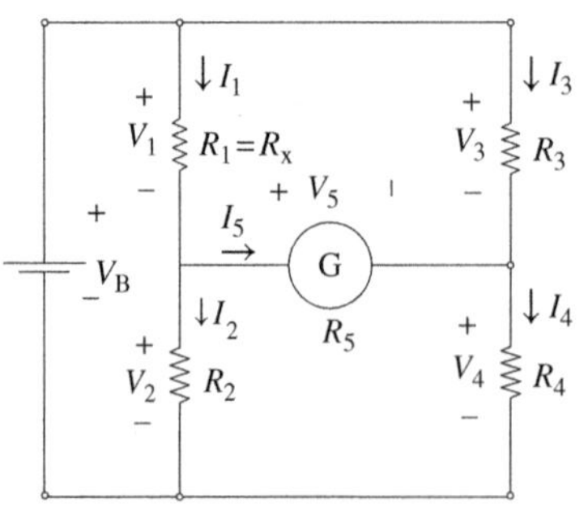

Figure 2.23 Wheatstone bridge for measuring the resistance $R_1 = R_x$ with galvanometer resistance R_5.

or

$$R_x = R_2(R_3/R_4). \tag{2.50}$$

The advantage of this resistance measuring method is that the DC voltage V_B does not influence the result for $R_1 = R_x$. For measuring the zero current $I_5 = 0$, a high-sensitivity moving-coil galvanometer G or current meter with a small internal resistance R_5 is used.

Application Example 2.7 Application of circuit laws to unbalanced Wheatstone bridge

Derive an expression for the current $I_5 = f(V_B, R_1, R_2, R_3, R_4, R_5)$ through the moving-coil galvanometer G of Figure 2.23 if the Wheatstone bridge is not balanced, that is, $I_5 \neq 0$.

Solution

In this application example most above-introduced circuit laws (e.g. Ohm's law, Kirchhoff's laws, voltage and current divider laws, Thévenin equivalent circuit) will be applied.

Step #1: Calculation of open-circuit voltage V_{oc} (across G with resistance R_5) based on Thévenin equivalent circuit is shown in Figure E2.7.1.

From Figure E2.7.1 one obtains from Kirchhoff's second law $\left(\sum_{j=1}^{N} V_j = 0\right)$

$$V_{oc} = V_{R3} - V_{R1}, \tag{E2.7.1a}$$

$$V_B = V_{R1} + V_{R2} = V_{R3} + V_{R4}, \tag{E2.7.1b}$$

and from Ohm's law

$$V_{R1} = I_1 R_1, \tag{E2.7.2a}$$

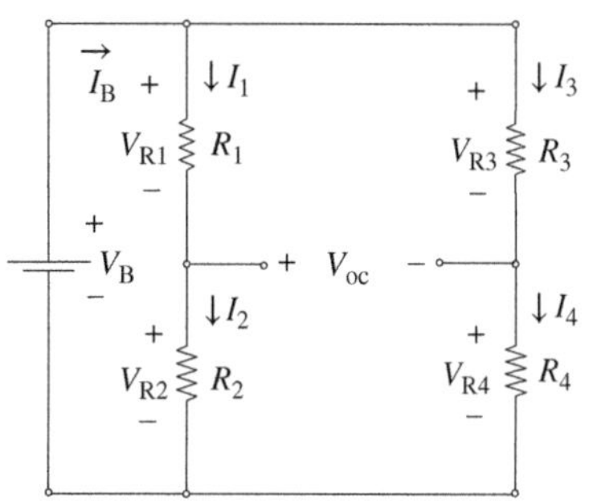

Figure E2.7.1 Determination of open-circuit voltage V_{oc}.

$$V_{R2} = I_2 R_2, \tag{E2.7.2b}$$

$$V_{R4} = I_4 R_4, \tag{E2.7.2c}$$

$$V_{R3} = I_3 R_3, \tag{E2.7.2d}$$

$$I_1 = I_2 = V_B/(R_1 + R_2), \tag{E2.7.3a}$$

$$I_3 = I_4 = V_B/(R_3 + R_4). \tag{E2.7.3b}$$

Introducing Eqs. (E2.7.3a) and (E2.7.3b) into Eq. (E2.7.1a) yields

$$V_{oc} = V_B[R_3/(R_3 + R_4) - R_1/(R_1 + R_2)]. \tag{E2.7.4}$$

Step #2: Calculation of short-circuit current I_{sc} based on Thévenin equivalent circuit is depicted in Figure E2.7.2.

From Figure E2.7.2 Kirchhoff's first law $\sum_{j=1}^{N} I_j = 0$ yields

$$I_{sc} = I_1 - I_2. \tag{E2.7.5a}$$

With Eqs. (E2.7.2a) and (E2.7.2b) one can write

$$I_{sc} = V_{R1}/R_1 - V_{R2}/R_2; \tag{E2.7.5b}$$

applying the voltage divider rule, one obtains

$$V_{R1} = V_B[R_1/(R_1 + R_2)] = V_{R3} = V_B[R_3/(R_3 + R_4)], \tag{E2.7.5c}$$

$$V_{R2} = V_B[R_2/(R_1 + R_2)] = V_{R4} = V_B[R_4/(R_3 + R_4)]; \tag{E2.7.5d}$$

introducing Eqs. (E2.7.5c) and (E2.7.5d) into Eq. (E2.7.5b) results in

$$I_{sc} = \frac{V_B}{(R_1 + R_2)} - V_B\left(\frac{R_4}{R_3 + R_4}\right)\frac{1}{R_2}. \tag{E2.7.6}$$

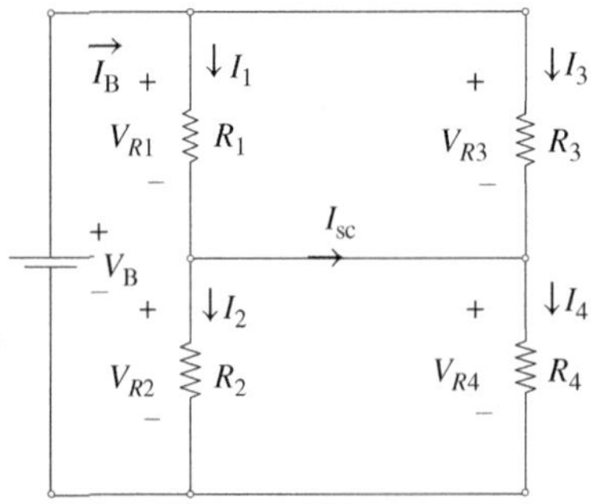

Figure E2.7.2 Determination of short-circuit current I_{sc} through galvanometer G.

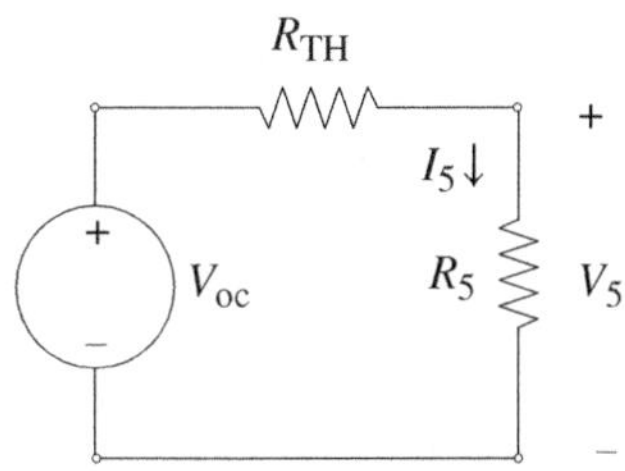

Figure E2.7.3 Determination of Thévenin resistance R_{TH}, where R_5 is the galvanometer (G) resistance.

Step #3: Determination of Thévenin resistance R_{TH} is illustrated in Figure E2.7.3. Dividing Eqs. (E2.7.4) by (E2.7.6) yields

$$R_{TH} = \frac{V_{oc}}{I_{sc}} = \frac{\dfrac{R_3}{(R_3 + R_4)} - \dfrac{R_1}{(R_1 + R_2)}}{\dfrac{1}{(R_1 + R_2)} - \left(\dfrac{R_4}{R_3 + R_4}\right)\dfrac{1}{R_2}}. \tag{E2.7.7}$$

The current I_5 through R_5 (galvanometer G) is now

$$I_5 = \frac{V_{oc}}{R_{TH} + R_5} = \frac{V_B\left[\dfrac{R_3}{R_3 + R_4} - \dfrac{R_1}{R_1 + R_2}\right]}{\dfrac{\dfrac{R_3}{(R_3 + R_4)} - \dfrac{R_1}{(R_1 + R_2)}}{\dfrac{1}{(R_1 + R_2)} - \left(\dfrac{R_4}{R_3 + R_4}\right)\dfrac{1}{R_2}} + R_5}. \tag{E2.7.8}$$

2.12 Summary

Electric circuit laws are valid for DC or time-independent unipolar currents and voltages, AC currents and voltages, and time-varying DC voltages and currents. Ohm's law is defined for the most general case of instantaneous voltages and currents, where voltage $v(t)$ is measured in volt (V) and current $i(t)$ in ampere (A) and the quotient of voltage and current is defined as resistance $R = v(t)/i(t)$ measured in ohm (Ω or V/A). Note that the term "resistor" refers to the physical component, while "resistance" is the ohmic value of a resistor.

The inverse of a resistance R is the conductance $G = 1/R$, measured in siemens (S). The instantaneous electric power $p(t) = v(t)i(t)$ is measured in watt (W or VA). This instantaneous definition is the most elementary definition of power, which will be expanded in later chapters to more specific power expressions.

Next, Kirchhoff's first law $\left(\sum_{j=1}^{N} I_j = 0\right)$, the current law, and Kirchhoff's second law $\left(\sum_{j=1}^{N} V_j = 0\right)$, the voltage law, are explained based on simple electric circuits defining the concept of an electric node and an electric mesh or loop. Voltage and current divider rules are addressed next. Then a single-node pair circuit and resistance combinations (series and parallel) are explained.

The reduction of resistance combinations is detailed based on the Δ-Y transformation. Nodal analysis, superposition, source exchange/transformation, Thévenin's theorem (equivalent impressed voltage source), and Norton's theorem (equivalent impressed current source) are explained and applied. The Wheatstone and Thomson bridge concepts and their applications are addressed last.

Problems

2.1 Application of Ohm's law $R = v(t)/i(t)$.

Apply Ohm's law to the three resistors and calculate the currents I_{R1}, I_{R2}, I_{R3} of the parallel resistive circuit of Figure P2.1.1 for the resistance values $R_1 = 5\,\Omega$, $R_2 = 12\,\Omega$, $R_3 = 130\,\Omega$, and DC voltage $v(t) = V_s = 120\,\text{V}$.

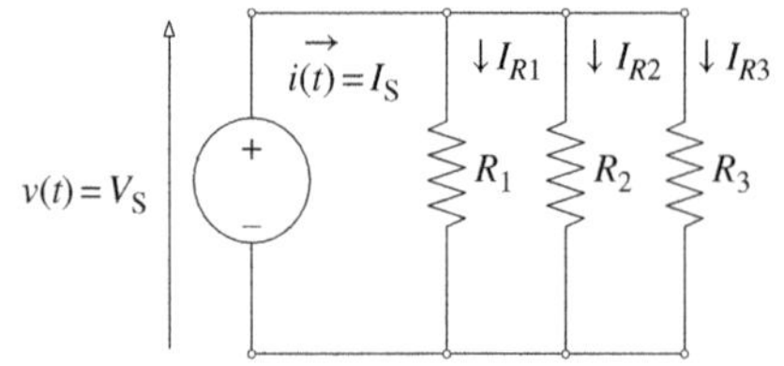

Figure P2.1.1 Parallel resistive circuit.

2.2 Calculation of instantaneous electric power $p(t) = v(t)i(t)$.

For the circuit of Figure P2.1.1, derive an expression for the instantaneous electric power and calculate it if $v(t) = V_s = 240\,\text{V}$ (where s stands for source) for $R_1 = 10\,\Omega$, $R_2 = 40\,\Omega$, and $R_3 = 5\,\Omega$. Note $1/\Omega = \text{S}$ or siemens.

2.3 Application of Kirchhoff's current (first) law (KCL) $\sum_{j=1}^{N} i_j(t) = 0$.

For the circuit of Figure P2.1.1, derive an expression for the current $i(t) = I_s$ and calculate it if $v(t) = V_s = 240\,\text{V}$ (where s stands for source) for $R_1 = 10\,\Omega$, $R_2 = 40\,\Omega$, and $R_3 = 5\,\Omega$.

2.4 Apply Kirchhoff's voltage (second) law (KVL) $\sum_{j=1}^{N} v_j(t) = 0$.

Find in Figure P2.4.1 for $V_s = 240\,\text{V}$, $R_1 = 5\,\Omega$, $R_2 = 20\,\Omega$, $R_3 = 1\,\Omega$ the voltages V_{R2} and V_{R3}.

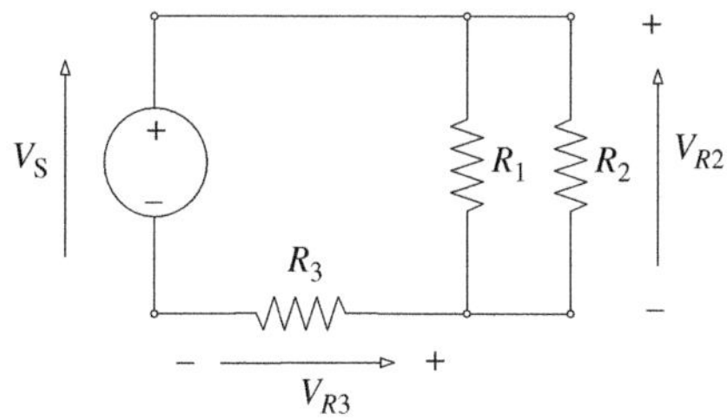

Figure P2.4.1 Application of Kirchhoff's second law.

2.5 Single-loop circuit.

Determine for Figure P2.5.1 the relation for I_s and compute its value for $V_{s1} = 240\,\text{V}$, $V_{s2} = 60\,\text{V}$, $R_1 = 4\,\Omega$, and $R_2 = 6\,\Omega$.

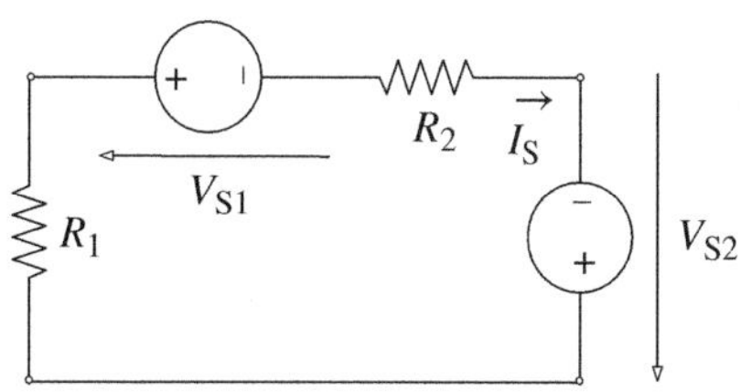

Figure P2.5.1 Single-loop circuit.

2.6 Single-node pair circuit.

Determine for Figure P2.6.1 the function $I_s = f(V_s, R_1, R_2, R_3)$ and calculate I_s for $V_s = 240\,\text{V}$, $R_1 = 11\,\Omega$, $R_2 = 70\,\Omega$, $R_3 = 5\,\Omega$.

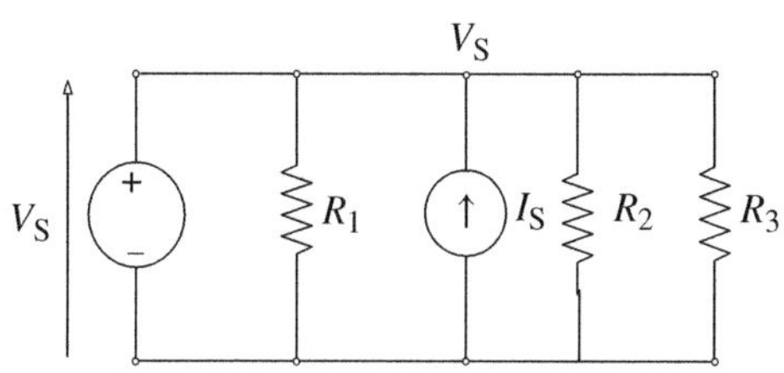

Figure P2.6.1 Single-node pair circuit.

2.7 Current division.

For Figure P2.7.1 determine for $V_s = 240$ V, $R_1 = 6\,\Omega$, $R_2 = 13\,\Omega$, $R_3 = 80\,\Omega$ the currents I_1, I_2, I_3, I_{12}, I_s, and explain the principle of current division.

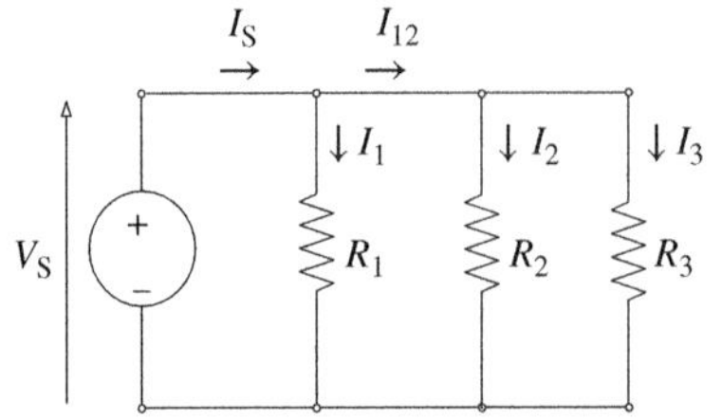

Figure P2.7.1 Current division and summation.

2.8 Voltage division.

For Figure P2.8.1 determine for $V_s = 240$ V, $R_1 = 6\,\Omega$, $R_2 = 13\,\Omega$, $R_3 = 80\,\Omega$, $R_4 = 8\,\Omega$ the voltages V_{R1}, V_{R2}, V_{R4}, and explain the principle of voltage division.

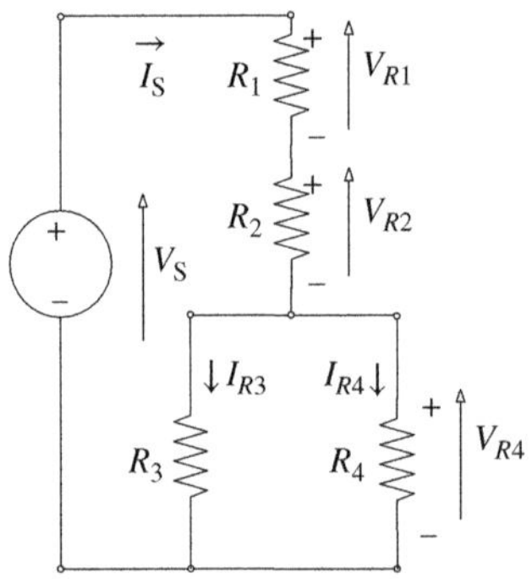

Figure P2.8.1 Voltage division and summation.

2.9 Resistors connected in series.

Derive an expression for the total series resistance R_{series} as indicated in Figure P2.9.1 and find its value for $R_1 = 7\,\Omega$, $R_2 = 14\,\Omega$, $R_3 = 21\,\Omega$.

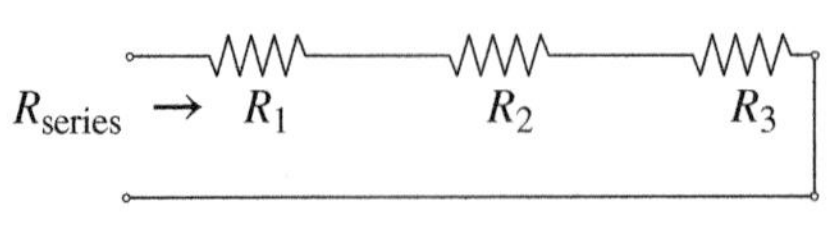

Figure P2.9.1 Series connection of resistors.

2.10 Resistors connected in parallel.

Derive an expression for the total parallel resistance R_{parallel} as indicated in Figure P2.10.1 and find its value for $R_1 = 7\,\Omega$, $R_2 = 14\,\Omega$, $R_3 = 21\,\Omega$, and $R_4 = 10\,\Omega$.

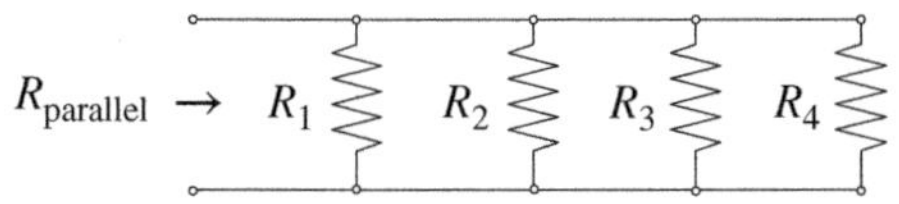

Figure P2.10.1 Parallel connection of resistors.

2.11 Nodal analysis.

For Figure P2.11.1 where $V_{s1} = 120\,\text{V}$, $V_{s2} = 12\,\text{V}$, $I_s = 11\,\text{A}$, $R_1 = 4\,\Omega$, $R_2 = 7\,\Omega$, $R_3 = 11\,\Omega$, derive analytical expressions for V_1 and V_2 and compute their numerical values using the nodal approach.

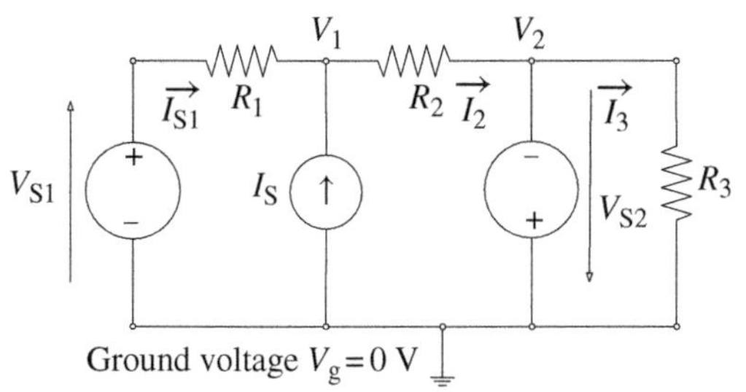

Figure P2.11.1 Nodal analysis circuit.

2.12 Loop/mesh analysis.

Determine the current I_o through resistor R_4 in Figure P2.12.1 using loop analysis for the quantities $R_1 = 1\,\Omega$, $R_2 = 2\,\Omega$, $R_3 = 3\,\Omega$, $R_4 = 4\,\Omega$, $I_s = 10\,\text{A}$, $V_s = 240\,\text{V}$.

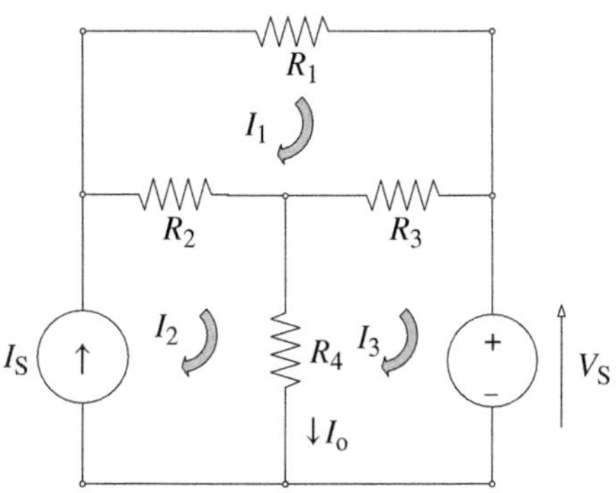

Figure P2.12.1 Circuit to be solved via loop/mesh analysis.

2.13 Superposition.

Determine the voltage V_{R1} in Figure P2.13.1 using superposition for the quantities $R_1 = 5\,\text{k}\Omega$, $R_2 = 8\,\text{k}\Omega$, $R_3 = 12\,\text{k}\Omega$, $I_s = 10\,\text{mA}$, $V_s = 24\,\text{V}$.

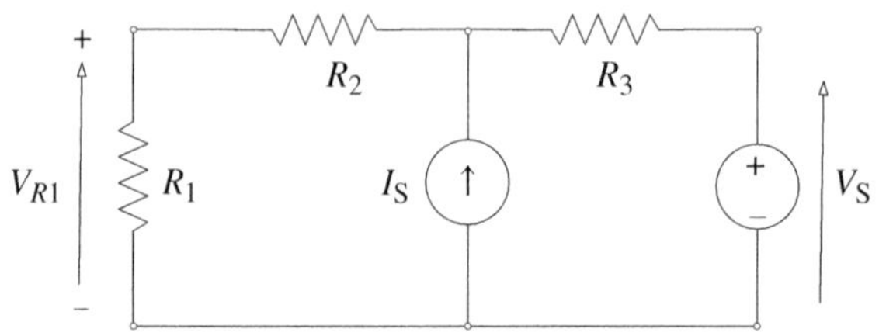

Figure P2.13.1 Circuit to be solved via the principle of superposition.

2.14 Source exchange/transformation.

Determine the voltage V_{R3} in Figure P2.14.1 using source transformation for the quantities $R_1 = 5\,\text{k}\Omega$, $R_2 = 8\,\text{k}\Omega$, $R_3 = 12\,\text{k}\Omega$, $R_4 = 7\,\text{k}\Omega$, $I_s = 10\,\text{mA}$, $V_s = 24\,\text{V}$.

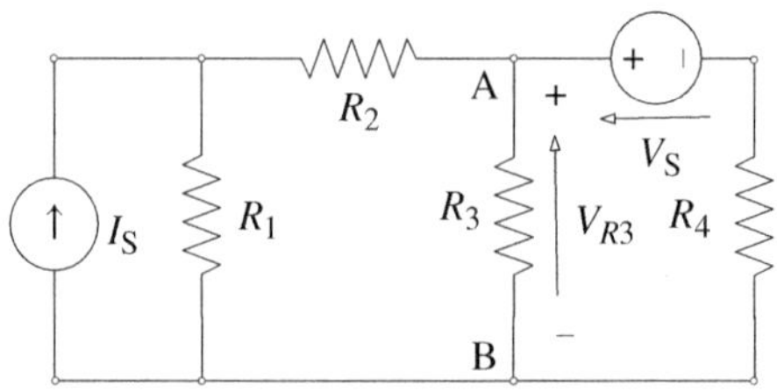

Figure P2.14.1 Circuit to be solved via source transformation.

2.15 Norton equivalent circuit.

Determine the voltage V_{R3} in Figure P2.14.1 using Norton's theorem for the quantities $R_1 = 5\,\text{k}\Omega$, $R_2 = 8\,\text{k}\Omega$, $R_3 = 12\,\text{k}\Omega$, $R_4 = 7\,\text{k}\Omega$, $I_s = 10\,\text{mA}$, $V_s = 24\,\text{V}$.

2.16 Thévenin equivalent circuit.

Determine the current I_o in Figure P2.16.1 using Thévenin's theorem for the quantities $R_1 = 5\,\text{k}\Omega$, $R_2 = 8\,\text{k}\Omega$, $R_3 = 12\,\text{k}\Omega$, $R_4 = 7\,\text{k}\Omega$, $I_{s1} = 10\,\text{mA}$, $I_{s2} = 24\,\text{mA}$.

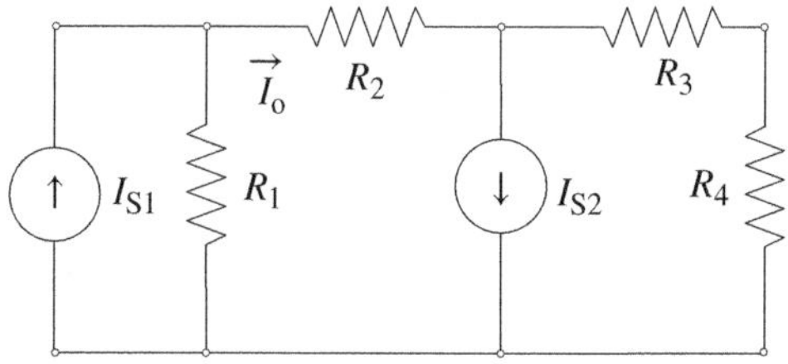

Figure P2.16.1 Circuit to be solved via Thévenin's theorem.

2.17 Application of source transformation.

Determine the voltage V_{R3} in Figure P2.17.1 using source transformation for the quantities $R_1 = 5\,\text{k}\Omega$, $R_2 = 8\,\text{k}\Omega$, $R_3 = 12\,\text{k}\Omega$, $I_s = 10\,\text{mA}$, $V_{s1} = 12\,\text{V}$, $V_{s2} = 24\,\text{V}$.

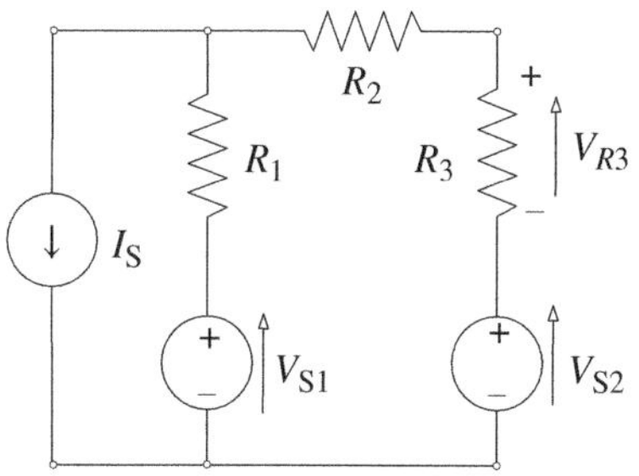

Figure P2.17.1 Circuit to be solved via source transformation.

2.18 Wheatstone bridge.

a) Compute the current $I_5 = f(V_B, R_1, R_2, R_3, R_4, R_5)$ as given by Eq. (E2.7.8) through the moving-coil galvanometer G of Figure 2.23 if the Wheatstone bridge is balanced for $R_1/R_2 = R_3/R_4$, where $R_1 = 1\,\Omega$, $R_2 = 2\,\Omega$, $R_3 = 3\,\Omega$, $R_4 = 6\,\Omega$, $V_B = 12\,\text{V}$ and the resistance of galvanometer G is $R_5 = 0.001\,\Omega$.

b) Compute the current $I_5 = f(R_4)$ for an unbalanced bridge, where $R_5 = 0.001\,\Omega$, $R_1 = 1\,\Omega$, $R_2 = 2\,\Omega$, $R_3 = 3\,\Omega$, $4\,\Omega < R_4 < 8\,\Omega$, and $V_B = 12\,\text{V}$. Plot $I_5 = f(R_4)$.

References

1 Hunt, I.W. (2020). Nikola Tesla. Encyclopædia Britannica, Inc.. https://www.britannica.com/biography/Nikola-Tesla (accessed 11 May 2020).

2 Conot, R.E. and Josephson, M. (2020). Thomas Edison. Encyclopædia Britannica, Inc. https://www.britannica.com/biography/Thomas-Edison (accessed 11 May 2020).

3 The Editors of Encyclopaedia Britannica (2009). Ohm. Encyclopædia Britannica, Inc. https://www.britannica.com/science/ohm (accessed 11 May 2020).

4 The Editors of Encyclopaedia Britannica (2014). Kirchhoff's rules. Encyclopædia Britannica, Inc. https://www.britannica.com/science/Kirchhoffs-rules (accessed 11 May 2020).

5 Kelly Bramble, Engineers Edge, LLC. Wheatstone bridge circuit equations and derivation, 4 pp. https://www.engineersedge.com/instrumentation/wheat_stone_bridge_operation.htm (accessed 11 May 2020).

6 Wenner, F. (1912). The four-terminal conductor and the Thomson Bridge. *Bulletin of Bureau of Standards, USA*, vol. 8, no. 3, pp. 559–610.

3

DC Circuit Transient Analysis

In this chapter, we examine the behavior of a DC network as a function of time after a sudden change in the network occurs due to opening or closing of switches. The DC network may contain resistors, inductors, and capacitors. First- and second-order ordinary differential equations will be analytically and numerically solved by well-known solution techniques and numerical methods such as PSPICE [1–4], MATLAB/Simulink [5], and Mathematica [6]. Note that there are standard resistor [7], capacitor [8], and inductor [9] values. In electric circuit designs, any (certain) resistor, capacitor, and inductor values might be required based on the desired design performance. These certain (e.g. resistance) values are not necessarily manufactured and sold, simplifying the manufacturing process of resistors: only well-defined resistance values of resistors are manufactured. For example, a design requires the resistance value of 25 Ω; then a series combination of two resistors with standard resistance values of 10 and 15 Ω must be used because $10\ \Omega + 15\ \Omega = 25\ \Omega$. For a 6 Ω resistor, the two resistors with standard values 10 and 15 Ω can be connected in parallel such that $\dfrac{10\,\Omega \cdot 15\,\Omega}{(10+15)\Omega} = 6\,\Omega$.

3.1 Capacitors

3.1.1 Energy Stored in a Capacitor

Ideal lossless (with no losses) capacitors consist of two conducting plates separated by a nonconducting, or dielectric, material. The simplest capacitor consists of two parallel plates as shown in Figure 3.1. Note the capacitor is the name of a device having the capacitance C. Any conducting or nonconducting material – which is

Introduction to Energy, Renewable Energy, and Electrical Engineering: Essentials for Engineering Science (STEM) Professionals and Students, First Edition. Ewald F. Fuchs and Heidi A. Fuchs.

Companion website: www.wiley.com/go/fuchs/renewableenergyandelectricalengineering

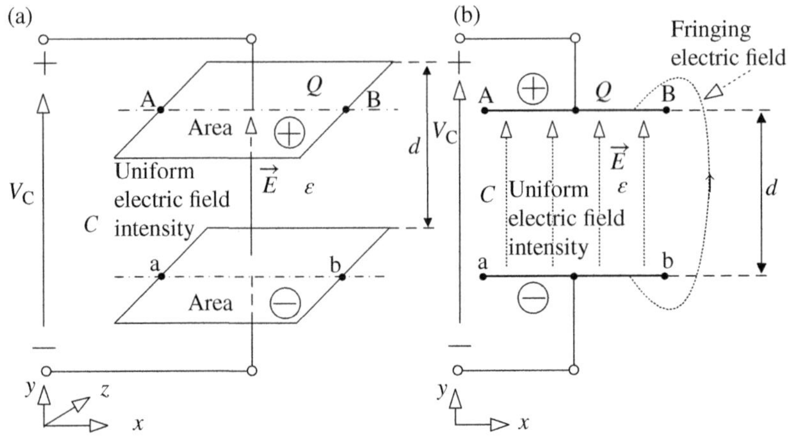

Figure 3.1 Parallel plate capacitor with capacitance *C*: (a) three-dimensional representation in the three-dimensional *y*, *x*, *z* space and (b) two-dimensional representation in the *y*, *x* space.

electrically isolated – can be charged with endowed (see Chapter 1, Appendix 1.B) electrons and has the property of capacitance *C*. Such a device is called a capacitor. For example, an isolated plastic ruler, rubbed on a fabric – such as a shirt sleeve – can collect free electrons residing on the sleeve and become endowed electrons due to the energy supplied through the rubbing process. These endowed electrons residing on the ruler produce a spark when the plastic ruler is approaching a conducting/grounded material. Another example is when a human with shoes – having rubber soles – walks on a carpet picking up free electrons residing on the carpet, which become endowed electrons through the rubbing process between shoes and carpet. This is evidenced through the electric shock experienced by the human touching a grounded object or when shaking hands with another person residing on a different electric potential.

$\vec{E} = iE_x + jE_y + kE_z$ is the electric field intensity vector that exists between negative and positive capacitor plates.

Figure 3.1a shows the capacitor plate configuration in three dimensions, while Figure 3.1b represents the capacitor in two dimensions. Between the positive plate indicated by a + sign in a circle and the negative plate defined by a – sign in a circle, the electric field intensity vector.

$\vec{E} = iE_x + jE_y + kE_z$ is shown. *i*, *j*, *k* are the unit vectors in three-dimensional space. Please do not mix up the unit vector *i* with the time-dependent current symbol usually denoted by *i*(*t*).

In Figure 3.1a and b the electric field intensity vector $\vec{E}$ is in y direction only; that is, $\vec{E} = E_y$ and E_x, E_z are zero, except at the left and right ends of the plate capacitor where fringing electric fields occur.

The electric field component E_y is defined as the gradient of the voltage V_c; that is, $E_y = V_c/dt$ or $V_c = E_y\ dt$. In general, the time-dependent integral relating voltage and electric field intensity $\vec{E}(t)$ in a capacitor C

$$v_c(t) = \int \vec{E}(t)\,dt \tag{3.1}$$

is valid. The unit of $\vec{E}(t)$ is volt/meter or V/m.

The capacitance of a parallel plate capacitor is

$$C = \frac{\varepsilon \cdot \text{area}}{d}\ [\text{farads or F}]. \tag{3.2}$$

The permittivity $\varepsilon = \varepsilon_r \cdot \varepsilon_0$ is a material constant, where the vacuum or free space permittivity is

$$\varepsilon_0 = 8.854\,187\,817 \cdot 10^{-12}\ \text{F/m}\ [\text{or farads per meter}], \tag{3.3}$$

and the relative permittivity ε_r – commonly known as dielectric constant – is the relative dielectric constant or permittivity of a material. Specifically, $\varepsilon_r = 1$ for free space (or approximately for air).

The quotient between charge Q and DC voltage V_c defines the capacitance

$$C = Q/V_c \tag{3.4}$$

The capacitance of a capacitor is usually measured in millifarad or mF = 10^{-3} F, microfarad or μF = 10^{-6} F, nanofarad or nF = 10^{-9} F, and picofarad or pF = 10^{-12} F, where the unit farad or F can be expressed as As/V.

The derivation of the energy stored in a capacitor is based on the circuit of Figure 3.2, where an ideal capacitor is connected to the time-dependent voltage $v_c(t)$.

The voltage $v_c(t)$ across a capacitor cannot change quickly, and therefore the voltage is called a state variable, characterizing the charging state of a capacitor. From Eq. (3.4) one can write for a time-dependent voltage $v_c(t)$, where $q(t)$ is the time-dependent charge of a capacitor:

$$q(t) = Cv_c(t). \tag{3.5}$$

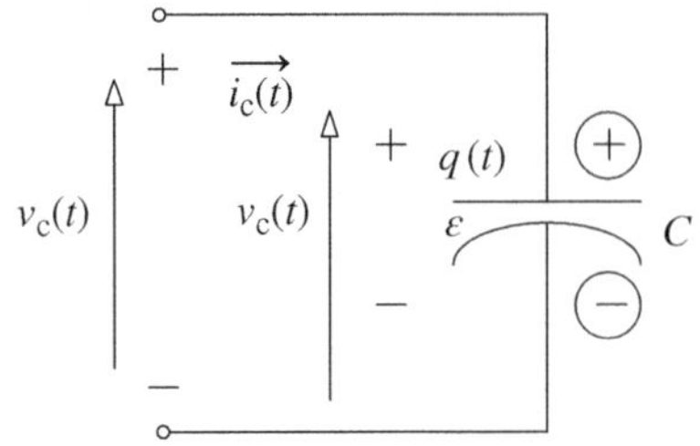

Figure 3.2 Derivation of energy storage of capacitor under ideal (lossless) conditions with capacitance C. ε is the dielectric constant, $q(t)$ the time-dependent charge, and $v_c(t)$ the applied terminal voltage.

The rate of change of a charge $q(t)$ is called current $i_c(t)$, which depends upon the capacitance C of a capacitor. Therefore,

$$i_c(t) = dq(t)/dt = C\,dv_c(t)/dt. \tag{3.6}$$

With

$$dv_c(t) = (1/C)i_c(t)dt \tag{3.7}$$

follows after integration

$$\begin{aligned} v_c(t) &= \frac{1}{C}\int_{t=-\infty}^{t>0} i_c(x)dx = \frac{1}{C}\int_{t=-\infty}^{t_0=0} i_c(x)dx + \frac{1}{C}\int_{t_0=0}^{t>0} i_c(x)dx \\ &= v_c(t_0) + \frac{1}{C}\int_{t=0}^{t} i_c(x)dx, \end{aligned} \tag{3.8}$$

where x is the variable of integration, $v_c(t_0)$ represents the initial capacitor voltage, and $v_c(t)$ is the voltage at any time for $t > 0$. $v_c(t)$ cannot quickly change because of the integration process, and $v_c(t)$ is therefore called a state variable as mentioned above, characterizing the charging state of a capacitor with capacitance C. The work or energy stored in a capacitor can be derived from the power delivered over time to it:

$$w_c(t) = \int_0^t p_c(t)dt = \int_0^t v_c(t)i_c(t)dt = \int_0^t v_c(t)\left[C\frac{dv_c(t)}{dt}\right]dt = \frac{1}{2}Cv_c^2(t), \tag{3.9}$$

measured either in joules or J.

The unit of joules is alternatively expressed as watt-seconds or Ws, which is identical to newton meters or Nm.

Application Example 3.1 Charging and discharging of an ideal capacitor having a capacitance of $C = 5\ \mu F$ with a sawtooth-type triangular voltage as illustrated in Figure E3.1.1a

Solution

The response of the capacitor to the sawtooth voltage waveshape of Figure E3.1.1a can be calculated by considering three time regions:

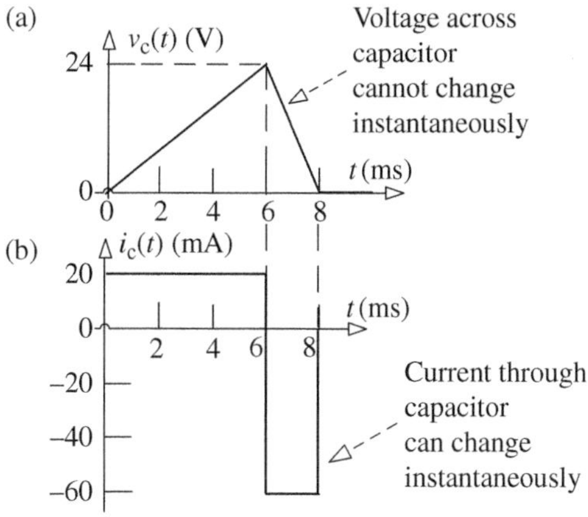

Figure E3.1.1 (a) Charging and discharging of a capacitor with the sawtooth function $v_c(t)$ and (b) charging current through capacitor $i_c(t)$ with a capacitance of 5 μF.

Time region 1: $0 \leq t \leq 6$ ms

The well-known slope m of a linear function

$$y = mx + b \tag{E3.1.1}$$

defined by x_1/y_1 and x_2/y_2 in an x, y coordinate system is

$$m = \frac{y - y_1}{x - x_1}, \tag{E3.1.2}$$

or

$$m = \frac{y - y_1}{x - x_1} = \frac{y_2 - y_1}{x_2 - x_1}. \tag{E3.1.3}$$

For the origin ($y_2 = 0$ V, $x_2 = 0$ ms) and for the maximum point of the $v_c(t)$ function ($y_1 = 24$ V, $x_1 = 6$ ms) follows the functional relationship

$$v_c(t) = \frac{0 - 24\,\text{V}}{0 - 6\,\text{ms}} t = 4k\left(\frac{\text{V}}{\text{s}}\right)t. \tag{E3.1.4}$$

With

$$i_c(t) = C\frac{dv_c(t)}{dt} \tag{E3.1.5}$$

follows for $C = 5\,\mu\text{F} = 5 \cdot 10^{-6}$ F

$$i_c(t) = 5 \cdot 10^{-6}\,\text{F} \cdot 4\,\text{kV/s} = 20 \cdot 10^{-3}(\text{V/s})(\text{As/V}) = 20\,\text{mA}. \tag{E3.1.6}$$

Figure E3.1.1b shows this computed value of $i_c(t) = 20$ mA.

Time region 2: $6\,\text{ms} \leq t \leq 8\,\text{ms}$

For the maximum point of the function ($y_2 = 24\,\text{V}, x_2 = 6\,\text{ms}$) and for the point on the abscissa ($y_1 = 0\,\text{V}, x_1 = 8\,\text{ms}$) follows

$$\frac{v_c(t) - 0\,\text{V}}{t - 8\,\text{ms}} = \frac{24\,\text{V} - 0\,\text{V}}{6\,\text{ms} - 8\,\text{ms}} = \frac{24\,\text{V}}{-2\,\text{ms}} = -12\,\text{kV/s}, \tag{E3.1.7}$$

or

$$v_c(t) - 0\,\text{V} = -12\,\text{k(V/s)}t + 96\,\text{V}. \tag{E3.1.8}$$

With

$$i_c(t) = C\frac{dv_c(t)}{dt} \tag{E3.1.9}$$

follows for $C = 5\,\mu\text{F} = 5 \cdot 10^{-6}\,\text{F}$

$$\begin{aligned} i_c(t) &= 5 \cdot 10^{-6}\,\text{F} \cdot (-12)\,\text{k(V/s)} \\ &= -60\,\text{k} \cdot 10^{-3}\,\text{(V/s)(As/V)} = -60\,\text{mA}. \end{aligned} \tag{E3.1.10}$$

Figure E3.1.1b shows this computed value of $i_c(t) = -60\,\text{mA}$.

Time region 3: $8\,\text{ms} \leq t$

$$v_c(t) = 0 \tag{E3.1.11}$$

with

$$i_c(t) = C\frac{dv_c(t)}{dt} \tag{E3.1.12}$$

follows for $C = 5\,\mu\text{F} = 5 \cdot 10^{-6}\,\text{F}$

$$i_c(t) = 5 \cdot 10^{-6}\,\text{F} \cdot 0 = 0\,\text{A}. \tag{E3.1.13}$$

Figure E3.1.1b shows this computed value of $i_c(t) = 0\,\text{A}$.

The analytically computed steady-state result of Figure E3.1.1b is now numerically computed based on PSPICE; see the program as listed in Table E3.1.1. Figure E3.1.2 represents the numerically calculated steady-state response. It shows that Figure E3.1.1b, the analytical result, confirms the numerically computed steady-state result.

Now we model a nonideal capacitor C including losses. A practical capacitor has a very large parallel (leakage) resistance R_C, which provides a parallel conduction path between the capacitor plates as illustrated in Figure 3.3.

Table E3.1.1 PSPICE program.

Application Example 3.1	C 1 0 5u
∗ piecewise linear waveform	∗Analysis request
vc 1 0 PWL(0 0 6m 24 8ms 0 10m 0)	.Tran 10us 10ms
∗capacitor as storage device	.Probe
	.end

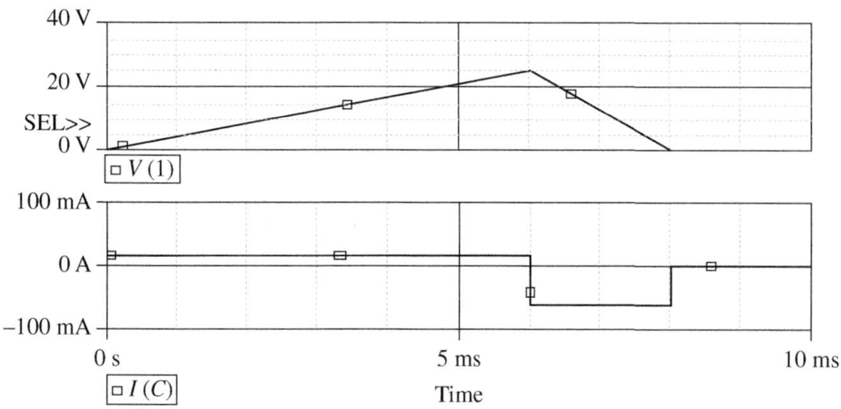

Figure E3.1.2 Numerically computed steady-state result $i_c(t) = I(c)$ for given $v_c(t) = V(1)$ based on PSPICE program as listed in Table E3.1.1.

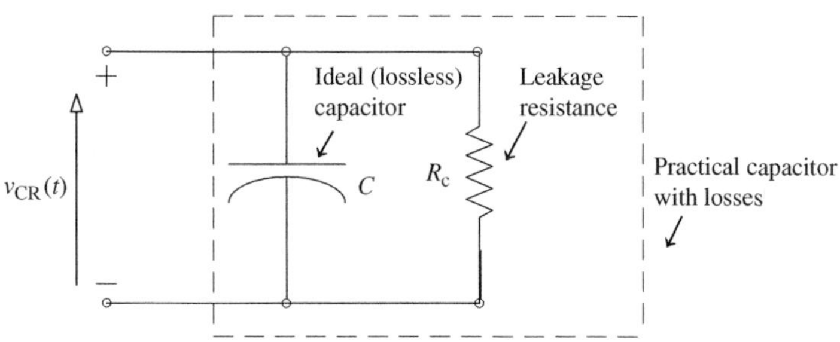

Figure 3.3 Practical capacitor with capacitance C including losses defined by either parallel (leakage) or series (not shown) resistor R_c and excited by terminal voltage $v_{CR}(t)$.

3.1.2 Capacitor Combination Formulas

Series-connected capacitors have the total capacitance of C_{series}, which is defined by the relation (Figure 3.4)

$$\frac{1}{C_{\text{series}}} = \frac{1}{C_1} + \frac{1}{C_2} + \cdots + \frac{1}{C_N}. \tag{3.10}$$

Parallel-connected capacitors have the total capacitance of C_{parallel}, which is defined by the relation (Figure 3.5)

$$C_{\text{parallel}} = C_1 + C_2 + \cdots + C_N. \tag{3.11}$$

Conclusions

The capacitances of series-connected capacitors combine like the resistances of resistors in parallel, and the capacitances of paralleled capacitors add up like the resistances of resistors connected in series. The analytically computed steady-state result of Figure E3.1.1b is confirmed by the numerically computed steady-state result of Figure E3.1.2.

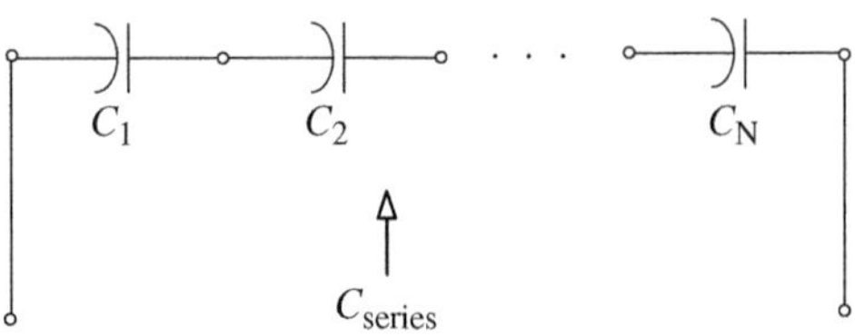

Figure 3.4 Series connection of capacitors C_1, C_2, ..., C_N.

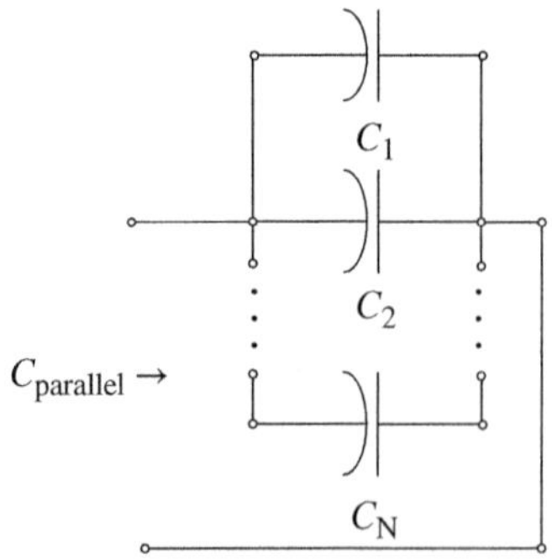

Figure 3.5 Parallel connection of capacitors C_1, C_2, ..., C_N.

3.2 Inductors

The ideal inductor has no losses and is a linear circuit element that consists of a conducting wire wound around either a magnetic or nonmagnetic core. Inductors form the basis of transformers discussed in Chapter 6. Figure 3.6 depicts the structure of a C-core inductor with an air gap g. The permeability of the core is $\mu = \mu_r\mu_0$, where $\mu_0 = 4\pi \cdot 10^{-7}$ H/m [henries/m or Vs/A] and μ_r is the relative permeability of the core material. For any (linear) inductor, the magnetic core must be equipped with an air gap g. Figure 3.7 represents a C-core inductor with a toroidal core and air-gap length g. N represents the number of turns of the winding.

The symbol for an ideal inductor is depicted in Figure 3.8, and the relation

$$v_L(t) = L\frac{di_L(t)}{dt} \tag{3.12}$$

governs its physical behavior. Equation (3.12) can be rewritten in integral form as

$$i_L(t) = \frac{1}{L}\int v_L(t)dt. \tag{3.13}$$

The above integral relation demonstrates that the current $i_L(t)$ is the result of integration, and therefore it cannot change quickly and therefore the current through an inductor $i_L(t)$ is a state variable.

The vector relation between the current $i_L(t)$ and the magnetic field intensities for the C-core inductor is either $\vec{H}_g$ or $\vec{H}_{core} = iH_x + jH_y + kH_z$ with x, y, z components, where i, j, k are unit vectors, and for the toroidal inductor is either $\vec{H}_g$ or $\vec{H}_{core} = iH_r + jH_\varphi + kH_z$ with r, φ, z components:

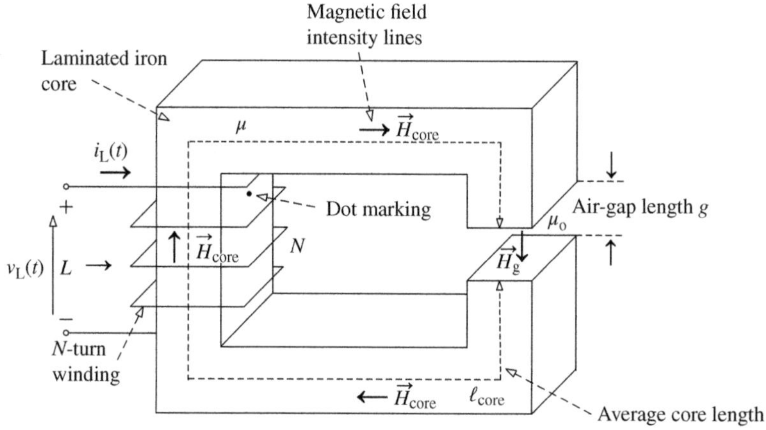

Figure 3.6 Inductor with C-core and air gap g. The open core is called a C-core because it looks like a C. L is the inductance of the inductor. ℓ_{core} is the average core length.

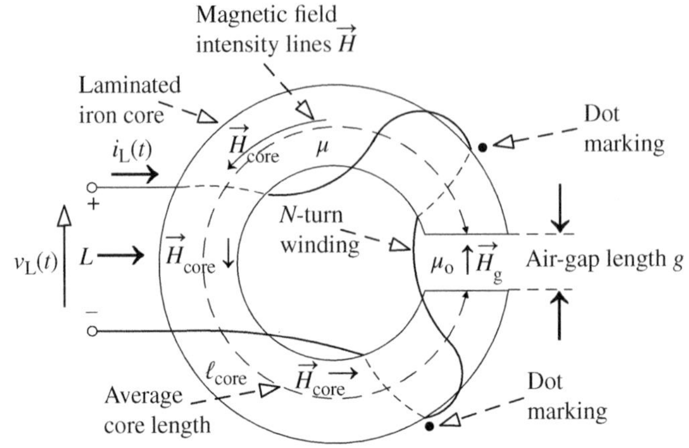

Figure 3.7 Inductor with toroidal C-core and air gap g. L is the inductance of the inductor. ℓ_{core} is the average core length.

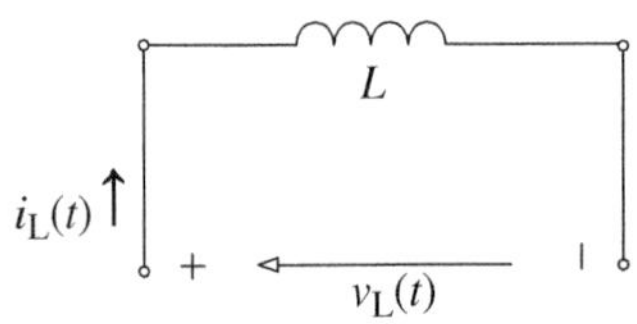

Figure 3.8 Symbol for an ideal inductor.

$$N \cdot i_L(t) = H_{core} \cdot \ell_{core} + H_g \cdot g, \tag{3.14}$$

where H_{core} and H_g are approximated by scalars. Because of the large difference in magnitudes of the relative permeability of the magnetic core $\mu_r = \mu_{r\_core}$ – such as 1000 – and the permeability of the air gap with $\mu_r = \mu_{r\_air\text{-}gap} = 1$, one obtains due to $H = \dfrac{B}{\mu_r}$ with B = constant the large inequality $H_{core} \ll H_g$, resulting in

$$N \cdot i_L(t) \approx H_g \cdot g. \tag{3.15}$$

3.2.1 Energy Stored in an Inductor

Following the development of the equations for a capacitor, one obtains for an inductor L from Eq. (3.13)

$$\begin{aligned} i_L(t) &= \frac{1}{L}\int_{t=-\infty}^{t} v_L(x)\mathrm{d}x = \frac{1}{L}\int_{t=-\infty}^{t=0} v_L(x)\mathrm{d}x + \frac{1}{L}\int_{t=0}^{t} v_L(x)\mathrm{d}x \\ &= i_L(t_0) + \frac{1}{L}\int_{t=0}^{t} v_L(x)\mathrm{d}x, \end{aligned} \tag{3.16}$$

where x is the variable of integration.

Figure 3.9 Practical inductor with inductance L including losses defined by series resistor R_L and excited by terminal voltage $v_{LR}(t)$.

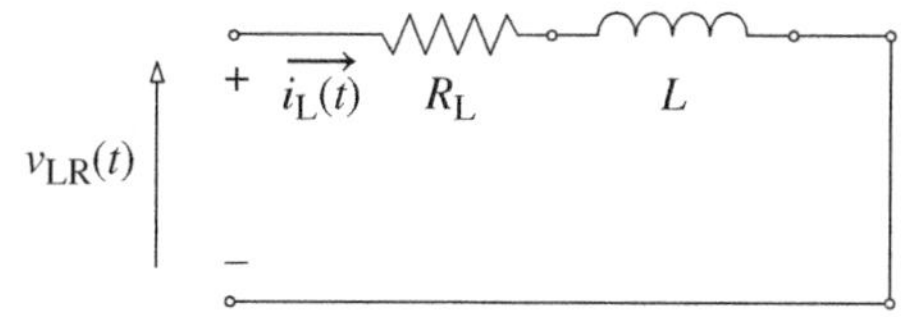

The instantaneous power delivered to an inductor is

$$p(t) = v_L(t)i_L(t) = [L\mathrm{d}i_L(t)/\mathrm{d}t]i_L(t) \tag{3.17}$$

By integrating the power of Eq. (3.17), one finds that the energy stored mainly in the magnetic air-gap field intensity H_g is

$$w_L(t) = \int_0^t p(x)\mathrm{d}x = \int_0^t \left[L\frac{\mathrm{d}i_L(x)}{\mathrm{d}x}\right] i_L(x)\mathrm{d}x = \frac{1}{2}Li_L^2(t) \text{ [joules or Ws].} \tag{3.18}$$

A practical inductor cannot store energy as well as a practical capacitor (e.g. supercapacitor, ultracapacitor) because there is always a winding resistance in the coil, which dissipates energy. Figure 3.9 shows the equivalent of a practical inductor. Just as the voltage across a capacitor cannot change instantaneously, the current in an inductor cannot change instantaneously because the current $i_L(t)$ is a result of integration, as shown in Eq. (3.16).

Application Example 3.2 Charging and discharging of an ideal inductor with inductance L = 10 mH with a triangular voltage as illustrated in Figure E3.2.1a

Figure E3.2.1 (a) Charging and discharging of an inductor with the triangular current $i_L(t)$ and (b) inductor voltage $v_L(t)$ with an inductance of 10 mH.

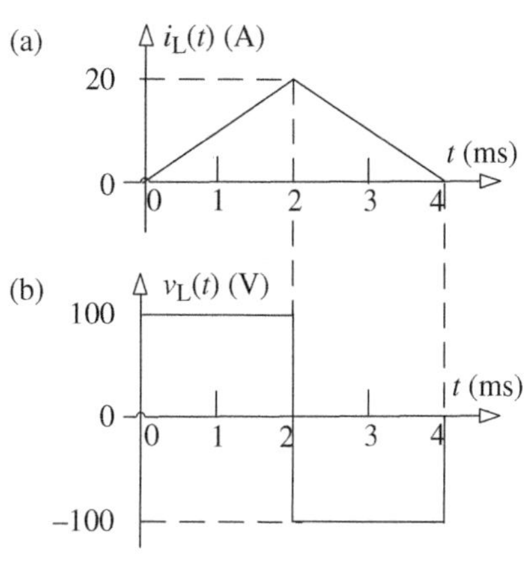

Solution

Time region 1: $0 \leq t \leq 2\,\text{ms}$

With Eq. (3.12)

$$v_{\text{L}}(t) = L\frac{\text{d}i_{\text{L}}(t)}{\text{d}t} = 10 \cdot 10^{-3}\frac{20}{2 \cdot 10^{-3}} = 100\,\text{V}. \tag{E3.2.1}$$

Time region 2: $2\,\text{ms} \leq t \leq 4\,\text{ms}$

$$v_{\text{L}}(t) = L\frac{\text{d}i_{\text{L}}(t)}{\text{d}t} = 10 \cdot 10^{-3}\frac{(-20)}{2 \cdot 10^{-3}} = -100\,\text{V}. \tag{E3.2.2}$$

Time region 3: $4\,\text{ms} \leq t$

$$v_{\text{L}}(t) = 0. \tag{E3.2.3}$$

The analytically computed result of Figure E3.2.1b is now numerically computed based on PSPICE; see the program as listed in Table E3.2.1. Figure E3.2.2 represents the numerically calculated response, and it shows that the analytical result of Figure E3.2.1b confirms the numerically computed result.

Table E3.2.1 PSPICE program.

```
*Application Example 3.2                L 1 0 10m
* piecewise linear waveform             *Analysis request
iL 1 0 PWL(0 0 2m 20 4ms 0 6m 0)        .Tran 10us 6ms
*inductor as storage device             .Probe
                                        .end
```

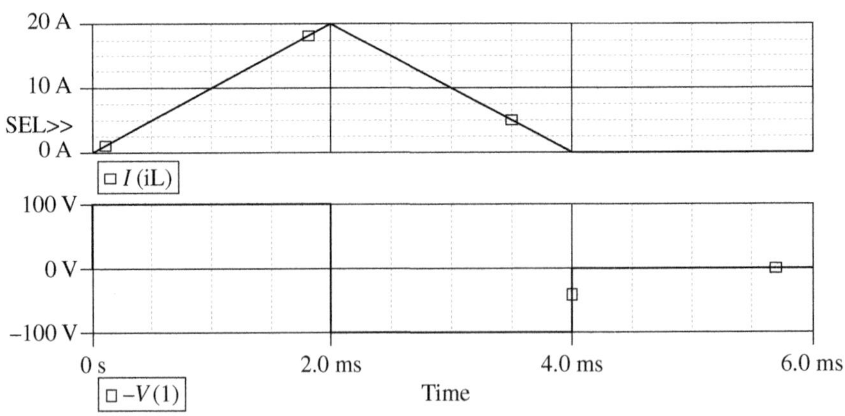

Figure E3.2.2 Numerically computed result for $v_{\text{L}}(t) = -V(1)$ as a function $i_{\text{L}}(t) = I(iL)$ based on PSPICE program as listed in Table E3.2.1.

Conclusions

The current of an inductor $i_L(t)$ cannot change quickly because it relates to the voltage $v_L(t)$ via an integral relation and $i_L(t)$ therefore called a state variable, while the voltage across an inductor can change quickly.

3.2.2 Inductor Combination Formulas

Series-connected inductors have the total inductance of L_{series}, which relates to the relation

$$L_{series} = L_1 + L_2 + \cdots + L_N, \tag{3.19}$$

that is, the inductances of series-connected inductors combine in exactly the same manner as the resistances of series-connected resistors (Figure 3.10).

Parallel-connected inductors have the total inductance of $L_{parallel}$, which relates to the relation

$$\frac{1}{L_{parallel}} = \frac{1}{L_1} + \frac{1}{L_2} + \ldots + \frac{1}{L_N}, \tag{3.20}$$

that is, the inductances of parallel-connected inductors combine in exactly the same manner as the resistances of parallel-connected resistors (Figure 3.11).

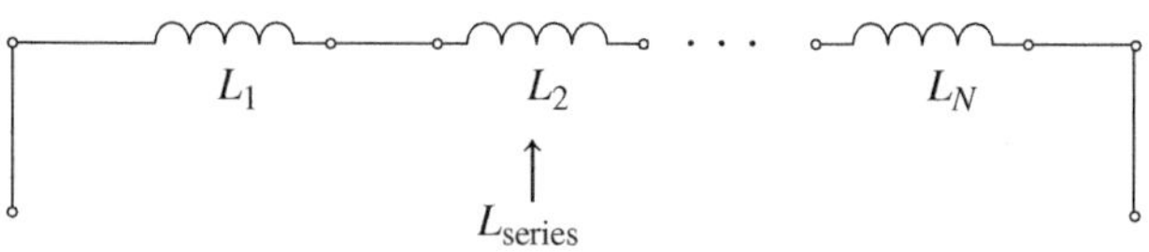

Figure 3.10 Series connection of inductances L_1, L_2, ..., L_N.

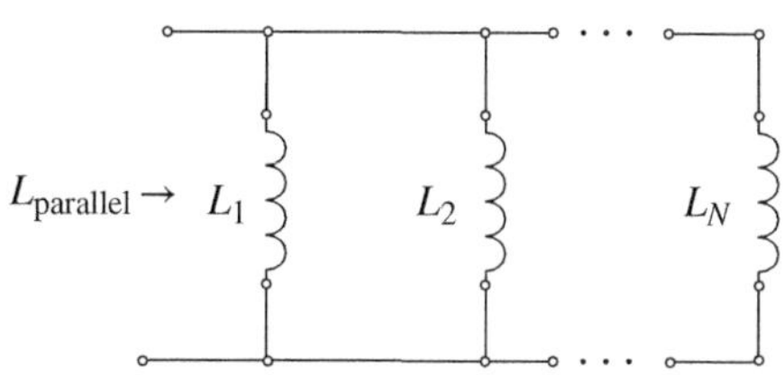

Figure 3.11 Parallel connection of inductances L_1, L_2, ..., L_N.

3.3 Transient Analysis Applied to Circuits Resulting in First-Order, Ordinary Differential Equations with Constant Coefficients

The solution of circuits with one storage element (e.g. C or L) must be formulated in terms of state variables: this means in an RC circuit a first-order ordinary differential equation with constant coefficients must be a function of $v_c(t)$ and the first-order ordinary differential equation of an RL circuit must be formulated in terms of $i_c(t)$.

3.3.1 RC Series Network and Time Constant τ_{RC}

The voltage response $v_c(t)$ of the RC series network of Figure 3.12 is to be analytically determined for $t \geq 0$.

The first-order ordinary differential equation with constant coefficients in terms of the state variable $v_c(t)$ is obtained by applying Kirchhoff's voltage law (KVL). If the switch is in position "1," then at $t = -0$ the initial capacitor voltage before switching is $v_c(t = -0) = 0$, and after switching at $t = +0$ in position "2," the capacitor voltage $v_c(t = +0) = 0$ starts to increase; at steady-state $v_c(t \rightarrow \infty) = V_s$.

$$V_s = v_R(t) + v_c(t) = Ri(t) + v_c(t), \tag{3.21}$$

with

$$i(t) = C\frac{dv_c}{dt} \tag{3.22}$$

introduced into Eq. (3.21) results in

$$V_s = RC\frac{dv_c(t)}{dt} + v_c(t), \tag{3.23}$$

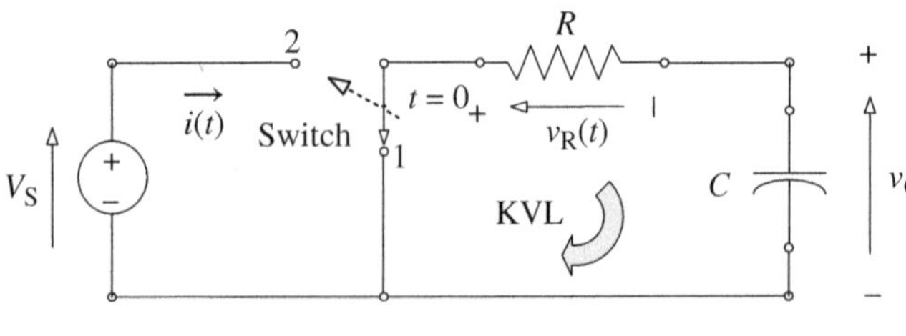

Figure 3.12 RC series circuit, where V_s is a source (s) DC voltage.

or the first-order ordinary differential equation with constant coefficients is

$$\frac{dv_c(t)}{dt} + \frac{v_c(t)}{RC} = \frac{V_s}{RC}, \tag{3.24}$$

where the time constant is $\tau_{RC} = RC$.

Application Example 3.3 Closed-form or analytical solution of a first-order ordinary differential equation with constant coefficients as described by Eq. (3.24)

Solution

The general form of a first-order ordinary differential equation with constant coefficients is in terms of the state variable x

$$\frac{dx(t)}{dt} + \alpha x(t) = A, \tag{E3.3.1}$$

where α is a damping factor and A is a forcing term.
Step 1: Solution of homogeneous equation or finding of natural transient response from the homogeneous first-order ordinary differential equation with constant coefficients:

$$\frac{dx(t)}{dt} + \alpha x(t) = 0. \tag{E3.3.2}$$

Assuming the homogeneous solution of the form

$$x(t)_{hom} = K_2 e^{-\left(\frac{t}{\tau}\right)}, \tag{E3.3.3}$$

differentiation yields

$$\frac{dx(t)}{dt} = \frac{(-K_2)}{\tau} e^{-\left(\frac{t}{\tau}\right)}. \tag{E3.3.4}$$

Introducing Eqs. (E3.3.3) and (E3.3.4) into Eq. (E3.3.2), one obtains

$$\left(\frac{-K_2}{\tau}\right) e^{-\left(\frac{t}{\tau}\right)} + \alpha K_2 e^{-\left(\frac{t}{\tau}\right)} = 0, \tag{E3.3.5}$$

where $K_2 \neq 0$; $\tau \neq 0$; $e^{-(t/\tau)} \neq 0$; as a consequence

$$-\frac{1}{\tau} + \alpha = 0,$$

or

$$\alpha = \frac{1}{\tau}, \tag{E3.3.6}$$

the solution of the homogeneous first-order ordinary differential equation is now

$$x(t)_{\text{hom}} = K_2 \mathrm{e}^{-\left(\frac{t}{\tau}\right)}. \tag{E3.3.7}$$

Step 2: Solution of particular solution of inhomogeneous (forced) first-order ordinary differential equation

Assume the solution to be of the form

$$x(t) = K_{1,} \tag{E3.3.8}$$

differentiation of Eq. (E3.3.8) yields

$$\mathrm{d}x(t)/\mathrm{d}t = 0, \tag{E3.3.9}$$

introducing Eqs. (E3.3.8) and (E3.3.9) into Eq. (E3.3.1) yields

$$\alpha K_1 = A, \tag{E3.3.10}$$

or

$$K_1 = A/\alpha. \tag{E3.3.11}$$

Step 3: General solution of first-order ordinary differential equation

The general solution is obtained by superimposing the homogeneous and the particular solutions as given by Eqs. (E3.3.7) and (E3.3.11), respectively:

$$x(t) = K_1 + K_2 \mathrm{e}^{-\left(\frac{t}{\tau}\right)}. \tag{E3.3.12}$$

The terms K_1, K_2, and τ are constants. K_1 is the steady-state solution because as $t \to \infty$ the term $K_2\mathrm{e}^{-(t/\tau)} \to 0$ representing the transient solution. The term τ is called the time constant of the network/circuit.

As given by Eq. (3.24) the differential equation of the RC network is

$$\frac{\mathrm{d}v_\mathrm{c}(t)}{\mathrm{d}t} + \frac{v_\mathrm{c}(t)}{\mathrm{RC}} = \frac{V_\mathrm{S}}{\mathrm{RC}}, \tag{3.25}$$

where the time constant is $\tau_{\mathrm{RC}} = \mathrm{RC}$.

According to Eq. (E3.3.12), the solution to Eq. (3.25) is

$$v_\mathrm{c}(t) = K_1 + K_2 \mathrm{e}^{-\left(\frac{t}{\tau_{\mathrm{RC}}}\right)}; \tag{3.26}$$

differentiation of Eq. (3.26) yields

$$\frac{\mathrm{d}v_\mathrm{c}(t)}{\mathrm{d}t} = \left(\frac{-K_2}{\tau}\right) \mathrm{e}^{-\left(\frac{t}{\tau_{\mathrm{RC}}}\right)}; \tag{3.27}$$

and introducing Eqs. (3.26) and (3.27) into Eq. (3.25) results in

$$\left(\frac{-K_2}{\tau}\right)e^{-\left(\frac{t}{\tau_{RC}}\right)} + \frac{1}{RC}\left[K_1 + K_2 e^{-\left(\frac{t}{\tau_{RC}}\right)}\right] = \frac{V_S}{RC}. \tag{3.28}$$

The constants K_1, K_2, and τ are obtained by satisfying the initial conditions of the storage element (e.g. capacitor C). Equating the constant terms and the coefficients of the exponential terms, one can write

$$\frac{1}{RC}K_1 = \frac{V_S}{RC}, \tag{3.29}$$

or

$$K_1 = V_S, \tag{3.30}$$

$$\frac{-K_2}{\tau_{RC}} + \frac{K_2}{RC} = 0, \tag{3.31}$$

$$\tau_{RC} = RC. \tag{3.32}$$

The solution of the first-order ordinary differential equation (3.24) is now

$$v_c(t) = V_s + K_2 e^{-\left(\frac{t}{RC}\right)}, \tag{3.33}$$

where V_s is the steady-state solution and $\tau_{RC} = RC$ is the network's time constant.
Step 4: Determination of the constant K_2 from the initial condition of the capacitor C

The only unknown in Eq. (3.33) is K_2, which can be determined from the initial condition of the capacitor C, which is here assumed to be $v_c(t) = 0$. Introducing $v_c(t) = 0$ into Eq. (3.33) leads to

$$0 = V_s + K_2 \tag{3.34}$$

or

$$K_2 = -V_{s.} \tag{3.35}$$

Introducing the results of Eqs. (3.30), (3.32), and (3.35) into Eq. (3.26) leads to the complete solution for $v_c(t)$, that is,

$$v_c(t) = V_s\left[1 - e^{-(t/RC)}\right]. \tag{3.36}$$

For $V_s = 120$ V, $R = 10\,\Omega$, and $C = 1$ Mf, one obtains $\tau_{RC} = RC = 10\,\Omega \cdot 1$ mF $= 10$ ms, and

$$v_c(t) = 120\left[1 - e^{-(t/10\text{ ms})}\right]\text{ V}. \tag{3.37}$$

The analytical solution, Eq. (3.37), is graphically displayed in Figure 3.13, and calculated values are listed in Table 3.1. The transient change due to one storage

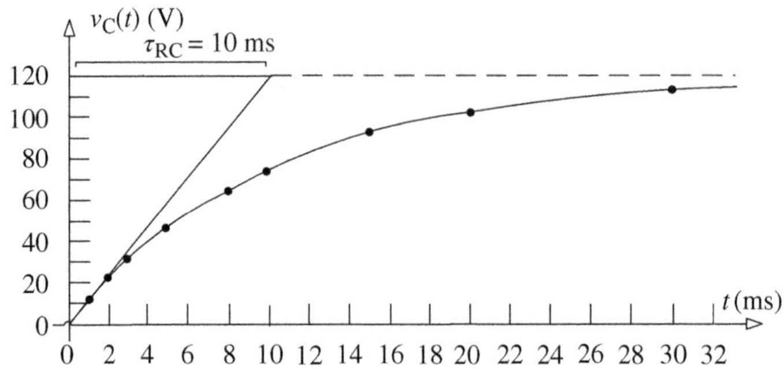

Figure 3.13 $v_c(t)$ as a function of time t, defined by Eq. (3.37), and displayed in the network of Figure 3.12.

Table 3.1 Calculated values for Figure 3.13.

t (ms)	0	1	2	3	5	8	10	15	20	30
$v_c(t)$ (V)	0	11.420	21.752	31.102	47.216	66.081	75.854	93.224	103.760	114.030

Table 3.2 PSPICE program.

```
RC network, Table 3.2
* transient analysis
Vs 1 0 120
*capacitor as storage device
*with resistor R in series
R 1 2 10
C 2 0 1m ic=0
*Analysis request
.Tran 10us 32ms
.Probe
.end
```

element (e.g. C, L) lasts about three times the time constant τ_{RC}: this is in Figure 3.13 during the time $3\tau_{RC} = 3 \cdot 10\text{ ms} = 30\text{ ms}$.

For a capacitor the voltage $v_c(t)$ is the state variable, which is continuous as a function of time t as can be seen in Figure 3.13.

The analytically computed results of Figure 3.13 are now numerically computed based on PSPICE; see the program as listed in Table 3.2. Figure 3.14 represents the numerically calculated transient response, and it shows that it confirms the analytically computed transient solution.

3.3.2 RL Series Network and Time Constant τ_{RL}

Figure 3.15 depicts the network/circuit with one nonideal storage element, an inductor L in series with a resistor R. The general solution is obtained by applying

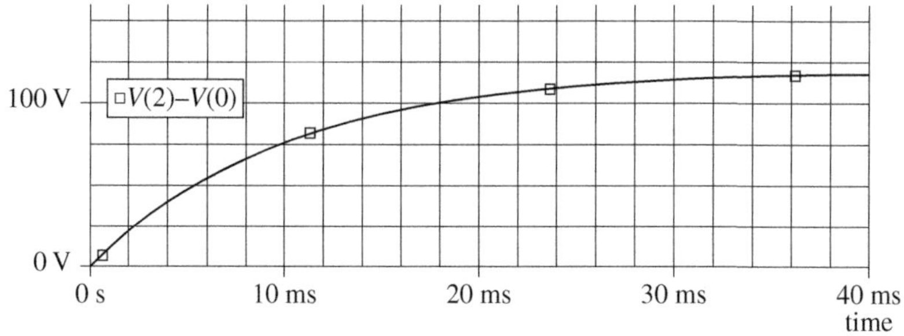

Figure 3.14 Numerically computed transient solution for $v_c(t) = V(2) - V(0)$ based on PSPICE program as listed in Table 3.2.

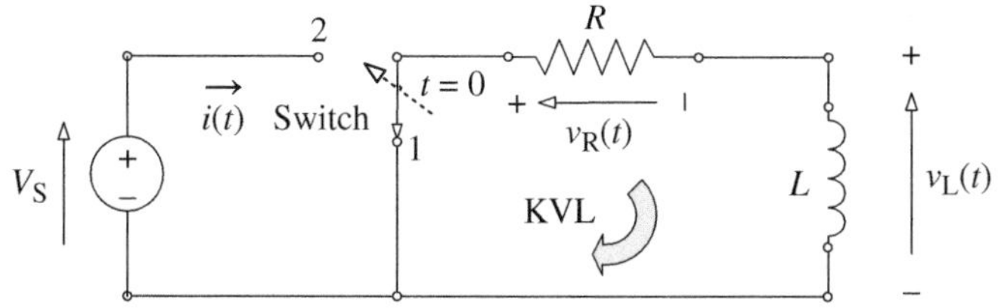

Figure 3.15 RL series circuit where V_s is a source (s) DC voltage.

KVL in a similar manner as has been done for the RC network of Figure 3.12. KVL results in

$$V_s = v_R(t) + v_L(t) = i(t)R + L\frac{di(t)}{dt},$$

or

$$\frac{di(t)}{dt} + \frac{R}{L}i(t) = \frac{V_s}{L}. \tag{3.38}$$

The first-order ordinary differential equation with constant coefficients in terms of the state variable $i(t) = i_L(t)$ is obtained by applying KVL. If the switch is in position "1," then at $t = -0$ the initial inductor current before switching is $i(t = -0) = 0$, and after switching at $t = +0$ in position "2," the inductor current $i(t = +0) = 0$ starts to increase; at steady-state $i(t \to \infty) = V_s/R$.

The general solution is according to Eq. (E3.3.12)

$$i(t) = K_1 + K_2 e^{-\left(\frac{t}{\tau_{RL}}\right)}. \tag{3.39}$$

With the derivative

$$\frac{di(t)}{dt} = \left(\frac{-1}{\tau}\right)K_2 \cdot e^{-\left(\frac{t}{\tau_{RL}}\right)} \tag{3.40}$$

follows with Eq. (3.38)

$$\left(\frac{-1}{\tau}\right)K_2 e^{-\left(\frac{t}{\tau_{RL}}\right)} + \frac{1}{\left(\frac{L}{R}\right)}\left[K_1 + K_2 \cdot e^{-\left(\frac{t}{\tau_{RL}}\right)}\right] = \frac{V_s}{L}. \tag{3.41}$$

Comparing the coefficients of the various terms results in $\frac{K_1}{L/R} = \frac{V_s}{L}$

or

$$K_1 = V_s/R \tag{3.42}$$

and

$$\left(\frac{-1}{\tau_{RL}}\right)K_2 - \frac{K_2}{\left(\frac{L}{R}\right)} = 0$$

or

$$\tau_{RL} = L/R. \tag{3.43}$$

The first-order ordinary differential equation is now

$$i(t) = \frac{V_s}{R} + K_2 e^{-\left(\frac{t}{\tau_{RL}}\right)}. \tag{3.44}$$

The constant K_2 can be determined from the initial condition: if one assumes $i(t = 0) = 0$, then $0 = V_s/R + K_2$ or $K_2 = -V_s/R$. The complete solution for $i(t)$ is

$$i(t) = \frac{V_s}{R}\left[1 - e^{-\left(\frac{t}{\tau_{RL}}\right)}\right]. \tag{3.45}$$

For $R = 10\,\Omega$, $L = 10$ mH, $V_s = 240$ V, one obtains with $\tau_{RL} = L/R = 1$ ms the equation

$$i(t) = 24\left[1 - e^{-\left(\frac{t}{1\,\text{ms}}\right)}\right]\text{ A}. \tag{3.46}$$

If the initial condition of the current is $i(t = 0) = I_o$, then $I_o = V_s/R + K_2$ or $K_2 = I_o - V_s/R$, resulting in the differential equation including initial nonzero current condition:

$$i(t) = \frac{V_s}{R}\left[1 - e^{-\left(\frac{t}{\tau_{RL}}\right)}\right] + I_o e^{-\left(\frac{t}{\tau_{RL}}\right)}. \tag{3.47}$$

The analytical solution, Eq. (3.46), is graphically displayed in Figure 3.16, and calculated values are listed in Table 3.3. The transient change due to one storage

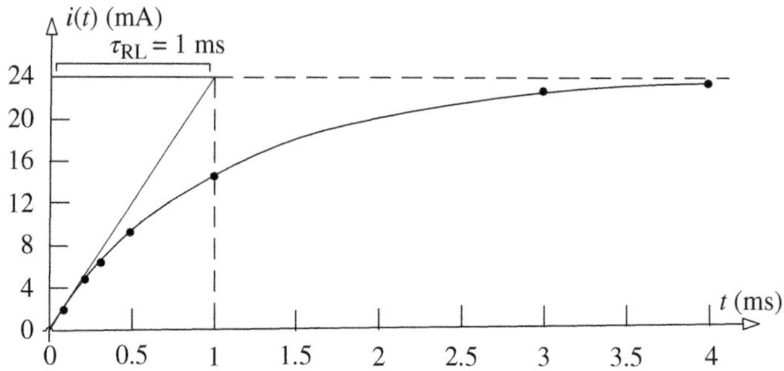

Figure 3.16 $i(t) = i_L(t)$ as a function of time t defined by Eq. (3.46) and displayed in the network of Figure 3.15.

Table 3.3 Calculated values for Figure 3.16.

t (ms)	**0**	**0.1**	**0.2**	**0.3**	**0.5**	**0.8**	**0.9**	**1**	**3**	**4**
$i(t)$ (A)	0	2.284	4.351	6.220	9.443	13.216	14.242	15.171	22.805	23.560

Table 3.4 PSPICE program.

RL network, Table 3.4	R 1 2 10
* transient analysis	L 2 0 10m ic=0
Vs 1 0 240	*Analysis request
*inductor as storage device	.Tran 10us 4ms
*with resistor R in series	.Probe
	.end

element (e.g. C, L) lasts about three times the time constant: this is in Figure 3.16 during the time $3\tau_{RL} = 3 \cdot 1\text{ ms} = 3\text{ ms}$.

For an inductor the current $i(t) = i_L(t)$ is the state variable, which is continuous as a function of time t as can be seen in Figure 3.16.

The analytically computed transient results of Figure 3.16 are now numerically computed based on PSPICE; see the program as listed in Table 3.4. Figure 3.17 represents the numerically calculated steady-state response, and it shows that it confirms the analytically computed steady-state result.

Performance of capacitor C and inductor L at steady-state operation: At steady-state condition, the capacitor voltage is constant $v_c(t) = V_s[1-e^{-(t/RC)}] = V_s$,

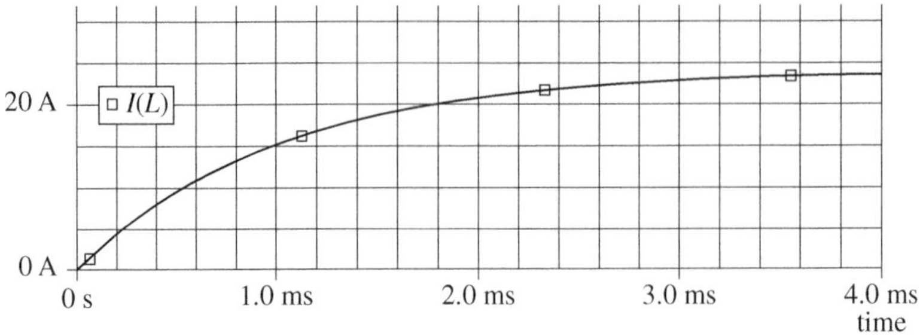

Figure 3.17 Numerically computed transient solution for $i(t) = I(L)$ based on PSPICE program as listed in Table 3.4.

because for $t \to \infty$ the exponential term is zero. The capacitor current is zero, and the capacitor represents an open circuit with a capacitor voltage $v_c(t) = V_S$.

At steady-state condition, the inductor current is constant $i(t) = \frac{V_s}{R}\left[1 - e^{-\left(\frac{t}{\tau_{RL}}\right)}\right] = \frac{V_s}{R}$, because for $t \to \infty$ the exponential term is zero. The inductor current is not zero, and the inductor represents a short circuit.

3.4 Transient Analysis Applied to Circuits Resulting in Second-Order, Ordinary Differential Equations with Constant Coefficients

We extend the transient analysis to the cases where capacitor C and inductor L are simultaneously present in the network. Figure 3.18 depicts the parallel RLC circuit generating a second-order ordinary differential equation where $v_c(t)$ is the state variable, common to both L and C.

The second-order ordinary differential equation with constant coefficients in terms of the state variable $v_c(t)$ is obtained by applying Kirchhoff's current law (KCL). If the switch is in position "1," then at $t = -0$ the initial capacitor voltage $v_c(t = -0) = 0$ and inductor current $i_L(t = -0) = 0$ before switching are zero, and after switching at $t = +0$ in position "2," the capacitor voltage $v_c(t = +0) = 0$ and the inductor current $i_L(t = +0) = 0$ start to increase; at steady-state $v_c(t \to \infty)$ reaches the constant value $V_{c\_steady\text{-}state}$.

From KCL one gets

$$I_s = i_R(t) + i_L(t) + i_c(t), \tag{3.48}$$

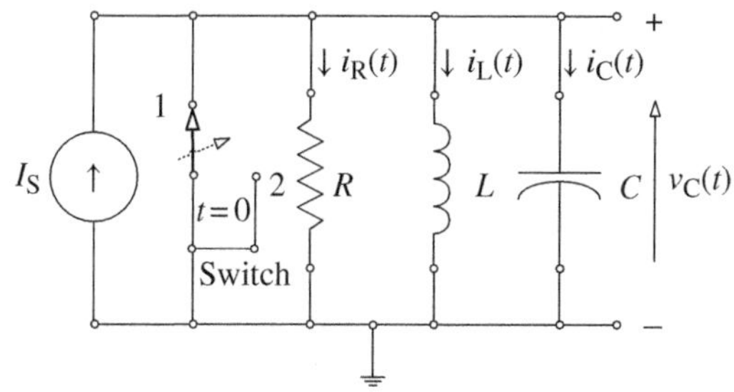

Figure 3.18 Two storage elements L and C in parallel with resistor R supplied by DC current source I_s.

with $i_R(t) = v_c(t)/R$, $i_L(t) = [1/L]\int v_c(t)\,dt$, and $i_c(t) = C[dv_c(t)/dt]$ leading to the second-order ordinary differential equation

$$I_s = \frac{v_c(t)}{R} + \frac{1}{L}\int v_c(t)dt + C\frac{dv_c(t)}{dt}, \tag{3.49}$$

the differentiation of which results in the final version of this second-order ordinary differential equation without integrals and with constant coefficients:

$$\frac{d^2v_c(t)}{dt^2} + \frac{1}{RC}\frac{dv_c(t)}{dt} + \frac{1}{LC}v_c(t) = 0. \tag{3.50}$$

Application Example 3.4 Analytic solution of second-order ordinary differential equation [10] with constant coefficients based on characteristic equation

Solution

Equation 3.50 can be expressed in the more general form as

$$\frac{d^2x(t)}{dt^2} + 2\alpha\frac{dx(t)}{dt} + \omega_0^2x(t) = 0, \tag{E3.4.1}$$

where $x(t)$ represents either $v_c(t)$ or $i(t)$. An important equation that is related to this differential equation is called the characteristic equation, where the derivatives are defined as $s^2 = \frac{d^2x(t)}{dt^2}$ and $s = \frac{dx(t)}{dt}$; thus

$$s^2 + 2\alpha s + \omega_0^2 = 0. \tag{E3.4.2}$$

Assuming the solution to be of the form

$$v_c(t) = Ae^{st}, \tag{E3.4.3}$$

where A and s are unknowns, the first and second derivatives are

$$\frac{dv_c(t)}{dt} = Ase^{st}, \tag{E3.4.4}$$

$$\frac{d^2 v_C(t)}{dt^2} = As^2 e^{st}, \tag{E3.4.5}$$

and introducing these derivatives into Eq. (3.50) yields

$$As^2 e^{st} + \left(\frac{1}{RC}\right) Ase^{st} + \left(\frac{1}{LC}\right) Ae^{st} = 0, \tag{E3.4.6}$$

for $A \neq 0$, $e^{st} \neq 0$ follows the characteristic equation

$$s^2 + \left(\frac{1}{RC}\right) s + \left(\frac{1}{LC}\right) = 0, \tag{E3.4.7}$$

where $\left(\frac{1}{RC}\right) = 2\alpha$, and $\left(\frac{1}{LC}\right) = \omega_0^2$.

The terms α and ω_0 are important factors: α is known as the exponential damping coefficient, and ω_0 is the undamped natural (angular) frequency (velocity).

The roots of the characteristic equation $s^2 + 2\alpha s + \omega_0^2 = 0$ are

$$s_1 = -\alpha + \sqrt{\alpha^2 - \omega_0^2}, \tag{E3.4.8}$$

$$s_2 = -\alpha - \sqrt{\alpha^2 - \omega_0^2}. \tag{E3.4.9}$$

s_1 and s_2 are called natural (angular) frequencies because they determine the natural or unforced response of the network. "Unforced" means the manner in which the network responds to an initial voltage on a capacitor or an initial current through an inductor. One can classify the response depending upon the values of α and ω_0 as follows:

Case 1: $\alpha > \omega_0$ results in an overdamped response where s_1 and s_2 are real and unequal generating the response

$$x(t) = K_1 e^{-s_2 t} + K_2 e^{-s_1 t},$$

or

$$x(t) = K_1 e^{-\left(\alpha - \sqrt{\alpha^2 - \omega_0^2}\right)t} + K_2 e^{-\left(\alpha + \sqrt{\alpha^2 - \omega_0^2}\right)t}, \tag{E3.4.10}$$

K_1, K_2 are constants and are determined from initial conditions $x(0)$ and $dx(0)/dt$.

Case 2: $\alpha < \omega_0$ results in an underdamped response where s_1 and s_2 are complex, generating the exponentially damped oscillation response

$$x(t) = e^{-\alpha t}\left(A_1 \cos\left[\sqrt{\omega_0^2 - \alpha^2}\right] t + A_2 \sin\left[\sqrt{\omega_0^2 - \alpha^2}\, t\right]\right), \tag{E3.4.11}$$

A_1, A_2 are constants and are evaluated based on $x(0)$ and $dx(0)/dt$.

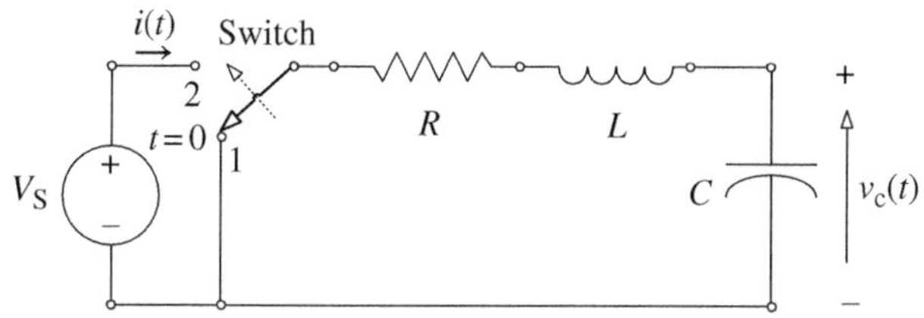

Figure E3.4.1 Two storage elements L and C in series with resistor R supplied by voltage source V_S.

Case 3: $\alpha = \omega_0$ results in a critically damped response where the characteristic equation has repeated roots s_1 and s_2 generating the response

$$x(t) = B_1 \mathrm{e}^{-\alpha t} + B_2 \mathrm{e}^{-\alpha t}. \tag{E3.4.12}$$

B_1 and B_2 are constants and found from initial conditions.

Figure E3.4.1 depicts the series RLC circuit generating a second-order ordinary differential equations where $v_c(t)$ is the state variable.

The second-order ordinary differential equation with constant coefficients in terms of the state variable $v_c(t)$ is obtained by applying Kirchhoff's current law (KVL). If the switch is in position "1," then at $t = -0$ the initial capacitor voltage $v_c(t = -0) = 0$ and inductor current $i_L(t = -0) = 0$ before switching are zero, and after switching at $t = +0$ in position "2," the capacitor voltage $v_c(t = +0) = 0$ and the inductor current $i_L(t = +0) = 0$ start to increase; at steady-state $v_c(t \to \infty)$ the capacitor voltage reaches the constant value V_S. Note the state variable $i(t)$ could be used as well.

The application of KVL generates

$$V_s = i(t)R + L\frac{\mathrm{d}i(t)}{\mathrm{d}t} + \frac{1}{C}\int i(t)\mathrm{d}t. \tag{E3.4.13}$$

Replacing in Eq. (E3.4.13) $v_c(t) = \frac{1}{C}\int i(t)\mathrm{d}t$ and $i(t) = C\frac{\mathrm{d}v_c(t)}{\mathrm{d}t}$ leads to

$$V_s = \mathrm{CR}\frac{\mathrm{d}v_c(t)}{\mathrm{d}t} + \mathrm{LC}\frac{\mathrm{d}^2 v_c(t)}{\mathrm{d}t^2} + v_c(t)$$

or

$$\frac{\mathrm{d}^2 v_c(t)}{\mathrm{d}t^2} + \left(\frac{R}{L}\right)\frac{\mathrm{d}v_c(t)}{\mathrm{d}t} + \frac{1}{\mathrm{LC}}v_c(t) = \frac{V_s}{LC}. \tag{E3.4.14}$$

Alternatively, differentiation of Eq. (E3.4.13) leads to the second-order ordinary differential equation without an integral and with constant coefficients whereby $i(t)$ is state variable:

$$0 = R\frac{\mathrm{d}i(t)}{\mathrm{d}t} + L\frac{\mathrm{d}^2 i(t)}{\mathrm{d}t^2} + \frac{1}{C}i(t), \tag{E3.4.15}$$

or

$$\frac{d^2 i(t)}{dt^2} + \left(\frac{R}{L}\right)\frac{di(t)}{dt} + \left(\frac{1}{LC}\right)i(t) = 0. \tag{E3.4.16}$$

Analytical solution for Eq. (E3.4.14) and Figure E3.4.1 at $V_s = 240\,V$, $R = 1\,\Omega$, $L = 10\,mH$, $C = 20\,mF$, and initial conditions $v_c(t = 0) = 0$, $v_c(t \to \infty) = 240\,V$, and $i(t = 0) = 0$ leads to the following differential equation:

$$\frac{d^2 v_c(t)}{dt^2} + 100\frac{dv_c(t)}{dt} + 5000\,v_c(t) = 1\,200\,000\,V, \tag{E3.4.17}$$

with the parameters $2\alpha = 1/(RC) = 50\cdot[1/\Omega F]$, $\alpha = 25\cdot[1/\Omega F]$, $\omega_0 = \sqrt{\frac{1}{LC}} = 70.711$ rad/s, and the characteristic equation $s^2 + s/RC + 1/LC = 0$ with the roots

$$s_1 = -\frac{1}{2RC} + \sqrt{\left(\frac{1}{2RC}\right)^2 - \left(\frac{1}{LC}\right)} = -25 + \sqrt{625 - 5000}$$
$$= (-25 + j66.144)\,[1/\Omega F],$$

$$s_2 = -\frac{1}{2RC} - \sqrt{\left(\frac{1}{2RC}\right)^2 - \left(\frac{1}{LC}\right)} = -25 - \sqrt{625 - 5000}$$
$$= (-25 - j66.144)\,[1/\Omega F].$$

This is *case 2* as stated in Eq. (E3.4.11), where $\alpha^2 < (\omega_0)^2$ results in an under-damped response with a slight overshoot because the characteristic equation has complex roots generating the response for the forcing voltage $V_s = A_3 = 240\,V$:

$$\begin{aligned} v_c(t) &= e^{-\alpha t}\left(A_1 \cos\left[\sqrt{\omega_0^2 - \alpha^2}\right]t + A_2 \sin\left[\sqrt{\omega_0^2 - \alpha^2}\,t\right]\right) + A_3 \\ &= e^{-25t}(A_1 \cos 66.144t + A_2 \sin 66.144t) + A_3. \end{aligned} \tag{E3.4.18}$$

A_1 and A_2 are constants and found from initial conditions $v_c(t = 0) = 0$, $i(t = 0) = 0$, and $v_c(t \to \infty)\ V_s = 240\,V$. Introducing the initial condition $v_c(t = 0) = 0$ into Eq. (E3.4.18) leads to the relation $v_c(t) = 0 = A_1 + A_3$, resulting in

$$A_1 = -A_3 = -240\,V. \tag{E3.4.19}$$

For the initial condition $i(t = 0) = 0$, the following expression based on the differentiation product rule $(uv)' = uv' + u'v$ is applied to Eq. (E3.4.18):

$$i(t) = C\frac{dv_c(t)}{dt} = C\{-25e^{-25t}A_1 \cos 66.144t - e^{-25t}A_1 66.144 \sin 66.144t$$
$$-25e^{-25t}A_2 \sin 66.144t + e^{-25t}A_2 66.144 \cos 66.144t\}, \quad \text{(E3.4.20)}$$
$$i(t=0) = 0 = C\{-25A_1 + 66.144A_2\};$$

with $C \neq 0$ follows

$$A_2 = 25A_1/66.144 = -240(25/66.144) = -90.711. \quad \text{(E3.4.21)}$$

The analytical solution is now

$$v_c(t) = -e^{-25t}240 \cos(66.144t)\text{V} - e^{-25t}90.711 \sin(66.177t)\text{ V} + 240\text{ V}. \quad \text{(E3.4.22)}$$

Table E3.4.1 presents calculated points for the analytically obtained solution Eq. (E3.4.22), and Figure E3.4.2 shows the underdamped response with an overshoot as defined by the parameters of Figure E3.4.1. Note the arguments of the trigonometric functions cos(50t) and sin(50t) must be in radians, not in degrees; that is, 25[rad/s] t[s] or [radians] and time t is in seconds.

A numerical PSPICE solution for the underdamped response (see Table E3.4.2) is illustrated in Figure E3.4.3. The parameters are $V_s = 240$ V, $R = 0.1\ \Omega$, $L = 0.5$ mH, $C = 2$ mF, and initial conditions $v_c(t=0) = 0$, $i(t=0) = 0$.

The numerical PSPICE solution for the overdamped response (see Table E3.4.3) is illustrated in Figure E3.4.4. The parameters are $V_s = 240$ V, $R = 1\ \Omega$, $L = 10$ mH, $C = 200$ mF, and initial conditions $v_c(t=0) = 0$, $i(t=0) = 0$.

Conclusions

First-order ordinary differential equations with constant coefficients are easily amenable to analytical solutions. Second-order ordinary differential equations are fairly amenable to the analytical approach, and it has been shown that their analytic solutions are identical with those of the numerical PSPICE approach. In conclusion, this text recommends the numerical approach for second-order and higher-order differential equations with constant and nonconstant coefficients. In control applications, an optimal response is not the overdamped or underdamped response but the critically damped response. The response of Figures E3.4.2 and E3.4.3 represents a fast control response, where a slight overshoot of the variable $v_c(t)$ is achieved.

Table E3.4.1 Analytical solution points of Eq. (E3.4.22).

***t* (ms)**	**0**	**10**	**20**	**30**	**50**	**60**	**80**	**100**	**150**	∞
$v_c(t)$ (V)	0	49.06	150.10	235.17	312.10	291.15	232.50	218.94	245.98	240

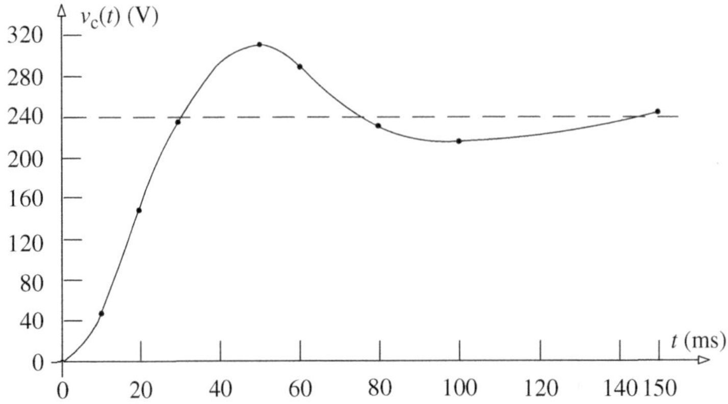

Figure E3.4.2 Analytical solution for the slightly underdamped response of $v_C(t)$ with overshoot as defined by the parameters of Figure E3.4.1.

Table E3.4.2 PSPICE program.

RLC network, Table E3.4.2	L 2 3 0.5m ic=0
* transient analysis, underdamped response	C 3 0 2m ic=0
Vs 1 0 240	*Analysis request
*capacitor and inductor as storage	.Tran 10us 70ms
*devices with resistor R in series	.Probe
R 1 2 0.1	.end

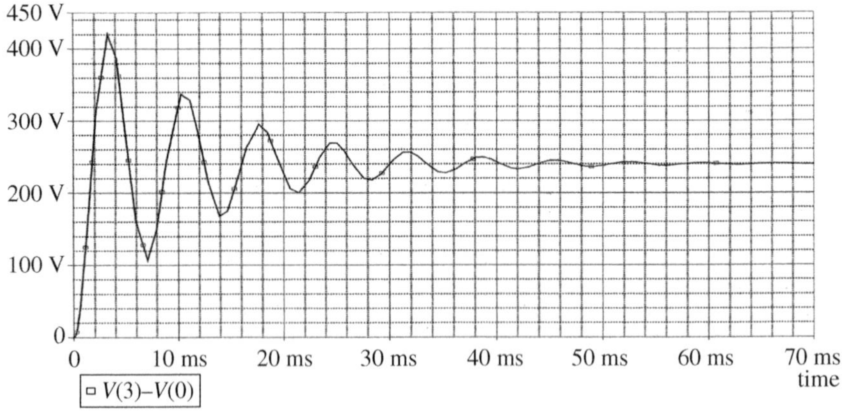

Figure E3.4.3 Numerical PSPICE solution for the underdamped response of $v_C(t) = V(3) - V(0)$ as defined by Figure E3.4.1.

Table E3.4.3 PSPICE program for calculation of Figure E3.4.4.

RLC network, Table E3.4.3	L 2 3 10m ic=0
* transient analysis, overdamped response	C 3 0 200m ic=0
Vs 1 0 240	*Analysis request
*capacitor and inductor as storage	.Tran 10us 900ms
*devices with resistor R in series	.Probe
R 1 2 1	.end

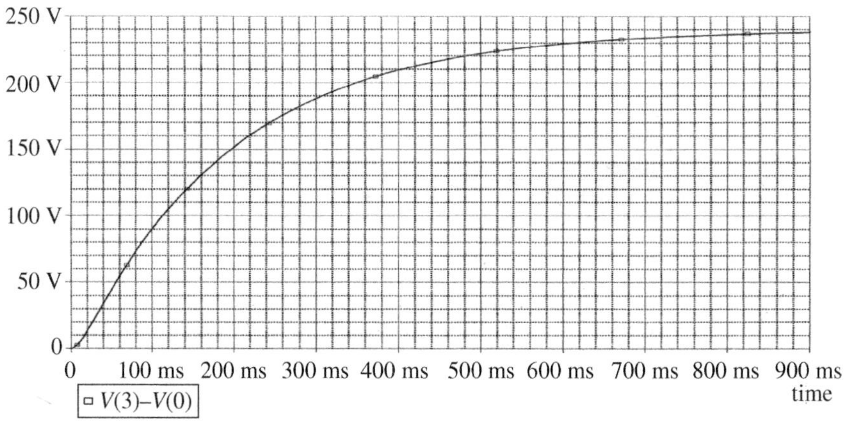

Figure E3.4.4 Numerical PSPICE solution for the overdamped response of $v_c(t) = V(3) - V(0)$ as defined by Figure E3.4.1.

3.5 Summary

Energy stored in a capacitor is derived based on the electric field intensity $\vec{E}$. The response of a capacitor C to a sawtooth voltage $v_c(t)$ is explained, and one notes that the voltage of a capacitor $v_c(t)$ cannot change quickly: it is a state function characterizing the charging state of a capacitor. The analytic solution is compared with the numerical solution obtained with PSPICE. Both methods generate the same current response. Series and parallel connections of several capacitors are explained.

Energy stored in an inductor L magnetized with magnetic field intensity $\vec{H}$ is found, and the solution response of an inductor to a triangular current $i_L(t)$ is demonstrated by analysis. In an inductor, the current $i_L(t)$ cannot change quickly, and the current is therefore denoted as a state function characterizing the charging state of an inductor. The analytic solution for the inductor voltage compares well with the numerical PSPICE solution. Inductor combinations (series, parallel) are addressed.

Transient analysis as applied to RC and RL circuits results in first-order ordinary differential equations with constant coefficients. Analytic solutions are compared with numerically obtained PSPICE responses. There is only one time constant for each circuit governing the response. Performance of capacitors and inductors under steady-state conditions is explained.

Transient analysis as applied to RLC circuits results in second-order ordinary differential equations with constant coefficients with two time constants. An analytical solution (Figure E3.4.2) is given for slightly underdamped response. Numerical solutions for underdamped (Figure E3.4.3) and overdamped (Figure E3.4.4) circuit conditions are included.

It is recommended to rely on numerical PSPICE solutions for circuits with more than two storage (e.g. C, L) elements because the analytic approach requires the solution of higher-order ordinary differential equations. Lastly, if any nonlinearity must be taken into account, the analytic approach becomes extremely difficult, if not nearly impossible. Further information can be found in references 10–13.

Problems

3.1 Calculation of the capacitor current waveform $i_c(t)$ for given voltage waveform $v_c(t)$ across a capacitor with capacitance $C = 88\ \mu F$ (Figure P3.1.1).

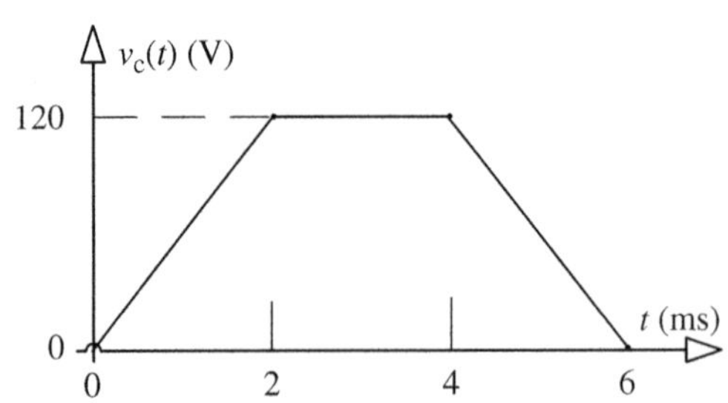

Figure P3.1.1 Calculation of capacitor current $i_c(t)$ for given capacitor voltage $v_c(t)$, which is the state variable.

3.2 Calculation of the capacitor current waveform $i_c(t)$ for given voltage waveform $v_c(t)$ across a capacitor with capacitance $C = 480\,\mu F$ (Figure P3.2.1).

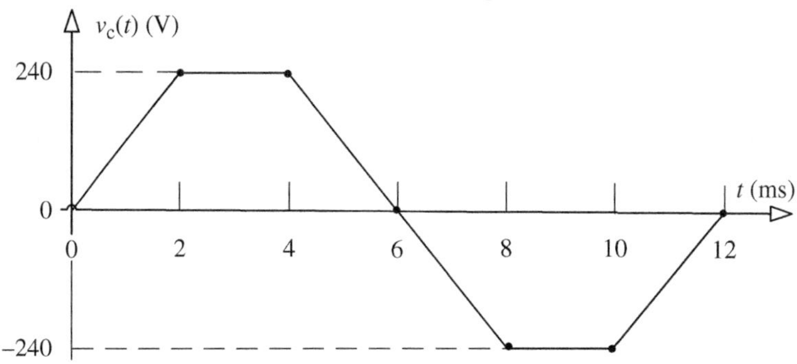

Figure P3.2.1 Calculation of capacitor current $i_c(t)$ for given capacitor voltage $v_c(t)$, which is a state variable.

3.3 Transient analysis of an RC circuit

The circuit of Figure P3.3.1 is given, assume the switch has been in position 1 for a long time and moves to position 2 at $t = 0$. Calculate the state variable voltage $v_c(t)$ for $t > 0$ for $R = 10\,\Omega$ and $C = 68\,\mu F$.

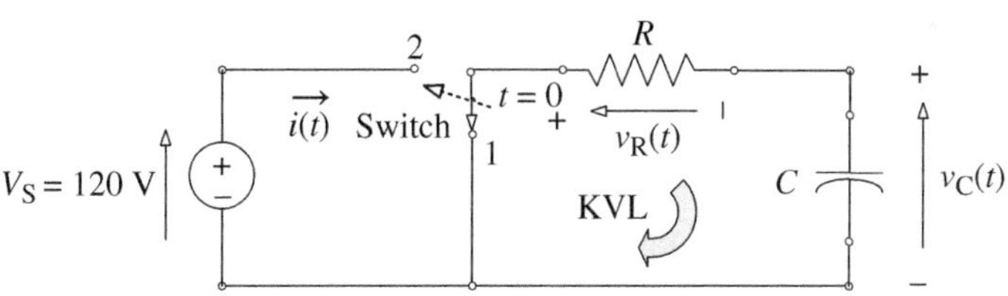

Figure P3.3.1 Calculation of transient voltage $v_c(t)$ for $t > 0$ and $V_s = 120$ V.

The first-order ordinary differential equation with constant coefficients in terms of the state variable $v_c(t)$ is obtained by applying Kirchhoff's voltage law (KVL). If the switch is in position "1," then at $t = -0$ the initial capacitor voltage before switching is $v_c(t = -0) = 0$, and after switching at $t = +0$ in position "2," the capacitor voltage $v_c(t = +0) = 0$ starts to increase; at steady-state $v_c(t \to \infty) = V_s$.

a Find the analytical solution by solving the governing first-order ordinary differential equation.

b Compute the numerical solution either using either PSPICE, Mathematica, or MATLAB.

3.4 Calculation of equivalent capacitance C_{AB}

On the one hand, standards permit the production of certain ratings of capacitances only, and on the other hand, design considerations require other than standard capacitance values. In order to satisfy both sets of constraints (e.g. standard, design), series and parallel connections of capacitors with standard capacitance values are required.

Determine the capacitance C_{AB} where $C_1 = 480\ \mu F$, $C_2 = 500\ \mu F$, $C_3 = 510\ \mu F$, $C_4 = 75\ \mu F$, $C_5 = 82\ \mu F$, $C_6 = 88\ \mu F$ (Figure P3.4.1).

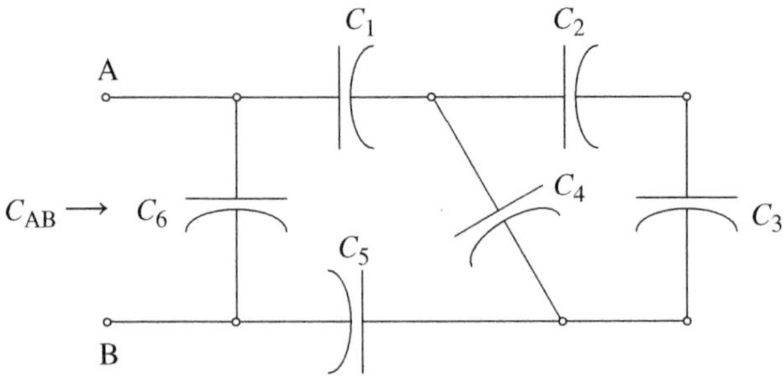

Figure P3.4.1 Calculation of equivalent capacitance C_{AB}.

3.5 Calculation of equivalent capacitance C_{AB}.

On the one hand, standards permit the production of certain ratings of capacitances only, and on the other hand, design considerations require other than standard capacitance values. In order to satisfy both sets of constraints (e.g. standard, design), series and parallel connections of capacitors with standard capacitance values are required.

Determine the capacitance C_{AB} where $C_1 = 0.33\ \mu F$, $C_2 = 0.47\ \mu F$, $C_3 = 68\ \mu F$, $C_4 = 72\ \mu F$, $C_5 = 75\ \mu F$, $C_6 = 82\ \mu F$ (Figure P3.5.1).

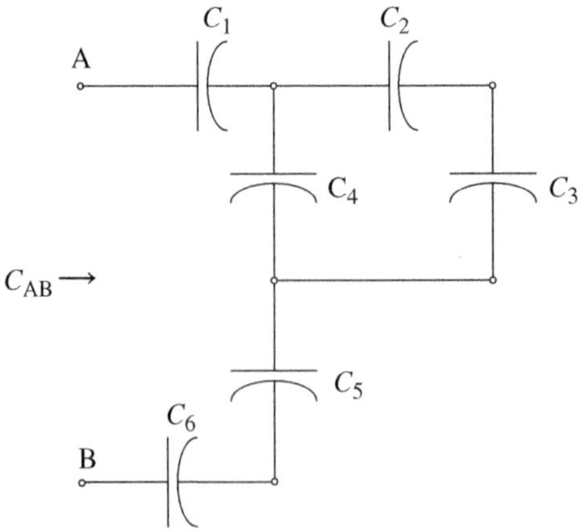

Figure P3.5.1 Calculation of equivalent capacitance C_{AB}.

3.6 Calculation of equivalent capacitance C_{AB}.

On the one hand, standards permit the production of certain ratings of capacitances only, and on the other hand, design considerations require other than standard capacitance values. In order to satisfy both sets of constraints (e.g. standard, design), series and parallel connections of capacitors with standard capacitance values are required.

Determine the capacitance C_{AB}, where $C_1 = 0.33\ \mu F$, $C_2 = 0.33\ \mu F$, $C_3 = 33\ \mu F$, $C_4 = 72\ \mu F$, $C_5 = 72\ \mu F$, $C_6 = 72\ \mu F$ (Figure P3.6.1).

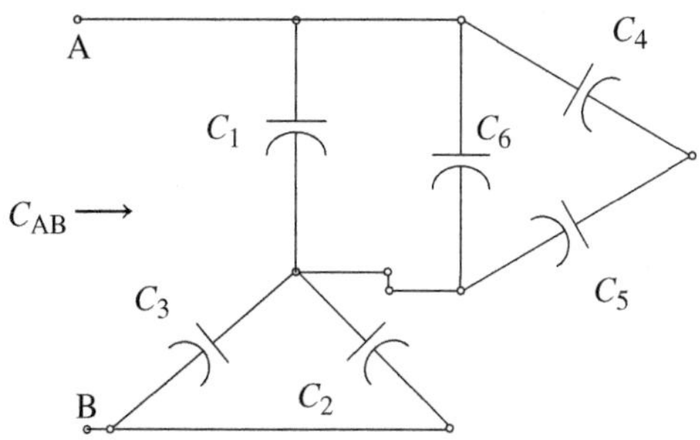

Figure P3.6.1 Calculation of equivalent capacitance C_{AB}.

3.7 Calculation of the inductor voltage waveform $v_L(t)$ for given current waveform $i_L(t)$ of an ideal inductor with inductance $L = 27\ \mu H$ (Figure P3.7.1).

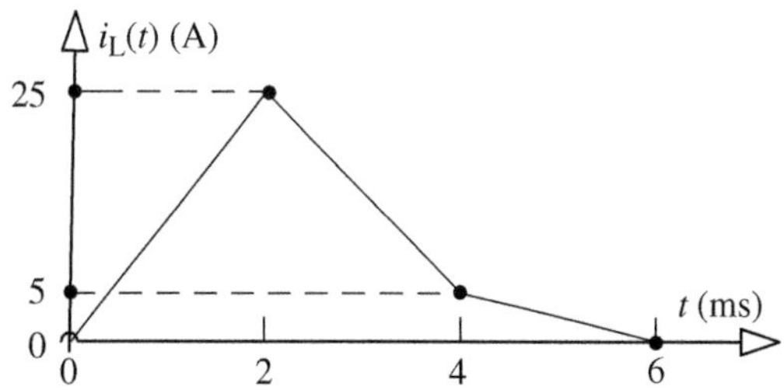

Figure P3.7.1 Calculation of inductor voltage $v_L(t)$ for given inductor current $i_L(t)$.

3.8 Calculation of the inductor voltage waveform $v_L(t)$ for given current waveform $i_L(t)$ of an ideal inductor with inductance $L = 240\ \mu H$ (Figure P3.8.1).

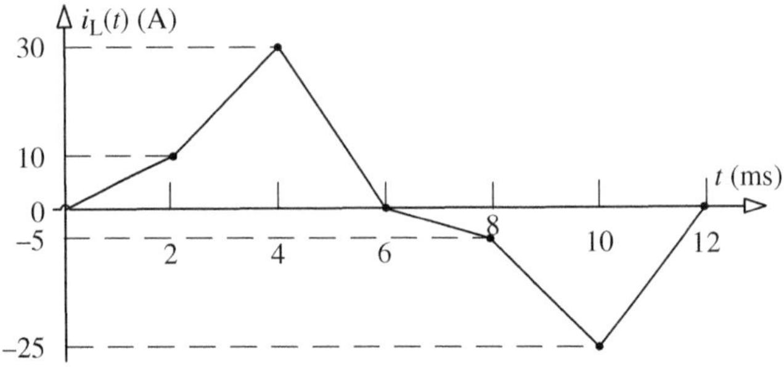

Figure P3.8.1 Calculation of inductor voltage $v_L(t)$ for given inductor current $i_L(t)$.

3.9 Transient analysis of an RL circuit.

The circuit of Figure P3.9.1 is given, assume the switch has been in position 1 for a long time and moves to position 2 at $t = 0$. Calculate the voltage $v_L(t)$ for $t > 0$ for $V_s = 240$ V, $R = 100\ \Omega$, and $L = 270$ mH.

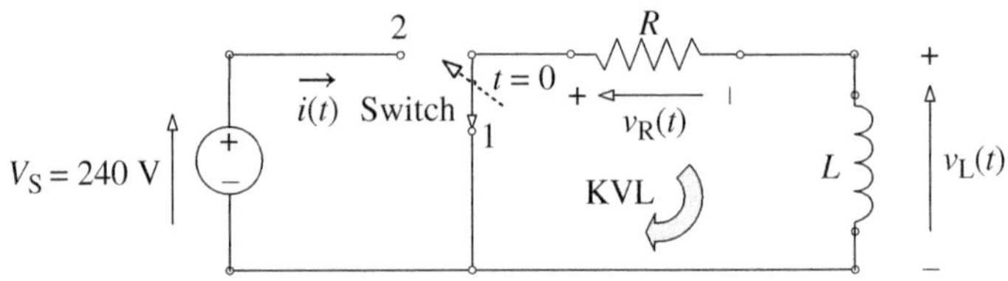

Figure P3.9.1 Calculation of transient current $i(t) = i_L(t)$ for $t > 0$.

The first-order ordinary differential equation with constant coefficients in terms of the state variable $i_L(t)$ is obtained by applying Kirchhoff's voltage law (KVL). If the switch is in position "1," then at $t = -0$ the initial inductor current before switching is $i(t = -0) = 0$, and after switching at $t = +0$ in position "2," the inductor current $i_L(t = +0) = i(t = -0) = 0$ starts to increase; at steady-state it is $i(t \to \infty) = 240 \cdot \text{V}/R = 240 \cdot \text{V}/100\ \Omega = 2.4$ A.

a Find the analytic solution by solving the governing first-order ordinary differential equation.

b Compute the numerical solution using either PSPICE [1–4], Mathematica [6], or MATLAB [5].

3.10 Calculation of equivalent inductance $L_{AB.}$

On the one hand, standards permit the production of inductors with certain ratings of inductances only, and on the other hand, design considerations require other than standard inductance values. In order to satisfy both sets of constraints (e.g. standard, design), series and parallel connections of inductors with standard inductance values are required.

Determine the inductance L_{AB} where $L_1 = 2.2\ \mu H$, $L_2 = 2.7\ \mu H$, $L_3 = 22\ \mu H$, $L_4 = 27\ \mu H$, $L_5 = 220\ \mu H$, $L_6 = 270\ \mu H$ (Figure P3.10.1).

Figure P3.10.1 Calculation of equivalent inductance L_{AB}.

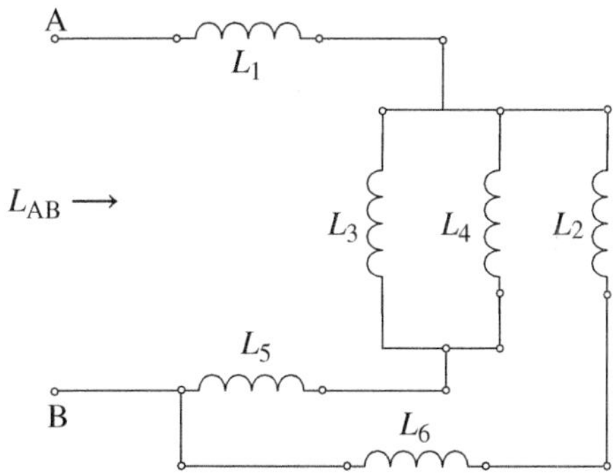

3.11 Calculation of equivalent inductance L_{AB}.

On the one hand, standards permit the production of inductors with certain ratings of inductances only, and on the other hand, design considerations require other than standard inductance values. In order to satisfy both sets of constraints (e.g. standard, design), series and parallel connections of inductors with standard inductance values are required.

Determine the inductance L_{AB} where $L_1 = 22\ \mu H$, $L_2 = 22\ \mu H$, $L_3 = 22\ \mu H$, $L_4 = 270\ \mu H$, $L_5 = 270\ \mu H$, $L_6 = 270\ \mu H$ (Figure P3.11.1).

Figure P3.11.1 Calculation of equivalent inductance L_{AB}.

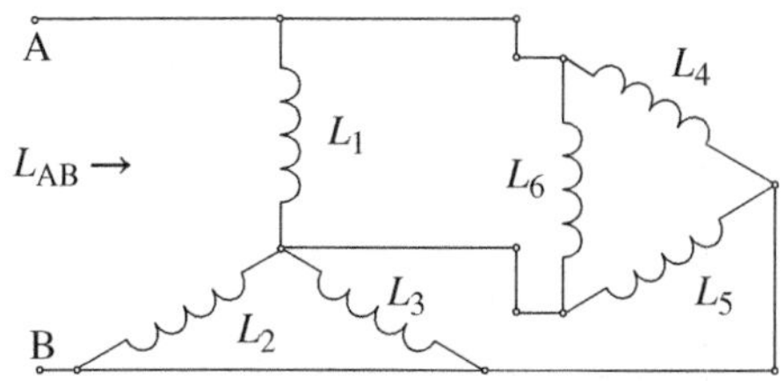

3.12 Calculation of equivalent inductance L_{AB}.

On the one hand, standards permit the production of inductors with certain ratings of inductances only, and on the other hand, design considerations require other than standard inductance values. In order to satisfy both sets of constraints (e.g. standard, design), series and parallel connections of inductors with standard inductance values are required.

Determine the inductance L_{AB}, where $L_1 = 0.1\ \mu H$, $L_2 = 0.15\ \mu H$, $L_3 = 0.22\ \mu H$, $L_4 = 0.33\ \mu H$, $L_5 = 0.47\ \mu H$, $L_6 = 68\ \mu H$ (Figure P3.12.1).

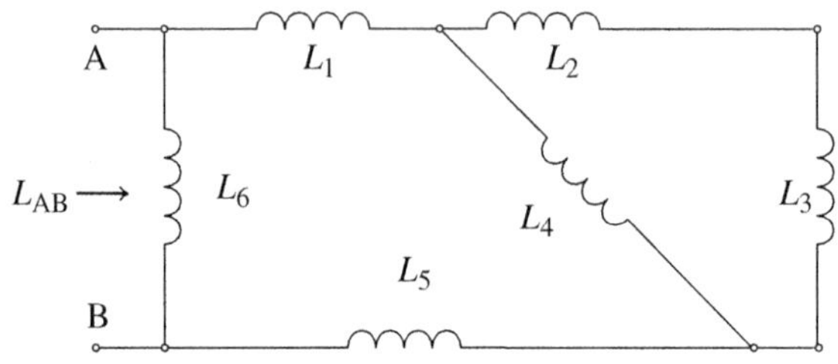

Figure P3.12.1 Calculation of equivalent inductance L_{AB}.

3.13 Transient analysis of an RLC circuit with two storage elements yielding a second-order ordinary differential equation.

Assume $R_1 = 5\ \Omega$, $R_2 = 0.005\ \Omega$, $R_3 = 0.2\ \Omega$, $L = 1$ mH, $C = 2$ mF, and $V_s = 60$ V. If the switch is in position "1," then at $t = -0$ the capacitor and inductor are discharged. The capacitor voltage before switching is $v_c(t = -0) = 0$, and after switching at $t = +0$ in position "2," the capacitor voltage $v_c(t = +0) = 0$ starts to increase. The inductor current $i_L(t)$ before switching is $i_L(t) = 0$, and after switching, $i_L(t)$ starts to increase (Figure P3.13.1).

Compute the numerical solution of $i(t)$ and $v_{R3L}(t)$ using either PSPICE [1–4], Mathematica [6], or MATLAB [5].

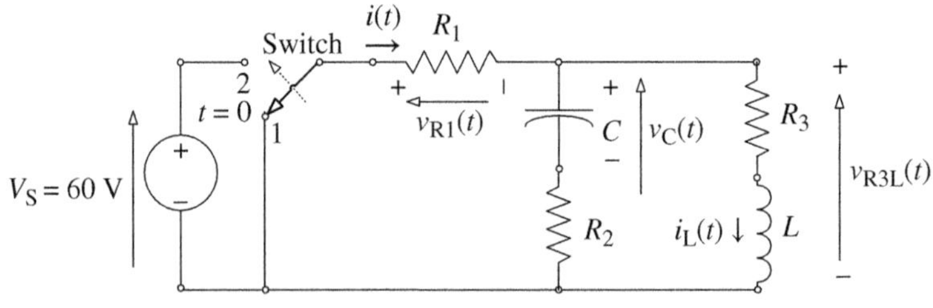

Figure P3.13.1 Calculation of inductor voltage $v_{R3L}(t)$ and current $i(t)$.

3.14 Charging of capacitor C through inductive R, L line based on transient analysis of an RLC circuit with two storage elements.

Assume $R = 5\ \Omega$, $L = 1$ mH, $C = 2$ mF, and $V_s = 120$ V, calculate capacitor voltage $v_c(t)$ and current $i(t)$ (Figure P3.14.1).

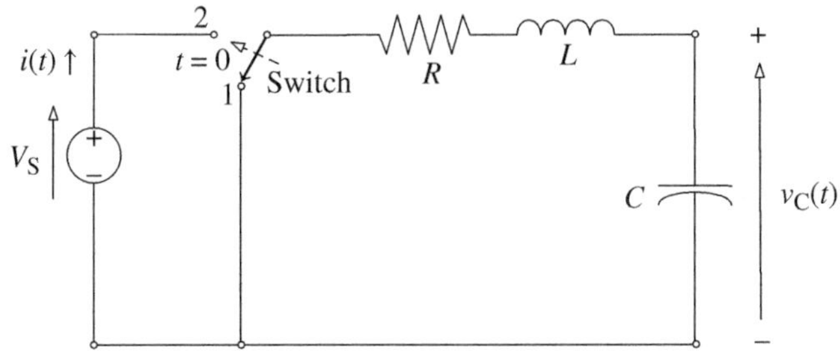

Figure P3.14.1 Calculation of capacitor voltage $v_c(t)$ and current $i(t)$.

The second-order ordinary differential equation with constant coefficients in terms of the state variable $v_c(t)$ is obtained by applying Kirchhoff's voltage law (KVL). If the switch is in position "1," then at $t = -0$ the initial capacitor voltage before switching is $v_c(t = -0) = 0$, and after switching at $t = +0$ in position "2," the capacitor voltage $v_c(t = +0) = 0$ starts to increase; at steady-state $v_c(t \to \infty) = V_S$.

Find the analytic solution by solving the governing second-order ordinary differential equation.

3.15 Discharging of capacitor C through inductive R, L line based on transient analysis of an RLC circuit with two storage elements.

Assume $R = 10\,\Omega$, $L = 100\,\text{mH}$, $C = 25\,\text{mF}$, and $V_s = 240\,\text{V}$ (Figure P3.15.1).

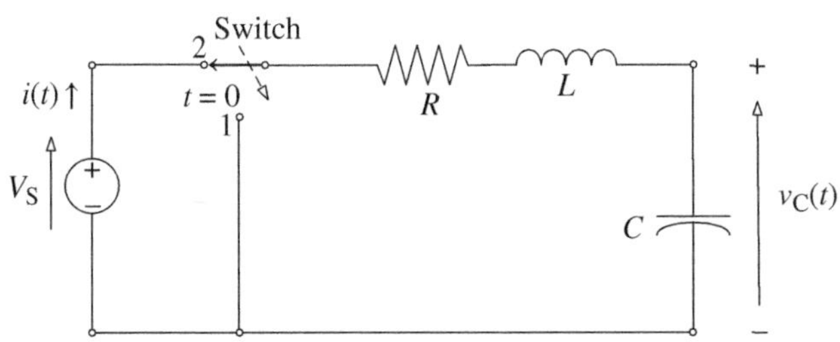

Figure P3.15.1 Calculation of capacitor voltage $v_c(t)$ and current $i(t)$.

The second-order ordinary differential equation with constant coefficients in terms of the state variable v_c(t) is obtained by applying Kirchhoff's voltage law (KVL). If the switch is in position "2," then at $t = -0$ the initial capacitor voltage before switching is $v_c(t = -0) = V_s = 240\,\text{V}$, and after switching at $t = +0$ in position "1," the capacitor voltage $v_c(t = +0) = 240$ V starts to decrease; at steady-state $v_c(t \to \infty) = 0$ V.

Find the analytic solution by solving the governing second-order ordinary differential equation.

3.16 Transient analysis of an RLC circuit acting as an oscillator (without damping) with more than two (e.g. 3) storage elements as shown in Figure P3.16.1.

Assume $R = 20\ \Omega$, $L_1 = 20$ mH, $L_2 = 1$ mH, and $C = 200\ \mu$F. This circuit demonstrates the principle of an LC oscillator where there is no damping in the parallel circuit of C and L_2.

Compute a numerical solution using either PSPICE [1–4], MATLAB/Simulink [5], or Mathematica [6].

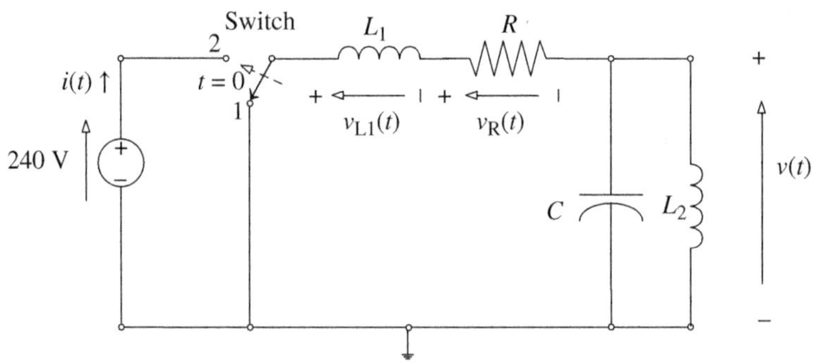

Figure P3.16.1 Calculation of transient charging current $i(t)$ and resonant/oscillatory voltage $v(t)$ for $t > 0$.

References

1 Tuinenga, P.W. (1992). *SPICE, A Guide to Circuit Simulation & Analysis Using PSpice*, 2e. Englewood Cliffs, NJ: Prentice Hall.

2 Roberts, G.W. and Sedra, A.S. (1996). *SPICE*, 2e. New York: Oxford University Press.

3 Banzhaf, W. (1992). *Computer-Aided Circuit Analysis Using PSpice*, 2e. Englewood Cliffs, NJ: Regents/Prentice Hall.

4 Rashid, M.H. (1992). *Spice for Power Electronics and Electric Power.* Englewood Cliffs, NJ: Prentice Hall.

5 MathWorks. MATLAB/SIMULINK. https://de.mathworks.com/products.html (accessed 11 May 2020).

6 Wolfram. Mathematica. http://www.wolfram.com/mathematica/new-in-11/?src=google&416 (accessed 11 May 2020).

7 Sawatzky, H. Design of electronic components. BOEKS GmbH. https://www.basictables.com/electronics/resistor/standard-resistor-values (accessed 11 May 2020).

8 Digi-Key Electronics. Common resistor and capacitor values. https://forum.digikey.com/t/common-resistor-and-capacitor-values/1899 (accessed 11 May 2020).

9 Marxhausen, P. and Grady, T. University of Nebraska-Lincoln. http://eeshop.unl.edu/standard_values.html (accessed 11 May 2020).

10 Nilsson, J.W. (1983). *Electric Circuits*. Reading, MA: Addison-Wesley Publishing Company.

11 Fuchs, E.F. and Masoum, M.A.S. (2011). *Power Conversion of Renewable Energy Systems*. New York, 2nd Printing: Springer, 692 pages.

12 Fuchs, E.F. (2010). ECEN 5787: power quality in power systems (Fall 2010). http://thiscourse.com/colorado/ecen5787/fa10 (accessed 11 May 2020).

13 Masoum, M.A.S. and Fuchs, E.F. (2015). *Power Quality in Power Systems and Electrical Machines*, 2e. London: Academic Press/Elsevier, 1123 pages.

4

Alternating Current (AC) Steady-State Analysis with Phasors

In Chapters 2 and 3, we have learned how to calculate both the transient (unforced, natural) and forced (steady-state) responses of circuits with DC forcing terms (e.g. voltages, currents). The transient response is simply a characteristic of the network and its initial conditions at time $t = 0$. In this chapter we concentrate on the steady-state or forced response of circuits with AC excitation, while the transient or natural response of alternating current (AC) circuits will be discussed in Chapter 7.

4.1 Sinusoidal and Cosinusoidal Functions

Figure 4.1 shows the general form of the sinusoid as

$$x(\omega t) = x_M \sin(\omega t + \theta), \tag{4.1}$$

where x is a variable (e.g. voltage, current, power, energy), x_M is amplitude or maximum excursion, $\omega = 2\pi f$ is the angular frequency or velocity measured in radians/second, $f = 1/T$ is the frequency with the unit Hz or 1/second, and T is the period of a periodic function. The constant pi is defined as $\pi = 3.141\,592\,653\,59\ldots$, and t is the time measured in seconds or s. The angle θ is the phase shift of the sinusoid with respect to the origin. If $\theta > 0$, then sinusoid leads the sinusoidal function $x(\omega t) = x_M \sin \omega t$, and if $\theta < 0$, then sinusoid lags the sinusoidal function.

The cosinusoidal function is

$$y(\omega t) = y_M \cos \omega t, \tag{4.2}$$

Introduction to Energy, Renewable Energy, and Electrical Engineering: Essentials for Engineering Science (STEM) Professionals and Students, First Edition. Ewald F. Fuchs and Heidi A. Fuchs.

Companion website: www.wiley.com/go/fuchs/renewableenergyandelectricalengineering

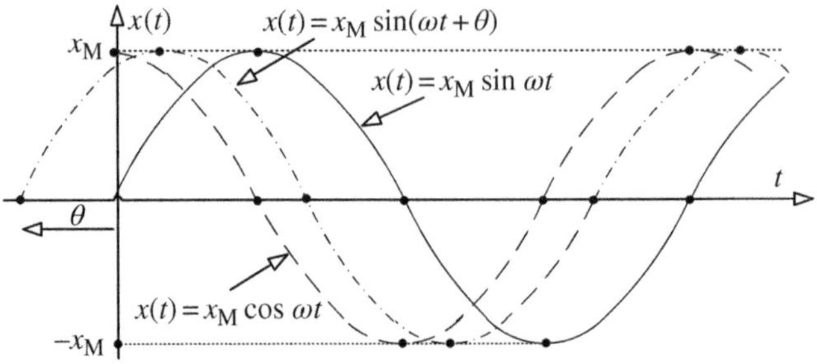

Figure 4.1 General form of a sinusoid in time domain.

4.2 Sinusoidal/Cosinusoidal and Complex Number Relations

The most general form of any analysis of an electric network is based on the time domain, for example, as shown in Figure 4.2, depicting an RL time-domain circuit energized by the terminal voltage $v(t)$.

The forcing function in Figure 4.2 is the terminal voltage $v(t)$ and has the form

$$v(t) = V_{\mathrm{M}} \cos(\omega t + \theta_{\mathrm{v}}). \tag{4.3}$$

If in Eq. (4.3) $\theta_{\mathrm{v}} = 0$, then this voltage relation is defined as real reference or cosinusoidal forcing function:

$$v(t) = V_{\mathrm{M}} \cos(\omega t). \tag{4.4}$$

"Real" means in a mathematical sense that it is neither complex nor imaginary, as will be later defined in this chapter.

In Figure 4.2 the current $i(t)$ is the real response function defined as

$$i(t) = I_{\mathrm{M}} \cos(\omega t + \theta_{\mathrm{i}}). \tag{4.5}$$

Note that we associate the phase shift angles θ_{v} and θ_{i} with voltages (v) and currents (i), respectively.

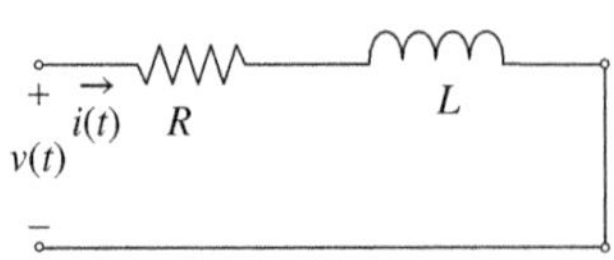

Figure 4.2 RL time domain network.

Time domain analysis is laborious if large electric networks are investigated, so phasor analysis is preferred, which is based on complex numbers. Note that phasor analysis is limited to steady-state considerations. Transient analyses must be performed in the time domain as, for example, done in Chapter 3.

4.2.1 Definition of Phasors

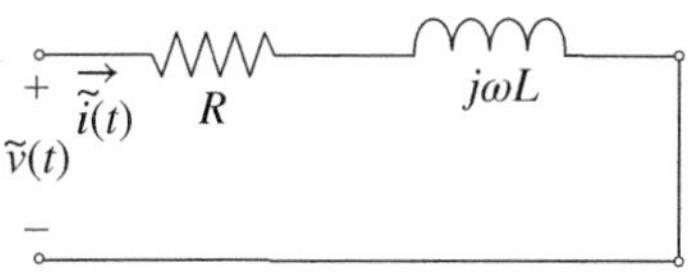

Figure 4.3 RL complex number domain network $\tilde{v}(t) = f\{\tilde{i}(t)\}$.

Figure 4.2 can be represented in the complex number domain as shown in Figure 4.3, where the tilde (~) signifies a complex quantity [e.g. voltage $\tilde{v}(t) = V_M(\cos \omega t + j \sin \omega t)$], which is similar to a complex number of the form $\tilde{z} = x + iy$. Note in mathematics the imaginary part "y" is indicated by an "i," while in electrical engineering a "j" is used in order to differentiate between imaginary identifier ("j") and current ("i").

In Figure 4.3 $\tilde{v}(t)$ is the complex voltage forcing function, and $\tilde{i}(t)$ is the corresponding complex response function.

Conclusions

- V_1 denotes a scalar (e.g. $V_1 = 240$ V).
- $\tilde{v}_1$ or $\widetilde{V_1}$ represents a complex number.
- $\tilde{v}_1(t)$ or $\widetilde{V_1}(t)$ is a complex number as a function of time t.
- As a convenience we use the same symbols in the complex number domain as those in the time domain, except that the time domain symbol $v_1(t)$ is provided with a tilde in the complex number domain $\tilde{v}_1(t)$.

The Swiss mathematician Leonhard Euler [1] invented the following identity relating the complex number "$e^{j\omega t}$" to the real part "$\cos \omega t$" and imaginary part "$\sin \omega t$":

$$e^{j\omega t} = \cos \omega t + j \sin \omega t. \tag{4.6}$$

The real (Re) part is identified by

$$\text{Re}\left(e^{j\omega t}\right) = \cos \omega t, \tag{4.7}$$

and the imaginary (Im) part is abbreviated by

$$\text{Im}\left(e^{j\omega t}\right) = \sin \omega t. \tag{4.8}$$

Complex numbers can be represented in the Gaussian [2] plane as depicted in Figure 4.4. Euler's equation provides a relation between rectangular and polar coordinates as shown in Figure 4.4.

We define the complex voltage forcing function as

$$\tilde{v}(t) = V_M e^{j(\omega t + \theta_v)}, \tag{4.9}$$

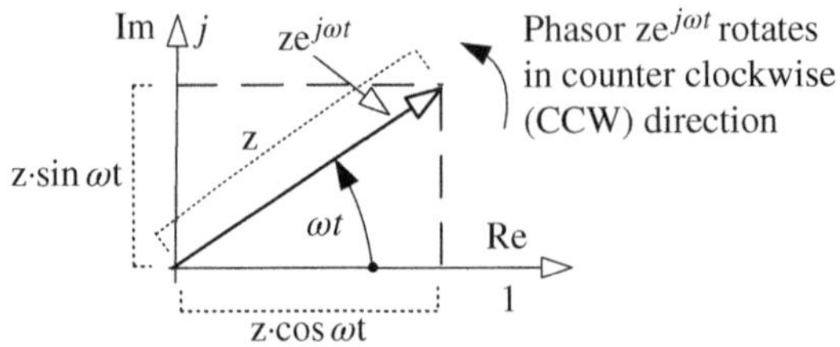

Figure 4.4 Complex number depicted in the Gaussian [2] plane relating rectangular ($x = z \cos \omega t$, $y = z \sin \omega t$) to polar ($|\tilde{z}| = z = \sqrt{x^2 + y^2}$, ωt) coordinates, resulting in the polar complex number $\tilde{z} = |\tilde{z}|e^{j\omega t} = ze^{j\omega t}$.

which becomes with $\theta_v = 0$ the complex reference voltage forcing function

$$\tilde{v}(t) = V_M\, e^{j\omega t}, \tag{4.10}$$

or relying on the Euler identity, one can write

$$\tilde{v}(t) = V_M \cos \omega t + jV_M \sin \omega t. \tag{4.11}$$

We can think of this complex reference forcing function $\tilde{v}(t)$ Eq. (4.11) as two real reference forcing functions $V_M \cos \omega t$ and $V_M \sin \omega t$. Because of assumed linearity [3], the principle of superposition [4] applies, and the complex current "i" response function is

$$\tilde{i}(t) = I_M \cos(\omega t + \theta_i) + jI_M \sin(\omega t + \theta_i); \tag{4.12}$$

where θ_i is the phase shift of the response current with respect to the forcing reference voltage $\tilde{v}(t)$. Equations (4.11) and (4.12) are related such that $I_M \cos(\omega t + \theta_i)$ is the response to $V_M \cos \omega t$ and $jI_M \sin(\omega t + \theta_i)$ is the response to $jV_M \sin \omega t$. Note that I_M and θ_i are unknowns.

Application Example 4.1 Solution of the current response $\tilde{i}(t)$ in the complex number domain for an RL circuit with the AC complex forcing voltage $\tilde{v}(t)$

Formulation in complex number domain:

Figure E4.1.1 shows the RL circuit with complex quantities. Note a voltage quantity consists of a value (e.g. $\sqrt{2} \cdot 240$) and a unit (e.g. V), that is, $V_M = \sqrt{2} \cdot 240$ V. In this example the complex forcing voltage is $\tilde{v}(t) = V_M e^{j\omega t} = V_M \cos \omega t + jV_M \sin \omega t$ [V], and the complex response current is $\tilde{i}(t) = I_M e^{j(\omega t + \theta_i)} = I_M \cos(\omega t + \theta_i) + jI_M \sin(\omega t + \theta_i)$ [A], where θ_i is the phase shift of the unknown complex response current $\tilde{i}(t)$ with respect to the known complex forcing voltage $\tilde{v}(t)$.

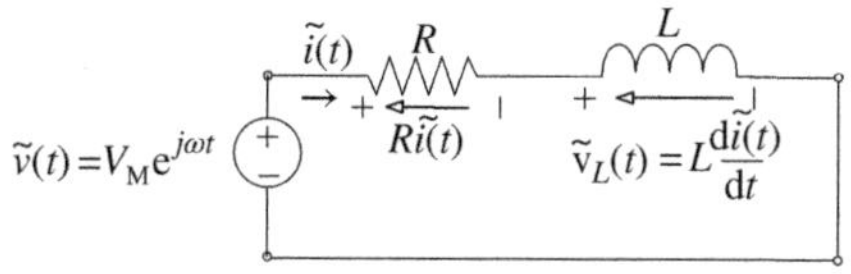

Figure E4.1.1 RL circuit response $\tilde{i}(t)$ solved in the complex domain for forcing reference function $\tilde{v}(t)$.

Solution

Applying Kirchhoff's voltage law, one gets

$$\tilde{v}(t) - R\tilde{i}(t) - L\frac{d\tilde{i}(t)}{dt} = 0, \tag{E4.1.1}$$

or

$$V_M\, e^{j\omega t} = R\tilde{i}(t) + L\frac{d\tilde{i}(t)}{dt}, \tag{E4.1.2}$$

with the unknown response current

$$\tilde{i}(t) = I_M\, e^{\,j(\omega t + \theta_i)}; \tag{E4.1.3}$$

one obtains by differentiation

$$\frac{d\tilde{i}(t)}{dt} = I_M\, j\omega e^{\,j(\omega t + \theta_i)}. \tag{E4.1.4}$$

Introducing Eqs. (E4.1.3) and (E4.1.4) into Eq. (E4.1.2) yields Kirchhoff's voltage equation in the complex domain

$$V_M\, e^{j\omega t} = RI_M e^{\,j(\omega t + \theta_i)} + LI_M\, j\omega e^{\,j(\omega t + \theta_i)} \tag{E4.1.5}$$

or

$$V_M e^{j\omega t} = RI_M\, e^{j\omega t} e^{\,j\theta_i} + LI_M\, j\omega e^{j\omega t} e^{\,j\theta_i}. \tag{E4.1.6}$$

Definition of phasors and analysis in phasor domain:

Equation (E4.1.6) can be divided by $e^{j\omega t} \neq 0$, which yields the reduced or simplified relation in mathematical notation

$$\tilde{V}_M = V_M\angle 0° = RI_M e^{\,j\theta_i} + (j\omega L)I_M e^{\,j\theta_i} \tag{E4.1.7}$$

or in electrical engineering notation with $\tilde{V}_M = V_M\angle 0°$, $\tilde{I}_M = I_M e^{\,j\theta_i}$ whereby $\angle 0°$ indicates a phase shift by 0 degrees (°) – also called the reference phase – Eq. (E4.1.7) is represented in Figure E4.1.2.

$$V_M\angle 0° = R\tilde{I}_M + j\omega L\tilde{I}_M. \tag{E4.1.8}$$

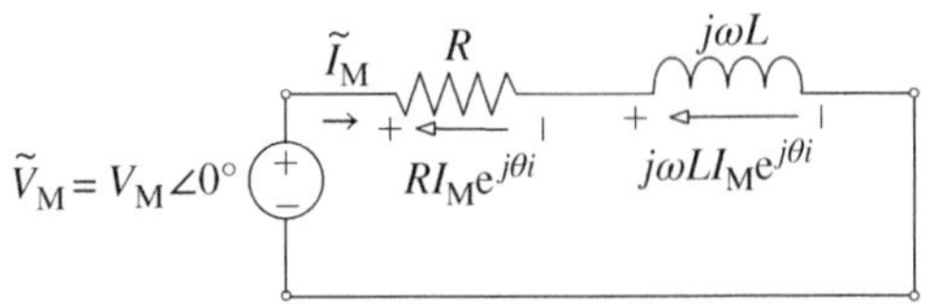

Figure E4.1.2 Representation of the complex quantities in Eq. (E4.1.7).

The quantities $V_M \angle 0°$, $\tilde{I}_M$ are called phasors and $j\omega L$ is the inductance expressed in the phasor domain, called complex (or more specific imaginary) impedance jX, where $X = \omega L$ is called reactance of an inductor. Note phasor analysis is only suitable for steady-state considerations and not for transient investigations.

Conclusions

A phasor consists of magnitude and phase shift, like a vector that consists of magnitude and direction. Use of complex number rules is permitted in phasor equations.

The response current is now obtained from Eq. (E4.1.8) as

$$\tilde{I}_M = \frac{V_M \angle 0°}{R + j\omega L} = \frac{\tilde{V}_M}{R + j\omega L} = \frac{V_M \angle 0°}{\sqrt{\left[R^2 + (\omega L)^2\right]} \angle \theta_i} = \frac{V_M \angle (0° - \theta_i)}{\sqrt{\left[R^2 + (\omega L)^2\right]}}. \tag{E4.1.9}$$

With the phase angle definition

$$\theta_i = \tan^{-1}\left(\frac{\omega L}{R}\right), \tag{E4.1.10}$$

one can rewrite the response current as

$$\tilde{I}_M = \frac{V_M \angle \left[0° - \tan^{-1}\left(\frac{\omega L}{R}\right)\right]}{\sqrt{\left[R^2 + (\omega L)^2\right]}}. \tag{E4.1.11}$$

The magnitude of the current is obtained from Eq. (E4.1.11) as

$$\left|\tilde{I}_M\right| = I_M = \frac{V_M}{\sqrt{\left[R^2 + (\omega L)^2\right]}}, \tag{E4.1.12}$$

and the phase shift of the response current with respect to the forcing voltage is

$$\theta_i = 0° - \tan^{-1}\left(\frac{\omega L}{R}\right). \tag{E4.1.13}$$

Back transformation of computed result in Eq. (E4.1.11) from phasor domain to time domain:

The reference forcing voltage in the time domain is $v(t) = V_M \cos \omega t$, and the magnitude of the response current is given by Eq. (E4.1.12), while the phase shift of the response current from the forcing reference voltage is stated in Eq. (E4.1.13).

Therefore the time domain expression of the response current is

$$i(t) = \frac{V_M}{\sqrt{R^2 + (\omega L)^2}} \cos\left[\omega t + 0^\circ - \tan^{-1}\left(\frac{\omega L}{R}\right)\right]$$

$$= \frac{V_M}{\sqrt{R^2 + (\omega L)^2}} \cos\left[\omega t - \tan^{-1}\left(\frac{\omega L}{R}\right)\right]$$

$$= I_M \cos\left[\omega t - \tan^{-1}\left(\frac{\omega L}{R}\right)\right] = \sqrt{2} I_{rms} \cos\left[\omega t - \tan^{-1}\left(\frac{\omega L}{R}\right)\right]. \quad \text{(E4.1.14)}$$

Root mean square (rms) or effective (eff) value of a sinusoidal/cosinusoidal function:

The rms value X_{rms} of the sinusoidal function $x(t) = X_M \sin(\omega t)$ with the period T is defined with $t_0 = 0$ as [5]

$$X_{rms} = \sqrt{\frac{1}{T}\int_{t_0}^{t_0+T} (X_M \sin \omega t)^2 dt} = \sqrt{\frac{1}{T} X_M^2 \left[\frac{x}{2} - \frac{1}{4}\sin 2x\right]_{t_0}^{t_0+T}}$$

$$= \sqrt{\frac{1}{T} X_M^2 \left[\frac{T}{2} - 0\right]} = \sqrt{\frac{X_M^2}{2}} = \frac{X_M}{\sqrt{2}} \quad \text{(E4.1.15)}$$

or

$$X_M = \sqrt{2} \cdot X_{rms} \quad \text{(E4.1.16)}$$

Conclusions

- Electric circuit or network is given in time domain.
- Analysis or calculation occurs in phasor domain.
- Response is formulated or expressed in time domain.
- The root mean square or effective value of a sinusoidal/cosinusoidal function is $X_{rms} = (X_M/\sqrt{2})$.

Application Example 4.2 Solution of RL circuit in phasor domain
Determine the solution of the current response $i(t)$ employing the phasor domain for an RL circuit with the AC forcing voltage $v(t) = V_M \cos \omega t$ [V] $= \sqrt{2} \cdot 240 \cos \omega t$ [V], with the resistance $R = 10\,\Omega$ and the inductance $L = 20$ mH at a power system frequency of $f = 60$ Hz. Calculate the root mean square (rms) value of the current $i(t)$ and its phase shift with respect to the forcing voltage $v(t)$.

Solution

Figure E4.2.1 shows the RL network under consideration.

The phasor reference voltage of the time function $v(t) = \sqrt{2} \cdot 240 \cos \omega t$ [V] is $\tilde{v}(t) = 240\,[\mathrm{V}]\angle 0°$ where by the voltage phasor represents its rms value, resulting in the phasor current

$$\tilde{i}(t) = \frac{\tilde{v}(t)}{R + j\omega L} = \frac{\tilde{v}(t)}{\sqrt{R^2 + (\omega L)^2}} \frac{1}{\angle \tan^{-1}\left(\frac{\omega L}{R}\right)} \tag{E4.2.1}$$

or in terms of rms values with $\omega = 2\pi f \approx 377$ rad/s

$$\tilde{i}_{\text{rms}} = \frac{240\angle 0°}{\sqrt{10^2 + (377 \cdot 20 \cdot 10^{-3})^2}} \frac{1}{\angle \tan^{-1}\left(\frac{377 \cdot 20 \cdot 10^{-3}}{10}\right)} \,[\mathrm{A}], \tag{E4.2.2}$$

$$\tilde{i}_{\text{rms}} = \frac{240\angle 0° - \tan^{-1}\left(\frac{377 \cdot 20 \cdot 10^{-3}}{10}\right)}{\sqrt{100 + 56.852}} = \frac{240\angle 0° - \tan^{-1}(0.754)}{\sqrt{100 + 56.852}} \,[\mathrm{A}], \tag{E4.2.3}$$

$$\tilde{i}_{\text{rms}} = 19.163\angle - 37.016°\,[\mathrm{A}]. \tag{E4.2.4}$$

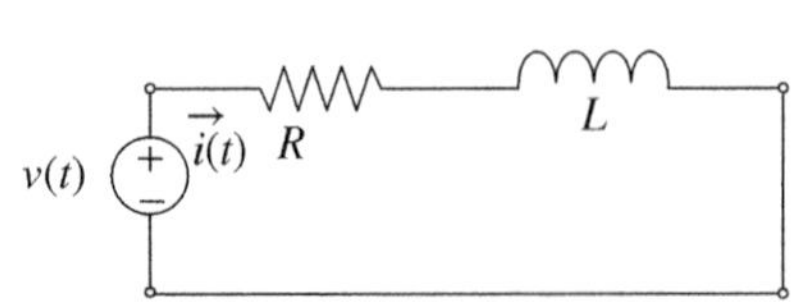

Figure E4.2.1 RL network with voltage $v(t)$ and current $i(t)$.

Conclusions

- Equation (E4.2.1) shows the equation in terms of general (either amplitude or rms) complex quantities.
- Equation (E4.2.2) shows the relation for the complex quantities in rms form where phasors are represented in terms of rms value and phase shifts in numerator and denominator.
- Equation (E4.2.3) shows the equation in terms of complex quantities in rms form where phasors are represented by rms value and phase shifts, whereby the phase shifts are combined in the numerator.
- Equation (E4.2.4) shows the final result for the current in rms form.

4.3 Phasor Relations for Circuit Elements such as Resistor, Inductor, and Capacitor

4.3.1 Resistor

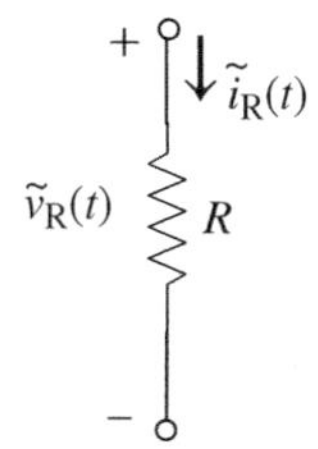

Figure 4.5 Resistor exposed to complex voltage $\tilde{v}_R(t) = \tilde{V}_R e^{j\omega t}$ resulting in complex response current $\tilde{i}_R(t) = \tilde{I}_R\, e^{j\omega t}$.

Figure 4.5 shows a resistor R where the forcing complex voltage $\tilde{v}_R(t) = \tilde{V}_R e^{j\omega t}$ is applied and results in the complex response current $\tilde{i}_R(t) = \tilde{I}_R e^{j\omega t}$. Note that $\tilde{V}_R$ and $\tilde{I}_R$ are by definition the associated phasor values.

Application of Ohm's law yields the complex number relation

$$\tilde{V}_R e^{j\omega t} = R \cdot \tilde{I}_R e^{j\omega t}. \tag{4.13}$$

Dividing this equation with $e^{j\omega t} \neq 0$ results in the phasor relation

$$\tilde{V}_R = R \cdot \tilde{I}_R. \tag{4.14}$$

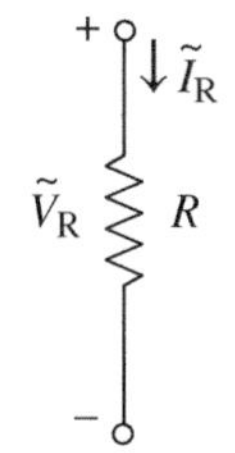

Figure 4.6 Phasor forcing voltage $\tilde{V}_R$ and phasor response current $\tilde{I}_R$ applied as to resistor R.

Figure 4.6 depicts Ohm's law for the resistor R where the forcing voltage $\tilde{V}_R$ and the response current $\tilde{I}_R$ are in phase, as shown in the graphical representation via the phasor diagram in Figure 4.7.

4.3.2 Inductor

Figure 4.8 shows an inductor L where the forcing complex voltage $\tilde{v}_L(t) = \tilde{V}_L e^{j\omega t}$ is applied and results in the complex

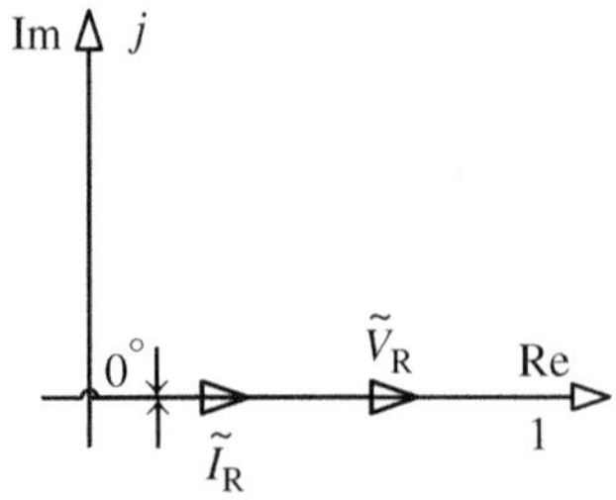

Figure 4.7 Graphical representation of forcing voltage $\tilde{V}_R$ and phasor response current $\tilde{I}_R$ in phasor diagram.

response current $\tilde{i}_L(t) = \tilde{I}_L e^{j\omega t}$. Note that $\tilde{V}_L$ and $\tilde{I}_L$ are by definition the associated phasor values.

Application of the voltage–current relationship for an inductor as discussed in Chapter 1 (Figure 1.22b) results in the complex differential equation

$$\tilde{v}_L(t) = L\frac{d\tilde{i}_L(t)}{dt}, \tag{4.15}$$

or

$$\tilde{V}_L e^{j\omega t} = L\frac{d\left(\tilde{I}_L e^{j\omega t}\right)}{dt}. \tag{4.16}$$

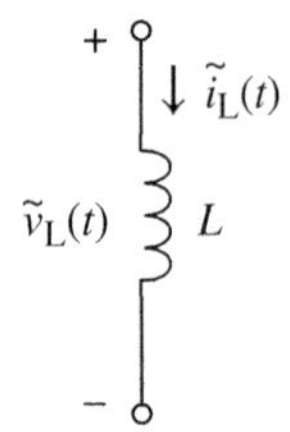

Figure 4.8 Inductor exposed to complex voltage $\tilde{v}_L(t) = \tilde{V}_L e^{j\omega t}$ resulting in complex response current $\tilde{i}_L(t) = \tilde{I}_L e^{j\omega t}$.

Differentiating Eq. (4.16) results in

$$\tilde{V}_L e^{j\omega t} = j\omega L\tilde{I}_L e^{j\omega t}. \tag{4.17}$$

Dividing this equation with $e^{j\omega t} \neq 0$ results in the phasor relation

$$\begin{aligned}\tilde{V}_L &= j\omega L \cdot \tilde{I}_L = \omega L \cdot \tilde{I}_L e^{j90°} \\ &= X_L \cdot \tilde{I}_L \angle 90°.\end{aligned} \tag{4.18}$$

where $\omega L = X_L$ is the reactance of an inductor.

Figure 4.9 depicts the forcing voltage and response current for the inductor L where the forcing voltage $\tilde{V}_L$ leads the response current $\tilde{I}_L$ by 90°, as is shown in the graphical representation via the phasor diagram in Figure 4.10.

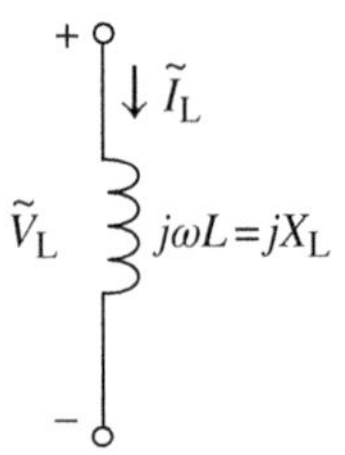

Figure 4.9 Phasor forcing voltage $\tilde{V}_L$ and phasor response current $\tilde{I}_L$ applied as to inductor L.

4.3.3 Capacitor

Figure 4.11 shows a capacitor C where the forcing complex voltage $\tilde{v}_c(t) = \tilde{V}_c e^{j\omega t}$ is applied and results in the complex response current $\tilde{i}_c(t) = \tilde{I}_c e^{j\omega t}$. Note that $\tilde{V}_c$ and $\tilde{I}_c$ are by definition the associated phasor values.

Application of the voltage–current relationship for an inductor as discussed in Chapter 1 (Figure 1.22b) results in the complex differential equation

$$\tilde{i}_c(t) = C\frac{d\tilde{v}_c(t)}{dt}, \tag{4.19}$$

or

$$\tilde{I}_c e^{j\omega t} = C \frac{d\left(\tilde{V}_c e^{j\omega t}\right)}{dt}. \quad (4.20)$$

Differentiating Eq. (4.20) results in

$$\tilde{I}_c e^{j\omega t} = j\omega C \tilde{V}_c e^{j\omega t}. \quad (4.21)$$

Dividing this equation with $e^{j\omega t} \neq 0$ results in the phasor relation

$$\begin{aligned} \tilde{V}_c &= \frac{1}{j\omega C} \cdot \tilde{I}_c = \frac{-j}{\omega C} \cdot \tilde{I}_c \\ &= \frac{1}{\omega C} \cdot \tilde{I}_c e^{-j90^\circ} = X_c \cdot \tilde{I}_c \angle -90^\circ \end{aligned} \quad (4.22)$$

where $\frac{1}{\omega C} = X_c$ is the reactance of a capacitor.

Figure 4.12 depicts the forcing voltage and response current for the capacitor C where the forcing voltage $\tilde{V}_c$ lags the response current $\tilde{I}_c$ by 90°, as is shown in the graphical representation via the phasor diagram in Figure 4.13.

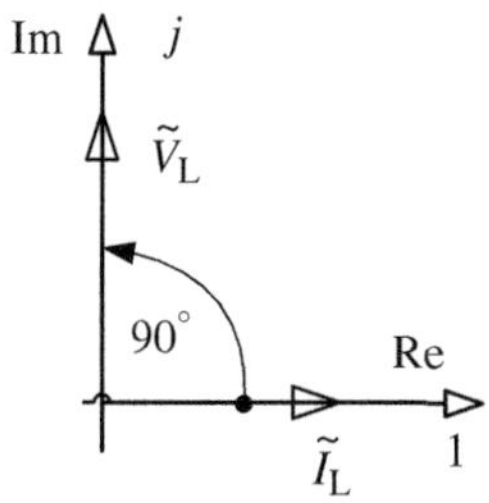

Figure 4.10 Graphical representation of forcing voltage $\tilde{V}_L$ and phasor response current $\tilde{I}_L$ in phasor diagram.

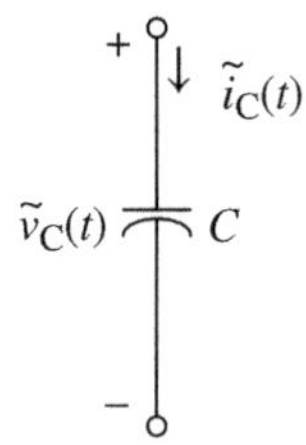

Figure 4.11 Capacitor exposed to complex voltage $\tilde{v}_c(t) = \tilde{V}_c e^{j\omega t}$ resulting in complex response current $\tilde{i}_c(t) = \tilde{I}_c e^{j\omega t}$.

4.3.4 Definition of Impedance $\tilde{Z}$ and Admittance $\tilde{Y}$

For phasors in the frequency domain, as shown in Figure 4.14, one can define the impedance as the ratio of the phasor voltage $\tilde{V}_M$ to the phasor current $\tilde{I}_M$:

$$\tilde{Z}_M = \frac{\tilde{V}_M}{\tilde{I}_M}, \quad (4.23)$$

where the subscript "M" refers to the amplitude or maximum (M) excursion of a sinusoidal/cosinusoidal function.

If the network of Figure 4.14 consists of

- the resistance R, then

$$\tilde{Z}_R = R, \quad (4.24)$$

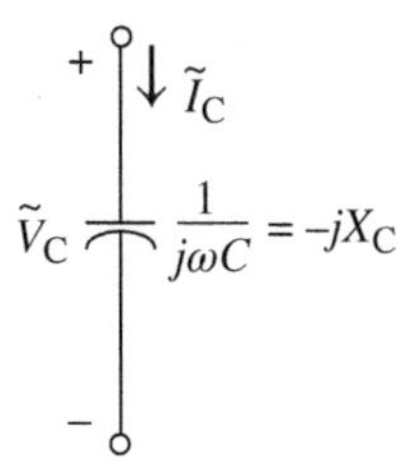

Figure 4.12 Phasor forcing voltage $\tilde{V}_c$ and phasor response current $\tilde{I}_c$ applied as to capacitor C.

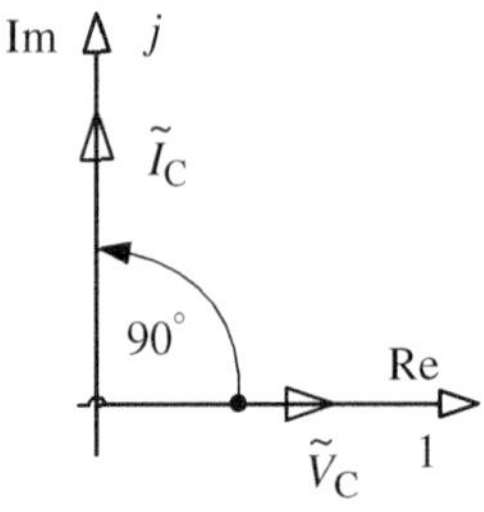

Figure 4.13 Graphical representation of forcing voltage $\tilde{V}_c$ and phasor response current $\tilde{I}_c$ in phasor diagram.

- the inductance L, then

$$\tilde{Z}_L = j\omega L = jX_L \text{ with } \omega L = X_L, \tag{4.25}$$

- the capacitance C, then

$$\tilde{Z}_c = \frac{1}{j\omega C} = \frac{-j}{\omega C} = -jX_c \text{ with } \frac{1}{\omega C} = X_c. \tag{4.26}$$

Having examined each of the passive circuit components in the frequency (phasor) domain (e.g. R, $j\omega L$, $1/j\omega C$), the input impedance is in polar form:

$$\tilde{Z}_M = \frac{\tilde{V}_M}{\tilde{I}_M} = \frac{|\tilde{V}_M|\angle\theta_v}{|\tilde{I}_M|\angle\theta_i} = |\tilde{Z}_M|\angle(\theta_v - \theta_i) = |\tilde{Z}_M|\angle\theta_z, \tag{4.27}$$

where the amplitude is

$$|\tilde{Z}_M| = \frac{|\tilde{V}_M|}{|\tilde{I}_M|} \tag{4.28a}$$

and the phase angle

$$\theta_z = \tan^{-1}\left\{\frac{\text{Im}\left(\tilde{Z}_M\right)}{\text{Re}\left(\tilde{Z}_M\right)}\right\}. \tag{4.28b}$$

The subscript "z" in Eqs. (4.27) and (4.28b) refers to the impedance $\tilde{Z}_M$. In rectangular form one writes

$$\tilde{Z}_M = (R + jX). \tag{4.29}$$

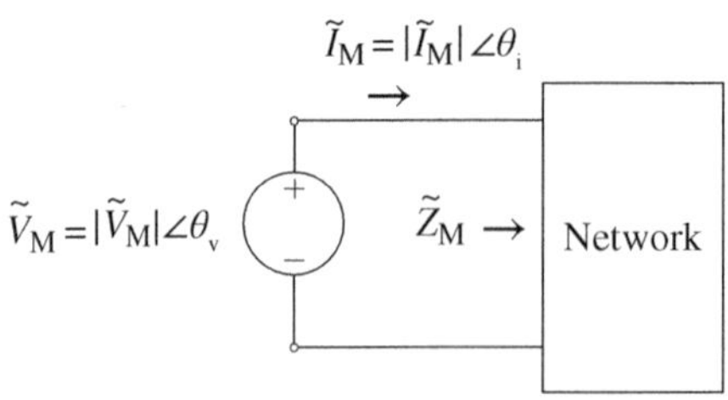

Figure 4.14 General network in a block diagram.

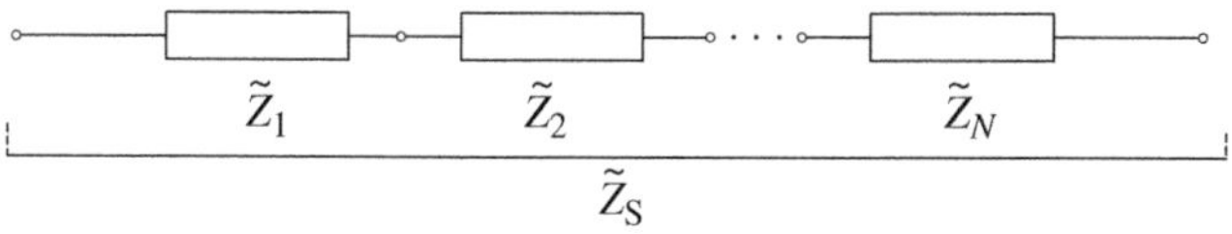

Figure 4.15 Series connection of impedances.

The series connection of impedances is similar to the series connection of resistors (Chapter 2, Eq. 2.28), as depicted in Figure 4.15 where the individual impedance is shown as a block.

The total series impedance is

$$\tilde{Z}_s = \tilde{Z}_1 + \tilde{Z}_2 + \tilde{Z}_3 + \ldots + \tilde{Z}_N, \tag{4.30}$$

and the total parallel connection is illustrated in Figure 4.16 with the associated formula in Eq. (4.31).

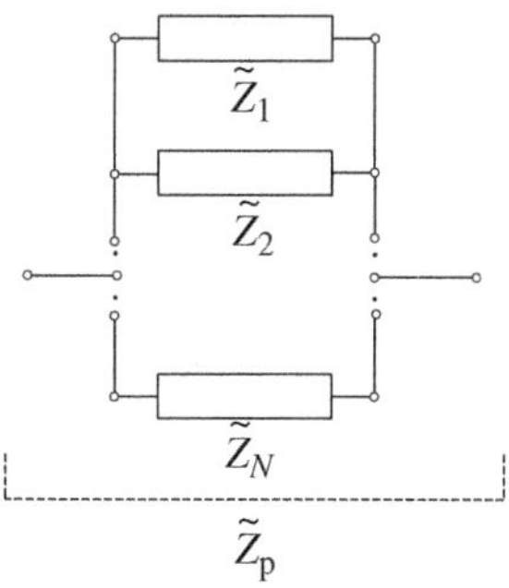

Figure 4.16 Parallel connection of impedances.

The total parallel impedance is

$$\frac{1}{\tilde{Z}_p} = \frac{1}{\tilde{Z}_1} + \frac{1}{\tilde{Z}_2} + \frac{1}{\tilde{Z}_3} + \ldots + \frac{1}{\tilde{Z}_N}. \tag{4.31}$$

Definition of admittance in polar form is

$$\tilde{Y}_M = \frac{1}{\tilde{Z}_M} = \frac{\tilde{I}_M}{\tilde{V}_M} = |\tilde{Y}_M| \angle\theta_y; \tag{4.32}$$

in Eq. (4.32) the subscript "y" refers to the admittance $\tilde{Y}_M$.

In rectangular form the admittance consists of conductance G, representing the real part of the admittance $\tilde{Y}_M$, and the susceptance B, representing the imaginary part of the admittance $\tilde{Y}_M$:

$$\tilde{Y} = G + jB = \frac{1}{R + jX} = \frac{(R - jX)}{(R + jX)(R - jX)} = \frac{(R - jX)}{(R^2 - jRX + jRX + X^2)}$$

$$= \frac{R}{R^2 + X^2} - \frac{jX}{R^2 + X^2}. \tag{4.33a}$$

The term $(R - jX)$ is the complex conjugate of $(R + jX)$ and

$$G = \frac{R}{R^2 + X^2}, \tag{4.33b}$$

$$B = -\left(\frac{X}{R^2 + X^2}\right). \tag{4.33c}$$

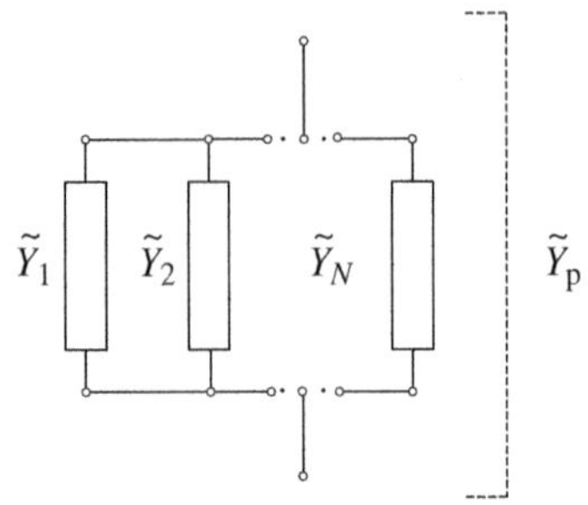

Figure 4.17 Parallel connection of admittances.

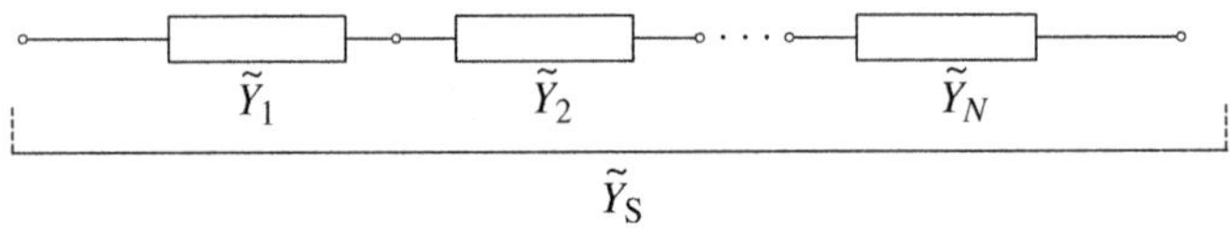

Figure 4.18 Series connection of admittances.

A parallel connection of admittances is shown in Figure 4.17, with its formula given in Eq. (4.34):

$$\widetilde{Y}_p = \widetilde{Y}_1 + \widetilde{Y}_2 + \widetilde{Y}_3 + \dots + \widetilde{Y}_N. \tag{4.34}$$

A series connection of admittances is shown in Figure 4.18, with its formula given in Eq. (4.35):

$$\frac{1}{\widetilde{Y}_s} = \frac{1}{\widetilde{Y}_1} + \frac{1}{\widetilde{Y}_2} + \frac{1}{\widetilde{Y}_3} + \dots + \frac{1}{\widetilde{Y}_N}. \tag{4.35}$$

Summary

- Resistor admittance: $\widetilde{Y}_R = \frac{1}{R} = G$.
- Inductor admittance: $\widetilde{Y}_L = \frac{1}{j\omega L} = -j\left(\frac{1}{\omega L}\right) = \frac{1}{\omega L\angle 90^\circ} = \frac{1\angle -90^\circ}{\omega L}$.
- Capacitor admittance: $\widetilde{Y}_c = j\omega C = \omega C\angle 90^\circ$.

Next, we apply to phasor calculations the concepts we have studied in Chapters 1 and 2 such as Ohm's law, Kirchhoff's current law, Kirchhoff's voltage law, Δ-Y transformation, superposition, Thévenin's theorem, Norton's theorem, nodal analysis, and mesh/loop analysis.

4.4 Delta-Wye Transformation

The Δ-Y and Y-Δ transformations presented in Chapter 2 for resistances are also valid for impedances and admittances in the phasor (frequency) domain.

4.5 Solution Based on Kirchhoff's Laws

Application Example 4.3 Solution of the currents and voltages in an RLC network for the forcing source (s) voltage $\widetilde{V}_s = 240\,\text{V}\angle 60^\circ$ employing the Kirchhoff's laws in the phasor domain

Figure E4.3.1 depicts the RLC network.

a) Determine the input impedance $\widetilde{Z}_{eq}$.
b) Calculate the input current $\widetilde{I}_1$.
c) Calculate the voltage $\widetilde{V}_1$ across the inductor impedance $j\omega L = j6\,\Omega$.
d) Find the branch currents $\widetilde{I}_2$ and $\widetilde{I}_3$.
e) Draw the phasors in the complex plane $(j - 1)$ or (Im – Re) plane for the above calculated voltages and currents including the forcing voltage $\widetilde{V}_s$.

Solution

a) The equivalent input impedance is

$$\widetilde{Z}_{eq} = 4 + \frac{j6(8-j4)}{(j6+8-j4)} = 4 + \frac{(24+j48)}{(8+j2)} = 4 + \frac{\sqrt{(24)^2+(48)^2}\angle\theta_N}{\sqrt{8^2+2^2}\angle\theta_D}$$

$$= \left\{4 + \frac{53.67\angle 1.1072\text{ rad}}{8.25\angle 0.2445\text{ rad}}\right\}\Omega, \tag{E4.3.1}$$

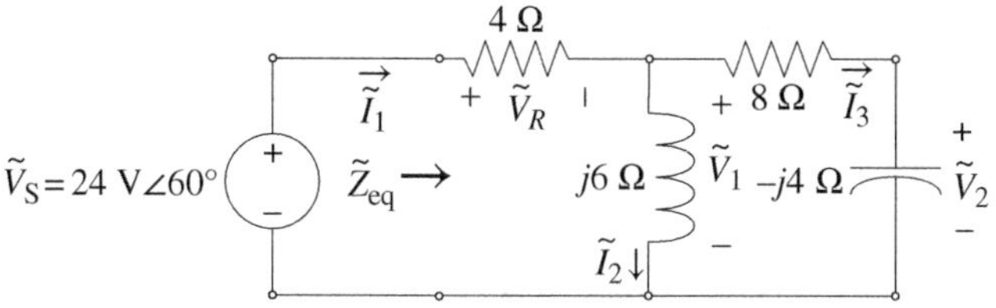

Figure E4.3.1 RLC circuit solved by applying Kirchhoff's voltage and current laws.

where subscript "N" stands for numerator and subscript "D" stands for denominator.

The phase angles are $\theta_N = \tan^{-1}\left(\frac{48}{24}\right) = 1.1072$ rad, $\theta_D = \tan^{-1}\left(\frac{2}{8}\right) = 0.2445$ rad. The unit "rad" stand for radians and relates to degrees (°) via the relation 360° corresponds to 2π rad or

$$1\ \text{rad} = 360°/2\pi, \tag{E4.3.2}$$

that is, $\theta_N = 1.1072 \cdot 360/2\pi = 63.44°$, and $\theta_D = 0.2445 \cdot 360/2\pi = 14.01°$.

Simplification results in

$$\begin{aligned}\widetilde{Z}_{eq} &= \{4 + 6.505\angle 0.8627\ \text{rad}\}\Omega = 4 + 6.505[\cos(0.8627) \\ &\quad + j\sin(0.8627)]\ \Omega = 4 + 4.231 + j4.94 = (8.231 + j4.941)\Omega \\ &= \sqrt{(8.231)^2 + (4.941)^2}\angle\theta = 9.60\ \Omega\angle 0.5406\ \text{rad},\end{aligned} \tag{E4.3.3a}$$

where the phase angle is

$$\theta = \tan^{-1}\left(\frac{4.941}{8.231}\right) = 0.5406\ \text{rad}. \tag{E4.3.3b}$$

b)

$$\begin{aligned}\widetilde{I}_1 &= \frac{\widetilde{V}_s}{\widetilde{Z}_{eq}} = \frac{240\ \text{V}\angle 60°}{9.6\ \Omega\angle(0.5406\ \text{rad})} = 25\ \text{A}\angle(1.0472\ \text{rad} - 0.5406\ \text{rad}) \\ &= 25\ \text{A}\angle(0.50657\ \text{rad or } 29.025°).\end{aligned} \tag{E4.3.4}$$

Application of KVL leads to

$$\widetilde{V}_s - 4\widetilde{I}_1 - \widetilde{V}_1 = 0 \tag{E4.3.5}$$

or

$$\begin{aligned}\widetilde{V}_1 &= \widetilde{V}_s - 4\widetilde{I}_1 = 240\ \text{V}\angle 60° - 4\ \Omega \cdot 25\ \text{A}\angle 0.506\,57\ \text{rad} = 240\ \text{V}[\cos 60° + j\sin 60°] \\ &\quad - 100\ \text{V}[\cos(0.506\,57\ \text{rad}) + j\sin(0.506\,57\ \text{rad})] \\ &= 120\ \text{V} + j207.8\ \text{V} - 87.441\ \text{V} - j48.518\ \text{V} = (32.559 + j159.30)\ \text{V}.\end{aligned} \tag{E4.3.6}$$

Note: One can either work with radians (rad) or degrees (°) or with both in the same equation. It is important not to mix up both concepts!

$$\widetilde{I}_2 = \frac{\widetilde{V}_1}{j6\ \Omega} = \frac{\sqrt{(32.56)^2 + (159.3)^2}\angle\theta_N}{6\ \Omega\angle 90°} = \frac{27.1\ A\angle 78.45°}{1\angle 90°} = 27.1\ \text{A}\angle -11.55°, \tag{E4.3.7a}$$

$$\theta_N = \tan^{-1}\left(\frac{159.3}{32.56}\right) = 1.369 \text{ rad} \equiv 78.45^\circ, \tag{E4.3.7b}$$

$$\widetilde{I}_3 = \frac{\widetilde{V}_1}{(8 - j4)\,\Omega} = \frac{\sqrt{(32.56)^2 + (159.3)^2}\angle\theta_N}{\sqrt{64 + 16}\angle\theta_D}, \tag{E4.3.8a}$$

where the angles in radians are

$$\theta_D = \tan^{-1}\left(\frac{-4}{8}\right) = -0.464 \text{ rad}, \tag{E4.3.8b}$$

$$\theta_N = \tan^{-1}\left(\frac{159.3}{32.56}\right) = 1.369 \text{ rad} \equiv 78.45^\circ, \tag{E4.3.8c}$$

$$\widetilde{I}_3 = \frac{162.5 \text{ V}\angle -0.464 \text{ rad}}{8.944\,\Omega\angle 1.368 \text{ rad}} = 18.18 \text{ A}\angle 1.833 \text{ rad}, \tag{E4.3.8d}$$

$$\widetilde{V}_2 = \widetilde{I}_3(-j4) = 18.18 \text{ A}\angle 1.833 \text{ rad} \cdot 4\,\Omega\angle -90^\circ = 72.72 \text{ V}\angle 15.03^\circ, \tag{E4.3.9}$$

$$\widetilde{V}_R = 4\,\Omega\widetilde{I}_1 = 100 \text{ V}\angle 29.025^\circ. \tag{E4.3.10}$$

Once the phasor voltages and currents are calculated, we can display them in a graphical fashion using a phasor diagram, as shown in Figure E4.3.2. In this diagram we simply plot the various phasors of interest, and in so doing we can immediately see the relationship among them.

It is important to realize that depending upon the location of a phasor within the complex plane (e.g. quadrants 1, 2, 3, and 4), the proper transformation of rectangular coordinates $\widetilde{z} = x + jy$ to polar coordinates $\widetilde{z} = |\widetilde{z}|e^{j\theta}$ is very important, as discussed in Appendix 4.A.

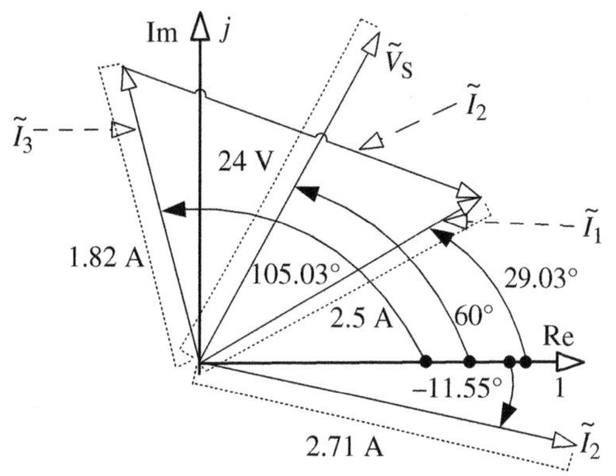

Figure E4.3.2 Phasor diagram in the complex plane of the above calculated voltage $\widetilde{V}_s$ and currents where $\widetilde{I}_1 = \widetilde{I}_2 + \widetilde{I}_3$.

4.6 Solution Using Nodal Analysis

Application Example 4.4 Solution of the currents and voltages in an RLC network for the forcing voltage source (s) $\tilde{V}_s$ = 120 V∠0° and the forcing current source $\tilde{I}_s$ = 20 A∠0° employing the nodal analysis

Figure E4.4.1 depicts the RLC network.

Calculate the node potentials/voltages $\tilde{V}_1$ and $\tilde{V}_2$ with Cramer's rule [6] (also called the method of determinants) and determine $\tilde{I}_o$.

Solution

Apply Kirchhoff's current law to node 1:

$$\tilde{I}_1 + \tilde{I}_3 + \tilde{I}_2 = 0 \tag{E4.4.1}$$

or

$$\frac{\tilde{V}_1 - 120\ V\angle 0^\circ}{R_1} + \frac{\tilde{V}_1}{j\Omega} + \frac{\tilde{V}_1 - \tilde{V}_1}{-j\Omega} = \tilde{V}_1 - 120\,\text{V}\angle 0^\circ - j\tilde{V}_1 + j\tilde{V}_1 - j\tilde{V}_2 = 0, \tag{E4.4.2}$$

or

$$\tilde{V}_1 - j\tilde{V}_2 = 120\,\text{V}\angle 0^\circ. \tag{E4.4.3}$$

Apply Kirchhoff's current law to node 2:

$$\tilde{I}_2 - \tilde{I}_o - \tilde{I}_s = 0 \tag{E4.4.4}$$

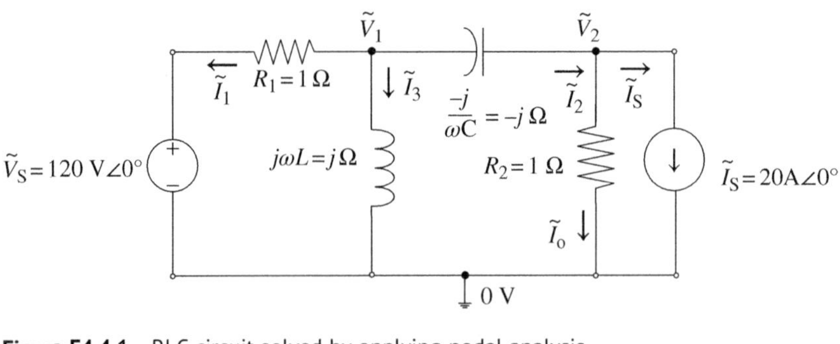

Figure E4.4.1 RLC circuit solved by applying nodal analysis.

or

$$\frac{\widetilde{V}_1 - \widetilde{V}_2}{-j\Omega} - \frac{\widetilde{V}_2}{1\,\Omega} - 20\,\text{A}\angle 0° = 0, \tag{E4.4.5}$$

or

$$j\widetilde{V}_1 - (1 + j)\widetilde{V}_2 = 20\,\text{V}\angle 0°. \tag{E4.4.6}$$

Solution of Eqs. (E4.4.3) and (E4.4.6) with Cramer's rule yields the determinant (D)

$$D = \begin{vmatrix} 1 & -j \\ j & -(1+j) \end{vmatrix} = -(1 + j) - 1 = -(2 + j); \tag{E4.4.7a}$$

the determinant instrumental for the calculation of $\widetilde{V}_1$ is

$$D_{V1} = \begin{vmatrix} 120 & -j \\ 20 & -(1+j) \end{vmatrix} = -120(1 + j) + j20 = -120 + 19j. \tag{E4.4.7b}$$

Therefore

$$\widetilde{V}_1 = \frac{D_{V1}}{D} = \frac{-120 + 19j}{-(2 + j)} = 54.332\,\text{V}\angle -35.562°. \tag{E4.4.7c}$$

The determinant for the calculation of $\widetilde{V}_2$ is

$$D_{V2} = \begin{vmatrix} 1 & 120 \\ j & 20 \end{vmatrix} = 20 - j120, \tag{E4.4.7d}$$

and thus

$$\widetilde{V}_2 = \frac{20 - 120j}{-(2 + j)} = 54.407\,\text{V}\angle -287.10°. \tag{E4.4.7e}$$

$$\check{I}_o = \frac{\widetilde{V}_2}{R} = \frac{54.407\,\text{V}\angle -287.10°}{1\,\Omega} = 54.407\,\text{A}\angle -287.10°. \tag{E4.4.7f}$$

4.7 Solution with Mesh and Loop Analysis by Applying Kirchhoff's and Ohm's Laws

Application Example 4.5 Solution of the currents and voltages in an RLC network for the forcing source (s) voltage $\tilde{V}_s = 120\,\text{V}\angle 0°$ and the forcing current source $\tilde{I}_s = 20\,\text{A}\angle 0°$ employing the mesh or loop analysis

Figure E4.5.1 depicts the RLC network. This example is the same as that of Application Example 4.4, but with a different solution method.

Solution

Calculate the current $\tilde{I}_s$ using Cramer's rule [6] (or method of determinants).

Mesh 3 yields

$$\tilde{I}_3 = 20\,\text{A}\angle 0°. \tag{E4.5.1}$$

Apply Kirchhoff's law to mesh 1:

$$120\,\text{V}\angle 0° - \tilde{I}_1 \cdot 1\,\Omega - \left(\tilde{I}_1 - \tilde{I}_2\right) j\,\Omega = 0, \tag{E4.5.2}$$

or

$$(1 + j)\,\Omega \cdot \tilde{I}_1 - j\,\Omega \cdot \tilde{I}_2 = 120\,\text{V}\angle 0°. \tag{E4.5.3}$$

Apply Kirchhoff's law to mesh 2:

$$-\left(\tilde{I}_1 - \tilde{I}_2\right) j\,\Omega + j\,\Omega \tilde{I}_2 - \left(\tilde{I}_2 - 20\,\text{A}\angle 0°\right) \cdot 1\,\Omega = 0, \tag{E4.5.4}$$

or

$$-j\,\Omega \cdot \tilde{I}_1 + 1\,\Omega \cdot \tilde{I}_2 = 20\,\text{A}\angle 0°. \tag{E4.5.5}$$

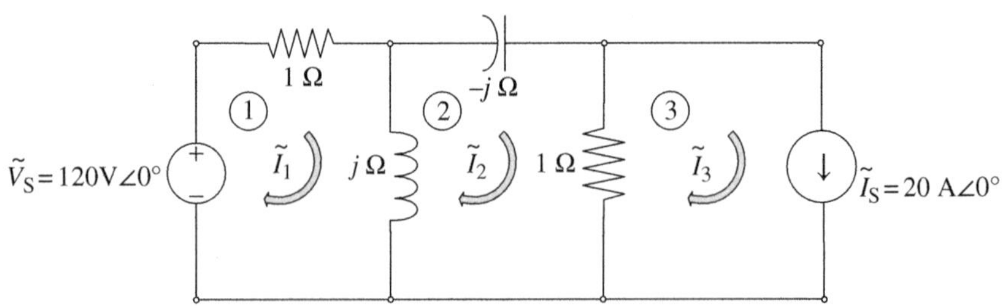

Figure E4.5.1 RLC circuit solved based on mesh/loop analysis.

Solution of Eqs. (E4.5.3) and (E4.5.5) with Cramer's rule yields the determinant (D)

$$D = \begin{vmatrix} (1+j) & -j \\ -j & 1 \end{vmatrix} = (1+j) + 1 = 2 + j; \tag{E4.5.6a}$$

the determinant instrumental for the calculation of $\widetilde{I}_1$ is

$$D_{I1} = \begin{vmatrix} 120 & -j \\ 20 & 1 \end{vmatrix} = 120 + j20; \tag{E4.5.6b}$$

therefore

$$\widetilde{I}_1 = \frac{D_{I1}}{D} = \frac{120 + j20}{(2+j)} = 54.405\ \text{A}\angle -17.103°. \tag{E4.5.6c}$$

The determinant for the calculation of $\widetilde{I}_2$ is

$$D_{I2} = \begin{vmatrix} (1+j) & 120 \\ -j & 20 \end{vmatrix} = 20(1+j) + j120 = 20 + 140j, \tag{E4.5.6d}$$

and thus

$$\widetilde{I}_2 = \frac{20 + 140j}{(2+j)} = 63.245\ \ \text{A}\angle 55.305°, \tag{E4.5.6e}$$

$$\begin{aligned} \widetilde{I}_o = \widetilde{I}_2 - \widetilde{I}_3 &= 63.245\ \ \text{A}\angle 55.305° - 20\ \text{A}\angle 0° = 63.245[\cos(55.305) \\ &+ j\sin(55.305)] - 20 = 63.245(0.569\,21 + j0.822\,19) - 20 = 36 \\ &+ j51.999 - 20 = 16 + j51.999 = 54.405\ \text{A}\angle 72.897°. \end{aligned} \tag{E4.5.6f}$$

4.8 Solution Based on Superposition

Application Example 4.6 Solution of the currents and voltages in an RLC network for the forcing voltage source (s) $\widetilde{V}_s = 120\,\text{V}\angle 0°$ and the forcing current source $\widetilde{I}_s = 20\,\text{A}\angle 0°$ employing the principle of superposition

Figure E4.6.1 depicts the RLC network. This example is the same as that of Application Examples 4.4 and 4.5, but with a different solution method.

a) Determine the input impedance $\widetilde{Z}_{\text{total\_v}}$, if the current source is zero.
b) Calculate the input current $\widetilde{I}_v$ and branch current $\widetilde{I}'_o$ when the current source is an open circuit.

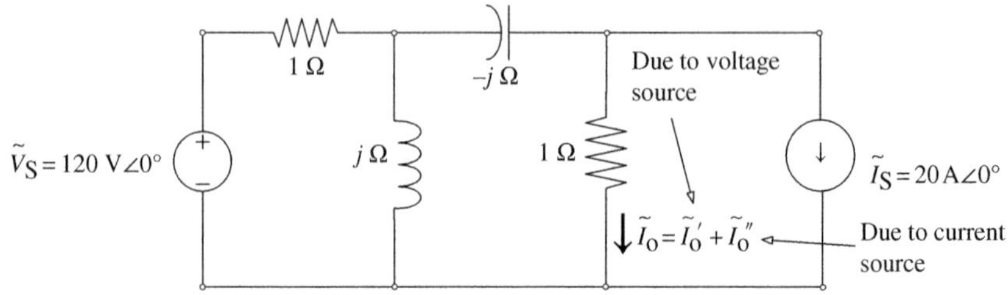

Figure E4.6.1 RLC network solved with the principle of superposition.

c) Determine the input impedance $\tilde{Z}_{\text{total\_i}}$, if the voltage source is zero.

d) Calculate the branch current $\tilde{I}_o''$ when the voltage source is a short circuit.

e) Use superposition to calculate $\tilde{I}_o = \tilde{I}_o' + \tilde{I}_o''$.

Solution

a) The input impedance is

$$\tilde{Z}_{\text{total\_v}} = 1 + \frac{j(1-j)}{j+1-j} = 1 + \frac{1+j}{1} = (2+j)\ \Omega. \quad \text{(E4.6.1)}$$

b) The input current is based on Figure E4.6.2 where the current source $\tilde{I}_s$ is an open circuit:

$$\tilde{I}_v = \frac{\tilde{V}_s}{\tilde{Z}_{\text{total\_v}}} = \frac{120\ \text{V}\angle 0°}{(2+j)\Omega} = \frac{120}{(2+j)}\ \text{A}. \quad \text{(E4.6.2)}$$

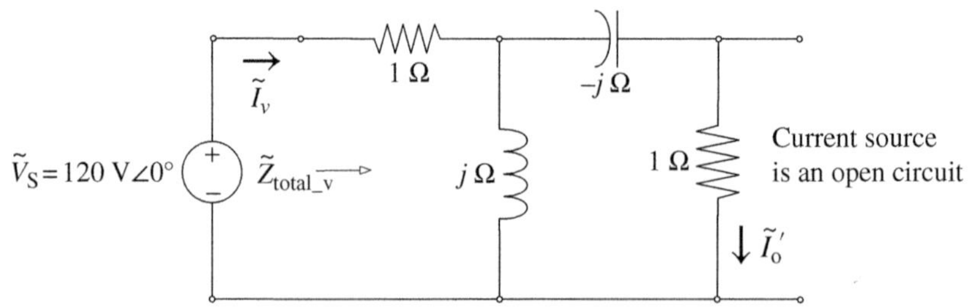

Figure E4.6.2 RLC circuit without current source $\tilde{I}_s$ (open circuit).

With the application of current division, one obtains

$$\tilde{I}'_o = \frac{j}{(1-j+j)}\tilde{J}_v = j\frac{120\angle 0°}{(2+j)} = \frac{120\,\mathrm{V}\,j}{(2+j)} = \frac{120j(2-j)}{5} = \frac{120}{5}(1+2j)$$

$$= \frac{(120+240j)}{5}\,\mathrm{A}. \tag{E4.6.3}$$

c) Calculation of the current caused by the current source by setting the voltage source to zero (short circuit), as depicted in Figure E4.6.3.
The input impedance is

$$\tilde{Z}_{\text{total\_i}} = \left(-j + \frac{j}{1+j}\right)\Omega = \frac{1-j}{2}\Omega. \tag{E4.6.4}$$

d) With the application of current division, one obtains

$$\tilde{I}''_o = -\frac{\frac{1-j}{2}}{\frac{1-j}{2}+1}20 = -\frac{(1-j)}{(1-j+2)}20 = -\frac{(1-j)}{(3-j)}20\,\mathrm{A}. \tag{E4.6.5}$$

e) Superposition

$$\tilde{I}_o = \tilde{I}'_o + \tilde{I}''_o = \frac{120+j240}{5} - \frac{1-j}{3-j}20 = \frac{120}{5} + j\frac{240}{5} - \frac{(1-j)(3+j)}{10}20, \tag{E4.6.6}$$

$$\tilde{I}_o = \frac{120}{5} + j\frac{240}{5} - \frac{(60+20j-j60+20)}{10} = \frac{120}{5} + j\frac{240}{5} - \frac{80}{10} + j\frac{40}{10}$$

$$= \left(\frac{80}{5} + j\frac{260}{5}\right)\mathrm{A} = (16+j52)\,\mathrm{A} = 54.406\,\mathrm{A}\angle 72.897°. \tag{E4.6.7}$$

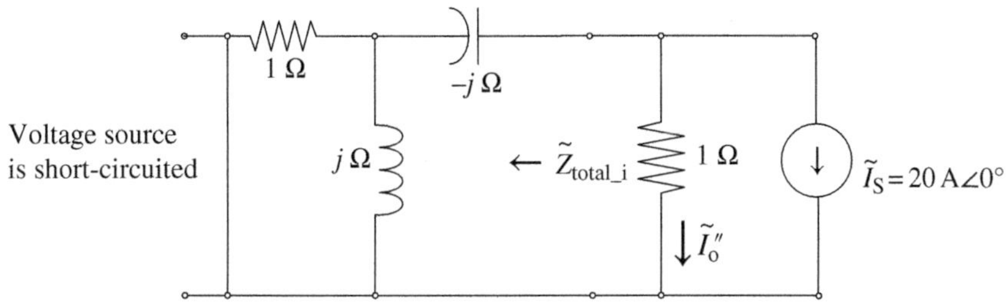

Figure E4.6.3 RLC circuit without voltage source $\tilde{V}_s$ (short circuit).

4.9 Solution with Source Transformation/Exchange

Application Example 4.7 Solution of the currents and voltages in an RLC network of Figure E4.7.1 for the forcing source (s) voltage $\tilde{V}_s$ = 120 V∠0° and the forcing current source $\tilde{I}_s$ = 20 A∠0° employing the source transformation/exchange

This example is the same as that of Application Examples 4.4–4.6, but with a different solution method.

Solution

We begin the network simplification by transforming the $\tilde{V}_s = 120\,\text{V}\angle 0°$ voltage source into a

$$\tilde{I}_1 = \frac{\tilde{V}_s}{1\,\Omega} = 120\,\text{A}\angle 0° \tag{E4.7.1}$$

current source in parallel with a 1 Ω resistor as shown in Figure E4.7.2.

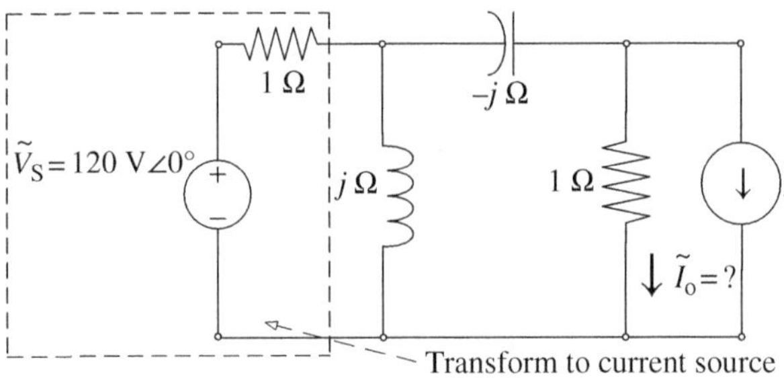

Figure E4.7.1 Original RLC network.

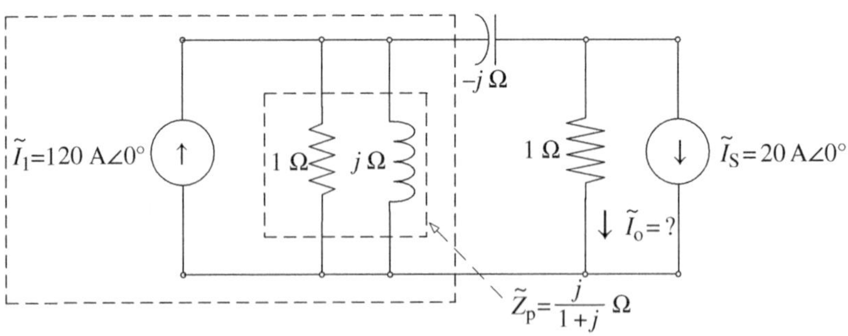

Figure E4.7.2 First transformation of RLC network.

The current source $\widetilde{I}_1 = \frac{\widetilde{V}_s}{1\,\Omega} = 120\ \text{A}\angle 0°$ in parallel with the 1 Ω resistor and the $j\ \Omega$ reactance (see Figure E4.7.2) is now transformed into an equivalent voltage source $\widetilde{V}_1$ by multiplying the parallel (p) impedance

$$\widetilde{Z}_\text{p} = \frac{j}{(1+j)} = \left(\frac{1}{2} + \frac{j}{2}\right)\Omega \tag{E4.7.2}$$

with the current $\widetilde{I}_1 = \frac{\widetilde{V}_s}{1\,\Omega} = 120\ \text{A}\angle 0°$

$$\widetilde{V}_1 = \widetilde{I}_1\widetilde{Z}_\text{p} = 120\angle 0°\frac{j}{1+j} = 120j\frac{(1-j)}{2} = 60j(1-j) = (60 + j60)\ \text{V}. \tag{E4.7.3}$$

Now we transform the network as shown in Figure E4.7.3 a second time, where the voltage source $\widetilde{V}_1$ is transformed back to a current source $\widetilde{I}_2$ as follows.

The series (s) impedance is

$$\widetilde{Z}_\text{s} = \frac{1}{2} + \frac{j}{2} - j = \left(\frac{1}{2} - \frac{j}{2}\right)\Omega, \tag{E4.7.4}$$

resulting in the equivalent circuit of Figure E4.7.4 with the current source $\widetilde{I}_2$ and the parallel impedance $\widetilde{Z}_\text{s}$.

The current source is

$$\widetilde{I}_2 = \frac{\widetilde{V}_1}{\widetilde{Z}_\text{s}} = \frac{60 + j60}{\frac{1}{2} - \frac{j}{2}} = \frac{120 + 120j}{1-j} = \frac{(120 + 120j)(1+j)}{2}$$
$$= (60 + 60j)(1+j) = j120\ \text{A}. \tag{E4.7.5}$$

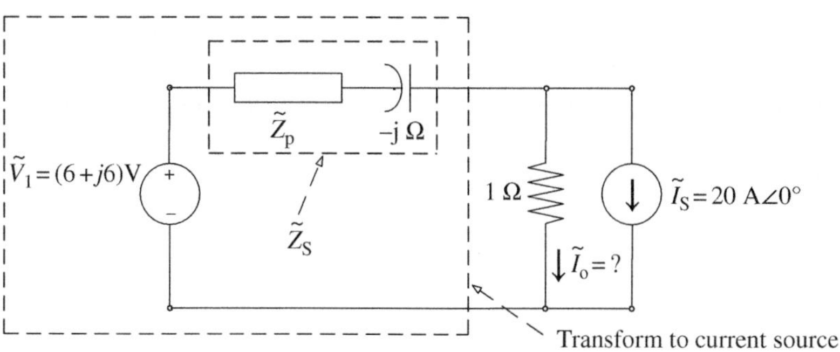

Figure E4.7.3 Second transformation of RLC network.

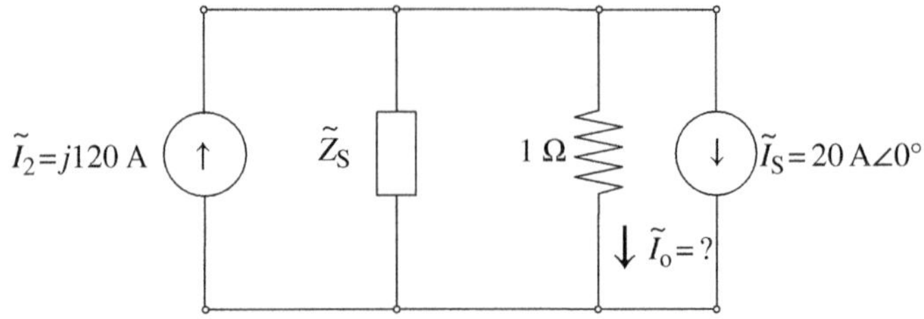

Figure E4.7.4 Final transformed RLC network.

Note that the two current sources are in parallel, resulting in the total current where the current $\tilde{I}_2$ is counted as positive and $\tilde{I}_s$ is negative, leading to the total current of

$$\tilde{I}_{\text{total}} = \left(\tilde{I}_2 - \tilde{I}_s\right) = (j120 - 20\angle 0°) = (120j - 20)\text{A}. \tag{E4.7.6}$$

Applying the current division rule yields the desired current through the 1 Ω resistor in parallel to the $\tilde{I}_s$ current source (see Figure E4.7.4):

$$\tilde{I}_o = \frac{\tilde{Z}_S}{1 + \tilde{Z}_S}\tilde{Z}_{\text{total}} = \frac{\frac{1-j}{2}}{1 + \frac{1-j}{2}}(120j - 20)\text{A} = \left(\frac{80}{5} + j\frac{260}{5}\right)\text{A}$$
$$= (16 + 52j)\text{A} = 54.406\ \text{A}\angle - 287.1°. \tag{E4.7.7}$$

4.10 Solutions Employing Thévenin's and Norton's Theorems and Source Transformations

Application Example 4.8 Solution of the currents and voltages in an RLC network for the forcing source (s) voltage $\tilde{V}_s = 120\,\text{V}\angle 0°$ and forcing source current $\tilde{I}_s = 20\,\text{A}\angle 0°$ employing Thévenin's and Norton's theorems in the phasor domain

This example is the same as that of Application Examples 4.4–4.7, but with a different solution method.

Solution

The given circuit is shown in Figure E4.8.1, and the goal of the simplification is the reduced circuit represented in Figure E4.8.2.

For finding the open-circuit voltage $\tilde{V}_{oc}$ across terminals A, B, rely on Figures E4.8.3 and E4.8.4.

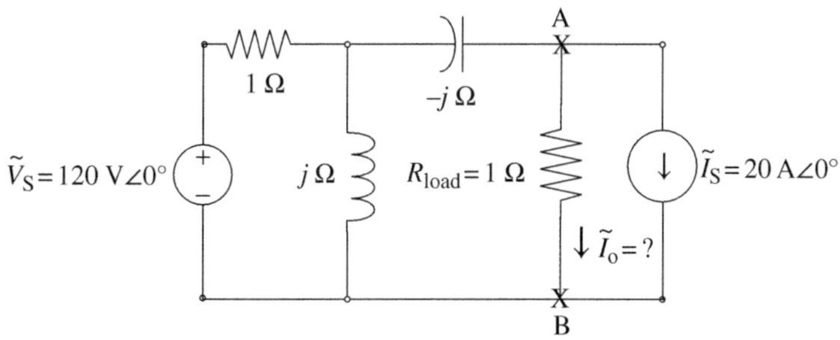

Figure E4.8.1 Original RLC circuit.

Figure E4.8.2 Desired reduced RLC circuit by applying Thévenin's theorem.

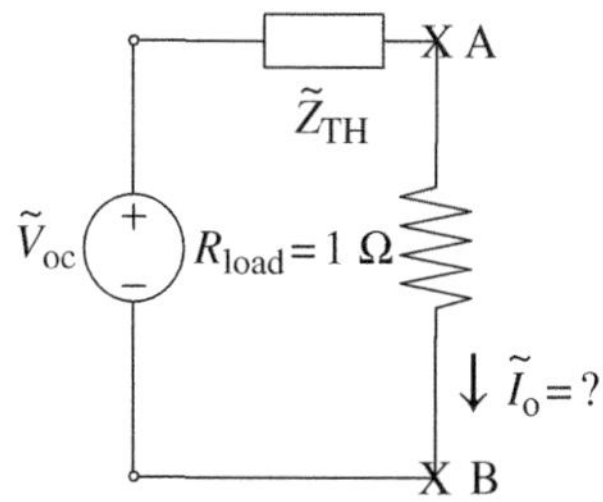

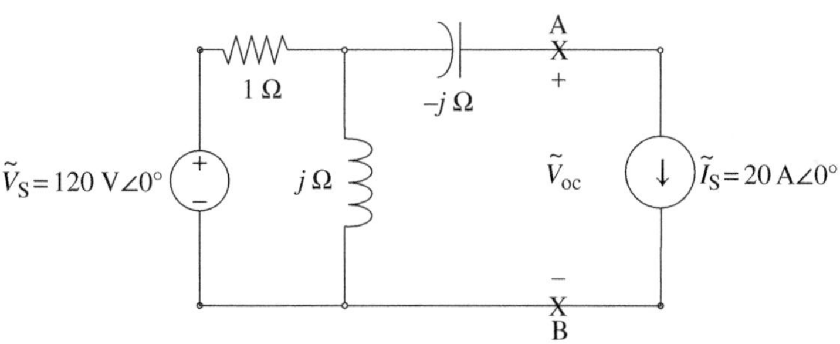

Figure E4.8.3 Circuit for finding the open-circuit voltage $\tilde{V}_{oc}$.

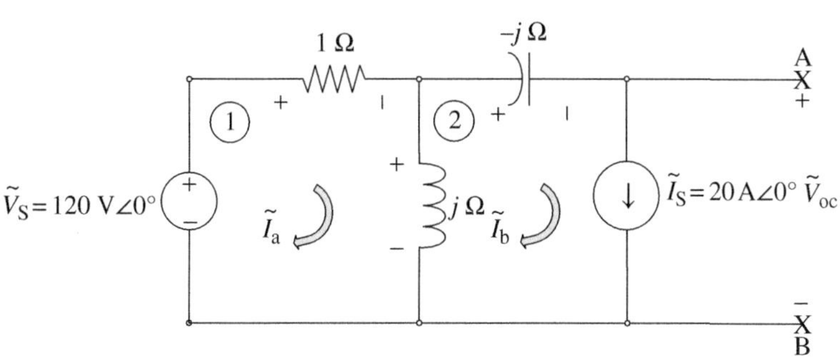

Figure E4.8.4 Redrawn circuit for finding the open-circuit voltage $\tilde{V}_{oc}$.

Note that in Figure E4.8.4

$$\widetilde{I}_b = 20\ \mathrm{A}\angle 0^\circ. \tag{E4.8.1}$$

For mesh/loop 1 one can write Kirchhoff's voltage law:

$$120\ \mathrm{V}\angle 0^\circ - \widetilde{I}_a 1\ \Omega - \left(\widetilde{I}_a - \widetilde{I}_b\right) j\ \Omega = 0. \tag{E4.8.2}$$

For mesh/loop 2 one can write Kirchhoff's voltage law:

$$-\left(\widetilde{I}_b - \widetilde{I}_a\right) j\ \Omega + \widetilde{I}_b j\ \Omega - \widetilde{V}_{oc} = 0. \tag{E4.8.3}$$

From Eq. (E4.8.2) follows

$$120\ \mathrm{V} - (1+j)\Omega \widetilde{I}_a + j\ \Omega \widetilde{I}_b = 0, \tag{E4.8.4}$$

or

$$\widetilde{I}_a = \frac{j\ \Omega \widetilde{I}_b + 120\ \mathrm{V}}{(1+j)\Omega} = \frac{120\ \mathrm{V} + 20\ \mathrm{A} j\ \Omega}{(1+j)\ \Omega}. \tag{E4.8.5}$$

From Eq. (E4.8.3) follows

$$\widetilde{V}_{oc} = j\ \Omega \widetilde{I}_b - \left(\widetilde{I}_b - \widetilde{I}_a\right) j\ \Omega = \widetilde{I}_a j\ \Omega; \tag{E4.8.6}$$

introducing Eq. (E4.8.5) into (E4.8.6) yields

$$\widetilde{V}_{oc} = \frac{120\ \mathrm{V} + 20\ \mathrm{A} j\ \Omega}{(1+j)\Omega} j\Omega = \frac{120\ \mathrm{V} j - 20\ \mathrm{A}\Omega}{(1+j)\Omega} = \frac{(120\ \mathrm{V} j - 20\ \mathrm{A}\Omega)(1-j)\Omega}{2\ \Omega}$$
$$= (50 + 70j)\mathrm{V}. \tag{E4.8.7}$$

The Thévenin (TH) impedance is found from Figure E4.8.5:

$$\widetilde{Z}_{TH} = -j\ \Omega + \frac{j\ \Omega}{(1+j)} = -j\ \Omega + \frac{j\ \Omega(1-j)}{2} = \left(\frac{1}{2} - \frac{j}{2}\right)\Omega. \tag{E4.8.8}$$

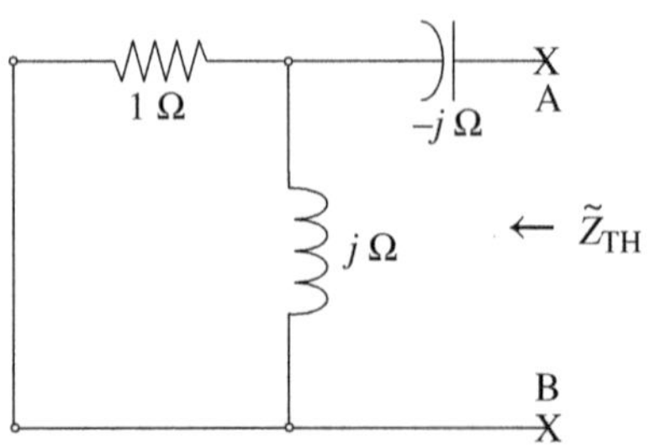

Figure E4.8.5 Definition of Thévenin impedance $\widetilde{Z}_{TH}$.

The current $\tilde{I}_o$ is now with $R_{load} = 1\,\Omega$ as indicated in the original circuit of Figure E4.8.1:

$$\tilde{I}_o = \frac{\tilde{V}_{oc}}{\tilde{Z}_{TH} + R_{load}} = \frac{(50 + 70j)\text{V}}{(\frac{1}{2} - \frac{j}{2})\Omega + 1\,\Omega} = \frac{(100 + 140j)\text{V}}{(3-j)\Omega} = (16 + 52j)\text{A}$$
$$= 54.406\ \text{A}\angle - 287.1°. \quad \text{(E4.8.9)}$$

Application Example 4.9 Solution of the currents and voltages in an RLC network for the forcing source (s) voltage $\tilde{V}_s = 120\ \text{V}\angle 0°$ and the forcing source current $\tilde{I}_s = 20\ \text{A}\angle 0°$, employing Thévenin's and Norton's theorems in the phasor domain and source transformations

This example is the same as that of Application Examples 4.4–4.8, but with a different solution method.

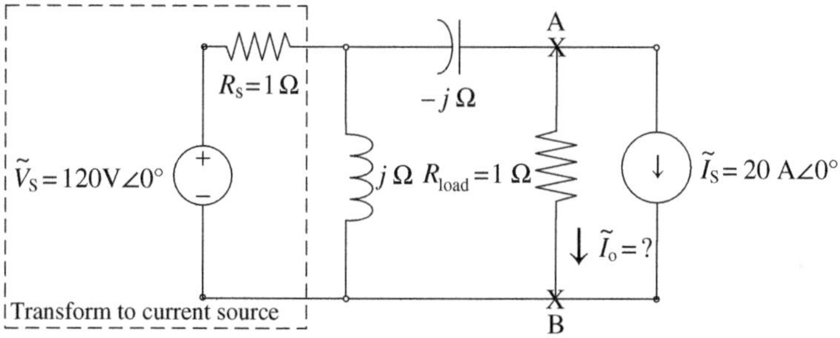

Figure E4.9.1 Original RLC circuit, where the voltage source $\tilde{V}_s = 120\ \text{V}\angle 0°$ is transformed to a current source resulting in Figure E4.9.2 with current $\tilde{I}_{sc1} = \frac{\tilde{V}_s}{R_s} = 120\ \text{A}\angle 0°$.

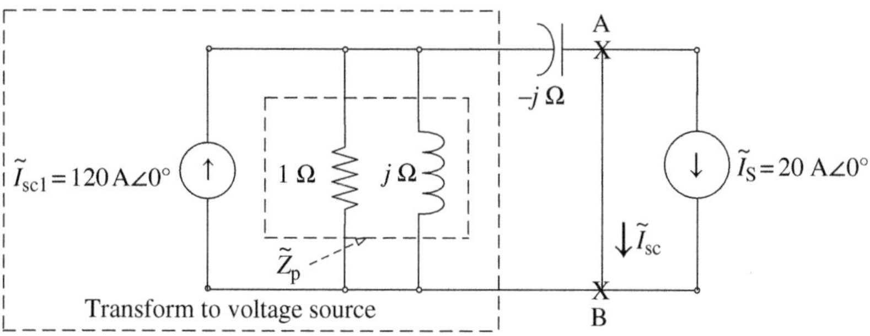

Figure E4.9.2 Circuit with the short-circuit current $\tilde{I}_{sc1}$ and parallel impedance $\tilde{Z}_p = \frac{j}{(1+j)}\ \Omega$.

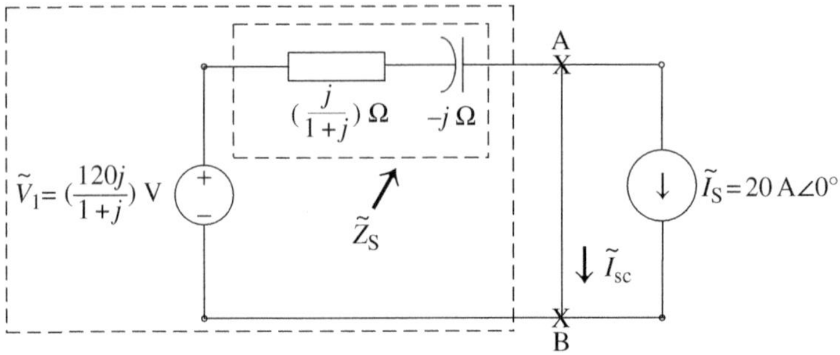

Figure E4.9.3 Circuit where the short-circuit current $\tilde{I}_{sc1}$ is replaced by voltage source $\tilde{V}_1 = \tilde{I}_{sc1}\tilde{Z}_p = \frac{120j}{(1+j)}$ V and series impedance $\tilde{Z}_s = \left[\frac{j}{(1+j)} - j\right]\Omega = \frac{1}{(1+j)}\Omega$, resulting in Figure E4.9.4.

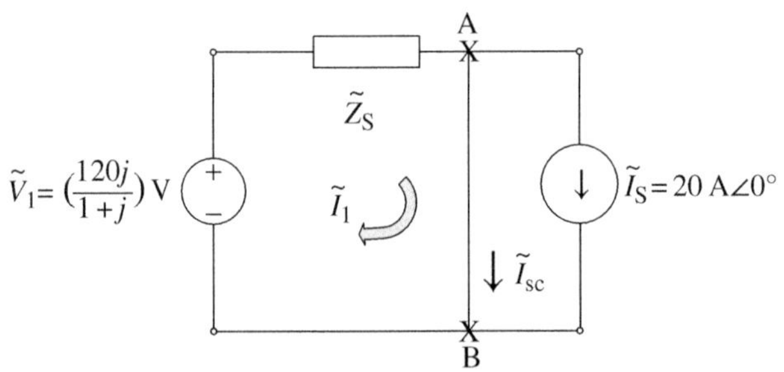

Figure E4.9.4 Reduced circuit.

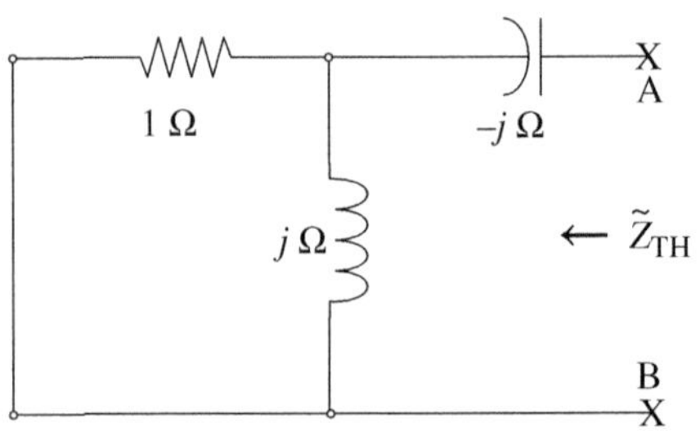

Figure E4.9.5 Definition of Thévenin impedance.

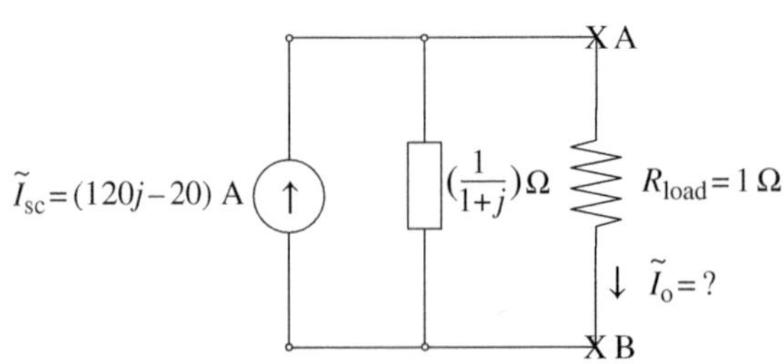

Figure E4.9.6 Circuit with current source in Eq. (E4.9.1) and $\tilde{I}_o$ defined in the original circuit in Figure E4.9.1.

Solution

The given circuit is shown in Figure E4.9.1 (identical with that of Figure E4.8.1), and the goal of the simplification is the reduced circuits represented in Figures E4.9.2–E4.9.6.

Note that in Figure E4.9.4

$$\widetilde{I}_{sc} = \widetilde{I}_1 - 20\angle 0°\text{A} = \frac{\widetilde{V}_1}{\widetilde{Z}_S} - 20\angle 0°\text{A} = \frac{\frac{120j}{(1+j)}}{\frac{1}{(1+j)}}\text{A} - 20\angle 0°\text{A} = (120j - 20)\text{A}. \tag{E4.9.1}$$

The Thévenin impedance is defined in Figure E4.9.5 and is

$$\widetilde{Z}_{TH} = -j\,\Omega + \frac{j}{(1+j)}\Omega = \frac{1}{(1+j)}\Omega. \tag{E4.9.2}$$

The final circuit for calculating the current $\widetilde{I}_0$ is given in Figure E4.9.6 whereby current division is applied.

$$\widetilde{I}_o = \frac{\frac{1}{(1+j)}}{1 + \frac{1}{(1+j)}}(120j - 20)\text{A} = \frac{(120j-20)}{(1+j+1)}\text{A} = \frac{(120j-20)}{(2+j)}$$
$$= \frac{(120j-20)(2-j)}{5} = \left(\frac{260j}{5} + \frac{80}{5}\right)\text{A} = (16 + 52j)\text{A} = 54.406\ \text{A}\angle -287.1°. \tag{E4.9.3}$$

4.11 Nonsinusoidal Steady-State Response

So far the voltages and currents we have considered were either (unipolar) DC or (sinusoids) AC with a specific frequency f. However, in electrical engineering, we encounter a variety of periodic signals, many of which are nonsinusoidal; one example is the human heart beat pulses as mentioned in Chapter 1.

Periodic functions obey the relationship

$$f(t) = f(t + nT), \tag{4.36}$$

where $n = \pm 1, = \pm 2, = \pm 3, \ldots$

Original or Rectangular/Trigonometric Form of Fourier Series

According to Fourier [5, 7] (1768–1830), we can decompose a nonsinusoidal, periodic signal/function $f(t)$ (e.g. heart beat) in an infinite series (sum) of signals with integer radial frequencies $n\omega_0$:

$$f(t) = a_0 + \sum_{n=1}^{\infty}(a_n \cos n\omega_0 t + b_n \sin n\omega_0 t), \tag{4.37}$$

where

$\omega_0 = 2\pi f = 2\pi/T$ with f the frequency and T the period of the fundamental ($n = 1$) sinusoid, a_0 is the average of the waveform over one period T.

The coefficients of Eq. (4.37) are defined as follows:

$$a_0 = \frac{1}{T}\int_{t_0}^{t_0+T} f(t)\,\mathrm{d}t, \tag{4.38}$$

$$a_\mathrm{k} = \frac{2}{T}\int_{t_0}^{t_0+T} f(t)\cos k\omega_0 t\,\mathrm{d}t, \tag{4.39}$$

$$b_\mathrm{k} = \frac{2}{T}\int_{t_0}^{t_0+T} f(t)\sin k\omega_0 t\,\mathrm{d}t. \tag{4.40}$$

In Eqs. (4.39) and (4.40) the subscript k is used to indicate the kth coefficient in the integer sequence 1, 2, 3, ….

An alternative trigonometric form of Fourier series is

$$f(t) = a_0 + \sum_{n=1}^{\infty} D_n \cos(n\omega_0 t + \theta_n). \tag{4.41}$$

Exponential Form of Fourier Series

Since the function $\cos(n\omega_0 t + \theta_n)$ can be written in exponential form using Euler's identity [1]

$$\mathrm{e}^{jx} = \cos x + j\sin x, \tag{4.42}$$

the periodic function $f(t)$ can be written as

$$f(t) = a_0 + \sum_{\substack{n=-\infty \\ n\neq 0}}^{\infty} \tilde{C}_n \mathrm{e}^{jn\omega_0 t}, \tag{4.43}$$

or

$$f(t) = \sum_{n=-\infty}^{\infty} \tilde{C}_n \mathrm{e}^{jn\omega_0 t}, \tag{4.44}$$

where the coefficient in this equation is

$$\tilde{C}_n = \frac{1}{T}\int_{t_0}^{t_0+T} f(t)\,\mathrm{e}^{-jn\omega_0 t}\mathrm{d}t. \tag{4.45}$$

Symmetry conditions that result in a simplification of Fourier analysis are discussed in Nilsson [5].

Application Example 4.10 Rectangular or trigonometric form of Fourier series development without applying certain symmetry conditions, representing the general case

Find the Fourier series of the sawtooth function of Figure E4.10.1.

The sawtooth function is frequently applied in electrical engineering apparatuses such as oscilloscopes, where a periodic function must be repetitively displayed.

Solution

The analytical expression for the function from the origin (0/0) to the point (T/V_m) is according to $y = mx$, where m is the slope $m = V_m/T$; thus one obtains for the voltage

$$v(t) = (V_m/T)t. \tag{E4.10.1}$$

Using Eq. (4.38) one can write

$$a_0 = \frac{1}{T}\int_0^T \left(\frac{V_m}{T}\right) t\,dt = \left(\frac{1}{2}\right)V_m. \tag{E4.10.2}$$

With Eq. (4.39) follows

$$a_k = \frac{2}{T}\int_0^T \left(\frac{V_m}{T}\right) t \cos k\omega_0\,dt = \frac{2V_m}{T^2}\left(\frac{1}{k^2\omega_0^2}\cos k\omega_0 t + \frac{t}{k\omega_0}\sin k\omega_0 t\Big|_0^T\right) \tag{E4.10.3}$$

or

$$a_k = \frac{2V_m}{T^2}\left[\frac{1}{k^2\omega_0^2}(\cos 2\pi k - 1)\right] = 0, \quad \text{for all } k. \tag{E4.10.4}$$

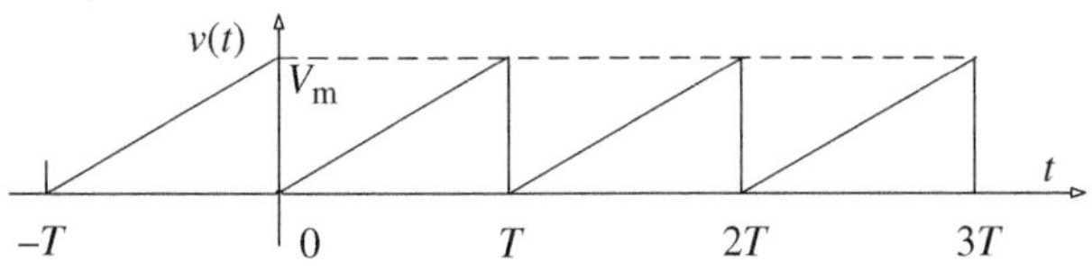

Figure E4.10.1 Asymmetric sawtooth function.

With Eq. (4.40) follows

$$b_k = \frac{2}{T}\int_0^T \left(\frac{V_\mathrm{m}}{T}\right) t \sin k\omega_0 \,\mathrm{d}t = \frac{2V_\mathrm{m}}{T^2}\left(\frac{1}{k^2\omega_0^2}\sin k\omega_0 t + \frac{t}{k\omega_0}\cos k\omega_0 t\Big|_0^T\right) \tag{E4.10.5}$$

or

$$b_k = \frac{2V_\mathrm{m}}{T^2}\left(0 - \frac{T}{k\omega_0}\cos 2\pi k\right) = \frac{-V_\mathrm{m}}{\pi k}. \tag{E4.10.6}$$

The Fourier series for $v(t)$ is

$$v(t) = \frac{V_\mathrm{m}}{2} - \frac{V_\mathrm{m}}{\pi}\sum_{n=1}^{\infty}\frac{1}{n}\sin n\omega_0 t = \frac{V_\mathrm{m}}{2} - \frac{V_\mathrm{m}}{\pi}\sin \omega_0 t - \frac{V_\mathrm{m}}{2\pi}\sin 2\omega_0 t - \frac{V_\mathrm{m}}{3\pi}\sin 3\omega_0 t.... \tag{E4.10.7}$$

Application Example 4.11 Exponential form of Fourier series development without applying certain symmetry conditions, representing the general case

Find the Fourier series of the pulse function of Figure E4.11.1.

The pulse function is frequently applied in electrical engineering apparatuses such as computers, where a periodic pulse functions are relied on for number manipulations such as addition and subtraction.

Solution

For the function shown in Figure E4.11.1, one can write with Eq. (4.45)

$$\tilde{C}_n = \frac{1}{T}\int_{-\Delta T}^{\Delta T} V_\mathrm{m}\mathrm{e}^{-jn\omega_0 t}\mathrm{d}t = \frac{V_\mathrm{m}}{T}\left(\frac{\mathrm{e}^{-jn\omega_0 t}}{-jn\omega_0}\right)\Bigg|_{-\Delta T}^{\Delta T} = \frac{jV_\mathrm{m}}{n\omega_0 T}\left(\mathrm{e}^{-jn\omega_0\Delta T} - \mathrm{e}^{jn\omega_0\Delta T}\right)$$

$$= \frac{2V_\mathrm{m}}{n\omega_0 T}\sin n\omega_0\Delta T = \frac{2V_\mathrm{m}\Delta T}{T}\left(\frac{\sin(n\omega_0\Delta T)}{n\omega_0\Delta T}\right). \tag{E4.11.1}$$

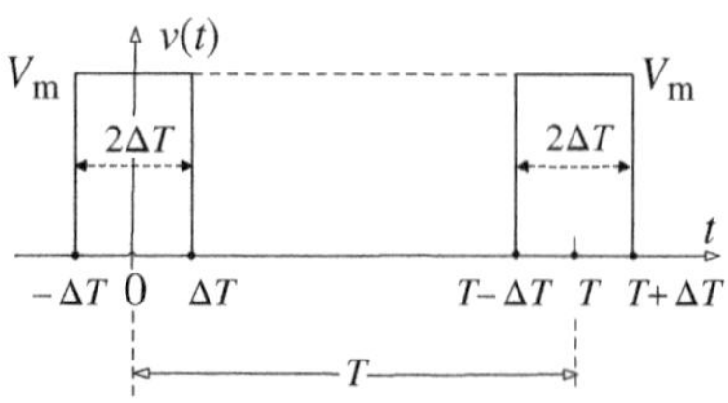

Figure E4.11.1 Pulse function.

The exponential Fourier series is

$$v(t) = \sum_{n=-\infty}^{n=\infty} \left(\frac{2V_{\mathrm{m}}\Delta T}{T}\right) \frac{\sin(n\omega_0\Delta T)}{n\omega_0\Delta T} \mathrm{e}^{jn\omega_0 t} = \left(\frac{2V_{\mathrm{m}}\Delta T}{T}\right) \sum_{n=-\infty}^{n=\infty} \frac{\sin(n\omega_0\Delta T)}{n\omega_0\Delta T} \mathrm{e}^{jn\omega_0 t}. \tag{E4.11.2}$$

4.12 Summary

This chapter presents the foundations for steady-state AC analysis introducing phasors based on complex number theory and represents a prerequisite for the design of power-related circuits for single- and three-phase electric apparatuses such as transformers, electric machines, and distribution power systems.

After introducing sinusoids (e.g. sinusoidal and cosinusoidal functions), the concept of a phasor is based on complex number relations. Indeed, a phasor represents a simplified complex number by ignoring the term $\mathrm{e}^{j\omega t}$. Phasor relations for resistor, inductor, and capacitor and the concept of impedance as well as admittance are specified. Also presented are Δ-Y transformation, solutions based on Kirchhoff's laws, nodal analysis, mesh/loop analysis, superposition, source transformation/exchange, Thévenin's (principle of constant voltage source) and Norton's (principle of constant current source) theorems, and steady-state power analysis.

Various power definitions are reviewed, such as instantaneous power $p(t)$, average or real power P, imaginary power Q, apparent power S, and complex power $\tilde{S}$. Amplitude excursions of voltages (V_{M}) and current (I_{M}) are defined, as well as root mean square (rms) error and effective (eff) values of signals.

Lastly, nonsinusoidal steady-state responses are treated with Fourier analysis, with its trigonometric and exponential forms.

Problems

4.1 Complex numbers: conversion from rectangular to polar form and vice versa.

1 Given three complex numbers: $\tilde{z} = 6 - j2$, $\tilde{w} = 7\angle 15°$, and $\tilde{u} = 10^{\,j2.2\ \mathrm{rad}}$. Note 1 radian ≡ 57.2958° or 1° ≡ π/180 radians = 0.017 45 rad.

a Find $|\tilde{z}|$.

b Find $\tilde{u}^*$ in rectangular form, where $*$ denotes the complex conjugate.

c Evaluate $\frac{\tilde{z}}{(\tilde{w}-\tilde{u})}$ and place the results in polar form.

d Solve for $\tilde{s}$ if $\tilde{z}(\tilde{s} - \tilde{w}) = \tilde{u}$ and express $\tilde{s}$ in polar form.

2 Evaluate the following expressions and put your answer in polar form.

a $(1.4 + j6) \cdot 4\angle 18^\circ + 6\angle 12^\circ$,

b $\dfrac{0.2 - j0.5}{13 + j7} \cdot \dfrac{1}{10\angle 66^\circ} - 10^{-3}$,

c $\dfrac{1}{\dfrac{1}{2 + j2} + \dfrac{1}{3\angle 40^\circ}}$.

3 What is the rms phasor for $v(t) = 5.2$ V$(100t - 38^\circ)$?

4 What time function is represented by the phasor $\widetilde{I} = (6 + j9)$A, if the frequency is $f = 400$ Hz?

5 Determine the frequency of the following two currents and the phase angle between them:

$$i_1(t) = 4\cos(754t - 60^\circ)\text{A}, i_2(t) = 3\cos(754t + 20^\circ)\text{A}.$$

6 In the circuit of Figure P4.1.1, $i_s(t) = 2\cos(377t + 60^\circ)$A find $\widetilde{V}_s$, where "s" stands for source.

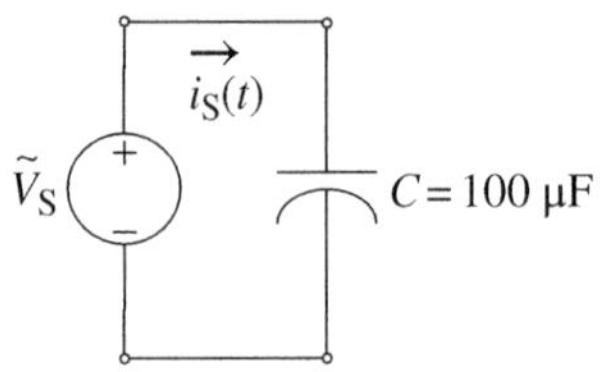

Figure P4.1.1 Capacitor supplied by current $i_s(t)$.

7 Find the equivalent impedance $\widetilde{Z}_{eq}$ shown in Figure P4.1.2, where the angular frequency is $\omega = 377$ rad/s.

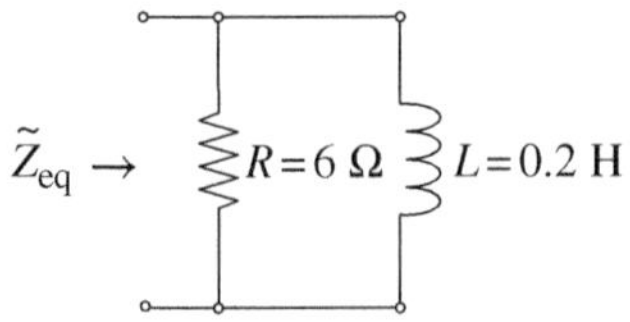

Figure P4.1.2 Impedance calculation $\widetilde{Z}_{eq}$.

8 Find the equivalent impedance $\tilde{Z}_{eq}$ shown in the circuit in Figure P4.1.3.

Figure P4.1.3 Equivalent impedance $\tilde{Z}_{eq}$.

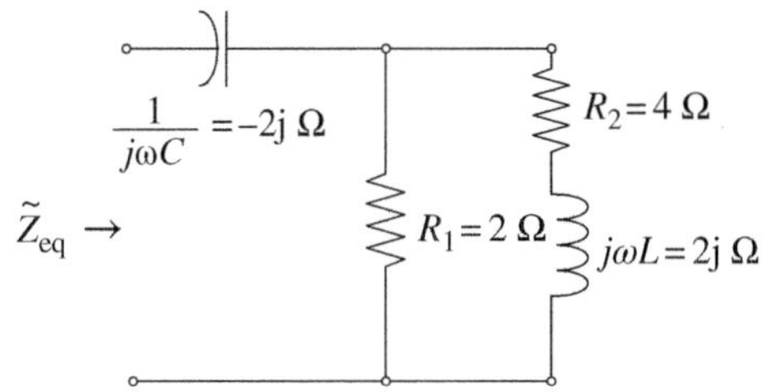

4.2 Calculate the equivalent admittance $\tilde{Y}_{eq}$ shown in the circuit in Figure P4.2.1. Note the "S" stands for the unit siemens = 1/ohms, that is, $S = 1/\Omega$.

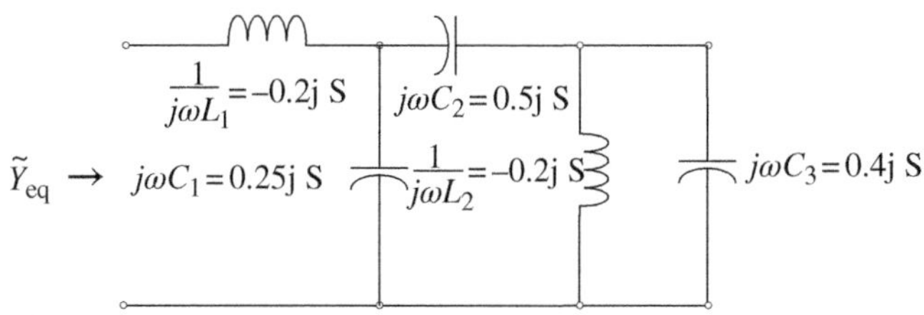

Figure P4.2.1 Equivalent admittance $\tilde{Y}_{eq}$.

4.3 Determine the equivalent admittance $\tilde{Y}_{eq}$ for the network of Figure P4.3.1 and use it to calculate the current $\tilde{I}_s$ for the voltage $\tilde{V}_s = 60\,\text{V}\angle 45°$, where "s" stands for source.

Figure P4.3.1 Equivalent admittance $\tilde{Y}_{eq}$ and current $\tilde{I}_s$.

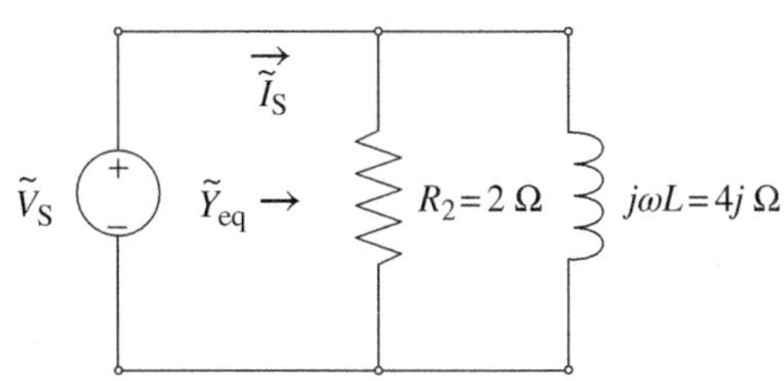

4.4 Find the expression for the equivalent impedance $\tilde{Z}_{eq}$ shown in the bridge-type circuit of Figure P4.4.1.

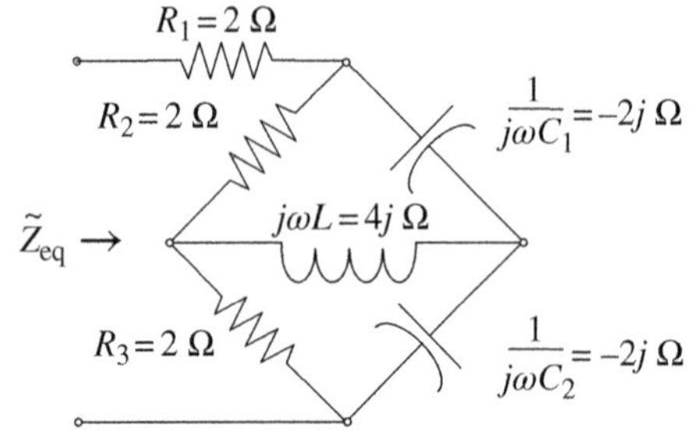

Figure P4.4.1 Bridge-type circuit.

4.5 For the circuit of Figure P4.5.1, determine $\tilde{V}_o$, where "o" stands for output and "s" stands for source.

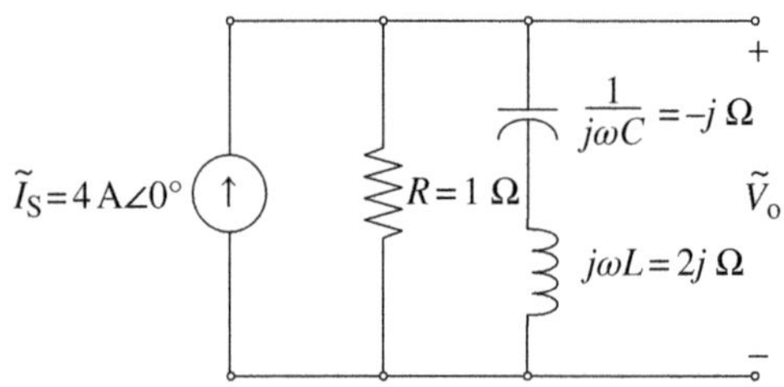

Figure P4.5.1 Voltage as a function of given current source.

4.6 For the circuit of Figure P4.6.1, determine $\tilde{I}_o$, where "o" stands for output and "s" stands for source.

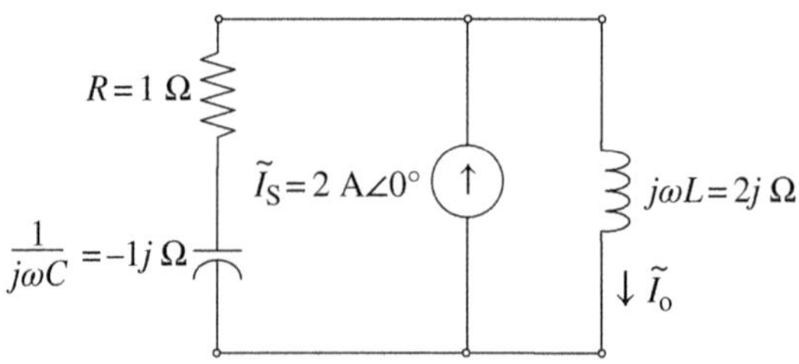

Figure P4.6.1 Current as a function of given current source.

4.7 Solution based on Kirchhoff's laws.

Determine the currents $\tilde{I}_1$, $\tilde{I}_2$, and $\tilde{I}_3$ in the network of Figure P4.7.1 based on Kirchhoff's voltage and current laws, KVL and KCL, respectively, where "o" stands for output and "s" stands for source.

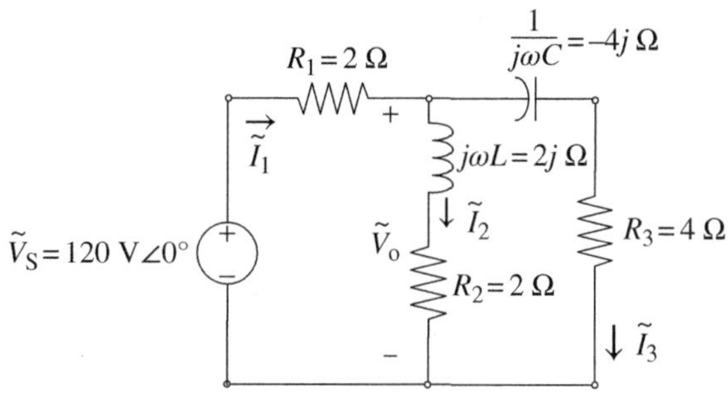

Figure P4.7.1 Application of KVL and KCL.

4.8 Solution using nodal analysis.

Find $\tilde{I}_o$ by using nodal analysis in the network of Figure P4.8.1, where "o" stands for output and "s" stands for source.

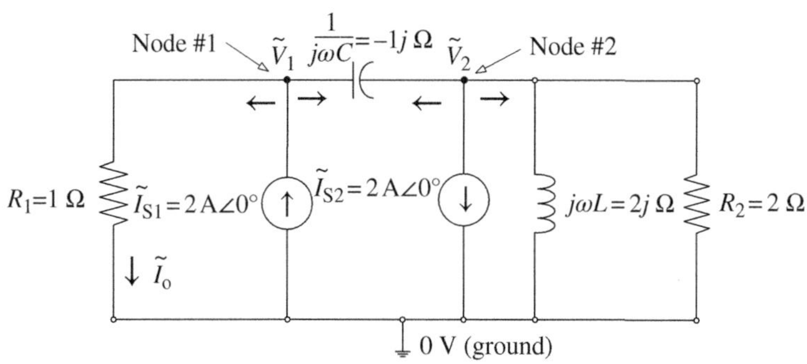

Figure P4.8.1 Application of nodal analysis.

4.9 Solution relying on mesh analysis.

For calculating the voltage $\tilde{V}_o$ of Figure P4.9.1, rely on mesh analysis, where "o" stands for output and "s" stands for source.

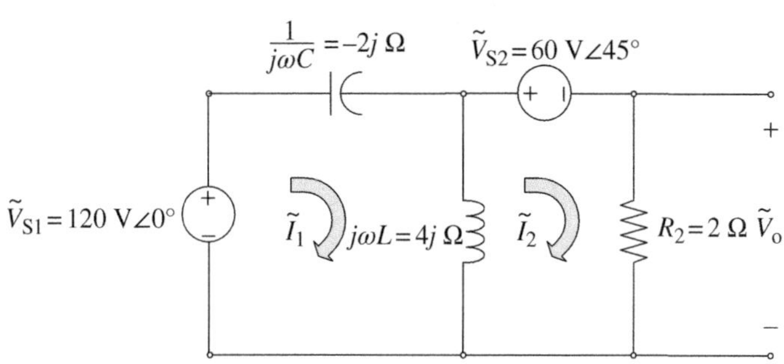

Figure P4.9.1 Mesh analysis.

4.10 Solution relying on loop/mesh analysis.

For calculating the current $\tilde{I}_o$ of Figure P4.10.1, rely on loop/mesh analysis, where "o" stands for output and "s" stands for source.

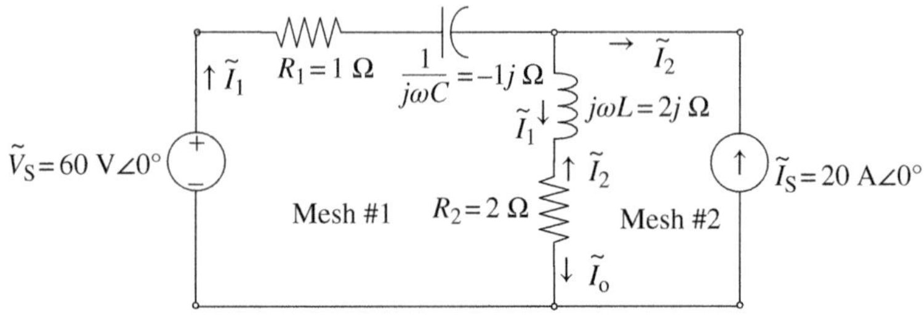

Figure P4.10.1 Loop/mesh analysis.

4.11 Solution employing principle of superposition.

Determine in circuit of Figure P4.11.1 $\tilde{V}_o$ using the principle of superposition, where "o" stands for output and "s" stands for source.

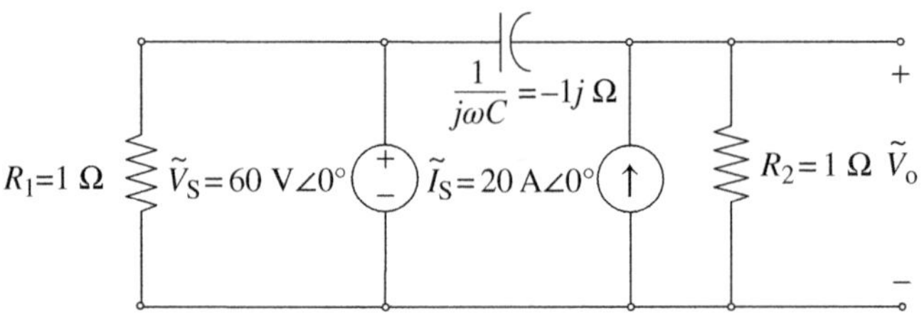

Figure P4.11.1 Analysis based on the principle of superposition.

4.12 Solution with source transformation/exchange.

Calculate $\tilde{V}_o$ in the circuit of Figure P4.12.1 using source transformation, where "o" stands for output and "s" stands for source.

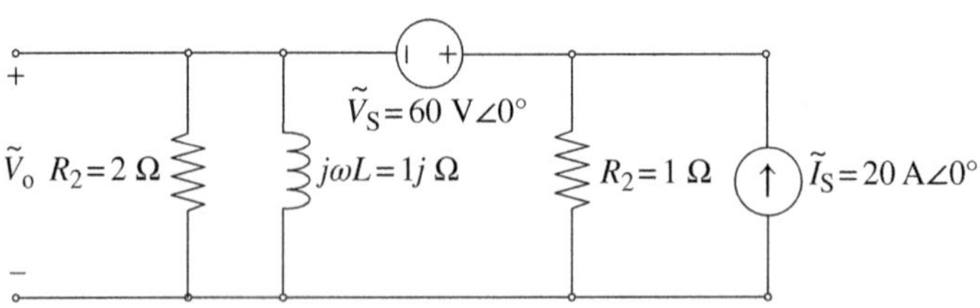

Figure P4.12.1 Analysis using source transformation/exchange.

4.13 Solution relying on Thévenin's theorem.

Find $\tilde{I}_o$ in the circuit of Figure P4.13.1 based on Thévenin's theorem, also called impressed voltage source approach, where "o" stands for output and "s" stands for source.

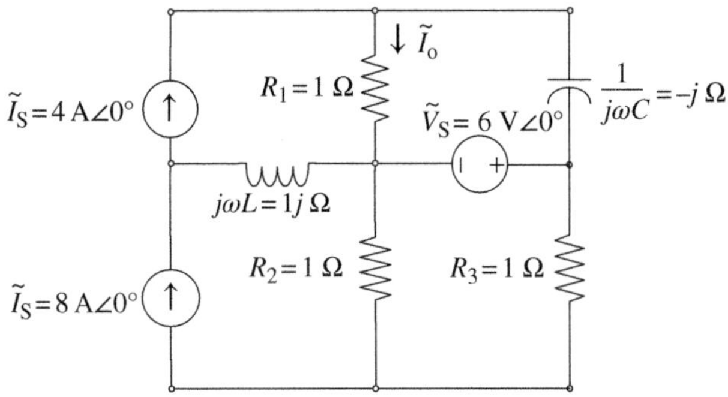

Figure P4.13.1 Application of Thévenin's theorem (impressed voltage source).

4.14 Solution based on Norton's theorem.

Find $\tilde{V}_o$ in the circuit of Figure P4.14.1 based on Norton's theorem, also called impressed current source approach, where "o" stands for output and "s" stands for source.

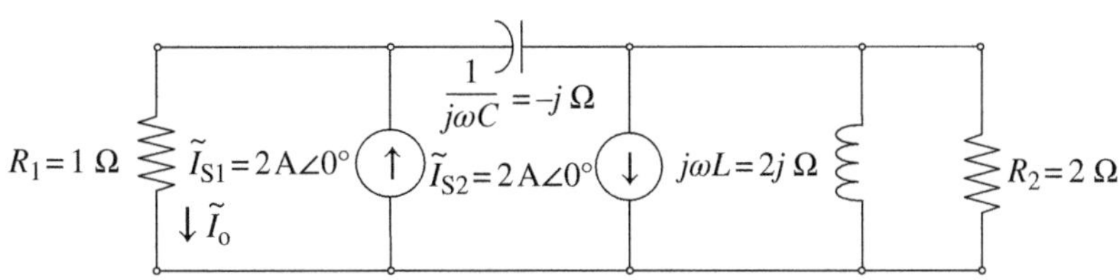

Figure P4.14.1 Application of Norton's theorem (impressed current source).

4.15 Fourier analysis employing rectangular/trigonometric form.

a Find the Fourier coefficients a_0, a_1, a_2, and a_3 as well as b_1, b_2, and b_3 in rectangular/trigonometric form for the symmetric waveform of Figure P4.15.1. Note that for symmetric waveforms certain Fourier coefficients are zero [5], simplifying the analysis.

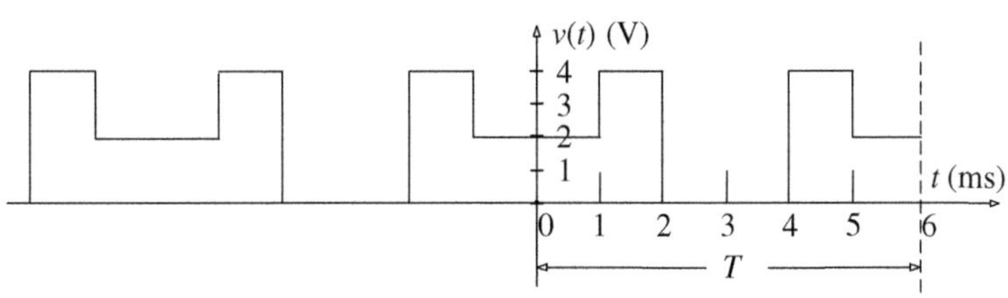

Figure P4.15.1 Symmetric waveform, rectangular/trigonometric analysis with period T.

b Plot the result of the above requested Fourier approximation and compare it with the given function of Figure P4.15.1.

4.16 Fourier transformation employing exponential form.

Find the Fourier coefficients in exponential form for the symmetric waveform of Figure P4.16.1. Note that for symmetric waveforms certain Fourier coefficients are zero [5], simplifying the analysis. Hint: For the calculation of the complex Fourier coefficient C with tilde (~) above C for $n = 0$ the application of the rule of de l'Hôspital [8] is required.

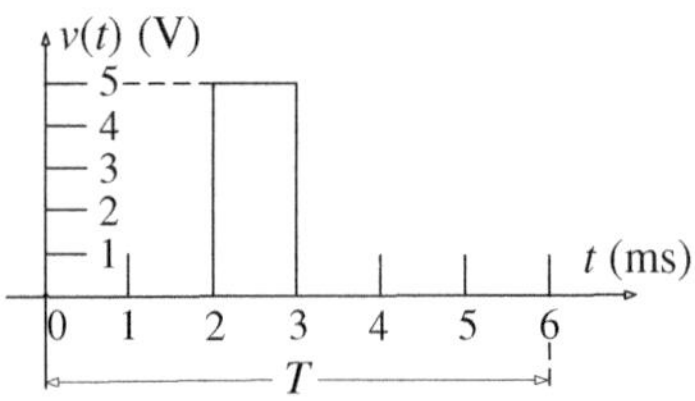

Figure P4.16.1 Symmetric waveform, exponential analysis with period *T*.

References

1 Smith, J.O. III (2007). *Mathematics of the Discrete Fourier Transform (DFT), with Audio Applications*, 2e. W3K Publishing.

2 Weisstein, Eric W. Complex plane. From MathWorld – A Wolfram Web Resource. https://mathworld.wolfram.com/ComplexPlane.html (accessed 12 May 12).

3 Smith, J.O. III (2007). *Introduction to Digital Filters with Audio Applications*. W3K Publishing.

4 Nave R. Superposition theorem. http://hyperphysics.phy-astr.gsu.edu/hbase/electric/suppos.html (accessed 12 May 2020).

5 Nilsson, J.W. (1986). *Electric Circuits*, 2e. Reading, MA: Addison-Wesley Publishing Company.

6 Stapel, Elizabeth. Cramer's rule. Purplemath. https://www.purplemath.com/modules/cramers.htm (accessed 13 May 2020).

7 The Wolfram Centre. Fourier series. http://reference.wolfram.com/language/ref/FourierSeries.html (accessed 12 May 2020).

8 Tutorial.math.lamar.edu/classes/calci/lhospitalsrule.aspx

Appendix 4.A Conversion of Phasors from Rectangular to Polar Form

The transformation from polar to rectangular and vice versa complex quantities in the four quadrants of the complex plane represents an important task, where mistakes frequently occur. To prevent such conversion mistakes, the transformation issues will be addressed for all four quadrants.

Quadrant 1: Transform the complex number with the unit of meter denoted as [m]

$$\tilde{z}_1 = (3 + j2)\ [\text{m}], \tag{4.A.1}$$

given in rectangular form to the polar form:

$$\tilde{z}_1 = |\tilde{z}_1| \text{e}^{\,j\theta_1}\ [\text{m}]. \tag{4.A.2}$$

Figure 4.A.1 shows this complex number in the Gaussian (complex) plane. The magnitude of $\tilde{z}_1$ is

$$|\tilde{z}_1| = \sqrt{3^2 + 2^2} = 3.6055\ [\text{m}], \tag{4.A.3}$$

and its angle θ_1 is

$$\theta_1 = \tan^{-1}\left(\frac{2}{3}\right) = 0.588 \text{ rad or } 33.69°. \tag{4.A.4}$$

The polar form is now

$$\tilde{z}_1 = |\tilde{z}_1| \angle\theta_1 = 3.6055\text{e}^{\,j0.588\text{ rad}}\ [\text{m}] = 3.6055\text{e}^{\,j33.69°}\ [\text{m}]. \tag{4.A.5}$$

Note the unit of the angular velocity or frequency is radians/s (rad/s), while the unit of the angle θ can be expressed either in radians (rad) or in degrees (°).

Quadrant 2: Transform the complex number

$$\tilde{z}_2 = (-3 + j2)\ [\text{m}] = -(3 - j2)\ [\text{m}], \tag{4.A.6}$$

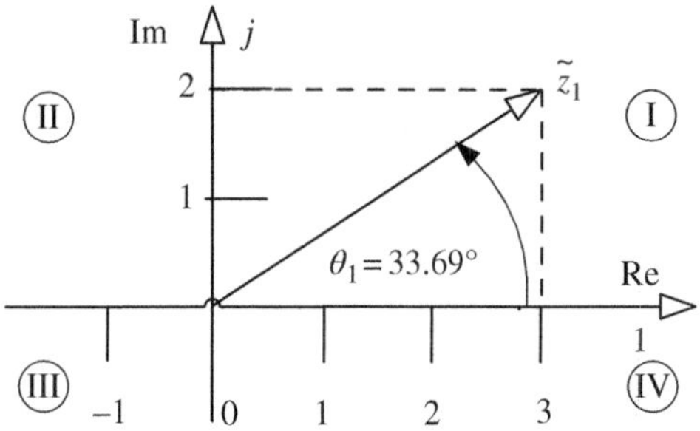

Figure 4.A.1 Complex number $\tilde{z}_1 = (3 + j2)$ [m] in Gaussian plane.

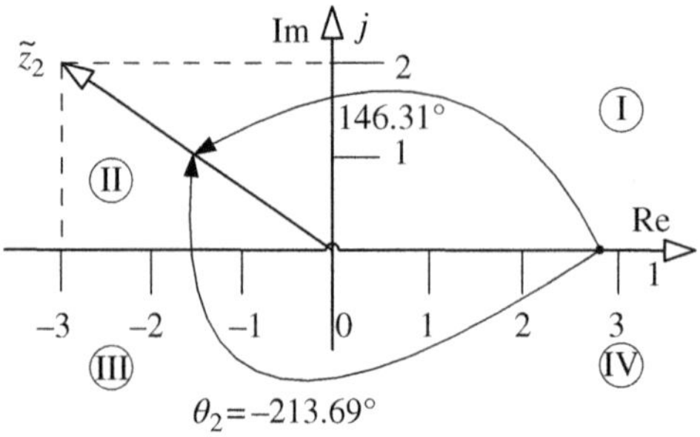

Figure 4.A.2 Complex number $\tilde{z}_2 = (-3 + j2)$ [m] = $-(3 - j2)$ [m] in Gaussian plane.

given in rectangular form to the polar form:

$$\tilde{z}_2 = |\tilde{z}_2| \mathrm{e}^{j\theta_2} \text{ [m]}. \tag{4.A.7}$$

Figure 4.A.2 shows this complex number in the Gaussian (complex) plane. The magnitude of $\tilde{z}_2$ is

$$|\tilde{z}_2| = \sqrt{3^2 + 2^2} = 3.6055 \text{ [m]}, \tag{4.A.8}$$

and its angle θ_2 is

$$\theta_2 = -\tan^{-1}\left(\frac{-2}{3}\right) = -\pi - 0.588 \text{ rad or} - 180° - 33.69°, \tag{4.A.9}$$

or

$$\theta_2 = -\tan^{-1}\left(\frac{-2}{3}\right) = -3.7295 \text{ rad or} - 213.69°. \tag{4.A.10}$$

The polar form is now

$$\tilde{z}_2 = |\tilde{z}_2| \angle\theta_2 = 3.6055\mathrm{e}^{-j3.729 \text{ rad}} \text{ [m]} = 3.6055\mathrm{e}^{-j213.69°} \text{ [m]}. \tag{4.A.11}$$

Quadrant 3: Transform the complex number

$$\tilde{z}_3 = (-3 - j2) \text{ [m]} = -(3 + j2) \text{ [m]}, \tag{4.A.12}$$

given in rectangular form to the polar form:

$$\tilde{z}_3 = |\tilde{z}_3| \mathrm{e}^{j\theta_3} \text{ [m]}. \tag{4.A.13}$$

Figure 4.A.3 shows this complex number in the Gaussian (complex) plane. The magnitude of $\tilde{z}_3$ is

$$|\tilde{z}_3| = \sqrt{3^2 + 2^2} = 3.6055 \text{ [m]}, \tag{4.A.14}$$

and its angle θ_3 is

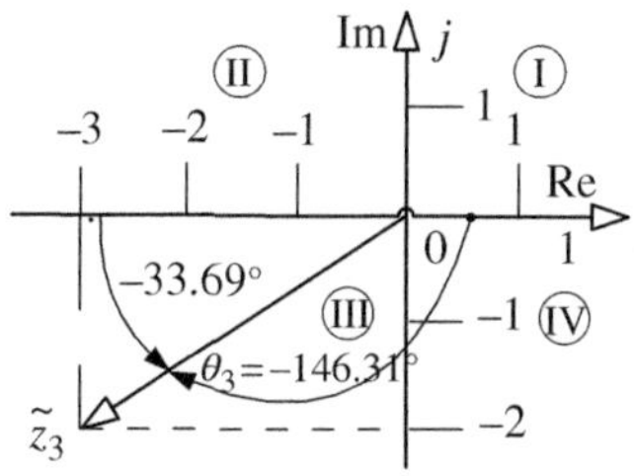

Figure 4.A.3 Complex number $\tilde{z}_3 = (-3 - j2)$ [m] $= -(3 + j2)$ [m] in Gaussian plane.

$$\theta_3 = -\tan^{-1}\left(\frac{2}{3}\right) = -\pi + 0.588 \text{ rad or } -180° + 33.69°, \quad (4.A.15)$$

or

$$\theta_3 = -\tan^{-1}\left(\frac{2}{3}\right) = -2.554 \text{ rad or } -146.31°. \quad (4.A.16)$$

The polar form is now

$$\tilde{z}_3 = |\tilde{z}_3|\angle\theta_3 = 3.6055e^{-j2.554 \text{ rad}} \text{ [m]} = 3.6055e^{-j146.31°} \text{ [m]}. \quad (4.A.17)$$

Quadrant 4: Transform the complex number

$$\tilde{z}_4 = (3 - j2) \text{ [m]}, \quad (4.A.18)$$

given in rectangular form to the polar form:

$$\tilde{z}_4 = |\tilde{z}_4|e^{j\theta_4} \text{ [m]}. \quad (4.A.19)$$

Figure 4.A.4 shows this complex number in the Gaussian (complex) plane. The magnitude of $\tilde{z}_4$ is

$$|\tilde{z}_4| = \sqrt{3^2 + 2^2} = 3.6055 \text{ [m]}, \quad (4.A.20)$$

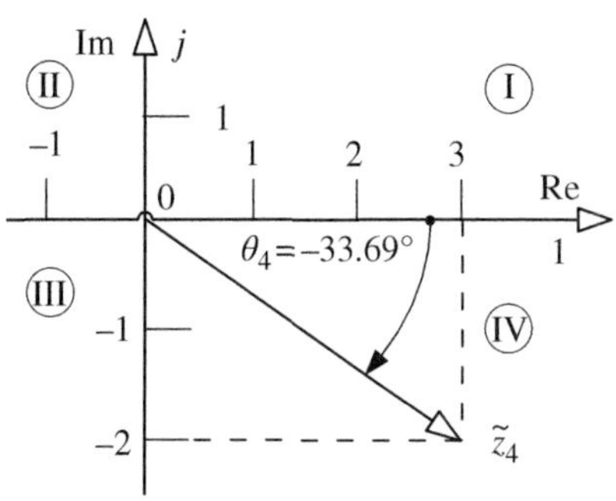

Figure 4.A.4 Complex number $\tilde{z}_4 = (3 - j2)$ [m] in Gaussian plane.

and its angle θ_4 is

$$\theta_4 = \tan^{-1}\left(\frac{-2}{3}\right) = -0.588 \text{ rad or} -33.69°. \tag{4.A.21}$$

The polar form is now

$$\widetilde{z}_4 = |\widetilde{z}_4|\angle\theta_4 = 3.6055e^{-j0.588 \text{ rad}}\,[\text{m}] = 3.6055e^{-j33.69°}\,[\text{m}]. \tag{4.A.22}$$

5

Instantaneous and Steady-State Power Analysis

5.1 Introduction

In this chapter, first we will determine the steady-state powers of single-phase AC circuits and then those of balanced three-phase AC networks. There are five different types of power:

- Instantaneous power $p(t)$, measured in watts [W],
- Average or real power P, measured in watts [W],
- Reactive or imaginary power Q, measured in volt-ampere reactive [VAr],
- Apparent power $S = \left|\widetilde{S}\right| = \sqrt{P^2 + Q^2}$, measured in volt-ampere [VA],
- Complex power $\widetilde{S} = P + jQ$, measured in volt-ampere [VA].

5.2 Instantaneous Power $p(t)$

A typical example for instantaneous power are the voltage and current of a lightning strike consisting of transient current $i_{\text{lightning}}(t)$ and voltage $v_{\text{lightning}}(t)$ with a very short duration in the range of a few hundred microseconds as shown in Figure 5.1a and b. A lightning discharge has incredibly large voltage and current signatures – up to 30 million volts at 100 thousand amperes – but is of very short duration; hence lightning power and energy cannot be harnessed or used; indeed it is frequently very destructive by setting buildings on fire or causing death for people without protection through a so-called Faraday cage. Problem P5.16 deals with the calculation/estimation of the power $P_{\text{lightning}}$ and energy $W_{\text{lightning}}$ contained/transmitted by a lightning strike.

Introduction to Energy, Renewable Energy, and Electrical Engineering: Essentials for Engineering Science (STEM) Professionals and Students, First Edition. Ewald F. Fuchs and Heidi A. Fuchs.

Companion website: www.wiley.com/go/fuchs/renewableenergyandelectricalengineering

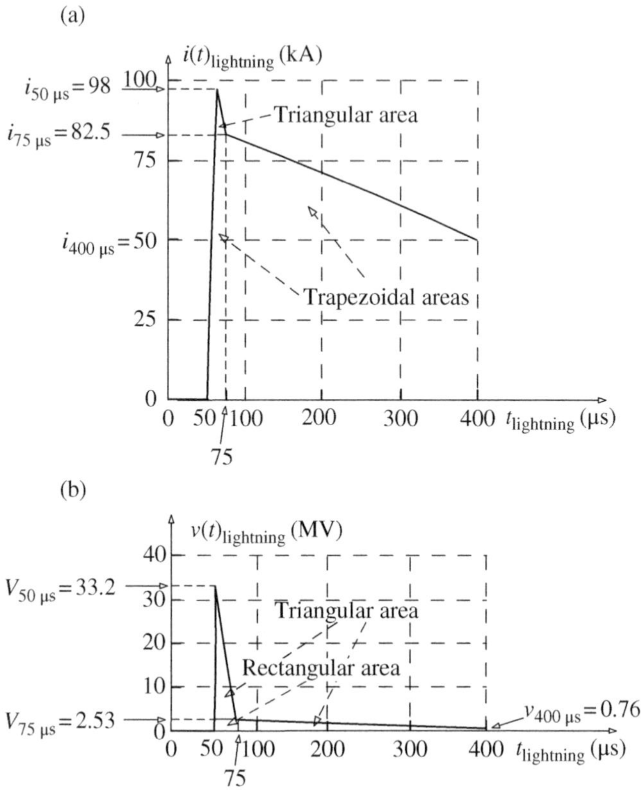

Figure 5.1 (a) Lightning instantaneous current $i_{\text{lightning}}(t)$. (b) Lighting instantaneous voltage $v_{\text{lightning}}(t)$.

Defining the instantaneous voltage and current in the time domain for useful electric engineering applications, one can write in terms of sinusoids for the source (s) voltage with the amplitude V_{M}.

$$v_{\text{s}}(t) = V_{\text{M}} \cos(\omega t + \theta_{\text{v}}), \tag{5.1}$$

and for the response current with the amplitude I_{M}

$$i(t) = I_{\text{M}} \cos(\omega t + \theta_{\text{i}}). \tag{5.2}$$

Equations (5.1) and (5.2) are represented in Figure 5.2a for any passive (R, L, C) load network. The angular frequency, also called angular velocity, is

$$\omega = 2\pi f, \tag{5.3}$$

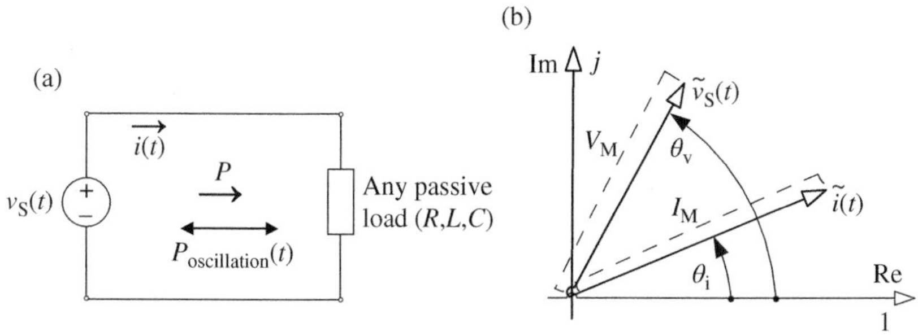

Figure 5.2 (a) Passive R, L, C load network/circuit supplied by source voltage $v_s(t)$ and the response current $i(t)$ including real power P and oscillating power $p_{oscillation}(t)$. (b) Phasor diagram showing voltage Eq. (5.4) and current (Eq. (5.5)) phasors.

where in most English-speaking countries the frequency $f = 60$ Hz and in most other countries $f = 50$ Hz. For $f = 60$ Hz one obtains $\omega = 2\pi 60$ rad/s $= 376.98$ rad/s ≈ 377 rad/s. For $f = 50$ Hz one gets $\omega = 2\pi 50$ rad/s $= 300$ rad/s.

The time functions of Eqs. (5.1) and (5.2) can be represented in the phasor domain as

$$\tilde{v}_s(t) = V_M \angle\theta_v, \tag{5.4}$$

$$\tilde{i}(t) = I_M \angle\theta_i. \tag{5.5}$$

Phasors (Eqs. (5.4) and (5.5)) are graphically depicted in the complex plane of Figure 5.2a and b, where Re stands for real axis and Im for imaginary axis of the complex plane. The real axis is taken as reference direction for all phasor diagrams. Also, in order to keep identification of variables consistent, we identify a phasor by attaching a tilde (~) to the time variable $x(t)$; for example, the time variable $v_s(t)$ becomes the phasor variable $\tilde{v}_s(t)$.

The instantaneous power is defined as

$$p(t) = v_s(t)i(t) = V_M I_M \cos(\omega t + \theta_v)\cos(\omega t + \theta_i); \tag{5.6}$$

using the trigonometric identity

$$\cos\Phi_1 \cos\Phi_2 = (1/2)[\cos(\Phi_1 - \Phi_2) + \cos(\Phi_1 + \Phi_2)] \tag{5.7}$$

results in

$$p(t) = \frac{V_M I_M}{2}[\cos(\theta_v - \theta_i) + (2\omega t + \theta_v + \theta_i)]. \tag{5.8}$$

The first term of Eq. (5.8) is independent of time t representing the useful average or real power:

$$P = \frac{V_M I_M}{2} \cos(\theta_v - \theta_i), \tag{5.9}$$

and the second term of Eq. (5.8) is time dependent and represents a useless power:

$$P_{\text{oscillation}}(t) = \frac{V_M I_M}{2} \cos(2\omega t + \theta_v + \theta_i). \tag{5.10}$$

$P_{\text{oscillation}}(t)$ varies or oscillates with twice the angular excitation frequency or angular velocity ω.

5.3 Average (Real) Power *P*

The average, or real, power is useful power, representing the capacity of the electricity to perform work, and is obtained from the integration over one period T:

$$\begin{aligned} P &= \frac{1}{T}\int_{t_0}^{t_0+T} p(t)\mathrm{d}t = \frac{1}{T}\int_{t_0}^{t_0+T} v_s(t)i(t)\mathrm{d}t \\ &= \frac{1}{T}\int_{t_0}^{t_0+T} V_M I_M \cos(\omega t + \theta_v)\cos(\omega t + \theta_i)\mathrm{d}\omega t, \end{aligned} \tag{5.11}$$

or with Eq. (5.8)

$$P = \frac{1}{T}\int_{t_0}^{t_0+T} \frac{V_M I_M}{2}[\cos(\theta_v - \theta_i) + \cos(2\omega t + \theta_v + \theta_i)]\mathrm{d}\omega t. \tag{5.12}$$

The second term of above integral is zero, and the first term, after integration, is

$$P = \frac{1}{T}\left[\frac{V_M I_M}{2}\cos(\theta_v - \theta_i)\right]\Bigg|_{t_0}^{t_0+T} = \frac{V_M I_M}{2}\cos(\theta_v - \theta_i). \tag{5.13}$$

Conclusion

For purely resistive circuits one obtains with $\theta_v = \theta_i$ when current and voltage are in phase

$$P = \frac{V_M I_M}{2}\cos(\theta_v - \theta_i) = \frac{V_M I_M}{2} = V_{\text{rms}} \cdot I_{\text{rms}} = V_{\text{eff}} \cdot I_{\text{eff}}. \tag{5.14}$$

In this equation the $V_{\text{rms}} = V_{\text{eff}}$ and $I_{\text{rms}} = I_{\text{eff}}$ represent root-mean-square (rms) or effective (eff) values, that is,

$$V_{\text{rms}} = V_{\text{eff}} = \frac{V_M}{\sqrt{2}} \tag{5.15a}$$

and

$$I_{\mathrm{rms}} = I_{\mathrm{eff}} = \frac{I_{\mathrm{M}}}{\sqrt{2}}, \tag{5.15b}$$

as shown in Section 5.4.

The voltage–current relationship is shown in the complex plane in Figure 5.3.

For purely inductive circuits, one obtains with $(\theta_{\mathrm{v}} - \theta_{\mathrm{i}}) = 90^{\circ}$ when the voltage leads the current by 90°

$$P = \frac{V_{\mathrm{M}} I_{\mathrm{M}}}{2} \cos(\theta_{\mathrm{v}} - \theta_{\mathrm{i}}) = 0. \tag{5.16}$$

The voltage–current relationship is shown in the complex plane in Figure 5.4.

For purely capacitive circuits, one obtains with $(\theta_{\mathrm{i}} - \theta_{\mathrm{v}}) = 90^{\circ}$ when the current leads the voltage by 90°

$$P = \frac{V_{\mathrm{M}} I_{\mathrm{M}}}{2} \cos(\theta_{\mathrm{v}} - \theta_{\mathrm{i}}) = 0. \tag{5.17}$$

The voltage–current relationship is shown in the complex plane in Figure 5.5.

Figure 5.3 Phasor voltage $\tilde{v}_{\mathrm{s}}(t) = V_{\mathrm{M}} \angle \theta_{\mathrm{v}}$ and phasor current $\tilde{i}(t) = I_{\mathrm{M}} \angle \theta_{\mathrm{i}}$ for a resistor R.

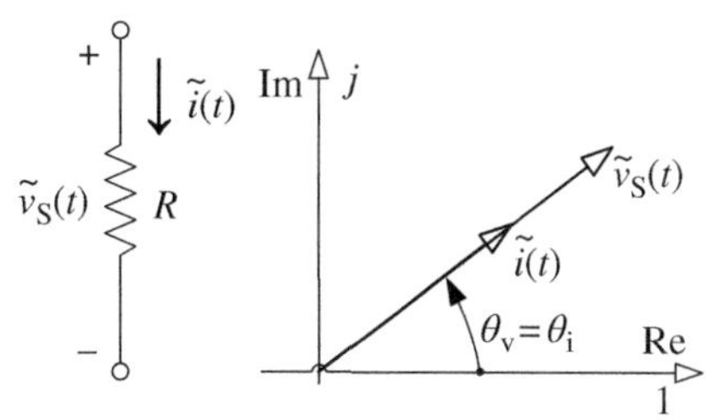

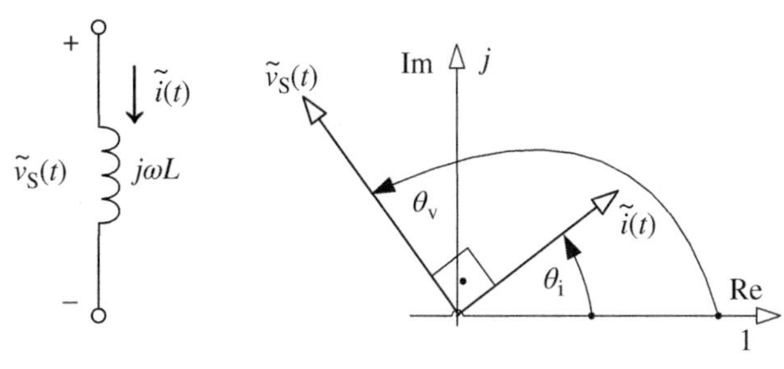

Figure 5.4 Phasor voltage $\tilde{v}_{\mathrm{s}}(t) = V_{\mathrm{M}} \angle \theta_{\mathrm{v}}$ and phasor current $\tilde{i}(t) = I_{\mathrm{M}} \angle \theta_{\mathrm{i}}$ for an inductor L or $j\omega L$.

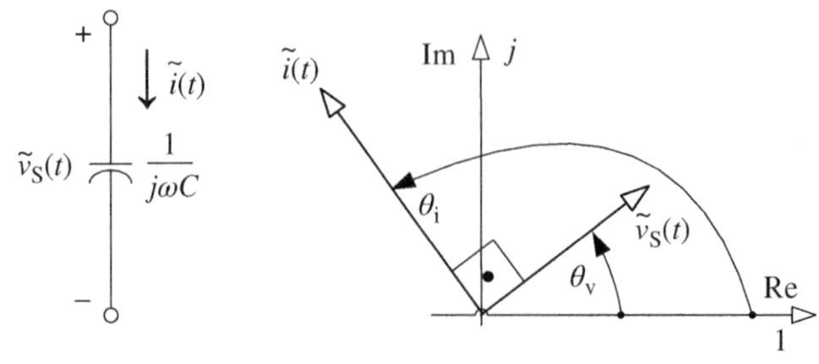

Figure 5.5 Phasor voltage $\tilde{v}_s(t) = V_M\angle\theta_v$ and phasor current $\tilde{i}(t) = I_M\angle\theta_i$ for a capacitor C or $1/j\omega C$.

5.4 Relation Between Root-Mean-Square (rms) or Effective (eff) Value and Amplitude

Relation between amplitude X_M and its rms value for the function $x(t) = X_M \cos \omega t$ is based on the integral $\int \cos^2 x \mathrm{d}x = \frac{x}{2} + \frac{1}{4}\sin 2x$. Note the variable X stands for any voltage or current:

$$
\begin{aligned}
X_{\mathrm{rms}} = X_{\mathrm{eff}} &= \sqrt{\frac{1}{T}\int_{t_0}^{t_0+T} [x(t)]^2 \mathrm{d}t} \\
&= \sqrt{\frac{1}{T}\int_{t_0}^{t_0+T} X_M^2[\cos(\omega t)]^2 \mathrm{d}\omega t}\sqrt{\frac{X_M^2}{T}\left(\frac{\omega t}{2} + \frac{1}{4}\sin 2\omega t\right)}\Bigg|_0^{T=\omega t} \\
&= \sqrt{\frac{X_M^2}{T}\left(\frac{T}{2} + \frac{1}{4}\sin 2T\right)} = \frac{X_M}{\sqrt{2}}.
\end{aligned}
\tag{5.18}
$$

Application Example 5.1

Average or real power supplied is equal to the average or real power absorbed: $P_{\text{supplied}} = P_{\text{absorbed}}$

Solution

The principle of conservation of energy (see Chapter 1), based on the first law of thermodynamics, states that

$$
\text{Energy or work} = \int p(t)\mathrm{d}t = \text{constant}, \tag{E5.1.1}
$$

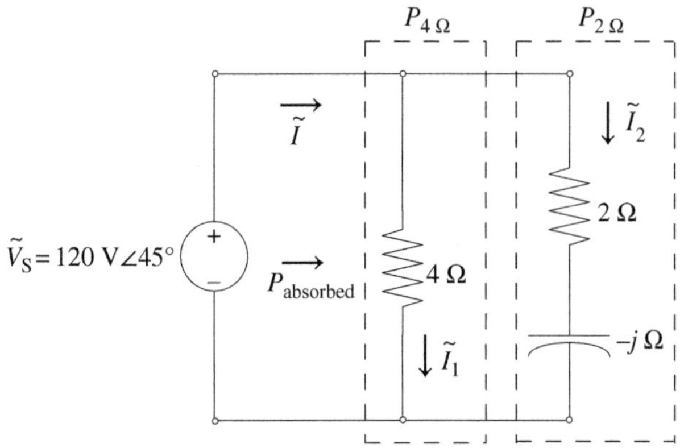

Figure E5.1.1 Circuit for calculating the power balance.

which implies that the source (input) and load (output) powers in a circuit/network must be balanced. For the circuit of Figure E5.1.1, the power balance will be demonstrated by computing the power delivered by the source (input) P_s and the load (output) power dissipated in the resistors $P_{4\Omega}$ and $P_{2\Omega}$.

$$\tilde{I}_1 = \frac{\tilde{V}_s}{4\,\Omega} = \frac{120\,V\angle 45°}{4\,\Omega} = 30\,A\angle 45°, \tag{E5.1.2}$$

$$\tilde{I}_2 = \frac{\tilde{V}_s}{(2-j)\,\Omega} = \frac{120\,V\angle 45°}{(2-j)\,\Omega} = \frac{120\,A\angle 45°}{\sqrt{5}\angle\theta_D}; \tag{E5.1.3}$$

with

$$\theta_D = \tan^{-1}\left(\frac{-1}{2}\right) = -26.565\,8°\text{or} - 0.463\,65\text{ rad}, \tag{E5.1.4}$$

follows

$$\tilde{I}_2 = \frac{120\,V\angle(45° + 26.5658°)}{\sqrt{5}\,\Omega} = 53.67\,A\angle 71.566°. \tag{E5.1.5}$$

The total input current is then

$$\begin{aligned}\tilde{I} &= \tilde{I}_1 + \tilde{I}_2 = (30\angle 45° + 53.67\angle 71.566°)A \\ &= 30[\cos(45°) + j\sin(45°)]A + 53.67[\cos(71.566°) + j\sin(71.566°)]\,A \\ &= [21.21 + j21.21 + 16.973 + j50.916]A = (38.183 + j72.13)A\end{aligned} \tag{E5.1.6}$$

or

$$\widetilde{I} = \sqrt{(38.183)^2 + (72.13)^2} \angle\theta_{\mathrm{i}}\ \mathrm{A} = 81.615\ \mathrm{A}\angle 62.107°.$$

The complex power supplied by the source (s) is with $\widetilde{V}_{\mathrm{s}} = V_{\mathrm{M}}\angle\theta_{\mathrm{v}}$ and $\widetilde{I} = I_{\mathrm{M}}\angle\theta_{\mathrm{i}}$

$$\begin{aligned}\widetilde{S}_{\mathrm{s}} &= \widetilde{V}_{\mathrm{s}}\widetilde{I} = V_{\mathrm{M}}\angle\theta_{\mathrm{v}} I_{\mathrm{M}}\angle\theta_{\mathrm{i}} = 120\mathrm{e}^{\,j45°}81.615\mathrm{e}^{\,j62.107°}\\ &= 120 \cdot 81.615\mathrm{e}^{\,j107.11°}\ \mathrm{VA},\end{aligned} \tag{E5.1.7}$$

resulting in the source (input) power

$$\begin{aligned}P_{\mathrm{s}} &= \frac{V_{\mathrm{M}}I_{\mathrm{M}}}{2}\cos(\theta_{\mathrm{v}} - \theta_{\mathrm{i}}) = \frac{V_{\mathrm{M}}I_{\mathrm{M}}}{2}\cos(107.11°)\\ &= \frac{9793.8}{2}\cos(107.11°) = 4896.9 \cdot 0.9566 = 4.68\ \mathrm{kW}.\end{aligned} \tag{E5.1.8}$$

Average power absorbed in 4 Ω resistor is according to Eq. (5.14)

$$P_{4\Omega} = \frac{V_{\mathrm{M}}I_{\mathrm{M}}}{2} = \frac{1}{2}120 \cdot 30\ \mathrm{W} = 1800\ \mathrm{W}. \tag{E5.1.9}$$

Average power absorbed in 2 Ω resistor is then

$$P_{2\Omega} = \frac{1}{2}(I_2)^2 R = \frac{1}{2}(53.67) \cdot 2\ \mathrm{W} = 2881\ \mathrm{W}, \tag{E5.1.10}$$

resulting in the total absorbed (load) power of

$$P_{\mathrm{absorbed}} = P_{4\Omega} + P_{2\Omega} = 1800\ \mathrm{W} + 2881\ \mathrm{W} = 4.68\ \mathrm{kW} = P_{\mathrm{s}}. \tag{E5.1.11}$$

5.5 Fundamental or Displacement Power Factor

The fundamental or displacement power factor (PF) is a very important quantity, stemming in part from the economic impact it has on users and distribution/transmission lines, including generation (e.g. synchronous and induction generators) and transformation (e.g. transformers, rectifiers, inverters) equipment. The PF is the ratio of the real power P absorbed by the load or delivered by the source to the apparent power $\left|\widetilde{S}\right|$ flowing in the circuit. If the fundamental PF is low (e.g. PF = 0.5 pu), the utilization of the voltage V_{rms} and the current I_{rms} is low, and if the fundamental PF is high (e.g. PF ≤ 1 pu), the utilization of voltage and current is high. The fundamental or displacement PF [1] relates to fundamental voltages and currents, while the total PF [2] not discussed here includes both the effects of fundamental and harmonic voltages and currents. Note the PF is dimensionless, which is indicated per unit or pu. If the fundamental PF is small, then a PF correction must be performed whereby the reactive power must be compensated as discussed in Section 5.7. Figure 5.6 shows an electric circuit consisting of applied fixed rms source voltage V_{rms}, distribution/transmission line impedance $\tilde{Z}_{line}$, and load or user/consumer impedance $\tilde{Z}_{load}$ whereby the responding rms current $\tilde{I}_{rms}$ is unknown and must be calculated.

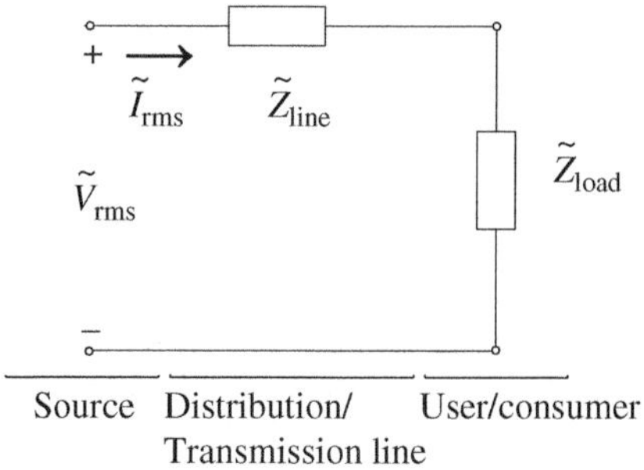

Figure 5.6 Single-phase source–load circuit with line impedance.

The average or real power with the unit watts [W] was defined in Eq. (5.13) as

$$P = \frac{V_M I_M}{2}\cos(\theta_v - \theta_i) = \frac{V_M}{\sqrt{2}} \cdot \frac{I_M}{\sqrt{2}}\cos(\theta_v - \theta_i) = V_{rms} I_{rms}\cos(\theta_v - \theta_i) \tag{5.19}$$

with V_{rms} (or V_{eff}) with unit volts [V] and I_{rms} (or I_{eff}) with unit amperes [A].

We define the apparent power with the unit volt-amperes [VA] as

$$\left|\tilde{S}\right| = V_{rms} I_{rms}. \tag{5.20}$$

Dividing the real power (Eq. (5.19)) by the apparent power (5.20) yields the fundamental PF:

$$\mathrm{PF} = \frac{P}{\left|\tilde{S}\right|} = \frac{V_{rms} I_{rms}\cos(\theta_v - \theta_i)}{V_{rms} I_{rms}} = \cos(\theta_v - \theta_i), \tag{5.21}$$

where the angle difference $(\theta_v - \theta_i)$ is referred to as the fundamental PF angle. Figure 5.7 depicts a load $\tilde{Z}_{load}$, where a given voltage $\tilde{V}_{load}$ is applied and the responding current is $\tilde{I}_{load}$.

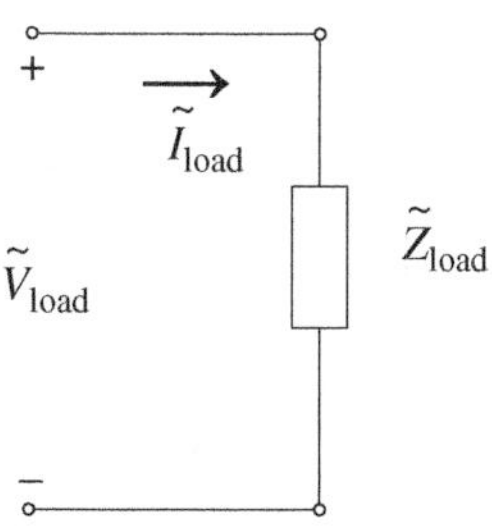

Figure 5.7 Given load voltage and resulting load current resulting in the impedance $\tilde{Z}_{load}$.

Assuming that the load voltage $\tilde{V}_{load} = \left|\tilde{V}_{load}\right| \angle \theta_v$ and the load current $\tilde{I}_{load} = \left|\tilde{I}_{load}\right| \angle \theta_v$ are known, then the load impedance is calculated as

$$\widetilde{Z}_{\text{load}} = \frac{\widetilde{V}_{\text{load}}}{\widetilde{I}_{\text{load}}} = \frac{|\widetilde{V}_{\text{load}}|\angle\theta_{\text{v}}}{|\widetilde{I}_{\text{load}}|\angle\theta_{\text{i}}} = |\widetilde{Z}_{\text{load}}|\frac{\angle\theta_{\text{v}}}{\angle\theta_{\text{i}}} = |\widetilde{Z}_{\text{load}}|\angle(\theta_{\text{v}} - \theta_{\text{i}}). \tag{5.22}$$

Conclusions

1) For purely resistive loads (R; see Figure 5.8a), $\widetilde{V}_{\text{load}}$ and $\widetilde{I}_{\text{load}}$ are in phase, that is, $(\theta_{\text{v}} - \theta_{\text{i}}) = 0$ or $\theta_{\text{v}} = \theta_{\text{i}}$. Figure 5.8b shows the corresponding phasor diagram. Therefore,

$$\widetilde{Z}_{\text{load}} = \frac{|\widetilde{V}_{\text{load}}|\angle\theta_{\text{v}}}{|\widetilde{I}_{\text{load}}|\angle\theta_{\text{i}}} = |\widetilde{Z}_{\text{load}}|\angle(\theta_{\text{v}} - \theta_{\text{i}}) = R\angle 0°\,\Omega. \tag{5.23a}$$

The power factor is

$$\text{PF} = \cos(\theta_{\text{v}} - \theta_{\text{i}}) = \cos(0) = 1. \tag{5.23b}$$

2) For purely inductive reactive loads ($j\omega L$; see Figure 5.9a), $\widetilde{V}_{\text{load}}$ leads $\widetilde{I}_{\text{load}}$ by 90°, that is, $(\theta_{\text{v}} - \theta_{\text{i}}) = 90°$. Figure 5.9b shows the corresponding phasor diagram. Therefore,

$$\widetilde{Z}_{\text{load}} = \frac{|\widetilde{V}_{\text{load}}|\angle\theta_{\text{v}}}{|\widetilde{I}_{\text{load}}|\angle\theta_{\text{i}}} = |\widetilde{Z}_{\text{load}}|\angle(\theta_{\text{v}} - \theta_{\text{i}}) = \omega L\angle 90°\,\Omega. \tag{5.24a}$$

The power factor is

$$\text{PF} = \cos(\theta_{\text{v}} - \theta_{\text{i}}) = \cos(90) = 0. \tag{5.24b}$$

3) For purely capacitive reactive loads ($1/j\omega C = -j/\omega C$; see Figure 5.10a), $\widetilde{V}_{\text{load}}$ lags $\widetilde{I}_{\text{load}}$ by 90°, that is, $(\theta_{\text{v}} - \theta_{\text{i}}) = -90°$. Figure 5.10b shows the corresponding phasor diagram.

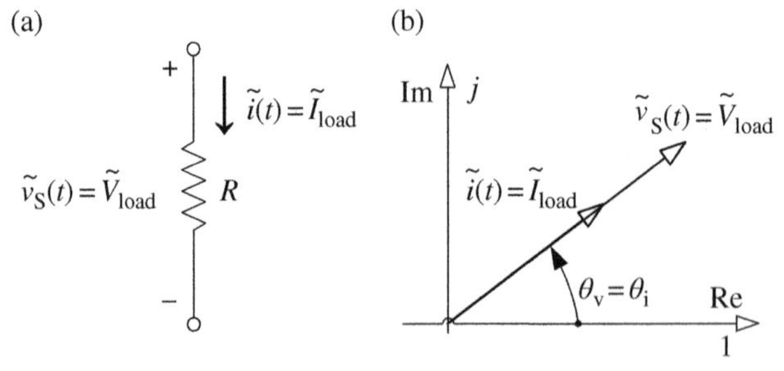

Figure 5.8 (a) Resistive load R. (b) Phasor diagram for resistive load R.

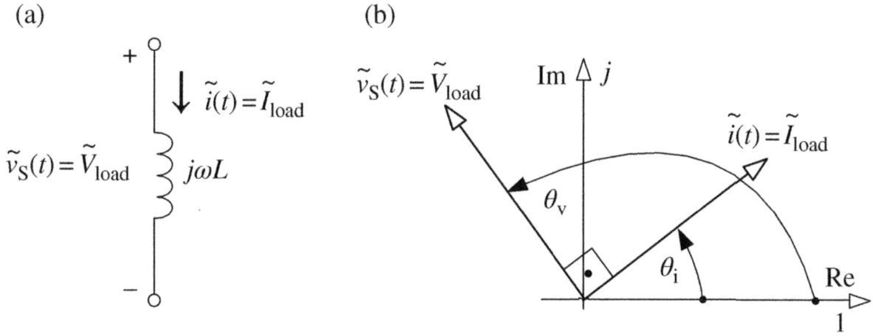

Figure 5.9 (a) Inductive reactive load $j\omega L$. (b) Phasor diagram for inductive reactive load $j\omega L$.

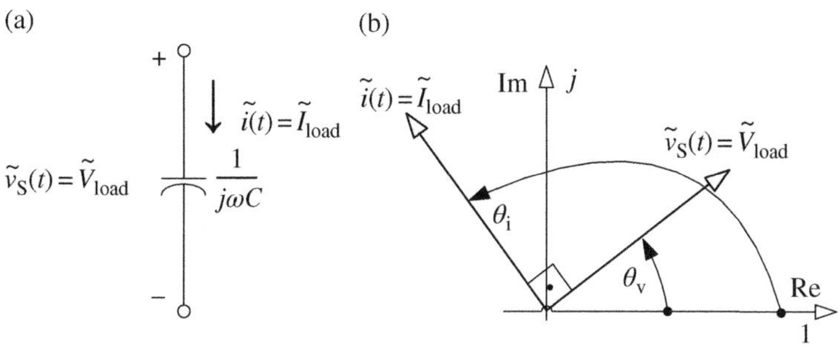

Figure 5.10 (a) Capacitive reactive load $1/j\omega C = -j/\omega C$. (b) Phasor diagram for capacitive reactive load $1/j\omega C = (-j/\omega C)$.

Therefore,

$$\tilde{Z}_{load} = \frac{|\tilde{V}_{load}|\angle\theta_v}{|\tilde{I}_{load}|\angle\theta_i} = |\tilde{Z}_{load}|\angle(\theta_v - \theta_i) = (1/\omega C)\angle-90°\Omega. \quad (5.25a)$$

The power factor is

$$\text{PF} = \cos(\theta_v - \theta_i) = \cos(-90) = 0. \quad (5.25b)$$

4) For resistive–inductive loads (RL) (see Figure 5.11a), $\tilde{V}_{load}$ leads $\tilde{I}_{load}$ by less than 90°, that is, $(\theta_v - \theta_i) < 90^\circ$.

Therefore,

$$\tilde{Z}_{load} = \frac{|\tilde{V}_{load}|\angle\theta_v}{|\tilde{I}_{load}|\angle\theta_i} = |\tilde{Z}_{load}|\angle(\theta_v - \theta_i). \quad (5.26a)$$

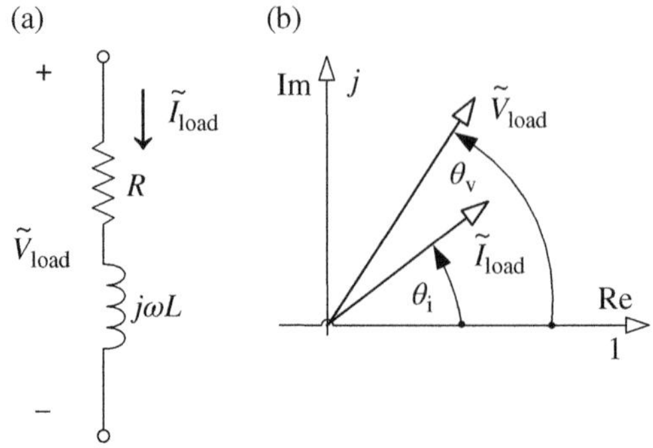

Figure 5.11 (a) Resistive–inductive load ($R + j\omega L$). (b) Phasor diagram for resistive–inductive load ($R + j\omega L$).

The power factor is

$$0 < \cos(\theta_v - \theta_i) < 1.0. \tag{5.26b}$$

Figure 5.11b shows the corresponding phasor diagram.

5) For resistive–capacitive loads (RC) (see Figure 5.12a), $\widetilde{V}_{load}$ lags $\widetilde{I}_{load}$ by less than 90°, that is, $(\theta_i - \theta_v) < 90°$.

Therefore,

$$\widetilde{Z}_{load} = \frac{|\widetilde{V}_{load}|\angle\theta_v}{|\widetilde{I}_{load}|\angle\theta_i} = |\widetilde{Z}_{load}|\angle(\theta_v - \theta_i). \tag{5.27a}$$

The power factor is

$$0 < \cos(\theta_i - \theta_v) < 1.0. \tag{5.27b}$$

Figure 5.12b shows the corresponding phasor diagram.

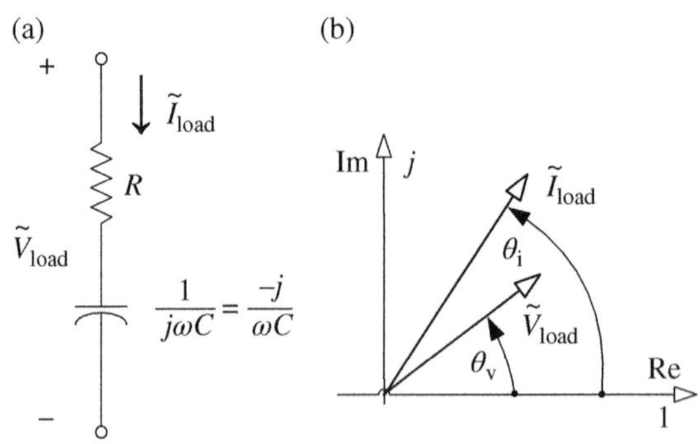

Figure 5.12 (a) Resistive–capacitive load ($R + 1/j\omega C$). (b) Phasor diagram for resistive–capacitive load ($R + 1/j\omega C$).

Application Example 5.2 Calculation of line current, line loss, and real power delivered by a source through a transmission line for different load fundamental power factors

a) An industrial single-phase load (see Figure E5.2.1) consumes $P_{\text{load}} = 88$ kW at a fundamental load power factor $\text{PF}_{\text{load}} = 0.707$ lagging at a load voltage magnitude $|\tilde{V}_{\text{rms load}}| = 480$ V supplied from a voltage source (s) $\tilde{V}_s$. The transmission line resistance is $R_{\text{line}} = 0.08\ \Omega$. Determine the real (average) power that must be supplied by the source with source voltage $\tilde{V}_s$.
b) If the fundamental load power factor is changed (see Application Example 5.4, where PF is increased by a capacitor in parallel to load) to $\text{PF}_{\text{load}} = 0.90$ lagging, repeat the analysis.

Solution

a) $\text{PF}_{\text{load}} = 0.707$ lagging

The real power delivered by the source voltage is

$$P_s = P_{\text{load}} + R_{\text{line}}\left|\tilde{I}_{\text{rms load}}\right|^2, \tag{E5.2.1}$$

where $|\tilde{I}_{\text{rms load}}|$ is calculated from $P_{\text{load}} = |\tilde{V}_{\text{rms load}}||\tilde{I}_{\text{rms load}}| \cos(\theta_v - \theta_i)$ at $\text{PF}_{\text{load}} = \cos(\theta_v - \theta_i) = 0.707$ pu, resulting in

$$\left|\tilde{I}_{\text{rms load}}\right| = \frac{P_{\text{load}}}{\left|\tilde{V}_{\text{rms load}}\right| \cos(\theta_v - \theta_i)} = \frac{88\ \text{kW}}{480\ \text{V} \cdot 0.707} = 259.31\ \text{A}. \tag{E5.2.2}$$

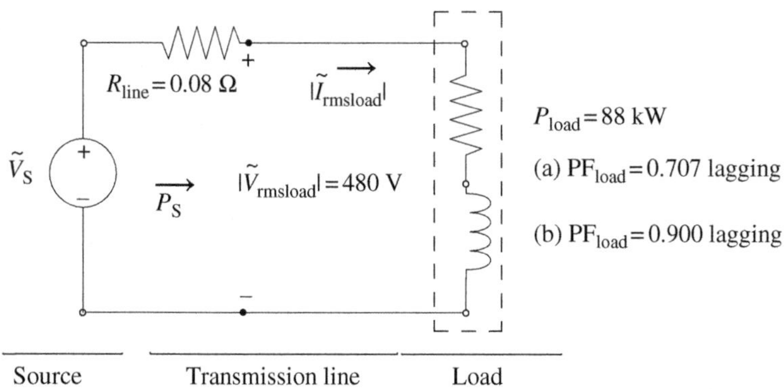

Figure E5.2.1 Transmission line supplying real power to load with different load fundamental power factors.

The source or sending end real (average) power is now

$$P_s = 88\ \text{kW} + 0.08(259.31)^2\ \text{kW} = 93.38\ \text{kW}, \tag{E5.2.3}$$

with the line loss

$$P_{\text{line}} = P_s - P_{\text{load}} = 93.38\ \text{kW} - 88\ \text{kW} = 5.38\ \text{kW}. \tag{E5.2.4}$$

b) $\text{PF}_{\text{load}} = 0.90$ lagging (increased by connecting a capacitor across load; see Application Example 5.4)

$$\left|\tilde{I}_{\text{rms load}}\right| = \frac{P_{\text{load}}}{\left|\tilde{V}_{\text{rms load}}\right| \cos(\theta_v - \theta_i)} = \frac{88\ \text{kW}}{480\ \text{V} \cdot 0.90} = 203.7\ \text{A}, \tag{E5.2.5}$$

$$P_s = 88\ \text{kW} + 0.08(203.7)^2\ \text{kW} = 91.32\ \text{kW}, \tag{E5.2.6}$$

$$P_{\text{line}} = P_s - P_{\text{load}} = 91.32\ \text{kW} - 88\ \text{kW} = 3.32\ \text{kW}. \tag{E5.2.7}$$

Conclusion

By increasing the fundamental power factor of the load PF_{load} from 0.707 to 0.90, the transmission line losses can be reduced significantly.

5.6 Complex Power

It is convenient to introduce the concept of complex power $\tilde{S}$. It represents the addition of the real power P and the reactive power Q in the complex plane spanned by the imaginary and real part axes as it will be explained below. All power calculations within electric machines and power systems are based on the complex power.

$$\tilde{S} = \tilde{V}_{\text{rms}} \tilde{I}^*_{\text{rms}}, \tag{5.28}$$

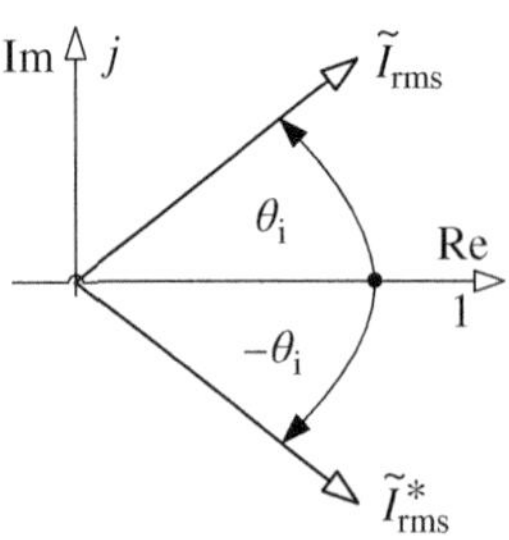

Figure 5.13 $\tilde{I}^*_{\text{rms}}$ is the complex conjugate of $\tilde{I}_{\text{rms}}$.

where $\tilde{I}^*_{\text{rms}}$ is the complex conjugate of $\tilde{I}_{\text{rms}}$. Figure 5.13 illustrates these two phasors in the complex plane, where $\tilde{I}_{\text{rms}} = |\tilde{I}_{\text{rms}}|\angle\theta_i$ and $\tilde{I}^*_{\text{rms}} = |\tilde{I}_{\text{rms}}|\angle -\theta_i$.

The complex power is now

$$\tilde{S} = \tilde{V}_{\text{rms}} \tilde{I}^*_{\text{rms}} = \left(|\tilde{V}_{\text{rms}}|\angle\theta_v\right)\left(|\tilde{I}_{\text{rms}}|\angle -\theta_i\right) = |\tilde{V}_{\text{rms}}||\tilde{I}_{\text{rms}}|\angle(\theta_v - \theta_i). \tag{5.29}$$

Figure 5.14 shows the phasors $\tilde{V}_{\text{rms}}$ and $\tilde{I}_{\text{rms}}$ in the complex plane.

Applying Euler's relation to Eq. (5.29), one gets in terms of cosine and sine terms

$$\widetilde{S} = |\widetilde{V}_{rms}||\widetilde{I}_{rms}|[\cos(\theta_v - \theta_i) + j\sin(\theta_v - \theta_i)]$$
$$= P + jQ \text{ with the unit volt-ampere (VA),} \quad (5.30)$$

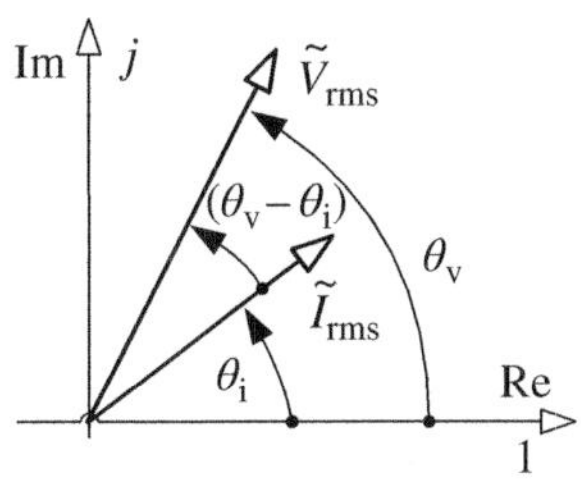

Figure 5.14 $\widetilde{V}_{rms}$ and $\widetilde{I}_{rms}$ in complex plane.

in Eq. (5.30) the real (average) power is

$$P = \text{Re}\{\widetilde{S}\} = |\widetilde{V}_{rms}||\widetilde{I}_{rms}|\cos(\theta_v - \theta_i) \text{ with the unit of watts (W),} \quad (5.31)$$

and the imaginary or reactive power is

$$Q = \text{Im}\{\widetilde{S}\} = |\widetilde{V}_{rms}||\widetilde{I}_{rms}|\sin(\theta_v - \theta_i) \text{ with the unit of volt-ampere reactive (VAr).} \quad (5.32)$$

The following relations are important as well:

$$P = |\widetilde{I}_{rms}|^2 \text{Re}\{\widetilde{Z}_{load}\}, \text{with the unit watts (W),} \quad (5.33a)$$

$$Q = |\widetilde{I}_{rms}|^2 \text{Im}\{\widetilde{Z}_{load}\}, \text{with the unit volt-ampere reactive (VAr),} \quad (5.33b)$$

$$|\widetilde{S}| = |\widetilde{I}_{rms}|^2 |\widetilde{Z}_{load}|, \text{with the unit volt-ampere (VA).} \quad (5.34a)$$

Conclusions

Complex ($\widetilde{S}$), real (P), and reactive (Q) powers can be represented in the complex plane as follows:

1) A resistive load as depicted in Figure 5.15a and b results with Eqs. (5.29) and (5.30) and $(\theta_v - \theta_i) = 0$ or $\theta_v = \theta_i = 38^\circ$ with $\widetilde{V}_{rms} = |\widetilde{V}_{rms}|\angle\theta_v = |\widetilde{V}_{rms}|\angle 38^\circ$, $\widetilde{I}_{rms} = |\widetilde{I}_{rms}|\angle\theta_i = |\widetilde{I}_{rms}|\angle 38^\circ$ in the real (resistive R) power

$$\widetilde{S} = |\widetilde{V}_{rms}||\widetilde{I}_{rms}|\angle(\theta_v - \theta_i) = |\widetilde{V}_{rms}||\widetilde{I}_{rms}|\angle 0^\circ = |\widetilde{V}_{rms}||\widetilde{I}_{rms}| = P_R, \quad (5.34b)$$

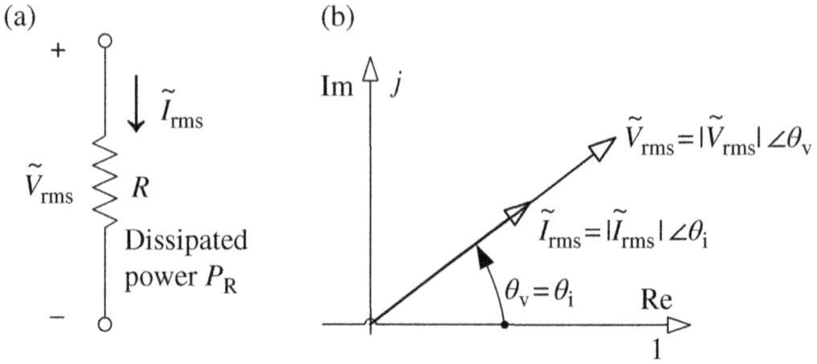

Figure 5.15 (a) Resistive load (R) with $\tilde{V}_{rms}$ and $\tilde{I}_{rms}$ and dissipated power P_R (b) $\tilde{V}_{rms}$ and $\tilde{I}_{rms}$ in complex plane.

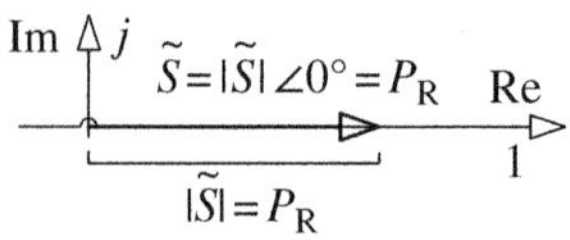

Figure 5.16 Real (average, resistive) power phasor $\tilde{S} = P_R$ in complex plane.

whereby the reactive power $Q_R = 0$. Figure 5.16 shows the real (average) power phasor $\tilde{S} = P_R$ in complex plane. Note that any angle $\theta_v = \theta_i$ (e.g. 38°) can be assumed as shown in Figure 5.15b.

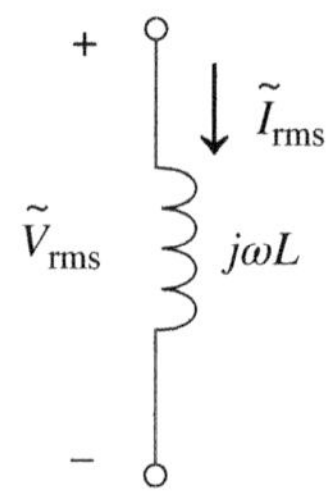

Figure 5.17 Inductive load $j\omega L$ with $\tilde{V}_{rms}$ and $\tilde{I}_{rms}$.

Conclusion

A resistive load absorbs no (zero) reactive power.

2) Inductive load ($j\omega L$) as depicted in Figure 5.17 results with Eqs. (5.29) and (5.30) and $(\theta_v - \theta_i) = 90°$ for $\theta_v = 0°$ in $\theta_i = -90°$; therefore $\tilde{V}_{rms} = |\tilde{V}_{rms}| \angle \theta_v = |\tilde{V}_{rms}| \angle 0°$, $\tilde{I}_{rms} = |\tilde{I}_{rms}| \angle \theta_i = |\tilde{I}_{rms}| \angle -90°$, generating the reactive inductive (L) power:

$$\begin{aligned}\tilde{S} &= |\tilde{V}_{rms}||\tilde{I}_{rms}| \angle (\theta_v - \theta_i) = |\tilde{V}_{rms}||\tilde{I}_{rms}| \angle -90° = |\tilde{S}| \angle -90° \\ &= |\tilde{S}| [\cos(-90°) + j \sin(-90°)] = |\tilde{S}| [0 - j] = -j|\tilde{S}| = -jQ_L,\end{aligned} \tag{5.35}$$

whereby the real power $P_L = 0$. Figures 5.18 and 5.19 show the voltage and current phasors and the reactive inductive power phasor $\tilde{S} = -jQ_L$ in complex plane, respectively.

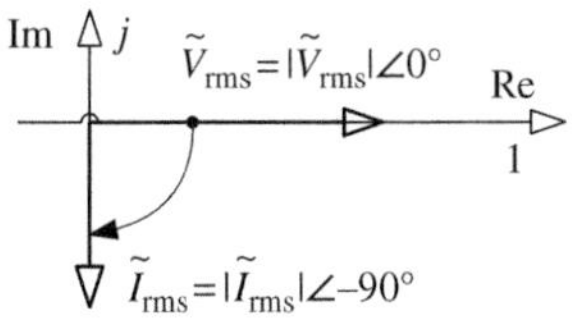

Figure 5.18 Voltage and current phasors for inductive load.

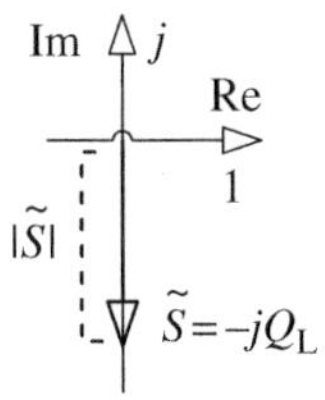

Figure 5.19 Reactive inductive power phasor $\tilde{S} = -jQ_L$ in complex plane.

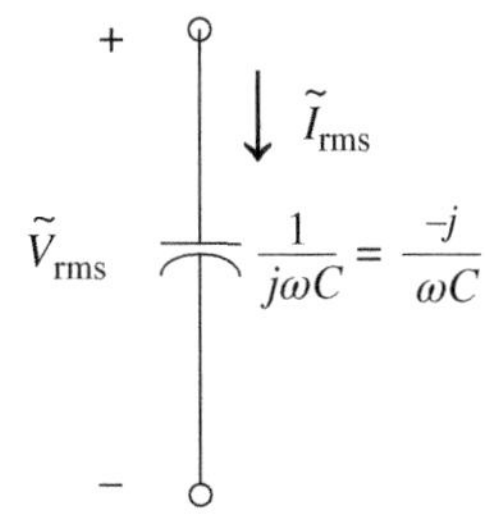

Figure 5.20 Capacitive load $1/j\omega C = -j/\omega C$ with $\tilde{V}_{rms}$ and $\tilde{I}_{rms}$.

Conclusion

An inductive load absorbs (negative) reactive power.

3) Capacitive load ($1/j\omega C = -j/\omega C$) as depicted in Figure 5.20 results with Eqs. (5.29) and (5.30) and $(\theta_v - \theta_i) = -90°$ for $\theta_v = 0°$ in $\theta_i = 90°$; therefore $\tilde{V}_{rms} = |\tilde{V}_{rms}|\angle\theta_v = |\tilde{V}_{rms}|\angle 0°$, $\tilde{I}_{rms} = |\tilde{I}_{rms}|\angle\theta_i = |\tilde{I}_{rms}|\angle 90°$, generating the reactive capacitive (C) power:

$$\begin{aligned}\tilde{S} &= |\tilde{V}_{rms}||\tilde{I}_{rms}|\angle(\theta_v - \theta_i)\\ &= |\tilde{V}_{rms}||\tilde{I}_{rms}|\angle 90° = |\tilde{S}|\angle 90°\\ &= |\tilde{S}|[\cos(90°) + j\sin(90°)]\\ &= |\tilde{S}|[0 + j] = j|\tilde{S}| = jQ_C,\end{aligned} \tag{5.36a}$$

whereby the real power $P_C = 0$. Figures 5.21 and 5.22 show the voltage and current phasors and the reactive capacitive power phasor $\tilde{S} = jQ_C$ in complex plane, respectively.

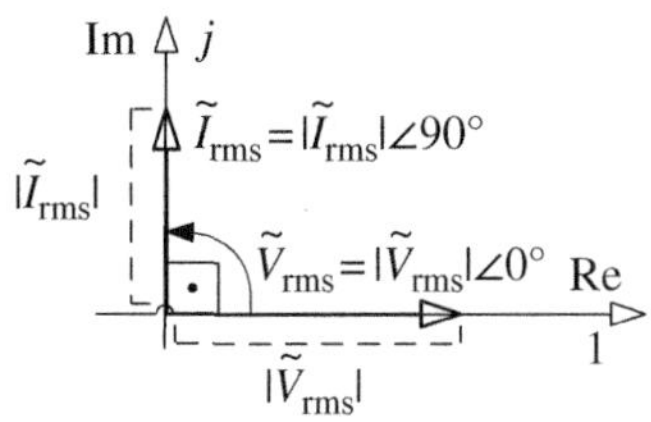

Figure 5.21 Voltage and current phasors for capacitive load.

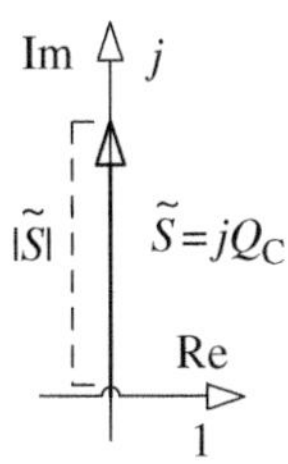

Figure 5.22 Reactive capacitive power phasor $\tilde{S} = jQ_C$ in complex plane.

Conclusion

A capacitive load supplies (positive) reactive power.

4) A resistive–inductive load $(R + j\omega L)$ as depicted in Figure 5.23 results with Eqs. (5.29) and (5.30) and, for example, $(\theta_v - \theta_i)$=45° for $\theta_v = 0°$ in $\theta_i = -45°$; therefore $\widetilde{V}_{rms} = |\widetilde{V}_{rms}|\angle\theta_v = |\widetilde{V}_{rms}|\angle 0°$, $\widetilde{I}_{rms} = |\widetilde{I}_{rms}|\angle\theta_i = |\widetilde{I}_{rms}|\angle -45°$, generating the resistive–inductive (RL) power:

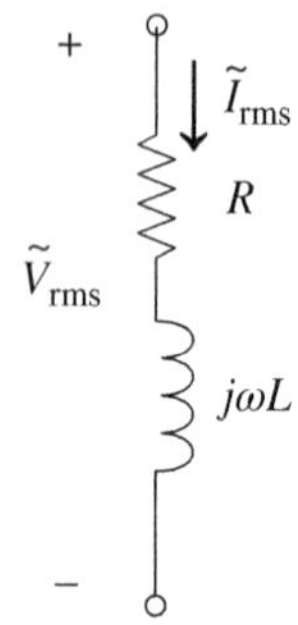

Figure 5.23 Resistive–inductive load $(R + j\omega L)$ with $\widetilde{V}_{rms}$ and $\widetilde{I}_{rms}$.

$$\begin{aligned}\widetilde{S} &= |\widetilde{V}_{rms}||\widetilde{I}_{rms}|\angle(\theta_v - \theta_i)\\ &= |\widetilde{V}_{rms}||\widetilde{I}_{rms}|\angle -45°\\ &= |\widetilde{S}|\angle -45°\\ &= |\widetilde{S}|[\cos(-45°) + j\sin(-45°)]\\ &= P_{RL} - jQ_{RL},\end{aligned} \tag{5.36b}$$

whereby the real power $P_{RL} = |\widetilde{S}|\cos(-45°)$. Figures 5.24 and 5.25 show the voltage and current phasors and the resistive–inductive (RL) power phasor $Q_{RL} = |\widetilde{S}|\sin(-45°)$ in complex plane, respectively.

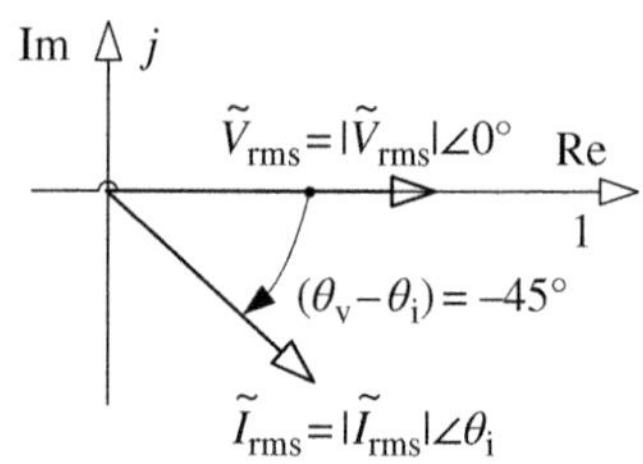

Figure 5.24 Voltage and current phasors for resistive–inductive load.

Conclusion

A resistive–inductive load absorbs (negative) reactive power.

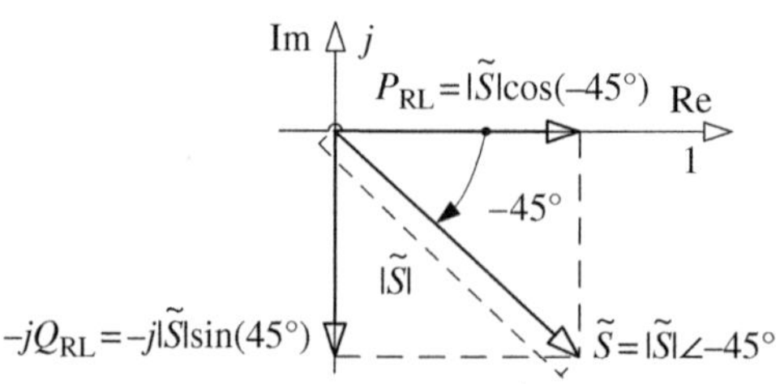

Figure 5.25 Reactive resistive–inductive power phasor $\widetilde{S} = P_{RL} - jQ_{RL}$ in complex plane.

5) A resistive–capacitive load ($R + 1/j\omega C = R - j/\omega C$) as depicted in Figure 5.26 results with Eqs. (5.29) and (5.30) and, for example, $(\theta_v - \theta_i) = -45°$ for $\theta_v = 0°$ in $\theta_i = 45°$; therefore $\tilde{V}_{rms} = |\tilde{V}_{rms}|\angle\theta_v = |\tilde{V}_{rms}|\angle 0°$, $\tilde{I}_{rms} = |\tilde{I}_{rms}|\angle\theta_i = |\tilde{I}_{rms}|\angle 45°$, generating the resistive–capacitive power:

$$\begin{aligned}\tilde{S} &= |\tilde{V}_{rms}||\tilde{I}_{rms}|\angle(\theta_v - \theta_i)\\ &= |\tilde{V}_{rms}||\tilde{I}_{rms}|\angle 45°\\ &= |\tilde{S}|\angle 45°\\ &= |\tilde{S}|[\cos(45°) + j\sin(45°)]\\ &= P_{RC} + jQ_{RC}\end{aligned} \tag{5.36c}$$

whereby the real power $P_{RC} = |\tilde{S}|\cos(45°)$. Figures 5.27 and 5.28 show the voltage and current phasors and the resistive–capacitive (RC) power phasor $Q_{RC} = |\tilde{S}|\sin(45°)$ in complex plane, respectively.

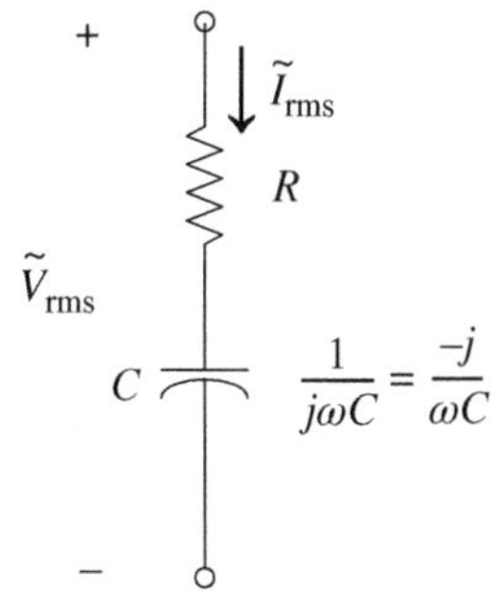

Figure 5.26 Capacitive load ($R + 1/j\omega C$) = ($R - j/\omega C$) with $\tilde{V}_{rms}$ and $\tilde{I}_{rms}$.

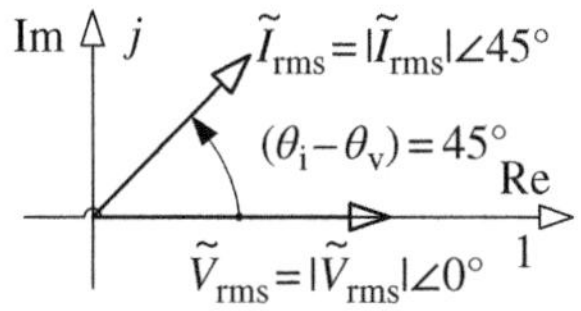

Figure 5.27 Voltage and current phasors for resistive–capacitive load.

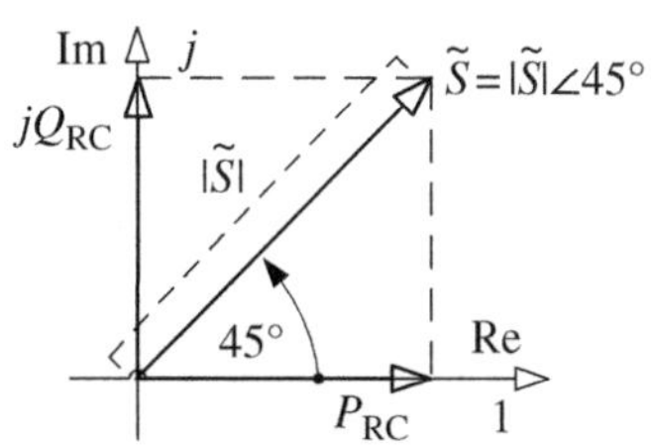

Figure 5.28 Reactive resistive–capacitive power phasor $\tilde{S} = P_{RC} + jQ_{RC}$ in complex plane.

Conclusion

A resistive–capacitive load supplies (positive) reactive power.

Application Example 5.3 Analysis of a transmission/distribution line feeder (Figure E5.3.1)

Solution

a) Complex power approach

The fundamental power factor at the load is $PF_{load} = \cos(\theta_v - \theta_i) = 0.85$ pu or its power factor angle is $(\theta_v - \theta_i) = \cos^{-1}(0.85) = 31.789°$. Selecting the load voltage $\tilde{V}_{rms\ load} = |\tilde{V}_{rms\ load}|\angle\theta_v = 480\ V\angle 0°$ as reference phasor ($\theta_v = 0$), then $\theta_i = -31.789°$. The minus sign indicates a lagging current; in other words, the load absorbs/consumes reactive power Q_{load}. Figure E5.3.2 shows the load voltage and its load current, and

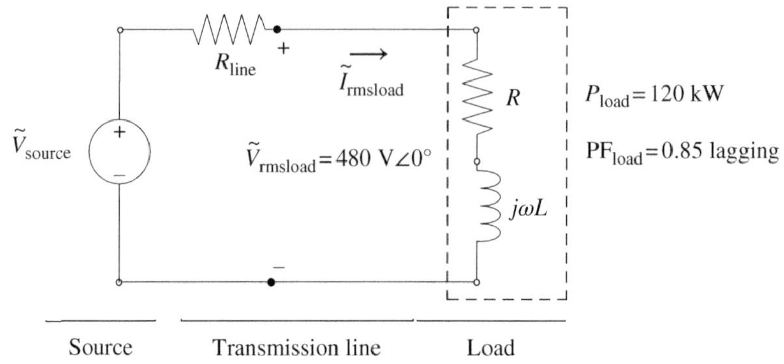

Figure E5.3.1 Transmission/distribution line feeder.

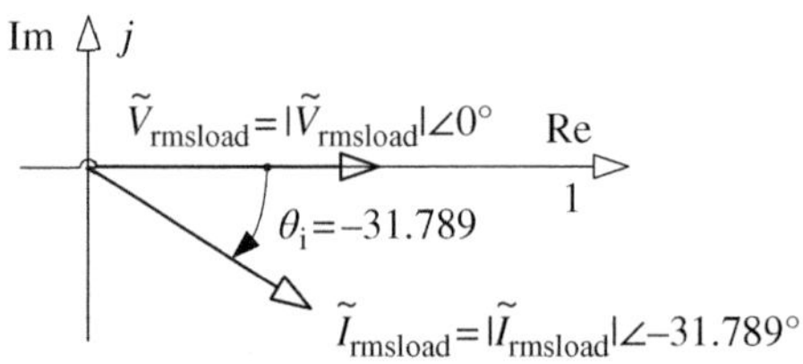

Figure E5.3.2 Load voltage and load current.

Figure E5.3.3 depicts the real and reactive load powers P_{load} and Q_{load}, respectively.

Note the reactive power phasor lags the real power phasor, or, in other words, the reactive power is absorbed/consumed by the load.

From Figure E5.3.3 one gets $\tan(-31.789°) = \frac{Q_{\text{load}}}{P_{\text{load}}}$ or

$$Q_{\text{load}} = P_{\text{load}} \cdot \tan(-31.789°) = -120\,\text{kW} \cdot 0.619\,74 = -74.369\,\text{kVAr}. \tag{E5.3.1}$$

With Eq. (5.30) one obtains

$$\tilde{S}_{\text{load}} = P_{\text{load}} + jQ_{\text{load}} = 120\,\text{kW} - j74.369\,\text{kVAr} = 141.18\,\text{kVA}\angle -31.789°. \tag{E5.3.2}$$

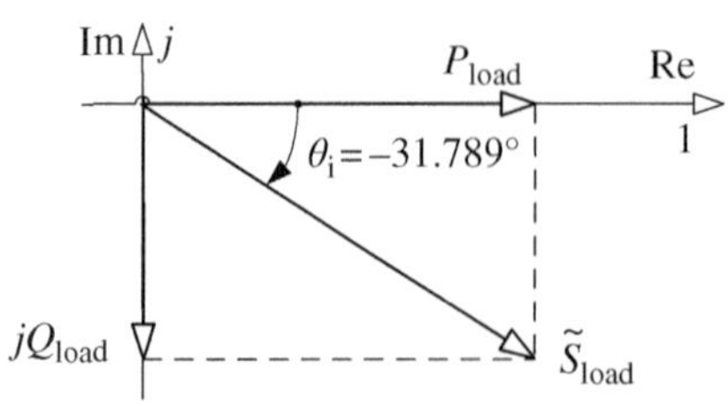

Figure E5.3.3 Load real and reactive powers P_{load} and Q_{load}, respectively, as well as the complex load power.

The magnitude of the load current is

$$\left|\widetilde{I}_{\text{rms load}}\right| = \frac{\left|\widetilde{S}_{\text{load}}\right|}{\left|\widetilde{V}_{\text{rms load}}\right|} = \frac{141.18\,\text{kVA}}{480\,\text{V}} = 294.13\,\text{A}. \qquad \text{(E5.3.3)}$$

The real power loss in the transmission/distribution line is

$$P_{\text{line}} = \left|\widetilde{I}_{\text{rms load}}\right|^2 R_{\text{line}} = (294.13)^2 0.1 = 8.65\,\text{kW}, \qquad \text{(E5.3.4)}$$

and the real power that must be supplied by the source is

$$P_{\text{s}} = P_{\text{load}} + P_{\text{line}} = 120\,\text{kW} + 8.65\,\text{kW} = 128.65\,\text{kW}. \qquad \text{(E5.3.5)}$$

Since the transmission line does not absorb reactive power $Q_{\text{line}} = 0$, the complex power delivered by the source is $Q_{\text{s}} = Q_{\text{load}}$:

$$\widetilde{S}_{\text{s}} = P_{\text{s}} + jQ_{\text{s}} = 128.65\,\text{kW} - j74.369\,\text{kVAr} = 148.60\angle - 30.03°\text{kVA}. \qquad \text{(E5.3.6)}$$

The magnitude of the source voltage is

$$\left|\widetilde{V}_{\text{s}}\right| = \frac{\left|\widetilde{S}_{\text{s}}\right|}{\left|\widetilde{I}_{\text{rms load}}\right|} = \frac{148.60\,\text{kVA}}{294.13\,\text{A}} = 505.22\,\text{V} \qquad \text{(E5.3.7)}$$

at a fundamental power factor at the source (see Eq. (E5.3.6)):

$$\text{PF}_{\text{s}} = \cos(-30.03°) = 0.8658\ \text{lagging}. \qquad \text{(E5.3.8)}$$

b) From part (a) use the results of Eqs. (E5.3.2) and (E5.3.3) and apply the KVL to the circuit of Figure E5.3.1, which yields $\widetilde{V}_{\text{s}} - \widetilde{I}_{\text{rms load}} R_{\text{line}} - \widetilde{V}_{\text{rms load}} = 0$ and results in

$$\begin{aligned}\widetilde{V}_{\text{s}} &= \widetilde{V}_{\text{rms load}} + \widetilde{I}_{\text{rms load}} R_{\text{line}} \\ &= 480\angle 0°\text{V} + (294.13\angle - 31.789°)0.1\,\text{V} = 505.22\angle - 1.76°\text{V}.\end{aligned} \qquad \text{(E5.3.9)}$$

The fundamental power factor angle of the source voltage is −1.76°, and the fundamental power factor angle of the current is −31.789°; therefore the fundamental power factor angle between source voltage and current is the difference 1.76° + (−31.789°) = −30.029°:

$$\text{PF}_{\text{s}} = \cos(-30.029°) = 0.866\ \text{lagging}. \qquad \text{(E5.3.10)}$$

Figure E5.3.4 shows the phasor voltages of the load and the source as well as the load current.

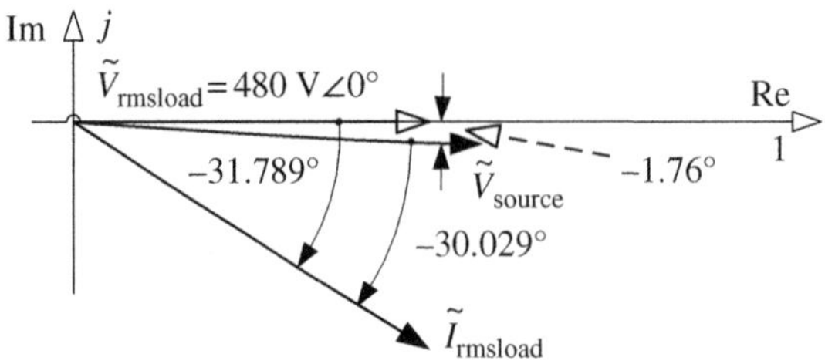

Figure E5.3.4 Phasor rms voltages of the load and the source as well as the load rms current.

5.7 Fundamental Power Factor Correction

Because of induction motor drives (about 60% of total load within an interconnected power system), industrial loads normally have lagging fundamental power factors; that is, the current $\tilde{I}_{\text{rms load}}$ lags the load voltage $\tilde{V}_{\text{rms load}}$. Note that the total installed electric power capacity within the United States is about $P_{total} = 1200$ GW $= 1.2 \cdot 10^{15}$ W, with a spinning reserve of about 100 GW. In order to appreciate the amount of electrical power generated, one can state that 1200 GW consumed by 340 million people within the United States provides 1200 GW/340 million $\approx$ 3.5 kW per person. Realizing that one person can generate about 75 W on a steady-state basis, then 3.5 kW/75 W $\approx$ 47, so-called energy slaves are required for each person's electrical power needs.

In view of the results obtained in Application Example 5.3, we are naturally led to ask if there is any convenient technique for raising the fundamental power factor of a load to make the power system more efficient. Usually a load power factor in the range 0.8 pu lagging $\leq PF_{load} \leq 1$ pu is desirable. The common method of raising the fundamental power factor is through the connection of parallel capacitors to the load impedance, as shown in Figure 5.29.

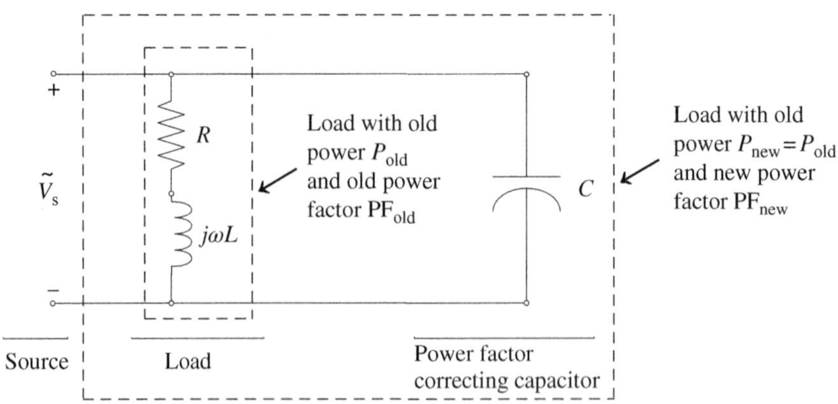

Figure 5.29 Circuit for power factor correction.

The above circuit illustrates a typical single-phase ($R + j\omega L$) load (e.g. single-phase representation of three-phase induction motors; see Chapter 12). In parallel with this load, we have connected a capacitor C as proposed in Application Example 5.2. Assume that the original complex power (without the capacitor C) for the load $\tilde{Z}_{\text{load}} = (R + j\omega L)$ is denoted as (see Figure 5.30)

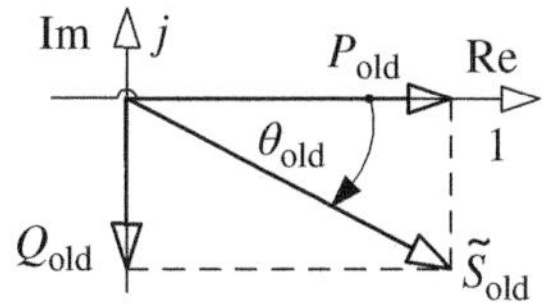

Figure 5.30 Power relationships before the capacitor C is connected. Note $P_{\text{old}} > 0$ and $Q_{\text{old}} < 0$.

$$\tilde{S}_{\text{old}} = P_{\text{old}} + jQ_{\text{old}} = \left|\tilde{S}_{\text{old}}\right| \angle\theta_{\text{old}}, \tag{5.37}$$

where

$$Q_{\text{old}} = P_{\text{old}} \cdot \tan(\theta_{\text{old}}). \tag{5.38}$$

The new complex power that results from adding a capacitor C is

$$\tilde{S}_{\text{new}} = P_{\text{new}} + jQ_{\text{new}} = \left|\tilde{S}_{\text{new}}\right| \angle\theta_{\text{new}}, \tag{5.39}$$

where

$$Q_{\text{new}} = P_{\text{old}} \cdot \tan(\theta_{\text{new}}). \tag{5.40}$$

Note that θ_{new} is specified by the required (new) fundamental power factor. The difference between the new and old complex power is caused by the addition of the capacitor C. The required complex power of the capacitor C is

$$\begin{aligned} \tilde{S}_{\text{capacitor}} &= \tilde{S}_{\text{new}} - \tilde{S}_{\text{old}} = (P_{\text{old}} + jQ_{\text{new}}) - (P_{\text{old}} + jQ_{\text{old}}) \\ &= j(Q_{\text{new}} - Q_{\text{old}}) = jQ_{\text{capacitor}} = j\omega C\left|\tilde{V}_{\text{rms}}\right|^2. \end{aligned} \tag{5.41}$$

Figure 5.31 illustrates the fundamental power phasors for the old (uncompensated) and new (compensated) reactive power conditions including the required reactive power generated by the capacitor C.

Figure 5.31 Power phasors for old (uncompensated) and new (compensated) conditions.

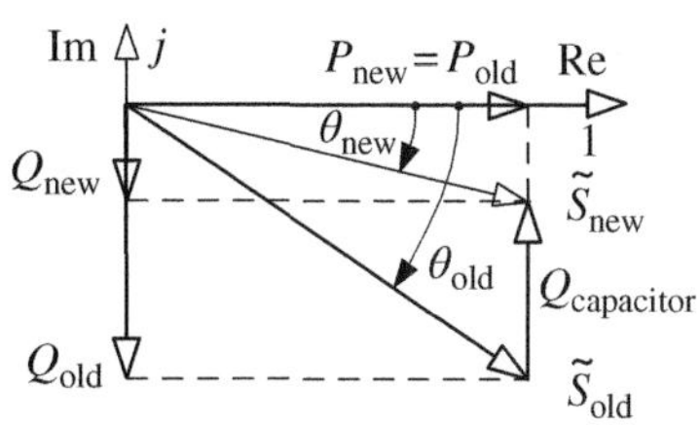

Application Example 5.4 Power factor compensation (increase of fundamental power factor) of an industrial load

An industrial load consisting of a bank of (single-phase) induction motors consumes $P_{\text{load old}} = 50\,\text{kW}$ at $\text{PF}_{\text{load old}} = \cos(\theta_{\text{old}}) = 0.8$ lagging. We wish to raise the fundamental power factor to $\text{PF}_{\text{load new}} = \cos(\theta_{\text{new}}) = 0.95$ lagging by placing a bank of capacitors C in parallel with the load as shown in Figure E5.4.1. Note that the source voltage is $\tilde{V}_{\text{source}} = 220\,\text{V}\angle 0°$ at a grid frequency of $f = 60\,\text{Hz}$. It is important to note that any fundamental power factor compensation must not produce leading power factors because of possible oscillations and instabilities within the power system.

Figure E5.4.2 depicts the real power ($P_{\text{load old}}$) and inductive reactive power ($Q_{\text{load old}}$) phasors in the complex plane.

Solution

In Figure E5.4.2 the old lagging (minus sign) fundamental power factor angle is

$$\Theta_{\text{old}} = \cos^{-1}(0.8) = -36.87°; \tag{E5.4.1}$$

the associated old inductive reactive power is then

$$\begin{aligned} Q_{\text{load old}} &= P_{\text{load old}} \cdot \tan(\theta_{\text{old}}) \\ &= 50 \cdot 10^3(-0.75) = -37.5\,\text{kVAr}. \end{aligned} \tag{E5.4.2}$$

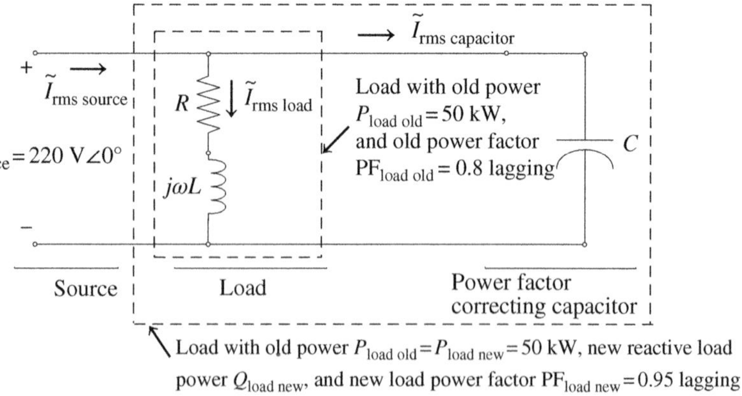

Figure E5.4.1 Power factor increase of a $(R + j\omega L)$ load with parallel capacitor bank C.

Hence, the complex old load power becomes

$$\tilde{S}_{\text{load old}} = P_{\text{load old}} + jQ_{\text{load old}} = 50\,\text{kW} - j37.5\,\text{kVAr}. \tag{E5.4.3}$$

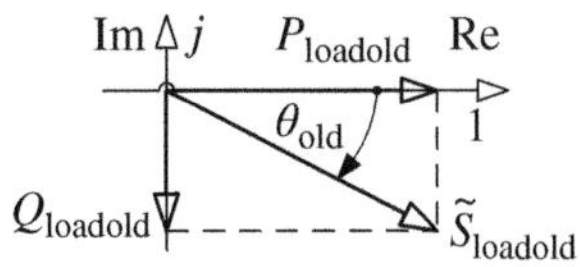

Figure E5.4.2 Real power ($P_{\text{load old}}$) and inductive reactive power ($Q_{\text{load old}}$) phasors in the complex plane.

The required new power factor including the capacitor bank C is

$$\text{PF}_{\text{load new}} = \cos(\theta_{\text{new}}) = 0.95 \text{ lagging}, \tag{E5.4.4}$$

resulting in the new power factor angle $\theta_{\text{new}} = -18.19^\circ$. The minus sign indicates lagging fundamental power factor (that is, the load current lags the load voltage). The new complex load power is defined as

$$\begin{aligned}\tilde{S}_{\text{load new}} &= P_{\text{load new}} + jQ_{\text{load new}} = P_{\text{load old}} + jP_{\text{load old}} \cdot \tan(\theta_{\text{new}}) \\ &= 50\,\text{kW} + 50\,\text{kW}\ \tan(-18.19^\circ) = 50\,\text{kW} - j16.43\,\text{kVAr};\end{aligned} \tag{E5.4.5}$$

therefore, the complex power delivered by the capacitor bank is

$$\begin{aligned}\tilde{S}_{\text{capacitor}} &= \tilde{S}_{\text{load new}} - \tilde{S}_{\text{load old}} = j(Q_{\text{load new}} - Q_{\text{load old}}) \\ &= jQ_{\text{capacitor}} = j[(-16.43\,\text{kVAr}) - (-37.5\,\text{kVAr})] \\ &= j21.07\,\text{kVAr}.\end{aligned} \tag{E5.4.6}$$

The capacitor reactive power is the reactive power delivered by the capacitor C:

$$\begin{aligned}\tilde{S}_{\text{capacitor}} &= \tilde{V}_{\text{rms capacitor}}\tilde{I}_{\text{rms capacitor}} \\ &= \left|\tilde{V}_{\text{rms source}}\right| \frac{\left|\tilde{V}_{\text{rms source}}\right|}{\dfrac{1}{j\omega C}} \\ &= \left|\tilde{V}_{\text{rms source}}\right|^2 j\omega C;\end{aligned} \tag{E5.4.7a}$$

the required capacitive apparent power is

$$S_{\text{capacitor}} = \left|\tilde{S}_{\text{capacitor}}\right| = \omega C \left|\tilde{V}_{\text{rms source}}\right|^2, \tag{E5.4.7b}$$

resulting in the capacitance C of the capacitor bank to be connected in parallel to the inductive load $(R + j\omega L)$:

$$C = \frac{\left|\tilde{S}_{\text{capacitor}}\right|}{\omega\left|\tilde{V}_{\text{rms source}}\right|^2} = \frac{21\,079\,\text{VA}}{377 \cdot (220)^2} = 1.15\,\text{mF} = 1150\,\mu\text{F}. \tag{E5.4.8}$$

5.8 Residential Single-Phase AC Power Circuits in the United States

The National Electrical Code (NEC) [3] and the National Electrical Safety Code [4] (NESC) are important documents for the installation and operation of residential AC power circuits. In the United States, the residential power system consists of a single-phase pole transformer providing two voltages to a residence where 120 V is used for lighting equipment and 240 V for appliances. Figure 5.32 shows such a single-phase circuit with a center-tapped pole transformer, residing on a pole or on the ground, supplying about 10 residences with electric power.

A three-phase substation (sub) transformer of $S_{sub} = 5$ MVA can supply about 200 pole transformers, each with an apparent power of $S_{pole} = 25$ kVA. In order to allow for residential growth, usually much fewer pole transformers (e.g. 100) are initially served by one 5 MVA substation transformer at a residential subdivision. Figure 5.33 shows the Δ-Y-connected substation transformer where one phase of the secondary supplies energy to about (5 MVA/3)/25 kVA = 66 pole transformers. The primary line-to-line voltage of the Δ-connected substation transformer is mostly $V_{\text{L-Lprimary\_rms}} = 138$ kV, while the secondary Y-connected phase voltage of the substation transformer is $V_{\text{L-Nsecondary\_rms}} = 7.2$ kV, resulting in a line-to-line voltage of $V_{\text{L-Lsecondary\_rms}} = 7.2\text{ kV} \cdot \sqrt{3} = 12.47$ kV.

Note that in the United States the distribution lines are mostly overhead lines because cities cover a large area but with low density of buildings (e.g. Colorado Springs) and underground distribution lines are not possible because cable capacitances resonate with substation transformer inductances, producing ferroresonance [1, 2] above secondary substation transformer voltages of $V_{\text{L-Lsecondary\_rms}} = 20$ kV.

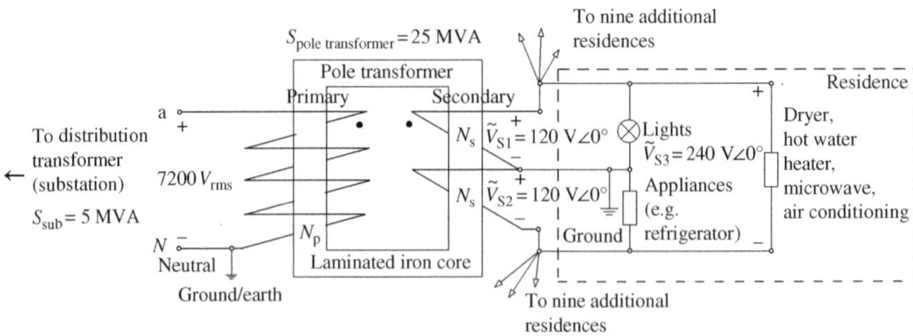

Figure 5.32 Residential power circuit consisting of a center-tapped $S_{pole} = 25$ kVA pole transformer supplying electric power to lighting at 120 V and appliances at 240 V.

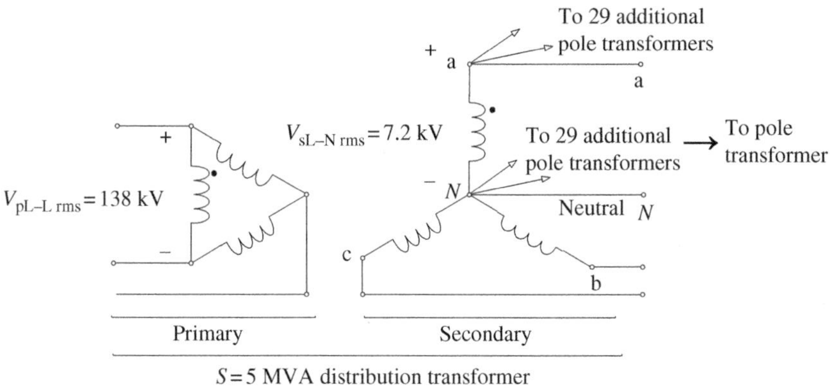

Figure 5.33 Residential Δ-Y-connected substation transformer, allowing for load growth up to about 100 pole transformers.

5.8.1 Power Requirements for Lighting Equipment

In the past, incandescent lightbulbs [5] were used. In recent years, builders rely on compact fluorescent lightbulbs [6] (CFLs) and light-emitting diodes [7] (LEDs). The "Système International [8] (SI)" in French or in English "International System of Units" defines the units of a lighting system such as:

1) lumen (lm), the luminous flux representing a measure of the perceived power of light,
2) candela (cd), the base unit of luminous intensity such that 1 lm = 1 cd·steradian = 1 lx·m^2,
3) lux (lx), the unit of illuminance,
4) kelvin (K), the unit for light color [9] (color of light, luminous color).

Application Example 5.5 Comparison of incandescent, CFL, and LED lightbulbs

An incandescent lightbulb consuming $P_{\text{incandescent}} = 60$ W in real power provides 600–850 lumens at a warm light with 2700 K at a fundamental power factor of $\cos\theta_{\text{incandescent}} = 1.0$, and no harmonic filters [2] are required. A CFL lightbulb with the same lumen and kelvin numbers consumes $P_{\text{CFL}} = 13$ W at a fundamental power factor of $\cos\theta_{\text{CFL}} \approx 0.75$ pu. This means the CFL lightbulb consumes 78% less real power than an incandescent one. However, it consumes an additional nonzero reactive power, and additionally it generates current harmonics, which must be reduced by filters; otherwise, the total power factor including harmonics must be considered.

A LED with similar properties (806 lumen, 2700 kelvin) consumes $P_{LED} = 9.5$ W at 110–120 V and input current of 0.106 A, resulting in a power factor of $\cos\theta_{LED} = 9.5\ \text{W}/(120\ \text{V}\cdot 0.106\ \text{A}) = 0.747$ pu. In addition a LED generates harmonic currents that must be reduced by appropriate filters [2]; otherwise, the total power factor including harmonics must be considered. Although a power reduction of 84% is achieved with the LED, the power quality problems [2] cannot be disregarded.

Note that the unit of kelvin [K] is used either for temperature measurements or the luminous color or light color of an illumination device.

Application Example 5.6 Analysis of a residential power system

Three residential power circuits are composed in Figure 5.32 of a $V_{rms1} = 120$ V lighting branch, an appliance branch $V_{rms2} = 120$ V, and a $V_{rms3} = 240$ V branch consisting of an appliance requiring larger power such as an electric dryer, as depicted in Figure E5.6.1.

The lighting branch must illuminate an area of 640 square feet [ft^2], the appliance branch serves 10 electrical receptacles (outlets), and the dryer is rated at $S_{dryer} = 5$ kVA.

a) Determine the circuit breaker required for each individual circuit.
b) Determine the currents in all parts of the power network.
c) Determine the monthly cost at $0.12/kWh for operating lights, appliances, and dryer 40, 8, and 3% of the time, respectively.

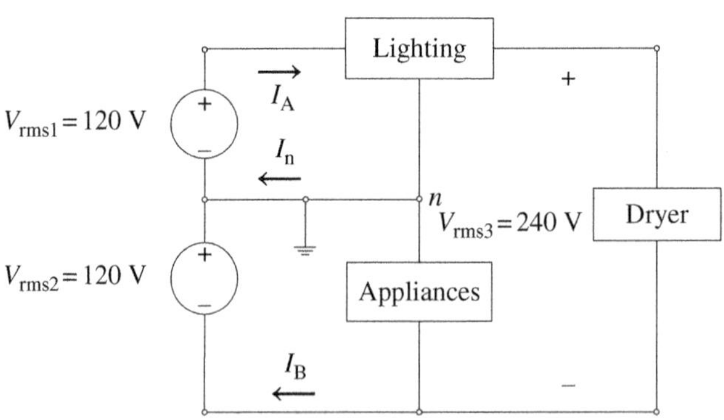

Figure E5.6.1 Typical three-branch circuit of a residence.

Solution

a) The general rule in the past for residential illumination was 3 W required per ft^2. In view of CFL and LED lighting, this power requirement can be reduced, say, to $0.6\,W/ft^2$, resulting in $P_{lighting} = [0.6W/(ft)^2]\cdot[640(ft)^2] = 384\,W$; operating approximately at a power factor of $\cos\theta_{lighting} \approx 0.75$ lagging results in an apparent power of $S_{lighting} = 384\,W/0.75 = 512\,VA$. At a voltage of $V_{rms1} = 120\,V$ the current drawn is $I_{lighting\_rms} = (512\,VA)/(120\,V) = 4.3\,A$. For this current an American wire gauge [1] (AWG) at a current density of $j = 1.4\,A/mm^2$ of AWG#12 is recommended. For the rated current of 4.3 A, a circuit breaker of 5.5 A must be used. Note that the current density for electrical apparatus without forced air cooling is usually from 1 to $2\,A/mm^2$ as listed in Table 4.2 of reference 2.

For the appliance branch $S_{receptacle} = 180\,VA$ per receptacle is usually recommended. This gives for 10 receptacles $S_{appliance} = 10 \cdot S_{receptacle} = 10 \cdot 180 = 1800\,VA$. At a voltage of $V_{rms2} = 120\,V$, a current of $I_{appliance\_rms} = 1800/120\,A = 15\,A$ is drawn from the grid requiring a wire size of AWG#7 with $j = 1.52\,A/mm^2$ and a circuit breaker of 20 A rating.

The electric dryer requires $S_{dryer} = 5\,kVA$ at $V_{rms3} = 240\,V$ drawing from the distribution system $I_{dryer\_rms} = 5000 \cdot VA/240\,V = 20.83\,A$ requiring an AWG#5 with $j = 1.34\,A/mm^2$ and a circuit breaker rating of 30 A.

b) The currents in all parts of the network of Figure E5.6.1 are with the definitions of apparent power $S = V_{rms}I_{rms}$ and real power $P = V_{rms}I_{rms}\cos\theta = S\cos\theta$, where $\cos\theta = 1$ pu is assumed because it is not known, resulting in total current drawn from transformer: I_A = lighting current + dryer current = 4.3 A + 20.83 A = 25.13 A, total current returns to transformer: I_B = appliance current + dryer current = 15 A + 20.83 A = 35.83 A, and total current fed into neutral (n): $I_n = I_A - I_B = 25.13\,A - 35.83\,A = (-)10.70\,A$.

It would be desirable to reduce the neutral current by reconfiguring lighting equipment, appliances, and the dryer in order to thermally balance the pole transformer.

c) The electricity cost per month assuming 30 days = $24 \cdot 30 = 720$ h is (electricity cost)/month = \$0.12/kWh[288 h · 0.512 kVA + 57.6 h · 1.8 kVA + 21.6 h · 5.0 kVA] = \$0.12/kWh[147.46 kWh + 103.68 kWh + 108 kWh] = \$43.096, which corresponds to 359.14 kWh used per month. This is small compared with the average electric energy used [10] in the United States of $E_{residential\_average}$ = 901 kWh per month; however, it is large compared with a highly insulated house [2] in Germany (see Chapter 1, Application Example 1.6, Table E1.6.1) with the energy used for lighting and appliances $E_{residential\_insulated\_house}$ = 601 kWh/year or 50.08 kWh/month, which does not include the energy supplied by the PV system. Overall estimate for $E_{residential\_insulated\_house}$ amounts to about 60–80 kWh/month.

5.9 Three-Phase Distribution and Transmission Networks

Up to this point we have dealt with single-phase circuits. Now we extend the analysis techniques to three-phase circuits, that is, circuits containing three voltage sources that are one-third of an $f = 50$ or 60 Hz cycle apart in time. Three-phase or polyphase circuits are important because they are technically better suited and more economical to transmit electric energy than single-phase networks. All high-voltage transmission lines [1] ranging from line-to-line voltages of $V_{L\text{-}L} = 230–765$ kV have three-phase wires and one ground wire. Figure 5.34 shows a typical three-phase transmission tower whereby two parallel three-phase systems are suspended and where for the reduction of magnetic fields on the ground a reverse-phased, double-circuit three-phase line [11] is employed.

In order to estimate the line-to-line voltage of a transmission line, one can estimate the length of the insulators as shown in Figure 5.34. For each 100 kV about 1 m length of the insulators is required. For example, if the insulator length is 4 m, then the transmission line has a line-to-line voltage of 380 kV. Figure 5.35 illustrates the reverse phasing. In order to make the inductances of the three-phase wires about the same, a transposition of the three-phase wires must take place about every 10 miles (~16 km).

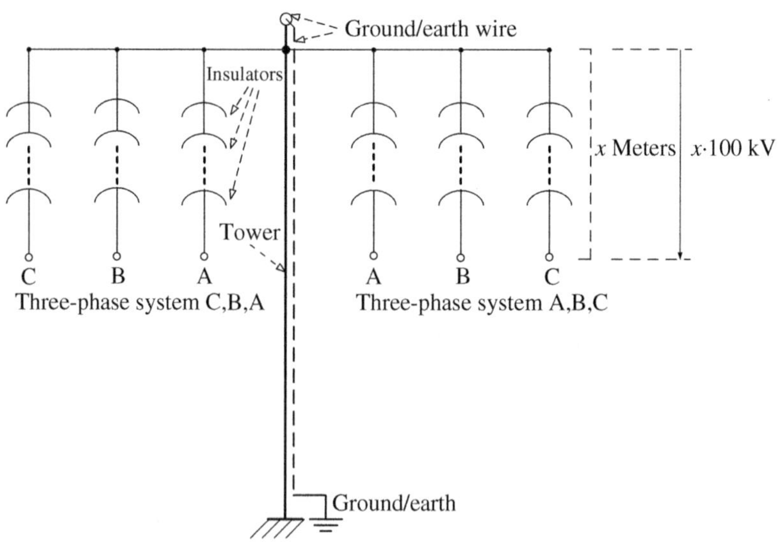

Figure 5.34 Two three-phase systems suspended on transmission tower.

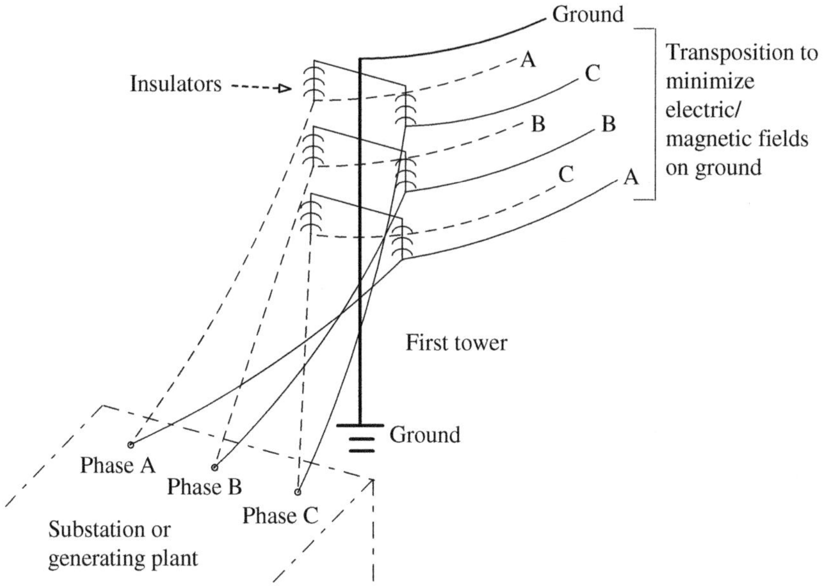

Figure 5.35 Reduction of magnetic fields on the ground by employing a reverse-phased, double-circuit three-phase line [11].

If the three sinusoidal voltages have the same amplitude and each is 120° out of phase with the other two, the voltages are said to be balanced. If the loads are such that the currents produced by the voltages are also balanced, the entire circuit is referred to as balanced three-phase circuit. A balanced set of three-phase voltages as depicted in Figure 5.36 can be formulated in the time domain as defined in Eqs. (5.42a)–5.42c and can be represented in the time domain as indicated in Figure 5.37.

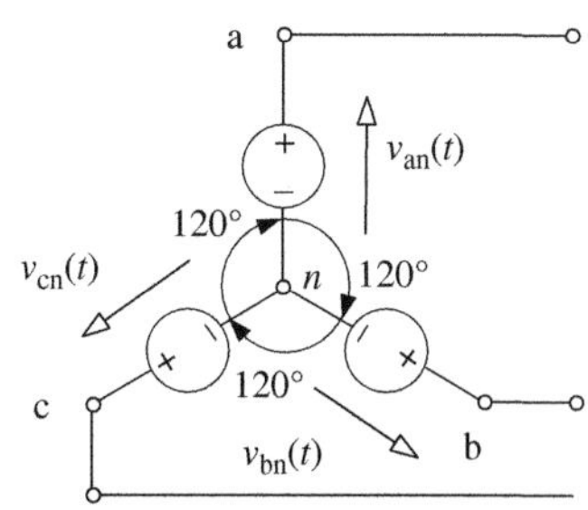

Figure 5.36 Time-domain representation of balanced voltages $v_{an}(t)$, $v_{bn}(t)$, $v_{cn}(t)$ of Y-connection.

$$v_{an}(t) = V_{anM}\cos(\omega t)\ [\mathrm{V}], \tag{5.42a}$$

$$v_{bn}(t) = V_{anM}\cos(\omega t - 120°)\ [\mathrm{V}], \tag{5.42b}$$

$$v_{cn}(\mathrm{t}) = V_{anM}\cos(\omega t - 240°)\ [\mathrm{V}]. \tag{5.42c}$$

The Y-connected rms voltages are defined in the phasor domain as follows, exhibiting positive-phase sequence (0°, −120°, −240°, rotating in clockwise direction):

$$\widetilde{V}_{an} = V_{an\ rms}\angle 0°, \tag{5.43a}$$

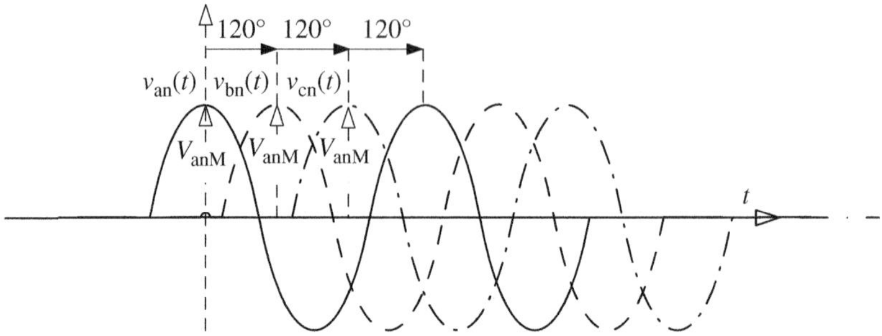

Figure 5.37 Three-phase Y-connected voltage sources, for example, in an electric generator.

$$\tilde{V}_{bn} = V_{an\ rms}\angle -120°, \tag{5.43b}$$

$$\tilde{V}_{cn} = V_{an\ rms}\angle -240° = V_{an\ rms}\angle +120°. \tag{5.43c}$$

If the load is balanced, the Y-connected currents produced by the source are for positive-phase sequence

$$i_a(t) = I_M \cos(\omega t + \theta_i)\ [\text{A}], \tag{5.44a}$$

$$i_b(t) = I_M \cos(\omega t - 120° + \theta_i)\ [\text{A}], \tag{5.44b}$$

$$i_c(t) = I_M \cos(\omega t - 240° + \theta_i)\ [\text{A}], \tag{5.44c}$$

where θ_i is either leading ($\theta_i > 0$) or lagging ($\theta_i < 0$) phase angle between voltage and current of each and every phase. Figure 5.38 shows the three-phase phasor voltages $\tilde{V}_{an}$, $\tilde{V}_{bn}$, $\tilde{V}_{cn}$ in the complex plane.

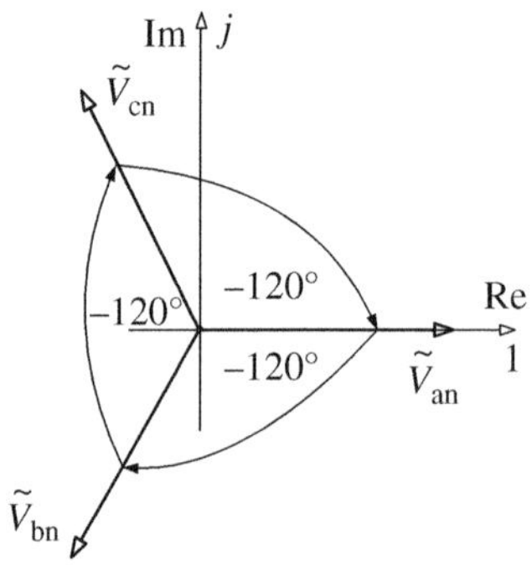

Figure 5.38 Three-phase Y-connected phasor voltages $\tilde{V}_{an}$, $\tilde{V}_{bn}$, $\tilde{V}_{cn}$ in the complex plane.

The instantaneous power of the three-phase system is now

$$\begin{aligned} p(t) &= p_a(t) + p_b(t) + p_c(t) \\ &= V_M I_M[\cos\omega t \cdot \cos(\omega t + \theta_i) + \cos(\omega t - 120°) \\ &\quad \cdot \cos(\omega t - 120° + \theta_i) + \cos(\omega t - 240°) \cdot \cos(\omega t - 240° + \theta_i)]; \end{aligned} \tag{5.45}$$

using the trigonometric relation $\cos\alpha\ \cos\beta = (1/2)[\cos(\alpha - \beta) + \cos(\alpha + \beta)$ and $[\cos\phi + \cos(\phi - 120°) + \cos(\phi - 240°)] = 0$ leads to

$$p(t) = P = 3\frac{V_M I_M}{2}\cos\theta_i = 3\frac{V_M}{\sqrt{2}}\frac{I_M}{\sqrt{2}}\cos\theta_i = 3V_{rms}I_{rms}\cos\theta_i. \tag{5.46}$$

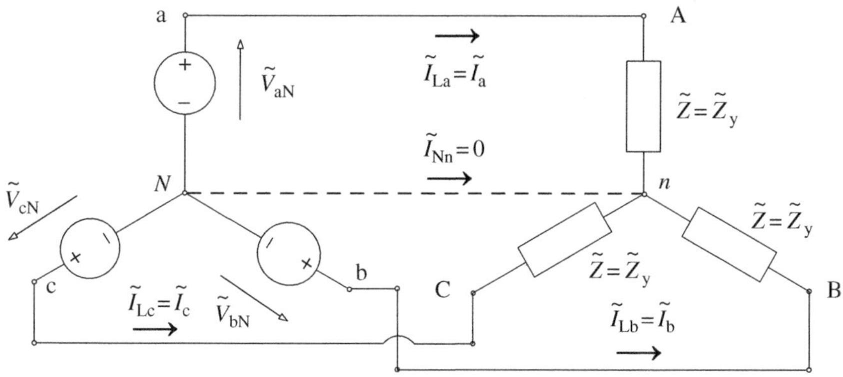

Figure 5.39 Y-Y configuration.

Equation (5.46) is not a function of time, but rather is a function of rms voltage V_{rms} and current I_{rms}, including fundamental or displacement or fundamental power factor PF $= \cos\theta_i$. This means that in contrast to the single-phase system where the power $p(t)$ is pulsating with twice the excitation frequency $f = \omega/(2\pi)$, in a balanced three-phase or polyphase system, the real power P or the resulting torque τ is always constant, which reduces wear on rotating machines or generators: a good reason to use or generate power in balanced three-phase or polyphase form.

The source and the load can be connected in either Y or Δ. Therefore one obtains four different source–load configurations. The Y-Δ and Δ-Y configurations eliminate the so-called zero-sequence [2] harmonics and are therefore mostly applied in power system networks. For convenient calculation, the Δ source or load is transformed into an equivalent Y as indicated in Figures 5.40–5.42 through dashed lines.

Y-Y configuration:

Figure 5.39 shows the Y-Y configuration with its associated voltages and currents.

Y-Δ configuration:

Figure 5.40 shows the Y-Δ configuration with its associated voltages and currents. This configuration is the most used one in distribution/transmission systems.

Δ-Y configuration:

Figure 5.41 shows the Δ-Y configuration with its associated voltages and currents.

Δ-Δ configuration:

Figure 5.42 shows the Δ-Δ configuration with its associated voltages and currents.

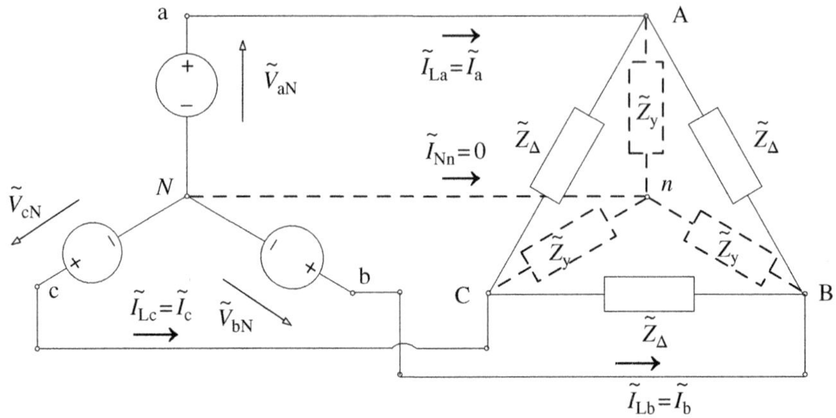

Figure 5.40 Y-Δ configuration.

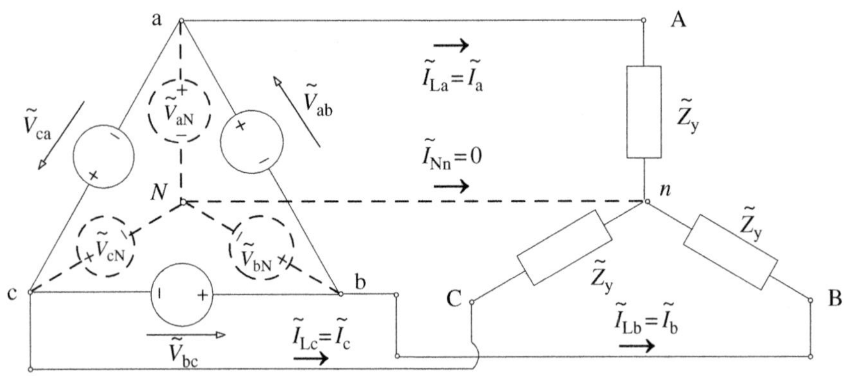

Figure 5.41 Δ-Y configuration.

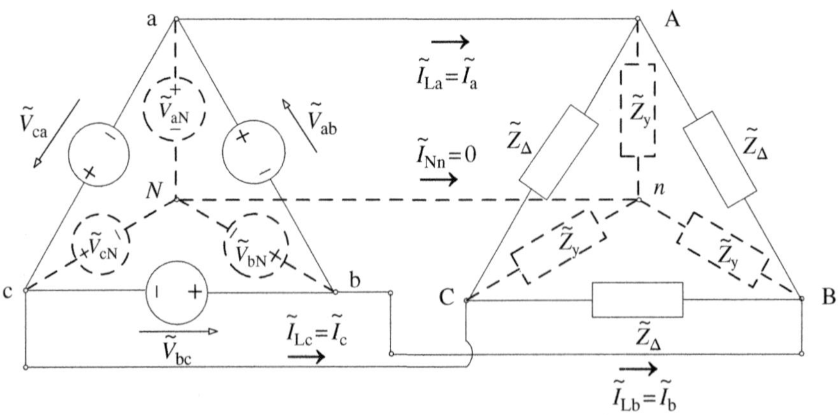

Figure 5.42 Δ-Δ configuration.

5.9.1 Balanced Wye (Y) Source-Wye (Y) Load Connection

Figure 5.43a represents the balanced Y-Y connection at the source and at the load, respectively.

The phase voltages of the Y-connected source in the complex plane are defined as

$$\tilde{V}_{\text{aN}} = \left|\tilde{V}_{\text{aN rms}}\right| \angle 0°, \tag{5.47a}$$

$$\tilde{V}_{\text{bN}} = \left|\tilde{V}_{\text{aN rms}}\right| \angle -120°, \tag{5.47b}$$

$$\tilde{V}_{\text{cN}} = \left|\tilde{V}_{\text{aN rms}}\right| \angle -240° = \left|\tilde{V}_{\text{aN rms}}\right| \angle +120°. \tag{5.47c}$$

The phase currents of the Y-connected source are with the phase impedance of the y-connected load $\tilde{Z}_{\text{y}}$:

$$\tilde{I}_{\text{a}} = \frac{\tilde{V}_{\text{aN}}}{\tilde{Z}_{\text{y}}} = \frac{\left|\tilde{V}_{\text{aN rms}}\right| \angle 0°}{\tilde{Z}_{\text{y}}}, \tag{5.48a}$$

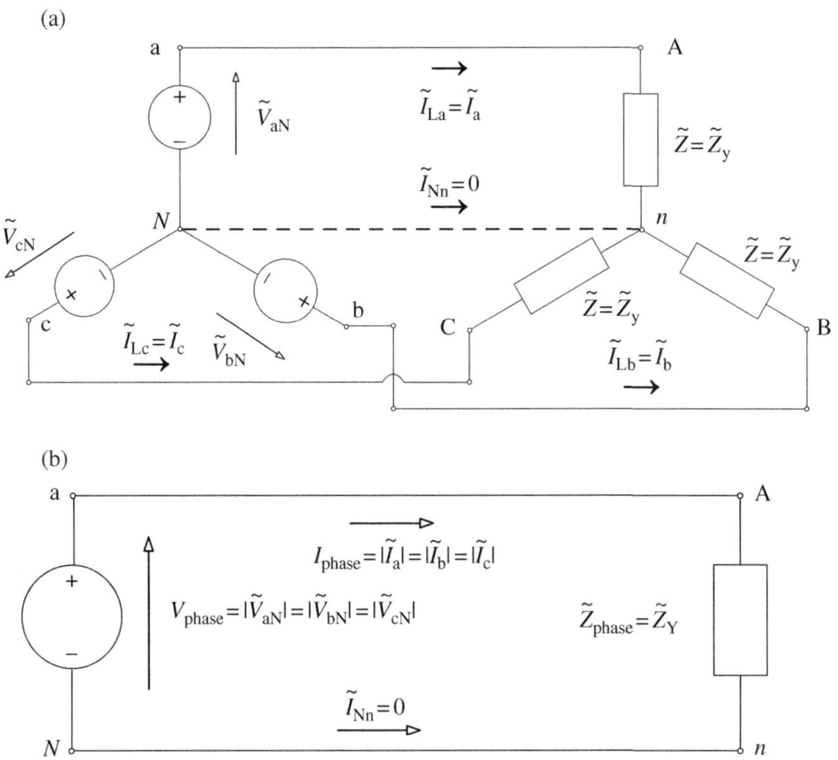

Figure 5.43 (a) Balanced Y-Y connections at source and at load, respectively. (b) Single-phase equivalent circuit for balanced three-phase network of (a).

$$\tilde{I}_b = \frac{\tilde{V}_{bN}}{\tilde{Z}_y} = \frac{|\tilde{V}_{aN\ rms}|\angle -120^\circ}{\tilde{Z}_y}, \tag{5.48b}$$

$$\tilde{I}_c = \frac{\tilde{V}_{cN}}{\tilde{Z}_y} = \frac{|\tilde{V}_{aN\ rms}|\angle -240^\circ}{\tilde{Z}_y} = \frac{|\tilde{V}_{aN\ rms}|\angle 120^\circ}{\tilde{Z}_y}. \tag{5.48c}$$

The sum of the phase currents of Figure 5.43a is $\tilde{I}_{Nn} = \tilde{I}_a + \tilde{I}_b + \tilde{I}_c = 0$, where N and n stand for neutrals (N, n).

The line voltage $\tilde{V}_{ab}$ of the Y-connected source is a function of the phase voltages:

$$\begin{aligned}
\tilde{V}_{ab} &= \tilde{V}_{aN} - \tilde{V}_{bN} = |\tilde{V}_{aN\ rms}|\angle 0^\circ - |\tilde{V}_{aN\ rms}|\angle -120^\circ \\
&= |\tilde{V}_{aN\ rms}| - |\tilde{V}_{aN\ rms}|\left[-\frac{1}{2} - j\frac{\sqrt{3}}{2}\right] \\
&= |\tilde{V}_{aN\ rms}|\left[\frac{3}{2} + j\frac{\sqrt{3}}{2}\right] \\
&= |\tilde{V}_{aN\ rms}|\sqrt{3}\left[\frac{\sqrt{3}}{2} + j\frac{1}{2}\right] \\
&= \sqrt{3}|\tilde{V}_{aN\ rms}|\sqrt{\left(\frac{\sqrt{3}}{2}\right)^2 + \left(\frac{1}{2}\right)^2}\angle\theta_v \\
&= \sqrt{3}|\tilde{V}_{aN\ rms}| \cdot 1\angle\theta_v,
\end{aligned}$$

where $\theta_v = \tan^{-1}\left(\frac{\frac{1}{2}}{\frac{\sqrt{3}}{2}}\right) = \tan^{-1}\left(\frac{1}{\sqrt{3}}\right) = 30^\circ$, resulting in

$$\tilde{V}_{ab} = \sqrt{3}|\tilde{V}_{aN\ rms}|\angle 30^\circ, \tag{5.49a}$$

correspondingly because of balanced three-phase system

$$\tilde{V}_{bc} = \sqrt{3}|\tilde{V}_{aN\ rms}|\angle -90^\circ. \tag{5.49b}$$

$$\tilde{V}_{ca} = \sqrt{3}|\tilde{V}_{aN\ rms}|\angle -210^\circ = \sqrt{3}|\tilde{V}_{aN\ rms}|\angle +150^\circ. \tag{5.49c}$$

The goal of this analysis is the reduction of a balanced three-phase to a single-phase network as depicted in Figure 5.43a and b.

Define

$$V_{phase} = |\tilde{V}_{aN}| = |\tilde{V}_{bN}| = |\tilde{V}_{cN}|, \tag{5.50}$$

$$V_{L-L} = |\tilde{V}_{ab}| = |\tilde{V}_{bc}| = |\tilde{V}_{ca}|, \tag{5.51}$$

$$I_{phase} = |\tilde{I}_a| = |\tilde{I}_b| = |\tilde{I}_c|. \tag{5.52}$$

The relation of the phase (phase) voltage magnitude to the line-to-line (L-L) voltage magnitude is

$$V_{\mathrm{L-L}} = \sqrt{3}V_{\mathrm{phase}}, \tag{5.53}$$

and the relation of the line (L) current magnitude to the phase (phase) current magnitude is

$$I_{\mathrm{L}} = I_{\mathrm{phase}}. \tag{5.54}$$

5.9.2 Balanced Wye (Y) Source-Delta (Δ) Load Connection

Figure 5.44 represents the balanced Y-Δ connection at the source and at the load, respectively.

The phase voltages of the Y-connected source in the complex plane are defined by Eqs. (5.47a–c); the line-to-line (L-L) voltages at the Y-connected source are defined in Eqs. (5.49a–c); and the line-to-line voltages at the Δ-connected load are in the absence of line impedance $\widetilde{Z}_{\mathrm{line}}$:

$$\widetilde{V}_{\mathrm{ab}} = \sqrt{3}\left|\widetilde{V}_{\mathrm{aN\ rms}}\right|\angle 30°, \tag{5.55a}$$

$$\widetilde{V}_{\mathrm{bc}} = \sqrt{3}\left|\widetilde{V}_{\mathrm{aN\ rms}}\right|\angle -90°. \tag{5.55b}$$

$$\widetilde{V}_{\mathrm{ca}} = \sqrt{3}\left|\widetilde{V}_{\mathrm{aN\ rms}}\right|\angle -210°. \tag{5.55c}$$

The line currents from source (node a) to load (node A) are $I_{\mathrm{La}} = \left|\widetilde{I}_{\mathrm{a}}\right| = \left|\widetilde{I}_{\mathrm{b}}\right| = \left|\widetilde{I}_{\mathrm{c}}\right|$.

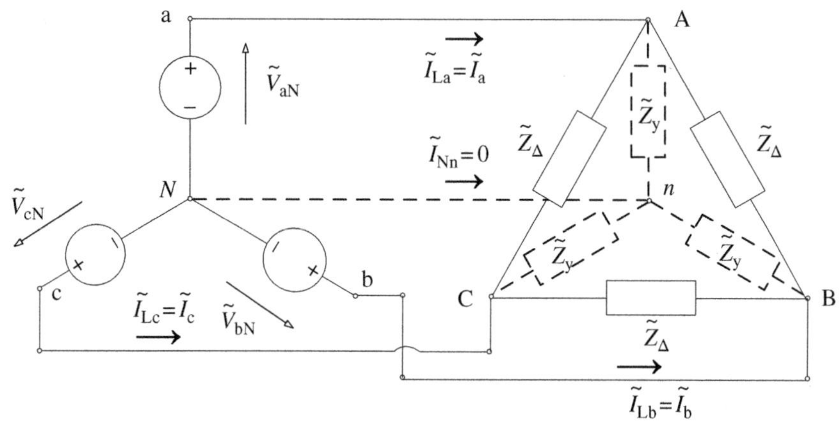

Figure 5.44 Balanced Y-Δ connections at source and at load, respectively.

The equivalent Y-connection impedances of the balanced Δ-load configuration are according to Chapter 2, Eq. (E2.1.4a) resulting in

$$\tilde{Z}_y = \frac{\tilde{Z}_\Delta}{3}. \tag{5.56}$$

5.9.3 Treatment of Delta (Δ)-Connected Source

The voltages of the Δ-connected source are given in Figure 5.45a as

$$\tilde{V}_{ab} = V_{L-L}\angle 0°, \tag{5.57a}$$

$$\tilde{V}_{bc} = V_{L-L}\angle - 120°, \tag{5.57b}$$

$$\tilde{V}_{ca} = V_{L-L}\angle - 240° = V_{L-L}\angle + 120°. \tag{5.57c}$$

The equivalent phase voltages are then shown in Figure 5.45b:

$$\tilde{V}_{an} = \left(\frac{V_{L-L}}{\sqrt{3}}\right)\angle - 30°, \tag{5.57d}$$

$$\tilde{V}_{bn} = \left(\frac{V_{L-L}}{\sqrt{3}}\right)\angle - 150°, \tag{5.57e}$$

$$\tilde{V}_{cn} = \left(\frac{V_{L-L}}{\sqrt{3}}\right)\angle - 270°. \tag{5.57f}$$

Relations between phase and line-to-line (L-L) voltages and line currents are in general for balanced Y-connection (see Figure 5.46a and b):

$$\tilde{V}_{an\ rms} = \left|\tilde{V}_{an\ rms}\right|\angle 0°, \tag{5.58a}$$

$$\tilde{V}_{bn\ rms} = \left|\tilde{V}_{an\ rms}\right|\angle - 120°, \tag{5.58b}$$

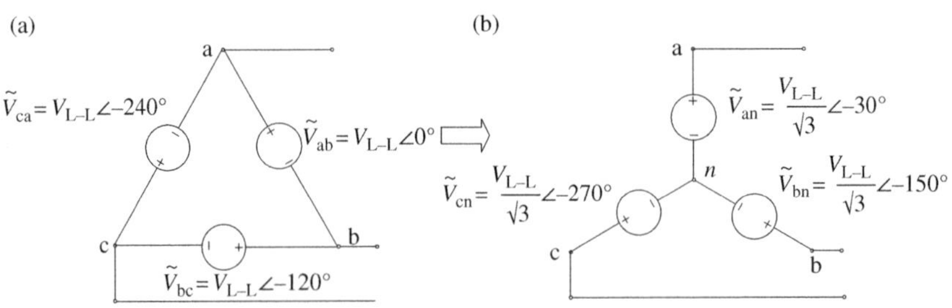

Figure 5.45 (a) Line-to-line (L-L) voltage definition for Δ-connected balanced three-phase source. (b) Phase voltage definition for equivalent Y-connected balanced three-phase source.

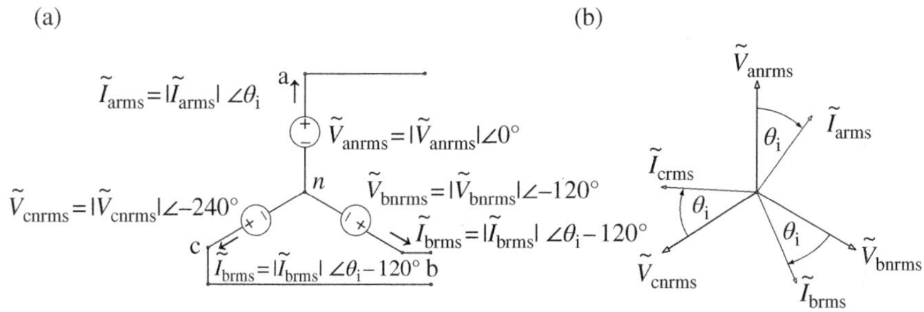

Figure 5.46 (a) Balanced Y-connection (either source or load) definition of voltages and currents. (b) Phasor diagram of balanced Y-connection.

$$\tilde{V}_{\text{cn rms}} = \left|\tilde{V}_{\text{an rms}}\right| \angle -240°. \tag{5.58c}$$

$$\tilde{V}_{\text{ab rms}} = \sqrt{3}\left|\tilde{V}_{\text{an rms}}\right| \angle 30° = \sqrt{3}\left|\tilde{V}_{\text{phase rms}}\right| \angle 30° = \left|\tilde{V}_{\text{L}-\text{L rms}}\right| \angle 30°, \tag{5.59a}$$

$$\tilde{V}_{\text{bc rms}} = \sqrt{3}\left|\tilde{V}_{\text{bn rms}}\right| \angle -90° = \sqrt{3}\left|\tilde{V}_{\text{phase rms}}\right| \angle -90° = \left|\tilde{V}_{\text{L}-\text{L rms}}\right| \angle -90°, \tag{5.59b}$$

$$\tilde{V}_{\text{ca rms}} = \sqrt{3}\left|\tilde{V}_{\text{cn rms}}\right| \angle -210° = \sqrt{3}\left|\tilde{V}_{\text{phase rms}}\right| \angle -210° = \left|\tilde{V}_{\text{L}-\text{L rms}}\right| \angle -210°, \tag{5.59c}$$

with rms magnitudes only $V_{\text{L-L rms}} = \sqrt{3} \cdot V_{\text{phase rms}}$.

The phase currents are

$$\tilde{I}_{\text{a}\ \text{rms}} = \left|\tilde{I}_{\text{a rms}}\right| \angle\theta_i = \left|\tilde{I}_{\text{phase rms}}\right| \angle\theta_i, \tag{5.60a}$$

$$\tilde{I}_{\text{b}\ \text{rms}} = \left|\tilde{I}_{\text{b rms}}\right| \angle(\theta_i - 120°) = \left|\tilde{I}_{\text{phase rms}}\right| \angle(\theta_i - 120°), \tag{5.60b}$$

$$\tilde{I}_{\text{c}\ \text{rms}} = \left|\tilde{I}_{\text{c rms}}\right| \angle(\theta_i - 240°) = \left|\tilde{I}_{\text{phase rms}}\right| \angle(\theta_i - 240°), \tag{5.60c}$$

with rms magnitudes only $I_{\text{a rms}} = I_{\text{phase rms}}$.

Relations between phase current and line current are in general for balanced Δ-connection (see Figure 5.47a and b):

$$\tilde{I}_{\text{ba}} = \tilde{I}_{\text{ba}\ \text{rms}} = \left|\tilde{I}_{\text{ba rms}}\right| \angle\theta_i, \tag{5.61a}$$

$$\tilde{I}_{\text{cb}} = \tilde{I}_{\text{cb}\ \text{rms}} = \left|\tilde{I}_{\text{cb rms}}\right| \angle\left(\theta_i - 120^{\circ}\right), \tag{5.61b}$$

$$\tilde{I}_{\text{ac}} = \tilde{I}_{\text{ac}\ \text{rms}} = \left|\tilde{I}_{\text{ac rms}}\right| \angle\left(\theta_i - 240^{\circ}\right), \tag{5.61c}$$

$$\tilde{I}_{\text{aA rms}} = \tilde{I}_{\text{ba rms}} - \tilde{I}_{\text{ac rms}} = \left|\tilde{I}_{\text{ba rms}}\right| \angle\theta_i - \left|\tilde{I}_{\text{ac rms}}\right| \angle(\theta_i - 240°), \tag{5.62a}$$

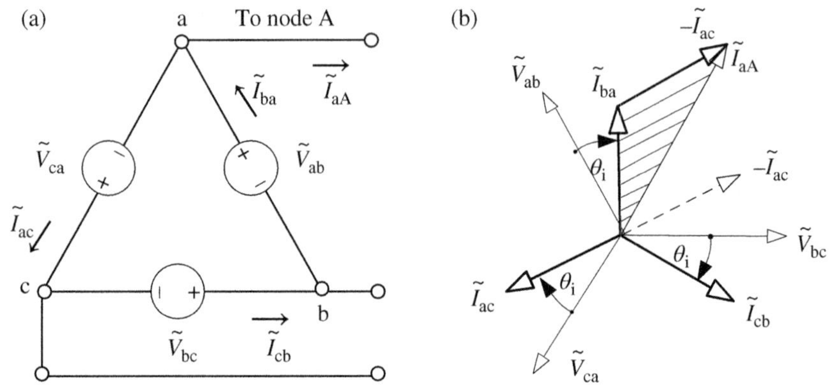

Figure 5.47 (a) Balanced Δ connection (either source or load). (b) Phasor diagram of balanced Δ connection.

with $I_{\text{phase rms}} = \left|\tilde{I}_{\text{ba rms}}\right| = \left|\tilde{I}_{\text{ac rms}}\right|$ based on trigonometric relations applied to the hatched triangle of Figure 5.47b

$$I_L = \left|\tilde{I}_{\text{aA rms}}\right| = \sqrt{3} \cdot I_{\text{phase rms}} = \sqrt{3} \cdot I_{\text{phase}}, \tag{5.62b}$$

or the line current is $\sqrt{3}$ times the phase current: $I_L = \sqrt{3} \cdot I_{\text{phase}}$.

5.9.4 Power Relationships for Three-Phase Balanced Systems

For a single-phase (single) system, the complex power is $\tilde{S}_{\text{single}} = V_{\text{rms}} I_{\text{rms}} e^{j(\theta_v - \theta_i)} = V_{\text{rms}} I_{\text{rms}} \angle(\theta_v - \theta_i)$, and for a balanced three-phase (three) system, one obtains the total complex power.

$$\begin{aligned} \tilde{S}_{\text{three}} &= 3\tilde{S}_{\text{single}} = 3V_{\text{phase rms}} I_{\text{phase rms}} e^{j(\theta_v - \theta_i)} \\ &= 3V_{\text{phase rms}} I_{\text{phase rms}} \angle(\theta_v - \theta_i) \\ &= \sqrt{3} V_{\text{L-L rms}} I_{\text{phase rms}} \angle(\theta_v - \theta_i) \\ &= \sqrt{3} V_{\text{phase rms}} I_{\text{L rms}} \angle(\theta_v - \theta_i) = P_{\text{three}} + jQ_{\text{three}}, \end{aligned} \tag{5.63}$$

where θ_v and θ_i are the phase shift angles of the phase voltage $\tilde{V}_{\text{phase rms}} = V_{\text{phase rms}} e^{j\theta_v} = V_{\text{phase rms}} \angle\theta_v$ and the phase current $\tilde{I}_{\text{phase rms}} = I_{\text{phase rms}} e^{j\theta_i} = I_{\text{phase rms}} \angle\theta_i$, respectively.

The real power is

$$P_{\text{three}} = \sqrt{3} V_{\text{L-L rms}} I_{\text{phase rms}} \cos(\theta_v - \theta_i) = \sqrt{3} \cdot V_{\text{phase rms}} I_{\text{L rms}} \cos(\theta_v - \theta_i), \tag{5.64}$$

the reactive power is

$$\begin{aligned} Q_{\text{three}} &= \sqrt{3} \cdot V_{\text{L}-\text{L rms}} I_{\text{phase rms}} \sin(\theta_{\text{v}} - \theta_{\text{i}}) \\ &= \sqrt{3} \cdot V_{\text{phase rms}} I_{\text{L rms}} \sin(\theta_{\text{v}} - \theta_{\text{i}}), \end{aligned} \tag{5.65}$$

and the apparent power is

$$\begin{aligned} S_{\text{three}} &= \left|\tilde{S}_{\text{three}}\right| = \sqrt{(P_{\text{three}})^2 + (Q_{\text{three}})^2} \\ &= \sqrt{3}\, V_{\text{L}-\text{L rms}} I_{\text{phase rms}} = \sqrt{3}\, V_{\text{phase rms}} I_{\text{L rms}} \end{aligned} \tag{5.66}$$

5.10 Summary

Starting from instantaneous power $p(t)$, the average or real power P and the reactive power Q are defined. The concept of the rms or eff value of a quantity is introduced. There are two types of power factors: the fundamental or the displacement power factor and the total power factor, which takes into account all harmonics present, reducing the displacement power factor. Next, the complex power $\tilde{S}$ and the apparent power S are defined. For efficient energy transmission, the displacement or fundamental power factor should be in the range $0.8 \leq \text{PF} \leq 1.0$ lagging, that is, the current lags the voltage. Leading power factors should be avoided because of instabilities in the power system. If low power factor exists, a power factor correction must be performed: this is mainly done by connecting capacitor banks C in parallel to an inductive load $(R + j\omega L)$.

The basic structure of the residential single-phase AC power circuits [1] in the United States is explained. Note that due to widespread towns and cities, a fairly high (e.g. larger than 20 kV) distribution system voltage is used to reduce losses, which are mainly a function of the current within the distribution network. It is well known that above 20 kV the distribution lines must be overhead in order to prevent ferroresonance [2], as it occurs when underground cable capacitances C resonate with transformer inductances L.

Also, large power ($P > 50$ kW) cannot be transmitted with single-phase circuits, and therefore three-phase circuits are used. This is also true for electric machines and drives. This chapter covers the basic four different three-phase transformer connections: Y-Y, Y-Δ, Δ-Y, and Δ-Δ transformers. The Y-Δ and Δ-Y types are the most popular because triplen (e.g. 3, 6, 9, ...) harmonics can be eliminated. Of course there are more sophisticated three-phase transformers such as the zigzag [1, 2] type, which are relied on for special applications. Finally, the power relations [1, 2] as they apply to three-phase systems (transformers, distribution and transmission circuits, electric machines such as generators and motors) are summarized in a simplified manner.

Problems

5.1 Calculation of instantaneous current $i_s(t)$ and power $p(t)$

In the circuit of Figure P5.1.1 $v_s(t) = 339.36 \cdot V\cos(377 \cdot t + 20°)$, determine the equations for the instantaneous current $i_s(t)$ and instantaneous power $p(t)$ as a function of time.

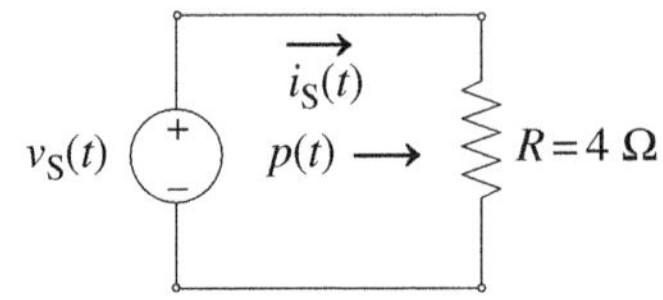

Figure P5.1.1 Calculation of instantaneous current $i_s(t)$ and instantaneous power $p(t)$.

5.2 Determination of average (real) power P and reactive power Q of resistive–inductive circuit

Calculate the average power P and reactive power Q that are supplied by the current source $\tilde{I}_s$ and absorbed by the two loads $(R_1 + j\omega L)$ and R_2 of the network of Figure P5.2.1.

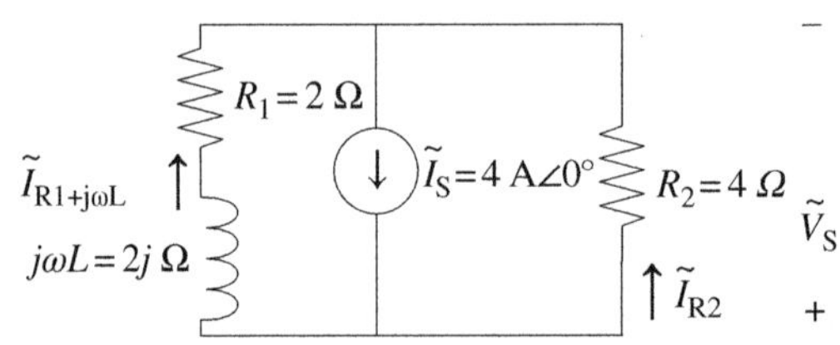

Figure P5.2.1 Calculation of average power P and reactive power Q absorbed by loads $(R_1 + j\omega L)$ and R_2 and supplied by current source $\tilde{I}_s$.

5.3 Determination of average (real) power P and reactive (im) power Q of resistive–capacitive circuit

Compute the average power P and reactive power Q absorbed by $(R_2 + 1/j\omega C)$ and R_1 of the network in Figure P5.3.1 and supplied by the two voltage sources $\tilde{V}_1$ and $\tilde{V}_2$.

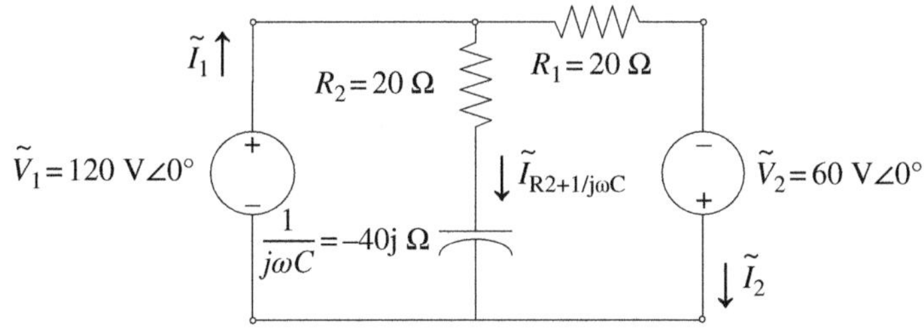

Figure P5.3.1 Computation of average power P and reactive power Q absorbed by loads ($R_2 + 1/j\omega C$) and R_1 and supplied by the two voltage sources $\tilde{V}_1$ and $\tilde{V}_2$.

5.4 Determination of average (real) power P and reactive (im) power Q of resistive–inductive–capacitive circuit

Show that the conservation of real (P) power holds for the circuit of Figure P5.4.1.

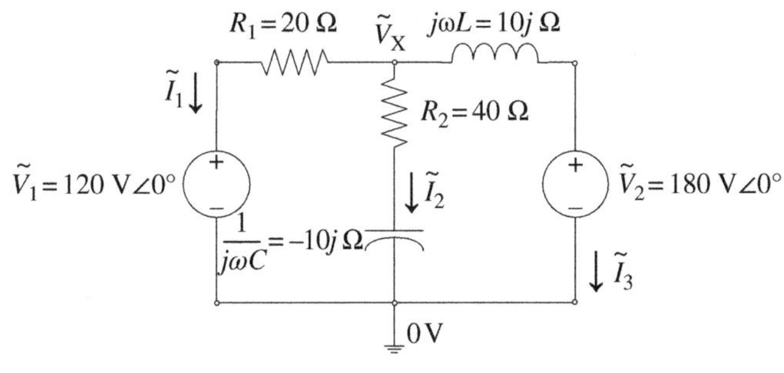

Figure P5.4.1 Conservation of real (P) power.

5.5 Root-mean-square (rms) value of voltage

Compute the rms value of the voltage waveform given in Figure P5.5.1a and b.

5.6 Transmission line losses

A single-phase load has a fundamental or displacement power factor $\text{PF}_{\text{load}} = \cos\theta_{\text{load}} = 0.8$ lagging. The power delivered to the load is $P_{\text{load}} = 40\ \text{kW}$ from a $\tilde{V}_{\text{load rms}} = 220\ \text{V}\angle 0°$, $f = 60\ \text{Hz}$ line. If the transmission line resistance is $R_{\text{line}} = 0.09\ \Omega$, determine the real power P_s that must be generated at the supply or source as shown in Figure P5.6.1.

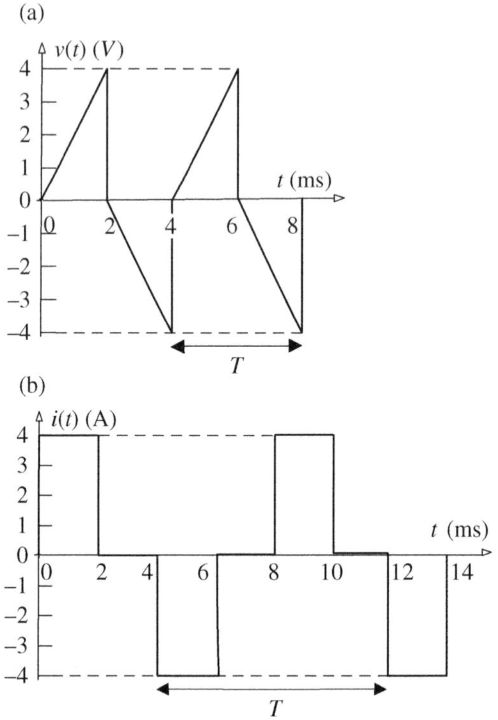

Figure P5.5.1 (a) Root-mean-square value of (asymmetric) sawtooth voltage. (b) Root-mean-square value of (symmetric) rectangular pulse voltage.

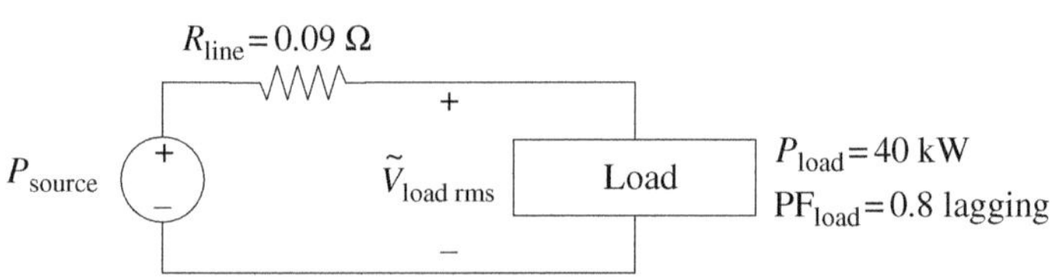

Figure P5.6.1 Transmission line with load and source.

5.7 Determination of fundamental or displacement power factor (PF) $\cos\theta$

An industrial plant at the end of a transmission line consumes $P_{\text{plant}} = 144\,\text{kW}$ at $\cos\theta_{\text{plant}} = 0.9$ lagging. The line voltage at the plant is $\tilde{V}_{\text{plant rms}} = 480\,\text{V}\angle 0°$. The resistance of the transmission line is $R_{\text{line}} = 0.15\,\Omega$ as given in Figure P5.7.1.

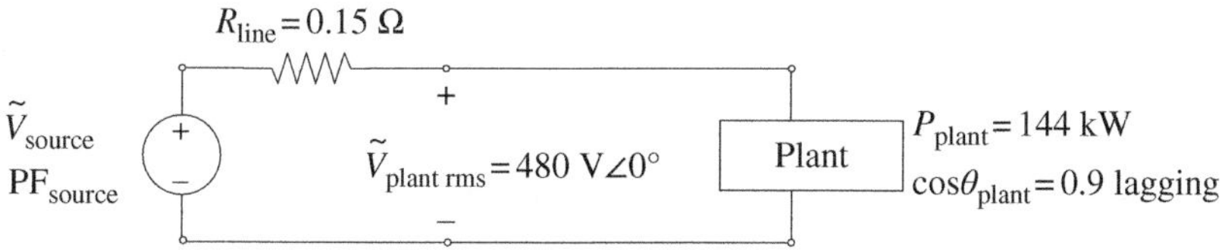

Figure P5.7.1 Voltage and power factor to be computed at the sending end (source) of transmission line.

a Determine the voltage and the power factor at the transmission line input/source $\tilde{V}_{source}$ and $PF_s = \cos\theta_s$.

b Draw the (not-to-scale) phasor diagram of current $\tilde{I}_{rms}$ and all voltages.

5.8 Complex power $\tilde{S}$

An industrial plant as shown in Figure P5.8.1, representing the load on a single-phase transmission line, consumes a real power of $P_{plant} = 110$ kW at a fundamental power factor of $PF_{plant} = \cos\theta_{plant} = 0.88$ lagging. The load voltage is $\tilde{V}_{plant\ rms} = 480\ V\angle 0°$.

a For the transmission line resistance of $R_{line} = 0.15\ \Omega$, determine the complex power at both ends of the line $\tilde{S}_s$ and $\tilde{S}_{plant}$.

b Draw the (not-to-scale) phasor diagram of current $\tilde{I}_{rms}$ and all voltages.

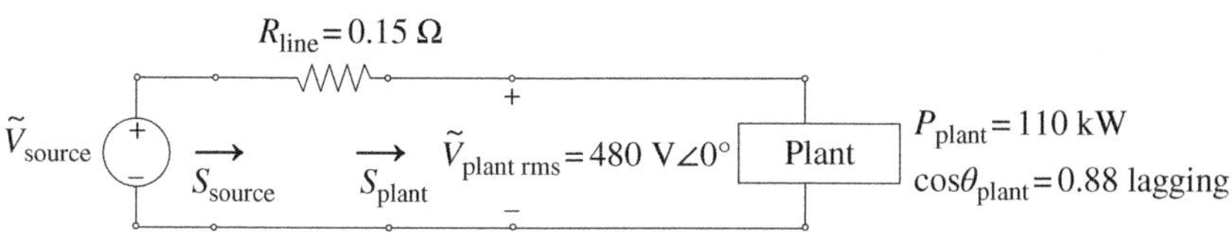

Figure P5.8.1 Calculation of the complex powers at the source and at the load $\tilde{S}_s$ and $\tilde{S}_{plant}$, respectively.

5.9 Transmission/distribution line feeder with impedance $\tilde{Z}_{line}$

An industrial plant as shown in Figure P5.9.1 consumes $P_{plant} = 160$ kW at $\cos\theta_{plant} = 0.9$ PF lagging. The line voltage at the plant is $\tilde{V}_{plant\ rms} = 480\ V\angle 0°$. The single-phase transmission line feeding the plant is 2 miles long.

a For the impedance of the line is $\tilde{Z}_{line} = R_{line} + jX_{line} = (0.1 + 0.2j)\Omega$, find the fundamental power factor at the sending end of the line $PF_{send} = \cos\theta_{send}$.

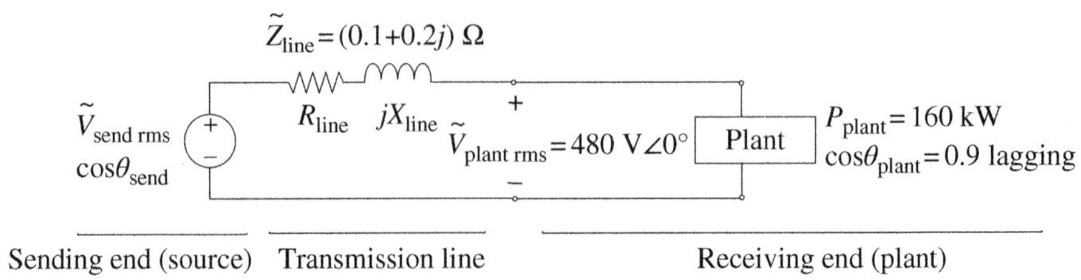

Figure P5.9.1 Single-phase transmission line with impedance $\tilde{Z}_{line} = R_{line} + jX_{line} = (0.1 + 0.2j)\Omega$.

b Draw the (not-to-scale) phasor diagram for current $\tilde{I}_{rms}$, voltages $\tilde{V}_{plantrms}$, $\tilde{V}_{Rline} = R_{line}\tilde{I}_{rms}$, $\tilde{V}_{Xline} = jX_{line}\tilde{I}_{rms}$, and $\tilde{V}_{sendrms}$.

5.10 Fundamental or displacement power factor correction/compensation

A small plant as shown in Figure P5.10.1 has a bank (several connected in parallel) of single-phase induction motors that consume $P_{motors} = 64$ kW at a fundamental power factor $PF_{fundamental} = \cos\theta_{motors\text{-}old} = 0.68$ lagging. The $f = 60$ Hz line voltage across the parallel-connected motors is $\tilde{V}_{motors\ rms} = 220\ V\angle 0°$. The power company has instructed the plant owner to raise the PF at the motors to $\cos\theta_{motors\text{-}new} = 0.92$ lagging. What value of the parallel-connected capacitor (C) across the motor bank is required (see Figure P5.10.1)?

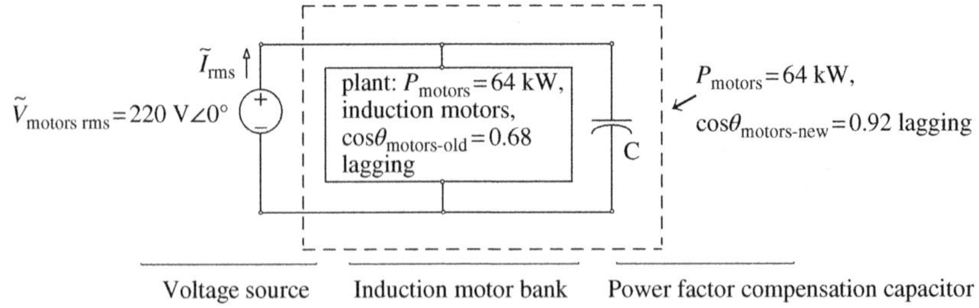

Figure P5.10.1 Fundamental power factor correction as applied to a bank of single-phase induction motors.

5.11 Residential single-phase power network

Residential power circuits are composed in Figure E5.6.1 of a $V_{rms1} = 120$ V lighting branch, an appliance branch $V_{rms2} = 120$ V, and a $V_{rms3} = 240$ V branch consisting of an electric dryer.

The lighting branch must illuminate an area of 1200 square feet (ft^2), the appliance branch serves 20 electrical receptacles (outlets), and the dryer is rated at $S_{dryer} = 8$ kVA.

a Determine the circuit breaker required for each individual circuit.

b Calculate the currents in all parts of the power network.

c Compute the monthly cost at \$0.12/kWh for operating the lights, appliances, and dryer 30, 12, and 5% of the time, respectively.

5.12 Three-phase distribution/transmission line network: Y-Y configuration

An abc-sequence balanced wye-connected source supplies a balanced wye-connected load as illustrated in Figure P5.12.1. If the per-phase line impedance is $\tilde{Z}_{line} = (1 + 1j)\Omega$, the load impedance is $\tilde{Z}_{Y\_load} = (12 + 8j)\Omega$, and the load voltage $\tilde{V}_{An\ rms} = 109.46\ V\angle 18.99°$, determine the source voltage $\tilde{V}_{aN\ rms}$.

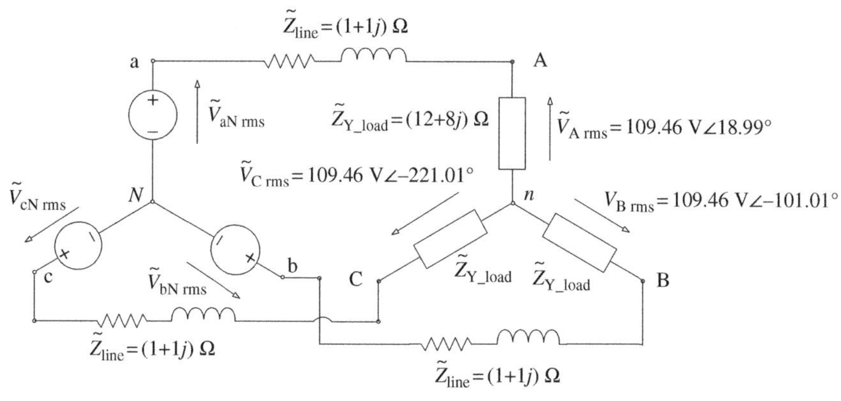

Figure P5.12.1 Y-Y configuration with given transmission line and load data.

5.13 Three-phase distribution/transmission line network: Y-Δ configuration

In a balanced three-phase wye–delta system (Figure P5.13.1), the (per-phase) line current is $\tilde{I}_{aA\ rms} = 8\ A\angle 13.13°$. If the impedance per phase of the delta load is $\tilde{Z}_{\Delta} = (36 + j27)\Omega$, find the source voltage $\tilde{V}_{aN\ rms}$ if the line impedance $\tilde{Z}_{line}$ can be ignored.

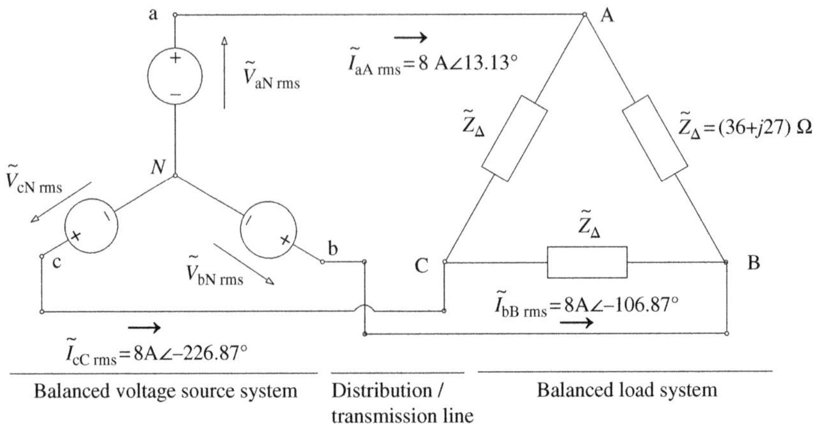

Figure P5.13.1 Y-Δ configuration with given line current and load data.

5.14 Three-phase distribution/transmission line network: Δ-Y configuration

In a balanced three-phase delta–wye system of Figure P5.14.1, compute the magnitude (rms value) of the line-to-line voltages $\tilde{V}_{AB\ rms}$, $\tilde{V}_{BC\ rms}$, $\tilde{V}_{CA\ rms}$ at the load and the magnitude of the phase currents in the delta-connected source, $\tilde{I}_{ab\ rms}$, $\tilde{I}_{bc\ rms}$, $\tilde{I}_{ca\ rms}$, provided the impedance per phase of the transmission line is $\tilde{Z}_{line} = (0.1 + 0.3j)\Omega$ and that of the Y-connected load is $\tilde{Z}_Y = (10 + 4j)\Omega$.

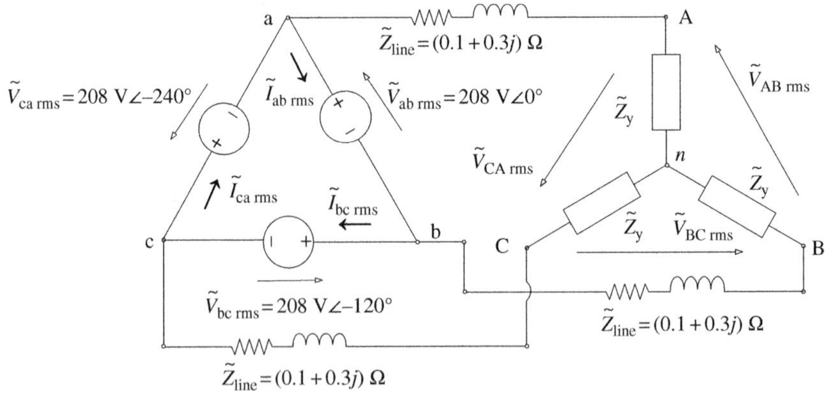

Figure P5.14.1 Δ-Y configuration with given source voltages, load, and line impedance data.

5.15 Three-phase distribution/transmission line network: Δ-Δ configuration

A balanced delta–delta three-phase system (Figure P5.15.1) has the following parameters:

$\tilde{V}_{ab\,rms} = 207.84\,\text{V}\angle -20°$, $\tilde{V}_{bc\,rms} = 207.84\,\text{V}\angle -140°$, $\tilde{V}_{ca\,rms} = 207.84\,\text{V}\angle -260°$, $\tilde{Z}_{line} = (1+2j)\Omega$, and $\tilde{Z}_{\Delta} = (18+12j)\Omega$. Compute the line currents $\tilde{I}_{aA\,line}$, $\tilde{I}_{bB\,line}$, and $\tilde{I}_{cC\,line}$.

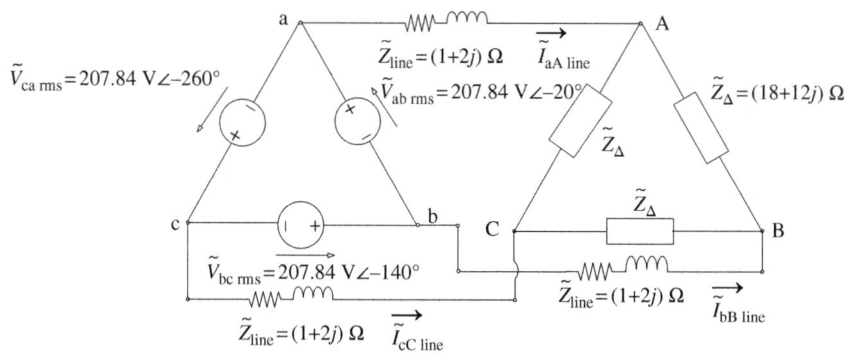

Figure P5.15.1 Δ-Δ configuration with given source voltages, load, and line impedance data.

5.16 Estimation of the energy content of a 100 kA, 40 MV lightning strike

A typical example for instantaneous power is the voltage and current of a lightning strike consisting of transient current $i_{lightning}(t)$ and voltage $v_{lightning}(t)$ with a very short duration in the range of a few hundred microseconds as shown in Figure 5.1a and b and Figure P5.16.1a and b. This lightning discharge has incredibly large voltage and current signatures – up to 30 million volts at 100 thousand amperes – but is of very short duration; hence lightning power and energy cannot be harnessed or used; indeed it is frequently very destructive by setting buildings on fire or causing death for people without protection through a so-called Faraday cage.

a Calculate/estimate the power $P_{lightning}$ and energy $W_{lightning}$ contained/transmitted by the lightning strike described by Figure P5.16.1a and b.

b Compare this lightning strike energy with the DC energy generated by the PV system of Application Example 1.6 during one year (07/14/2014–07/14/2015).

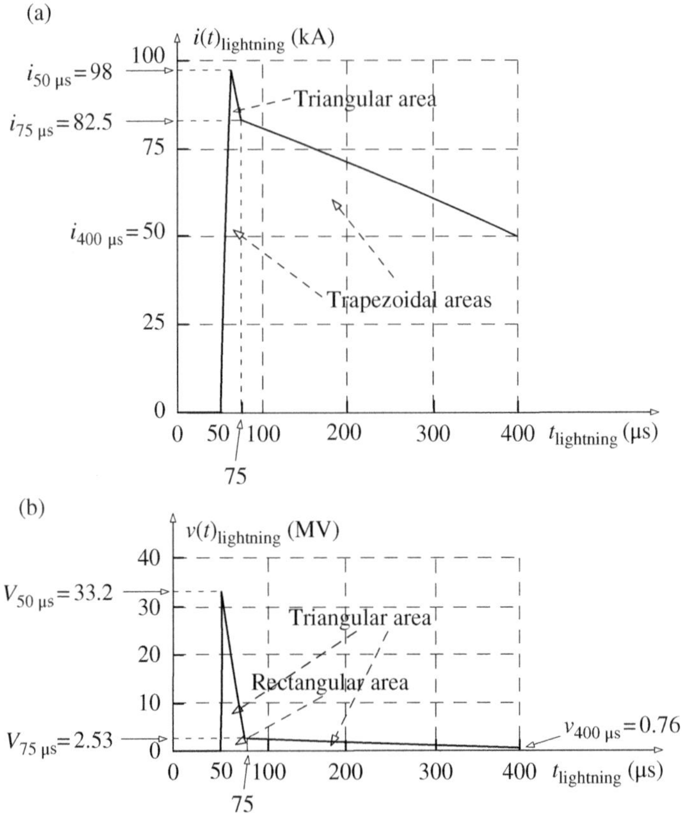

Figure P5.16.1 (a) Lightning instantaneous current $i_{\text{lightning}}(t)$. (b) Lighting instantaneous voltage $v_{\text{lightning}}(t)$.

References

1 Fuchs, E.F. and Masoum, M.A.S. (2012). *Power Conversion of Renewable Energy Systems*, 1e, 2nd Printing. USA: Springer, 692 pp.

2 Masoum, M.A.S. and Fuchs, E.F. (2015). *Power Quality in Power Systems and Electrical Machines*, 2e. Elsevier/Academic Press, 1123 pp.

3 Mark W. Earley, Christopher D. Coache, Mark Cloutier, Gil Moniz, Derek Vigstol (2016) *National Electrical Code Handbook 2017 (International Electrical Code)*, 14. Cengage Learning. https://www.amazon.de/National-Electrical-Code-2017-Handbook/dp/1455912840 (accessed 11 May 2020).

4 Clapp, A.L. (2019). National electrical safety code ANSI C2. http://ethw.org/National_Electrical_Safety_Code_ANSI_C2 (accessed 11 May 2020).

5 Jennifer Latson (2014). How Edison invented the light bulb — and lots of myths about himself. *Time* (21 October 2014) http://time.com/3517011/thomas-edison (accessed 11 May 2020).

6 Elizabeth Palermo (2017) Who invented the light bulb? (17 August 2017) © Future US, Inc. https://www.livescience.com/43424-who-invented-the-light-bulb.html (accessed 11 May 2020).

7 Palermo, E. (2017). Who invented the light bulb? US – IFL Science LLC. http://www.iflscience.com/physics/nobel-prize-physics-awarded-scientists-who-invented-blue-led (accessed 11 May 2020).

8 International Bureau of Weights and Measures (BIPM) (2019) SI brochure: the International System of Units (SI). 9. https://www.bipm.org/en/publications/si-brochure/ (accessed 11 May 2020).

9 Ledmarkt24 UG (2018) Was ist die Farbtemperatur? Die Farbtemperatur bei LED Lampen. http://www.ledmarkt24.de/Was-ist-die-Farbtemperatur (accessed 11 May 2020).

10 U.S. Energy Information Administration (2019) How much electricity does an American home use? Last updated: 2 October 2019 https://www.eia.gov/tools/faqs/faq.php?id=97&t=3 (accessed 11 May 2020).

11 Leeper, E.A. (2001). *Silencing the Fields*. Symmetry Books.

6

Coupled Magnetic Circuits, Single- and Three-Phase Transformers

6.1 Introduction

Magnetic circuits are the building blocks of transformers and electric machines. To understand the concepts of magnetic fields such as magnetic flux Φ, magnetic field intensity $\vec{H}$, magnetic flux density $\vec{B}$, induced voltage $e(t)$, magnetomotive force (mmf) $\mathcal{F}$, reluctance $\Re$, and the resulting (mechanical) forces $\vec{F}$, simple permanent-magnet configurations are considered first, where the flux Φ cannot be controlled, but is given by the magnetization of the "green" or not yet magnetized magnet material. Magnetization of the green magnet material occurs by exposing it to a very high flux density $B(t)$ for a very short time. A controllable time-dependent flux $\Phi(t)$ is achieved by generating the flux via winding coils excited by a controllable time-dependent current $i(t)$. From the arrangement of windings or coils in a certain manner, one can construct transformers and electric machines serving as either generator or motor – or for certain applications, interchangeably used as generator and motor. Single- and three- or multiphase transformers are elementary building blocks of renewable power/energy systems. This chapter will present some practical applications of coupled magnetic circuits and single- and three-phase transformers.

6.2 Magnetic Circuits

Magnetic field flux Φ, magnetic field intensity vector $\vec{H}$, and magnetic flux density vector $\vec{B}$ are identified by considering a single permanent magnet as shown in Figure 6.1. The resulting (mechanical) force vector $\vec{F}$ generated by the interaction

Introduction to Energy, Renewable Energy, and Electrical Engineering: Essentials for Engineering Science (STEM) Professionals and Students, First Edition. Ewald F. Fuchs and Heidi A. Fuchs.

Companion website: www.wiley.com/go/fuchs/renewableenergyandelectricalengineering

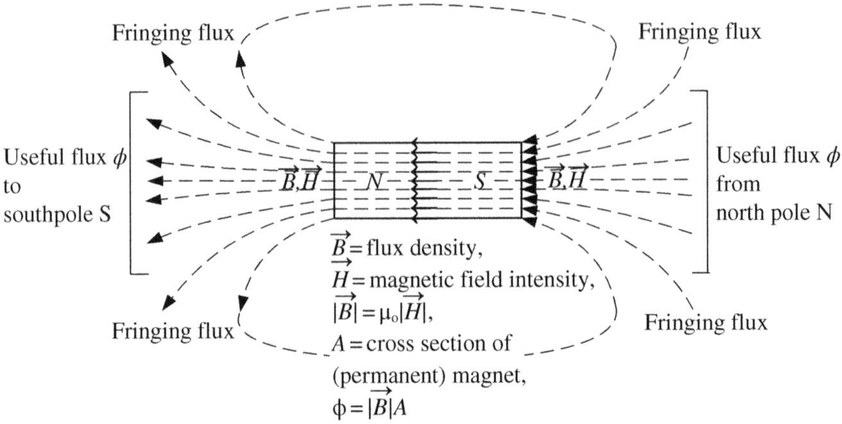

Figure 6.1 Single permanent magnet and associated magnetic flux Φ, magnetic field strength $\vec{H}$, and magnetic flux density $\vec{B}$. Identification of useful and fringing (not useful) fluxes.

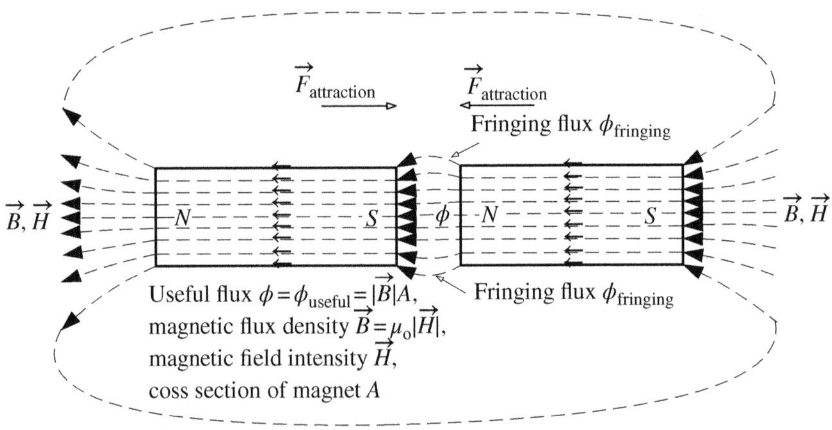

Figure 6.2 Pair of permanent magnets and associated magnetic fields resulting in unlike-pole attraction with (mechanical) magnetic force $\vec{F}_{attraction}$. Identification of useful Φ_{useful} and fringing (not useful) $\Phi_{fringing}$ fluxes.

of the abovementioned magnetic fields of two permanent magnets is best explained by the interaction of a pair of permanent magnets resulting in unlike-pole attraction as shown in Figure 6.2 and a pair of like-pole repulsion as depicted in Figure 6.3. The calculation of forces due to magnetic origins is addressed in Chapter 12.

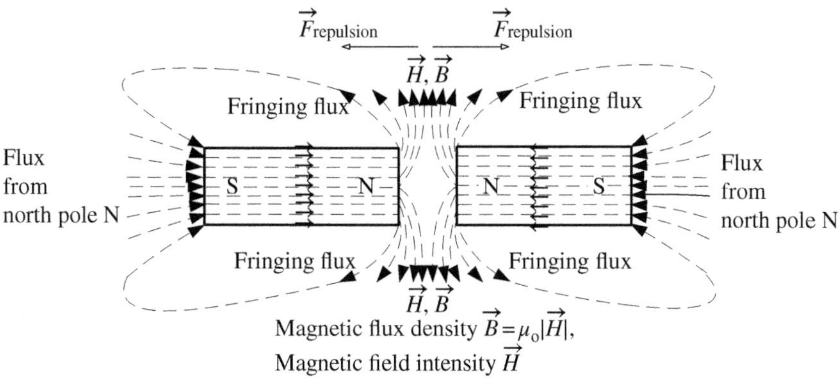

Figure 6.3 Pair of permanent magnets and associated magnetic fields resulting in like-pole repulsion with (mechanical) magnetic force $\vec{F}_{\text{repulsion}}$.

One distinguishes between useful flux Φ_{useful} and fringing (not useful) flux Φ_{fringing} as illustrated in Figures 6.1–6.3. Fringing fluxes do not contribute to the cross product for either force or torque production:

$$\vec{T} = \vec{R} \times \vec{F}, \tag{6.1}$$

where $\vec{R}$ is the radius, where the mechanical force $\vec{F}$ acts on a rotating member, e.g. rotor of electric machine.

The useful flux density is defined as the scalar

$$B = \frac{\Phi_{\text{useful}}}{A}, \tag{6.2}$$

where A is the area through which the useful flux flows as defined in Figure 6.1. The useful magnetic field intensity is defined by the scalar magnetic field intensity H as

$$H = \frac{B}{\mu_r \mu_0}, \tag{6.3}$$

where μ_r is the relative permeability of a material and μ_0 is the permeability of free space.

Permanent-magnet materials are applied in renewable energy systems such as electric machine (both generator and motor) drives and variable-speed wind power plants as discussed in Chapter 10. In order to control the flux $\Phi(t)$ in these applications, inverters – electronic commutating devices – are employed, as demonstrated in Chapter 10.

For certain applications such as coupled circuits as used in electronic and transformer applications, permanent magnets are not suitable because it is desirable to control the flux $\Phi(t)$ as a function of the exciting coil/winding current $i(t)$. Such applications call for winding/coil arrangements – also called inductors – which can be magnetically coupled.

A simple relation for the magnetic field intensity $\vec{H}$ as a function of the exciting current for a long wire is showcased in Figure 6.4a and b as defined by Ampere's law [1, 2] relating a surface integral (LHS = left hand side) to a line integral (RHS = right-hand side):

$$\text{LHS} = \overset{\text{surface}}{\int_S} \vec{J} \cdot \vec{dS} = \overset{\text{line}}{\oint_C} \vec{H} \cdot \vec{d\ell} = \text{RHS} \tag{6.4}$$

Equation (6.4) can be illustrated based on the so-called right-hand rule [3]: this right-hand rule illustrated in Figure 6.5 shows how a current-carrying wire with current I generates a magnetic field H_ϕ. If you point your thumb in the direction of the current I and let your fingers assume a curved position, the magnetic field $\vec{H}$ with component H_ϕ circling around this (see Figure 6.4a) or those wires (see Figure 6.4b) flow in the direction in which your four fingers point.

Evaluating Eq. (6.4) for an infinitely long single wire carrying current I leads with Eq. (6.4) and Figure 6.4a to the simple relation $\left|\vec{H}\right| = H_\Phi$, that is, for the magnetic field intensity in ϕ direction of a (r, ϕ, z) coordinate system, defined by the magnetic field intensity vector $\vec{H} = H_r + H_\phi + H_z$

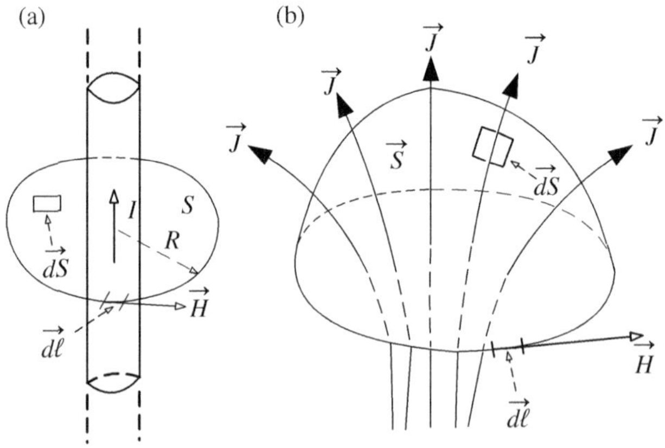

Figure 6.4 (a) Application of Ampere's law to a uniform long wire located in free space. (b) Relation of a surface integral (LHS) and a line integral (RHS) as defined by Eq. (6.4).

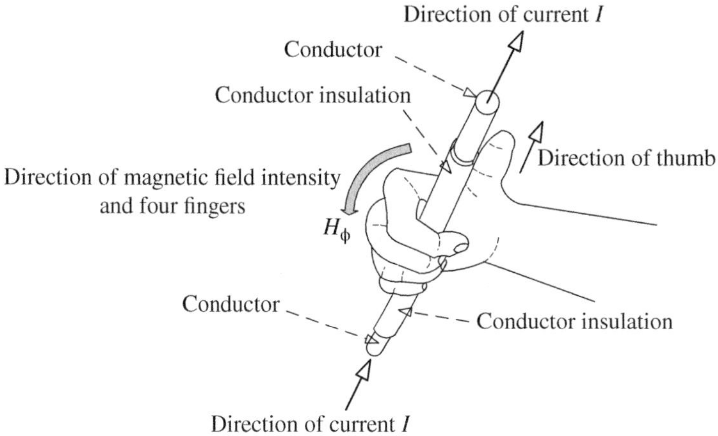

Figure 6.5 Right-hand rule [3], where current I generates H_Φ, which is a component of $\vec{H} = H_r + H_\phi + H_z$.

$$I = H_\phi 2\pi R, \tag{6.5}$$

$$H_\phi = I/(2\pi R), \tag{6.6}$$

where R is the radius as indicated in Figure 6.4a and the direction of H_ϕ is obtained based on the right-hand rule.

Application Example 6.1 Calculate magnetic field intensity H_Φ and magnetic field density B_Φ inside (H_{Φ_in}, B_{Φ_in}) and outside (H_{Φ_out}, B_{Φ_out}) of a single infinitely long conductor with radius $R = 3$ mm conducting a DC current of $I = 50$ A in a uniform manner as shown in Figure E6.1.1

a) Compute the current density J inside the round wire conductor.
b) Determine the magnetic field intensity inside the conductor H_{Φ_in} and the magnetic field density B_{Φ_in} inside the conductor as a function of the radius r_{in} as defined in Figure E6.1.1.
c) Calculate the magnetic field intensity outside the conductor H_{Φ_out} and the magnetic density B_{Φ_out} outside the conductor as a function of the radius R.
d) The DC earth magnetic field [1] density is about (depending upon location on earth) $B_{earth} = 0.7$ gauss. Determine the percentage of $(B_{\Phi\_out}/B_{earth}) \cdot 100\%$. Is there a health problem due to B_{Φ_out}?

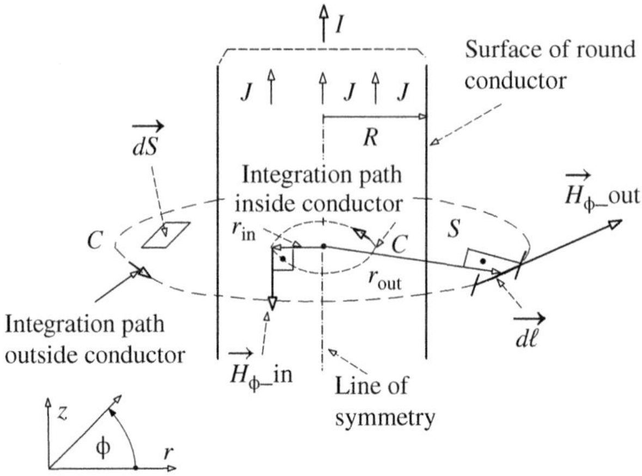

Figure E6.1.1 Infinitely long single wire conducting a DC current of I = 50 A in a uniform manner (side view).

Solution

a) The current density within the round conductor/wire is $J = I/(\pi R^2) = 50\ \text{A}/[\pi \cdot (3\ \text{mm})^2] = 1.768\ \text{A/mm}^2$ representing a value that requires no special cooling [1, 2, 4]. Such current densities are employed for electric installation in residences and factories.

b) Magnetic field density inside round conductor

According to Ampere's law, Eq. (6.4)

$$\text{LHS} = \int_S^{\text{surface}} \vec{J} \cdot \vec{dS} = \oint_C^{\text{line}} \vec{H} \cdot \vec{d\ell} = \text{RHS}, \tag{E6.1.1}$$

one obtains for the magnetic field intensity inside the conductor as shown in Figures E6.1.1 and E6.1.2 $H_{\Phi\_in}(2\pi r_{in}) = I_{in,}$ where H_{Φ_in} is the magnitude of the vector $\vec{H}_{\Phi\_in}$; therefore

$$H_{\Phi\_in} = I_{in}/(2\pi r_{in}), \tag{E6.1.2}$$

resulting in

$$I_{in} = I\left(\frac{\pi \cdot r_{in}^2}{\pi R^2}\right) \tag{E6.1.3}$$

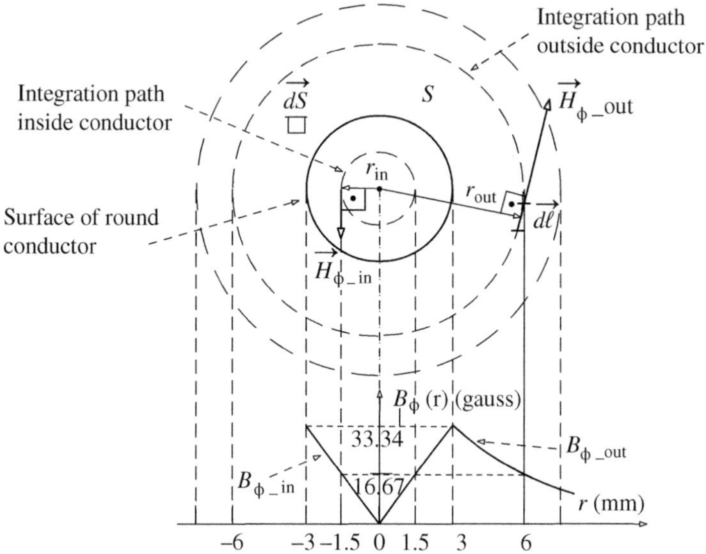

Figure E6.1.2 Magnetic field densities inside and outside of a single conductor B_{Φ_in} and B_{Φ_out}, respectively, with radius R (top view).

in

$$H_{\Phi\_in} = I\left[r_{in}/\left(2\pi R^2\right)\right]. \tag{E6.1.4}$$

The magnetic field density inside the conductor is now

$$B_{\Phi\_in} = \mu_0 H_{\Phi\_in} = \mu_0 I\left[r_{in}/\left(2\pi R^2\right)\right]. \tag{E6.1.5}$$

One concludes that the magnetic field density inside of the round conductor is a linear function of r_{in}.

For the chosen radius $r_{in} = 1.5\text{ mm} = 1.5 \cdot 10^{-3}$ m inside the conductor, one gets the numerical value

$$B_{\Phi\_in}(1.5\text{ mm}) = \frac{4\pi \cdot 10^{-7}\frac{\text{H}}{\text{m}} \cdot 50\text{ A} \cdot 1.5 \cdot 10^{-3}\text{m}}{2\pi(3 \cdot 10^{-3})^2\text{m}^2} = 1.667 \cdot 10^{-3}\text{ tesla}. \tag{E6.1.6}$$

c) Magnetic field density outside round conductor

The magnetic field intensity outside the conductor is

$$H_{\Phi\_out} = I/(2\pi R) = 50\text{ A}/(2\pi R). \tag{E6.1.7}$$

For the chosen radius $r_{in} = 6\ \text{mm} = 6 \cdot 10^{-3}$ m outside the conductor, one gets the numerical value

$$B_{\Phi\_out}(6\ \text{mm}) = \frac{4\pi \cdot 10^{-7}\frac{\text{H}}{\text{m}} \cdot 50\ \text{A}}{2\pi(6 \cdot 10^{-3})\ \text{m}} = 1.667 \cdot 10^{-3}\ \text{tesla}. \tag{E6.1.8}$$

The magnetic field density $B_\Phi(r)$ is shown in Figure E6.1.2. With the identity 1 tesla = 10^4 gauss, one obtains $B_{\Phi\_out} = 1.667 \cdot 10^{-3}$ tesla = $1.667 \cdot 10^{-3} \cdot 10^4$ gauss = 16.67 gauss. On the surface of the round conductor, the magnetic field density is B_Φ (3 mm) = 33.34 gauss, which is 33.34/0.7 · 100% = 4763% of the DC earth field. Because the magnetic fields decay in inverse proportion to the radius r, the B_Φ (1 m) = 14.29% of the earth field. DC and low-frequency (e.g. 0–5 kHz) magnetic/electric fields are not impacting human health.

The magnetic fields obtained by a permanent magnet (see Chapter 12) can be also generated by an inductor (see Chapters 3 and 4) consisting of a laminated iron core with permeability $\mu = \mu_r\mu_0$ with an air gap g and a winding/coil of N turns as shown in Figure 6.6. The relative permeability μ_r is a property of the chosen laminated iron core, and $\mu_0 = 4\pi \cdot 10^{-7}$ H/m is the permeability of free space. The relation between the vectors $\vec{B}$ and $\vec{H}$ is defined by the constituent relation

$$\vec{B} = \mu_r\mu_0\vec{H}, \tag{6.7}$$

where the relative permeability of free space is $\mu_r = 1$ and that of an iron core, for example, $\mu_r = 10\,000$. Note that iron cores exhibit the phenomenon of saturation discussed later in this chapter, that is, μ_r changes from a very high value at low saturation, e.g. $\mu_r = 100\,000$, to a very low value at high saturation, e.g. $\mu_r \geq 1.0$. In Figure 6.6 the current $i(t)$ represents the source of flux Φ.

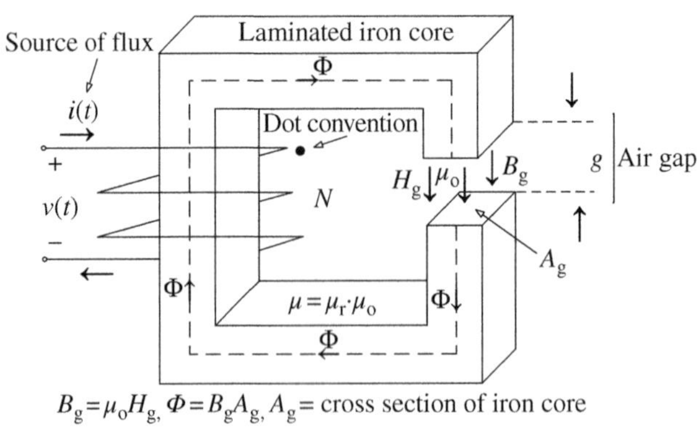

Figure 6.6 C-type iron core with one winding/coil having N turns and an air gap of length g.

The mmf of the C-core of Figure 6.6 is

$$\mathcal{F} = N \quad i(t). \tag{6.8}$$

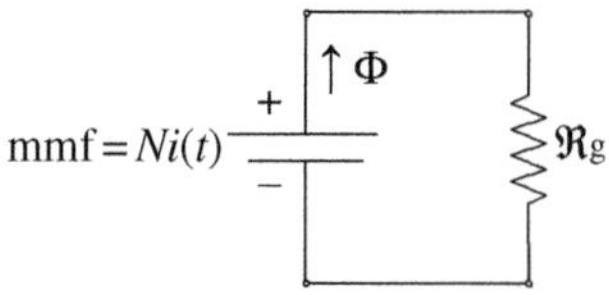

Figure 6.7 Equivalent circuit for Figure 6.6 with mmf = $\mathcal{F}$ = $N\, i(t)$, flux Φ, and reluctance $\Re_g$.

Note that the term "magnetomotive force" is not a mechanical force due to magnetics. It is really a misnomer. In electrical engineering, physical configurations are frequently represented by equivalent circuits. Figure 6.7 represents the equivalent circuit for Figure 6.6, where the source of flux Φ is the mmf = $\mathcal{F} = N \;\; i(t)$ and the load experienced by the flux is the reluctance $\Re_g$.

In analogy to Ohm's law, we can write with g as the gap length and H_g the magnetic field intensity within the gap g the equation for the reluctance – also called magnetic resistance – ($\mathfrak{R}_g$) of the gap

$$\mathfrak{R}_g = \frac{g}{(\mu_0 H_g)}, \tag{6.9}$$

resulting in Ohm's law for a magnetic circuit

$$\text{mmf} = \mathcal{F} = \Phi \mathfrak{R}_g, \tag{6.10}$$

where $\mathcal{F}$ is the source of flux, $\mathfrak{R}_g$ represents the load similar to a resistance, and Φ takes the place of the current as used in an electric circuit. The reluctance of the iron core is defined as

$$\mathfrak{R}_g = \ell_c/(\mu_c H_c), \tag{6.11}$$

where ℓ_c is the iron-core length, $\mu_c = \mu_r \mu_0$ is its permeability, and H_c is the magnetic field intensity in the iron core. The induced voltage across the winding turns of Figure 6.6 is according to Faraday's law [1, 2]:

$$v(t) = L\mathrm{d}i(t)/\mathrm{d}t, \tag{6.12}$$

where L is called the inductance and d$i(t)$/dt is the rate of change of current $i(t)$ with respect to time t. The inductance L is defined as

$$L = \frac{N^2}{(\Re_g + \Re_c)}, \tag{6.13}$$

and the flux linkages are

$$\lambda = N\Phi. \tag{6.14}$$

Single inductors are mostly used for filters [4–6]. In order to convert power/energy from one voltage or one current to another, one must employ several windings/coils as discussed in Sections 6.3–6.7.

Application Example 6.2 Calculation of reluctances $\Re_g$ and $\Re_c$, flux linkages λ, and magnetic stored energy (or work) in the air gap and laminated core W_{gap}, W_{core}, respectively, in a magnetic circuit representing an inductor with one winding using the MKSAK system of units

The MKSAK system of units relies on the units of meter (m), kilogram (k), seconds (s), ampere (A), and kelvin (K). An inductor with one winding used for an electronic circuit is shown in Figure E6.2.1. It has the iron-stack depth $D = 0.01$ m, a uniform core width $W = 0.01$ m, and air-gap length of $g = 0.02$ m. Assume that the laminated iron core [1, 2] has an infinitely large core (c) permeability $\mu_c = \mu_r\mu_0$ with $\mu_r \rightarrow \infty$, and neglect the effects of magnetic leakage and flux fringing (see Figure E6.2.1).

a) Calculate the reluctance of the air gap $\Re_g$ and that of the iron core $\Re_c$ for an air-gap length of $g = 0.02$ m and an iron-core length of $\ell_c = 0.1$ m. The cross section of the core at the air gap is $A_g = D \cdot W = 0.01\ \text{m} \cdot 0.01\ \text{m} = 10^{-4}\ \text{m}^2$.
b) For $N = 200$ turns calculate the flux Φ in Vs, self-inductance L in henries or the unit Vs/A of the inductor provided the flux density in the air gap is $B_g = 0.8$ tesla or Vs/m^2.
c) Determine the DC current I_{DC} measured in amperes or A required.
d) Determine the flux linkages λ measured in Vs·turns of the coil, where the term "turns" is not a unit.
e) Calculate the stored energy/work within the air gap W_{gap} and within the iron core W_{core}.

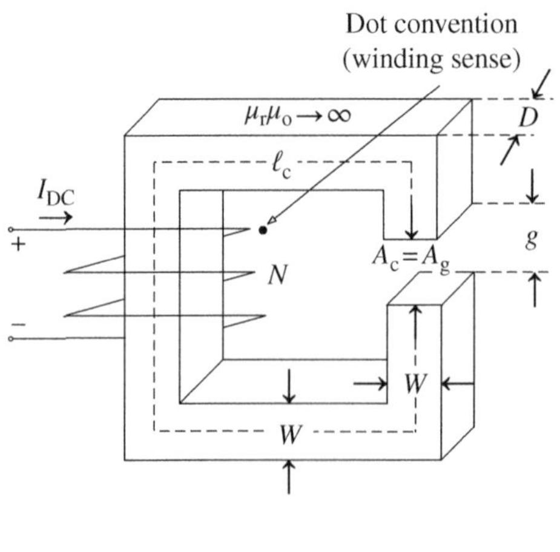

Figure E6.2.1 C-type iron-core magnetic circuit with gap $g = 0.02$ m and relative permeability $\mu_r \rightarrow \infty$.

Solution

a) The reluctance of the air gap is

$$\Re_g = g/(\mu_0 A_g) = 0.02\ \text{m}/[4\pi \cdot 10^{-7}\ \text{H/m} \cdot 1 \cdot 10^{-4}\ \text{m}^2] = 159.16 \cdot 10^6\ \text{A/Vs}, \quad \text{(E6.2.1)}$$

with $\mu_c = \mu_r \mu_0 \rightarrow \infty$ follows

$$\Re_c = \ell_c/(\mu_c A_c) \rightarrow 0. \quad \text{(E6.2.2)}$$

b) The flux is

$$\Phi = B_g A_g = B_g D \cdot W = 0.8\ \text{T} \cdot 0.1\ \text{m} \cdot 0.1\ \text{m} = 80 \cdot 10^{-6}\ \text{Wb or Vs}, \quad \text{(E6.2.3)}$$

and the inductance

$$L = N^2/(\Re_g + \Re_c) = 200^2/[159.16 \cdot 10^6 + 0] = 0.251\ \text{mH}. \quad \text{(E6.2.4)}$$

c) The magnetomotive force is with $H_c = 0$

$$\begin{aligned} Ni(t) = N \cdot I_{DC} &= H_g \cdot g + H_c \cdot \ell_c = (B_g/\mu_0) \cdot g = (0.8\ \text{T} \cdot 0.02\ \text{m})/(4\pi \cdot 10^{-7}\ \text{H/m}) \\ &= (0.8\ \text{Vs/m}^2 \cdot 0.02\ \text{m})/(4\pi \cdot 10^{-7}\ \text{Vs/Am}) = 1.27310^4\ \text{A turns} \end{aligned} \quad \text{(E6.2.5)}$$

or

$$I_{DC} = 1.2733 \cdot 10^4/N = 1.2733 \cdot 10^4/200 = 63.665\ \text{A}. \quad \text{(E6.2.6)}$$

d) The flux linkages are

$$\lambda = N\Phi = 200\ \text{turns} \cdot 80 \cdot 10^{-6}\ \text{Vs} = 16\ \text{mWb turns or mVs turns}. \quad \text{(E6.2.7)}$$

e) The energy/work stored in the gap is

$$\begin{aligned} W_{\text{gap}} = W_{\text{magetic}} &= \frac{1}{2} \int_{\text{volume}}^{\text{air pap}} B_g H_g dV = \frac{B_g B_g}{2\mu_o} g \cdot W \cdot D \\ &= \frac{0.8\left(\frac{\text{Vs}}{\text{m}^2}\right)}{2 \cdot 4\pi \cdot 10^{-7} \cdot \frac{\text{H}}{\text{m}}} 0.02\ \text{m} \cdot 0.01\ \text{m} \cdot 0.01\ \text{m} \\ &= 63.664 \cdot \frac{10^{-2}\left(\frac{\text{Vs}}{\text{m}^2}\right)^2 \text{m}^3}{\frac{\text{Vs}}{\text{mA}}} = 0.637\ \text{VAs or Ws}, \end{aligned} \quad \text{(E6.2.8)}$$

and the energy stored in the iron core is

$$W_{\text{core}} = 0, \text{because of } H_c = B_g/(\mu_r \mu_0) = B_g/\mu_c \text{ and } \mu_c \rightarrow \infty. \quad \text{(E6.2.9)}$$

6.3 Magnetically Coupled Circuits, Definition of Self- and Mutual Inductances

Figure 6.8 illustrates the coupling between winding 1 with turns N_1 and winding 2 with turns N_2. Without an iron core, the coupling between two windings is very small. To have a really good coupling coefficient k (see Figure 6.10), an iron core on which the two windings reside is used, resulting in a magnetically coupled circuit.

Winding 1 with N_1 turns is excited by the voltage $v_1(t)$, producing in the iron core the total flux Φ_{11} (not shown) and in the surrounding air regions the leakage (leak) or fringing flux Φ_{leak1}. Correspondingly winding 2 with N_2 turns is excited by the voltage $v_2(t)$, producing in the iron core the useful flux Φ_{22} and in the surrounding air regions the leakage or fringing flux Φ_{leak2}. The mutual fluxes that link winding 1 and winding 2 are denoted as Φ_{12} and Φ_{21} (not shown), respectively, whereby the first subscript denotes the winding by which the flux is received/linked and the second subscript denotes the winding that generates the flux. Windings 1 and 2 are supplied by the two voltage sources with the currents $i_1(t)$ and $i_2(t)$, respectively.

Based on the addition of fluxes, one can write for the total flux generated by coil 1

$$\Phi_{11} = \Phi_{21} + \Phi_{\mathrm{leak1}} \tag{6.15}$$

and for the total flux generated by coil 2

$$\Phi_{22} = \Phi_{12} + \Phi_{\mathrm{leak2}}. \tag{6.16}$$

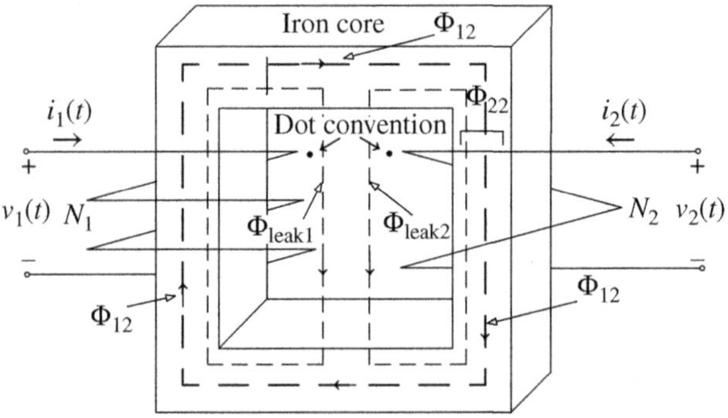

Figure 6.8 Magnetically coupled circuit with two windings. The dot (•) convention indicates the winding sense of a coil/winding residing on an iron core with leakage (leak) fluxes Φ_{leak1} and Φ_{leak2}. The total fluxes linked with windings 1 and 2, Φ_1 and Φ_2, respectively, are not shown in order to simplify the drawing; see for their definitions Eqs. (6.17) and (6.18).

The total flux linked with coil/winding 1 is

$$\Phi_1 = \Phi_{11} + \Phi_{12} = \Phi_{21} + \Phi_{\text{leak1}} + \Phi_{12}, \tag{6.17}$$

and the total flux linked with coil/winding 2 is

$$\Phi_2 = \Phi_{22} + \Phi_{21} = \Phi_{12} + \Phi_{\text{leak2}} + \Phi_{21}. \tag{6.18}$$

According to Faraday's law, one obtains for the voltage $v_1(t)$ of winding 1

$$v_1(t) = N_1 \mathrm{d}\Phi_1/\mathrm{d}t = N_1 \mathrm{d}\Phi_{11}/\mathrm{d}t + N_1 \mathrm{d}\Phi_{12}/\mathrm{d}t \tag{6.19}$$

for a linear transformer; where the flux is proportional to the current, one can write $\Phi_{11} = c_1 i_1$ and $\Phi_{12} = c_2 i_2$, and Eq. (6.19) becomes now with the self-inductance of winding 1 L_{11} and the mutual inductance between windings 1 and 2 L_{12}

$$v_1(t) = L_{11} \mathrm{d}i_1/\mathrm{d}t + L_{12} \mathrm{d}i_2/\mathrm{d}t; \tag{6.20}$$

correspondingly, one can write with the mutual inductance L_{21} between winding 2 and winding 1 and self-inductance L_{22} of winding 2

$$v_2(t) = L_{21} \mathrm{d}i_1/\mathrm{d}t + L_{22} \mathrm{d}i_2/\mathrm{d}t. \tag{6.21}$$

Based on the law of reciprocity, one can write for the mutual (M) inductance $L_{12} = L_{21} = M$ and with $L_1 = L_{11}$, $L_2 = L_{22}$

$$v_1(t) = L_1 \mathrm{d}i_1/\mathrm{d}t + M \mathrm{d}i_2/\mathrm{d}t, \tag{6.22}$$

$$v_2(t) = M \mathrm{d}i_1/\mathrm{d}t + L_2 \mathrm{d}i_2/\mathrm{d}t. \tag{6.23}$$

The magnetic arrangement of the two windings and the iron core is shown in Figure 6.9 with the dot (•) convention.

Figure 6.10 depicts the electric equivalent circuit of the two-winding coupled circuit of Figure 6.9.

For sinusoidal/cosinusoidal steady-state analysis, we use phasor notation $v_1(t) \to \widetilde{V}_1$, $i_1(t) \to \widetilde{I}_1$ and $L\frac{\mathrm{d}()}{\mathrm{d}t} \to j\omega L$, resulting in the (complex) phasor equations

$$\widetilde{V}_1 = j\omega L_1 \widetilde{I}_1 + j\omega M \widetilde{I}_2, \tag{6.24}$$

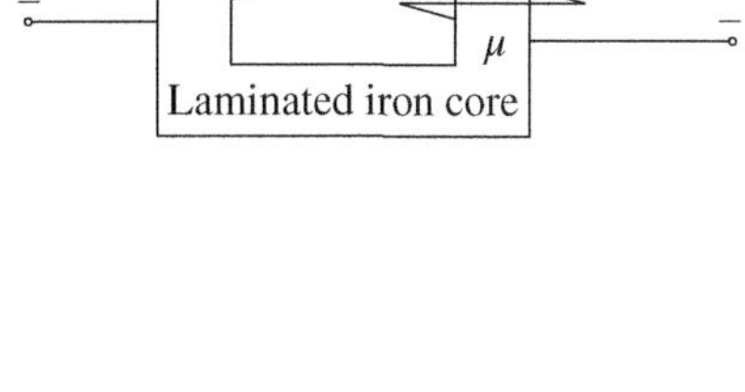

Figure 6.9 Magnetic arrangement of two coupled windings residing on iron core, where μ is constant.

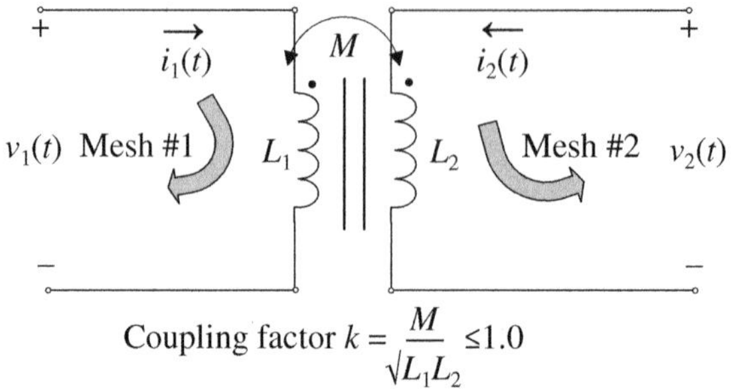

Figure 6.10 Electric equivalent circuit of a coupled two-winding system residing on iron core. The two perpendicular bars indicate symbolically the iron core coupling the two windings.

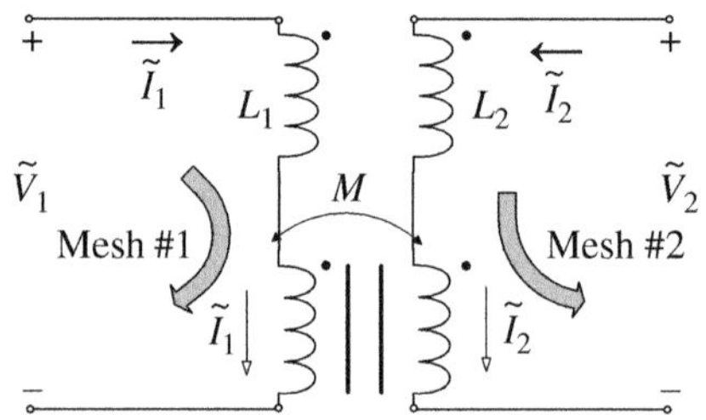

Figure 6.11 Detailed electric equivalent circuit with zero leakage fluxes; L_1 and L_2 are self-inductances, and M is the mutual inductance.

$$\tilde{V}_2 = j\omega M\tilde{I}_1 + j\omega L_2\tilde{I}_2. \tag{6.25}$$

A detailed equivalent circuit reflecting Eqs. (6.24) and (6.25) based on Kirchhoff's voltage law is represented in Figure 6.11, where the coupling coefficient k varies between 0 and 1.0:

$$k = \frac{M}{\sqrt{L_1L_2}}, \tag{6.26}$$

$$0 \leq k \leq 1.0. \tag{6.27}$$

For $k = 1.0$ then $M = \sqrt{L_1L_2}$, there is perfect coupling between the two windings, and the leakage/fringing fluxes $\Phi_{\text{leak1}} = \Phi_{\text{leak2}} = 0$.

6.4 Unsaturated or Linear Single-Phase Transformer

A transformer is a coupling device exploiting the properties of two magnetically coupled coils residing on an iron core without any air gap. Note that the previously discussed inductances are mostly residing on an iron core as well but have an air gap g in which the stored energy [1, 2] resides: the air gap g makes the difference between a two-winding coupled inductor and a two-winding transformer.

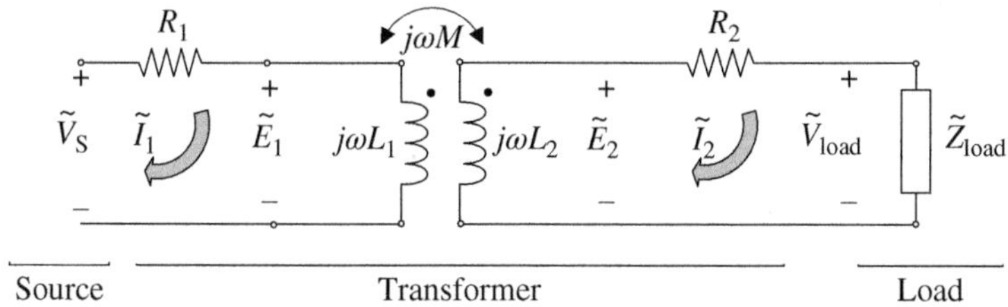

Figure 6.12 Single-phase transformer with voltage source $\tilde{V}_s$, load impedance $\tilde{Z}_{load}$, induced voltages $\tilde{E}_1$ and $\tilde{E}_2$, self-inductances L_1 and L_2, mutual inductance M, and winding resistances R_1 and R_2.

A typical two-winding or single-phase transformer is shown in Figure 6.12. All windings except superconducting ones, which are not discussed here, have resistances R_1 and R_2 as indicated in Figure 6.12.

Applying Kirchhoff's voltage law yields for the primary and secondary meshes/loops indicated in Figure 6.12 by the loop currents by $\tilde{I}_1$ and $\tilde{I}_2$

$$\tilde{V}_s = (R_1 + j\omega L_1)\tilde{I}_1 - j\omega M\tilde{I}_2, \tag{6.28}$$

$$0 = -j\omega M\tilde{I}_1 + \left(R_2 + j\omega L_2 + \tilde{Z}_{load}\right)\tilde{I}_2. \tag{6.29}$$

Solving Eq. (6.29) for $\tilde{I}_2$ and substituting in Eq. (6.28) yield

$$\tilde{V}_s = \left(R_1 + j\omega L_1 + \frac{\omega^2 M^2}{\left(R_2 + j\omega L_2 + \tilde{Z}_{load}\right)}\right)\tilde{I}_1. \tag{6.30}$$

The input impedance is obtained from

$$\tilde{Z}_{in} = \frac{\tilde{V}_s}{\tilde{I}_1} = \left(R_1 + j\omega L_1 + \frac{\omega^2 M^2}{\left(R_2 + j\omega L_2 + \tilde{Z}_{load}\right)}\right) = R_1 + j\omega L_1 + \tilde{Z}_{reflect}, \tag{6.31}$$

where $\tilde{Z}_{reflect}$ is the reflected impedance of the secondary circuit as viewed from the primary winding of the transformer; with the load impedance $\tilde{Z}_{load} = R_{load} + jX_{load}$, one obtains

$$\tilde{Z}_{reflect} = \frac{\omega^2 M^2}{\left(R_2 + j\omega L_2 + \tilde{Z}_{load}\right)} = \frac{\omega^2 M^2}{[R_2 + R_{load} + j(\omega L_2 + X_{load})]}$$

$$= \frac{\omega^2 M^2\{(R_2 + R_{load}) - j(\omega L_2 + X_{load})\}}{(R_2 + R_{load})^2 + (\omega L_2 + X_{load})^2}. \tag{6.32}$$

If $(\omega L_2 + X_{load}) = 0$, that is, there exists resonance on the secondary side, then $\widetilde{Z}_{reflect}$ is purely resistive:

$$\widetilde{Z}_{reflect} = \frac{\omega^2 M^2}{(R_2 + R_{load})}. \tag{6.33}$$

Application Example 6.3 Circuit with transformer represented by two coupled inductors (see Figure 6.12) and load impedance $\widetilde{Z}_{load}$

For $\widetilde{V}_s = 240\,\text{V}\angle 0°$, $R_1 = 1\,\Omega$, $\omega = 2\pi f$ where $f = 60$ Hz, $L_1 = 2$ mH, $L_2 = 3$ mH, $M = 2.5$ mH, $R_2 = 1\,\Omega$, and $\widetilde{Z}_{load} = R_{load} + jX_{load} = (3 + j10)\,\Omega$, calculate $\widetilde{I}_1$, $\widetilde{Z}_{reflect}$ and $\widetilde{Z}_{in}$ for non-resonant conditions.

Solution

According to Eq. (6.30)

$$\widetilde{I}_1 = \frac{\widetilde{V}_s}{R_1 + j\omega L_1 + \dfrac{\omega^2 M^2}{R_2 + j\omega L_2 + \widetilde{Z}_{load}}}\,\text{A} = \frac{240\,\text{V}\angle 0°}{\left[1 + j0.754 + \dfrac{6.25 \cdot 10^{-6}}{1 + j1.131 + 3 + j10}\right]\Omega}$$

$$= \frac{240\,\text{V}\angle 0°}{\left[1 + j0.754 + \dfrac{6.25 \cdot 10^{-6}(4 - j11.131)}{(4 + j11.131)(4 - j11.131)}\right]\Omega}$$

$$= \frac{240\,\text{V}\angle 0°}{\left[1 + j0.754 + \dfrac{25 \cdot 10^{-6} - j69.569 \cdot 10^{-6}}{16 + 123.9}\right]\Omega}$$

$$= \frac{240\,\text{V}\angle 0°}{[1 + j0.754 + (0.114\,37 - j0.497\,28)10^{-6}]\Omega} = \frac{240\,\text{V}\angle 0°}{(1 + j0.754)\Omega}$$

$$= \frac{240\,\text{V}\angle(0 - 37.016)°}{\sqrt{1 + 0.568\,52)}\,\Omega}$$

$$= 191.63\,\text{A}\angle - 37.016°. \tag{E6.3.1}$$

From Eq. (6.32) one gets

$$\widetilde{Z}_{reflect} = \frac{0.888\,31 \cdot 10^6\,\Omega^2}{[1 + 3 + j(1.131 + 10)]\,\Omega} = \frac{0.888\,31 \cdot 10^6\,\Omega}{4 + j11.131}$$

$$= \frac{0.888\,31 \cdot 10^6\,\Omega(4 - j11.131)}{(4 + j11.131)(4 - j11.131)}$$

$$= \frac{0.888\,31 \cdot 10^6\,\Omega(4 - j11.131)}{139.90} = \frac{(3.5532 - j9.8878) \cdot 10^6}{139.90}$$

$$= (25.398 - j70.678)\,\text{k}\Omega. \tag{E6.3.2}$$

Using Eq. (6.31) one obtains

$$\widetilde{Z}_{\text{in}} = (1 + j0.745)\ \Omega + (25.398 - j70.678)\ \text{k}\Omega \approx (25.399 - j69.924)\ \text{k}\Omega. \tag{E6.3.3}$$

6.5 Ideal Transformer

An ideal transformer [1, 2] is a linear device for which $L_1 = L_2 = M$, that is, $k = 1.0$ (there is no leakage flux and perfect coupling between primary and secondary windings exists, $M \rightarrow \infty$), and $R_1 = R_2 = 0$. Figure 6.13 shows the magnetic circuit, where the primary and secondary windings reside on an iron core with very high permeability $\mu = \mu_r\mu_0 \rightarrow \infty$.

According to Faraday's law [1, 2], one can write

$$v_1(t) = N_1 \frac{\mathrm{d}\Phi}{\mathrm{d}t}, \tag{6.34}$$

$$v_2(t) = N_2 \frac{\mathrm{d}\Phi}{\mathrm{d}t}. \tag{6.35}$$

Dividing Eq. (6.34) by Eq. (6.35), one obtains

$$\frac{v_1(t)}{v_2(t)} = \frac{N_1}{N_2}. \tag{6.36}$$

According to Ampere's law [1, 2] (Eq. 6.4), one obtains for $H = 0$ within the iron core because of $\mu = \mu_r\mu_0 \rightarrow \infty$ the mmf (Eq. 6.8) balance relation, where the current $i_1(t)$ produces a flux or mmf$_1$ in clockwise direction and $i_2(t)$ a flux or mmf$_2$ in counterclockwise direction:

$$N_1 i_1(t) - N_2 i_2(t) = 0, \tag{6.37}$$

or

$$\frac{i_1(t)}{i_2(t)} = \frac{N_2}{N_1}. \tag{6.38}$$

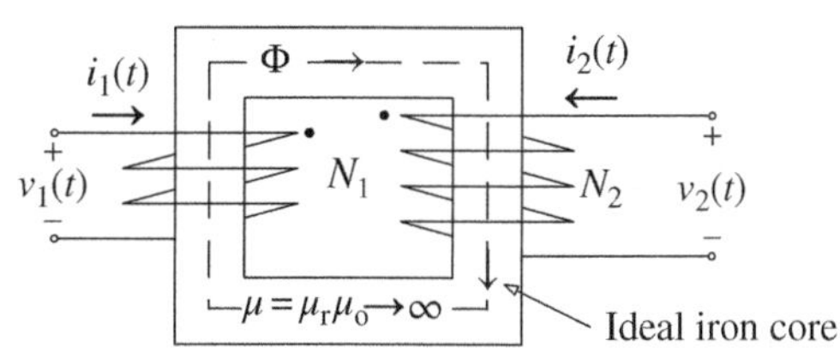

Figure 6.13 Magnetic circuit of ideal transformer.

An ideal transformer is lossless; that is, the winding and iron core incur no losses, and therefore the input power is identical to the output power

$$p_{in}(t) = v_1(t)i_1(t) = v_2(t)i_2(t) = p_{out}(t). \tag{6.39}$$

The input impedance is with the output or load impedance $Z_{out} = v_2(t)/i_2(t)$

$$Z_{in} = \frac{v_1(t)}{i_1(t)} = \frac{\frac{N_1}{N_2}v_2(t)}{\frac{N_2}{N_1}i_2(t)} = \frac{N_1^2}{N_2^2}\frac{v_2(t)}{i_2(t)} = \frac{N_1^2}{N_2^2}Z_{out} = Z'_{load}. \tag{6.40}$$

The output or load impedance referred to the primary winding is indicated by a prime (′)

$$Z'_{load} = Z'_{out} = \frac{N_1^2}{N_2^2}Z_{out}; \tag{6.41}$$

in other words, it is the load impedance as seen from the primary winding.

The instantaneous input power is

$$p_{in} = v_1(t)i_1(t), \tag{6.42}$$

and the instantaneous output power is

$$p_{out} = v_2(t)i_2(t). \tag{6.43}$$

Figure 6.14 represents the circuit symbol of an ideal transformer with ideal coupling between primary and secondary, with instantaneous input power P_{in}, instantaneous output/load power P_{out}, input impedance $Z_{in} = v_1(t)/i_1(t)$, and output impedance $Z_{out} = v_2(t)/i_2(t)$.

In phasor notation the above relations are

$$\frac{\tilde{V}_1}{\tilde{V}_2} = \frac{N_1}{N_2}, \tag{6.44}$$

$$\frac{\tilde{I}_1}{\tilde{I}_2} = \frac{N_2}{N_1}, \tag{6.45}$$

$$\tilde{Z}_{in} = \frac{\tilde{V}_1}{\tilde{I}_1} = \frac{\frac{N_1}{N_2}\tilde{V}_2}{\frac{N_2}{N_1}\tilde{I}_2} = \frac{N_1^2}{N_2^2}\frac{\tilde{V}_2}{\tilde{I}_2} = \frac{N_1^2}{N_2^2}\tilde{Z}_{out} = \tilde{Z}'_{load}. \tag{6.46}$$

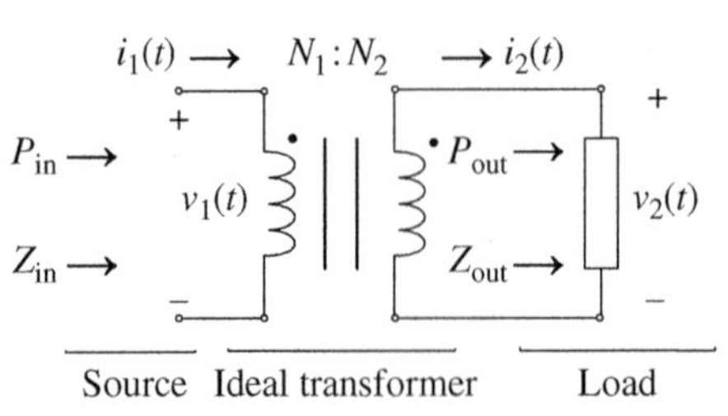

Figure 6.14 Circuit symbol for ideal (single-phase) transformer. The two parallel bars between the two coupled windings signify the ideal iron core with $\mu \rightarrow \infty$.

Application Example 6.4 Analysis of circuit with two coupled inductors

For the network of Figure E6.4.1, the measured output voltage is $\tilde{V}_{out} = 120\,\text{V}\angle 30°$. Find the required current source $\tilde{I}_s$.

Solution

The secondary transformer current is

$$\tilde{I}_2 = \frac{\tilde{V}_{out}}{R_2} = \frac{120\,\text{V}\angle 30°}{40}\Omega = 3\,\text{A}\angle 30°. \tag{E6.4.1}$$

According to the turns ratio of the ideal transformer, one obtains

$$\frac{\tilde{I}_1}{\tilde{I}_2} = \frac{N_2}{N_1} = \frac{4}{1} = 4, \tag{E6.4.2}$$

resulting in

$$\tilde{I}_1 = 4\tilde{I}_2 = 12\angle 30°\text{A}. \tag{E6.4.3}$$

The secondary transformer voltage is by applying Kirchhoff's voltage law (KVL)

$$\tilde{V}_2 = \tilde{I}_2\left(R_2 - \frac{j}{\omega C_2}\right) = 3\,\text{A}\angle 30°(40 - j10)\,\Omega = 3\,\text{A}(\cos\,30° + j\sin\,(30°))(40 - j10)\,\Omega$$
$$= 3\,\text{A}(0.866 + j0.5)(40 - j10)\,\Omega = (103.92 - j25.98 + j60 + 15) = (118.92 + j34.02)\,\text{V}. \tag{E6.4.4}$$

The voltage relation at an ideal transformer is

$$\frac{\tilde{V}_1}{\tilde{V}_2} = \frac{N_1}{N_2} = 1/4; \tag{E6.4.5}$$

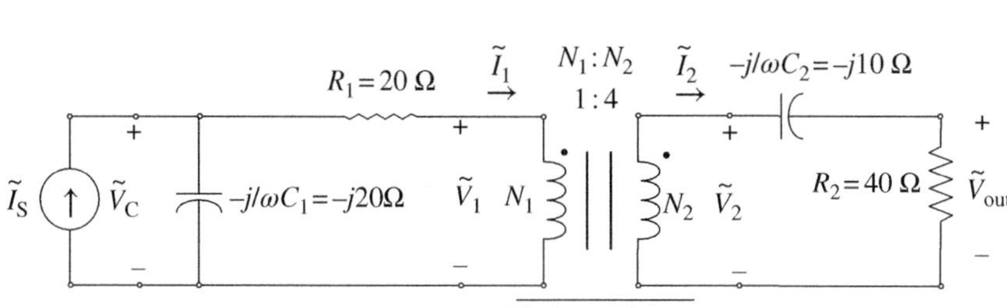

Figure E6.4.1 Electric circuit with two coupled inductors supplied by current source $\tilde{I}_s$.

therefore

$$\tilde{V}_1 = \left(\frac{1}{4}\right)(118.92 + j34.02)\ \text{V} = (29.73 + j8.505)\ \text{V}, \tag{E6.4.6}$$

yielding the capacitor voltage by applying the KVL:

$$\tilde{V}_c = 20\tilde{I}_1 + \tilde{V}_1 = 20\ \Omega \cdot 12\ \text{A}\angle 30^\circ + (29.73 + j8.505)\ \text{V} = 240(0.866 + j0.5)\ \text{V} + (29.73 + j8.505)\ \text{V} = (207.84 + j120 + 29.73 + j8.505)\ \text{V} = (237.57 + j128.51)\ \text{V}. \tag{E6.4.7}$$

The capacitor current is now

$$\tilde{I}_c = \frac{\tilde{V}_c}{(-j20)\ \Omega} = \frac{(237.57 + \text{j}128.51)\ \text{V}}{(-j20)\ \Omega} = (-6.426 + j11.879)\ \text{A}, \tag{E6.4.8}$$

and the source (s) current by applying KCL in rectangular form

$$\tilde{I}_s = \tilde{I}_1 + \tilde{I}_c = [12\angle 30^\circ - 6.426 + j11.879]\ \text{A} = [12(0.866 + j0.5) - 6.426 + j11.879]\ \text{A} = [10.392 + j6 - 6.426 + j11.879]\ \text{A} = [3.966 + j17.879]\ \text{A} \tag{E6.4.9}$$

or in polar form

$$\tilde{I}_s = \sqrt{(3.9665)^2 + (17.879)^2}\ \text{A}\angle 77.493^\circ = 18.314\ \text{A}\angle 77.493^\circ. \tag{E6.4.10}$$

Application Example 6.5 Circuit with ideal (iron-core permeability $\mu \to \infty$) single-phase transformer and known voltage source $\tilde{V}_s$ with resistive load R_{load}

An ideal single-phase transformer with turns ratio $N_1/N_2 = 4/2$ is energized by the voltage source $\tilde{V}_s = 480\ \text{V}\angle 0^\circ$ and loaded by the resistive load $R_{load} = 10\ \Omega$ as shown in Figure E6.5.1. Calculate the voltages $\tilde{V}_2$, $\tilde{V}_{load}$, currents $\tilde{I}_1$, $\tilde{I}_2$, the output real power P_{out}, the input real power P_{in}, and the impedances $\tilde{Z}_{out}$, $\tilde{Z}_{in}$.

Solution

The secondary voltage of the transformer determined by the relation is $\frac{|\tilde{V}_1|}{|\tilde{V}_2|} = \frac{4}{2} = 2$ or

$$\tilde{V}_2 = \frac{\tilde{V}_1}{2} = 240\ \text{V}\angle 0^\circ = \tilde{V}_{load}; \tag{E6.5.1}$$

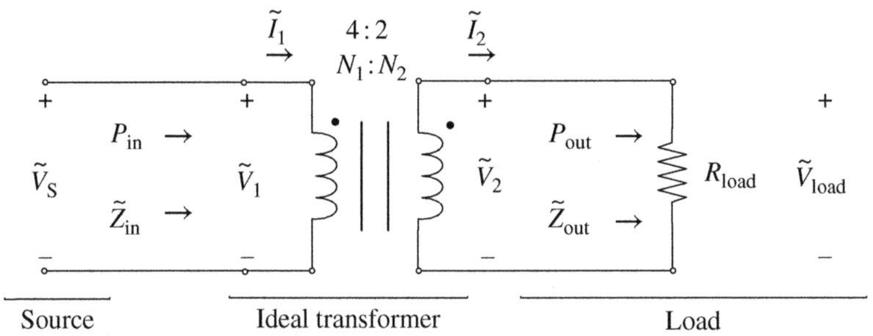

Figure E6.5.1 Electric circuit with ideal single-phase transformer and given voltage source $\tilde{V}_s$ with resistive load R_{load}.

thus the secondary current is

$$\tilde{I}_2 = \frac{\tilde{V}_{load}}{R_{load}} = 240\,\text{V}\angle 0^\circ / 10\,\Omega = 24\,\text{A}\angle 0^\circ. \tag{E6.5.2}$$

The primary (input) current to the transformer is $\dfrac{|\tilde{I}_1|}{|\tilde{I}_2|} = \dfrac{2}{4} = 1/2$ or

$$\tilde{I}_1 = \left(\frac{1}{2}\right)\tilde{I}_2 = 12\,\text{A}\angle 0^\circ. \tag{E6.5.3}$$

The output real power is

$$P_{out} = |\tilde{V}_2||\tilde{I}_2| = 240\,\text{V} \cdot 24\,\text{A} = 5.67\,\text{kW}, \tag{E6.5.4}$$

and the input real power

$$P_{in} = |\tilde{V}_1||\tilde{I}_1| = 480\,\text{V} \cdot 12\,\text{A} = 5.67\,\text{kW}. \tag{E6.5.5}$$

Note that $P_{out} = P_{in}$ because a lossless transformer, which is an ideal transformer, is used. The output impedance is

$$\tilde{Z}_{out} = \frac{\tilde{V}_2}{\tilde{I}_2} = \frac{240\,\text{V}\angle 0^\circ}{24\,\text{A}\angle 0^\circ} = 10\,\Omega = R_{load}, \tag{E6.5.6}$$

and the input impedance is

$$\tilde{Z}_{in} = \frac{\tilde{V}_1}{\tilde{I}_1} = \frac{480\,\text{V}\angle 0^\circ}{12\,\text{A}\angle 0^\circ} = 40\,\Omega = R'_{load}. \tag{E6.5.7}$$

The result of Eq. (E6.5.7) is confirmed by Eq. (6.46)

$$\tilde{Z}_{in} = \frac{\tilde{V}_1}{\tilde{I}_1} = \frac{\frac{N_1}{N_2}\tilde{V}_2}{\frac{N_2}{N_1}\tilde{I}_2} = \frac{N_1^2 \tilde{V}_2}{N_2^2 \tilde{I}_2} = \frac{N_1^2}{N_2^2} R_{load} = \left(\frac{4}{2}\right)^2 R_{load} = R'_{load} = 4 \cdot 10\,\Omega = 40\,\Omega. \tag{E6.5.8}$$

Application Example 6.6 Circuit with ideal (iron-core permeability $\mu \to \infty$) single-phase transformer and known current source $\tilde{I}_s$ with inductive load ($R_{load} + j\omega L_{load}$) at $f = 60$ Hz

An ideal single-phase transformer with turns ratio $N_1/N_2 = 6/2$ is energized by the current source $\tilde{I}_s = 20\ \text{A}\angle 0°$ and loaded by the resistive–inductive load $\tilde{Z}_{load} = (R_{load} + j\omega L_{load}) = (1 + j0.5)\ \Omega$ as shown in Figure E6.6.1. Calculate the currents $\tilde{I}_1, \tilde{I}_2$, voltages $\tilde{V}_2, \tilde{V}_{load}, \tilde{V}_1$, output and input real powers P_{out}, P_{in}, respectively, and the impedances $\tilde{Z}_{out}, \tilde{Z}_{in}$ of the ideal transformer.

Solution

The primary current of the ideal transformer is $\tilde{I}_1 = \tilde{I}_s = 20$ A.

The secondary (output) current to the transformer is $\frac{|\tilde{I}_1|}{|\tilde{I}_2|} = \frac{N_2}{N_1} = \frac{2}{6} = 1/3$ or

$$\tilde{I}_2 = 3\tilde{I}_1 = 60\ \text{A}\angle 0°. \tag{E6.6.1}$$

The secondary voltage of the transformer is determined by Ohm's law

$$\tilde{V}_2 = \tilde{V}_{load} = \tilde{I}_2 \tilde{Z}_{load} = \tilde{I}_2(R_{load} + j\omega L_{load}) = 60\ \text{A}\angle 0°(1 + j0.5)\ \Omega = (60 + j30)\ \text{V}. \tag{E6.6.2}$$

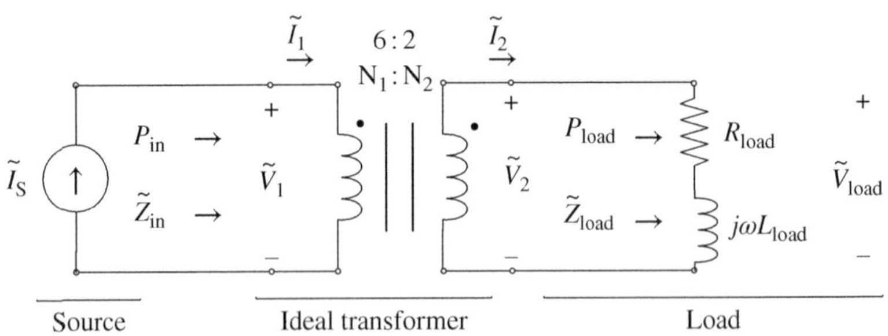

Figure E6.6.1 Ideal transformer fed by current source $\tilde{I}_s$ supplying resistive–inductive load $\tilde{Z}_{load} = R_{load} + j\omega L_{load}$, where $\omega = 2\pi f$ and $f = 60$ Hz.

The relation is $\frac{\left|\widetilde{V}_1\right|}{\left|\widetilde{V}_2\right|} = \frac{N_1}{N_2} = \frac{6}{2} = 3$, resulting in the primary voltage

$$\widetilde{V}_1 = 3\widetilde{V}_2 = 3(60 + j30)\ \text{V} = (180 + j90)\ \text{V}. \quad \text{(E6.6.3)}$$

The output complex power is

$$\widetilde{S}_{\text{out}} = \widetilde{I}_2^*\widetilde{V}_2 = 60\ \text{A}\angle 0°(60 + j30)\ \text{V} = (3600 + j1800)\ \text{VA}, \quad \text{(E6.6.4)}$$

resulting in the real output power $P_{\text{out}} = 3.6$ kW, and because the ideal transformer is lossless, the real input power is

$$P_{\text{in}} = P_{\text{out}} = 3.6\ \text{kW}, \quad \text{(E6.6.5)}$$

which is confirmed by the relation

$$P_{\text{out}} = \widetilde{I}_2^2 R_{\text{load}} = (60)^2\text{A}^2 \cdot 1\ \Omega = 3.6\ \text{kW}. \quad \text{(E6.6.6)}$$

The output impedance is

$$\widetilde{Z}_{\text{out}} = \widetilde{Z}_{\text{load}} = (1 + j0.5)\ \Omega, \quad \text{(E6.6.7)}$$

and the input impedance is given by

$$\widetilde{Z}_{\text{in}} = \frac{\widetilde{V}_1}{\widetilde{I}_1} = \frac{\frac{N_1}{N_2}\widetilde{V}_2}{\frac{N_2}{N_1}\widetilde{I}_2} = \frac{N_1^2\,\widetilde{V}_2}{N_2^2\,\widetilde{I}_2} = \frac{N_1^2}{N_2^2}(R_{\text{load}} + j\omega L_{\text{load}})$$

$$= \left(\frac{6}{2}\right)^2 (1 + j0.5)\ \Omega = (9 + j4.5)\ \Omega, \quad \text{(E6.6.8)}$$

which is confirmed by Ohm's law

$$\widetilde{Z}_{\text{in}} = \frac{\widetilde{V}_1}{\widetilde{I}_1} = \frac{(180 + j90)\ \text{V}}{20\angle 0°\text{A}} = (9 + j4.5)\ \Omega. \quad \text{(E6.6.9)}$$

Application Example 6.7 Circuit with ideal (iron-core permeability $\mu \to \infty$) single-phase transformer and known current source $\widetilde{I}_s$ with inductive load $\widetilde{Z}_{\text{load}} = (R_{\text{load}} + j\omega L_{\text{load}})$ in parallel with a power factor correction capacitor C

An ideal single-phase transformer with turns ratio $N_1/N_2 = 6/2$ is energized by the current source $\widetilde{I}_s = 20\ \text{A}\angle 0°$ and loaded by the resistive–inductive load $\widetilde{Z}_{\text{load}} = (R_{\text{load}} + j\omega L_{\text{load}}) = (1 + j0.5)\ \Omega$, where $\omega = 2\pi f$ and $f = 60$ Hz including a power factor correction capacitor $C = 10\ \mu\text{F}$, as shown in Figure E6.7.1. Calculate the currents $\widetilde{I}_1$, $\widetilde{I}_2$, voltages $\widetilde{V}_2$, $\widetilde{V}_{\text{load}}$, $\widetilde{V}_1$, the output real power P_{out}, the input real power P_{in}, and the impedances $\widetilde{Z}_{\text{out}}$, $\widetilde{Z}_{\text{in}}$.

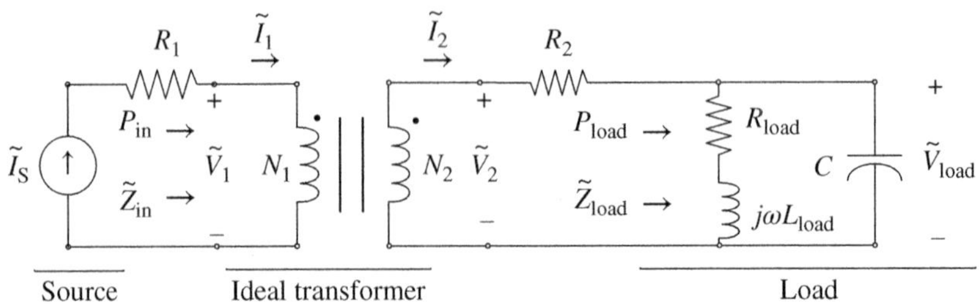

Figure E6.7.1 Electric circuit with ideal single-phase transformer and given current source $\tilde{I}_s$ with inductive load $\tilde{Z}_{load}$ = (R_{load} + $j\omega L_{load}$), N_1/N_2 = 6/2 and power factor correction with capacitor C = 10 μF.

Solution

The primary current of the ideal transformer is $\tilde{I}_1 = \tilde{I}_s = 20$ A.

The secondary (output) current to the transformer is $\dfrac{|\tilde{I}_1|}{|\tilde{I}_2|} = \dfrac{N_2}{N_1} = \dfrac{2}{6} = 1/3$ or

$$\tilde{I}_2 = 3\tilde{I}_1 = 60\ \text{A}\angle 0°. \tag{E6.7.1}$$

The total secondary impedance is the parallel connection of $\tilde{Z}_{load}$ with the capacitor $C = 10\ \mu\text{F}$, yielding the reactive impedance at $f = 60$ Hz:

$$jX_C = -j/\omega C = -j/\left(377 \cdot 10 \cdot 10^{-6}\right) = -j265.27\ \Omega, \tag{E6.7.2}$$

$$\tilde{Z}_{total} = \frac{\tilde{Z}_{load}\left(-\frac{j}{\omega C}\right)}{\tilde{Z}_{load} + \left(-\frac{j}{\omega C}\right)} = \frac{(1 + j0.5)(-j265.27)}{1 + j0.5 - j265.27} = \frac{(-j265.27 + 132.63)}{(1 - j264.77)}$$

$$= \frac{(132.63 - j265.27)(1 + j264.27)}{(1 + 70\,103)} = \frac{132.63 - j265.27 + j35\,116 + 70\,236}{70\,104}$$

$$= \left[\frac{70\,368 + j34\,851}{70\,104}\right]\Omega = (1.004 + j0.497)\ \Omega. \tag{E6.7.3}$$

The secondary voltage of the transformer is determined by Ohm's law:

$$\tilde{V}_2 = \tilde{V}_{load} = \tilde{I}_2\tilde{Z}_{total} = 60\ \text{A}\angle 0°(1.004 + j0.497)\ \Omega = (60.24 + j29.82)\text{V}. \tag{E6.7.4}$$

The relation $\dfrac{|\tilde{V}_1|}{|\tilde{V}_2|} = \dfrac{N_1}{N_2} = \dfrac{6}{2} = 3$ results in the primary voltage

$$\tilde{V}_1 = 3\tilde{V}_2 = 3(60.24 + j29.82)\ \mathrm{V} = (180.72 + j89.46)\ \mathrm{V}. \tag{E6.7.5}$$

The complex output power is

$$\tilde{S}_{\text{out}} = \tilde{I}_2^*\tilde{V}_2 = 60\ \mathrm{A}\angle 0°(60.24 + j29.82)\ \mathrm{V} = (3614.4 + j1789.2)\ \mathrm{VA}, \tag{E6.7.6}$$

resulting in the real output power $P_{\text{out}} = 3.614\ \mathrm{kW}$, and because the ideal transformer is lossless, the real input power is

$$P_{\text{in}} = P_{\text{out}} = 3.614\ \mathrm{kW}, \tag{E6.7.7}$$

the output impedance is

$$\tilde{Z}_{\text{out}} = \tilde{Z}_{\text{total}} = (1.004 + j0.497)\ \Omega, \tag{E6.7.8}$$

and the input impedance is given by

$$\tilde{Z}_{\text{in}} = \frac{\tilde{V}_1}{\tilde{I}_1} = \frac{\frac{N_1}{N_2}\tilde{V}_2}{\frac{N_2}{N_1}\tilde{I}_2} = \frac{N_1^2\tilde{V}_2}{N_2^2\tilde{I}_2} = \frac{N_1^2}{N_2^2}\tilde{Z}_{\text{total}} = \frac{6^2}{2^2}(1.004 + j0.497)\ \Omega = (9.04 + j4.47)\ \Omega. \tag{E6.7.9}$$

Conclusions

The addition of the parallel capacitor of 10 μF decreases the reactive part of the output impedance from 0.5 to 0.497 Ω. One concludes that the capacitance of the parallel capacitor can be increased (e.g. $C = 100\ \mu\mathrm{F}$) in order to significantly reduce the reactive part of the output impedance of 0.497 Ω, thus increasing the power factor at the output and input of this circuit.

6.6 Applications of Single-Phase Power Transformers

There are two types of single-phase transformers: dry-type ones for low power ratings (e.g. 20 kW) and liquid-filled transformers from 10 kW to 333 MW [1, 2, 4–14]. Single-phase units of 25 kW are used in distribution systems and the latter ones in high-voltage transmission systems as part of a three-phase transformer bank with ratings of 999 MW.

Figure 6.15 shows voltages, currents, and powers of a single-phase transformer in the time and phasor domains. The iron core and primary/secondary-winding

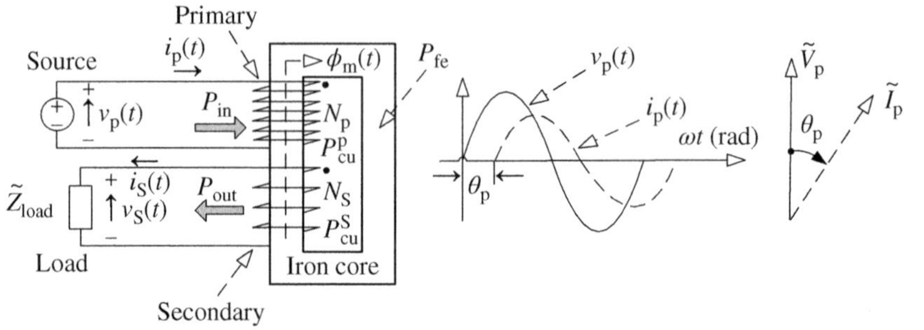

Figure 6.15 Single-phase transformer with input power P_{in}, output power P_{out} and losses in primary and secondary windings P^p_{cu}, P^s_{cu}, and iron-core loss P_{fe} are indicated. Voltage and current time functions $v_p(t)$ and $i_p(t)$, respectively, are shown, as well as phasor quantities $\tilde{V}_p$ and $\tilde{I}_p$.

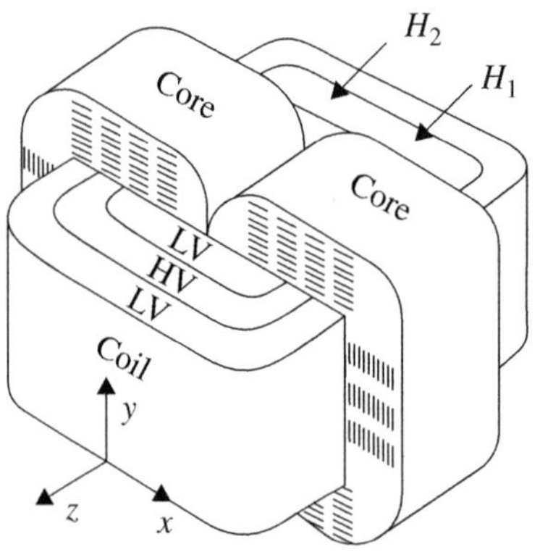

Figure 6.16 Shell-form, single-phase transformer with interleaved primary and secondary windings wrapped around two wound iron cores [14].

assembly of a shell-type, single-phase transformer is detailed in Figure 6.16, while Figures 6.17a–c demonstrate the assembly of an isotropic (due to rolling direction), laminated core with associated inter-lamination and butt-to-butt air gaps. The winding arrangement of a single-phase transformer is shown in Figure 6.18, where the secondary, low-voltage winding consists of thin aluminum sheets and the primary high-voltage winding of copper conductors. The linear T-equivalent circuit is given in Figure 6.19. It is called a T-circuit because the equivalent circuit without load impedance looks like a T. Figure 6.20 represents the phasor diagram of voltages and currents in the complex (1–j) Gaussian plane for the single-phase equivalent circuit of Figure 6.19. Note the superscripts from (1) to (15) attached to the

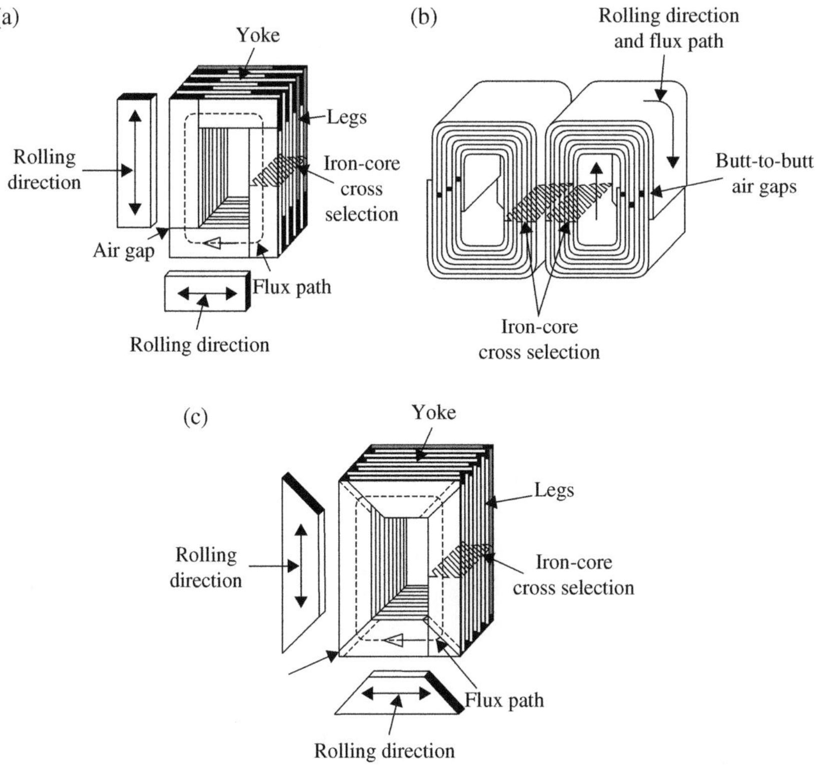

Figure 6.17 (a) Butt-to-butt, (b) wound, and (c) mitered iron cores: for either shell-form or core-form transformers [15].

phasor voltages and currents indicate the sequence of how the phasor diagram is logically constructed.

As indicated in the above equivalent circuit and phasor diagram, the reactive circuit elements are given in terms of reactance values for leakage (e.g. $X_{\ell p}, X'_{\ell s}$) in the series branch, and in the magnetizing or cross-branch, the core loss in terms of the core-loss resistance R_c or the core-(c) loss conductance G_c, where $G_c = 1/R_c$. The magnetizing reactance jX_m is frequently replaced by the magnetizing susceptance $-jB_m$, where $1/jX_m = -j(1/X_m) = -jB_m$. In the following application examples, the inductance (e.g. $X_{\ell p}, L_m$) and the corresponding reactance (e.g. X_m) values are given together with the core-loss conductance G_c. The susceptance ($-jB_m$) is not used as is done in other [1, 2] texts.

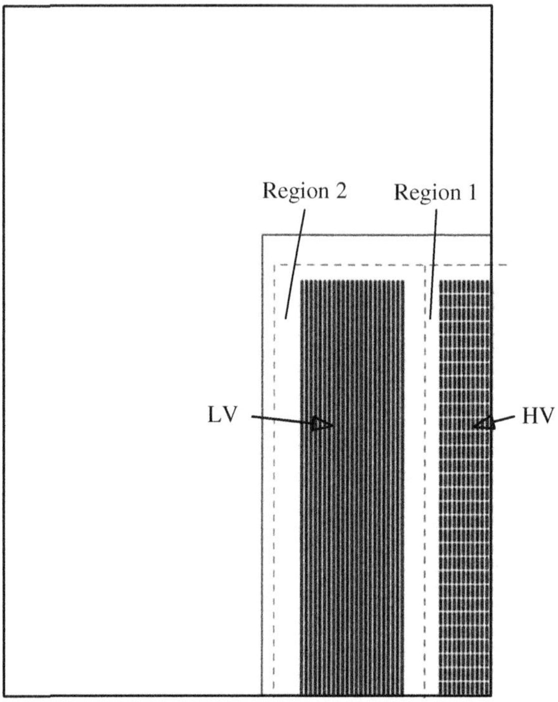

Figure 6.18 Winding arrangement of one-eighth of an $S = 25$ kVA oil-cooled single-phase pole transformer. LV indicates the low-voltage winding consisting (in this case) of aluminum sheets and HV the high-voltage copper winding [14].

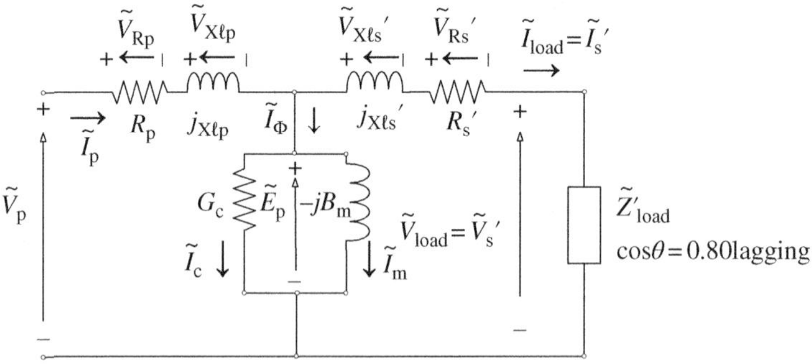

Figure 6.19 Linear *T*-equivalent circuit of two-winding, single-phase transformer, including series and excitation branches. The conductance G_c and the susceptance B_m are assumed to be linear, that is, operating point independent and constant. The same applies to the leakage reactances $X_{\ell p}$ and $X'_{\ell s}$, where subscript "ℓ" stands for leakage.

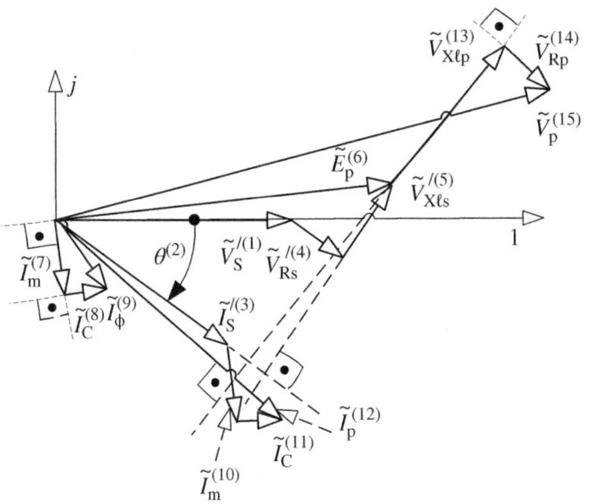

Figure 6.20 Phasor diagram of two-winding single-phase transformer derived from Figure 6.19. The numbers indicate the construction sequence of the phasor diagram within the complex plane indicated by "j" and "1"; that is, one starts with $\tilde{V}_s^{/(1)}$ and ends with $\tilde{V}_p^{(15)}$, where (1)…(15) indicate the sequence numbers.

Application Example 6.8

Circuit with linear single-phase transformer at resistive load R_{load}

An apparent power $S = 50$ kVA, $|\tilde{V}_p| = 2400$ V, $|\tilde{V}_s| = |\tilde{V}_{\text{load}}| = 240$ V, $f = 60$ Hz distribution transformer has the leakage impedance of $\tilde{Z}_p = (R_p + j\omega L_{\ell p}) = (R_p + jX_{\ell p}) = (0.72 + j0.92)\ \Omega$ in the high-voltage primary (p) winding and $\tilde{Z}_s = (R_s + j\omega L_{\ell s}) = (R_s + jX_{\ell s}) = (0.007 + j0.009)\ \Omega$ referred to the low-voltage secondary (s) winding. Referred to the high-voltage winding, the secondary impedance $\tilde{Z}_s$ is

$$\tilde{Z}_s' = R_s' + j\omega L_{\ell s}' = R_s' + jX_{\ell s}' = \left(\frac{2400}{240}\right)^2 (R_s + jX_{\ell s}) = 100(0.007 + j0.009)$$
$$= (0.7 + j0.9)\ \Omega.$$

At rated voltage and frequency, the cross-branch admittance is $\tilde{Y}_\phi = G_c - jB_m = (0.324 - j2.24)$ mS. This corresponds to an impedance of

$$\tilde{Z}_\phi = \frac{1}{\tilde{Y}_\phi} = \frac{1}{(G_c - jB_m)} = \frac{1}{(0.324 - j2.24)\ \text{mS}} = (63.25 + j437.28)\ \Omega = (R_c + jX_m).$$

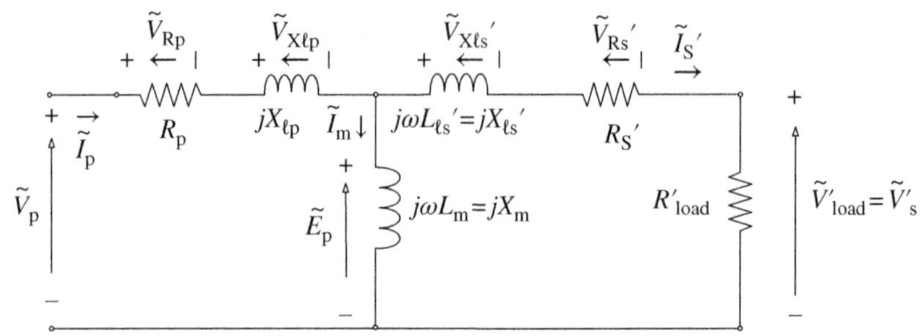

Figure E6.8.1 Linear single-phase transformer at resistive load R_{load} = 1.152 Ω or R'_{load} = 115.2 Ω.

The inductance parameters of the equivalent circuit of Figure E6.8.1 are at a frequency of $f = 60$ Hz: $L_{\ell p} = 2.44$ mH, $L_m = 1.16$ H, $L'_{\ell s} = 2.39$ H.

The associated reactance parameters are as given above $X_{\ell p} = 0.92\,\Omega$, $X_m = 437.28\,\Omega$, $X'_{ls} = 0.9\,\Omega$, and the resistive components are $R_p = 0.72\,\Omega$, $R'_s = 0.7\,\Omega$. The core-loss resistance is in this case assumed to be $R_c \to \infty$ in order to simplify the solution: $\tilde{I}_{load} = \frac{S}{|\tilde{V}_{load}|} = 50\,000\,\frac{\text{VA}}{240\text{ V}} = 208.33$ A, $R_{load} = 240$ V/208.33 A = 1.152 Ω, or R_{load} referred to primary side $R'_{load} = \left(\frac{|\tilde{V}_p|}{|\tilde{V}_s|}\right)^2 \cdot R_{load} = (2400/240)^2\ 1.152\,\Omega = 115.2\,\Omega$. $\tilde{V}'_{load} = \left(\frac{2400}{240}\right) 240\text{ V} = 2400$ V. It is assumed that the transformer delivers rated real power $S = P = 50$ kVA to the load resistance R_{load}, resulting in the secondary current referred to the primary $|\tilde{I}'_s| = \frac{S}{|\tilde{V}'_{load}|} = \frac{50\,000}{2400}$ = 20.833 A.

a) Draw the phasor diagram (not to scale).
b) Calculate the primary current $\tilde{I}_p$, primary voltage $\tilde{V}_p$, and the (fundamental) power factor at the primary side $\cos(\theta_{Vp} - \theta_{Ip}) = \cos\theta_{Vp-Ip}$.

Solution

a) The not-to-scale phasor diagram for the circuit of Figure E6.8.1 is shown in Figure E6.8.2 where the load voltage $\tilde{V}'_{load}$ is identical with $\tilde{V}'_s$, where $\tilde{V}'_s$ and $\tilde{I}'_s$ are in phase, that is, $\theta_{V'load} - \theta_{Is'} = 0$.
b) The primary rated current is

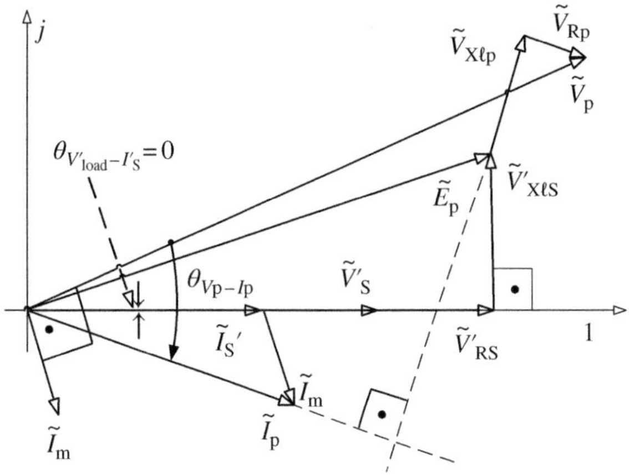

Figure E6.8.2 Not-to-scale phasor diagram for the linear transformer equivalent circuit (Figure E6.8.1), where the load voltage $\tilde{V}'_{load}$ is identical with $\tilde{V}'_s$, where $\tilde{V}'_s$ and $\tilde{I}'_s$ are in phase, that is, $\theta_{V'load} - \theta_{Is'} = 0$.

$$|\tilde{I}_p| = \frac{S}{|\tilde{V}_p|} = \frac{50\,000\text{ VA}}{2400\text{ V}} = 20.833\text{ A}, \tag{E6.8.1}$$

and the rated secondary current is

$$|\tilde{I}_s| = |\tilde{I}_{load}| = \frac{S}{|\tilde{V}_s|} = \frac{50\,000\text{ VA}}{240\text{ V}} = 208.33\text{ A}; \tag{E6.8.2}$$

by referring the secondary current $\tilde{I}_s$ to the primary side, one gets

$$\frac{|\tilde{I}'_{load}|}{|\tilde{I}_{load}|} = \frac{N_2}{N_1} = \frac{240}{2400} = 0.1, \tag{E6.8.3}$$

or

$$|\tilde{I}'_{load}| = |\tilde{I}'_s| = 0.1|\tilde{I}_{load}| = 20.833\text{ A}. \tag{E6.8.4}$$

By choosing $\tilde{V}_{load} = 240\text{ V}\angle 0°$ as reference, then the secondary voltage, referred to the primary, is

$$\frac{|\tilde{V}'_{load}|}{|\tilde{V}_{load}|} = \frac{N_1}{N_2} = \frac{2400}{240} = 10, \tag{E6.8.5}$$

or

$$\left|\widetilde{V}'_{\text{load}}\right| = \frac{N_1}{N_2}\left|\widetilde{V}_{\text{load}}\right| = \left(\frac{2400}{240}\right)240\text{ V} = 2400\text{ V}, \tag{E6.8.6}$$

resulting due to the purely resistive load in the two phasors

$$\widetilde{V}'_{\text{load}} = \widetilde{V}'_{\text{s}} = 2400\text{ V}\angle 0°, \tag{E6.8.7}$$

$$\widetilde{I}'_{\text{load}} = \widetilde{I}'_{\text{s}} = 20.833\text{ A}\angle 0°. \tag{E6.8.8}$$

The induced voltage across the mutual reactance jX_{m} is obtained by applying the KVL:

$$\begin{aligned}\widetilde{E}_{\text{p}} &= \widetilde{I}'_{\text{s}}(\,jX'_{\ell\text{s}} + R'_{\text{s}} + R'_{\text{load}}) = 20.833\text{ A}\angle 0°(\,j0.9 + 0.7 + 115.2)\,\Omega\\ &= (2414.5 + j18.75)\text{ V}.\end{aligned} \tag{E6.8.9}$$

The magnetizing current through jX_{m} is

$$\widetilde{I}_{\text{m}} = \frac{\widetilde{E}_{\text{p}}}{jX_{\text{m}}} = \frac{(2414.5 + j18.75)}{j437.28} = (0.043 - j5.5216)\text{ A}, \tag{E6.8.10}$$

and the resulting actual primary current becomes

$$\begin{aligned}\widetilde{I}_{\text{p}} &= \widetilde{I}'_{\text{s}} + \widetilde{I}_{\text{m}} = 20.833\text{ A}\angle 0° + (0.043 - j5.5216)\text{ A} = (20.876 - j5.5216)\text{ A}\\ &= \sqrt{(20.876)^2 + (5.5216)^2}\angle\phi_{I\text{p}} = 21.594\text{ A}\angle 14.815°.\end{aligned} \tag{E6.8.11}$$

The voltage drop across the primary leakage reactance is

$$\widetilde{V}_{X\ell\text{p}} = jX_{\ell\text{p}}\widetilde{I}_{\text{p}} = j0.92\,\Omega(20.876 - j5.5216)\text{ A} = (5.08 + j19.206)\text{ V}, \tag{E6.8.12}$$

and the voltage drop across the primary resistance

$$\widetilde{V}_{R\text{p}} = R_{\text{p}}\widetilde{I}_{\text{p}} = 0.72\,\Omega(20.876 - j5.5216)\text{ A} = (15.031 - j3.9756)\text{ V}, \tag{E6.8.13}$$

yielding the primary actual voltage by applying the KVL:

$$\begin{aligned}\widetilde{V}_{\text{p}} &= \widetilde{V}_{R\text{p}} + \widetilde{V}_{X\ell\text{p}} + \widetilde{E}_{\text{p}} = 15.031 - j3.9756 + 5.08 + j19.206 + 2414.5 + j18.75\\ &= (24346 + j33.98)\text{ V} = \sqrt{(24346)^2 + (33.98)^2}\angle\theta_{V\text{p}} = 24\,346\text{ V}\angle 0.08°.\end{aligned} \tag{E6.8.14}$$

The fundamental power factor at the primary side of the transformer is

$$\cos\theta_{V\mathrm{p}-I\mathrm{p}} = \cos(0.08 - 14.815)^\circ = 0.967 \text{ lagging,} \tag{E6.8.15}$$

that is, the current lags the voltage.

Application Example 6.9 Circuit with linear single-phase transformer at resistive–inductive load $\widetilde{Z}_{\mathrm{load}} = (R_{\mathrm{load}} + j\omega L_{\mathrm{load}})$ with lagging power factor (PF) (current lags the voltage)

Most loads are of resistive–inductive nature, for example, induction motors. In this example a resistive–inductive load is considered, resulting in a lagging power factor, that is, the current lags the voltage by a relatively large angle.

An apparent power $S = 50$ kVA, $|\widetilde{V}_{\mathrm{p}}| = 2400$ V, $|\widetilde{V}_{\mathrm{s}}| = |\widetilde{V}_{\mathrm{load}}| = 240$ V, $f = 60$ Hz distribution transformer has the leakage impedance of $\widetilde{Z}_{\mathrm{p}} = R_{\mathrm{p}} + j\omega L_{\ell\mathrm{p}} = R_{\mathrm{p}} + jX_{\ell\mathrm{p}} = (0.72 + j0.92)\ \Omega$ in the high-voltage primary (p) winding and $\widetilde{Z}_{\mathrm{s}} = R_{\mathrm{s}} + j\omega L_{\ell\mathrm{s}} = R_{\mathrm{s}} + jX_{\ell\mathrm{s}} = (0.007 + j0.009)\ \Omega$ referred to the low-voltage secondary (s) winding. Referred to the high-voltage winding, the secondary impedance is

$$\widetilde{Z}'_{\mathrm{s}} = R'_{\mathrm{s}} + j\omega L'_{\ell\mathrm{s}} = R'_{\mathrm{s}} + jX'_{\ell\mathrm{s}} = \left(\frac{2400}{240}\right)^2 (R_{\mathrm{s}} + jX_{\ell\mathrm{s}}) = 100(0.007 + j0.009)$$
$$= (0.7 + j0.9)\ \Omega.$$

The core-loss resistance is in this case assumed to be very large and can be neglected ($R_{\mathrm{c}} = 100\ \mathrm{k}\Omega$, $G_{\mathrm{c}} \approx 0.0$ S). At rated voltage and frequency, the cross-branch admittance is $\widetilde{Y}_{\phi} = G_{\mathrm{c}} + jB_{\mathrm{m}} = (0.0 - j2.24)$ mS $= -j2.24$ mS, which corresponds to an impedance of

$$\widetilde{Z}_{\phi} = \frac{1}{\widetilde{Y}_{\phi}} = \frac{1}{G_{\mathrm{c}} - jB_{\mathrm{m}}} = \frac{1}{(0.0 - j2.24)\ \mathrm{mS}} = (\ j446.43)\ \Omega = jX_{\mathrm{m}}.$$

The inductance parameters of the equivalent circuit of Figure E6.9.1 are at a frequency of $f = 60$ Hz:

$$L_{\ell\mathrm{p}} = 2.44\ \mathrm{mH}, L_{\mathrm{m}} = 1.16\ \mathrm{H}, L'_{\ell\mathrm{s}} = 2.39\ \mathrm{mH}.$$

The associated reactance parameters are as given above $X_{\ell\mathrm{p}} = 0.92\ \Omega$, $X_{\mathrm{m}} = 437.28\ \Omega$, $X'_{\ell\mathrm{s}} = 0.9\ \Omega$, and the resistive components are $R_{\mathrm{p}} = 0.72\ \Omega$, $R'_{\mathrm{s}} = 0.7\ \Omega$. The core-loss resistance is in this case assumed to be $R_{\mathrm{c}} \to \infty$ in order to simplify the solution: $\widetilde{I}_{\mathrm{load}} = \frac{S}{|\widetilde{V}_{\mathrm{load}}|} = 50\,000\,\frac{\mathrm{VA}}{240\ \mathrm{V}} = 208.33$ A , $R_{\mathrm{load}} + j\omega L_{\mathrm{load}} = 240\ \mathrm{V}/208.33$ A $\angle\theta_{V'\mathrm{load}} - \theta_{Is'} = 1.152\ \Omega\angle 45^\circ = (0.815 + 0.815)\ \Omega$ or referred to primary side

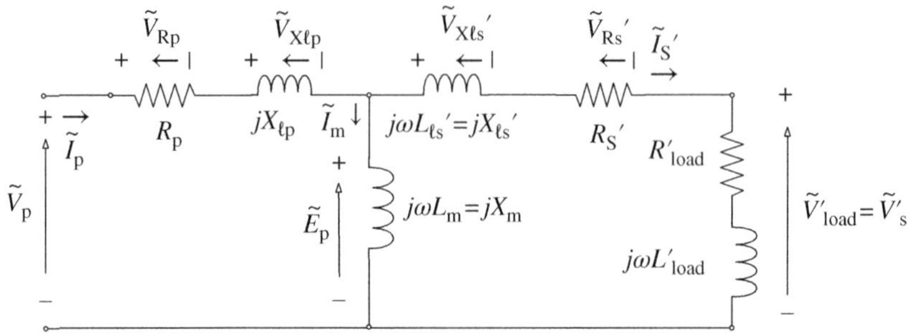

Figure E6.9.1 Linear single-phase transformer at resistive–inductive load $\tilde{Z}_{load} = (R_{load} + j\omega L_{load})$ with lagging power factor (PF) (secondary current lags the secondary voltage by 45°).

$$\tilde{Z}'_{load} = (R'_{load} + j\omega L'_{load}) = \left(\frac{|\tilde{V}_p|}{|\tilde{V}_s|}\right)^2 (R_{load} + j\omega L_{load})$$

$$= \left(\frac{2400}{240}\right)^2 (0.815 + j0.815)\,\Omega = (\,81.5 + j81.5)\,\Omega = 115.24\,\Omega\angle 45°.$$

$\tilde{V}'_{load} = \left(\frac{2400}{240}\right) 240\text{ V} = 2400\text{ V}$. The secondary or load current is

$$\tilde{I}_s = \frac{\tilde{V}_{load}}{R_{load} + j\omega L_{load}} = \frac{240\text{ V}}{(0.815 + j0.815)\,\Omega} = \frac{240\text{ V}}{1.152\,\Omega\angle 45°} = 208.33\text{ A}\angle -45°,$$ or

referred to the primary by multiplying it by $\left(\frac{\tilde{V}_s}{\tilde{V}_p}\right) = \left(\frac{\tilde{V}_{load}}{\tilde{V}_p}\right) = \frac{N_s}{N_p} = \frac{240}{2400}$

yields $\tilde{I}'_s = \left(\frac{240}{2400}\right) 208.33\text{ A}\angle -45° = 20.833\text{ A}\angle -45°$. It is assumed that the transformer delivers rated apparent power $S = 50\text{ kVA}$ to the load impedance $\tilde{Z}_{load} = (R_{load} + j\omega L_{load})$, resulting in the secondary current referred to the primary $\left|\tilde{I}'_s\right| = \frac{S}{\left|\tilde{V}'_{load}\right|} = \frac{50\,000}{2400} = 20.833\text{ A}.$

a) Draw the phasor diagram (not to scale).

b) Calculate the primary current $\tilde{I}_p$, primary voltage $\tilde{V}_p$, and the power factor angle at the primary side $\cos(\theta_{Vp-Ip})$.

Solution

a) Figure E6.9.2 is a not-to-scale phasor diagram.
b) The rated primary current is

$$\left|\tilde{I}_{\mathrm{p}}\right| = \frac{S}{\left|\tilde{V}_{\mathrm{p}}\right|} = \frac{50\,000\ \mathrm{VA}}{2400\ \mathrm{V}} = 20.833\ \mathrm{A}, \tag{E6.9.1}$$

and the rated secondary current is

$$\left|\tilde{I}_{\mathrm{s}}\right| = \left|\tilde{I}_{\mathrm{load}}\right| = \frac{S}{\left|\tilde{V}_{\mathrm{s}}\right|} = \frac{50\,000\ \mathrm{VA}}{240\ \mathrm{V}} = 208.33\ \mathrm{A}; \tag{E6.9.2}$$

by referring the secondary current $\tilde{I}_{\mathrm{s}}$ to the primary side, one gets

$$\frac{\left|\tilde{I}'_{\mathrm{load}}\right|}{\left|\tilde{I}_{\mathrm{load}}\right|} = \frac{N_2}{N_1} = \frac{N_{\mathrm{s}}}{N_{\mathrm{p}}} = \frac{240}{2400} = 0.1, \tag{E6.9.3}$$

or

$$\left|\tilde{I}'_{\mathrm{load}}\right| = \left|\tilde{I}'_{\mathrm{s}}\right| = 0.1\left|\tilde{I}_{\mathrm{load}}\right| = 20.833\ \mathrm{A}. \tag{E6.9.4}$$

By choosing $\tilde{V}_{\mathrm{load}} = 240\ \mathrm{V}\angle 0°$ as reference, then the secondary voltage, referred to the primary, is

$$\frac{\left|\tilde{V}'_{\mathrm{load}}\right|}{\left|\tilde{V}_{\mathrm{load}}\right|} = \frac{N_{\mathrm{p}}}{N_{\mathrm{s}}} = \frac{2400}{240} = 10, \tag{E6.9.5}$$

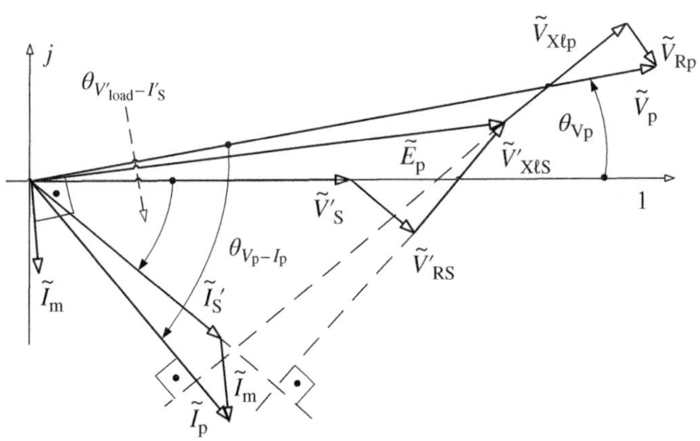

Figure E6.9.2 Not-to-scale phasor diagram for the linear transformer equivalent circuit at resistive–inductive load with $\tilde{R}'_{\mathrm{load}} = 81.5\ \Omega$ and $j\omega L'_{\mathrm{load}} = j81.5\ \Omega$

or

$$\left|\tilde{V}'_{\text{load}}\right| = \frac{N_{\text{p}}}{N_{\text{s}}}\left|\tilde{V}_{\text{load}}\right| = \left(\frac{2400}{240}\right)240\text{ V} = 2400\text{ V}, \tag{E6.9.6}$$

resulting due to the resistive–inductive load, where it is assumed that the secondary current lags the secondary voltage:

$$\tilde{V}'_{\text{load}} = \tilde{V}'_{\text{s}} = 2400\text{ V}\angle 0°, \tag{E6.9.7}$$

$$\tilde{I}'_{\text{load}} = \tilde{I}'_{\text{s}} = 20.833\text{ A}\angle -45°. \tag{E6.9.8}$$

The induced voltage across the mutual reactance jX_{m} is obtained by applying the KVL:

$$\begin{aligned}\tilde{E}_{\text{p}} &= \tilde{I}'_{\text{s}}(jX'_{\ell s} + R'_{\text{s}} + R'_{\text{load}} + j\omega L'_{\text{load}}) = 20.833\text{ A}\angle -45°(\,j0.9 + 0.7 + 81.5 + j81.5)\ \Omega\\ &= 20.833[\cos(-45°) - j\sin(45°)](82.2 + 82.4)\\ &= 20.833(0.7071 - j0.7071)[82.2 + j82.4]\\ &= [14.731 - j14.731][82.2 + j82.4] = [1210.9 + j1213.8 - j1210.9 + 1213.8]\\ &= (2424.7 + j2.9)\text{ V}.\end{aligned} \tag{E6.9.9}$$

The magnetizing current through jX_{m} is

$$\tilde{I}_{\text{m}} = \frac{\tilde{E}_{\text{p}}}{jX_{\text{m}}} = \frac{(2424.7 + j2.9)}{j446.43} = (0.0065 - j5.431)\text{ A}, \tag{E6.9.10}$$

and the resulting actual primary current becomes

$$\begin{aligned}\tilde{I}_{\text{p}} &= \tilde{I}'_{\text{s}} + \tilde{I}_{\text{m}} = 20.833\text{ A}\angle -45° + (0.0065 - j5.431)\text{ A}\\ &= 20.833(0.7071 - j0.7071) + 0.0065 - j5.431\\ &= (14.738 - j20.126)\text{ A} = 25\text{ A}\angle -53.83°.\end{aligned} \tag{E6.9.11}$$

The voltage drop across the primary leakage reactance is

$$\tilde{V}_{X\ell\text{p}} = jX_{\ell\text{p}}\tilde{I}_{\text{p}} = j0.92\ \Omega(25\text{ A}\angle -53.83°) = 23.054\text{ V}\angle 36.01° = (18.649 + j13.664)\text{ V}, \tag{E6.9.12}$$

and the voltage drop across the primary resistance

$$\tilde{V}_{R\text{p}} = R_{\text{p}}\tilde{I}_{\text{p}} = 0.72\ \Omega(25\text{ A}\angle -53.83°) = 18.04\text{ V}\angle -53.99° = (10.606 - j14.189)\text{ V}, \tag{E6.9.13}$$

yielding the primary actual voltage by applying the KVL:

$$\tilde{V}_p = \tilde{V}_{Rp} + \tilde{V}_{X\ell p} + \tilde{E}_p = 10.606 - j14.189 + 18.649 + j13.554 + 2424.7 + j2.9$$
$$= (2454 + j2.265)\,\text{V} = \sqrt{(2454)^2 + (2.265)^2}\angle\theta_{Vp} = 2454\,\text{V}\angle 0.053°. \quad \text{(E6.9.14)}$$

The fundamental power factor (PF) at the primary side of the transformer is

$$\text{PF} = \cos\theta_{Vp-Ip} = \cos(0.053° - 53.99°) = 0.588 \text{ lagging}, \quad \text{(E6.9.15)}$$

that is, the current lags the voltage by 53.94°.

Application Example 6.10 Circuit with linear single-phase transformer at resistive–inductive load with parallel capacitor C = 2 mF, resulting in lagging but increased power factor (PF) (current lags voltage)

In power system applications, leading power factors cause oscillations and are therefore to be avoided. Desirable lagging power factors range from 0.8 to somewhat below unity power factor (1.00). To improve the power factor obtained in Application Example 6.9, a parallel capacitor $C = 2$ mF is connected to the resistive–inductive load as shown in Figure E6.10.1.

An apparent power $S = 50$ kVA, $|\tilde{V}_p| = 2400$ V, $|\tilde{V}_s| = |\tilde{V}_{load}| = 240$ V, $f = 60$ Hz distribution transformer has the leakage impedance of $\tilde{Z}_p = R_p + j\omega L_{\ell p} = R_p + jX_{\ell p} = (0.72 + j0.92)\,\Omega$ in the high-voltage primary (p) winding and $\tilde{Z}_s = R_s + j\omega L_{\ell s} = R_s + jX_{\ell s} = (0.007 + j0.009)\,\Omega$ referred to the low-voltage

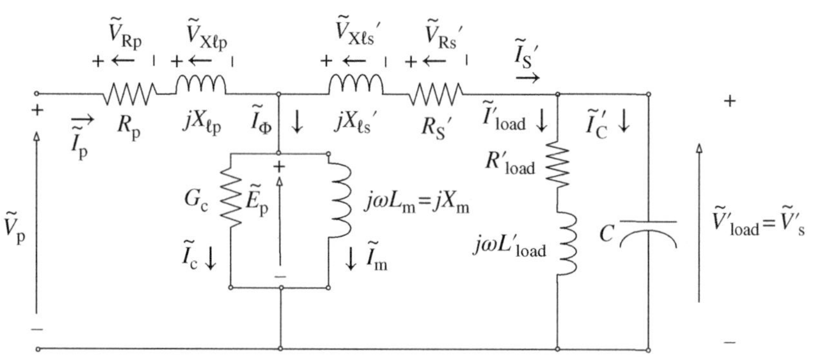

Figure E6.10.1 Linear single-phase transformer at resistive–inductive load with parallel capacitor C = 2 mF, resulting in improved but lagging power factor (PF) (current lags the voltage).

secondary (s) winding. Referred to the high-voltage winding, the secondary impedance is

$$\widetilde{Z}_s' = R_s' + j\omega L_{\ell s}' = R_s' + jX_{\ell s}' = \left(\frac{2400}{240}\right)^2 (R_s + jX_{\ell s}) = 100(0.007 + j0.009)$$

$$= (0.7 + j0.9)\ \Omega.$$

At rated voltage and frequency, the cross-branch admittance is $\widetilde{Y}_\phi = G_c - jB_m = (0.324 - j2.24)$ mS, which corresponds to an impedance of

$$\widetilde{Z}_\phi = \frac{1}{\widetilde{Y}_\phi} = \frac{1}{G_c - jB_m} = \frac{1}{(0.324 - j2.24)\ \text{mS}} = (63.25 + j437.28)\ \Omega = (R_c + jX_m).$$

The inductance parameters of the equivalent circuit of Figure E6.10.1 are at a frequency of $f = 60$ Hz:

$$L_{\ell p} = 2.44\ \text{mH}, L_m = 1.16\ \text{H}, L_{\ell s}' = 2.39\ \text{mH}.$$

The associated reactance parameters are as given above $X_{\ell p} = 0.92\ \Omega$, $X_m = 437.28\ \Omega$, $X_{\ell s}' = 0.9\ \Omega$, and the resistive components are $R_p = 0.72\ \Omega$, $R_s' = 0.7\ \Omega$. The core-loss resistance is in this case assumed to be $R_c = 1000\ \Omega$ in order to have a complete linear model for the transformer:

$$\widetilde{I}_{\text{load}} = \frac{S}{\left|\widetilde{V}_{\text{load}}\right|} = 50\,000\frac{\text{VA}}{240\ \text{V}} = 208.33\ \text{A},$$

$R_{\text{load}} + j\omega L_{\text{load}} = 240\ \text{V}/208.33\ \text{A}\ \angle\phi_{I_s'} = 1.152\ \Omega\angle 45° = (0.815 + 0.815)\ \Omega$ or referred to primary side

$$\widetilde{Z}_{\text{load}}' = \left(R_{\text{load}}' + j\omega L_{\text{load}}'\right) = \left(\frac{|\widetilde{V}_p|}{|\widetilde{V}_s|}\right)^2 (R_{\text{load}} + j\omega L_{\text{load}}) = \left(\frac{2400}{240}\right)^2 (0.815 + j0.815)\ \Omega$$

$$= (81.5 + j81.5)\ \Omega = 115.24\ \Omega\angle 45°.$$

$$\widetilde{V}_{\text{load}}' = \left(\frac{2400}{240}\right) 240\ \text{V} = 2400\ \text{V}.$$

The load impedance in parallel with the capacitance $C = 2$ mF is

$$\frac{(R_{\text{load}} + j\omega L_{\text{load}})\dfrac{1}{j\omega C}}{R_{\text{load}} + j\omega L_{\text{load}} + \dfrac{1}{j\omega C}} = \frac{(0.815 + j0.815)\left(\dfrac{-j}{377 \cdot 2 \cdot 10^{-3}}\right)}{0.815 + j0.815 - \dfrac{j}{377 \cdot 2 \cdot 10^{-3}}} = \frac{1.081 - j1.081}{0.815 - j0.5113}$$

$$= (1.548 - j0.328)\ \Omega = 1.588\ \Omega\angle - 12.96°$$

$$\tilde{I}_s = \frac{\tilde{V}_{load}}{\frac{(R_{load}+j\omega L_{load})\left(\frac{-j}{\omega C}\right)}{R_{load}+j\omega L_{load}-\frac{j}{\omega C}}} = \frac{240\ \text{V}}{1.589\angle -12.96°} = 151.04\ \text{A}\angle -12.96°$$

or referred to the primary by multiplying it by $\left(\frac{|\tilde{V}_s|}{|\tilde{V}_p|}\right) = \left(\frac{|\tilde{V}_{load}|}{|\tilde{V}_p|}\right) = \frac{N_s}{N_p} = 240/2400$ yields

$$\tilde{I}_s' = \frac{240}{2400} 151.04\ \text{A}\angle -12.96° = 15.104\ \text{A}\angle -12.96°.$$

a) Draw the phasor diagram (not to scale).
b) Calculate the primary current $\tilde{I}_p$, primary voltage $\tilde{V}_p$, and the power factor angle at the primary side $\cos(\Phi_{Vp-Ip})$.

Solution

a) Not-to-scale phasor diagram for equivalent circuit of Figure E6.10.1 is shown in Figure E6.10.2.
b) The rated primary current is

$$|\tilde{I}_p| = \frac{S}{|\tilde{V}_p|} = \frac{50\,000\ \text{VA}}{2400\ \text{V}} = 20.833\ \text{A}, \tag{E6.10.1}$$

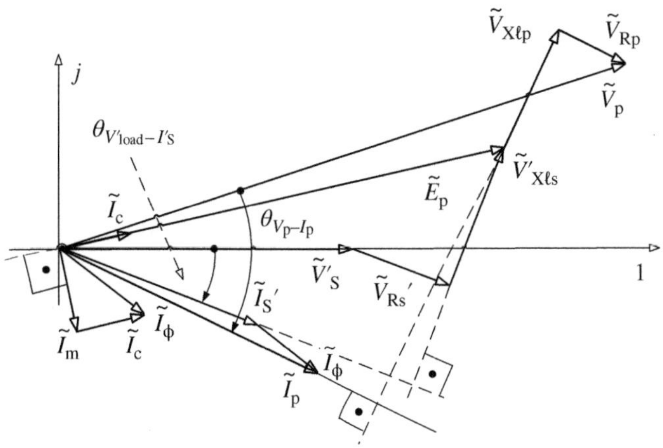

Figure E6.10.2 Not-to-scale phasor diagram for the linear transformer equivalent circuit at resistive–inductive load with R'_{load} = 81.5 Ω and $j\omega L'_{load}$ = j81.5 Ω and a parallel capacitor C = 2 mF.

and the rated secondary current is

$$\left|\widetilde{I}_{\text{load}}\right| = \frac{S}{\left|\widetilde{V}_{\text{load}}\right|} = \frac{50\,000\ \text{VA}}{2400\ \text{V}} = 208.33\ \text{A}, \tag{E6.10.2}$$

by referring the secondary rated current $\widetilde{I}_s$ to the primary side, one gets

$$\frac{\left|\widetilde{I}'_{\text{load}}\right|}{\left|\widetilde{I}_{\text{load}}\right|} = \frac{N_s}{N_p} = \frac{240}{2400} = 0.1, \tag{E6.10.3}$$

or

$$\left|\widetilde{I}'_{\text{load}}\right| = \left|\widetilde{I}'_s\right| = 0.1\left|\widetilde{I}_{\text{load}}\right| = 20.833\ \text{A}. \tag{E6.10.4}$$

By choosing $\widetilde{V}_{\text{load}} = 240\ \text{V}\angle 0°$ as reference, then the secondary voltage, referred to the primary, is

$$\frac{\left|\widetilde{V}'_{\text{load}}\right|}{\left|\widetilde{V}_{\text{load}}\right|} = \frac{N_p}{N_s} = \frac{2400}{240} = 10, \tag{E6.10.5}$$

or

$$\left|\widetilde{V}'_{\text{load}}\right| = \frac{N_p}{N_s}\left|\widetilde{V}_{\text{load}}\right| = (2400/240)240\ \text{V} = 2400\ \text{V}, \tag{E6.10.6}$$

resulting due to the resistive–inductive load with parallel capacitor $C = 2\ \text{mF}$, where it is assumed that the secondary current lags somewhat the secondary voltage (see problem definition above):

$$\widetilde{V}'_{\text{load}} = \widetilde{V}'_s = 2400\ \text{V}\angle 0°, \tag{E6.10.7}$$

$$\widetilde{I}'_s = \frac{240}{2400} 151.04\ \text{A}\angle -12.96° = 15.104\ \text{A}\angle -12.96°. \tag{E6.10.8}$$

The induced voltage across the mutual reactance jX_m is obtained by applying the KVL:

$$\begin{aligned}\widetilde{E}_p &= \widetilde{I}'_s\left(jX'_{\ell s} + R'_s + R'_{\text{load}} + j\omega L'_{\text{load}}\right) = 15.104\ \text{A}\angle -12.96°(\ j0.9 + 0.7 + 81.5 + j81.5)\ \Omega\\ &= 15.104[\cos(-12.96°) - j\sin(12.96°)][82.2 + j82.4]\\ &= 15.104(0.9745 - j0.2243)[82.2 + j82.4]\\ &= [14.719 - j3.387][82.2 + j82.4] = [1209.9 + j1212.8 - j278.41 + 279.09]\\ &= (1489 + j934.39)\ \text{V}.\end{aligned} \tag{E6.10.9}$$

The magnetizing current through jX_m is

$$\tilde{I}_m = \frac{\tilde{E}_p}{jX_m} = \frac{1489 + j934.39}{437.28} = (3.405 - j2.237)\ \text{A}. \tag{E6.10.10}$$

The core-loss current through $R_c = 1\ \text{k}\Omega$ is

$$\tilde{I}_c = \frac{\tilde{E}_p}{R_c} = \frac{(1489 + j934.39)\ \text{V}}{1000\ \Omega} = (1.489 + j0.934), \tag{E6.10.11}$$

and the resulting actual primary current becomes

$$\begin{aligned}\tilde{I}_p &= \tilde{I}_s' + \tilde{I}_m + \tilde{I}_c = 15.104\ \text{A}\angle -12.96^\circ + (3.405 - j2.237)\ \text{A} + (1.489 + j0.934)\ \text{A}\\ &= 15.104[\cos(-12.96^\circ) - j\sin(12.96^\circ)] + (3.405 - j2.237)\ \text{A} + (1.489 + j0.934)\ \text{A}\\ &= 14.719 - j3.387 + 3.405 - j2.237 + 1.489 + j0.934 = 19.613 - j4.69\\ &= 20.166\ \text{A}\angle -13.44^\circ.\end{aligned} \tag{E6.10.12}$$

The voltage drop across the primary leakage reactance is

$$\begin{aligned}\tilde{V}_{X\ell p} &= jX_{\ell p}\tilde{I}_p = j0.92\ \Omega \cdot 20.166\ \text{A}\angle -13.44^\circ = 18.553\ \text{V}\angle 76.56^\circ\\ &= (4.312 + j18.045)\ \text{V},\end{aligned} \tag{E6.10.13}$$

and the voltage drop across the primary resistance

$$\begin{aligned}\tilde{V}_{Rp} &= R_p\tilde{I}_p = 0.72\ \Omega \cdot 20.166\ \text{A}\angle -13.44^\circ = 14.52\ \text{V}\angle -13.44^\circ\\ &= (14.122 - j3.375)\ \text{A},\end{aligned} \tag{E6.10.14}$$

yielding the primary actual voltage by applying the KVL:

$$\begin{aligned}\tilde{V}_p &= \tilde{V}_{Rp} + \tilde{V}_{X\ell p} + \tilde{E}_p = 14.122 - j3.375 + 4.312 + j18.045 + 1489 + j934.39\\ &= (1507.4 + j949.06)\ \text{V} = \sqrt{(1507.4)^2 + (949.06)^2}\angle\theta_{Vp} = 1781.3\ \text{V}\angle 32.2^\circ\end{aligned} \tag{E6.10.15}$$

The fundamental power factor (PF) at the primary side of the transformer is

$$\text{PF} = \cos\theta_{Vp-Ip} = \cos(32.2 - (-13.44))^\circ = 0.6992\ \text{lagging}, \tag{E6.10.16}$$

that is, the current $\tilde{I}_p$ lags the voltage $\tilde{V}_p$ by 45.64°.

Conclusions

1) The fundamental power factor at the transformer primary has improved due to $C = 2\ \text{mF}$ from 0.588 lagging in Application Example 6.9 to 0.6992 lagging in Application Example 6.10. Since a power factor of about 0.8 is desirable, the capacitance of $C = 2\ \text{mF}$ should be increased.

2) The required primary voltage decreased with increased power factor, and therefore the load of the transformer should be decreased by increasing the load impedance ($R_{\text{load}} + j\omega L_{\text{load}}$).

The grid structure of a 25 kVA single-phase transformer for numerical field calculation [9] is shown in Figure 6.21a where grain directions are indicated. Figure 6.21b illustrates a sample strip [9, 14, 16, 17] for mixed cross grain and with cross grain for the calculation of representative (B–H) characteristic. As a result of the numerical field calculation [9], one obtains Figures 6.22–6.24.

6.7 Three-Phase Power Transformers

Three-phase transformers are indispensable components within conventional and future power systems dealing with renewable energy sources. This is due firstly to grounding (e.g. Y-connected transformers) issues and voltage and current transformations, including efficiency considerations where high voltages (e.g. 230–765 kV) and low currents are required for long-distance energy transmission. The following section discusses some of the very basic transformer configurations. Some transformer connections [1, 4, 5, 8] serve as filters (e.g. Δ connection represents a trap for triplen harmonics) and symmetrizing load conditions (e.g. zigzag connection).

Figure 6.25a shows the primary (p) three-phase voltages with the period $T = 1/f$ (e.g. $f = 60$ Hz) and displaced by 120° as applied on the input of three-phase transformers, where the resulting primary current set (i_A^p, i_B^p, i_C^p) is lagging the voltage set (v_A^p, v_B^p, v_C^p) due to the load on the secondary side of the transformer. The two phasor voltage and current sets are shown in Figure 6.25b, where the current set lags the voltage set by θ^p degrees. Three single-phase transformers are arranged in Figure 6.25c in a symmetric manner – forming a bank of three transformers – so that the total flux Φ_{total} is zero due to the trigonometric relation where three cosine time functions are displaced in time by 120° from one another:

$$\cos \omega t + \cos(\omega t - 120°) + \cos(\omega t - 240°) = 0. \tag{6.47}$$

The symmetrical arrangement of three single-phase transformers, where the adjacent three limbs or legs do not carry any flux because the three fluxes are canceling at symmetric three-phase operation as documented by Eq. (6.47) and Figure 6.25c, leads to the three-phase transformer of Figure 6.25d, which has somewhat asymmetric limbs due to their different geometric lengths; as can be seen, the two outside limbs are longer than the center limb. This small asymmetry does not significantly influence transformer operation. As an

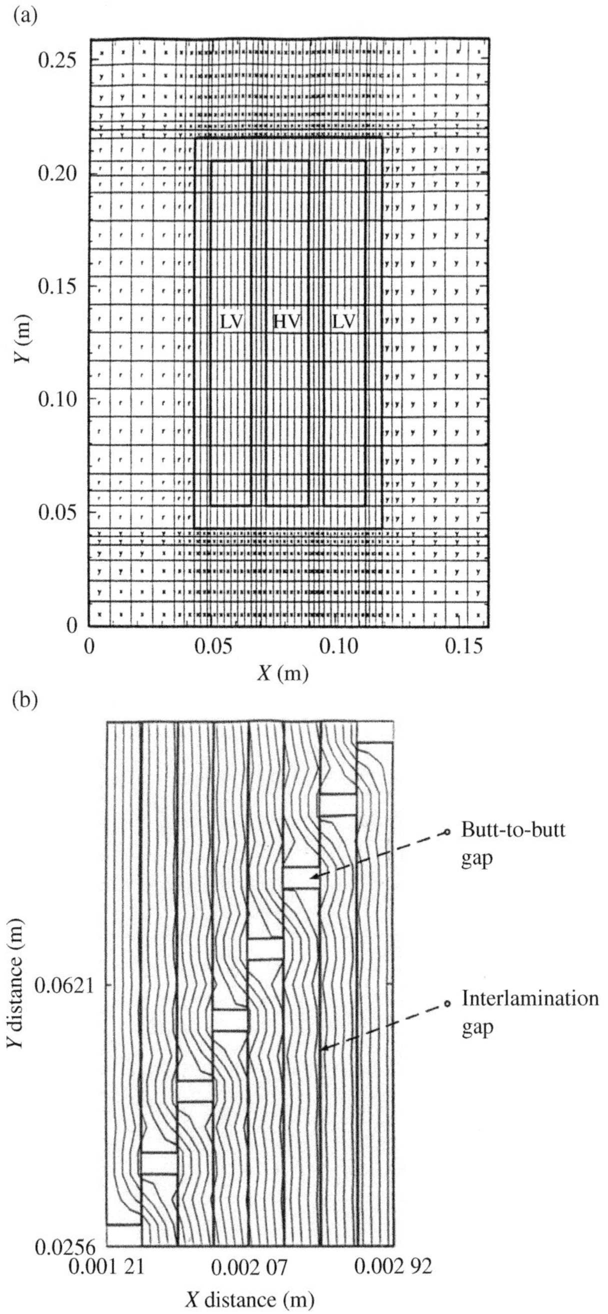

Figure 6.21 (a) Grid structure for numerical field calculation [9] (see Chapter 11) in (x–y) plane for an oil-cooled, two-winding, single-phase 25 kVA pole transformer. Grain directions are indicated by x and y; the overlap joints are identified by r. LV = low-voltage winding, HV = high-voltage winding. (b) Sample strip [9, 14, 16, 17] with mixed cross grain for calculation of representative (B–H) characteristic. Flux does not cross the butt-to-butt gaps, but rather the interlamination gaps, minimizing the equivalent reluctance of the laminated (wound) core.

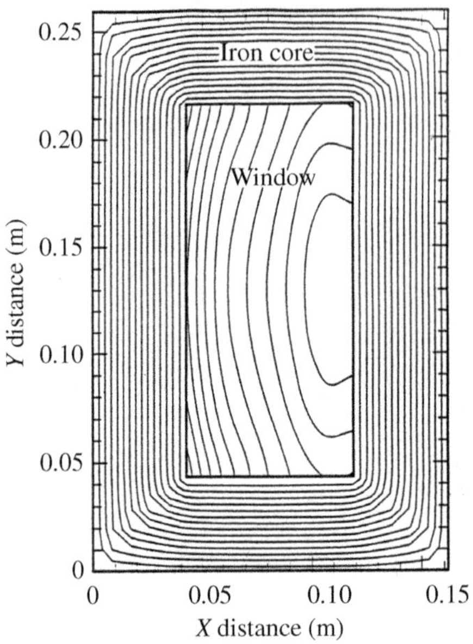

Figure 6.22 Magnetostatic flux distribution in (x–y) plane [9]. The flux tubes within the window contain less flux than within the core.

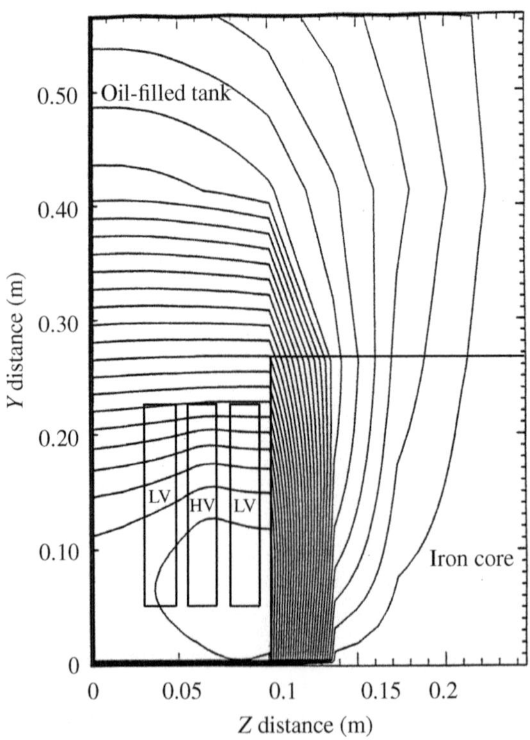

Figure 6.23 Magnetostatic flux distribution in y–z plane [9]. Flux tubes do not contain equal flux. LV = low-voltage winding, HV = high-voltage winding.

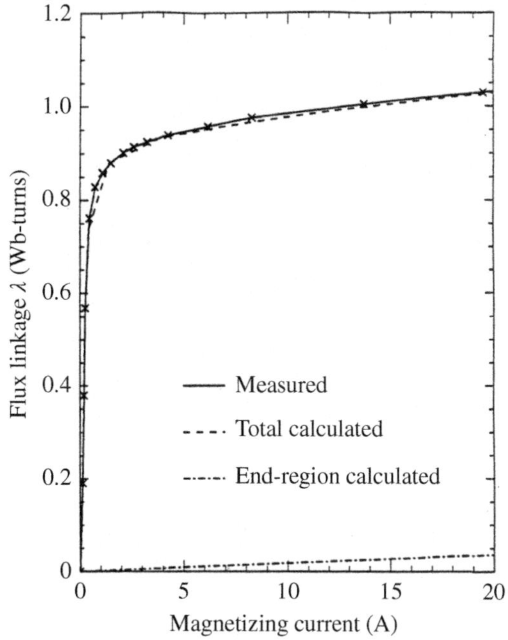

Figure 6.24 Measured and calculated nonlinear ($\lambda - i$) characteristics [9]. Total characteristic of the end region (y–z; see Figure 6.23) plus that of the iron-core region (x–y; see Figure 6.22).

advantage, a three-phase transformer of the type of Figure 6.25d costs less than three single-phase transformers depicted in Figure 6.25c. One significant advantage of the configuration of Figure 6.25c is that in case of transformer failure, only one transformer must be replaced, and this can be done on short notice so that the transformer outage is minimized.

Note the flux in the three center limbs/legs of Figure 6.25c is $\Phi_{total} = 0$ of the voltage sets v_A^p, v_B^p, v_C^p and current sets i_A^p, i_B^p, i_C^p are symmetric with respect to amplitude and phase and have identical waveforms and fundamental frequency. N_p and N_s are the winding turns of primary and secondary windings residing on each single-phase transformer. In the following treatment, ideal transformer conditions (Figures E6.5.1, E6.6.1, E6.7.1) are assumed, that is, $V^p/V^s = V_1/V_2 = N_p/N_s = N_1/N_2$, and $P_1 = P_{in} = P_2 = P_{out} = P_{load}$.

Note that the three limbs [1, 4, 5, 7, 8] are somewhat asymmetric because the two outer limbs are longer than the center limb. This asymmetry is of minor importance and in most cases acceptable. The voltage sets v_A^p, v_B^p, v_C^p and current sets i_A^p, i_B^p, i_C^p are symmetric with respect to amplitude and phase and have identical waveforms and fundamental frequency. N_p and N_s are the winding turns of primary and secondary windings residing on each limb of the three-phase transformer.

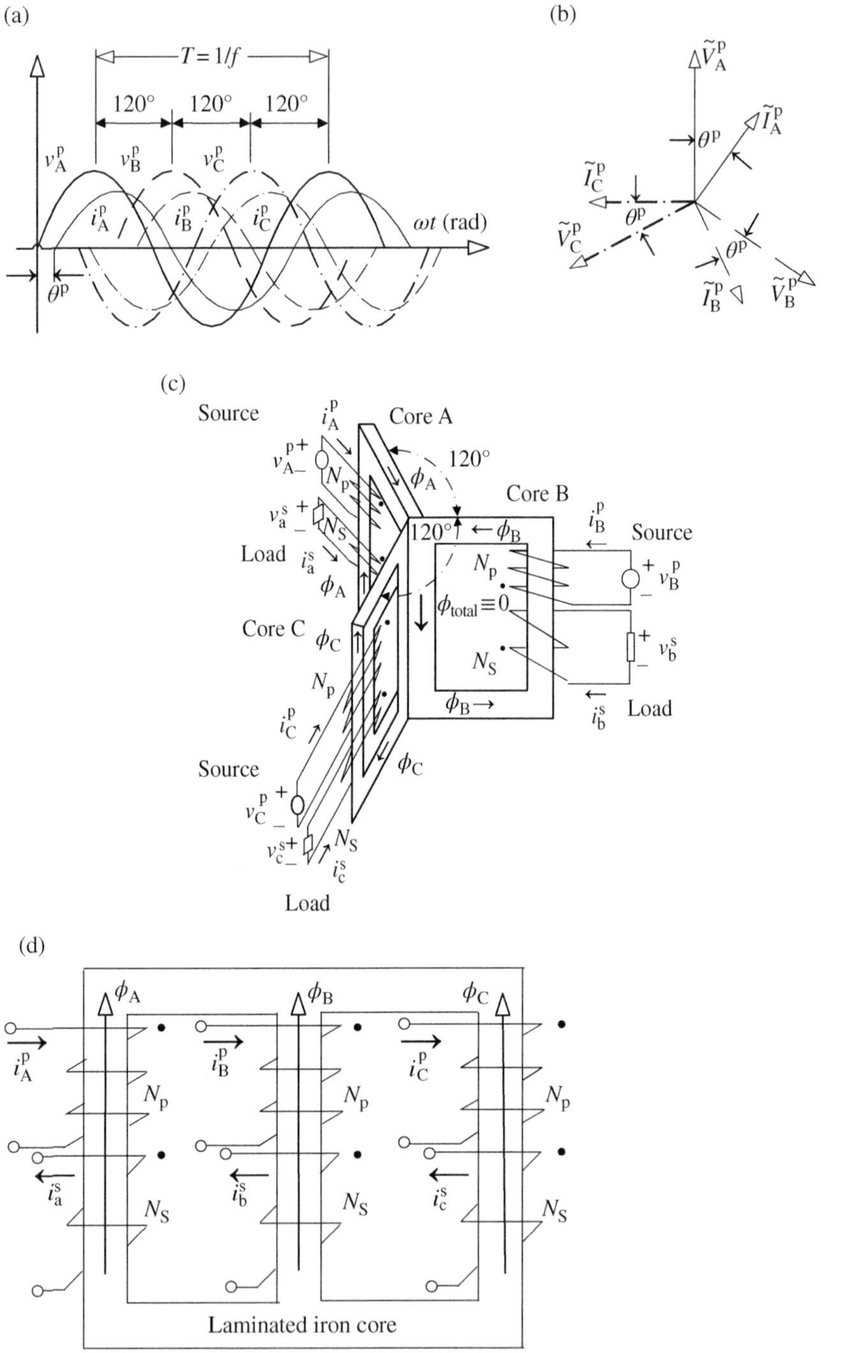

Figure 6.25 (a) Balanced instantaneous three-phase voltages (v_A^p, v_B^p, v_C^p) leading the balanced instantaneous currents (i_A^p, i_B^p, i_C^p) by θ^p degrees during one period T. (b) Balanced three-phase phasor voltages $\widetilde{V}_A^p, \widetilde{V}_B^p, \widetilde{V}_C^p$ leading the balanced phasor currents $\widetilde{I}_A^p, \widetilde{I}_B^p, \widetilde{I}_C^p$ by θ^p degrees. (c) Bank of three single-phase transformers arranged in a symmetric (displaced by 120°) manner to form a three-phase transformer. (d) Elimination of the center limb/leg of (c) leads to three-limb [1, 4, 5, 7, 8] three-phase transformer.

There are quite a few different types of (either single unit or a bank of three single-phase units) three-phase transformers and their applications: the Δ-Y configuration is a very popular one because it permits the trapping of triplen (e.g. 3rd, 6th, 9th, 12th, ...) harmonics due to the Δ-transformer winding and the possibility of grounding due to the Y-winding. Other configurations are Δ–Δ, Y-Y, zigzag, Δ/Y-Y six-phase configurations, and even higher phase/pulse arrangements as used for three-phase rectifiers. In the following application examples, a few three-phase transformers are analyzed in a very simple manner to foster the basic understanding of such devices.

Application Example 6.11 Circuit with linear Y-Y three-phase transformer at Y-connected load resistances R_{load}, that is, $\tilde{Z}_{load} = (R_{load} + j0)$ and line impedances $\tilde{Z}_{line} = (R_{line} + j\omega L_{line}) = (R_{line} + jX_{line})$

This transformer is frequently used as an isolation unit with different reference grounding points n_p, n_s. The root-mean-square (rms) values of the primary balanced voltages are $|\tilde{V}_{Anp}| = |\tilde{V}_{Bnp}| = |\tilde{V}_{Cnp}| = 480\,\text{V}$, and those of the secondary balanced voltages $|\tilde{V}_{ans}| = |\tilde{V}_{bns}| = |\tilde{V}_{cns}| = 277\,\text{V}$. $R_{load} = 50.0\,\Omega$, and $\tilde{Z}_{line} = (0.05 + j0.2)\,\Omega$. The resistive load occurs very rarely because any wire has an inductance due to the magnetic field intensity around it. Also the (distribution, transmission) line has a reactance that is mostly larger than its resistance $X_{line} > R_{line}$.

a) Sketch the phasor diagram of the primary and secondary transformer voltages.
b) Draw the single-phase equivalent circuit corresponding to Figure E6.11.1.
c) Calculate the current through the load $\tilde{I}_{load}$ and the apparent power $S = 3|\tilde{V}_{ans}||\tilde{I}_{load}|$.

Solution

a) Figure E6.11.2 shows the not-to-scale phasor diagram of the sets of primary and secondary voltages together with the orientation of the complex plane indicated by the real (Re) and imaginary (Im) axes. By assuming clockwise (cw) phase rotation, one can write for the primary phase voltages

$$\tilde{V}_{Anp} = 480\,\text{V}\angle 0°, \tilde{V}_{Bnp} = 480\,\text{V}\angle -120°, \tilde{V}_{Cnp} = 480\,\text{V}\angle -240°$$

and the secondary phase voltages

$$\tilde{V}_{ans} = 277\,\text{V}\angle 0°, \tilde{V}_{bns} = 277\,\text{V}\angle -120°, \tilde{V}_{cns} = 277\,\text{V}\angle -240°.$$

b) Figure E6.11.3 depicts the single-phase equivalent circuit of Figure E6.11.1 where the source voltage is the secondary phase voltage $\tilde{V}_{ans} = 277\,\text{V}\angle 0°$.

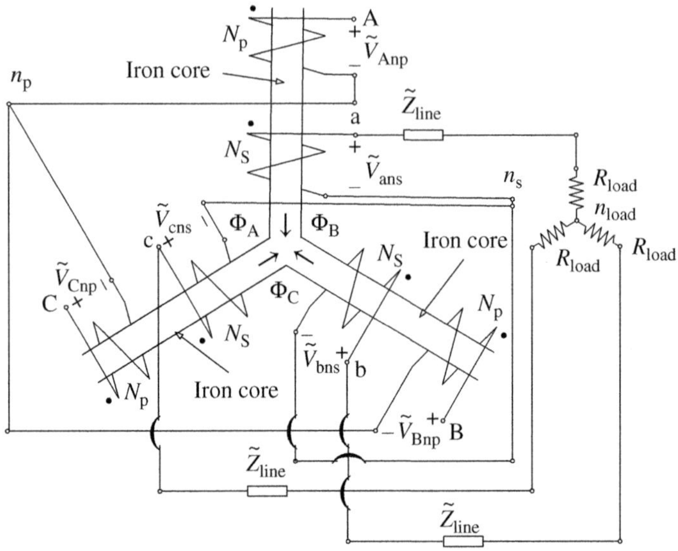

Figure E6.11.1 Y-Y three-phase transformer at Y-connected resistive load, resulting in load fundamental power factor (PF) $\cos\theta = 1.0$. The primary neutral is identified by n_p and the secondary neutral by n_s. In the following treatment, ideal transformer conditions (Figures E6.5.1, E6.6.1, E6.7.1) are assumed, that is, $V_{Anp}/V_{ans} = N_p/N_s$.

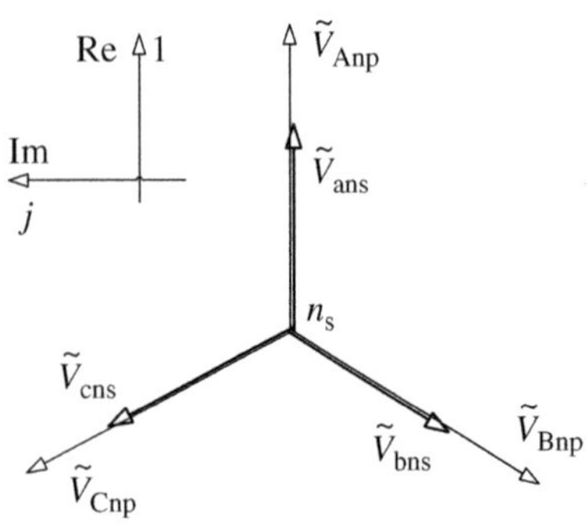

Figure E6.11.2 Not-to-scale phasor diagram of primary and secondary phase voltages of Figure E6.11.1.

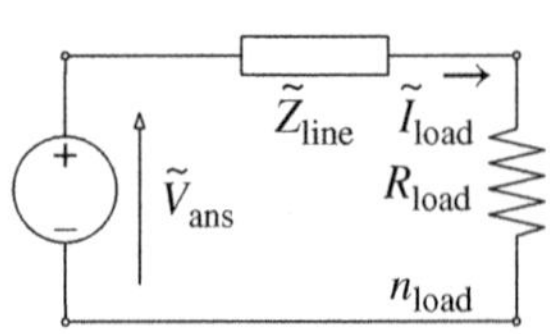

Figure E6.11.3 Single-phase equivalent circuit of Figure E6.11.1. In the following treatment, ideal transformer conditions (Figures E6.5.1, E6.6.1, E6.7.1) are assumed, that is, $V_{Anp}/V_{ans} = N_p/N_s$.

c) The load current is

$$\tilde{I}_{\text{load}} = \frac{\tilde{V}_{\text{ans}}}{\tilde{Z}_{\text{line}} + R_{\text{load}}} = \frac{277\,\text{V}\angle 0^\circ}{(0.05 + j0.2 + 50)\,\Omega} = \frac{277\,\text{V}}{(50.05 + j0.2)\Omega}$$

$$= \frac{277\,\text{V}(50.05 - j0.2)}{(2505 + 0.04)\Omega} = (5.534 - j0.022)\,\text{A} = 5.534\,\text{A}\angle -0.227^\circ,$$

and the apparent power is $S = 3 \cdot 277 \cdot 5.534\,\text{VA} = 4.6\,\text{kVA}$.

Application Example 6.12 Circuit with linear Y-Y three-phase transformer at Δ-connected resistive–inductive load $\tilde{Z}^{\Delta}_{\text{load}} = R^{\Delta}_{\text{load}} + j\omega L^{\Delta}_{\text{load}}$ and line impedances $\tilde{Z}_{\text{line}}$ with lagging power factor (PF) $\cos\theta$ (load current lags the load voltage)

Due to the Δ-Y transformation (Chapter 2), the equivalent Y-connected load per phase is $\tilde{Z}_{\text{load}} = \frac{\tilde{Z}^{\Delta}_{\text{load}}}{3} = R_{\text{load}} + j\omega L_{\text{load}}$. This transformer is frequently used to establish separate primary (p) and secondary (s) grounding points n_p, n_s supplying power to three-phase loads in industrial settings. The primary balanced phase voltages are $\tilde{V}_{\text{Anp}} = 7200\,\text{V}\angle 0^\circ$, $\tilde{V}_{\text{Bnp}} = 7200\,\text{V}\angle -120^\circ$, $\tilde{V}_{\text{Cnp}} = 7200\,\text{V}\angle -240^\circ$, and the secondary balanced phase voltages $\tilde{V}_{\text{ans}} = 240\,\text{V}\angle 0^\circ$, $\tilde{V}_{\text{bns}} = 240\,\text{V}\angle -120^\circ$, $\tilde{V}_{\text{cns}} = 240\,\text{V}\angle -240^\circ$.

The load impedance is $\tilde{Z}^{\Delta}_{\text{load}} = (1.0 + j0.5)\,\Omega$, $\tilde{Z}_{\text{line}} = (0.05 + j0.1)\,\Omega$.

a) Sketch the phasor diagram of the primary and secondary transformer voltages.
b) Draw the single-phase equivalent circuit corresponding to Figure E6.12.1.
c) Calculate the current through the load $\tilde{I}_{\text{load}}$ and the apparent power $S = 3|\tilde{V}_{\text{ans}}||\tilde{I}_{\text{load}}|$.

Solution

a) Figure E6.12.2 shows the not-to-scale phasor diagram of the sets of primary and secondary voltages together with the orientation of the complex plane indicated by the real (Re) and imaginary (Im) axes. For clockwise (cw) phase rotation, the primary phase voltages are given as $\tilde{V}_{\text{Anp}} = 7200\,\text{V}\angle 0^\circ$, $\tilde{V}_{\text{Bnp}} = 7200\,\text{V}\angle -120^\circ$, $\tilde{V}_{\text{Cnp}} = 7200\,\text{V}\angle -240^\circ$ and the secondary balanced phase voltages $\tilde{V}_{\text{ans}} = 240\,\text{V}\angle 0^\circ$, $\tilde{V}_{\text{bns}} = 240\,\text{V}\angle -120^\circ$, $\tilde{V}_{\text{cns}} = 240\,\text{V}\angle -240^\circ$.

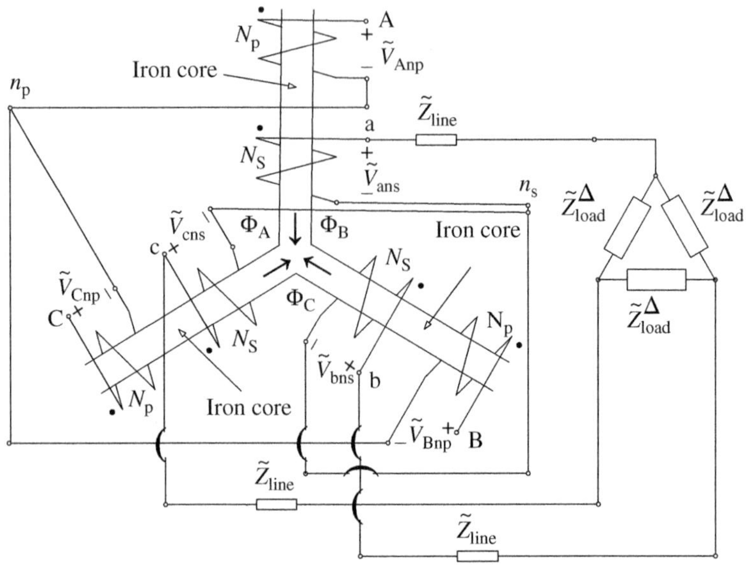

Figure E6.12.1 Y-Y three-phase transformer at Δ-connected resistive–inductive load with lagging fundamental power factor (PF) cos θ (load current lags the load voltage). The primary neutral is identified by n_p and the secondary neutral by n_s.

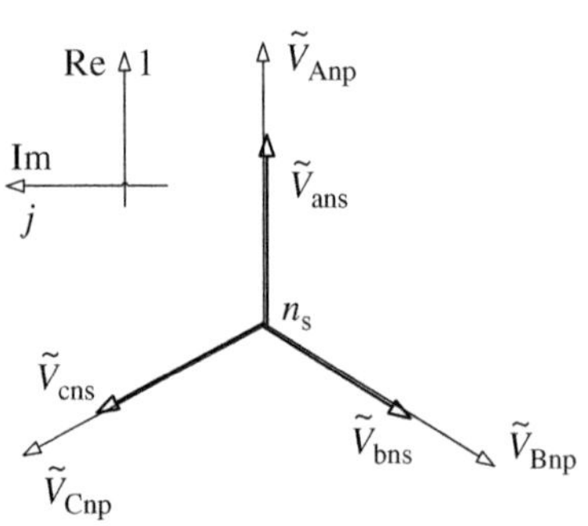

Figure E6.12.2 Not-to-scale phasor diagram of primary and secondary phase voltages of Figure E6.12.1.

b) Figure E6.12.3 depicts the single-phase equivalent circuit of Figure E6.12.1 where the source voltage is the secondary phase voltage $\tilde{V}_{ans} = 240\,\text{V}\angle 0°$.

c) The load current is

Figure E6.12.3 Single-phase equivalent circuit of Figure E6.12.1. In the following treatment ideal transformer conditions (Figures E6.5.1, E6.6.1, E6.7.1) are assumed, that is, $V_{Anp}/V_{ans} = N_p/N_s$.

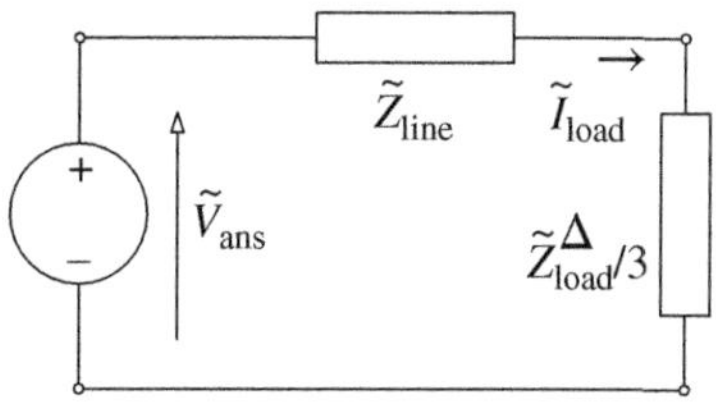

$$\tilde{I}_{load} = \frac{\tilde{V}_{ans}}{\tilde{Z}_{line} + \frac{\tilde{Z}^{\Delta}_{load}}{3}} = \frac{240\,\text{V}\angle 0^\circ}{(0.05 + j0.1 + 0.333 + j0.167)\,\Omega} = \frac{240\,\text{V}}{(0.383 + j0.267)}$$

$$= \frac{240\,\text{V}(0.383 - j0.267)}{(0.147 + 0.071)\Omega} = (421.65 - j293.94)\,\text{A} = 514\,\text{A}\angle -34.88^\circ,$$

and the apparent power is $S = 3 \cdot 240 \cdot 514\,\text{VA} = 370\,\text{kVA}$.

Application Example 6.13 Circuit with linear Δ-Y three-phase transformer at Y-connected resistive–inductive load $\tilde{Z}_{load} = R_{load} + j\omega L_{load}$ and line impedances $\tilde{Z}_{load}$ with lagging power factor (PF) cos θ (load current lags the load voltage)

This unit is frequently used as a $S = 5$ MVA substation transformer. The rms values of the primary balanced voltages are $|\tilde{V}_{AB}| = |\tilde{V}_{BC}| = |\tilde{V}_{CA}| = 135\,\text{kV}$, and those of the secondary balanced voltages $|\tilde{V}_{ans}| = |\tilde{V}_{bns}| = |\tilde{V}_{cns}| = 7.2\,\text{kV}$. $\tilde{Z}_{load} = (30 + j15)\,\Omega$ and $\tilde{Z}_{line} = (1 + j5)\,\Omega$.

a) Sketch the phasor diagram of the primary and secondary transformer voltages.
b) Draw the single-phase equivalent circuit corresponding to Figure E6.13.1.
c) Calculate the current through the load $\tilde{I}_{load}$ and the apparent power $S = 3|\tilde{V}_{ans}||\tilde{I}_{load}|$.

Solution

a) Figure E6.13.2 shows the not-to-scale phasor diagram of the sets of primary and secondary voltages together with the orientation of the complex plane indicated by the real (Re) and imaginary (Im) axes. By assuming clockwise (cw) phase rotation, one can write for the line-to-line voltages $\tilde{V}_{AB} = 135\,\text{kV}\angle 30^\circ$, $\tilde{V}_{BC} = 135\,\text{kV}\angle -90^\circ$, $\tilde{V}_{CA} = 135\,\text{kV}\angle -210^\circ$ and for the phase voltages $\tilde{V}_{ans} = 7.2\,\text{kV}\angle 0^\circ$, $\tilde{V}_{bns} = 7.2\,\text{kV}\angle -120^\circ$, $\tilde{V}_{cns} = 7.2\,\text{kV}\angle -240^\circ$.
b) Figure E6.13.3 depicts the single-phase equivalent circuit of Figure E6.13.1 where the source voltage is the secondary phase voltage $\tilde{V}_{ans} = 7.2\,\text{kV}\angle 0^\circ$.

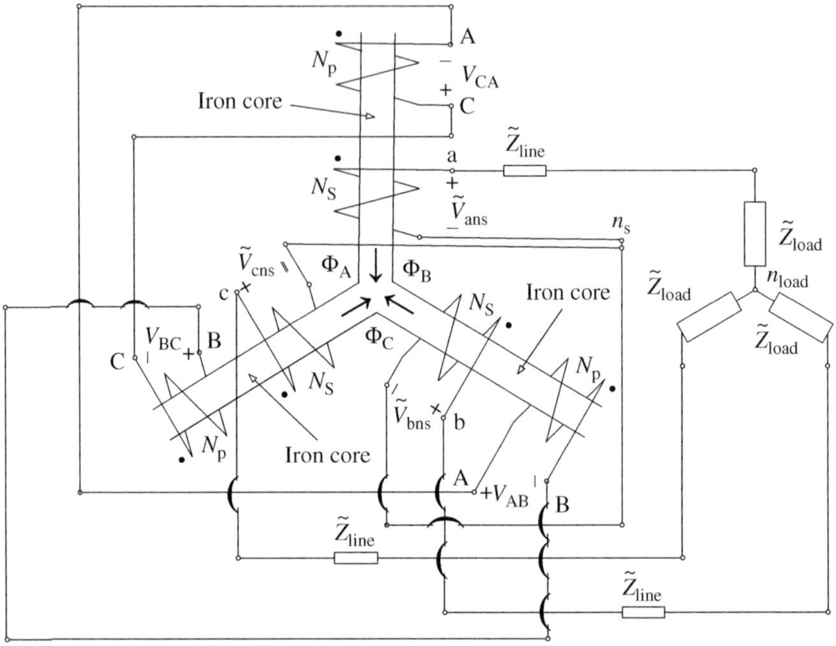

Figure E6.13.1 Δ-Y three-phase transformer at Y-connected resistive–inductive load with lagging power factor (PF) cos θ (load current lags the load voltage). The secondary neutral is identified by n_s. In the following treatment ideal transformer conditions (Figures E6.5.1, E6.6.1, E6.7.1) are assumed, that is, $V_{CA}/V_{ans} = N_p/N_s$.

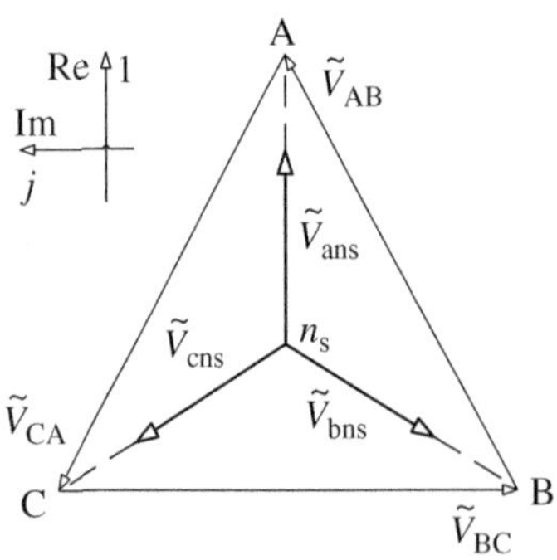

Figure E6.13.2 Not-to-scale phasor diagram of primary line-to-line and secondary phase voltages of Figure E6.13.1.

The load current is

$$\tilde{I}_{load} = \frac{\tilde{V}_{ans}}{\tilde{Z}_{line} + \tilde{Z}_{load}} = \frac{7200\,\text{V}\angle 0^\circ}{(1 + j5 + 30 + j15)\Omega} = \frac{7200}{(31 + j20)} = \frac{7200(31 - j20)}{(961 + 400)},$$
$$= (164 - j105.8)\,\text{A} = 195.17\,\text{A}\angle - 32.83^\circ$$

and the apparent power is $S = 3 \cdot 7200 \cdot 195.17\,\text{VA} = 4.22\,\text{MVA}$.

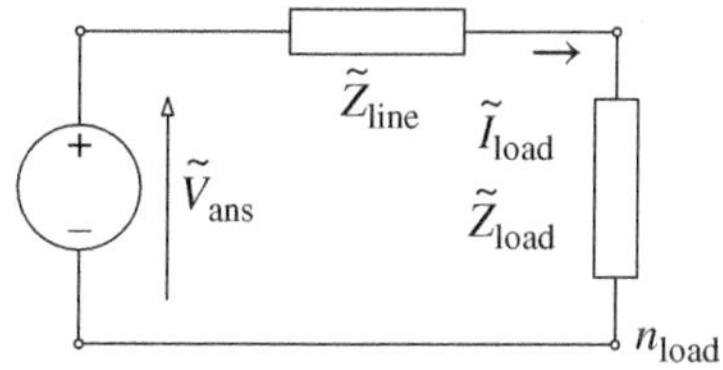

Figure E6.13.3 Single-phase equivalent circuit of Figure E6.13.1.

Conclusions

When transformers are designed and installed, they are overdesigned such that overload conditions in emergencies can be accommodated and a power reserve added for future load growth when residential neighborhoods expand.

Application Example 6.14 Circuit with linear Δ-Y three-phase transformer at Y-connected resistive–inductive load $\tilde{Z}_{load} = (2 + j0.3)\ \Omega$ with power factor (PF) correction employing a Δ-connected capacitor bank C leading to lagging but increased PF $\cos\theta$ (phase load current lags the phase load voltage). The secondary neutral is identified by n_s

This unit is frequently used as a sub-transmission transformer. The rms values of the primary balanced voltages are $|\tilde{V}_{AB}| = |\tilde{V}_{BC}| = |\tilde{V}_{CA}| = 115\ \text{kV}$, and those of the secondary balanced voltages $|\tilde{V}_{ans}| = |\tilde{V}_{bns}| = |\tilde{V}_{cns}| = 7.97\ \text{kV}$. $\tilde{Z}_{load} = (2 + j0.3)\ \Omega$ and $\tilde{Z}_{line} = (0.05 + j0.1)\ \Omega$.

a) Sketch the phasor diagram of the primary and secondary transformer voltages.
b) Draw the single-phase equivalent circuit corresponding to Figure E6.14.1.
c) Calculate the current through the load $\tilde{I}_{load}$ and the apparent power $S = 3|\tilde{V}_{ans}||\tilde{I}_{load}|$.

Solution

a) Figure E6.14.2 shows the not-to-scale phasor diagram of the sets of primary and secondary voltages together with the orientation of the complex plane indicated by the real (Re) and imaginary (Im) axes. By assuming clockwise (cw) phase rotation, one can write for the line-to-line voltages $\tilde{V}_{AB} = 115\ \text{kV}\angle 30°$, $\tilde{V}_{BC} = 115\ \text{kV}\angle -90°$, $\tilde{V}_{CA} = 115\ \text{kV}\angle -210°$ and for the phase voltages $\tilde{V}_{ans} = 7.97\ \text{kV}\angle 0°$, $\tilde{V}_{bns} = 7.97\ \text{kV}\angle -120°$, $\tilde{V}_{cns} = 7.97\ \text{kV}\angle -240°$.
b) Figure E6.14.3 depicts the single-phase equivalent circuit of Figure E6.14.1 where the source voltage is the secondary phase voltage $\tilde{V}_{ans} = 7.97\ \text{kV}\angle 0°$ and the Y-connected load impedance $\tilde{Z}_{load}$ is paralleled with a Δ-connected capacitor bank $C = 133\ \mu\text{F}$ at $f = 60\ \text{Hz}$. Due to the Δ-Y transformation

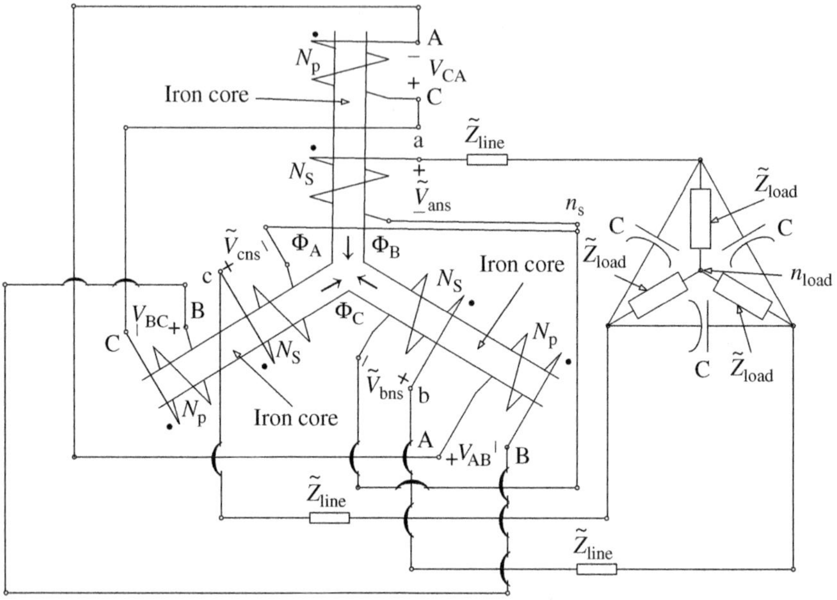

Figure E6.14.1 Δ-Y three-phase transformer at Y-connected resistive–inductive load with parallel connected power factor (PF) correction capacitors C (phase load current lags the phase load voltage). The secondary neutral is identified by n_s. In the following treatment, ideal transformer conditions (Figures E6.5.1, E6.6.1, E6.7.1) are assumed, that is, $V_{CA}/V_{ans} = N_p/N_s$.

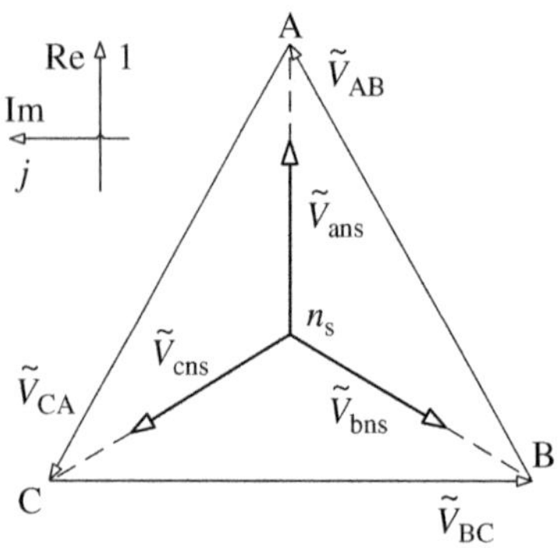

Figure E6.14.2 Not-to-scale phasor diagram of primary line-to-line and secondary phase voltages of Figure E6.14.1.

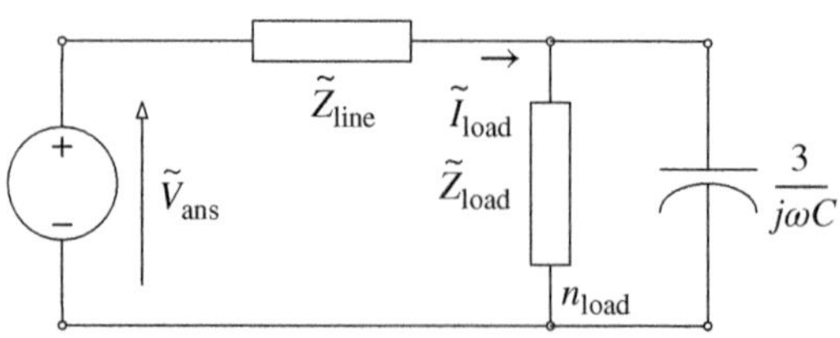

Figure E6.14.3 Single-phase equivalent circuit of Figure E6.14.1.

(Chapter 2), the capacitor's impedance becomes on a per-phase basis $\left(\frac{1}{\frac{j\omega C}{3}}\right) = \frac{3}{j\omega C} = -\frac{j3}{0.05} = -j60\,\Omega$.

c) The load current is

$$\tilde{I}_{\text{load}} = \frac{\tilde{V}_{\text{ans}}}{\tilde{Z}_{\text{line}} + \frac{\tilde{Z}_{\text{load}}\left(\frac{3}{j\omega C}\right)}{\left(\tilde{Z}_{\text{load}} + \frac{3}{j\omega C}\right)}} = \frac{7970\,\text{V}\angle 0^{\circ}}{\left[(0.05 + j0.1) + \frac{(2 + j0.3)\cdot(-j60)}{(2 + j0.3 - j60)}\right]\Omega}$$

$$= \frac{7970\,\text{V}\angle 0^{\circ}}{\left[(0.05 + j0.1) + \frac{(18 - j120)}{(2 - j59.7)}\right]\Omega} = (3753 - j606.1)\,\text{A} = 3801.6\,\text{A}\angle -9.17^{\circ},$$

and the apparent power is $S = 3 \cdot 7970 \cdot 3801.6\,\text{VA} = 90.9\,\text{MVA}$.

Conclusions

When transformers are designed and installed, they are overdesigned such that overload conditions in emergencies can be accommodated and a power reserve added for future load growth when residential neighborhoods expand.

6.8 To Ground or Not to Ground? That Is the Question

Transformer grounding is very important for personnel safety, equipment protection, and power continuity to electrical loads. Electric power systems require two types of grounding: system grounding and equipment grounding; each has its own functions. The IEEE Green Book called *IEEE Recommended Practice for Grounding of Industrial and Commercial Power Systems, ANSI/IEEE C114.1, Std.142*, is the authoritative reference on grounding techniques. There are very special cases where grounding is not advisable. For example, a submarine electric system is not grounded in order to continue its operation if a ground fault occurs, and the ship can reach the harbor with one ground fault before a second ground fault occurs, disabling the ship's operation [4, 5].

6.9 Results Obtained Through More Accurate Calculation and Measurement Methods

Linear methods for the calculation of the circuit parameters generate generally approximate values for inductances and transformers only. Most textbooks neglect the influence of harmonics of electric signals, and derating due to temperature rise is ignored. Failure mechanisms and protection of circuits are mostly addressed in advanced texts, which discuss the following advances in safe and reliable circuit designs.

1) Accurate calculation methods based on the finite-difference and finite-element methods (see Chapter 11, Appendices 11.A and 11.B) permit iron-core modeling taking into account butt-to-butt and interlamination gaps including saturation. New measurement methods permit loss measurements where voltage and current differences are calibrated as compared with the well-established methods where primary and secondary voltages and current are calibrated. Improved efficiency and derating through measurements and calculation are proposed [18–26].
2) The influence of voltage and current harmonics [27–30] on maximum temperature including cooling (hot spot calculation) is taken into account.
3) Transformer derating due to temperature increases is measured and compared with calculations. Aging [4, 5, 31–33] of transformers is addressed.
4) Voltage and current harmonic limits [34] for transformers are proposed as a function of occurring harmonics.
5) Protection [35–37] limits are addressed for power system operation.
6) The performance of DC magnetization [38–41] is discussed.
7) IEEE guidelines [42–46] summarize basic principles for power system and transformer operation.
8) Advanced textbooks complement the issues with respect to transformer operation [47].

Similar advanced analyses are required for other power apparatuses such as electronic circuits (e.g. rectifiers, inverters) and electric machines and other devices instrumental for the design of renewable energy systems.

6.10 Summary

Magnetic materials occur naturally in rocks and have been known to humans for a long time. Certain materials (e.g. neodymium–iron–boron [NdFeB]) can be magnetized through exposing them to a very large magnetic field intensity $\vec{H}$ generated by a large instantaneous current $i(t)$ through a winding with N turns. The magnetic field intensity in turn entails a magnetic flux density $\vec{B} = \mu\vec{H}$, resulting in the flux $\Phi = \int \vec{B} \cdot \overrightarrow{dS}$ when $\vec{B}$ is integrated over the surface S, where the permeability $\mu = \mu_r\mu_0$ is the product of the relative (r) permeability of a material and that of free space. As a consequence of the flux, an electric force $\vec{F}$ is produced that can be used in form of a torque $\vec{T} = (\vec{r} \times \vec{F})$ for propulsion applications; $\vec{r}$

represents the radius where the force $\vec{F}$ acts. One distinguishes between useful flux Φ_{useful} and fringing or leakage flux $\Phi_{fringing}$, which is not useful. A pair of permanent magnets and associated magnetic fields can result in unlike-pole attraction with (mechanical) magnetic force $\vec{F}_{attraction}$. In contrast, a pair of permanent magnets and associated magnetic fields can result in like-pole repulsion with (mechanical) magnetic force $\vec{F}_{repulsion}$.

The application of Ampere's law leads to the right-hand rule [3] for the computation of $\vec{H}(t)$ as a function of the current $i(t)$ within the turns N of a winding/coil. Magnetic and electric fields of low frequency ($0 < f < 20$ kHz) are useful for energy production and transformation. One prominent magnetic field is that of the earth at frequency $f = 0$ Hz – also called a DC field – which amounts to $B_{earth} = 0.7$ gauss, varying per the geographic location and altitude. Shielding magnetic fields of electric apparatus is important to prevent magnetic interferences. Starting from a C-type iron core carrying one winding with N turns, Ohm's law applied to this circuit leads to the definition of a magnetic resistance – also called reluctance – of the gap (g) and core (c) $\Re_g$ and $\Re_c$, respectively.

The flux linkages are defined as $\lambda = N\Phi$, which leads to the definition of the inductance L and the magnetic energy/work stored in the gap (energy$_g$) and in the core (energy$_c$): it turns out that energy$_c \ll$ energy$_g$. Introducing the dot (•) convention permits us to indicate the winding sense of a coil residing on either a magnetic or nonmagnetic core. Introducing at least two windings on an (iron) core leads to a coupled magnetic circuit with its coupling coefficient k. Next, the transition to an unsaturated or linear single-phase transformer is made with self-(s) and mutual (m) inductances L_s and L_m, respectively. To simplify the application, an ideal transformer circuit is specified that can be relied on for an approximate analysis. Varying loads of transformers are considered, and the concept of (fundamental) factor correction is introduced. There are butt-to-butt, wound, and mitered iron cores used for transformer construction. The linear T-equivalent circuit and its associated phasor diagram permit a good prediction of single-phase power transformer performance. A more advanced analysis is based on numerical magnetic fields analyses in the x–y and y–z planes confirmed by measurements: nonlinear flux-linkage (λ versus current i) characteristics are very important for harmonic performance and aging due to temperature increases.

Three-phase power transformers with their three-phase voltage and current excitations are analyzed in the time and phasor domain concentrating on a bank of three-single phase transformers and three-phase units of the types Y-Y with Y-load, Y-Y with Δ-load, Δ-Y with Y-load, and Δ-Y with Y-load and Δ-connected capacitor bank for (fundamental) power factor correction.

Grounding of transformers is briefly addressed, and pertinent references are given, as well as more accurate and advanced measurement and calculation methods.

Problems

6.1 Find the root-mean-square (rms) phasor currents $\tilde{I}_1$ and $\tilde{I}_2$ in the network of Figure P6.1.1 based on the mesh/loop current method.

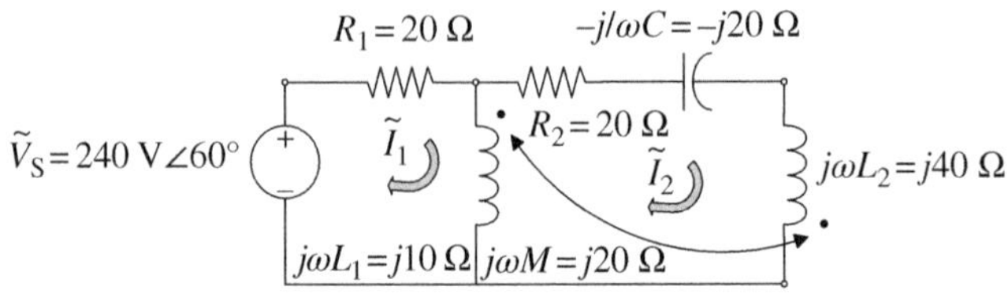

Figure P6.1.1 Network with two coupled linear (without any iron core) inductors, with self-inductances L_1, L_2, resistances R_1, R_2, capacitance C, and mutual inductance M.

6.2 Determine the rms phasor currents $\tilde{I}_1$ and $\tilde{I}_2$ in the circuit of Figure P6.2.1.

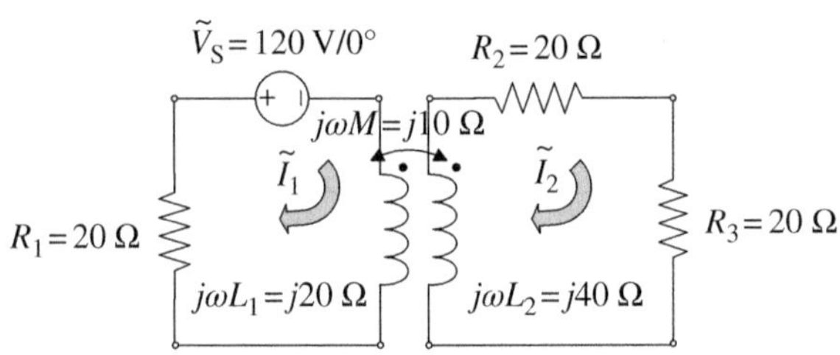

Figure P6.2.1 Network with two coupled linear (without any iron core) inductors, with self-inductances L_1, L_2, resistances R_1, R_2, and R_3, and mutual inductance M.

6.3 Find the rms voltage $\tilde{V}_x$ in the network of Figure P6.3.1.

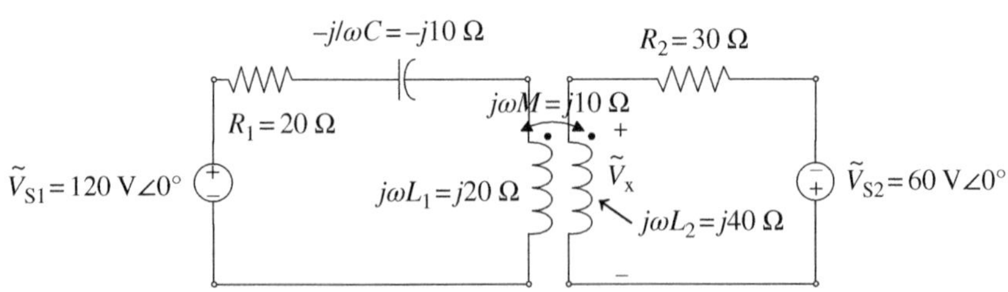

Figure P6.3.1 Network with two coupled linear (without any iron core) inductors, with self-inductances L_1, L_2, resistances R_1, R_2, capacitance C, and mutual inductance M.

6.4 Calculate the rms phasor current $\tilde{I}_2$ for the circuit of Figure P6.4.1.

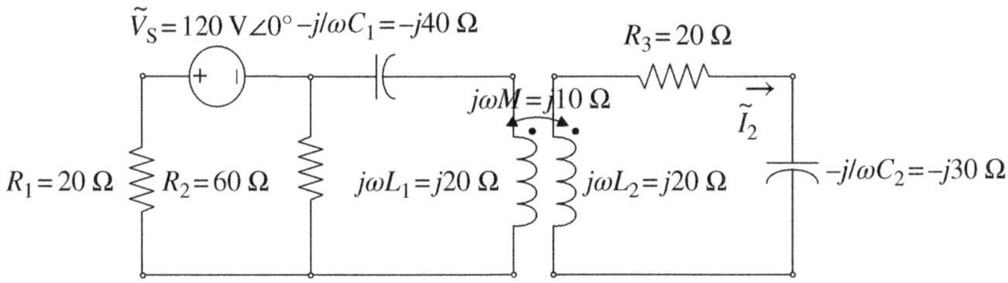

Figure P6.4.1 Network with two coupled linear (without any iron core) inductors, with self-inductances L_1, L_2, resistances R_1, R_2, and R_3, capacitances C_1, C_2, and mutual inductance M.

6.5 Determine the impedance $\tilde{Z}_{in}$ seen by the rms phasor voltage source $\tilde{V}_s$ as given in Figure P6.5.1.

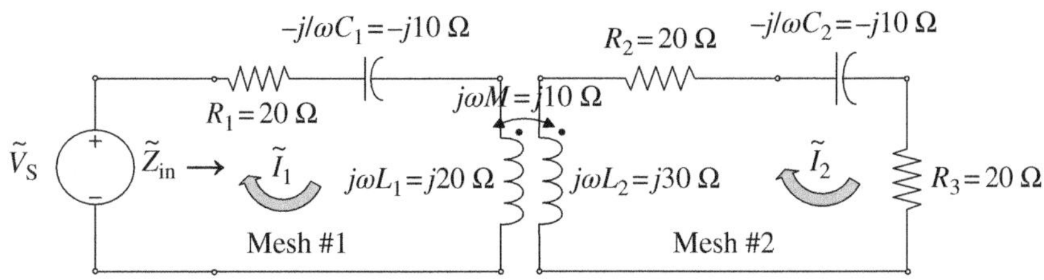

Figure P6.5.1 Network with two coupled linear (without any iron core) inductors, with inductances L_1, L_2, resistances R_1, R_2, R_3, capacitances C_1, C_2, and mutual inductance M.

6.6 Find the rms phasor currents $\tilde{I}_1$ and $\tilde{I}_2$ in the circuit shown in Figure P6.6.1 for $\tilde{V}_s = 60$ V∠0°.

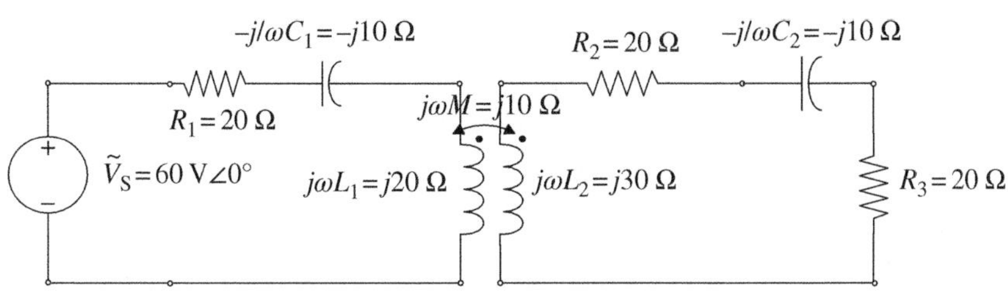

Figure P6.6.1 Network with two coupled linear (without any iron core) inductors, with inductances L_1, L_2, resistances R_1, R_2, R_3, capacitances C_1, C_2, and mutual inductance M.

6.7 In the network of Figure P6.7.1, the rms phasor voltage $\tilde{V}_o = 120\,\text{V}\angle 0°$. Find the output current $\tilde{I}_2$, the input current $\tilde{I}_1$, the impedance of secondary circuit $\tilde{Z}_2 = \left(R_2 + \frac{1}{j\omega C}\right)$ referred to the primary side $\tilde{Z}_2'$, and the input voltage $\tilde{V}_{s\ \text{rms}}$.

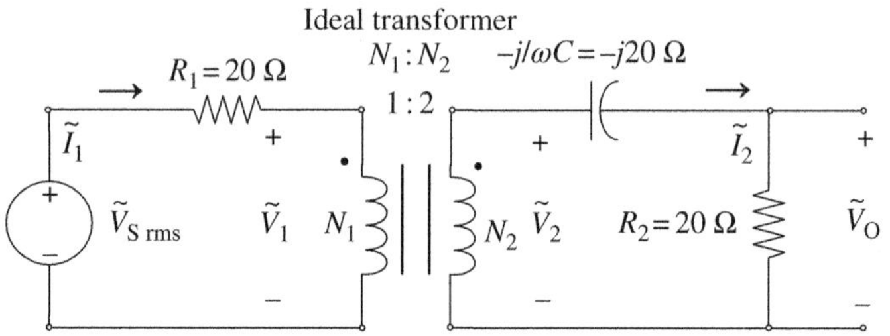

Figure P6.7.1 Network with ideal (step-up) transformer with $N_1 : N_2$ winding ratio, resistances, and capacitance C.

6.8 If the load voltage in the network shown in Figure P6.8.1 is $\tilde{V}_{\text{load\_rms}} = 240\,\text{V}\angle 0°$, determine output current $\tilde{I}_2$, the input current $\tilde{I}_1$, and the rms input phasor voltage $\tilde{V}_{\text{s\_rms}}$.

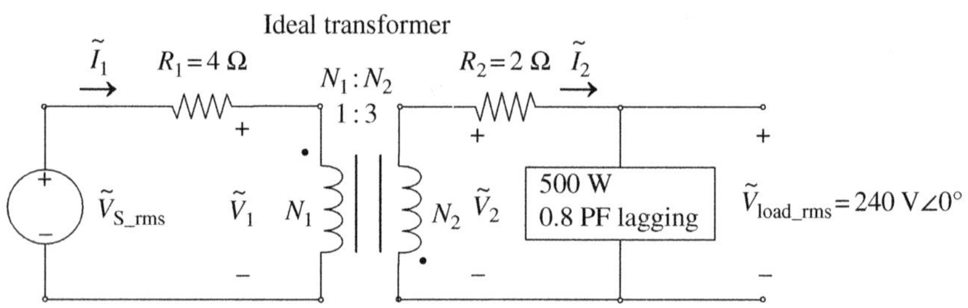

Figure P6.8.1 Network with ideal (step-up) transformer with $N_1 : N_2$ winding ratio, resistances, and lagging power factor (PF) load at $\cos\theta = 0.80$ lagging and $P = 500$ W.

6.9 Circuit with linear single-phase transformer at resistive (unity power factor) load of Figure P6.9.1 with power factor (PF) $\cos\theta = 1.0$ (load voltage and load current are in phase). Calculate for $R = 2\,\Omega$ and $\tilde{V}_1 = 120\,\text{V}\angle 0°$ the secondary transformer voltage $\tilde{V}_2$, the impedance of the secondary circuit $\tilde{Z}_2$, the impedance of the secondary circuit referred to primary side $\tilde{Z}_2'$, and the currents $\tilde{I}_1$, $\tilde{I}_2$.

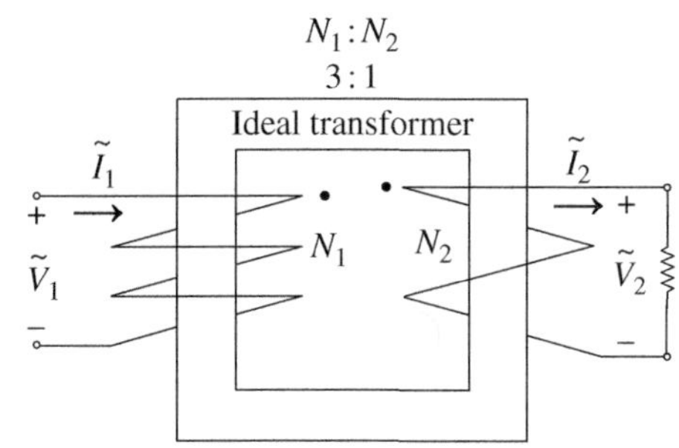

Figure P6.9.1 Network with ideal (step-down) transformer with $N_1 : N_2$ winding ratio at resistive load (unity power factor).

6.10 Figure P6.10.1 represents a circuit with linear single-phase transformer at resistive–inductive load (load current lags the load voltage). Calculate for $(R + j\omega L) = (1 + j2)\ \Omega$ and $\tilde{V}_1 = 240\ \text{V}\angle 0°$ the secondary transformer voltage $\tilde{V}_2$, the impedance of the secondary circuit $\tilde{Z}_2$, the impedance of the secondary circuit referred to primary side $\tilde{Z}_2'$, and the currents $\tilde{I}_1$, $\tilde{I}_2$.

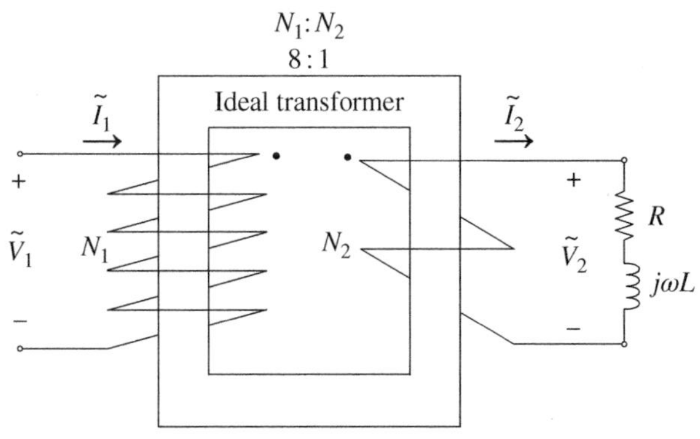

Figure P6.10.1 Network with ideal (step-down) transformer with $N_1 : N_2$ winding ratio at resistive–inductive load (lagging power factor; voltage leads the current).

6.11 Figure P6.11.1 represents a circuit with linear single-phase transformer at resistive–inductive load with power factor correction using a parallel capacitor C, resulting in (PF) $\cos\theta = 0.951$ lagging (load current lags the load voltage). Calculate for $(R + j\omega L) = (1 + j2)\ \Omega$, $\frac{1}{j\omega C} = \frac{-j}{\omega C} = -3j\ \Omega$, $\tilde{V}_1 = 240\ \text{V}\angle 0°$ the secondary transformer voltage $\tilde{V}_2$, the impedance of the secondary circuit $\tilde{Z}_2$, the impedance of the secondary circuit referred to primary side $\tilde{Z}_2'$, and the currents $\tilde{I}_1$, $\tilde{I}_2$.

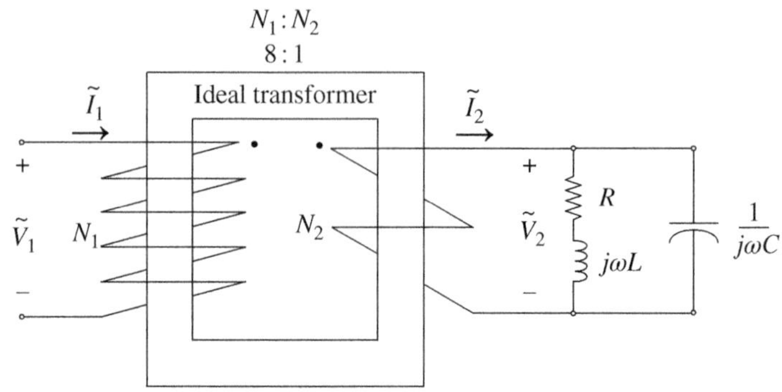

Figure P6.11.1 Network with ideal (step-down) transformer with $N_1 : N_2$ winding ratio at resistive–inductive load with power factor correction capacitor C, resulting in $\cos\theta = 0.951$ lagging (load current lags the load voltage).

6.12 Circuit with Y-Δ three-phase distribution transformer with two parallel Δ-connected resistive–inductive loads $\tilde{Z}^{\Delta}_{\text{load1}}$, $\tilde{Z}^{\Delta}_{\text{load2}}$, resulting in lagging power factor (PF) (load current lags the load voltage).

The line impedances are $\tilde{Z}_{\text{line}} = (0.1 + j0.5)\ \Omega$. Due to the Δ-Y transformation (Chapter 2), the equivalent Y-connected load per phase is $\tilde{Z}^{Y}_{\text{load}i} = \tilde{Z}^{\Delta}_{\text{load}i}/3$. This transformer is frequently used to trap in the secondary Δ-winding triplen (e.g. 3rd, 6th, 9th, ...) harmonic currents whereby the primary (p) Y is grounded at n_p.

The rms values of the primary balanced phase voltages are $|\tilde{V}_{\text{Anp}}| = |\tilde{V}_{\text{Bnp}}| = |\tilde{V}_{\text{Cnp}}| = 24\,000\ \text{V}$, and those of the secondary balanced line-to-line voltages $|\tilde{V}_{\text{aa}'}| = |\tilde{V}_{\text{bb}'}| = |\tilde{V}_{\text{cc}'}| = 2400\ \text{V}$. The rated apparent power is $S_{\text{rat}} = 650\ \text{kVA}$. The load impedances $\tilde{Z}^{\Delta}_{\text{load1}} = R^{\Delta}_{\text{load1}} + j\omega L^{\Delta}_{\text{load1}} = (40 + j20)\ \Omega$, $\tilde{Z}^{\Delta}_{\text{load2}} = R^{\Delta}_{\text{load2}} + j\omega L^{\Delta}_{\text{load2}} = (60 + j30)\ \Omega$. From these follows via Δ-Y transformation $\tilde{Z}^{Y}_{\text{load1}} = R^{Y}_{\text{load1}} + j\omega L^{Y}_{\text{load1}} = (13.33 + j6.66)\ \Omega$, $\tilde{Z}^{Y}_{\text{load2}} = R^{Y}_{\text{load2}} + j\omega L^{Y}_{\text{load2}} = (20 + j10)\ \Omega$.

a Sketch the phasor diagram of the primary and secondary transformer voltages.

b Draw the single-phase equivalent circuit corresponding to Figure P6.12.1.

c Calculate the current through the load $\tilde{I}_{\text{load}}$, the actual apparent power $S_{\text{actual}} = 3|\tilde{V}_{\text{ans}}||\tilde{I}_{\text{load}}|$, and the actual real power $P_{\text{actual}} = S_{\text{actual}} \cos\theta$.

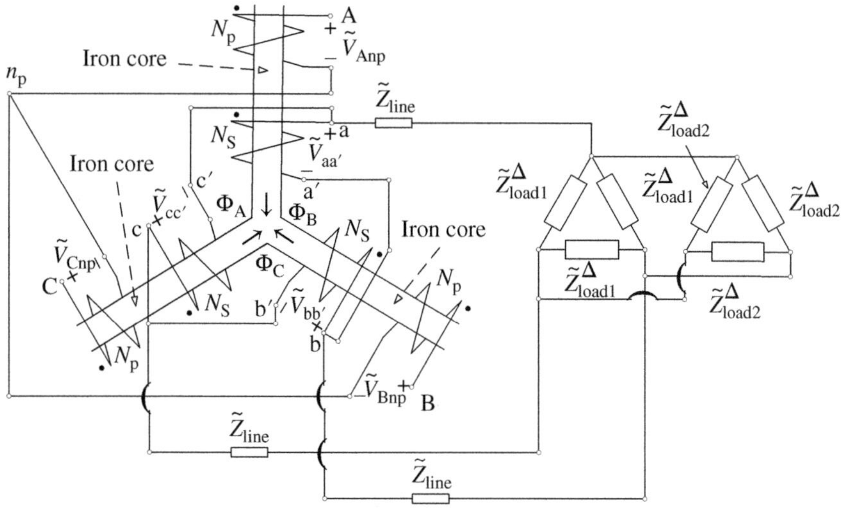

Figure P6.12.1 Circuit with Y-Δ three-phase distribution transformer with two parallel Δ-connected resistive–inductive loads leading to lagging power factor (PF) (load current lags the load voltage). Ideal transformer conditions are assumed with $V_{Anp}/V_{aa'} = N_p/N_S$.

6.13 Circuit with Y-Δ three-phase distribution transformer at Δ-connected resistive–inductive load in parallel with Δ-connected capacitor bank C for power factor (PF) correction, leading to increased power factor but lagging $\cos\theta$ (load current lags the load voltage).

Consider a circuit with Y-Δ three-phase distribution transformer with Δ-connected resistive–inductive load $\tilde{Z}_{load}$ with Δ-connected capacitor bank with capacitance $C = 200\ \mu F$, resulting in increased but lagging power factor (PF) (load current lags the load voltage). The line impedances are $\tilde{Z}_{line} = (0.1 + j0.5)\ \Omega$. Due to the Δ-Y transformation (Chapter 2), the equivalent Y-connected load per phase is $\tilde{Z}^{Y}_{load} = \tilde{Z}^{\Delta}_{load}/3$. This transformer is frequently used to trap in the secondary Δ-winding triplen (e.g. 3rd, 6th, 9th, ...) harmonic currents whereby the primary (p) Y is grounded at n_p.

The rms values of the primary balanced phase voltages are $|\tilde{V}_{Anp}| = |\tilde{V}_{Bnp}| = |\tilde{V}_{Cnp}| = 24\,000$ V, and those of the secondary balanced line-to-line voltages $|\tilde{V}_{aa'}| = |\tilde{V}_{bb'}| = |\tilde{V}_{cc'}| = 2400$ V. The rated apparent power is $S_{rat} = 700$ kVA. The load impedance is $\tilde{Z}^{\Delta}_{load} = R^{\Delta}_{load} + j\omega L^{\Delta}_{load} = (20 + j10)\ \Omega$. Via Δ-Y transformation one gets $\tilde{Z}^{Y}_{load} = R^{Y}_{load} + j\omega L^{Y}_{load} = (6.66 + j3.33)\ \Omega$.

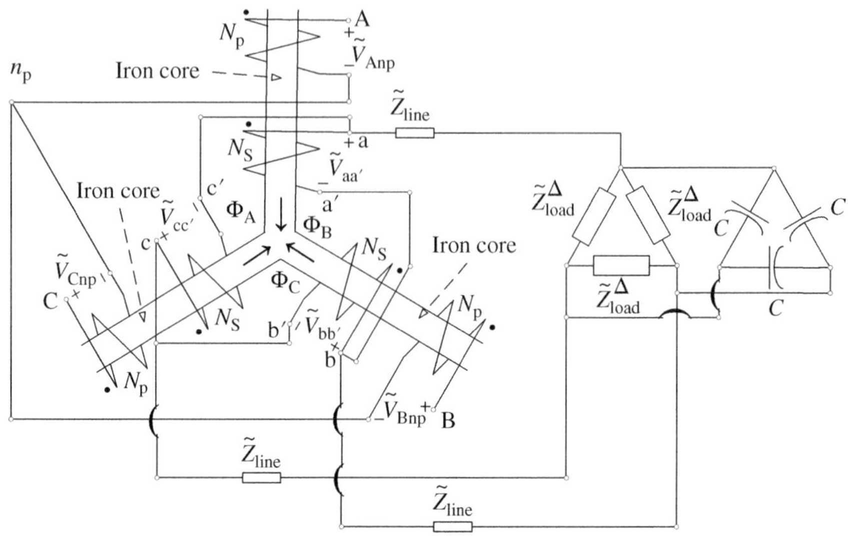

Figure P6.13.1 Circuit with Y-Δ three-phase distribution transformer with Δ-connected resistive–inductive loads and with Δ-connected capacitor bank, leading to lagging power factor (PF) (load current lags the load voltage). Ideal transformer conditions are assumed with $V_{Anp}/V_{aa'} = N_p/N_S$.

a Sketch the phasor diagram of the primary and secondary transformer voltages.

b Draw the single-phase equivalent circuit corresponding to Figure P6.13.1.

c Calculate the current through the load $\tilde{I}_{load}$, the actual apparent power $S_{actual} = 3|\tilde{V}_{ans}||\tilde{I}_{load}|$, and the actual real power $P_{actual} = S_{actual}\cos\theta$.

6.14 Circuit with linear Δ-Δ three-phase sub-transmission transformer at Y-connected resistive–inductive load with lagging (fundamental) power factor (PF) cos θ (load current lags the load voltage).

Consider a circuit with Δ–Δ three-phase sub-transmission transformer with Y-connected resistive–inductive load $\tilde{Z}_{load}$, resulting in lagging fundamental power factor (PF) (load current lags the load voltage). The three line impedances are $\tilde{Z}_{load} = (0.2 + j0.5)\ \Omega$. The rms values of the primary balanced line-to-line voltages are $|\tilde{V}_{AA'}| = |\tilde{V}_{BB'}| = |\tilde{V}_{CC'}| = 225\,000$ V, and those of the secondary balanced line-to-line voltages $|\tilde{V}_{aa'}| = |\tilde{V}_{bb'}| = |\tilde{V}_{cc'}| = 26\,400$ V. The rated apparent power is S_{rat} = 80 MVA. Via Δ-Y transformation one gets for the secondary phase voltage

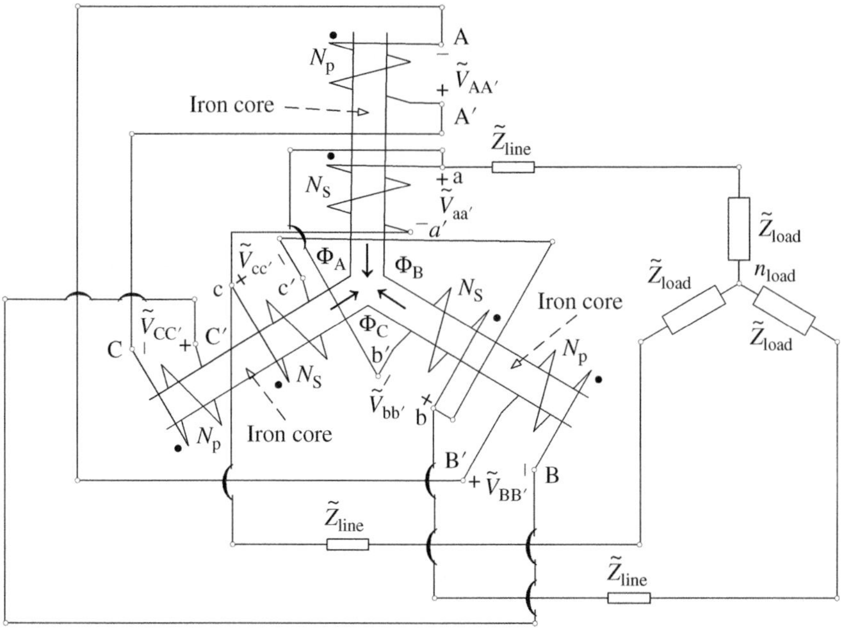

Figure P6.14.1 Circuit with Δ–Δ three-phase, sub-transmission transformer with Y-connected resistive–inductive load leading to lagging power factor (PF) (load current lags the load voltage). Ideal transformer conditions are assumed with $V_{AA'}/V_{aa'} = N_p/N_s$.

$\tilde{V}_{ans} = \left(\frac{|\tilde{V}_{aa'}|}{\sqrt{3}}\right)\angle 0° = \left(\frac{26\,400}{\sqrt{3}}\right)\text{V} = 15\,242\ \text{V}\angle 0°$, and

$\tilde{Z}_{load} = R_{load} + j\omega L_{load} = (7.11 + j4.5)\ \Omega$.

a Sketch the phasor diagram of the primary and secondary transformer voltages.

b Draw the single-phase equivalent circuit corresponding to Figure P6.14.1.

c Calculate the current through the load $\tilde{I}_{load}$, the actual apparent power $S_{actual} = 3|\tilde{V}_{ans}||\tilde{I}_{load}|$, and the actual real power $P_{actual} = S_{actual}\cos\theta$.

6.15 Circuit with Δ–Δ three-phase sub-transmission transformer with two parallel Y-connected resistive–inductive loads with lagging power factor (PF) $\cos\theta$ (load current lags the load voltage).

Ideal transformer conditions are assumed with $V_{BA}/V_{ba} = N_p/N_S$. Consider a circuit with Δ-Δ three-phase sub-transmission transformer with two parallel Y-connected resistive–inductive loads $\tilde{Z}_{load}$, resulting in lagging fundamental power factor (PF) (load current lags the load voltage). The three line impedances are $\tilde{Z}_{line} = (0.2 + j0.5)\ \Omega$. The rms values of the primary balanced line-to-line voltages are $|\tilde{V}_{BA}| = |\tilde{V}_{CB}| = |\tilde{V}_{AC}| = 225\,000$ V, and those of the secondary balanced line-to-line voltages $|\tilde{V}_{ba}| = |\tilde{V}_{cb}| = |\tilde{V}_{ac}| = 26\,400$ V. The rated apparent power is $S_{rat} = 100$ MVA. Via Δ-Y transformation one gets for the secondary phase voltage $\tilde{V}_{ans} = \left(\frac{|\tilde{V}_{ba}|}{\sqrt{3}}\right)\angle 0° = \left(\frac{26\,400}{\sqrt{3}}\right)\text{V}\angle 0° = 15\,242\,\text{V}\angle 0°$, and

$\tilde{Z}_{load1} = R_{load1} + j\omega L_{load1} = (10 + j5)\ \Omega$,
$\tilde{Z}_{load2} = R_{load2} + j\omega L_{load2} = (15 + j10)\ \Omega$, resulting in

$$\tilde{Z}_{load} = \frac{\tilde{Z}_{load1}\tilde{Z}_{load2}}{\tilde{Z}_{load1} + \tilde{Z}_{load2}} = \frac{(10 + j5)(15 + j10)}{(10 + j5 + 15 + j10)}$$
$$= \frac{(150 + j75 + j100 - 50)}{(25 + j15)} = \frac{(100 + j175)(25 - j15)}{625 + 225}$$
$$= \frac{(2500 + j4375 - j1500 + 2625)}{850} = \frac{(5125 + j2875)}{850}$$
$$= (6.029 + j3.382)\ \Omega = R_{load} + j\omega L_{load}.$$

a Sketch the phasor diagram of the primary and secondary transformer voltages.

b Draw the single-phase equivalent circuit corresponding to Figure P6.15.1.

c Calculate the current through the load $\tilde{I}_{load}$, the actual apparent power $S_{actual} = 3|\tilde{V}_{ans}||\tilde{I}_{load}|$, and the actual real power $P_{actual} = S_{actual}\cos\theta$.

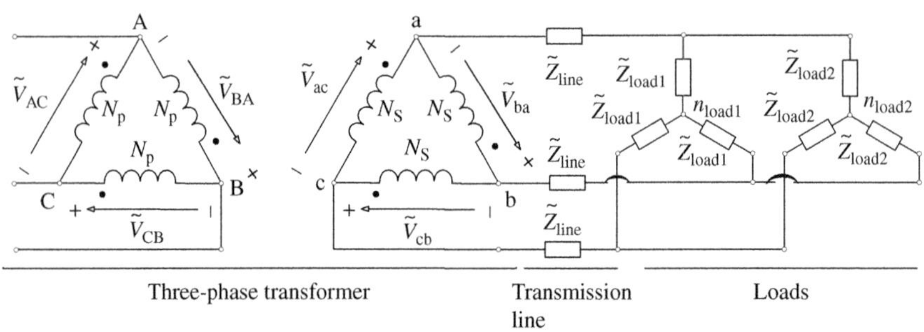

Figure P6.15.1 Circuit with Δ–Δ three-phase sub-transmission transformer with two parallel Y-connected resistive–inductive loads with lagging power factor (PF) (load current lags the load voltage). Ideal transformer conditions are assumed with $V_{BA}/V_{ba} = N_p/N_s$.

6.16 Circuit with Δ–Δ three-phase transmission transformer with two parallel Y-connected resistive–inductive loads with lagging power factor (PF) cos θ, and power factor correction capacitor Δ-connected bank $C = 1$ mF (load current lags the load voltage).

Ideal transformer conditions are assumed with $V_{BA}/V_{ba} = N_p/N_S$. Consider a circuit with Δ–Δ three-phase sub-transmission transformer with two parallel Y-connected resistive–inductive loads $\tilde{Z}_{load}$, resulting in lagging fundamental power factor (PF) (load current lags the load voltage) and a Δ-connected capacitor bank $C = 1$ mF for power factor correction. The line impedances are $\tilde{Z}_{line} = (0.1 + j0.3)\ \Omega$. The rms values of the primary balanced line-to-line voltages are $|\tilde{V}_{BA}| = |\tilde{V}_{CB}| = |\tilde{V}_{AC}| = 345\,000$ V, and those of the secondary balanced line-to-line voltages $|\tilde{V}_{ba}| = |\tilde{V}_{cb}| = |\tilde{V}_{ac}| = 24\,500$ V. The rated apparent power is $S_{rat} = 250$ MVA. Via Δ-Y transformation one gets for the secondary phase voltage $\tilde{V}_{ans} = \left(\frac{|\tilde{V}_{ba}|}{\sqrt{3}}\right)\angle 0° = (24\,500/\sqrt{3})\text{V}\angle 0° = 14\,145\ \text{V}\angle 0°$, and

$$\tilde{Z}_{load1} = R_{load1} + j\omega L_{load1} = (5 + j5)\ \Omega,$$

$$\tilde{Z}_{load2} = R_{load2} + j\omega L_{load2} = (3 + j2)\ \Omega, \text{ resulting in}$$

$$\tilde{Z}_{load} = R_{load} + j\omega L_{load} = \frac{\tilde{Z}_{load1} \cdot \tilde{Z}_{load2}}{\tilde{Z}_{load1} + \tilde{Z}_{load2}} = \frac{(5 + j5)\,(3 + j2)}{(5 + j5 + 3 + j2\,)}\Omega$$

$$= \frac{15 + j15 + j10 - 10}{8 + j7}\Omega = \frac{(5 + j25)(8 - j7)}{64 + 49}\Omega = (1.92 + j1.46)\ \Omega.$$

a Sketch the phasor diagram of the primary and secondary transformer voltages.

b Draw the single-phase equivalent circuit corresponding to Figure P6.16.1.

c Calculate the current through the load $\tilde{I}_{load}$, the actual apparent power $S_{actual} = 3|\tilde{V}_{ans}||\tilde{I}_{load}|$, and the actual real power $P_{actual} = S_{actual}\cos\theta$.

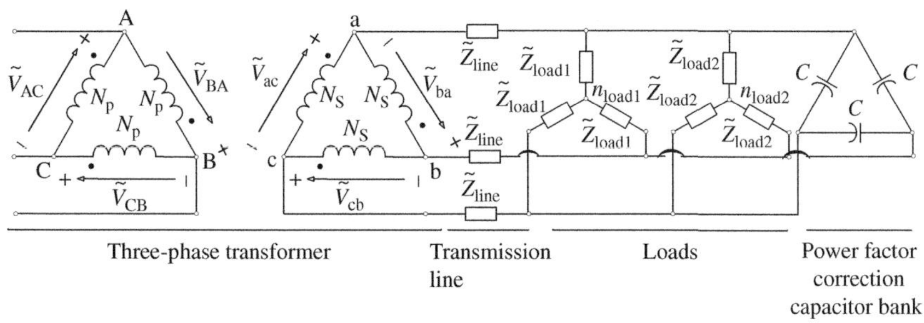

Figure P6.16.1 Circuit with Δ–Δ three-phase transmission transformer with two parallel Y-connected resistive–inductive loads leading to lagging power factor (PF) and power factor correction capacitor Δ-connected bank $C = 1$ mF (load current lags the load voltage). Ideal transformer conditions are assumed with $V_{BA}/V_{ba} = N_p/N_S$.

References

1 Fuchs, E.F. and Masoum, M.A.S. (2012). *Power Conversion of Renewable Energy Systems*, 1e, 2nd Printing. USA: Springer, 692 pp.

2 Fuchs, E.F. and Masoum, M.A.S. (2012). *Instructor Manual for Power Conversion of Renewable Energy Systems*. USA: Springer, 378 pp.

3 Fleming, J.A. (2019). Right and left hand rules. The National High Magnetic Field Laboratory. Lat modified 17 June 2019. https://nationalmaglab.org/education/magnet-academy/watch-play/interactive/right-and-left-hand-rules (accessed 15 May 2020).

4 Masoum, M.A.S. and Fuchs, E.F. (2015). *Power Quality in Power Systems and Electrical Machines*, 2e. Elsevier/Academic Press, 1123 pp.

5 Fuchs, E.F. and Masoum, M.A.S. (2008). *Power Quality in Power Systems and Electrical Machines*, 1e. Elsevier/Academic Press, 638 pp.

6 Nilsson, J.W. (1983). *Electric Circuits*. Reading, MA: Addison-Wesley Publishing Company, Inc., 789 pp.

7 Fuchs, E. F. (2017). Liquid-filled transformers: design, optimization, and monitoring. In: *Wiley Encyclopedia of Electrical and Electronics Engineering* (ed. J. Webster). Hoboken, NJ: Wiley. https://onlinelibrary.wiley.com/action/doSearch?AllField=Fuchs,%20Ewald (accessed 15 May 2020).

8 Fuchs, E. F. (1999). Transformers, liquid filled. In: *Wiley Encyclopedia of Electrical and Electronics Engineering* (ed. J. Webster). New York, NY: Wiley. https://onlinelibrary.wiley.com/doi/abs/10.1002/047134608X.W6132 (accessed 15 May 2020).

9 Stensland, T.D. (1995). Effects of voltage harmonics on single-phase transformers and induction machines including pre-processing for power flow. MS thesis. University of Colorado at Boulder.

10 Stensland, T.D., Fuchs, E.F., Grady, W.M., and Doyle, M. (1997). Modeling of magnetizing and core-loss currents in single-phase transformers with voltage harmonics for use in power flow. *IEEE Trans. Power Deliv.* 12 (2): 768–774.

11 Fuchs, E.F., Masoum, M.A.S., and Roesler, D.J. (1991). Large signal nonlinear model of anisotropic transformers for nonsinusoidal operation, part I: ($\lambda - i$) characteristic. *IEEE Trans. Power Deliv.* TPWRD-6: 174–186.

12 Fuchs, E.F., Masoum, M.A.S., and Roesler, D.J. (1991). Large signal nonlinear model of anisotropic transformers for nonsinusoidal operation, part I: ($\lambda - i$) characteristic. *IEEE Power Eng. Rev.* 11 (1): 47.

13 Masoum, M.A.S., Fuchs, E.F., and Roesler, D.J. (1991). Large signal nonlinear model of anisotropic transformers for nonsinusoidal operation, part II: magnetizing and core-loss currents. *IEEE Trans. Power Deliv.* TPWRD-6: 1509–1516.

14 Batan, T. (1999). Real-time monitoring and calculation of the derating of single-phase transformers under non-sinusoidal operation. PhD thesis. University of Colorado at Boulder (April 1999).

15 Shultz, G.P. (1989). *Transformers and Motors*. Indianapolis: SAMS.

16 Masoum, M. A. S. (1991). Generation and propagation of harmonics in power system feeders containing nonlinear transformers and loads. PhD thesis. University of Colorado at Boulder.

17 Fuchs, E.F., Yildirim, D., and Batan, T. (1999). Innovative procedure for measurement of losses of transformers supplying nonsinusoidal loads. *IEE Proc. - Gener., Transm. Distrib.* 146 (6): 617–625. IEE Proceedings Online No. 19990684.

18 Fuchs, E.F. and Fei, R. (1996). A new computer-aided method for the efficiency measurement of low-loss transformers and inductors under nonsinusoidal operation. *IEEE Trans. Power Deliv.* 11 (1): 292–304.

19 Fuchs, E.F., Yildirim, D., and Grady, W.M. (2000). Measurement of Eddy-current loss coefficient P_{EC-R}, derating of single-phase transformers, and comparison with K-factor approach. *IEEE Trans. Power Deliv.* 15 (1): 148–154.

20 Fuchs, E.F., Yildirim, D., and Grady, W.M. (2000). Corrections to "measurement of Eddy-current loss coefficient P_{EC-R}, derating of single-phase transformers, and comparison with K-factor approach". *IEEE Trans. Power Deliv.* 15 (4): 1357–1357.

21 Fuchs, E.F., Yildirim, D., and Grady, W.M. (2000). Closure of "measurement of Eddy-current loss coefficient P_{EC-R}, derating of single-phase transformers, and comparison with K-factor approach". *IEEE Trans. Power Deliv.* 15 (4): 1331–1333.

22 Yildirim, D. and Fuchs, E.F. (2000). Measured transformer derating and comparison with harmonic loss factor F_{HL} approach. *IEEE Trans. Power Deliv.* 15 (1): 186–191.

23 Yildirim, D. and Fuchs, E.F. (2000). Corrections to "measured transformer derating and comparison with harmonic loss factor F_{HL} approach". *IEEE Trans. Power Deliv.* 15 (4): 1357–1357.

24 Yildirim, D. and Fuchs, E.F. (2000). Closure of "measured transformer derating and comparison with harmonic loss factor F_{HL} approach". *IEEE Trans. Power Deliv.* 15 (4): 1328–1329.

25 Fuchs, E.F., Lin, D., and Martynaitis, J. (2006). Measurement of three-phase transformer derating and reactive power demand under nonlinear loading conditions. *IEEE Trans. Power Deliv.* 21 (2): 665–672.

26 Lin, D. and Fuchs, E.F. (2006). Real-time monitoring of iron-core and copper losses of three-phase transformers under (non)sinusoidal operation. *IEEE Trans. Power Deliv.* 21 (3): 1333–1341.

27 Masoum, M.A.S., Fuchs, E.F., and Roesler, D.J. (1991). Impact of nonlinear loads on anisotropic transformers. *IEEE Trans. Power Deliv.* TPWRD-6: 1781–1788.

28 Chowdhury, A.H., Grady, W.M., and Fuchs, E.F. (1999). An investigation of harmonic characteristics of transformer excitation current under nonsinusoidal supply voltage. *IEEE Trans. Power Deliv.* 14 (2): 450–458.

29 Masoum, M.A.S. and Fuchs, E.F. (2003). Derating of anisotropic transformers under nonsinusoidal operating conditions. *Int. J. Electr. Power Energy Syst.* 25 (1): 1–12.

30 Masoum, M.A.S. and Fuchs, E.F. (1994). Transformer magnetizing current and core losses in harmonic power flow. *IEEE Trans. Power Deliv.* PWRD-9: 10–20.

31 Fuchs, E.F., Roesler, D.J., and Kovacs, K.P. (1986). Aging of electrical appliances due to harmonics of the power system's voltage. *IEEE Trans. Power Deliv.* TPWRD-1 (3): 301–307.

32 Fuchs, E.F., Roesler, D.J., and Alashhab, F.S. (1987). Sensitivity of electrical appliances to harmonics and fractional harmonics of the power system's voltage, part I. *IEEE Trans. Power Deliv.* TPWRD-2 (2): 437–444.

33 Fuchs, E.F., Roesler, D.J., and Kovacs, K.P. (1987). Sensitivity of electrical appliances to harmonics and fractional harmonics of the power system's voltage, part II. *IEEE Trans. Power Deliv.* TPWRD-2 (2): 445–453.

34 Fuchs, E.F., Roesler, D.J., and Masoum, M.A.S. (2004). Are harmonic recommendations according to IEEE and IEC too restrictive? *IEEE Trans. Power Deliv.* 19 (4): 1775–1786.

35 Fuller, J.F., Fuchs, E.F., and Roesler, D.J. (1988). Influence of harmonics on power system distribution protection. *IEEE Trans. Power Deliv.* TPWRD-3 (2): 546–554. (PES/IEEE Prize Paper Award 1989).

36 Fuller, J.F., Fuchs, E.F., and Roesler, D.J. (1989). Influence of harmonics on power system distribution protection. *Power Eng. Rev.* 9 (10): 11–12. (Abstract of PES/IEEE Prize Paper Award 1989).

37 Fuller, J.F., Fuchs, E.F., and Roesler, D.J. (1992). Influence of harmonics on power distribution system protection. In: *Protective Relaying for Power Systems II* (ed. S.H. Horowitz), 29–37. IEEE Press.

38 You, Y., Fuchs, E.F., Lin, D., and Barnes, P.R. (1996). Reactive power demand of transformers with DC bias. *IEEE Ind. Appl. Soc. Mag.* 2 (4): 45–52.

39 Fuchs, E.F., You, Y., and Roesler, D.J. (1999). Modeling, simulation and their validation of three-phase transformers with three legs under DC bias. *IEEE Trans. Power Deliv.* 14 (2): 443–449.

40 Fuchs, E.F. and You, Y. (2002). Measurement of ($\lambda - i$) characteristics of asymmetric three-phase transformers and their applications. *IEEE Trans. Power Deliv.* 17 (4): 983–990.

41 Fuchs, E.F. and You, Y. (2003). Closure of "measurement of ($\lambda - i$) characteristics of asymmetric three-phase transformers and their applications". *IEEE Trans. Power Deliv.* 18 (2): 642–643.

42 Briskman, R.D.; IEEE Standards Board (1974). *IEEE Recommended Practice for Electric Power Systems in Commercial Buildings.* IEEE Std. 241-1974. The Institute of Electrical and Electronics Engineers Inc.

43 Koepfinger, J.L.; IEEE Standards Board (1975). *IEEE Recommended Practice for Protection and Coordination of Industrial and Commercial Power Systems.* IEEE Std. 242-1975. The Institute of Electrical and Electronics Engineers Inc.

44 Koepfinger, J.L.; IEEE Standards Board (1980). *IEEE Recommended Practice for the Design of Reliable Industrial and Commercial Power Systems.* IEEE Std. 393-1980. The Institute of Electrical and Electronics Engineers Inc./Wiley-Interscience.

45 Koepfinger, J.L.; IEEE Standards Board (1980). *IEEE Recommended Practice for Emergency and Standby Power systems for Industrial and Commercial Applications.* IEEE Std. 446-1980. The Institute of Electrical and Electronics Engineers Inc./Wiley-Interscience.

46 Koepfinger, J.L.; IEEE Standards Board. *IEEE Recommended Practice for Power System Analysis.* IEEE Std. 399-1980. The Institute of Electrical and Electronics Engineers Inc./Wiley-Interscience.

47 Wildi, T. (2002). *Electrical Machines, Drives, and Power Systems*, 5e. Prentice Hall.

7

Frequency Characteristics of Electric Circuits

7.1 Introduction

Although DC voltages and currents are important in some electrical engineering analyses, AC quantities are very important for power generation, distribution and transmission, utilization, and control applications. While at steady-state constant frequency assumptions are sufficient, at transient operating conditions, the frequency characteristics are of high interest. In Chapter 1 a survey of applications shows that the entire frequency range from $f = (0\text{–}10^9)$ Hz is utilized (e.g. from DC transmission lines to wireless communication signals). Chapters 2 and 3 discuss steady-state and transient DC analysis, respectively. The circuits of Chapter 4 are limited to power frequencies from $f = (15\text{–}2000)$ Hz and are analyzed by phasors, a very useful tool for investigating voltage, current, and power relations for AC steady-state circuits. Chapters 5 and 6 build on the phasor method of Chapter 4 and apply it to single- and three-phase circuits, including transformers.

This chapter discusses the network performance including inductors, capacitors, and resistors as a function of angular frequency ($\omega = 2\pi f$) up to $\omega = 100$ krad/s. Concepts such as resonance, frequency selectivity, bandwidth, cutoff frequency, and quality factor are introduced for passive and resonating circuits such as first-order and second-order filters – low-pass, high-pass, and band elimination – which are very important for either passing useful frequencies or rejecting unwanted frequency components.

Note that inductors with inductance L have the frequency dependency $j\omega L$. For capacitors with the capacitance C, the frequency-dependent relation is $1/(j\omega C)$, and resistors with the resistance R are, as a first approximation, independent of frequency. Because of these frequency-dependent properties, the components R, L, and C lend themselves to passive filters.

Introduction to Energy, Renewable Energy, and Electrical Engineering: Essentials for Engineering Science (STEM) Professionals and Students, First Edition. Ewald F. Fuchs and Heidi A. Fuchs.

Companion website: www.wiley.com/go/fuchs/renewableenergyandelectricalengineering

Finally, the application of active filters based on RLC components and semiconductor switches is mentioned, where passive filters are neither useful nor applicable for solving power quality and control problems.

7.2 Sinusoidal/Cosinusoidal Frequency Analysis

Many natural dynamic systems have frequency-dependent sound, such a waterfall or the pulse of humans; that is, the frequency and associated sound changes as a function of time. In contrast, organ pipes are tuned to one frequency only. Analogous systems occurring in the technical world are the voltage–current frequency of a power system, which changes with load slightly around a nominal value of $f = 60$ Hz. Resonant circuits in electrical engineering are tuned to one frequency, similar to sound emanating from an organ pipe. In this chapter we expose circuits to sinusoidal/cosinusoidal voltages/currents at varying frequencies to examine their behavior as a function of frequency, which is called sinusoidal frequency response or simply frequency response of a circuit. In other words, the frequency response characterizes the behavior of an electric network as a function of frequency.

7.3 Passive Filters

Passive filters [1, 2] consist of the network components resistance R, inductance L resulting in the inductive impedance $\tilde{Z}_L = j\omega L\ (\Omega)$, and capacitance C, yielding the capacitive impedance $\tilde{Z}_c = \dfrac{1}{j\omega C}\ (\Omega)$. These reactive and capacitive impedances lend themselves to passive filters because at $\omega = 0$ results in $\tilde{Z}_L = 0$ and $\tilde{Z}_c \to \infty$, while at $\omega \to \infty$ yields $\tilde{Z}_L \to \infty$ and $\tilde{Z}_c \to 0$. For a first-order low-pass filter as shown in Figure E7.1.1, the voltage across a capacitor C is used at the output, while the phasor source voltage is $\tilde{V}_s(\omega) = 120\,\text{V}\angle 0°$ supplied to the capacitor C via a resistor R. This circuit is called a first-order low-pass filter because it contains only one storage component, the capacitor C. For a first-order high-pass filter as shown in Figure E7.2.1, the voltage across a resistor R is sensed at the output, while the phasor source voltage is $\tilde{V}_s(\omega) = 120\,\text{V}\angle 0°$ supplied to the resistor R via a capacitor C. This circuit is called a first-order high-pass filter because it contains one storage component only, the capacitor C; a high-pass filter permits the passing of high frequencies from the input to the output while low frequencies are suppressed. If two or more storage elements such as inductors and capacitors are employed in the circuit, we call the circuit of second or higher (third, fourth, etc.) order.

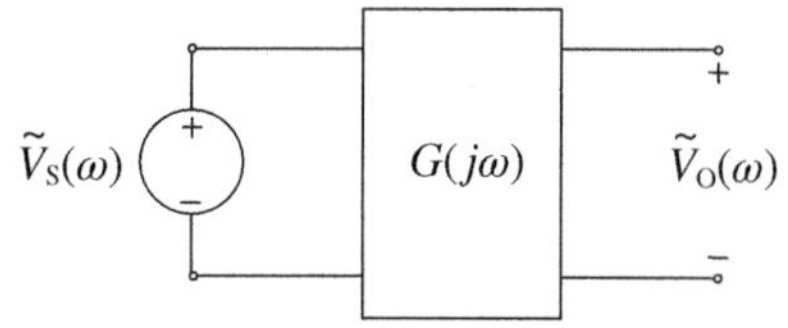

Figure 7.1 Definition of general transfer function $G(j\omega) = M(\omega)e^{j\phi(\omega)} = M(\omega) \angle \phi(\omega)$.

Higher-order filters are mostly replaced by active filters consisting of storage components and semiconductor elements such as switches.

Figure 7.1 shows the input–output relation for a passive filter defined by the so-called transfer function:

$$\frac{\widetilde{V}_o(\omega)}{\widetilde{V}_s(\omega)} = G(j\omega) = M(\omega)e^{j\phi(\omega)}, \tag{7.1}$$

where $\widetilde{V}_s(\omega)$ is the input filter source (s) voltage and $\widetilde{V}_o(\omega)$ is the output (o) filter voltage: while the input voltage may have undesirable harmonics (e.g. V_{harmonic}), the output voltage may have a reduced or almost eliminated V_{harmonic} magnitude.

7.3.1 Poles and Zeros of Transfer Function

The transfer function $G(j\omega)$ as defined by Eq. (7.1) can be expressed as the ratio of two polynomials in $(j\omega)$. Since the values of the circuit elements and sources are real numbers, the coefficients of the two polynomials are real numbers as well. If the transfer function (Eq. 7.1) is expressed as $N(j\omega)$ representing the numerator (N) polynomial of degree m and $D(j\omega)$ the denominator (D) polynomial of degree n, one can write

$$G(j\omega) = \frac{N(j\omega)}{D(j\omega)} = \frac{a_m(j\omega)^m + a_{m-1}(j\omega)^{m-1} + \cdots + a_1 j\omega + a_0}{b_n(j\omega)^n + b_{n-1}(j\omega)^{n-1} + \cdots + b_1 j\omega + b_0}, \tag{7.2}$$

or

$$G(j\omega) = K_0 \frac{(j\omega - \widetilde{z}_1)(j\omega - \widetilde{z}_2)\cdots(j\omega - \widetilde{z}_m)}{(j\omega - \widetilde{p}_1)(j\omega - \widetilde{p}_2)\cdots(j\omega - \widetilde{p}_n)}, \tag{7.3}$$

where K_0 is a constant and $\widetilde{z}_1 ... \widetilde{z}_m$ are the roots of $N(j\omega)$ and $\widetilde{p}_1 ... \widetilde{p}_n$ are the roots of $D(j\omega)$. If $j\omega = \widetilde{z}_1$ or $\widetilde{z}_2 ... \widetilde{z}_m$, then $G(j\omega)$ becomes zero, and therefore $\widetilde{z}_1 ... \widetilde{z}_m$ are called zeros of the transfer function. If $j\omega = \widetilde{p}_1$ or $\widetilde{p}_2 ... \widetilde{p}_n$, then $G(j\omega)$ becomes infinite, and therefore $\widetilde{p}_1 ... \widetilde{p}_n$ are called poles of the transfer function. The zeros and poles may be complex: if they are complex, they must be in conjugate pairs since the coefficients of the polynomial are real.

In the following, different passive filters will be explained in Application Examples 7.1–7.7.

7.3.2 First-Order RC Low-Pass Filter Circuit and Its Frequency Characteristics

Application Example 7.1 First-order low-pass filter with phasor voltage $\widetilde{V}_s(\omega) = 120\,\mathrm{V}\angle 0°$ and parameters $R = 5\,\Omega$, $C = 10\,\mu\mathrm{F}$

Figure E7.1.1 shows a first-order RC low-pass filter with the input voltage $\widetilde{V}_s(\omega) = 120\,\mathrm{V}\angle 0°$ and output voltage $\widetilde{V}_o(\omega)$.

Applying KVL to the circuit of Figure E7.1.1 yields

$$\widetilde{V}_s(\omega) = \widetilde{I}_s(\omega)\left(R + \frac{1}{j\omega C}\right), \tag{E7.1.1}$$

and with Ohm's law one can write

$$\widetilde{I}_s(\omega) = \widetilde{V}_o(\omega)/\left(\frac{1}{j\omega C}\right). \tag{E7.1.2}$$

Introducing Eq. (E7.1.2) into Eq. (E7.1.1), one gets for the output (o)–input/source (s) voltage ratio with the definition of the transfer function Eq. (7.1) and the time constant definition $\tau = RC$

$$\frac{\widetilde{V}_o(\omega)}{\widetilde{V}_s(\omega)} = \frac{1}{(1 + j\omega\tau)} = \frac{1}{\sqrt{1 + (\omega\tau)^2}\angle \tan^{-1}(\omega\tau)} = G(j\omega) = M(\omega)e^{j\phi(\omega)} = M(\omega)\angle\phi(\omega), \tag{E7.1.3}$$

where the magnitude is

$$M(\omega) = \frac{1}{\sqrt{1 + (\omega\tau)^2}} \tag{E7.1.4}$$

and the phase angle

$$\phi(\omega) = -\tan^{-1}(\omega\tau). \tag{E7.1.5}$$

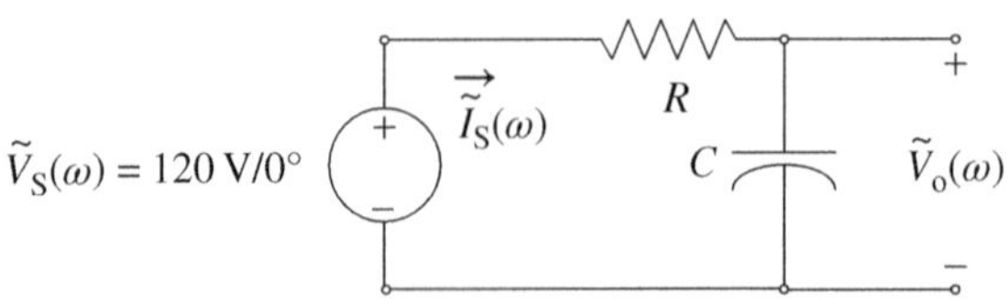

Figure E7.1.1 First-order RC low-pass filter circuit.

With $s = j\omega$, one can write for the transfer function $\frac{\widetilde{V}_o(\omega)}{\widetilde{V}_s(\omega)} = \frac{1}{(1 + s\tau)}$ where the subscript s stands for source and the variable s for $j\omega$. The denominator is represented by the pole $s = -(1/\tau)$, which becomes important for controllers of feedback systems as discussed in reference 3.

For $R = 5\,\Omega$ and $C = 10\,\mu\text{F}$, the time constant is $\tau = RC = 50 \cdot 10^{-6}$ s, the magnitude $M(\omega) = \frac{1}{\sqrt{1 + (\omega \cdot 50 \cdot 10^{-6})^2}}$, and the phase angle $\Phi(\omega) = -\tan^{-1}(\omega \cdot 50 \cdot 10^{-6})$. The break frequency is defined as $\omega_0\tau = 1$ or

$$\omega_0 = 1/\tau = 20\,\text{krad/s}. \tag{E7.1.6}$$

Table E7.1.1 represents the dependency of the magnitude $M(\omega)$ and the phase angle $\Phi(\omega)$ as a function of ω. The magnitude $M(\omega)$ and the phase angle $\Phi(\omega)$ are graphed in Figure E7.1.2.

Table E7.1.1 Magnitude $M(\omega)$ and phase angle $\Phi(\omega)$ for first-order RC low-pass filter, where the break angular velocity ω_0 = 20 krad/s is identified.

ω (krad/s)	0	5	10	12.5	15	17.5	$\omega_0 = 20$	22.5	25
$M(\omega)$ (pu)	1	0.970	0.894	0.848	0.800	0.753	$1/\sqrt{2} \approx 0.707$	0.664	0.625
$\Phi(\omega)$ (°)	0	−14.04	−26.57	−32.00	−36.87	−41.19	−45	−48.37	−51.34

ω (krad/s)	27.5	30	35	40	100
$M(\omega)$ (pu)	0.588	0.555	0.496	0.447	0.196
$\Phi(\omega)$ (°)	−53.97	−56.31	−60.26	−63.44	−78.69

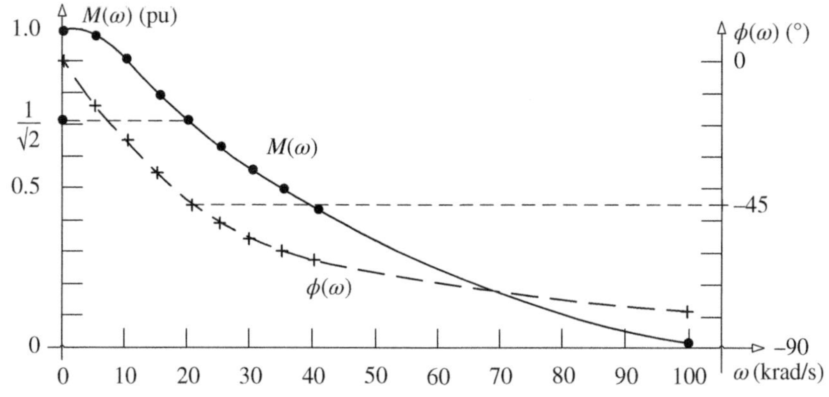

Figure E7.1.2 Magnitude $M(\omega)$ and the phase angle $\Phi(\omega)$ for first-order RC low-pass filter.

Specific value for output voltage as a function of input/source voltage at $\omega_0 = 20$ krad/s:

$$\widetilde{V}_o\left(20\frac{\text{krad}}{\text{s}}\right) = G\left(j20\frac{\text{krad}}{\text{s}}\right) = M\left(20\frac{\text{krad}}{\text{s}}\right)e^{j\phi\left(20\frac{\text{krad}}{\text{s}}\right)}$$
$$= 120\cdot\frac{1}{\sqrt{2}}\,\text{V}\angle -45^\circ = 84.87\,\text{V}\angle -45^\circ. \tag{E7.1.7}$$

7.3.3 First-Order RC High-Pass Filter Circuit and Its Frequency Characteristics

Application Example 7.2 First-order RC high-pass filter with phasor voltage $\widetilde{V}_s(\omega) = 120\,\text{V}\angle 0^\circ$ and parameters $R = 5\,\Omega$, $C = 10\,\mu\text{F}$

Figure E7.2.1 shows a first-order RC high-pass filter with the input voltage $\widetilde{V}_s(\omega) = 120\,\text{V}\angle 0^\circ$ and output voltage $\widetilde{V}_o(\omega)$. Applying KVL to the circuit of Figure E7.2.1 yields

$$\widetilde{V}_s(\omega) = \widetilde{I}_s(\omega)\left(R + \frac{1}{j\omega C}\right), \tag{E7.2.1}$$

and with Ohm's law one can write

$$\widetilde{I}_s(\omega) = \widetilde{V}_o(\omega)/R. \tag{E7.2.2}$$

Introducing Eq. (E7.2.2) into Eq. (E7.2.1), one gets for the output (o)–input/source (s) voltage ratio with the definition of the transfer function Eq. (7.1) and the time constant definition $\tau = RC$

$$\frac{\widetilde{V}_o(\omega)}{\widetilde{V}_s(\omega)} = \frac{j\omega\tau}{(1 + j\omega\tau)} = \frac{\omega\tau\angle 90^\circ}{\sqrt{1 + (\omega\tau)^2}\angle \tan^{-1}(\omega\tau)} = G(j\omega) = M(\omega)e^{j\phi(\omega)} = M(\omega)\angle\phi(\omega), \tag{E7.2.3}$$

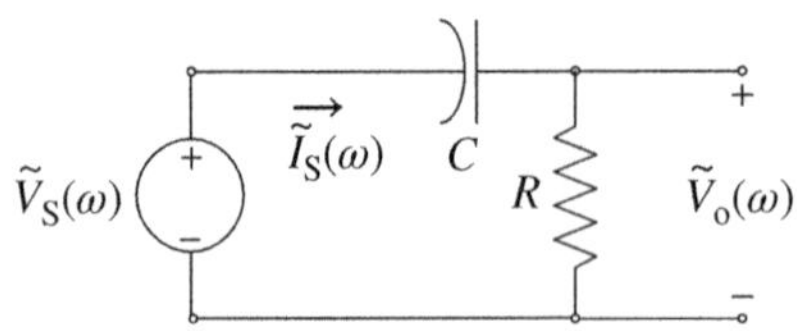

Figure E7.2.1 First-order RC high-pass filter circuit.

where the magnitude is

$$M(\omega) = \frac{1}{\sqrt{1 + (\omega\tau)^2}}, \tag{E7.2.4}$$

and the phase angle

$$\phi(\omega) = 90° - \tan^{-1}(\omega\tau). \tag{E7.2.5}$$

With $s = j\omega$, one can write for the transfer function $\frac{\widetilde{V}_o(\omega)}{\widetilde{V}_s(\omega)} = \frac{s\tau}{(1 + s\tau)}$ where the subscript s stands for source and s for $j\omega$. The denominator is represented by the pole $s = -(1/\tau)$, and the numerator is represented by the zero $s = 0$, which becomes important for controllers of feedback systems as discussed in reference 3.

For $R = 5\,\Omega$, $C = 10\,\mu\text{F}$ the time constant is $\tau = RC = 50 \cdot 10^{-6}$ s, the magnitude $M(\omega) = \frac{\omega \cdot 50 \cdot 10^{-6}}{\sqrt{1 + (\omega \cdot 50 \cdot 10^{-6})^2}}$, and the phase angle $\Phi(\omega) = 90° - \tan^{-1}(\omega \cdot 50 \cdot 10^{-6})$. The break frequency is defined as $\omega_0\tau = 1$ or

$$\omega_0 = 1/\tau = 20\,\text{krad/s}. \tag{E7.2.6}$$

Table E7.2.1 represents the dependency of the magnitude $M(\omega)$ and the phase angle $\Phi(\omega)$ as a function of ω. The magnitude $M(\omega)$ and the phase angle $\Phi(\omega)$ are graphed in Figure E7.2.2.

Table E7.2.1 Magnitude $M(\omega)$ and phase angle $\Phi(\omega)$ for first-order RC high-pass filter, where the break angular velocity ω_0 = 20 krad/s is identified.

ω (krad/s)	0	5	10	12.5	15	17.5	$\omega_0 = 20$	22.5
$M(\omega)$ (pu)	0	0.243	0.447	0.530	0.600	0.659	$1/\sqrt{2} = 0.707$	0.751
$\Phi(\omega)$ (°)	90	75.96	63.44	60	53.13	48.81	45	41.51

ω (krad/s)	25	27.5	30	35	40	100
$M(\omega)$ (pu)	0.781	0.809	0.832	0.868	0.894	0.981
$\Phi(\omega)$ (°)	38.66	36.03	33.69	29.75	26.57	11.31

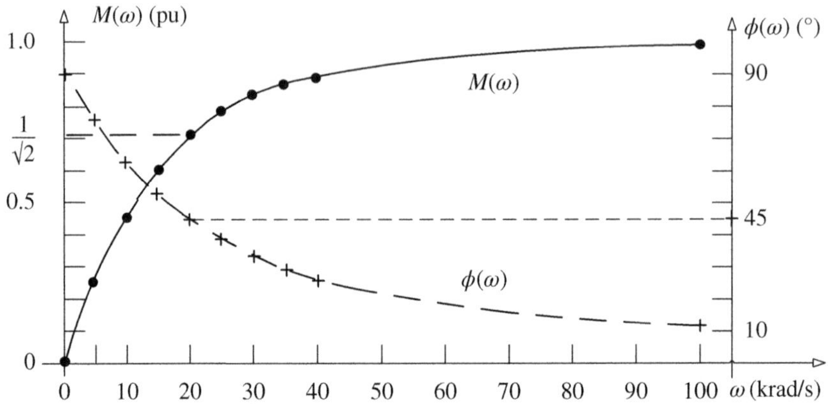

Figure E7.2.2 Magnitude $M(\omega)$ and the phase angle $\Phi(\omega)$ for first-order RC high-pass filter.

The specific value for output voltage as a function of input/source voltage at $\omega_0 = 20\,\text{krad/s}$:

$$\begin{aligned} \widetilde{V}_o\left(20\frac{\text{krad}}{\text{s}}\right) &= G\left(j20\frac{\text{krad}}{\text{s}}\right) = M\left(20\frac{\text{krad}}{\text{s}}\right)e^{j\phi\left(20\,\frac{\text{krad}}{\text{s}}\right)} \\ &= 120\cdot\frac{1}{\sqrt{2}}\text{V}\angle 45^\circ = 84.87\,\text{V}\angle 45^\circ. \end{aligned} \tag{E7.2.7}$$

7.3.4 Band-Pass and Band-Rejection (Second-Order) Filter Circuits and Their Frequency Characteristics

Electric signals such as voltages and currents sometimes contain unwanted DC components, harmonics, and subharmonics or fractional harmonics. To selectively filter these, band-pass and band-rejection circuits are designed. Typical amplitude $M(\omega)$ characteristics for simple band-pass and rejection filters will be analyzed in Application Examples 7.3 and 7.4, respectively. ω_0 is the resonance or center angular velocity/frequency of the pass or rejection band and the angular frequency at which the maximum or minimum amplitude occurs. ω_{LO} and ω_{HI} are the lower and upper (high) angular break or cutoff frequencies, where the amplitude is $1/\sqrt{2}$ of the maximum value. The width of the pass or rejection band is called the bandwidth $\text{BW} = \omega_{\text{HI}} - \omega_{\text{LO}}$.

Application Example 7.3 Band-pass (second-order) RLC filter with $\widetilde{V}_s(\omega) = 120\,\text{V}\angle 0^\circ$, $R = 5\,\Omega$, $L = 1\,\text{mH}$, $C = 10\,\mu\text{F}$

A band-pass filter is of the second-order type because it contains two storage elements L and C.

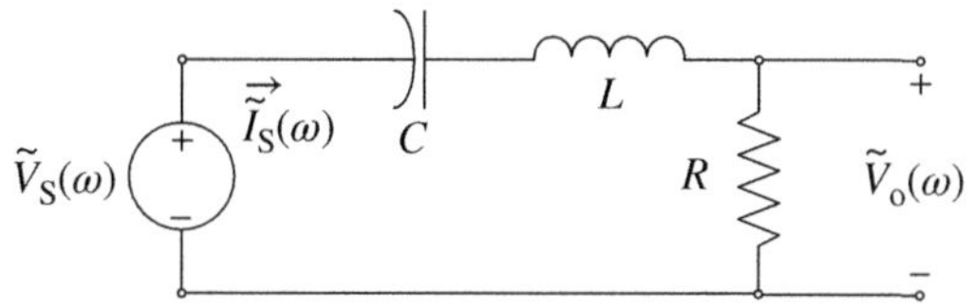

Figure E7.3.1 Band-pass (second-order) RLC filter circuit.

Applying KVL to the circuit of Figure E7.3.1 yields

$$\widetilde{V}_s(\omega) = \widetilde{I}_s(\omega)\left(R + j\omega L + \frac{1}{j\omega C}\right), \tag{E7.3.1}$$

and with Ohm's law one can write

$$\widetilde{I}_s(\omega) = \widetilde{V}_o(\omega)/R. \tag{E7.3.2}$$

Introducing Eq. (E7.3.2) into Eq. (E7.3.1), one gets for the output (o)–input/source (s) voltage ratio with the definition of the transfer function Eq. (7.1) and the time constant definition $\tau = RC$

$$\frac{\widetilde{V}_o(\omega)}{\widetilde{V}_s(\omega)} = \frac{R}{R + j\left(\omega L - \dfrac{1}{\omega C}\right)} = \frac{R}{\sqrt{R^2 + \left(\omega L - \dfrac{1}{\omega C}\right)^2}} \angle - \tan^{-1}\left[\frac{\omega L - \frac{1}{\omega C}}{R}\right] = G(j\omega)$$
$$= M(\omega)e^{j\phi(\omega)} = M(\omega)\angle\phi(\omega), \tag{E7.3.3}$$

where the magnitude is

$$M(\omega) = \frac{R}{\sqrt{R^2 + (\omega L - 1/\omega C)^2}}, \tag{E7.3.4}$$

and the phase angle

$$\Phi(\omega) = -\tan^{-1}[(\omega L - 1/\omega C)/R]. \tag{E7.3.5}$$

The center frequency or angular velocity is defined as

$$\omega_0 = 1/\sqrt{LC} = 10\,\text{krad/s}. \tag{E7.3.6}$$

With $s = j\omega$, one can write for the transfer function $\dfrac{\widetilde{V}_o(\omega)}{\widetilde{V}_s(\omega)} = \dfrac{sRC}{(sRC + 1 + s^2LC)}$ where the subscript s stands for source and the variable s for $j\omega$. The denominator is represented by the pole expression $(s^2LC + RCs + 1)$ and the numerator by $sRC = s\tau$, which becomes important for controllers of feedback systems as discussed in reference 3.

Table E7.3.1 represents the dependency of the magnitude $M(\omega)$ and the phase angle $\Phi(\omega)$ as a function of ω. The magnitude $M(\omega)$ and the phase angle $\Phi(\omega)$ are graphed in Figure E7.3.2.

Specific value for output voltage as a function of input/source voltage at center frequency or angular velocity $\omega_0 = 10\,\text{krad/s}$. $\omega_{\text{HI}} = 12.54\,\text{krad/s}$ and $\omega_{\text{LO}} = 8.85$ krad/s.

$$\widetilde{V}_o\left(10\frac{\text{krad}}{\text{s}}\right) = G\left(j10\frac{\text{krad}}{\text{s}}\right) = M\left(10\frac{\text{krad}}{\text{s}}\right)e^{j\phi\left(10\frac{\text{krad}}{\text{s}}\right)} \tag{E7.3.7}$$
$$= 120 \cdot 1\,\text{V}\angle 180^\circ = 120\,\text{V}\angle 180^\circ.$$

The bandwidth is defined as $\text{BW} = \omega_{\text{HI}} - \omega_{\text{LO}}$, where $M(\omega_{\text{HI}}) = M(\omega_{\text{LO}}) = 1/\sqrt{2} = 0.7071$. From Figure E7.3.2, one obtains $\omega_{\text{HI}} = 12.54\,\text{krad/s}$ and $\omega_{\text{LO}} = 8.85$ krad/s, resulting in the bandwidth $\text{BW} = \omega_{\text{HI}} - \omega_{\text{LO}} = 3.69\,\text{krad/s}$.

Table E7.3.1 Magnitude $M(\omega)$ and phase angle $\Phi(\omega)$ for (second-order) RLC band-pass filter, where the left angular velocity ω_0 = 10 krad/s is identified.

ω (krad/s)	2.5	5	7.5	$\omega_0 = 10$	12.5	15	17.5	20
$M(\omega)$ (pu)	0.132	0.316	0.651	1	0.743	0.515	0.391	0.316
$\Phi(\omega)$ (°)	82.41	71.57	49.40	0	−41.99	−59.04	−67.00	−71.57

ω (krad/s)	25	30	35	40	100
$M(\omega)$ (pu)	0.231	0.184	0.154	0.132	0.050
$\Phi(\omega)$ (°)	−78.69	−79.38	−81.16	−82.41	−87.11

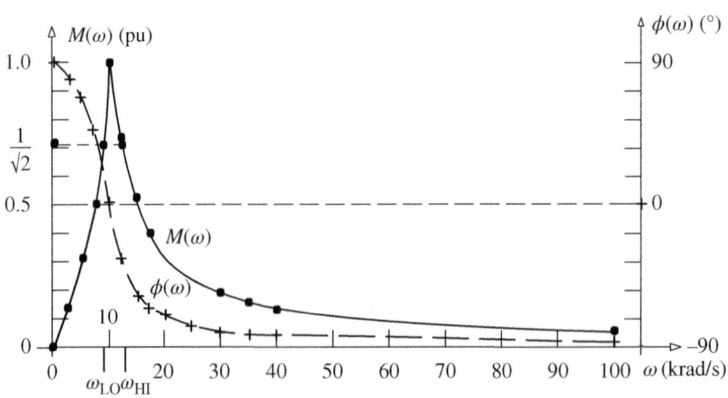

Figure E7.3.2 Magnitude $M(\omega)$ and phase angle $\Phi(\omega)$ for (second-order) RLC band-pass filter.

The high cutoff frequency ω_{HI} and the low cutoff frequency ω_{LO} are also called half-power frequencies.

Application Example 7.4 Band-rejection (second-order) filter with $\widetilde{V}_s(\omega) = 120\,\mathrm{V}\angle 0°$, $R = 5\,\Omega$, $L = 1\,\mathrm{mH}$, $C = 10\,\mu\mathrm{F}$

A band-rejection filter is of the second-order type because it contains two storage elements L and C.

Applying KVL to the circuit of Figure E7.4.1 yields

$$\widetilde{V}_s(\omega) = \widetilde{I}_s(\omega)\left[R + j\omega L + \frac{1}{j\omega C}\right], \tag{E7.4.1}$$

and with Ohm's law one can write

$$\widetilde{I}_s(\omega) = \frac{\widetilde{V}_o(\omega)}{\left(j\omega L + \dfrac{1}{j\omega C}\right)}. \tag{E7.4.2}$$

Introducing Eq. (E7.4.2) into Eq. (E7.4.1), one gets for the output (o)–input/source (s) voltage ratio with the definition of the transfer function Eq. (7.1) and the time constant definition $\tau = RC$

$$\frac{\widetilde{V}_o(\omega)}{\widetilde{V}_s(\omega)} = \frac{\left(j\omega L + \dfrac{1}{j\omega C}\right)}{R + j\left(\omega L - \dfrac{1}{\omega C}\right)} = \frac{\sqrt{\left(\omega L - \dfrac{1}{\omega C}\right)^2}}{\sqrt{R^2 + \left(\omega L - \dfrac{1}{\omega C}\right)^2}}\angle 90° - \tan^{-1}[(\omega L - 1/\omega C)/R]$$
$$= G(j\omega) = (\omega)e^{j\phi(\omega)} = M(\omega)\angle\phi(\omega), \tag{E7.4.3}$$

where the magnitude is

$$M(\omega) = \frac{\sqrt{\left(\omega L - \frac{1}{\omega C}\right)^2}}{\sqrt{R^2 + (\omega L - 1/\omega C)^2}}, \tag{E7.4.4}$$

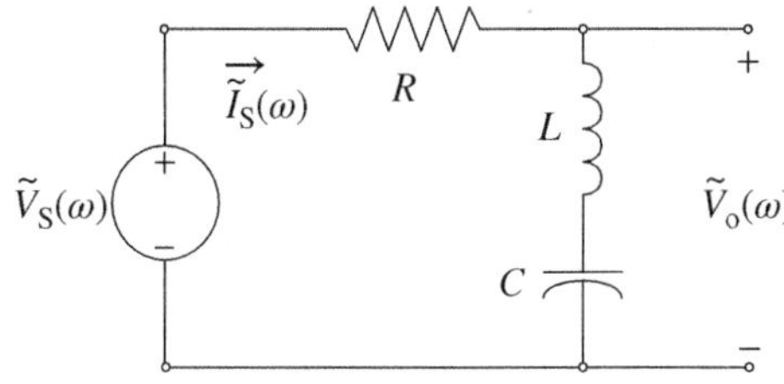

Figure E7.4.1 Band-rejection (second-order) RLC filter circuit.

and the phase angle

$$\Phi(\omega) = 90° - \tan^{-1}[(\omega L - 1/\omega C)/R]. \tag{E7.4.5}$$

The center frequency or angular velocity is defined as

$$\omega_0 = 1/\sqrt{LC} = 10\,\text{krad/s}. \tag{E7.4.6}$$

With $s = j\omega$ one can write for the transfer function $\dfrac{\widetilde{V}_o(\omega)}{\widetilde{V}_s(\omega)} = \dfrac{sL + 1/sC}{(R + sL + 1/sC)}$ where the subscript s stands for source and the variable s for $j\omega$. The denominator is represented by the pole expression $(R + sL + 1/sC)$ and the numerator by $(sL + 1/sC)$, which becomes important for controllers of feedback systems as discussed in reference 3. Table E7.4.1 represents the dependency of the magnitude $M(\omega)$ and the phase angle $\Phi(\omega)$ as a function of ω. The magnitude $M(\omega)$ and the phase angle $\Phi(\omega)$ are graphed in Figure E7.4.2.

Specific value for output voltage as a function of input/source voltage at center frequency or angular velocity $\omega_p = 10$ krad/s:

$$\widetilde{V}_o\left(10\frac{\text{krad}}{\text{s}}\right) = G\left(j10\frac{\text{krad}}{\text{s}}\right) = M\left(10\frac{\text{krad}}{\text{s}}\right)e^{j\phi\left(10\frac{\text{krad}}{\text{s}}\right)} \tag{E7.4.7}$$
$$= 120 \cdot 0\,\text{V}\angle 90° = 0\,\text{V}\angle 90°.$$

The bandwidth is defined as $\text{BW} = \omega_{HI} - \omega_{LO}$, where $M(\omega_{HI}) = M(\omega_{LO}) = 1/\sqrt{2} = 0.7071$. From Figure E7.4.2, one obtains $\omega_{HI} = 12.68$ krad/s and $\omega_{LO} = 7.32$ krad/s, resulting in the bandwidth $\text{BW} = \omega_{HI} - \omega_{LO} = 5.36$ krad/s.

Table E7.4.1 Magnitude $M(\omega)$ and phase angle $\Phi(\omega)$ for (second-order) RLC band-rejection filter, where the center angular velocity ω_0 = 10 krad/s is identified.

ω (krad/s)	0	2.5	5	7.5	$\omega_0 = 10$	12.5	15	17.5	20
$M(\omega)$ (pu)	1	0.991	0.949	0.652	0	0.669	0.857	0.920	0.949
$\Phi(\omega)$ (°)	180	172.41	161.57	139.40	90	48.01	30.97	22.99	18.44

ω (krad/s)	25	30	35	40	100
$M(\omega)$ (pu)	0.973	0.983	0.988	0.991	0.999
$\Phi(\omega)$ (°)	13.39	10.62	8.84	7.59	2.89

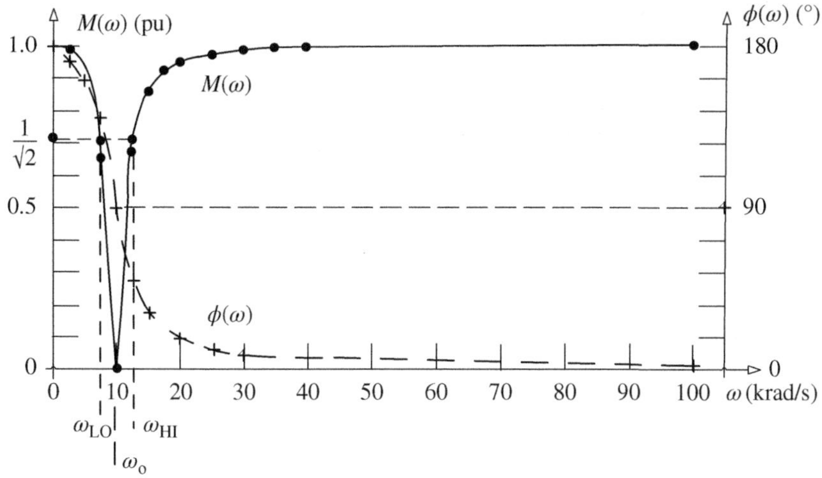

Figure E7.4.2 Magnitude $M(\omega)$ and phase angle $\Phi(\omega)$ for (second-order) RLC band-rejection filter. ω_{HI} = 12.68 krad/s and ω_{LO} = 7.32 krad/s, where the center angular velocity ω_0 = 10 krad/s is identified.

7.3.5 Series and Parallel Resonant RLC (Second-Order) Circuits

At resonance of an RLC circuit, the reactances of the inductor L and capacitor C are equal; that is, $j\omega L = 1/j\omega C$, resulting in the angular resonant frequency or velocity

$$\omega_0 = 1/\sqrt{LC}. \tag{7.4}$$

For this condition the voltage $\widetilde{V}_s(\omega_0)$ and the current $\widetilde{I}_s(\omega_0)$ are in phase, and therefore the phase angle $\widetilde{\phi}_{Is}(\omega_0) = 0$, resulting in a fundamental power factor of $\cos\theta = 1$. In the following two examples, the series- and parallel-connected resistor, inductor, and capacitor configurations are analyzed.

At resonance the voltage and current across and through resistance R are in phase, resulting in zero phase angle between current and voltage; that is, the power factor is unity. For the series connection of Figure E7.5.1, one defines the so-called quality factor, which is an important parameter for resonant circuits:

$$Q = \frac{\omega_0 L}{R} = \frac{1}{\omega_0 CR} = \frac{\sqrt{\text{LC}}}{\text{RC}} \tag{7.5}$$

Application Example 7.5 Series RLC (second-order) resonant circuit with the source voltage $\widetilde{V}_s = |\widetilde{V}_s(\omega)|\angle\phi_s = 120\,\text{V}\angle 0°$ and the circuit components $R = 5\,\Omega$, $L = 1\,\text{mH}$, $C = 10\,\mu\text{F}$

Figure E7.5.1 shows the circuit of the RLC series resonant circuit. In the series case the impedance is

$$\widetilde{Z}(\omega) = \left(R + j\omega L + \frac{1}{j\omega C}\right), \tag{E7.5.1}$$

and it reaches a minimum at resonance $\omega = \omega_0$

$$\widetilde{Z}(\omega_0) = R, \tag{E7.5.2}$$

and therefore the current $|\widetilde{I}_s(\omega_0)|$ is a maximum for a given voltage $\widetilde{V}_s(\omega_0)$. At low frequencies the impedance of Eq. (E7.5.1) is dominated by the capacitor term $1/j\omega C$, leading to leading current $\widetilde{I}_s(\omega)$ as compared with $\widetilde{V}_s(\omega)$, and at high frequencies the impedance $j\omega L$ of the inductive component of the series circuit prevails.

Applying Kirchhoff's voltage law to the circuit of Figure E7.5.1 leads to

$$\widetilde{I}_s(\omega) = \frac{\widetilde{V}_s(\omega)}{\widetilde{Z}_s(\omega)} = \frac{\widetilde{V}_s(\omega)}{\left(R + j\omega L + \frac{1}{j\omega C}\right)} = \frac{\widetilde{V}_s(\omega)}{R + j\left(\omega L - \frac{1}{\omega C}\right)} = |\widetilde{I}_s(\omega)|\angle\phi_{Is}(\omega), \tag{E7.5.3}$$

the magnitude is

$$|\widetilde{I}_s(\omega)| = \frac{120}{\sqrt{R^2 + \left(\omega L - \frac{1}{\omega C}\right)^2}}, \tag{E7.5.4}$$

and the phase angle

$$\phi_{Is}(\omega) = -\tan^{-1}\left[\frac{\left(\omega L - \frac{1}{\omega C}\right)}{R}\right]. \tag{E7.5.5}$$

With $s = j\omega$ one can write for the transfer function $\widetilde{I}_s(\omega) = \frac{\widetilde{V}_s(\omega)}{\left(R + sL + \frac{1}{sC}\right)}$ where the subscript s stands for source and variable s for $j\omega$. The denominator is

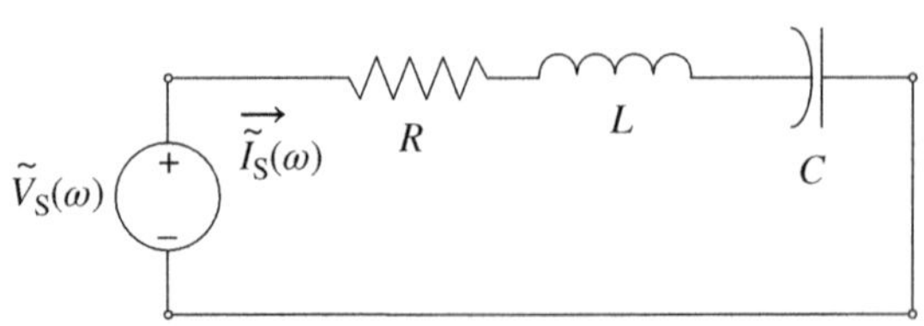

Figure E7.5.1 Series RLC (second-order) resonant circuit.

Table E7.5.1 Magnitude $\left|\tilde{I}_s(\omega_0)\right|$ and phase angle $\Phi_{Is}(\omega)$ for (second-order) RLC series resonant circuit, where the center angular velocity ω_0 = 10 krad/s is identified.

ω (krad/s)	2.5	5	7.5	$\omega_0 = 10$	12.5	15	17.5	20
$\left\|\tilde{I}_s(\omega)\right\|$ (A)	3.17	7.59	15.62	24	17.84	12.35	9.27	7.59
$\Phi_{Is}(\omega)$ (°)	82.41	71.57	49.40	0	−41.99	−59.03	−67.02	−71.57

represented by the pole expression ($R + sL + 1/sC$), which becomes important for controllers of feedback systems as discussed in reference 4. Table E7.5.1 lists the magnitude $\left|\tilde{I}_s(\omega_0)\right|$ and phase angle $\Phi_{Is}(\omega)$ the of the source current $\tilde{I}_s(\omega)$ as a function of the source voltage $\tilde{V}_s(\omega) = 120\,\text{V}\angle 0°$.

Figure E7.5.2 depicts the magnitude $\left|\tilde{I}_s(\omega_0)\right|$ of the source current $\tilde{I}_s(\omega)$ as a function of the source voltage $\tilde{V}_s(\omega) = \left|\tilde{V}_s(\omega)\right|\angle\phi_s = 120\,\text{V}\angle 0°$.

Another view of the (series) resonance is based on phasor diagrams: Figure E7.5.3 here depicts the phasors $\tilde{V}_s(\omega) = \left|\tilde{V}_s(\omega)\right|\angle\phi_s = 120\,\text{V}\angle 0°$, $\tilde{I}_s(\omega)$, $\tilde{V}_R(\omega)$, $\tilde{V}_L(\omega)$, and $V_c(\omega)$ for ω = 7.5 rad/s, 10 krad/s, and 12.5 krad/s. For ω

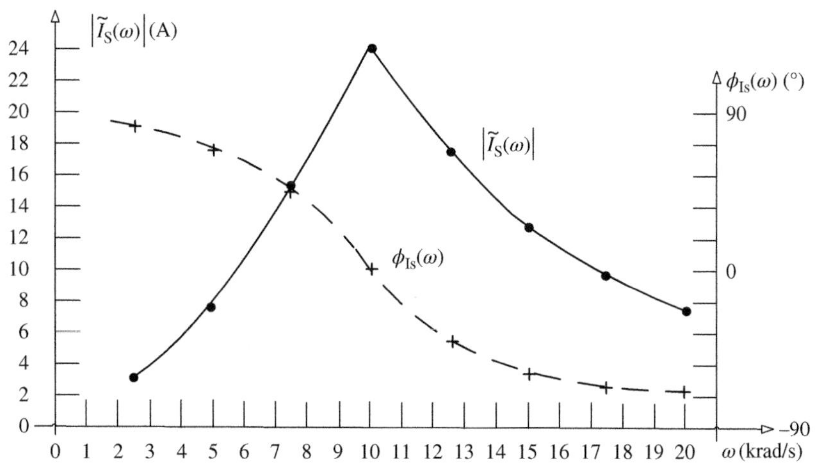

Figure E7.5.2 Magnitude $\left|\tilde{I}_s(\omega_0)\right|$ and phase angle $\Phi_{Is}(\omega)$ of the source current $\tilde{I}_s(\omega)$ as a function of the source voltage $\tilde{V}_s(\omega) = \left|\tilde{V}_s(\omega)\right|\angle\phi_s = 120\,\text{V}\angle 0°$ and of angular frequency ω for the series RLC (second-order) resonant circuit of Figure E7.5.1. The center angular velocity is ω_0 = 10 krad/s.

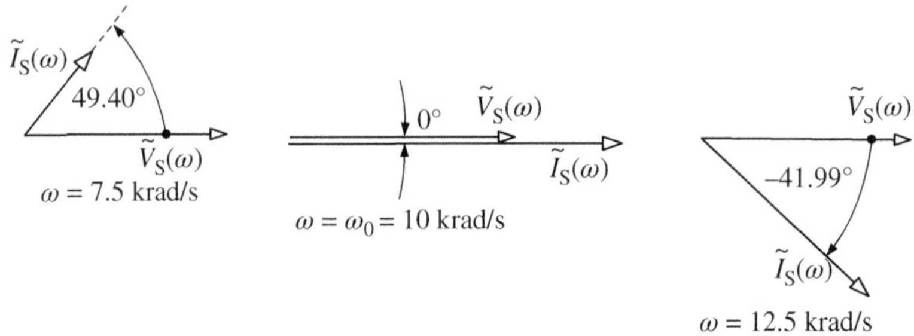

Figure E7.5.3 Phasors $\tilde{V}_s(\omega) = |\tilde{V}_s(\omega)|\angle 0°$ and $\tilde{I}_s(\omega)$. For 7.5 krad/s < ω_0 the current $\tilde{I}_s(\omega)$ leads $\tilde{V}_s(\omega)$, that is, the capacitor is dominant; for the center angular velocity 10 krad/s = ω_0 current $\tilde{I}_s(\omega)$ is in phase with $\tilde{V}_s(\omega)$, that is, resonance exists; and for 12.5 krad/s > ω_0 the current $\tilde{I}_s(\omega)$ lags $\tilde{V}_s(\omega)$, that is, the inductor dominates, whereby the voltage $\tilde{V}_s(\omega)$ is the reference phasor within the complex plane.

$< \omega_0$ the current $\tilde{I}_s(\omega)$ leads $\tilde{V}_s(\omega)$, for $\omega = \omega_0 = 10$ krad/s current $\tilde{I}_s(\omega)$ is in phase with $\tilde{V}_s(\omega)$, and for $\omega > \omega_0$ the current $\tilde{I}_s(\omega)$ lags $\tilde{V}_s(\omega)$, whereby the voltage $\tilde{V}_s(\omega)$ is the reference phasor within the complex plane.

Application Example 7.6 Parallel RLC (second-order) resonant circuit with the source voltage $\tilde{V}_s(\omega) = |\tilde{V}_s(\omega)|\angle\phi_s = 120\,\text{V}\angle 0°$ and the circuit components $R = 5\,\Omega$, $L = 1\,\text{mH}$, $C = 10\,\mu\text{F}$

Figure E7.6.1 here shows the circuit of the RLC parallel resonant circuit. The admittance is

$$\tilde{Y}(\omega) = \left(G + j\omega C + \frac{1}{j\omega L}\right), \tag{E7.6.1}$$

which reaches a minimum at $\omega = \omega_0 = 1/\sqrt{LC}$. In the parallel case at $\omega = \omega_0$, at resonance the admittance is $G = 1/R$ or

$$\tilde{Y}(\omega_0) = G, \tag{E7.6.2}$$

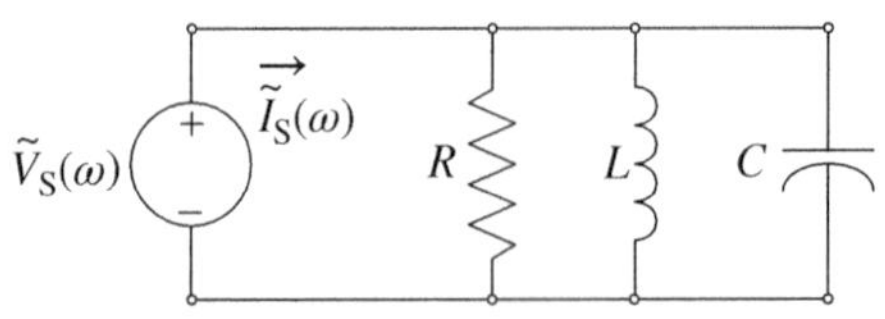

Figure E7.6.1 Parallel GLC (second-order) resonant circuit, where $G = 1/R$.

and, therefore, the current $|\widetilde{I}_s(\omega_0)|$ is a maximum for a given voltage $\widetilde{V}_s(\omega_0)$. At low frequencies the admittance of Eq. (E7.6.1) is dominated by the inductive term $1/j\omega L$, resulting in lagging current $\widetilde{I}_s(\omega)$ as compared with $\widetilde{V}_s(\omega)$, and at high frequencies the admittance $j\omega C$ of the capacitive component of the parallel circuit prevails.

Applying Kirchhoff's current law to the circuit of Figure E7.6.1 leads to

$$\widetilde{I}_s(\omega) = \widetilde{I}_G(\omega) + \widetilde{I}_L(\omega) + \widetilde{I}_c(\omega) = \widetilde{V}_s(\omega)\cdot G + \widetilde{V}_s(\omega)\cdot j\omega C + \frac{\widetilde{V}_s(\omega)}{j\omega L} = \widetilde{V}_s(\omega)\left[G + j\omega C - \frac{j}{\omega L}\right] = \widetilde{V}_s(\omega)\left[\frac{G\omega L + j(\omega^2 LC - 1)}{\omega L}\right] = \left|\widetilde{I}_s(\omega)\right| \angle\phi_{Is}(\omega), \tag{E7.6.3}$$

the magnitude is

$$\left|\widetilde{I}_s(\omega)\right| = \frac{\widetilde{V}_s(\omega)\sqrt{(G\omega L)^2 + (\omega^2 LC - 1)^2}}{(\omega L)^2}, \tag{E7.6.4}$$

and the phase angle

$$\phi_{Is}(\omega) = \tan^{-1}\left[\frac{(\omega^2 LC - 1)}{G\omega L}\right] = \tan^{-1}\left[\frac{(\omega C - \frac{1}{\omega L})}{G}\right]. \tag{E7.6.5}$$

With $s = j\omega$ one can write for the transfer function $\widetilde{I}_s(\omega) = \widetilde{V}_s(\omega)\left(G + sC + \frac{1}{sL}\right) = \widetilde{V}_s(\omega)\left[\frac{sLG + s^2CL + 1}{sL}\right]$ where the subscript s stands for source and variable s for $j\omega$. The denominator is represented by the pole expression sL and the numerator by $[s^2CL + (sLG + 1)]$, which become important for controllers of feedback systems as discussed in reference 3.

Table E7.6.1 here lists the magnitude $|\widetilde{I}_s(\omega)|$ and phase angle $\Phi_{Is}(\omega)$ the of the source current $\widetilde{I}_s(\omega)$ as a function of the source voltage $\widetilde{V}_s(\omega) = |\widetilde{V}_s(\omega)| \angle\phi_s = 120\,\text{V}\angle 0°$.

The magnitudes and phase angles are plotted up to $\omega = 20$ krad/s in Figure E7.6.2.

Table E7.6.1 Magnitude $|\widetilde{I}_s(\omega)|$ and phase angle $\Phi_{Is}(\omega)$ for (second-order) RLC parallel resonant circuit, where the center angular velocity ω_0 = 10 krad/s is identified.

ω (krad/s)	2.5	5	$\omega_0 = 10$	12.5	15	20	30
$\|\widetilde{I}_s(\omega)\|$ (A)	9.60	4.95	2.68	2.26	2.0	1.70	0.81
$\Phi_{Is}(\omega)$ (°)	−61.93	−36.87	0	12.68	22.62	36.87	53.13

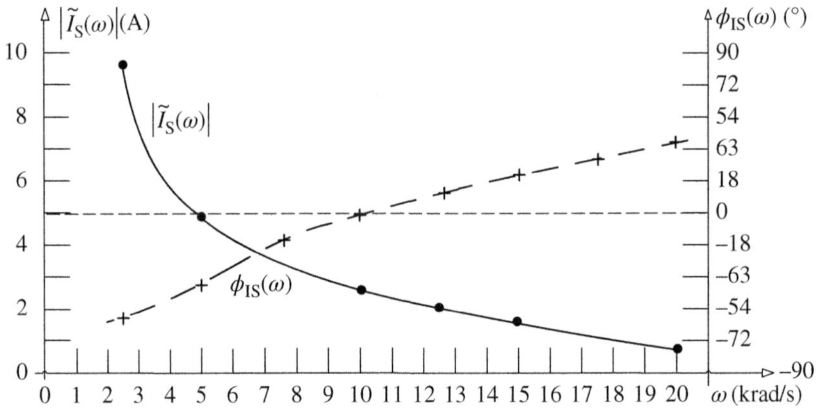

Figure E7.6.2 Magnitude $|\tilde{I}_s(\omega)|$ and phase angle $\Phi_{Is}(\omega)$ of the source current $\tilde{I}_s(\omega)$ as a function of the source voltage $\tilde{V}_s(\omega) = |\tilde{V}_s(\omega)| \angle\phi_s = 120\,\text{V}\angle 0°$ and of angular frequency ω for the parallel RLC (second-order) resonant circuit of Figure E7.6.1, where the center angular velocity is $\omega_0 = 10$ krad/s.

Depending upon the (angular) frequency omega the voltage phasor $\tilde{V}_s(\omega)$ leads the current phasor $\tilde{I}_s(\omega)$ at frequencies below the resonant frequency ω_0, both are in phase at ω_0, and at frequencies higher than ω_0 the current phasor leads the voltage phasor.

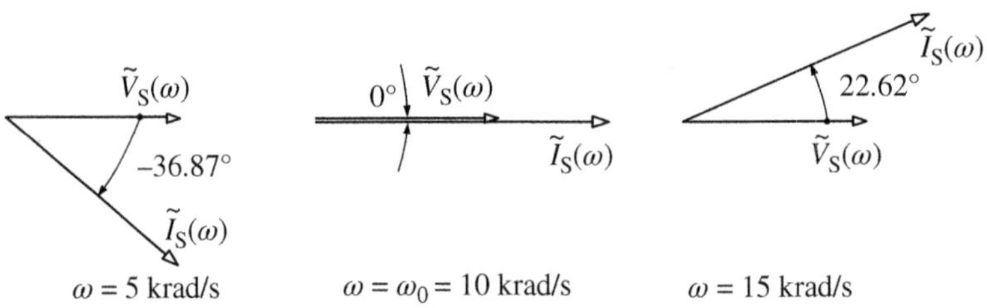

Figure E7.6.3 Phasors $\tilde{V}_s(\omega) = |\tilde{V}_s(\omega)| \angle 0°$ and $\tilde{I}_s(\omega)$. For 5 krad/s < ω_0 the current $\tilde{I}_s(\omega)$ lags $\tilde{V}_s(\omega)$, that is, the inductor is dominant; for 10 krad/s = ω_0 current $\tilde{I}_s(\omega)$ is in phase with $\tilde{V}_s(\omega)$, that is, resonance exists; and for 15 krad/s > ω_0 the current $\tilde{I}_s(\omega)$ leads $\tilde{V}_s(\omega)$, that is, the capacitor dominates, whereby the voltage $\tilde{V}_s(\omega)$ is the reference phasor within the complex plane.

Application Example 7.7 Calculation of bandwidth BW

Calculate the bandwidth BW and half-power angular frequencies ω_{HI}, ω_{LO} of the circuit of Figure E7.7.1 for $R = 5\,\Omega$, $L = 10\,\text{mH}$, $C = 50\,\mu\text{F}$.

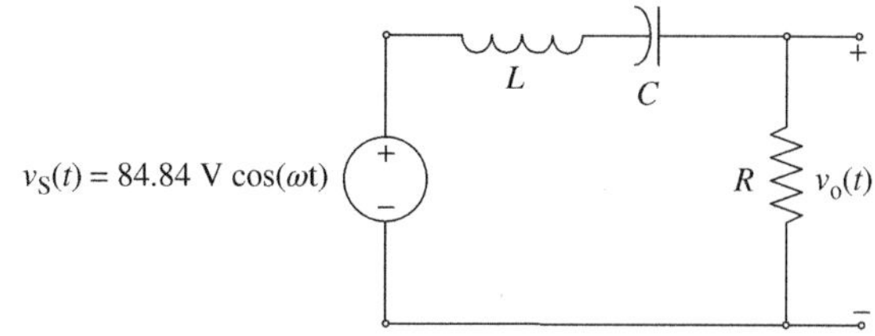

Figure E7.7.1 RLC series network.

Analysis is performed in the phasor domain with $\widetilde{V}_s(\omega) = 60\ \text{V} \angle 0°$ where $\widetilde{V}_o(\omega)$ is obtained based on voltage divider rule:

$$\frac{\widetilde{V}_o(\omega)}{\widetilde{V}_s(\omega)} = \frac{R}{R + j\omega L + \frac{1}{j\omega C}} = \frac{R}{R + j\left(\omega L - \frac{1}{\omega C}\right)} = \frac{RC\omega}{RC\omega + j(\omega^2 LC - 1)},$$

the magnitude $M(\omega)$ is at the half-power frequencies (ω_{HI} and ω_{LO}): $\frac{1}{\sqrt{2}} = \frac{RC\omega}{\sqrt{(RC\omega)^2 + (\omega^2 LC - 1)^2}}$, or $2(RC\omega)^2 = (RC\omega)^2 + (\omega^2 LC - 1)^2$; simplified one obtains $(RC\omega)^2 = (\omega^2 LC - 1)^2$, resulting in $\pm(RC\omega) = (\omega^2 LC - 1)$.

1) At lower cutoff or lower half-power frequency (ω_{LO}), the minus sign is valid: $-(RC\omega) = (\omega^2 LC - 1)$.

   Resonant frequency or angular velocity is $\omega_0 = \frac{1}{\sqrt{LC}} = \frac{1}{\sqrt{10 \cdot 10^{-3} \cdot 50 \cdot 10^{-6}}} = 1.414$ krad/s or $\omega_0^2 = 2\ (\text{krad/s})^2$. With $\omega^2 = \omega_0^2 = \frac{1}{LC}$ results in $C = \frac{1}{\omega_0^2 L}$ and one obtains $(\omega^2 LC - 1) + \frac{R\omega}{\omega_0^2 L} = 0$; simplified one gets $\omega^2 LC\omega_0^2 - \omega_0^2 + \frac{R}{L}\omega = 0$ or $\omega^2 + \frac{R}{L}\omega - \omega_0^2 = 0$.

   The solution of this quadratic equation leads for the low (LO) and high (HI) angular velocities to $\omega_{\text{LO,HI}} = \frac{-\frac{R}{L} \mp \sqrt{\left(\frac{R}{L}\right)^2 + 4(\omega_0^2)}}{2}$; the low angular velocity is $\omega_{\text{LO}} = \frac{-\frac{R}{L} + \sqrt{\left(\frac{R}{L}\right)^2 + 4(\omega_0^2)}}{2}$.

2) At higher cutoff or upper half-power frequency (ω_{HI}), the plus sign is valid: $+(RC\omega) = (\omega^2 LC - 1)$.With $\omega^2 = \omega_0^2$

$$= \frac{1}{LC} \text{ or } C = \frac{1}{\omega_0^2 L} \text{ one obtains } (\omega^2 LC - 1) + \frac{R\omega}{\omega_0^2 L} = 0$$

   or

$$\omega^2 - \frac{R}{L}\omega - \omega_0^2 = 0, \text{ resulting in } \omega_{\text{LO,HI}} = \frac{\frac{R}{L} \mp \sqrt{\left(\frac{R}{L}\right)^2 + 4(\omega_0^2)}}{2}.$$

For ω_{HI} one gets $\omega_{HI} = \dfrac{\frac{R}{L} + \sqrt{\left(\frac{R}{L}\right)^2 + 4(\omega_0^2)}}{2}$.

The bandwidth BW is defined as

$$BW = \omega_{HI} - \omega_{LO} = \frac{\frac{R}{L} + \sqrt{\left(\frac{R}{L}\right)^2 + 4(\omega_0^2)}}{2} - \frac{-\frac{R}{L} + \sqrt{\left(\frac{R}{L}\right)^2 + 4(\omega_0^2)}}{2} = \frac{R}{L}.$$

The lower cutoff or lower half-power frequency is $\omega_{LO} = \dfrac{-\frac{5000}{10} + \sqrt{25\,000\,000 + 4 \cdot (1414)^2}}{2} = 2.622\ \text{krad/s}$, and the higher cutoff or higher half-power frequency is

$$\omega_{HI} = \frac{+\frac{5000}{10} + \sqrt{25\,000\,000 + 4 \cdot (1414)^2}}{2} = 3.122\ \text{krad/s}.$$

The bandwidth is $BW = \omega_{HI} - \omega_{LO} = 500$ rad/s.

Filters of higher order than two: higher-order [1, 2, 3] or composite filters are sometimes used in power system applications, and their performance can be analyzed as in Application Examples 7.1–7.6.

7.4 Active Filters

Active filters [1, 2] consist of coupling transformers, capacitor and inductor banks, inverters [2, 3] – discussed in Chapters 10 and 12 – and their associated control circuits. The combination of passive and active filters is called hybrid filters. These filters are beyond the scope of this textbook because they require more advanced knowledge of (power) electronic circuits.

7.5 Summary

Voltage and current harmonics, including power quality problems, call for passive first and second up to the nth-order filters consisting of RLC components. Application examples are presented for first- and second-order filters, including series and parallel resonance circuits. All filters can be analyzed based on the phasor approach, where the magnitudes $M(\omega)$ and the phase angles $\Phi(\omega)$ can be computed as a function of the angular frequency ω. Active and hybrid filters rely on (power) electronic circuits [1, 2, 3] and their control and are therefore not discussed because they are beyond the scope of this introductory text. Half-power frequencies, bandwidth BW, quality factor Q, poles and zeros, and resonant angular frequency ω_0 are defined [4, 5].

Problems

7.1 First-order low-pass filter transfer function.

Derive based on phasor notation the first-order low-pass filter with the transfer function of Figure P7.1.1 $\frac{\widetilde{V}_o(\omega)}{\widetilde{V}_s(\omega)} = G(j\omega) = M(\omega)\angle\phi(\omega)$. Plot the magnitude $M(\omega)$ and the phase angle $\phi(\omega)$ as a function of the angular frequency ω for the range 0 rad/s < ω < 100 krad/s for $R_1 = 2\,\Omega$, $R_2 = 4\,\Omega$, $C = 100\,\mu\text{F}$.

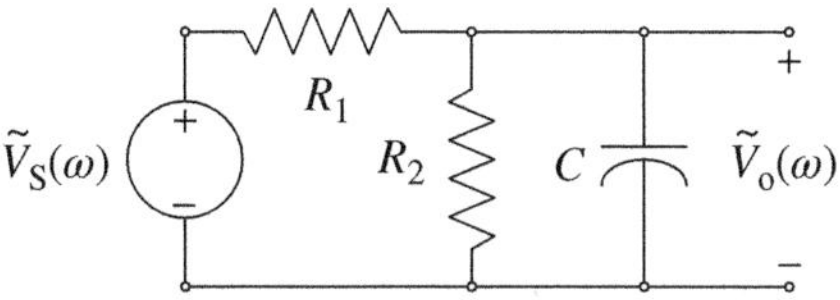

Figure P7.1.1 Network R_1, R_2, and C.

7.2 Poles and zeros of second-order transfer function.

Determine the poles and zeros of the second-order transfer function in Figure P7.2.1 $\frac{\widetilde{V}_o(\omega)}{\widetilde{V}_s(\omega)} = G(j\omega) = M(\omega)\angle\phi(\omega)$. Plot the magnitude $M(\omega)$ and the phase angle $\angle\phi(\omega)$ as a function of the angular frequency ω for the range 0 rad/s < ω < 2.5 krad/s for $R = 2\,\Omega$, $L = 10\,\text{mH}$, $C = 100\,\mu\text{F}$.

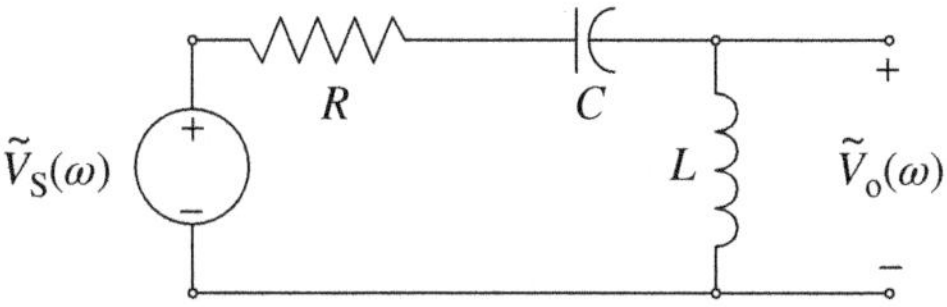

Figure P7.2.1 Network R, L, and C.

7.3 Poles and zeros of second-order transfer function.

Determine the poles and zeros of the second-order transfer function in Figure P7.3.1 $\frac{\widetilde{V}_o(\omega)}{\widetilde{V}_s(\omega)} = G(j\omega) = M(\omega)\angle\phi(\omega)$. Plot the magnitude $M(\omega)$ and the phase angle $\phi(\omega)$ as a function of the angular frequency ω for the range 0 rad/s < ω < 10 krad/s for $R_1 = 2\,\Omega$, $R_2 = 3\,\Omega$, $L = 10\,\text{mH}$, $C = 100\,\mu\text{F}$.

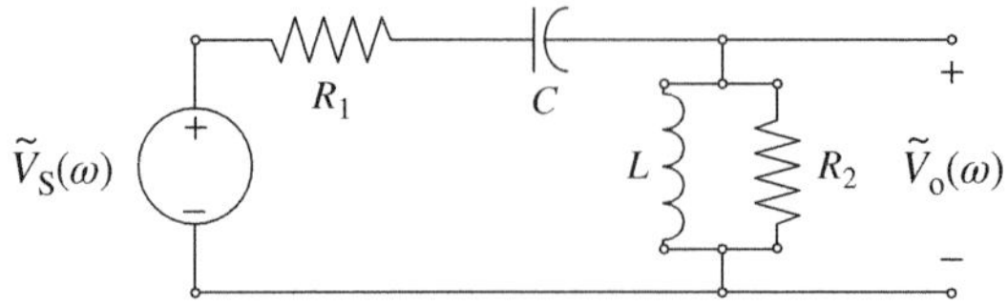

Figure P7.3.1 Network R_1, R_2, L, and C.

7.4 First-order high-pass transfer function.

Derive based on phasor notation the first-order high-pass filter with the transfer function of Figure P7.4.1 $\frac{\tilde{V}_o(\omega)}{\tilde{V}_s(\omega)} = G(j\omega) = M(\omega)\angle\phi(\omega)$. Plot the magnitude $M(\omega)$ and the phase angle $\phi(\omega)$ as a function of the angular frequency ω for the range 0 rad/s $< \omega <$ 20 krad/s for $R = 2\,\Omega$, $C = 100\,\mu\text{F}$.

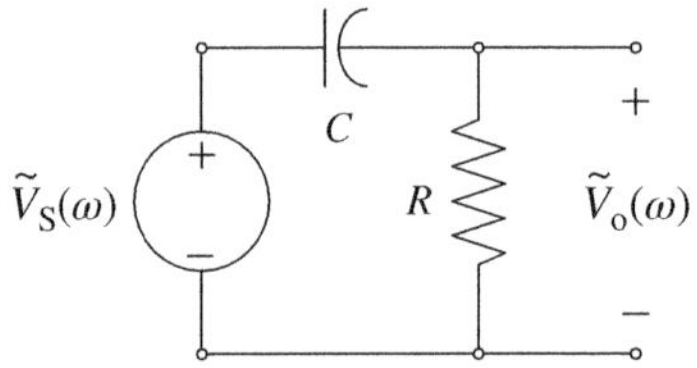

Figure P7.4.1 RC network.

7.5 First-order transfer function.

Derive based on phasor notation the transfer function of Figure P7.5.1 $\frac{\tilde{V}_o(\omega)}{\tilde{V}_s(\omega)} = G(j\omega) = M(\omega)\angle\phi(\omega)$. Plot the magnitude $M(\omega)$ and the phase angle $\phi(\omega)$ as a function of the angular frequency ω for the range 0 rad/s $< \omega <$ 40 krad/s for $R_1 = 2\,\Omega$, $R_2 = 5\,\Omega$, $C = 100\,\mu\text{F}$. What type of filter does it represent?

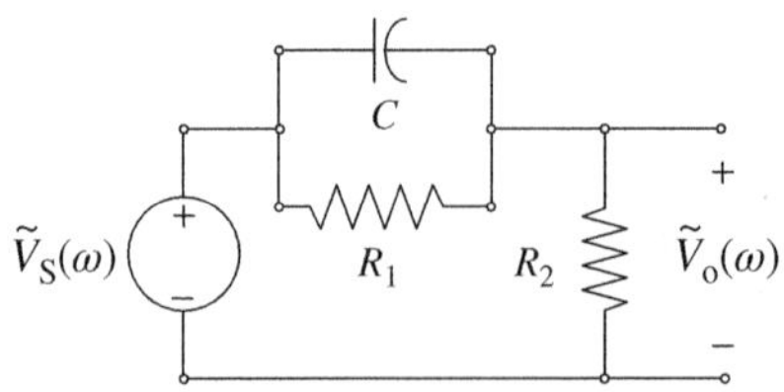

Figure P7.5.1 Network R_1, R_2, and C.

7.6 Selection of capacitor value.

Select for the variable capacitor C of the network of Figure P7.6.1 that will place the circuit in resonance at $\omega_0 = 2\,\text{krad/s}$ for $R = 4\,\Omega$, $L = 50\,\text{mH}$.

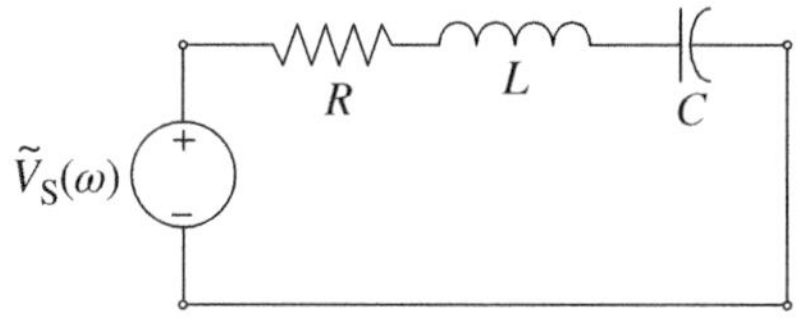

Figure P7.6.1 Series network *R*, *L*, and *C*.

7.7 Calculation of bandwidth BW.

Calculate the bandwidth BW and half-power angular frequencies ω of the circuit of Figure P7.7.1 for $R = 3\,\Omega$, $L = 100\,\text{mH}$, $C = 100\,\mu\text{F}$.

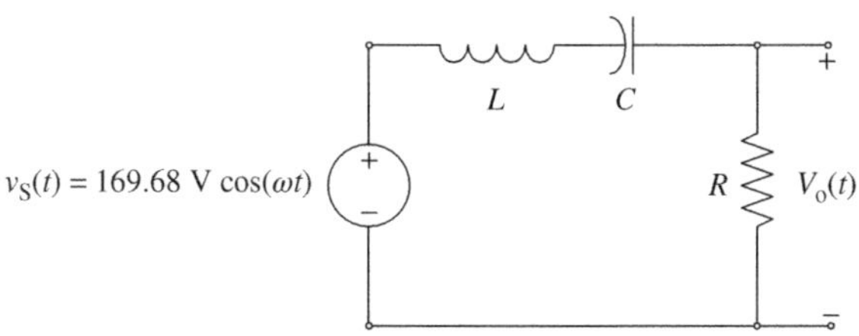

Figure P7.7.1 Series network *R*, *L*, and *C* with $v_s(t) = 169.68\,\text{V}\cos\omega t$ or rms phasor $\tilde{V}_s(\omega) = \dfrac{169.68\,\text{V}\angle 0^\circ}{\sqrt{2}} = 120\,\text{V}\angle 0^\circ$.

7.8 Voltage across capacitor is given.

For the network of Figure P7.7.1 at resonance, the magnitude of the voltage across the capacitor is 60 V_{rms}. Determine the value of the resistor R for $L = 40\,\text{mH}$, $C = 10\,\mu\text{F}$.

7.9 Quality factor Q.

The quality factor Q and resonant angular frequency ω_0 of the series RLC circuit of Figure P7.6.1 are known to be 50 and 7000 rad/s, respectively. For $R = 4\,\Omega$, determine the values of the inductance L and capacitance C.

7.10 Series resonance circuit.

The series RLC network of Figure P7.7.1 has a resistance at resonance of $R = 3.6\,\Omega$. The circuit has a bandwidth $\text{BW} = 200\,\text{rad/s}$ and a resonance angular frequency $\omega_0 = 2\,\text{krad/s}$. Determine the circuit parameters.

7.11 Parallel resonance circuit.

For the parallel RLC network of Figure P7.11.1, determine the resonant angular frequency ω_0 and the voltages across the circuit components at resonance for $R = 200\,\Omega$, $L = 160\,\text{mH}$, and $C = 480\,\mu\text{F}$.

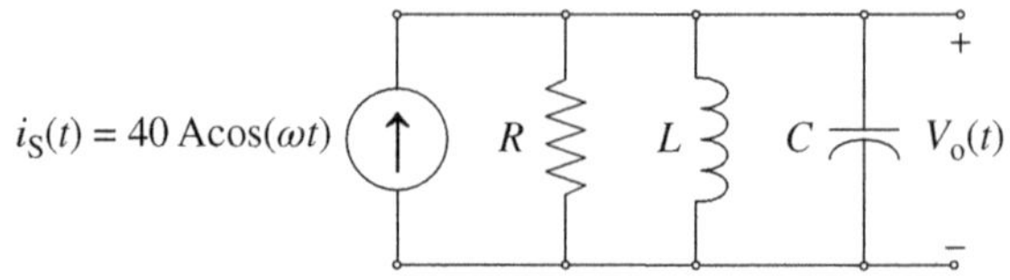

Figure P7.11.1 Parallel network.

7.12 Half-power frequencies, bandwidth, and quality factor.

For the parallel RLC network of Figure P7.12.1

a derive the expression for the resonant angular frequency ω_0, the half-power frequencies ω_{HI} and ω_{LO}, the bandwidth BW, and the quality factor Q for the transfer characteristic $\dfrac{\tilde{V}_o(\omega)}{\tilde{V}_s(\omega)} = G(j\omega) = M(\omega)\angle\phi(\omega)$ in terms of the circuit parameters R, L, C.

b Compute the quantities in part (a) if $R = (1/G) = 2\,\text{k}\Omega$, $L = 10\,\text{mH}$, and $C = 100\,\mu\text{F}$.

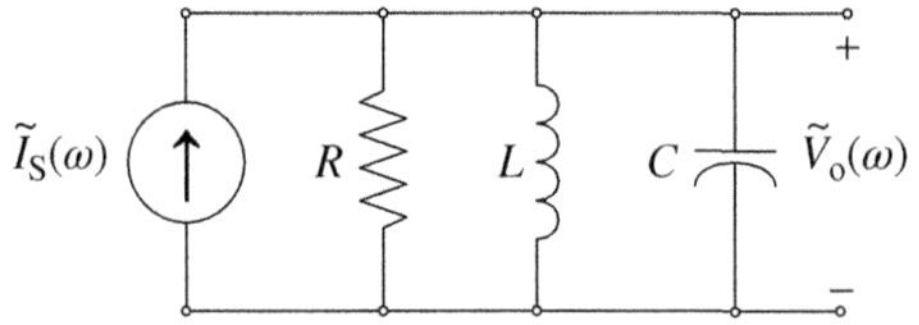

Figure P7.12.1 Parallel network R, L, and C.

7.13 Calculation of capacitor value C for network.

Calculate the value of C in the circuit shown in Figure P7.13.1 in order for the circuit to be in resonance. $R_1 = 6\,\Omega$, $R_2 = 8\,\Omega$, $L = 100\,\text{mH}$.

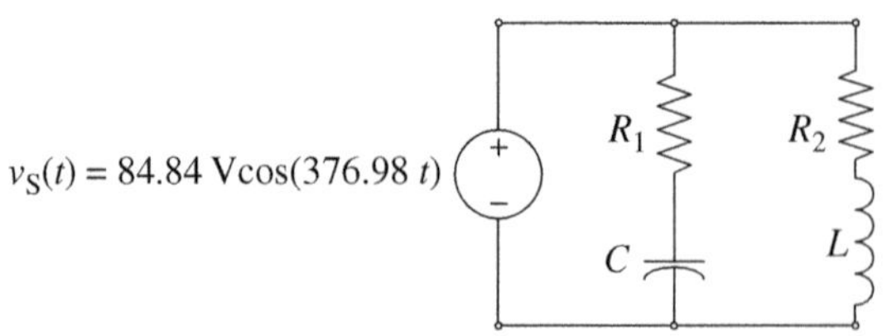

Figure P7.13.1 Network R_1, R_2, L, and C.

References

1 Fuchs, E.F. and Masoum, M.A.S. (2008). *Power Quality in Power Systems and Electrical Machines*, 1e. Elsevier/Academic Press, 638 pp.

2 Masoum, M.A.S. and Fuchs, E.F. (2015). *Power Quality in Power Systems and Electrical Machines*, 2e. Elsevier/Academic Press, 1123 pp.

3 Fuchs, E.F. and Masoum, M.A.S. (2012). *Power Conversion of Renewable Energy Systems*, 1e, 2nd Printing. USA: Springer, 692 pp.

4 Nilsson, J.W. (1983). *Electric Circuits*. Addison-Wesley Publishing Company.

5 Kerns, D.V. Jr. and Irwin, J.D. (2004). *Essentials of Electrical and Computer Engineering*. Pearson Prentice Hall.

8

Operational Amplifiers

8.1 Introduction

The operational (OP) amplifier is a basic building block for the design of analog electronic systems such as feedback control circuits [1], where the angular velocity $\omega_m(t)$, angular position $\theta_m(t) = \int \omega_m(t)dt$, current $i(t)$, and torque $T(t)$ must be controlled. An OP amplifier with no external circuitry is a voltage amplifier with a very high open-loop voltage gain that ideally approaches infinity $A_v \to \infty$, as shown in a symbolic manner in Figure 8.1. Note that input/output voltages can be either time varying or time independent, and they can be either periodic or nonperiodic.

OP amplifiers are integrated circuits (IC), which can be purchased off the shelf, and they contain transistors, diodes, resistors, and capacitors all fabricated on a single silicon chip. Inductors are bulky and are mostly avoided in OP amplifier designs. The analog electronic circuit of Figure 8.1 has two inputs: a negative (v_-) and a positive (v_+) input supplied by the voltages v_{in1} and v_{in2}, respectively, whose difference in voltage ($v_{in2} - v_{in1}$) produces an output signal voltage v_{out} that is a replica of the input voltage difference. The two-input voltage approach is useful because stray signals/electric noise that appear on both inputs are cancelled, and only the voltage difference is amplified. An OP amplifier is also called a differential amplifier because the output voltage is a voltage that is A_v times the voltage difference between the two inputs. During the past half century, OP amplifiers have been refined and improved to the extent that they have nearly ideal properties, and therefore we discuss in this textbook the performance of ideal OP amplifiers only. The understanding of nonideal properties of OP amplifiers requires detailed electronic knowledge [2].

Introduction to Energy, Renewable Energy, and Electrical Engineering: Essentials for Engineering Science (STEM) Professionals and Students, First Edition. Ewald F. Fuchs and Heidi A. Fuchs.

Companion website: www.wiley.com/go/fuchs/renewableenergyandelectricalengineering

8.2 Ideal Operational (OP) Amplifier

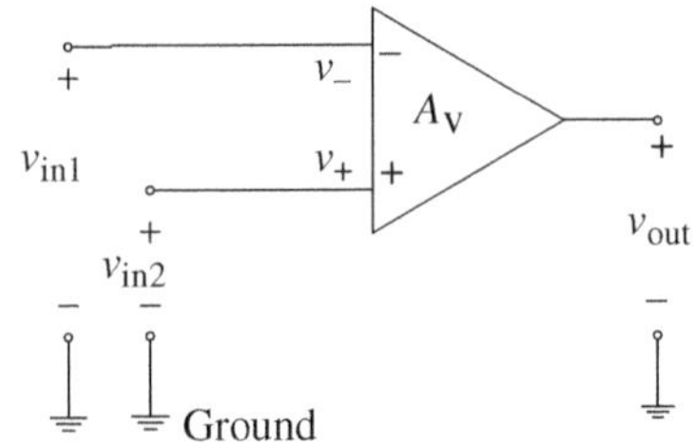

Figure 8.1 Linear ideal operational amplifier with input (v_{in1},v_{in2}) and output voltages v_{out}, respectively, with open-loop voltage gain $A_v \to \infty$.

As shown in Figure 8.1, the ideal OP amplifier has a positive (+) and a negative (−) input, the latter also called the inverting input, while the former is called the noninverting input. The voltage at the positive input is designated as v_+, and the voltage at the negative input is designated v_-. The input currents i_+ and i_- are zero because an ideal OP amplifier has infinite input resistance:

$$i_+ = i_- = 0. \tag{8.1}$$

The output voltage $v_{out}(t)$ is given by the open-loop amplifier gain A_v such that

$$v_{out} = A_v(v_+ - v_-) = A_v(v_{in2} - v_{in1}); \tag{8.2}$$

this identity holds because $i_+ = i_- = 0$. Negative feedback of an OP amplifier is used for many stable control circuits, such as speed, torque, and displacement control applications of variable-speed drives. Negative feedback speed control is achieved when the output terminal voltage v_{out} is connected through a network or circuit component (e.g. resistor R) to the inverting input v_- as illustrated in Figure 8.2. The negative feedback and the open-loop gain A_v approaching infinity allow us to make the following assumption: the potential at the negative input v_- terminal is following that of the positive input terminal v_+; that is, the two inputs are tracking each other and have the same potential or voltage:

$$v_- = v_+. \tag{8.3}$$

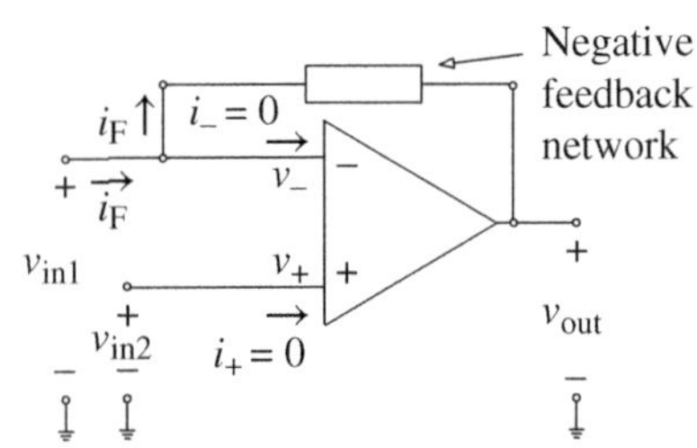

Figure 8.2 Linear ideal operational amplifier with input voltages v_{in2},v_{in1} and output voltage v_{out} with negative feedback network and negative feedback current i_F.

In addition, the output voltage v_{out} of an OP amplifier cannot exceed the power supply voltages; for example, $V_{DD} = +15$ V and $V_{SS} = -15$ V as shown in Figure 8.3. The subscripts DD and SS stem from the drain (D) and source (S) terminals of a field-effect transistor (see Chapter 9).

$$V_{SS} < v_{out} < V_{DD}. \tag{8.4}$$

8.3 Noninverting OP Amplifier

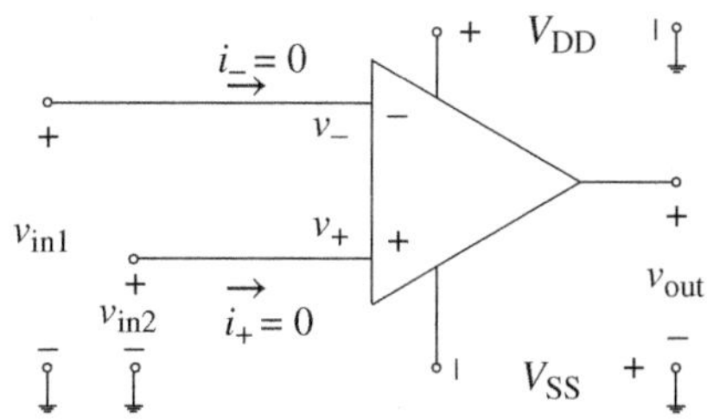

Figure 8.3 Power supply voltages V_{DD} and V_{SS} of linear, ideal operational (OP) amplifier with $V_{SS} < v_{out} < V_{DD}$.

Figure 8.4 shows the noninverting OP amplifier with

$$v_{in} = v_+ = v_-, \tag{8.5}$$

and

$$v_- = [R_1/(R_1 + R_2)]v_{out}; \tag{8.6}$$

it follows

$$v_{out} = [(R_1 + R_2)/R_1]v_{in}, \tag{8.7}$$

and the closed-loop voltage gain of the noninverting (NI) configuration is

$$G_V^{NI} = \frac{v_{out}}{v_{in}} = \frac{R_1 + R_2}{R_1} = 1 + \frac{R_2}{R_1}. \tag{8.8}$$

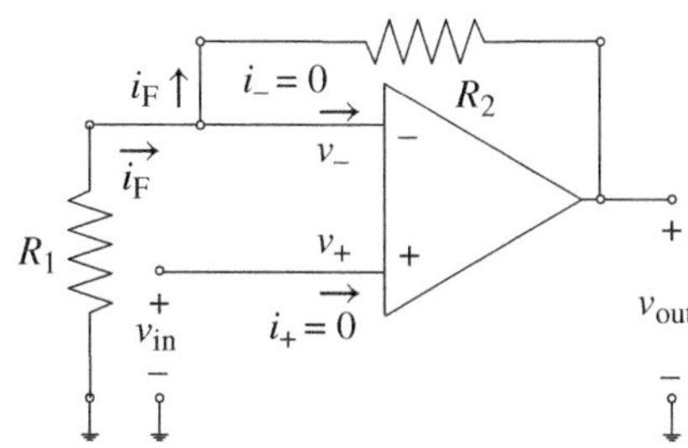

Figure 8.4 Noninverting OP amplifier with resistor R_2 as negative feedback network.

The voltage (subscript v) gain $G_V^{NI} = \frac{v_{out}}{v_{in}}$ is positive, and thus one obtains for a positive input voltage v_{in} a positive output voltage v_{out}. The magnitude of the voltage gain G_V^{NI} is governed by the ratio of the two external resistors R_1 and R_2. If $R_2 = 0$, then the closed-loop voltage gain is unity.

Application Example 8.1 Input and output voltages of a noninverting (NI) linear OP amplifier

Input and output voltages $v_{in}(t)$ and $v_{out}(t)$, respectively, of a noninverting linear OP amplifier with power supply voltages of $V_{DD} = 10$ V and $V_{SS} = -10$ V are shown in Figure E8.1.1a and b. Find the voltage gain G_V^{NI} of this linear amplifier.

Solution

The noninverting (NI) OP amplifier is a linear device; that is, $v_{out}(t) = G_V^{NI} v_{in}(t)$, or applying it to Figure E8.1.1a and b at $t = 3$ ms, the closed-loop voltage (v) gain is

$$G_V^{NI} = \frac{v_{out}(t = 3\,\text{ms})}{v_{in}(t = 3\,\text{ms})} = \frac{8\,\text{V}}{10\,\text{mV}} = 800\,\frac{\text{V}}{\text{V}} = 800\,\text{pu}. \tag{E8.1.1}$$

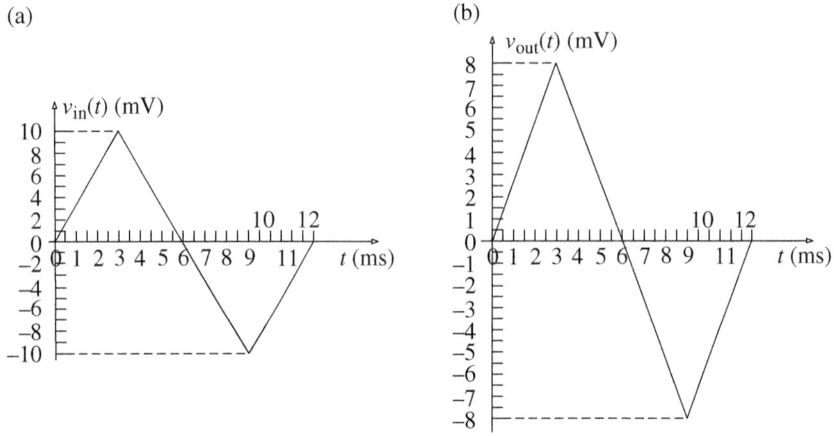

Figure E8.1.1 (a) Input and (b) output voltage of noninverting linear OP amplifier with power supply voltages of V_{DD} = 15 V and V_{SS} = −15 V.

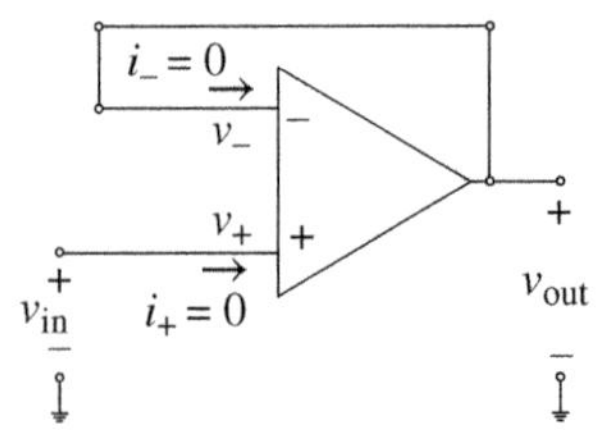

Figure 8.5 Unity-gain OP amplifier with zero (short-circuited) negative feedback network.

8.4 Unity-Gain OP Amplifier

Figure 8.5 illustrates the unity-gain (UG) OP amplifier. The closed-loop voltage gain of the UG buffer is one or unity; that is, the output voltage is the same as the input voltage:

$$v_{out} = v_{in}. \tag{8.9}$$

There is no need to employ a resistor R_1 as is done for the noninverting configuration. $R_2 = 0$ represents a short circuit between the output and the inverting input. The closed-loop voltage gain of the UG amplifier is based on the relations

$$v_{in} = v_+, \tag{8.10}$$

and

$$v_- = v_+, \tag{8.11}$$

because v_{out} is connected to v_-

$$v_{out} = v_-, \tag{8.12}$$

leading to $v_{out} = v_{in}$, or the closed-loop voltage gain for this configuration is

$$G_V^{UG} = \frac{v_{out}}{v_{in}} = 1. \tag{8.13}$$

This circuit represents an impedance converter or transformer that can receive a voltage from a high-impedance source and provides an output voltage at a very low output impedance, resulting in a substantial closed-loop current gain $G_V^{UG} = 1$with a high output current. Circuits of this type are used to send a voltage signal down a long wire or cable.

Application Example 8.2 Output voltage of a unity-gain linear operational amplifier

For the circuit shown in Figure E8.2.1, determine for the AC input voltage $v_{in}(t) = 5$ V $\cos \omega t$, the corresponding output voltage $v_{out}(t)$, and the associated current through the load resistor [3, 4] $R_L = 1\ \text{k}\Omega$.

Solution

The closed-loop voltage gain of the unity-gain amplifier is $G_V^{UG} = 1$; therefore $v_{out}(t) = G_V^{UG} \cdot v_{in}(t) = 5\,\text{V} \cos \omega t$. The current through $R_L = 1\,\text{k}\Omega$ is $i_L(t) = v_{out}(t)/R_L = 5\,\text{V} \cos \omega t/(1\,\text{k}\Omega) = 5\,\text{mA} \cos \omega t$.

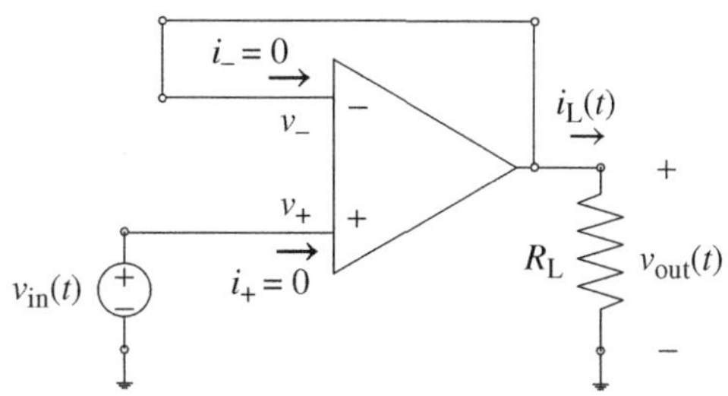

Figure E8.2.1 Unity-gain (UG) linear OP amplifier with power supply voltages of $V_{DD} = 10$ V and $V_{SS} = -10$ V.

8.5 Inverting OP Amplifier

Figure 8.6 represents an inverting (IN) linear OP amplifier that has a negative closed-loop voltage gain G_V^{IN} because the output voltage has the opposite sign of the input voltage. The positive input terminal is connected to ground, resulting in

$$v_+ = v_- = 0, \tag{8.14}$$

because both OP amplifier inputs are tracking each other. The current through resistor R_1 is

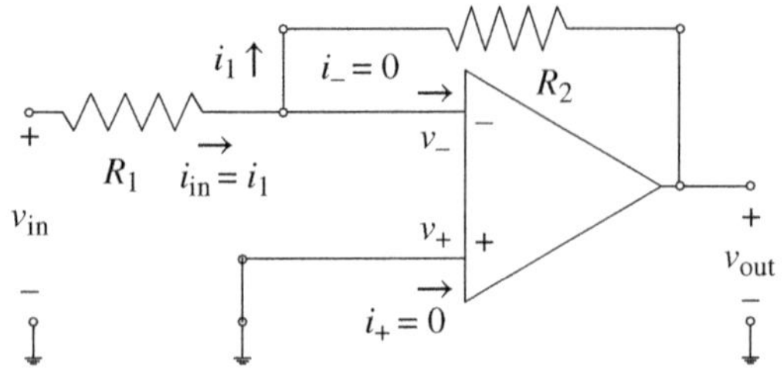

Figure 8.6 Inverting OP linear amplifier with resistor R_2 as negative feedback network.

$$i_{in} = i_1 = \frac{v_{in} - 0}{R_1} = \frac{v_{in}}{R_1}. \tag{8.15}$$

Due to $i_- = 0$ one obtains

$$i_1 = \frac{0 - v_{out}}{R_2}. \tag{8.16}$$

Equating Eqs. (8.15) and (8.16) results in

$$v_{out} = -\frac{R_2}{R_1} v_{in}, \tag{8.17}$$

and the closed-loop voltage gain of the inverting configuration is

$$G_V^{IN} = \frac{v_{out}}{v_{in}} = -\frac{R_2}{R_1}. \tag{8.18}$$

The gain $G_V^{IN} = \frac{v_{out}}{v_{in}} = -\frac{R_2}{R_1}$ is negative, and thus one obtains for a positive input voltage v_{in} a negative output voltage v_{out}. The magnitude of the gain G_V^{IN} is governed by the ratio of the two external resistors R_2 and R_1.

Application Example 8.3 Voltage gain of inverting linear operational amplifier

Determine the voltage gain G_V^{IN} for the circuit of Figure E8.3.1 assuming an ideal OP amplifier for the resistors [3, 4] $R_1 = 237\ \Omega$, $R_2 = 3.6\ \mathrm{k\Omega}$, $R_3 = 2\ \mathrm{k\Omega}$.

Solution

The parallel resistance of the feedback network is $R_F = R_2//R_3 = R_2R_3/(R_2 + R_3) = 3.6\ \mathrm{k} \cdot 2\ \mathrm{k}/(3.6\ \mathrm{k} + 2\ \mathrm{k}) = 1.286\ \mathrm{k\Omega}$, and therefore the closed-loop voltage gain of this inverting configuration is according to Eq. (8.18) $G_V^{IN} = \frac{v_{out}(t)}{v_{in}(t)}$

$$= -\frac{R_F}{R_1} = -\frac{1.286\ \mathrm{k\Omega}}{237\ \Omega} = -5.42\ \mathrm{pu}.$$

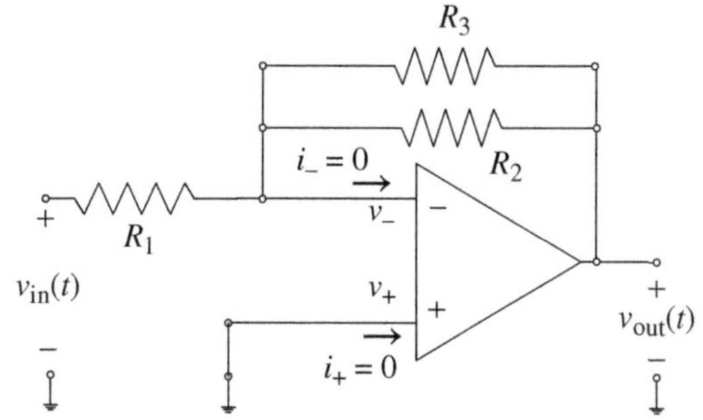

Figure E8.3.1 Inverting linear OP amplifier with power supply voltages of V_{DD} = 10 V and V_{SS} = −10 V.

8.6 Differential Amplifier

The OP amplifier of Figure 8.7 is a differential amplifier: the output voltage v_{out} is proportional to the difference of the two input voltages v_{in1} and v_{in2}. Assuming an ideal OP amplifier, one can write

$$v_+ = \frac{R_2}{R_1 + R_2} v_{in2}; \tag{8.19}$$

with $v_- = v_+$ one gets for the current through the feedback network

$$i_1 = \frac{v_{in1} - v_-}{R_1} = -\left(\frac{v_{out} - v_-}{R_2}\right). \tag{8.20}$$

Combining these two equations leads to

$$v_{out} = -\frac{R_2}{R_1}(v_{in1} - v_{in2}). \tag{8.21}$$

The differential amplifier produces an output voltage that is proportional to the difference of input voltages v_{in1} and v_{in2}. The same voltages applied to both inputs are rejected: interferences or electric noise from stray electric or magnetic fields

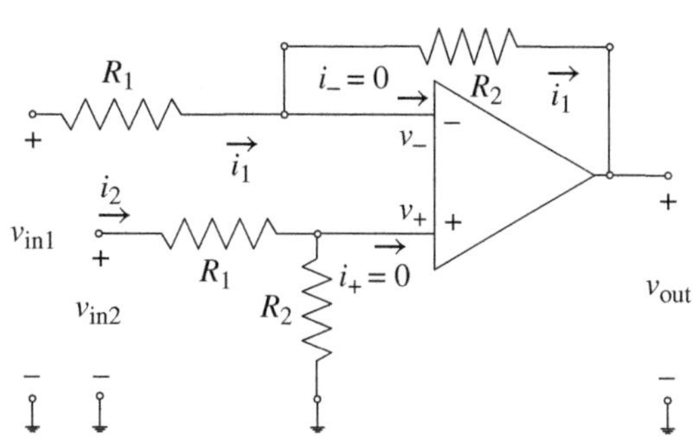

Figure 8.7 Linear ideal differential amplifier with input voltages v_{in1} and v_{in2} and output voltage v_{out}.

picked up on both input leads and common to both inputs is therefore cancelled. Differential amplifiers have the property of rejecting common-mode signals; this property is measured by the common-mode noise rejection ratio (CMRR) expressed in decibel (dB). The CMRR is the ratio of the differential (D) mode gain G_D to the common-mode gain G_{CMRR}; this ratio should be large:

$$\text{CMRR} = G_D/G_{CMRR}. \tag{8.22}$$

An important application of a differential amplifier is the measurement of the electrocardiogram also called "elektrokardiogram (EKG)" measuring the voltage difference generated by the heart of two people (see Chapter 1, Figures E1.9.1 and E1.9.2) of different age. Another application is electroencephalography (EEG), where the brain waves are measured by electrodes attached to the scalp at different locations.

Application Example 8.4 Output voltage $v_{out}(t)$ for a differential linear OP amplifier

Determine the output voltage $v_{out}(t)$ for the differential linear OP amplifier of Figure 8.7 provided the input voltages are $v_{in1}(t) = 10\,\text{V}\sin\omega t$, $v_{in2} = 5\,\text{V}\cos\omega t$, the resistors [3, 4] are $R_1 = 1\,\text{k}\Omega$, $R_2 = 562\,\Omega$, and power supply voltages are $V_{DD} = 15\,\text{V}$ and $V_{SS} = -15\,\text{V}$.

Solution

According to Eq. (8.21) one gets

$$v_{out}(t) = -\frac{R_2}{R_1}[v_{in1}(t) - v_{in2}(t)] = -\frac{562}{1\,\text{k}}[10\,\text{V}\sin\omega t - 5\,\text{V}\cos\omega t]$$
$$= -3.62\,\text{V}\sin\omega t + 1.81\,\text{V}\cos\omega t.$$

8.7 Summing Networks

The summing circuit of Figure 8.8 is similar to the inverting amplifier shown in Figure 8.6. The only difference is that Figure 8.8 has two input voltages v_{inA} and v_{inB} and two input resistors R_{1A} and R_{1B}. The output voltage v_{out} is the weighted sum of the values of the two input voltages. Assuming ideal OP amplifier with $v_+ = v_- = 0$ and $i_+ = i_- = 0$, one can write for the two input currents

$$i_A = v_{inA}/R_{1A} \tag{8.23a}$$
$$i_B = v_{inB}/R_{1B} \tag{8.23b}$$

and for the feedback (F) current

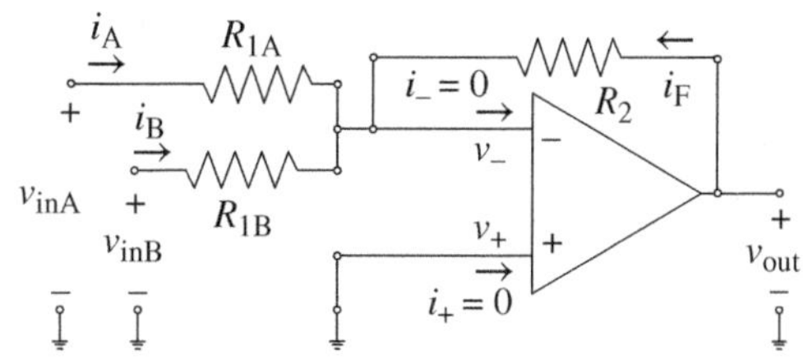

Figure 8.8 Summing network with two inputs.

$$i_F = v_{out}/R_2. \tag{8.24}$$

With $i_A + i_B = -i_F$ one can write

$$v_{inA}/R_{1A} + v_{inB}/R_{1B} = -v_{out}/R_2$$

or

$$v_{out} = -\left(\frac{R_2}{R_{1A}}v_{inA} + \frac{R_2}{R_{1B}}v_{inB}\right). \tag{8.25}$$

This method of summing the weighted input voltages can be applied to three, four, and a higher number of input voltages.

Application Example 8.5 Output voltage for summing network

Determine the output voltage $v_{out}(t)$ for the summing network of Figure 8.8 provided the input voltages are $v_{inA}(t) = 10\text{ V} \sin \omega t$, $v_{inB} = 5\text{ V} \cos \omega t$, the resistors [3, 4] are $R_{1A} = 1\text{ k}\Omega$, $R_{1B} = 422\,\Omega$, $R_2 = 562\,\Omega$, and the power supply voltages are $V_{DD} = 10\text{ V}$ and $V_{SS} = -10\text{ V}$.

Solution

According to Eq. (8.25) one gets

$$v_{out}(t) = -\left[\frac{R_2}{R_{1A}}v_{inA}(t) + \frac{R_2}{R_{1B}}v_{inB}(t)\right] = -[0.562 \cdot 10\text{ V} \sin \omega t + 1.332 \cdot 5\text{ V} \cos \omega t]$$
$$= -[5.62 \sin \omega t + 6.66 \cos \omega t]\ V.$$

8.8 Integrating and Differentiating Networks

Integrating and differentiating networks are frequently applied in control [1] networks. Integrating networks eliminate/reduce electric noise by smoothing the signal, while differentiating networks are applied to anticipate the change (increase or decrease) of the control variable, making the control faster as compared with the integrating network. However, the differentiating approach increases existing electric noise in the signal, while the integration reduces signal noise. In the circuit of Figure 8.9, a capacitor represents the feedback network of the integrating OP amplifier.

Applying the conditions of an ideal OP amplifier, one gets

$$i_F = C\frac{dv_{out}}{dt} = -i_{in} \tag{8.26}$$

and

$$i_{in} = v_{in}/R, \tag{8.27}$$

or combining these two equations leads to

$$\frac{v_{in}}{R} = -C\frac{dv_{out}}{dt} \tag{8.28}$$

and solving for the output voltage

$$v_{out} = -\frac{1}{RC}\int_0^t v_{in} dt + v_{out}(t=0). \tag{8.29}$$

The term $v_{out}(t = 0)$ is the output voltage $v_{out}(t)$ at time $t = 0$ resulting from the initial charge of the capacitor. This circuit will provide an output voltage proportional to the integral of the input voltage.

In the circuit of Figure 8.10, a capacitor together with input resistor is placed in the input network of the inverting OP amplifier acting as a differentiating circuit.

Applying the conditions of an ideal OP amplifier, one gets

$$i_F = v_{out}/R = -i_{in} \tag{8.30}$$

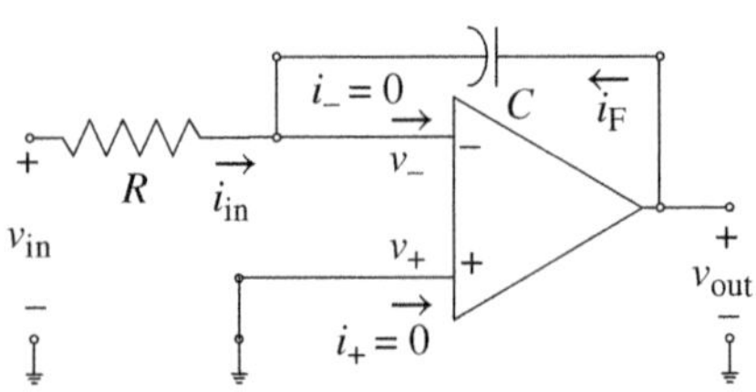

Figure 8.9 Integrating circuit.

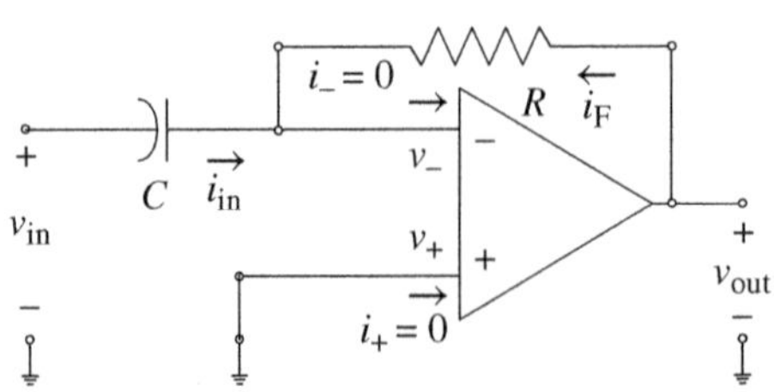

Figure 8.10 Differentiating circuit.

and

$$i_{in} = C\frac{dv_{in}}{dt}, \tag{8.31}$$

or combining these two equations leads to

$$v_{out} = -RC\frac{dv_{in}}{dt}. \tag{8.32}$$

This circuit will provide an output voltage v_{out} proportional to the differential of the input voltage v_{in}.

Application Example 8.6 Determination of output voltage of integrating circuit

For Figure E8.6.1a plot the output voltage $v_{out}(t)$, whereby the input voltage $v_{in}(t)$ is defined by the step function: from $t = 0$ s to $t = 2$ s, $v_{in} = 3$ V; from $t = 2$ s to $t = 5$ s, $v_{in} = 5$ V; from $t = 5$ s to $t = 8$ s, $v_{in} = -2$ V; from $t = 8$ s to $t = 12$ s, $v_{in} = 1$ V.

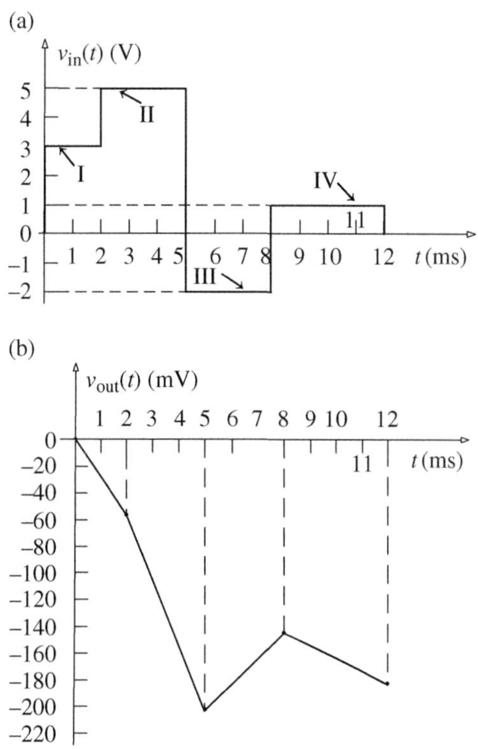

Figure E8.6.1 (a) Definition of input step voltage $v_{in}(t)$ via the functions I, II, III, and IV. (b) Computed output function $v_{out}(t)$ representing the integral of (a), the input voltage $v_{in}(t)$. Note that the inverting linear OP amplifier generates a sign inversion and the power supply voltages are $V_{DD} = 7.5$ V and $V_{SS} = -7.5$ V.

You may assume the initial condition $v_{out}(t = 0) = 0$. The resistor [3, 4] is $R = 4.7$ kΩ, the capacitor [4] is $C = 22\ \mu F$, and the power supply voltages are $V_{DD} = 10$ V and $V_{SS} = -10$ V.

Solution

According to Eq. (8.29) one can write for the output voltage

$$v_{out}(t) = -\frac{1}{RC}\int_0^t v_{in}(t)dt + v_{out}(t=0) = -9.67\frac{1}{s}\int_0^t v_{in}(t)dt + 0.$$

The application of this relation leads to the following:

- For the interval (0 ms < t < 2 ms) with origin (0/0) and the input voltage $v_{in}(t) = 3$ V with the initial condition $v_{out}(t = 0)_I = 0$ to the integral

$$v_{out}(t)_I = -9.67\frac{1}{s}\int_{t=0\ ms}^{t=2\ ms} v_{in}(t)dt + v_{out}(t=0)_I = -9.67\frac{1}{s}\cdot 3t|_{0\ ms}^{2\ ms} + 0$$

$$= -9.67\frac{1}{s}\cdot 3\ V\cdot(2-0)\ ms + 0 = -58.02\ mV \text{ at } t = 2\ ms;$$

- For the interval (2 ms < t < 5 ms) with origin (2 ms/−58.02 mV) and the input voltage $v_{in}(t) = 5$ V with the initial condition $v_{out}(t = 2\ ms)_{II} = -58.02$ mV to the integral

$$v_{out}(t)_{II} = -9.67\frac{1}{s}\int_{t=2\ ms}^{t=5\ ms} v_{in}(t)dt + v_{out}(t=2\ ms)_{II} = -9.67\frac{1}{s}\cdot 5t|_{2\ ms}^{5\ ms} - 58.02\ mV$$

$$= -9.67\frac{1}{s}\cdot 5\ V\cdot(5-2)\ ms - 58.02\ mV = -203.07\ mV \text{ at } t = 5\ ms;$$

- For the interval (5 ms < t < 8 ms) with origin (5 ms/−203.07 mV) and the input voltage $v_{in}(t) = -2$ V with the initial condition $v_{out}(t = 5\ ms)_{III} = -203.07$ mV to the integral

$$v_{out}(t)_{III} = -9.67\frac{1}{s}\int_{t=5\ ms}^{t=8\ ms} v_{in}(t)dt + v_{out}(t=5\ ms)_{III}$$

$$= -9.67\frac{1}{s}\cdot(-2t)|_{5\ ms}^{8\ ms} - 203.07\ mV = -9.67\frac{1}{s}\cdot(-2\ V)\cdot(8-5)\ ms$$

$$-203.07\ mV = -145.05\ mV \text{ at } t = 8\ ms;$$

- For the interval (8 ms < t < 12 ms) with origin (8 ms/−145.05 mV) and the input voltage $v_{in}(t) = 1$ V with the initial condition $v_{out}(t = 8\ ms)_{IV} = -145.05$ mV to the integral

$$v_{out}(t)_{IV} = -9.67\frac{1}{s}\int_{t=8\text{ ms}}^{t=12\text{ ms}} v_{in}(t)dt + v_{out}(t = 8\text{ ms})_{IV} = -9.67\frac{1}{s}\cdot t|_{8\text{ ms}}^{12\text{ ms}} - 145.05\text{ mV}$$

$$= -9.67\frac{1}{s}\cdot 1\text{ V}\cdot(12-8)\text{ ms} - 145.05\text{ mV} = -183.73\text{ mV at } t = 12\text{ ms}.$$

The integrated function $v_{out}(t)$ is shown in Figure E8.6.1b.

Application Example 8.7 Determine the output voltage of differentiating circuit

For Figure E8.7.1a, plot the output voltage $v_{out}(t)$, if the input voltage is defined through a polygonal function in the interval $0\ \mu s < t < 1\ \mu s$, where $v_{in}(t)_I$ changes linearly from 0 to 3 V; from $1\ \mu s < t < 3\ \mu s$ $v_{in}(t)_{II}$ changes linearly from 3 to 5 V; from $3\ \mu s < t < 4\ \mu s$ $v_{in}(t)_{III}$ changes linearly from 5 to −1 V; from $4\ \mu s < t < 6\ \mu s$ $v_{in}(t)_{IV}$ changes linearly from −1 to 1 V. Assume the resistor [3, 4] $R = 4.7\text{ k}\Omega$, and the capacitor [4] $C = 22\ \mu F$.

Solution

According to Eq. (8.32) one can write

$$v_{out}(t) = -RC\frac{dv_{in}(t)}{dt} = 4700\ \Omega\cdot 22\ \mu F\ \frac{dv_{in}(t)}{dt} = -0.1034\text{ s}\frac{dv_{in}(t)}{dt}.$$

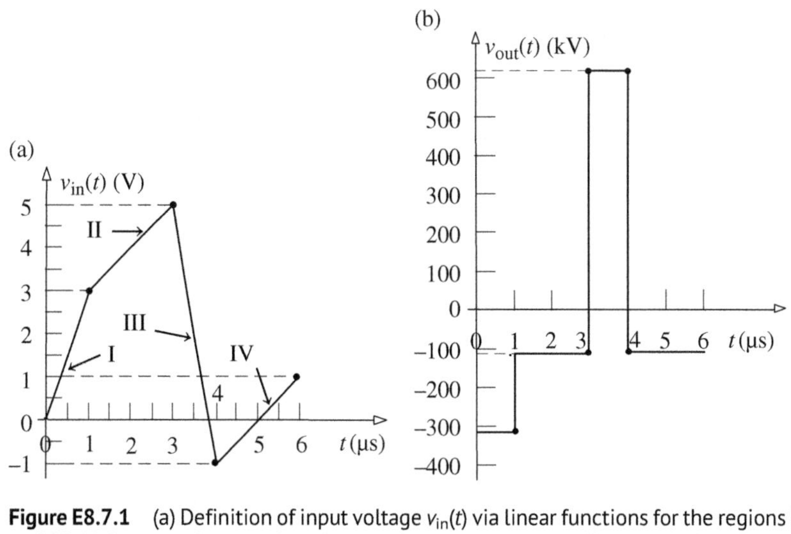

Figure E8.7.1 (a) Definition of input voltage $v_{in}(t)$ via linear functions for the regions I, II, III, and IV. (b) Computed output function $v_{out}(t)$ representing the differential of (a), which is the input voltage $v_{in}(t)$. Note the inverting linear OP amplifier generates a sign inversion.

- For the interval $(0\,\mu s < t < 1\,\mu s)$ with origin $(0\,\mu s/0\,V)$ is the input voltage $v_{in}(t) = 3t$ V, which results in the differential equation

$$v_{out}(t)_{I} = -0.1034\,s \frac{dv_{in}(t)}{dt}\bigg|_{I} \text{ whereby } \frac{dv_{in}(t)}{dt}\bigg|_{I} = \frac{3\,V}{1\,\mu s},$$

yielding

$$v_{out}(t)_{I} = -0.1034\,s \cdot \frac{3\,V}{1\,\mu s} = -310\,kV;$$

- For the interval $(1\,\mu s < t < 3\,\mu s)$ with origin $(1\,\mu s/3\,V)$ is the differential equation

$$v_{out}(t)_{II} = -0.1034\,s \frac{dv_{in}(t)}{dt}\bigg|_{II} \text{ whereby } \frac{dv_{in}(t)}{dt}\bigg|_{II} = \frac{5\,V - 3\,V}{2\,\mu s} = \frac{1\,V}{1\,\mu s},$$

resulting in

$$v_{out}(t)_{II} = -0.1034\,s \cdot \frac{1\,V}{1\,\mu s} = -103.4\,kV;$$

- For the interval $(3\,\mu s < t < 4\,\mu s)$ with origin $(3\,\mu s/5\,V)$ is the differential equation

$$v_{out}(t)_{III} = -0.1034\,s \frac{dv_{in}(t)}{dt}\bigg|_{III} \text{ whereby } \frac{dv_{in}(t)}{dt}\bigg|_{III} = \frac{-1\,V - 5\,V}{1\,\mu s} = \frac{(-6)\,V}{1\,\mu s},$$

producing

$$v_{out}(t)_{III} = -0.1034\,s \cdot \frac{(-6)\,V}{1\,\mu s} = 620.4\,kV;$$

- For the interval $(4\,\mu s < t < 6\,\mu s)$ with origin $(4\,\mu s/-1\,V)$ is the differential equation

$$v_{out}(t)_{IV} = -0.1034\,s \frac{dv_{in}(t)}{dt}\bigg|_{IV} \text{ whereby } \frac{dv_{in}(t)}{dt}\bigg|_{IV} = \frac{1\,V - (-1\,V)}{2\,\mu s} = \frac{1\,V}{1\,\mu s},$$

resulting in

$$v_{out}(t)_{IV} = -0.1034\,s \cdot \frac{1\,V}{1\,\mu s} = -103.4\,kV.$$

Conclusions

The output voltage $v_{out}(t)$ of Figure E8.7.1b is not realistic for solid-state circuits because linear OP amplifiers have usually maximum power supply voltages of $V_{DD} = 15\,V$ and $V_{SS} = -15\,V$. It is therefore necessary to change the resistance [3, 4] R and the capacitance [4] C to $47\,\Omega$ and $22\,nF$, respectively, resulting in the time constant $\tau = RC = 47\,\Omega \cdot 22 \cdot 10^{-9}\,As/V = 1.034\,\mu s$, yielding the maximum modified voltage excursion of $v_{out}(t)_{III_{modified}} = -1.034\,\mu s \cdot \frac{(-6)\,V}{1\,\mu s} = 6.204\,V$ suitable for a power supply voltages of $V_{DD} = 10\,V$ and $V_{SS} = -10\,V$.

Application Example 8.6 demonstrates that for a given input voltage $v_{in}(t)$, an integrating circuit reduces the voltage variations of the output voltage $v_{out}(t)$ as shown in Figure E8.6.1b, where the voltage is decreased from the volt range to the millivolt range, respectively. A differentiating circuit increases the voltage variations of the output voltage as depicted in Figure E8.7.1b, resulting from the volt range at the input to the kilovolt range at the output.

8.9 Active Filters

Chapter 7 discusses passive [5] filters, which have a certain frequency selectivity with a magnitude of the transfer function $M(\omega) \leq 1.0$: that is, there is no voltage gain. Passive filters use components such as capacitors, resistors, and inductors. Active filters rely on amplifiers (e.g. OP amplifiers) and other semiconductor components (e.g. inverters), which avoid the use of bulky inductors, resulting in lower cost and higher performance including closed-loop voltage gains of larger than 1.0. If only resistors with OP amplifiers are employed, then the closed-loop voltage gain G_v is real and frequency independent. With capacitors and/or inductors connected to an OP amplifier, the corresponding voltage gains are complex.

Consider the low-pass (LP) active filter of Figure 8.11 consisting of an inverting OP amplifier where the R_2C network in the feedback path is frequency selective.

The closed-loop complex voltage gain can be computed by taking the ratio of the impedance of the feedback branch to that of the input branch. In other words, the

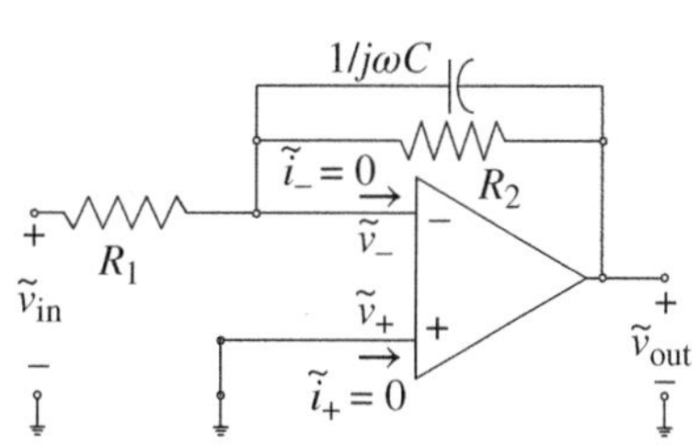

Figure 8.11 Low-pass (LP) active filter.

complex voltage gain is the ratio of the phasor output voltage to that of the input voltage:

$$G_V^{LP}(\omega) = \frac{\tilde{v}_{out}}{\tilde{v}_{in}} = \frac{\tilde{Z}_2(\omega)}{\tilde{Z}_1(\omega)}, \tag{8.33}$$

where

$$\tilde{Z}_2(\omega) = \frac{\frac{1}{j\omega C} \cdot R_2}{\frac{1}{j\omega C} + R_2} = \frac{R_2}{1 + j\omega C R_2}; \tag{8.34}$$

$$\tilde{Z}_1(\omega) = R_1; \tag{8.35}$$

therefore,

$$G_V^{LP}(\omega) = -\frac{R_2/R_1}{1 + j\omega C R_2}. \tag{8.36}$$

Note the negative (−) sign is there because the OP amplifier is operating in the inverting mode.

The magnitude of Eq. (8.36) is

$$\left|G_V^{LP}(\omega)\right| = \frac{R_2/R_1}{\sqrt{1 + (\omega C R_2)^2}}. \tag{8.37}$$

The frequency response of Eq. (8.37) is listed in Table 8.1 for $R_1 = 10\,\Omega$, $R_2 = 100\,\Omega$, and $C = 10\,\mu\text{F}$ and plotted in Figure 8.12. At low frequency, the gain (Eq. 8.37) is constant due to $R_2/R_1 = 10$.

For this circuit the half-power frequency (see Chapter 7)

$$\omega_{HI} = 1/R_2C = 1/\left(100\,\Omega \cdot 10 \cdot 10^{-6}\,\text{s}/\Omega\right) = 1\,\text{krad/s} \tag{8.38a}$$

is the magnitude if the closed-loop voltage gain

$$(R_2/R_1)/\sqrt{2} = 10/1.414 = 7.07\,\text{pu}. \tag{8.38b}$$

This active filter has at low frequencies a larger than 1.0 closed-loop gain due to feedback resistance R_2. Passive filters as discussed in Chapter 7 do not have such a property, where the voltage gain is increased to above 1.0.

Table 8.1 Low-pass active filter frequency response $\left|G_V^{LP}(\omega)\right|$ (pu) with $R_2/R_1 = 10$.

ω **(krad/s)**	**0.01**	**0.1**	ω_{HI} **= 1**	**2**	**3**	**4**	**5**
$\left\|G_V^{LP}(\omega)\right\|$ (pu)	9.99	9.95	$7.07 = 10/\sqrt{2}$	4.47	3.16	2.43	1.96

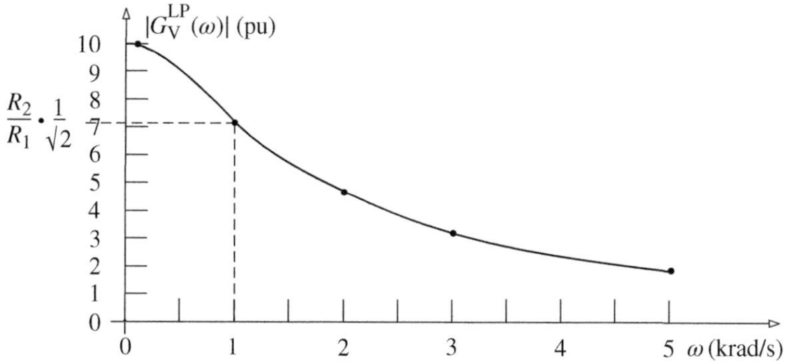

Figure 8.12 Low-pass (LP) active filter frequency response.

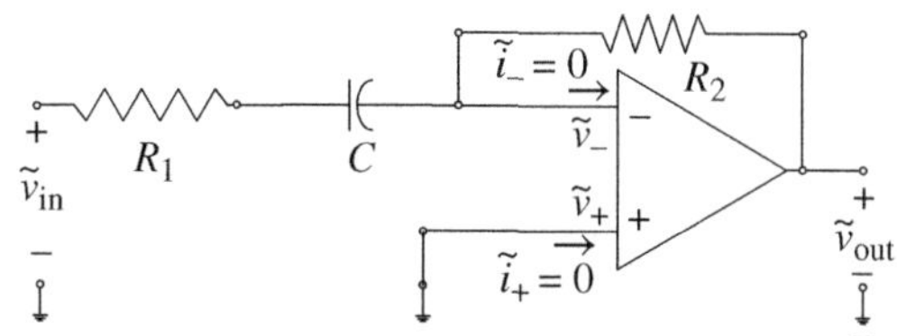

Figure 8.13 Frequency-selective high-pass (HP) active filter.

Consider the high-pass (HP) active filter of Figure 8.13, which is frequency selective consisting of an inverting OP amplifier, where the R_1C is in the input network and the resistor R_2 is in the feedback branch.

The closed-loop complex voltage gain can be computed by taking the ratio of the impedance of the feedback branch to that of the input branch. In other words, the complex voltage gain is the ratio of the phasor output voltage to that of the input voltage:

$$G_V^{HP}(\omega) = \frac{\tilde{V}_{out}}{\tilde{V}_{in}} = \frac{\tilde{Z}_2(\omega)}{\tilde{Z}_1(\omega)}, \tag{8.39}$$

where

$$\tilde{Z}_2(\omega) = R_2; \tag{8.40}$$

$$\tilde{Z}_1(\omega) = \frac{1}{j\omega C} + R_1, \tag{8.41}$$

therefore

$$G_V^{HP}(\omega) = -\frac{R_2}{R_1 + \frac{1}{j\omega C}} = -\frac{j\omega CR_2}{1 + j\omega CR_1}. \tag{8.42}$$

Table 8.2 High-pass active filter frequency response $|G_V^{HP}(\omega)|$ (pu).

ω (krad/s)	0.01	0.1	1	2	3	4	5
$\lvert G_V^{HP}(\omega)\rvert$ (pu)	0.01	0.1	0.995	1.96	2.87	3.72	4.47

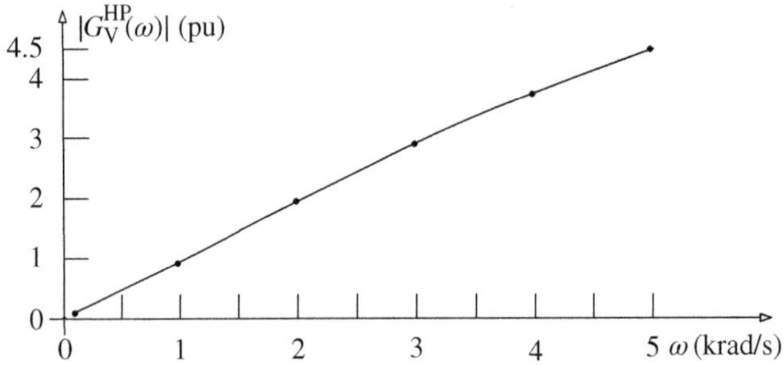

Figure 8.14 High-pass (HP) active filter frequency response $|G_V^{HP}(\omega)|$ (pu).

Note the negative (−) sign is there because the OP amplifier is operating in the inverting mode.

The magnitude of Eq. (8.42) is

$$|G_V^{HP}(\omega)| = \frac{\omega C R_2}{\sqrt{1 + (\omega C R_1)^2}}. \tag{8.43}$$

The frequency response of Eq. (8.43) is listed in Table 8.2 for $R_1 = 10\,\Omega$, $R_2 = 100\,\Omega$, and $C = 10\,\mu\text{F}$ and plotted in Figure 8.14.

This active filter has at high frequencies a larger than 1.0 closed-loop gain due to feedback resistance R_2. Passive filters as discussed in Chapter 7 do not have such a property, where the voltage gain is increased to above 1.0.

8.10 Current-to-Voltage Converter

For some applications it is required to convert a given current i to a proportional voltage v: examples are temperature measurements where the current source must be converted to an output voltage, a photodiode generates a current that is converted to a voltage, and on a large scale a photovoltaic array is a current source and delivers energy to the utility system, which is a voltage source. Figure 8.15

presents a circuit consisting of an OP amplifier operating in the inverting mode, whereby a resistor R_F is placed in the feedback branch. The input current i_{in} cannot flow into the ideal OP amplifier, but flows entirely through the resistor R_F.

Thus

$$v_{out} = -R_F \cdot i_{in}. \tag{8.44}$$

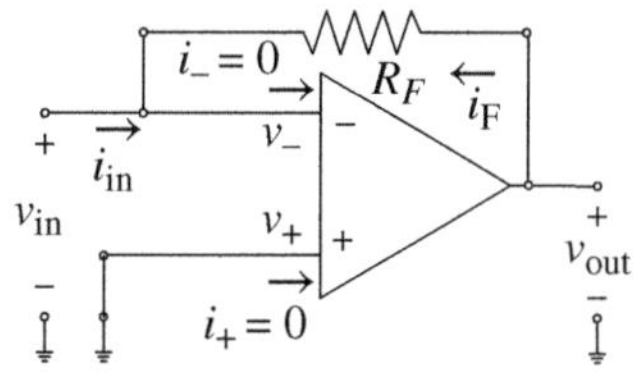

Figure 8.15 Current–voltage converter.

The parameter "transresistance" describes how much output voltage v_{out} is produced by the input current i_{in}. The word "trans" means that the current and voltage are not measured at the same location within the circuit. The transresistance is defined as

$$R_{trans} = v_{out}/i_{in} = -R_F. \tag{8.45}$$

Application Example 8.8 Calculation of transresistance of current-to-voltage converter

The circuit presented in Figure 8.15 is a current-to-voltage converter. Calculate the transresistance of this circuit by assuming an ideal linear OP amplifier, where the feedback (F) resistor is $R_F = 4.7\ \text{k}\Omega$.

Solution

According to Eq. (8.45) the transresistance is

$$R_{trans} = v_{out}(t)/i_{in}(t), \tag{E8.8.1}$$

and the output voltage

$$v_{out}(t) = -R_F i_{in}(t). \tag{E8.8.2}$$

Introducing Eq. (E8.8.2) into Eq. (E8.8.1) yields

$$R_{trans} = -(i_{in}(t)R_F)/i_{in}(t) = -R_F = -4.7\ \text{k}\Omega. \tag{E8.8.3}$$

8.11 Controllers for Electric Circuits

To control, for example, current $i(t)$, torque $T(t)$, angular velocity $\omega_m(t)$, and the angular displacement $\theta_m(t) = \int \omega_m(t)dt$ of electric drives (e.g. wind power plant, electric automobile, heat pumps), smart networks and sensors are required. Some of them are proportional (P) regulators also called P controllers, proportional–integral (PI) regulators or PI controllers, proportional–integral–differential (PID) regulators or PID controllers, and proportional–differential (PD) regulators or PD controllers.

1) A P controller has the greatest stability margin [1]; however, it has a gain-dependent, steady-state error: the smaller the gain of the P controller, the larger the steady-state deviation from the value of the desired control variable.
2) An I controller has a smaller stability margin [1, 5–7] than the P controller, and it is hardly used.
3) A PI controller is the most often [1, 5–7] used regulator: it has a zero steady-state error from the desired control variable, but has a somewhat reduced stability margin as compared with the P controller: in other words, it is less stable, and it takes longer time to achieve the desired control objective.
4) A D controller is not very useful because the gain is not variable and the differentiation increases existing electric noise so that no reliable control is possible.
5) A PID controller is sometimes relied on when the change of the control variable must be predicted/anticipated to make the control possible at all. One example is the suspension of a magnetic (e.g. iron) sphere/cylinder within a magnetic field generated by a current through a coil. This approach has been extensively researched in suspending electric trains without mechanical track contacts such as wheels.
6) A PD controller is not frequently employed except for special application such as controlling servomotors.

Of course, more specialized regulators exist, but the abovementioned ones are the most basic, and other regulators are in some manner related to them.

8.11.1 P Controller

A P controller consists of two OP amplifiers as shown in Figure 8.16a: the first one achieves the proportional (P) control relying on an inverting configuration such that

$$v_{\text{out1}} = -(R_2/R_1)\,v_{\text{in}}; \tag{8.46}$$

the second OP amplifier is called an inverter that inverts v_{out1}, resulting in

$$v_{\text{out2}} = -(R_4/R_3)v_{\text{out1}} = (R_2/R_1)(R_4/R_3)v_{\text{in}} = (R_2R_4/R_1R_3)v_{\text{in}}. \tag{8.47}$$

Note that the word "inverter" is used in a different manner here than in Chapters 10 and 12. In this chapter, the (voltage) signals are inverted (that is, changed

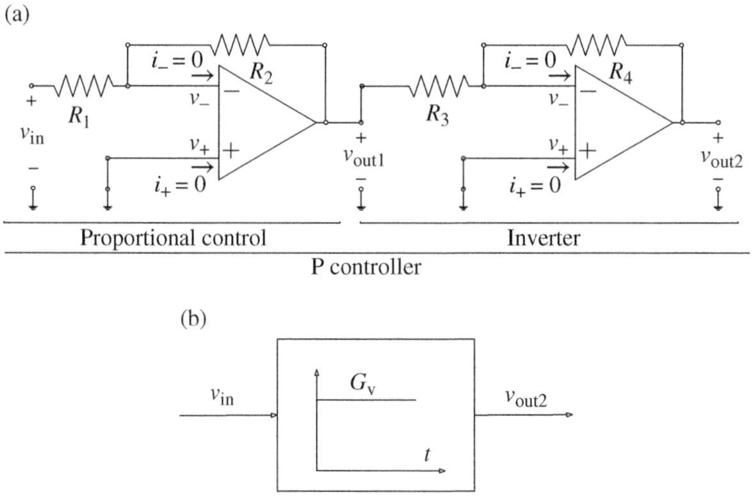

Figure 8.16 (a) P-controller network with closed-loop voltage gain $G_v = v_{out2}/v_{in}$. (b) Symbolic representation of P-controller network of (a) with closed-loop voltage gain $G_v = v_{out2}/v_{in}$.

from negative to positive), while in Chapters 10 and 12 DC voltages or currents are converted to AC through different types of inverters.

In general, proportional (P) control is the fastest regulation method, but the steady-state error $\varepsilon = 1/(1 + G_v)$ depends upon the overall gain [1, 6, 7] G_v: that is, for a large gain G_v the steady-state error is small, but control oscillations might be large; for small G_v values the steady-state error is large, and the control response is slow. For these reasons P control is seldom employed for control applications.

Application Example 8.9 Current $i(t)$ and angular velocity $\omega_m(t)$ control of variable-speed electric drives

The control of an electric variable-speed drive is important to guarantee flexible, smooth, and safe operation. Frequently, two independent control loops are used. The first one occurs where the stator current $i(t)$ is regulated and limited to safe magnitudes, and the second one is where the angular mechanical (m) velocity $\omega_m(t)$ is permitted to change up to a safe maximum limit so that the electric drive that may be part of an electric car can start at low speed or angular velocity with high torque and operate at high speed with sufficient torque so that an electric automobile can pass another within a short time. In most electric car drives, brushless DC machines – also called permanent-magnet machines supplied by an inverter – where frequency and magnitudes of voltages and current can be changed are very popular due to their high efficiency and light weight [8–11]. Another popular choice is the three-phase induction machine [10–21] supplied by a current-controlled voltage source inverter. An electric machine can be either used as a motor or a generator. This enables the automobile to regenerate [1, 22–25]; that is, to recharge the onboard battery during any braking action. One of the control algorithm is based on proportional (P) control with the overall gain G, where the steady-state deviation of the actual variable from the reference or nominal variable is given by the relation $\varepsilon = 1/(1+G)$, as will be demonstrated below for a DC machine electric drive.

DC Machine as Part of a Variable-Speed Drive
As an introduction to P and PI control, a variable-speed drive consisting of a separately excited DC machine (Chapter 11) with armature current $i_a(t)$ and angular velocity $\omega_m(t)$ control loops will be used to demonstrate P control. In this case, the armature current $i_a(t)$ supplied to the rotor (Chapter 11) and the angular velocity $\omega_m(t)$ are controlled with an armature current limiter so that the DC machine does not overheat under all driving conditions. Figure E8.9.1 shows the block diagram of a separately excited DC machine drive where the armature current $i_a(t)$ and the angular velocity $\omega_m(t)$ are controlled within a wide range as it applies to starting at zero speed and maximum (max) mechanical (m) speed $n_m|_{max} = \omega_m/2\pi|_{max}$. P control will be performed from 0 to 2 seconds and PI control from 2 to 5 seconds.

Solution

a) *P control of angular velocity ω_m from 0 to 2 seconds*

To understand the block diagram of Figure E8.9.1 for P control of ω_m, the components are explained as follows:

1) Difference of reference ω_m^* and actual ω_m mechanical (subscript m) angular velocities

$$\Delta\omega_m = \omega_m^* - \omega_m. \tag{E8.9.1}$$

Note the (1) above the circle with the summation ($\sum$) sign refers to the first block of Figure E8.9.1.

2) Due to P control of ω_m, the angular velocity gain $G_{\omega m}$ generates from the angular velocity difference $\Delta\omega_m$ the input current i_{in} to the current limiter:

$$i_{in} = G_{\omega m}\Delta\omega_m. \tag{E8.9.2}$$

Note the (2) refers to the second block of Figure E8.9.1.

3) Current limiter provides for input current i_{in} the output current i_a^*, which serves as input to the current difference node:

$$i_a^* = \pm\,\text{limit}(i_{in}). \tag{E8.9.3}$$

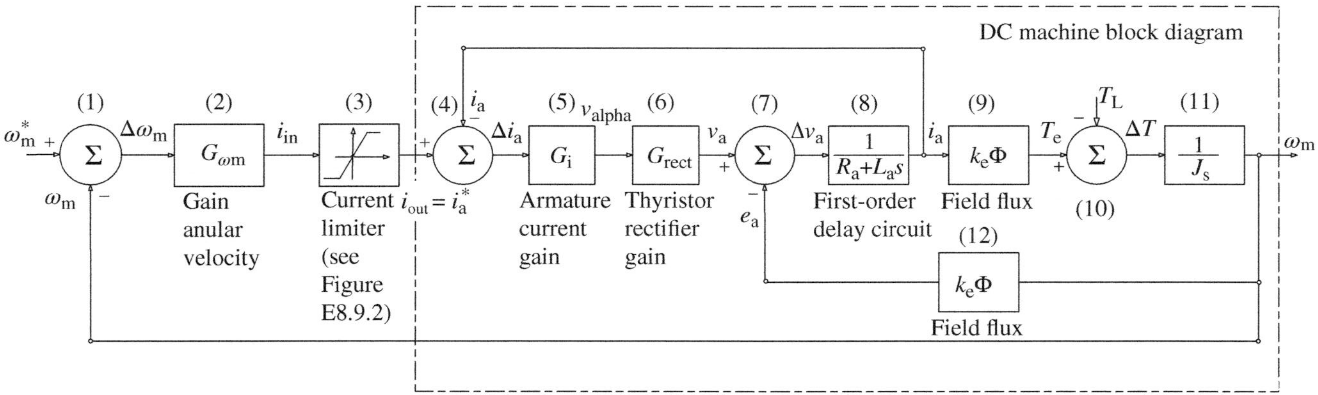

Figure E8.9.1 Block diagram of separately excited DC machine drive where the armature current $i_a(t)$ and the angular mechanical (m) velocity $\omega_m(t)$ are changed via P and PI control.

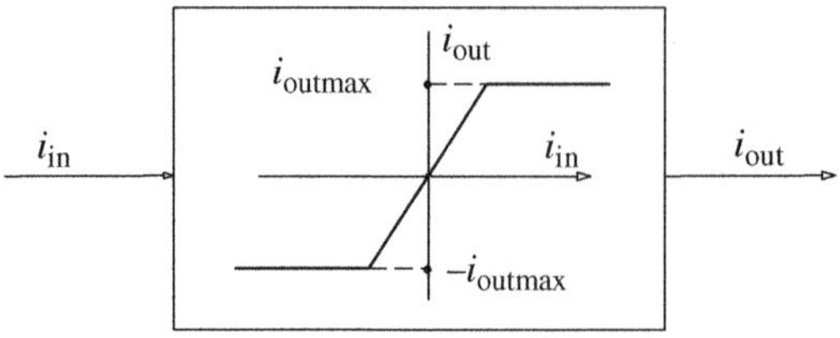

Figure E8.9.2 Definition of the parameter i_{outmax} of the current limiter.

Equation (E8.9.3) is explained in Figure E8.9.2 defining the limit of the output current i_{out} by $\pm i_{outmax}$.

4) Difference of reference i_a^* and actual i_a armature currents

$$\Delta i_a = i_a^* - i_a. \tag{E8.9.4}$$

5) Due to P-control current gain G_i generates from the current difference Δi_a the input control variable v_{alpha} to the thyristor rectifier [1, 14, 15, 26–27] controlled by the angle α:

$$v_{alpha} = G_i \Delta i_a. \tag{E8.9.5}$$

6) Thyristor rectifier with P control and gain G_{rect} produces from the input control variable v_{alpha} the thyristor rectifier approximate (with harmonics due to α) DC voltage v_a:

$$v_a = G_{rect} v_{alpha}. \tag{E8.9.6}$$

7) The difference of approximate DC voltage v_a and induced voltage e_a generates

$$\Delta v_a = v_a - e_a, \tag{E8.9.7}$$

where the induced voltage of the DC machine is

$$e_a = (k_e \Phi)\omega_m. \tag{E8.9.8}$$

8) The armature (subscript a) voltage–current relation of a DC machine is (Chapter 11)

$$v_a = R_a i_a + e_a + L_a \mathrm{d}i_a/\mathrm{d}t, \tag{E8.9.9}$$

or

$$\Delta v_a = v_a - e_a = R_a i_a + L_a s i_a = (R_a + L_a s) i_a, \tag{E8.9.10}$$

where $s = \mathrm{d}.../\mathrm{d}t$, and therefore

$$i_a = \left(\frac{1}{R_a + L_a s}\right)\Delta v_a. \tag{E8.9.11}$$

Equation (E8.9.11) is a first-order delay complex function [28] of the armature voltage Δv_a versus armature current i_a of a separately excited DC machine.

9) The electric (subscript e) torque T_e as a function of the armature current i_a is (Chapter 11)

$$T_e = k_e \Phi i_a. \tag{E8.9.12}$$

10) Difference between electric torque T_e and mechanical load (subscript L) torque T_L is

$$\Delta T = T_e - T_L, \tag{E8.9.13}$$

where the total load torque is

$$T_L = T_m + B\omega_m + T_c + c \cdot (\omega_m)^2 + T_s, \tag{E8.9.14}$$

T_m is the mechanical (subscript m) torque performing useful work, and the various frictional torques [1, 6] are $B\omega_m$, T_c, $c{\cdot}\omega_m{}^2$, and T_s, which do no useful work.

11) The angular velocity ω_m versus torque difference ΔT is based on the relation

$$\Delta T = J(\mathrm{d}\omega_m/\mathrm{d}t) \tag{E8.9.15}$$

or

$$\omega_m = \int \frac{\Delta T}{J} \mathrm{d}t, \tag{E8.9.16}$$

where J is the total mechanical inertia of the drive and $s = \mathrm{d}(\ldots)/\mathrm{d}t$. Note that Eqs. (E8.9. 1)–(E8.9.8) and (E8.9.11)–(E8.9.15) are represented in the block diagram (Figure E8.9.1).

12) The field flux $k_e\phi$ induces the voltage e_a as a function of angular velocity ω_m and provides a feedback path.

DC machines serve as a role model for control of angular velocity and torque. Their design parameters can be obtained from magnetic field calculations [29–31]. All graphs as shown in Figures E8.9.3–E8.9.11 are calculated with the Mathematica [32] program as listed in Table E8.9.1 for the parameters

$$\begin{gathered}\omega_m^* = 100\,\mathrm{rad/s}, G_{\omega m} = 20\,\mathrm{pu}, i_{outmax} = -200/200\,\mathrm{A}, G_i = 5\,\mathrm{pu}, G_{rect} = 15\,\mathrm{pu}, \\ R_a = 0.1\,\Omega, L_a = 0.005\,\mathrm{H}, k_e\Phi = 2.5\,\mathrm{V\,s/rad}, J = 0.6\,\mathrm{Nm\,s/rad}, T_m = 200\,\mathrm{Nm}, \\ T_c = 5\,\mathrm{Nm}, B = 0.1\,\mathrm{Nm\,s/rad}, c = 0.002\,\mathrm{Nm\,(s)^2/(rad)^2}, \text{and } T_s = 0\,\mathrm{Nm}.\end{gathered} \tag{E8.9.17}$$

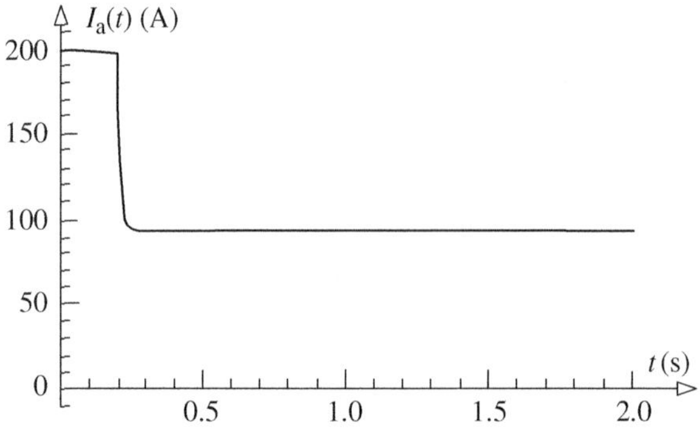

Figure E8.9.3 DC machine armature I_a (A) with current limiter of Figure E8.9.2 for the parameters as listed in Eq. (E8.9.17) from 0 to 2 s P control with rated angular velocity gain $G_{\omega m}$ = 20 pu.

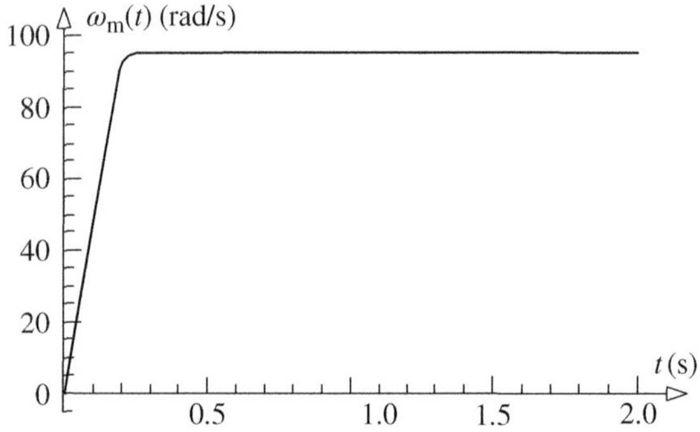

Figure E8.9.4 DC machine angular velocity ω_m (rad/s) with current limiter of Figure E8.9.2 for the parameters as listed in Eq. (E8.9.17) from 0 to 2 s P control with rated angular velocity gain $G_{\omega m}$ = 20 pu.

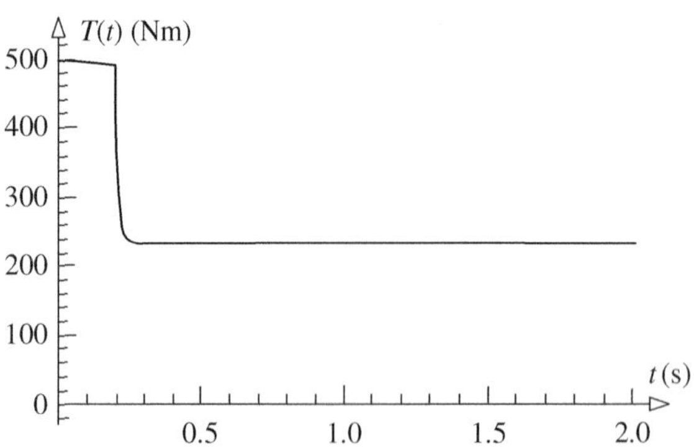

Figure E8.9.5 DC machine torque T (Nm) with current limiter of Figure E8.9.2 for the parameters as listed in Eq. (E8.9.17) from 0 to 2 s P control with rated angular velocity gain $G_{\omega m}$ = 20 pu.

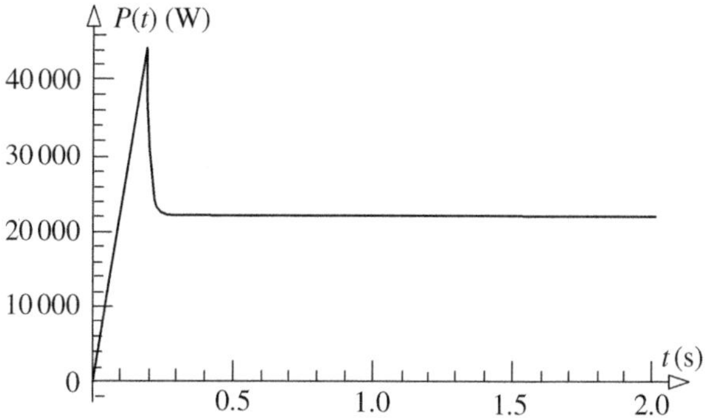

Figure E8.9.6 DC machine output power P (W) with current limiter of Figure E8.9.2 for the parameters as listed in Eq. (E8.9.17) from 0 to 2 s P control with rated angular velocity gain $G_{\omega m}$ = 20 pu.

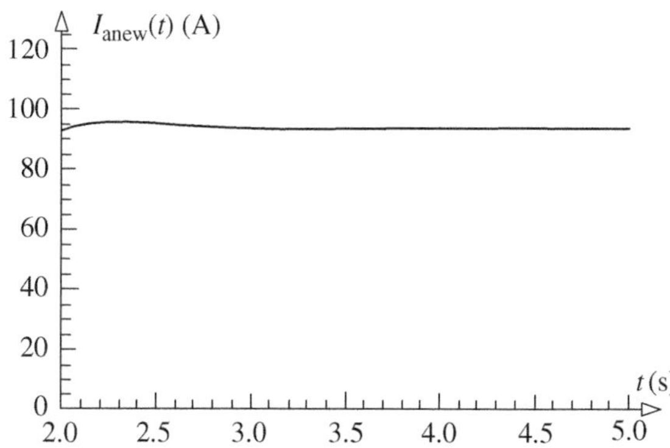

Figure E8.9.7 DC machine armature I_{anew} (A) with current limiter of Figure E8.9.2 for the parameters as listed in Eq. (E8.9.17) from 2 to 5 s PI control with rated angular velocity gain $G_{\omega m}$ = 20 pu.

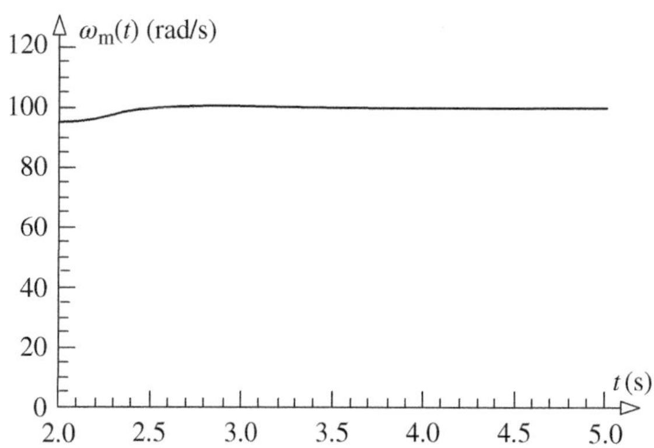

Figure E8.9.8 DC machine angular velocity ω_m (rad/s) with current limiter of Figure E8.9.2 for the parameters as listed in Eq. (E8.9.17) from 2 to 5 s PI control with rated angular velocity gain $G_{\omega m}$ = 20 pu.

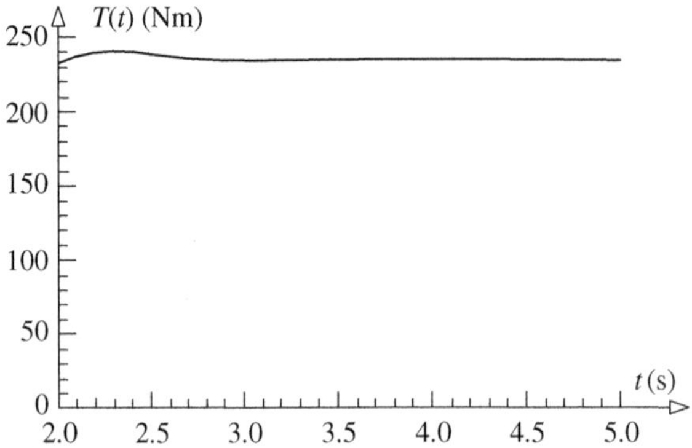

Figure E8.9.9 DC machine torque T (Nm) with current limiter of Figure E8.9.2 for the parameters as listed in Eq. (E8.9.17) from 2 to 5 s PI control with rated angular velocity gain $G_{\omega m}$ = 20 pu.

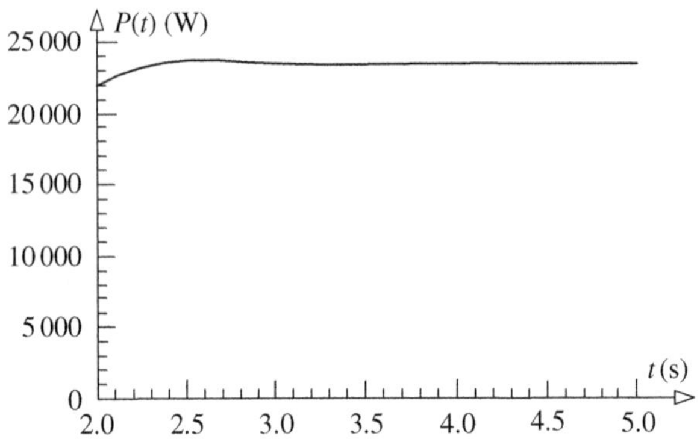

Figure E8.9.10 DC machine output power P (W) with current limiter of Figure E8.9.2 for the parameters as listed in Eq. (E8.9.17) from 2 to 5 s PI control with rated angular velocity gain $G_{\omega m}$ = 20 pu.

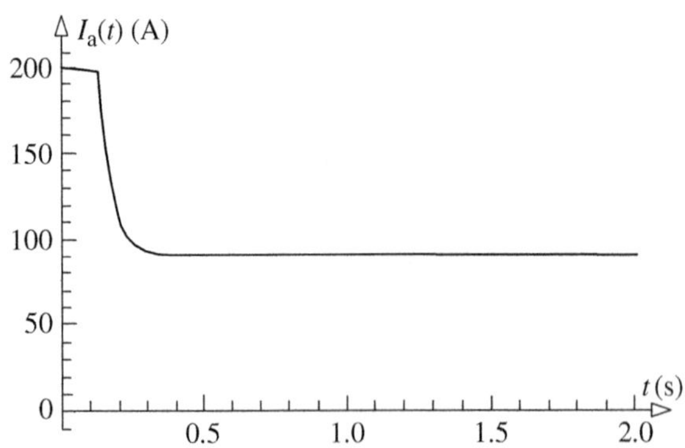

Figure E8.9.11 DC machine armature I_a (A) with current limiter of Figure E8.9.2 for the parameters as listed in Eq. (E8.9.17) from 0 to 2 s P control, but with reduced angular velocity gain $G_{\omega m}$ = 5 pu.

Table E8.9.1 Mathematica [32] program from $t = 0$ to 2 s for P control, and from 2 to 5 s for PI control.

```
Remove[T,Wmstar,eqn1,eqn2,eqn3,eqn4,
eqn5,ic1,ic2,ic3,ic4,ic5];
La=0.005;
Ra=0.1;
kphi=2.5;
J=0.6;
Tc=5;
B=0.1;
c=0.002;
Ts=0;
Grect=15;
Gi=5.0;
Gwm=20.0;
(*Gwm=10.0*)
(*Gwm=5.0*)
Tm=200;
Wmstar=100;
ic1=Ia[0]==0;
ic2=Wm[0]==0;
Valpha[t_]:=Gi*(Iastar[t]-Ia[t]);
Iastar[t_]:=Clip[Gwm*(Wmstar-Wm[t]),
{-200,200},{-200,200}];
Va[t_]:=Grect*Valpha[t];
T[t_]:=kphi*Ia[t];
P[t_]:=T[t]*Wm[t];
eqn1=Ia'[t]==(1/La)*(Va[t]-Ra*Ia[t]-
kphi*Wm[t]);
eqn2=Wm'[t]==(1/J)*(kphi*Ia[t]-Tm-
B*Wm[t]-Tc-c*(Wm[t]^2)-Ts);
sol1=NDSolve[{eqn1,eqn2,ic1,ic2},{Ia[t],
Wm[t]},{t,0,2},MaxSteps-> 10000,Method-
> {"ExplicitRungeKutta","StiffnessTest"->
False}];
Plot[Ia[t]/.sol1[[1]],{t,0,2},PlotRange-
>All,AxesLabel->{"t (s)","Ia (A)"}]
```

```
Plot[Wm[t]/.sol1[[1]],{t,0,2},PlotRange-
>All,AxesLabel->{"t (s)","Wm (rad/s)"}]
Plot[Evaluate[T[t]]/.sol1[[1]],{t,0,2},
PlotRange->All,AxesLabel->{"t (s)","T
(Nm)"}]
Plot[Evaluate[P[t]]/.sol1[[1]],{t,0,2},
PlotRange->All,AxesLabel->{"t (s)","P (W)"}]
Wmb=(Wm[t]/.sol1[[1]])/.t->2;
Iab=(Ia[t]/.sol1[[1]])/.t->2;
Iastarb=(Iastar[t]/.sol1[[1]])/.t->2;
ic3=Wm[2]==Wmb;
ic4=Ianew[2]==Iab;
ic5=Iastarnew[2]==Iastarb;
Va[t_]:=Grect*Valpha[t];
Valpha[t_]:=Gi*(Iastarnew[t]-Ianew[t]);
T[t_]:=kphi*Ianew[t];
P[t_]:=T[t]*Wm[t];
eqn3=Wm'[t]==(1/J)*(kphi*Ianew[t]-Tm-
B*Wm[t]-Tc-c*(Wm[t]^2)-Ts);
eqn4=Ianew'[t]==(1/La)*(Va[t]-Ra*Ianew
[t]-kphi*Wm[t]);
(*PI-control*)
eqn5=Iastarnew'[t]==(Wmstar-Wm[t])/0.2-
Wm'[t];
sol2=NDSolve[{eqn3,eqn4,eqn5,ic3,ic4,ic5},
{Wm[t],Ianew[t],Iastarnew[t]},{t,2,5},
MaxSteps->10000];
Plot[Ianew[t]/.sol2[[1]],{t,2,5},PlotRange->
{2,5} {0,120},AxesLabel->{"t(s)","Ianew(A)"}]
Plot[Wm[t]/. sol2[[1]], {t, 2, 5}, PlotRange
      → {2, 6}{0,120}, AxesLabel
      → {"t (s) " , " Wm (rad/s)"}]
Plot[Evaluate[T[t]/. sol2[[1]]], {t, 2, 5}, □
PlotRange → {2, 5}{0,250}, AxesLabel
      → {"t (s) " , " T (Nm)"}]
Plot[Evaluate[P[t]/. sol2[[1]]], {t, 2, 5},
PlotRange → {2, 5}{0, 25000}, AxesLabel
      → {"t (s) " , " P (W)"}]
```

Note: all calculation statements must be on one line and terminated with a semicolon (;).

b) *PI control of angular velocity ω_m from 2 to 5 seconds.*

Figures E8.9.12–E8.9.18 are computed for the parameters as given in Eq. (E8.9.17), but with $G_{\omega m} = 5$ pu. The above discussed P control for the angular velocity ω_m is now replaced in the second part of the Mathematica program by PI

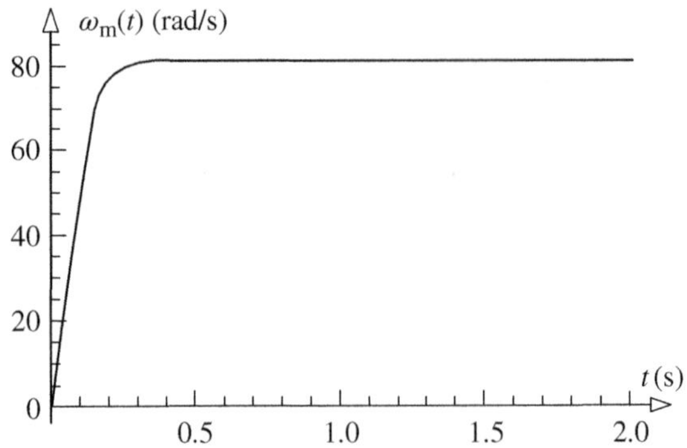

Figure E8.9.12 DC machine angular velocity ω_m (rad/s) with current limiter of Figure E8.9.2 for the parameters as listed in Eq. (E8.9.17) from 0 to 2 s P control, but with reduced angular velocity gain $G_{\omega m}$ = 5 pu.

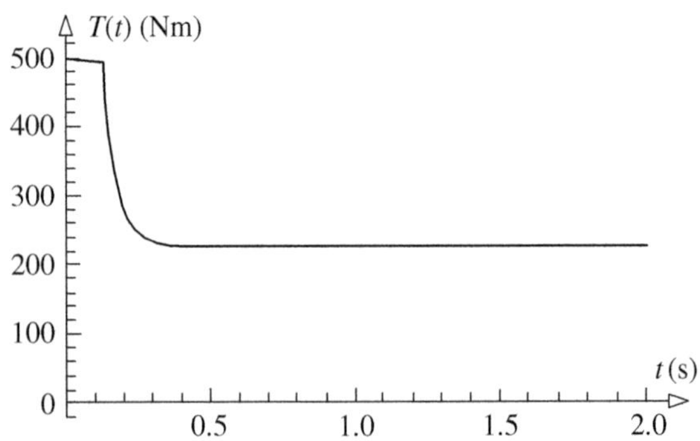

Figure E8.9.13 DC machine torque T (Nm) with current limiter of Figure E8.9.2 for the parameters as listed in Eq. (E8.9.17) from 0 to 2 s P control, but with reduced angular velocity gain $G_{\omega m}$ = 5 pu.

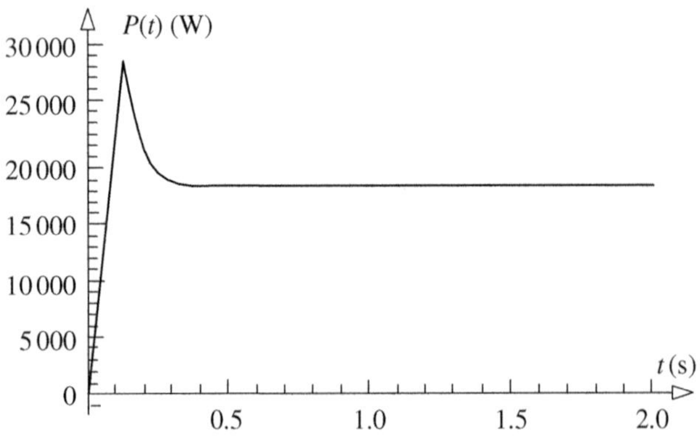

Figure E8.9.14 DC machine output power P (W) with current limiter of Figure E8.9.2 for the parameters as listed in Eq. (E8.9.17) from 0 to 2 s P control, but with reduced angular velocity gain $G_{\omega m}$ = 5 pu.

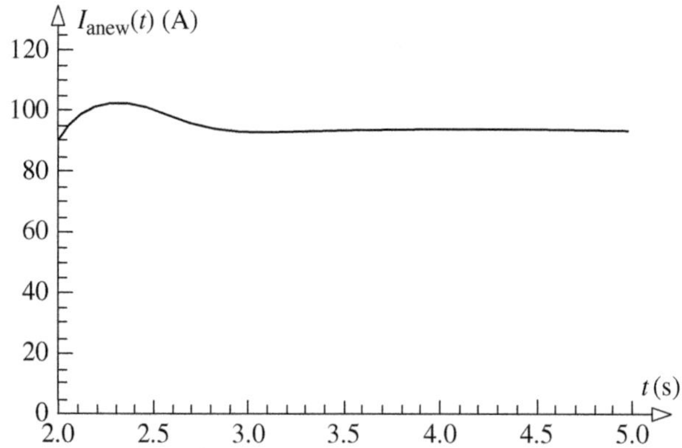

Figure E8.9.15 DC machine armature I_{anew} (A) with current limiter of Figure E8.9.2 for the parameters as listed in Eq. (E8.9.17) from 2 to 5 s PI control, but with reduced angular velocity gain $G_{\omega m} = 5$ pu.

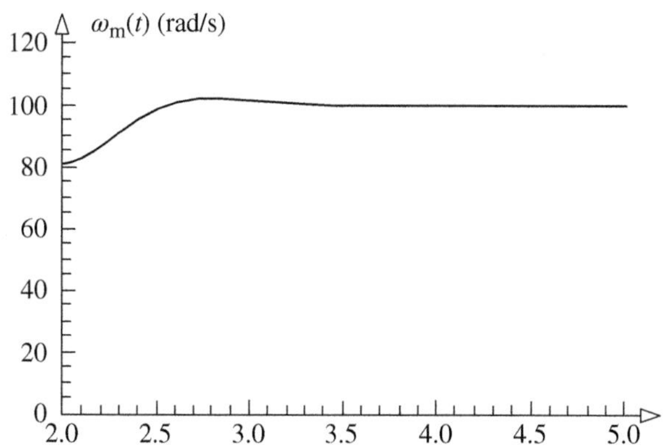

Figure E8.9.16 DC machine angular velocity ω_m (rad/s) with current limiter of Figure E8.9.2 for the parameters as listed in Eq. (E8.9.17) from 2 to 5 s PI control, but with reduced angular velocity gain $G_{\omega m} = 5$ pu.

control by employing the additional differential equation [1], which replaces the $G_{\omega m}$ block of Figure E8.9.1 by the ordinary differential equation

$$\frac{di_a^*}{dt} = \frac{(\omega_m^* - \omega_m)}{0.2} - \frac{d\omega_m}{dt}. \tag{E8.9.18}$$

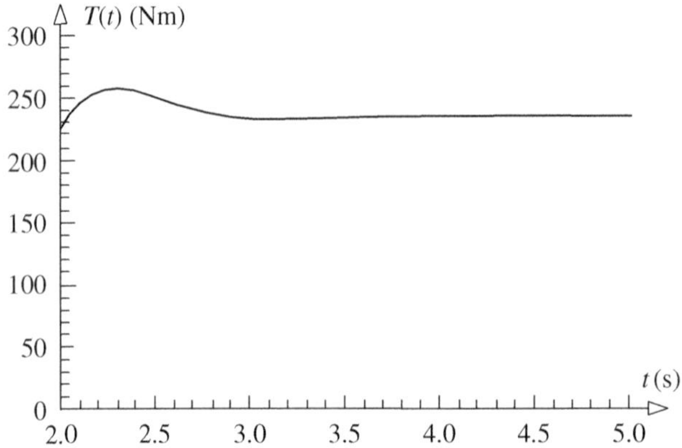

Figure E8.9.17 DC machine torque T (Nm) with current limiter of Figure E8.9.2 for the parameters as listed in Eq. (E8.9.17) from 2 to 5 s PI control, but with reduced angular velocity gain $G_{\omega m}$ = 5 pu.

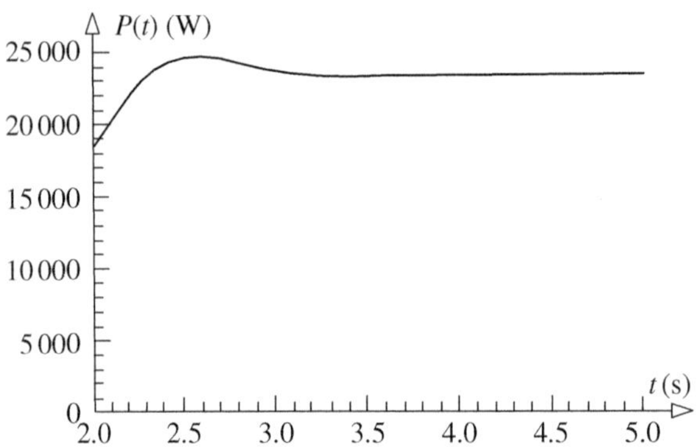

Figure E8.9.18 DC machine output power P (W) with current limiter of Figure E8.9.2 for the parameters as listed in Eq. (E8.9.17) from 2 to 5 s PI control, but with reduced angular velocity gain $G_{\omega m}$ = 5 pu.

c) *P control from 0 to 2 seconds, and PI control from 2 to 5 seconds, respectively.*

Combined results for P control of angular velocity ω_m (see first part of Mathematica program of Table E8.9.1) and PI control (see second part of Mathematica program of Table E8.9.1) are listed below (Mathematica program and graphs).

Table E8.9.1 lists the Mathematica [32] program. Note that Mathematica and MATLAB [33] can handle derivatives that occur in differential equations, while

PSPICE [34] is set up to mostly model algebraic equations. Note that when a Mathematica program is copied from the notebook (nb) file with extension .nb as listed in Table E8.9.1, the program list might not be preserved, and it is best to retype the entire program and not copy the program from either a document (.docx) or portable document file (.pdf).

How do we run the software Mathematica [32]?

1) It is important to realize that any software program is being continually improved; presently, the newest version to date is Mathematica 11.2.
2) All Mathematica files have the extension .nb or notebook (.nb).
3) When entering a line or equation, a single equal sign (=) sets the value of a parameter, whereas a double equal sign (= =) means "equal" and relates to equations in Mathematica.
4) All functions of t are defined by x[t_]:=, where the two parts are t_ and :=, which are required whenever defining a function.
5) To run the program, one can execute the program one instruction at a time or have the computer go through all instructions by going to Evaluation → Kernel Quit → Local → Evaluate Notebook.
6) Any red or pink messages from Mathematica signify an error.
7) Any blue messages represent comments, which are not part of the program to be executed.

This example demonstrates that the steady-state error $\varepsilon_{\omega m} = \left(\omega_m^* - \omega_m\right)/\omega_m^* = 1/(1 + G_{\omega m})$ of the control variable $\omega_m(t)$ depends upon the angular velocity gain $G_{\omega m}$ (see Figure E8.9.1) during the time from 0 to 2 seconds with P control, as shown in Figure E8.9.19.

Figures E8.9.8 and E8.9.16 illustrate that from 2 to 5 seconds for PI control, the steady-state error of the angular velocity $\omega_m(t)$ is $\varepsilon_{\omega m} = 0$, because $\omega_m = \omega_m^*$ at $t = 5$ s where steady-state operation is reached.

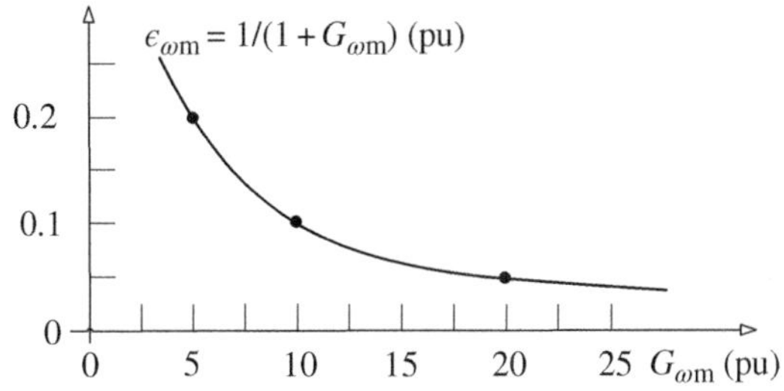

Figure E8.9.19 Steady-state error $\varepsilon_{\omega m} = \left(\omega_m^* - \omega_m\right)/\omega_m^* = 1/(1 + G_{\omega m})$ for angular velocity $\omega_m(t)$ for P control from 0 to 2 s as a function of the angular velocity gain $G_{\omega m}$.

Similar considerations apply to the rectifier G_{rect} and current G_i gains (see Figure E8.9.1) with respect to the steady-state errors ε_{rect} and ε_i, respectively.

Brushless DC Machine [1, 22–25] Used in a Variable-Speed Drive
Permanent-magnet machines can be designed (Chapter 12) with a high power density [8, 9] measured in kW/kg, where an output power density of more than 4 kW/kg can be reached. Brushless DC machines (either motor or generator) are three-phase permanent-magnet machines fed by a PWM converter operating at a high switching frequency f_s, whereby the machine speed n_m can be changed from very low values of $0.01n_{m\_rated}$ to high values of $3n_{m\_rated}$ by relying on flux weakening (FW) [22–25]. A compensation of flux weakening (CFW) [10, 11] is possible, which permits a further increase of the torque up to a maximum of $6n_{m\_rated}$.

Polyphase Induction Machine [1, 12–21] Employed in a Variable-Speed Drive
Compensation of flux weakening [10, 11] as proposed for brushless DC machines can be applied to three-phase induction machines as well. The advantage of three-phase induction machines together with current-controlled voltage-source inverters (Chapter 12) is that their controllability [1, 6, 7, 12–16] is similar to that of permanent-magnet machines and their design [8, 9, 22–25], but is less expensive and not as complicated as brushless DC machines.

8.11.2 I Controller

I controller consists of two OP amplifiers: the first one achieves the integral (I) control relying on an inverting configuration such that

$$v_{out1} = -\frac{1}{R_1 C}\int_0^t v_{in}\,dt, \tag{8.48}$$

that is, the output is the integral of the input voltage. The second OP amplifier is called an inverter, which inverts v_{out1}, resulting in

$$v_{out2} = -\frac{\left(\frac{R_3}{R_2}\right)}{R_1 C}\int_0^t v_{in}\,dt. \tag{8.49}$$

Therefore it is assumed that the initial capacitor charge is zero; that is, $v_{out1}(t = 0) = 0$, which is the voltage of the capacitor C at time $t = 0$. Note that the I controller as shown in Figure 8.17a is not frequently relied on because the gain of the controller is not variable. The block diagram of the I-controller of Figure 8.17a is represented in Figure 8.17b.

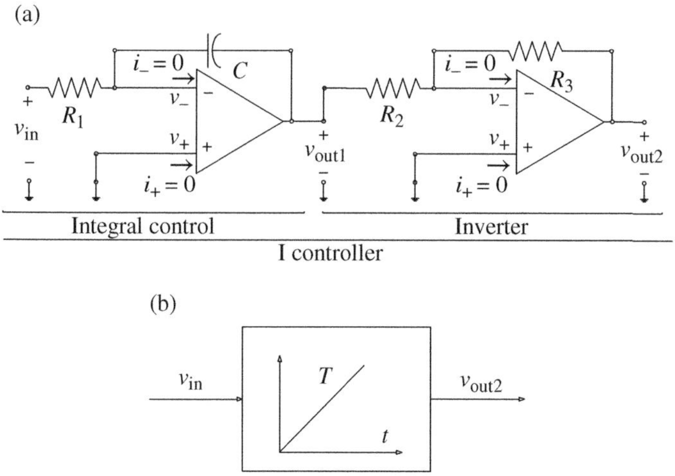

Figure 8.17 (a) I-controller network with time constant $T = R_1C$ and closed-loop voltage gain $G_v = v_{out2}/v_{in}$. (b) Symbolic representation of I-controller network of (a) with time constant $T = R_1C$ and closed-loop voltage gain $G_v = v_{out2}/v_{in}$.

8.11.3 PI Controller

The PI controller, see Figure 8.18a, consists of three OP amplifiers: the P controller in parallel to the I controller and a third one to invert the voltage signal. The first one provides the variable gain and the second one the integral property; both control networks rely on the inverting configuration such that the third OP amplifier is used to invert the output voltage of the two parallel-connected P and I controllers. A PI-controller with combined P- and I-networks is given in Figure 8.18b.

8.11.4 D Controller

The D controller as shown in Figure 8.19a consists of two OP amplifiers: the first one achieves the differential (D) control relying on an inverting configuration such that

$$v_{in} = (1/C)\int i_c dt + i_c R_1 \tag{8.50}$$

or

$$i_c = -v_{out1}/R_2 \tag{8.51}$$

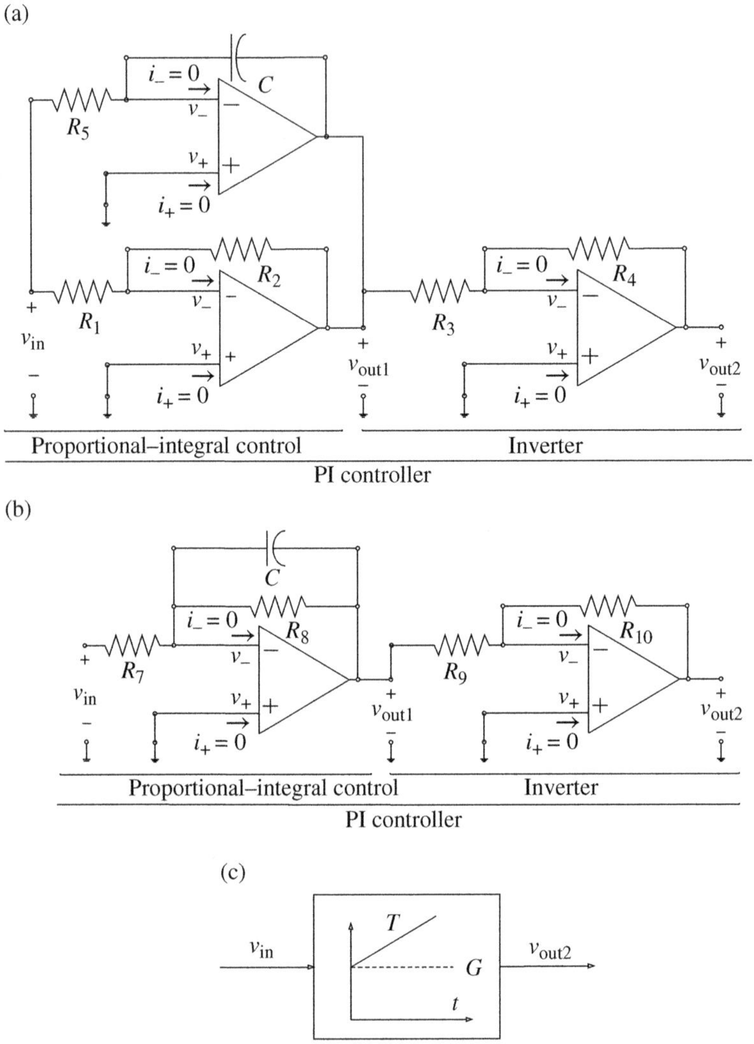

Figure 8.18 (a) PI-controller network with proportional–integral control connected in parallel using two OP amplifiers with closed-loop voltage gain $G_v = v_{out2}/v_{in}$ and time constant $T = R_5C$. (b) PI-controller network with combined P and I controllers relying on one OP amplifier with closed-loop voltage gain $G_v = v_{out2}/v_{in}$ and time constant $T = R_7C$. (c) Symbolic representation of IP-controller network of (b) with closed-loop voltage gain $G_v = v_{out2}/v_{in}$ and time constant $T = R_7C$.

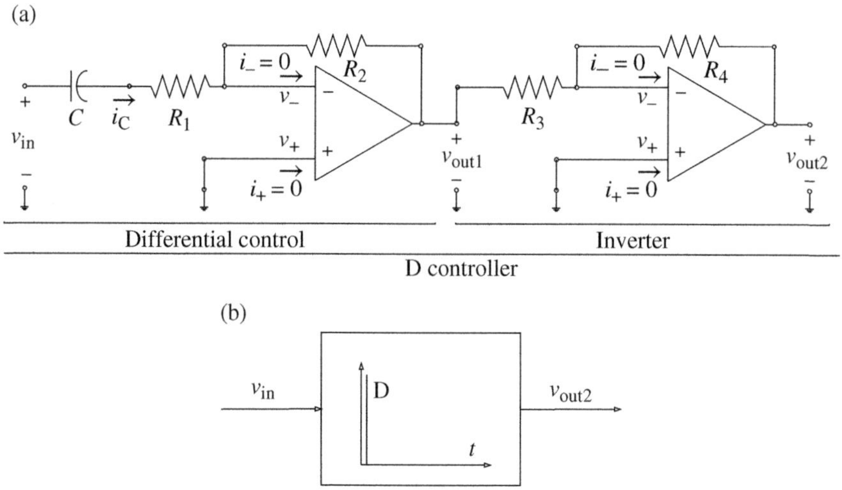

Figure 8.19 (a) D controller. (b) Symbolic representation of D controller of (a).

$$v_{in} = (1/C)\int i_c dt - R_1(v_{out1}/R_2) = -(1/C)\int (v_{out1}/R_2)dt - (R_1/R_2)v_{out1} \tag{8.52}$$

$$dv_{in}/dt = -(1/CR_2)v_{out1} - (R_1/R_2)dv_{out1}/dt \tag{8.53}$$

$$-v_{out1}/CR_2 = dv_{in}/dt + (R_1/R_2)dv_{out1}/dt \tag{8.54}$$

$$v_{out1} = -CR_2[dv_{in}/dt + (R_1/R_2)dv_{out1}/dt]. \tag{8.55}$$

Resistor $R_1 = R_c$ represents the capacitor loss resistance; for $R_c \to 0$ one obtains from Eq. (8.55)

$$v_{out1} \approx -CR_2 dv_{in}/dt; \tag{8.56}$$

that is, the output is the differential of the input voltage. The second OP amplifier is called an inverter, which inverts v_{out1}, resulting in

$$v_{out2} = \left(\frac{R_4}{R_3}\right) CR_2 \frac{dv_{in}}{dt}. \tag{8.57}$$

The block diagram of D-controller is shown in Figure 8.19b.

The D controller is not useful, as explained above; however, the D component will be relied on for PID control, discussed below.

8.11.5 PID Controller

The PID controller, see Figure 8.20a, consists of four OP amplifiers: the P controller in parallel to the I controller and D controller and a fourth one to invert the voltage

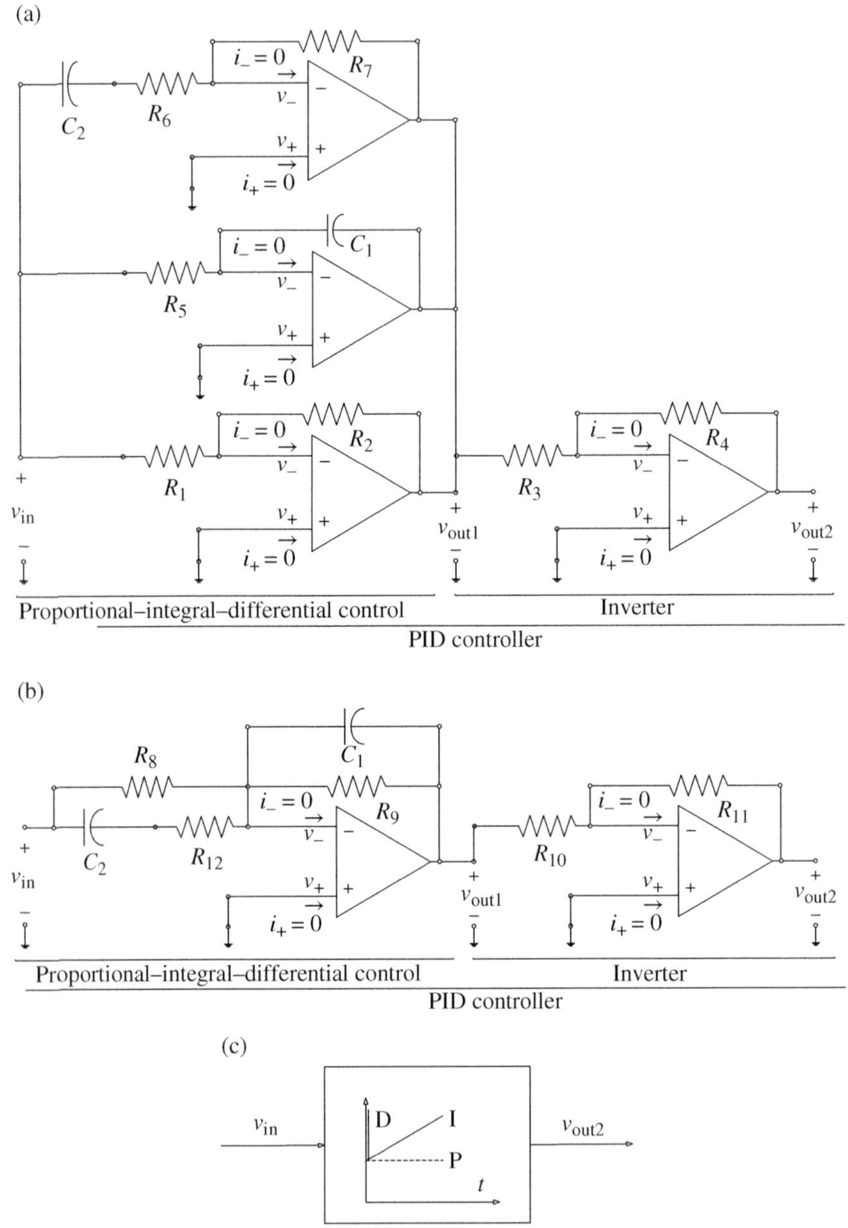

Figure 8.20 (a) PID controller with P, I, and D circuits connected in parallel using three OP amplifiers with one OP amplifier to invert the signal. $R_6 = R_c \rightarrow 0$ represents the capacitor loss resistor. (b) PID-controller network with combined P, I, and D circuits relying on one OP amplifier. $R_{12} = R_c \rightarrow 0$ represents the capacitor loss resistor. (c) Symbolic representation of PID controller of Figure 8.18a and b.

signal. The first one provides the variable gain and the second one the integral property; the third one anticipates the change/slope of the signal to be controlled. P-, I-, and D-control networks rely on the inverting configuration such that the fourth OP amplifier is used to invert the output voltage of the three parallel-connected P, I, and D controllers. A PID-controller with combined P-, I-, and D-networks is given in Figure 8.20b. The block diagram of the PID-controller of Figure 8.20a and b is represented in Figure 8.20c.

Application Example 8.10 PID control

The PID controller represents a frequently applied tool to control variables: P-control component provides through the gain a fast response, the I-control component achieves a steady-state error of zero, and the D-control component anticipates the change of the control variable by relying on its derivative so that the overall control response becomes faster. Figure E8.10.1 shows the schematic of such a control approach.

Solution

Table E8.10.1 represents the PSPICE [34] program for Figure E8.10.1, where P control through R_{p2} is possible; I control is possible through capacitor C_i, and D control is performed by capacitance C_d.

A controlled voltage v_{Rload} = V(17) response through $R_L = R_{load} = 100\,\Omega$ for $R_{p2} = 20\,k\Omega$, $C_i = 0.1\,\mu F$, $C_d = 10\,mF$ is shown in Figure E8.10.2 for input rectangular PULSE voltage ranging from −7.5 to 5 V {vs 1 0 PULSE(−7.5v 5v 0 0 0 100ms 200ms)}, where the steady-state error is small and transient oscillations are also small.

A controlled voltage v_{Rload} = V(17) response through $R_L = R_{load} = 100\,\Omega$ for $R_{p2} = 1\,k\Omega$, $C_i = 0.1\,\mu F$, $C_d = 0.1\,\mu F$ is shown in Figure E8.10.3 for input rectangular PULSE voltage ranging from −7.5 to 5 V {vs 1 0 PULSE(−7.5v 5v 0 0 0 100ms 200ms)}, where the steady-state error is large and transient oscillations are also large.

A controlled voltage v_{Rload} = V(17) response through $R_L = R_{load} = 100\,\Omega$ for $R_{p2} = 1\,k\Omega$, $C_i = 0.1\,\mu F$, $C_d = 10\,mF$ is shown in Figure E8.10.4 for input rectangular PULSE voltage ranging from −7.5 to 5 V {vs 1 0 PULSE(−7.5v 5v 0 0 0 100ms 200ms)}, where the steady-state error ε is large and transient oscillations are somewhat large.

A controlled voltage v_{Rload} = V(17) response through $R_L = R_{load} = 100\,\Omega$ for $R_{p2} = 1\,k\Omega$, $C_i = 10\,mF$, $C_d = 0.1\,\mu F$ is shown in Figure E8.10.5 for input rectangular PULSE voltage ranging from −7.5 to 5 V {vs 1 0 PULSE(−7.5v 5v 0 0 0 100ms 200ms)}, where the steady-state error ε is very large and transient oscillations are relatively small.

Table E8.10.1 PSPICE program for Figure E8.10.1.

```
*OP Amplifier based PID controller for
*inductive/capacitive (e.g., inductance is in
*parallel with capacitance), resistive load.
* Control variable is the load current through
* RL=Rload.
* Input rectangular PULSE voltage with -7.5 V
*and 5V magnitudes.
vs      1     0     PULSE(-7.5v 5v 0 0 0 +100ms
200ms)
*
* For Rp2=20k ohms the steady-state error
*is very small, and for Rp2=1k ohms the
*steady-state error is large.
*
*For Cd=10mF transient oscillation is smaller
*than for 0.1uF, but steady-state error is larger
*than for Rp2=20k ohms.
*
*For Ci=10mF transient oscillation is smallest,
*and steady-state error tends to become zero.
*
* Summation of positive input and negative
*feedback. Values for Fig. E8.10.5.
Rs1     1    3     5k
Rs2     19   3     5k
Rs3     3    4     5k
XOP1    0 3 4      OPAMP1
* P - proportional (p) controller.
Rp1     4    5     1k
*Rp2    5    6     20k
Rp2     5    6     1k
XOP2    0 5 6      OPAMP1
* D - differentiating (d) controller.
Cd 4 7 0.1uF IC=0
*Cd     4    7     10mF IC=0
Rc      7    8     200
Rd      8    9     1k
XOP3    0 8 9      OPAMP1
* I - integral (i) controller.
Ri      4      10     200
*Ci     10     11     0.1uF IC=0
Ci      10     11     10mF IC=0
XOP4    0 10 11       OPAMP1
* Summation of above three
*controller outputs.
Rs4     6    12    10k
Rs5     9    12    10k
Rs6     11     12     10k
Rs7     12     13     10k
Rs8     13     14     10k
XOP5    0 13 14       OPAMP1
* Signal inversion
Rin1    14     15     10k
Rin2    15     16     10k
XOP6    0 15 16       OPAMP1
*Load consisting of inductor,
*capacitor, and resistor.
Lload 16 17 6m
Cload  16   17 50uF IC=0
Rload  17   0  100
*Signal inversion from load.
Rin3 17 18 10k
Rin4 18 19 10k
XOP7 0 18 19 OPAMP1
* OP amplifier model for maximal
*15V.
* connections: non-inverting input
*                | inverting input
*                | | output
*                | | |
.SUBCKT OPAMP1       1 2 6
* Input impedance.
RIN     1     2      10MEG
* DC gain=100K and pole1=100Hz
* Unity gain = DCgain X pole1 =
*10MHz.
EGAIN   3  0   1  2   100k
RP1     3     4      100k
CP1     4     0      0.0159uF
* Zener Diode.
D1      4     7      DZ
D2      0     7      DZ
* Output buffer and resistance.
EBUFFER     5 0     4 0     1
ROUT    5     6      10
.model      DZ D(Is=0.05u Rs=0.1 +Bv=15
Ibv=0.05u)
.ENDS
* Transient analysis.
.TRAN       0.1mS 300mS
* Printing and plotting.
.PRINT TRAN      V(1),V(17)
.PROBE
.END
```

Note: All comment lines must be preceded with an asterisk (*) while continued calculation statements must be preceded with a plus sign (+).

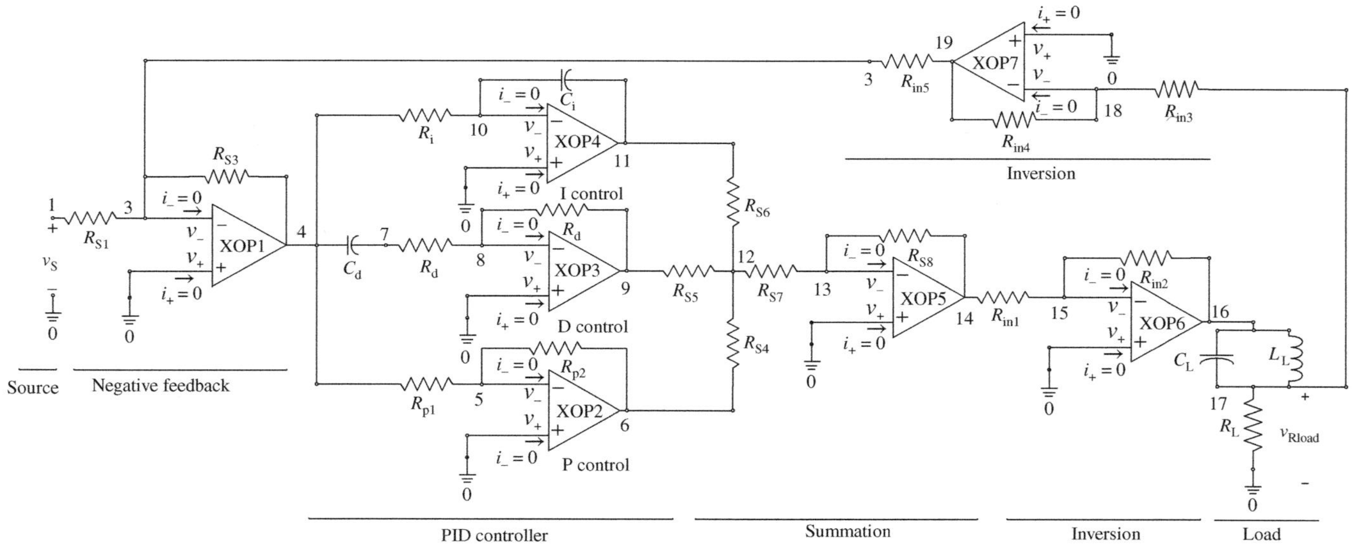

Voltage control of inductive/capacitive–resistive load for rectangular input source v_S

Figure E8.10.1 PID control of current through load resistor R_L with seven ideal linear OP amplifiers.

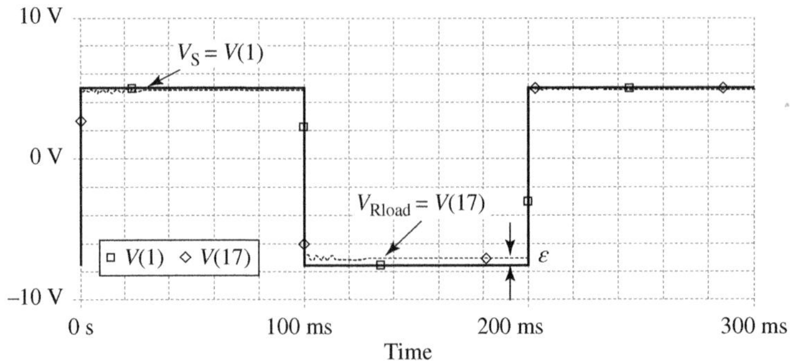

Figure E8.10.2 Controlled voltage response $v_{Rload} = V(17)$ across $R_L = R_{load} = 100\ \Omega$ for $R_{p2} = 20\ k\Omega$, $C_i = 0.1\ \mu F$, $C_d = 10\ mF$ for $v_s = V(1)$; v_{Rload} has a steady-state error $\varepsilon = 1/(1 + G)$ where G is the overall gain of the control network. The steady-state error ε is small, and transient oscillations are small.

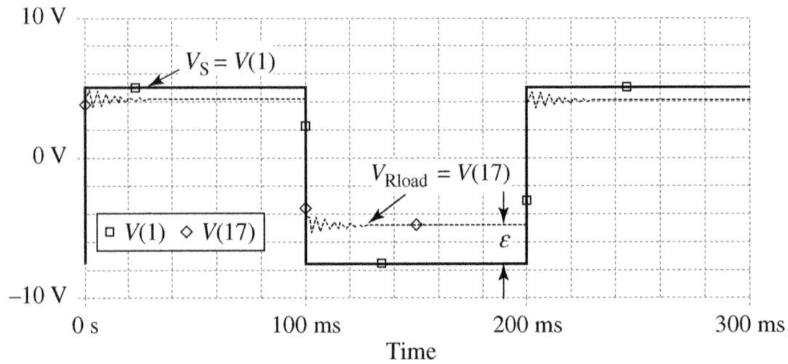

Figure E8.10.3 Controlled voltage $v_{Rload} = V(17)$ response through $R_L = R_{load} = 100\ \Omega$ for $R_{p2} = 1\ k\Omega$, $C_i = 0.1\ \mu F$, $C_d = 0.1\ \mu F$ for $v_s = V(1)$; v_{Rload} has a steady-state error $\varepsilon = 1/(1 + G)$ where G is the overall gain of the control network. Steady-state error ε is large, and transient oscillations are large.

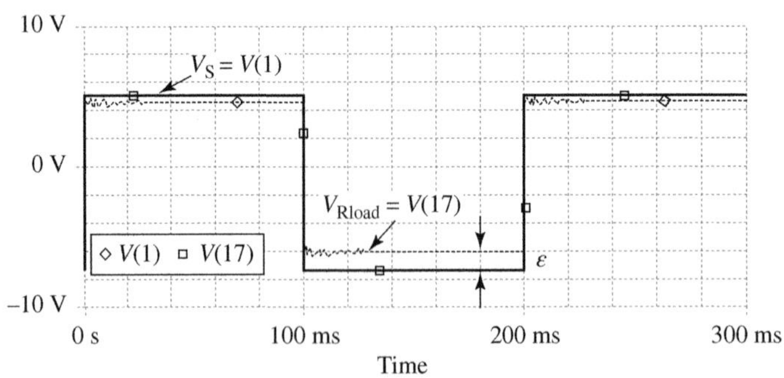

Figure E8.10.4 Controlled voltage $v_{Rload} = V(17)$ response through $R_L = R_{load} = 100\ \Omega$ for $R_{p2} = 1\ k\Omega$, $C_i = 0.1\ \mu F$, $C_d = 10\ mF$ for $v_s = V(1)$; v_{Rload} has a steady-state error $\varepsilon = 1/(1 + G)$ where G is the overall gain of the control network. Steady-state error ε is large, and transient oscillations are somewhat large.

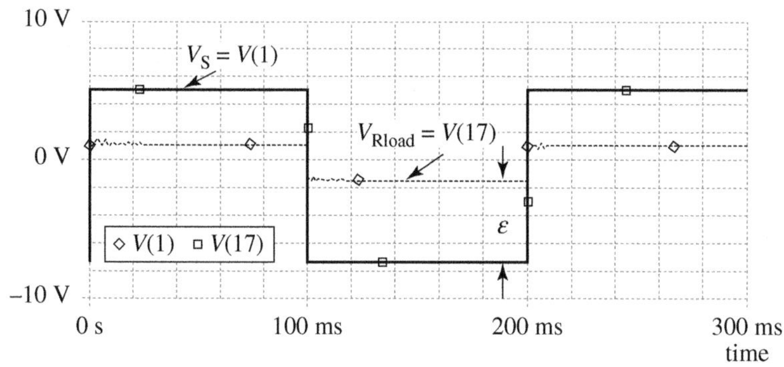

Figure E8.10.5 Controlled voltage $v_{Rload} = V(17)$ response through $R_L = R_{load} = 100\ \Omega$ for $R_{p2} = 1\ k\Omega$, $C_i = 10$ mF, $C_d = 0.1\ \mu F$ for $v_s = V(1)$; v_{Rload} has a steady-state error $\varepsilon = 1/(1 + G)$ where G is the overall gain of the control network. Steady-state error ε is large, and transient oscillations are relatively small.

Conclusions

In general, the larger the gain of the control variable is (e.g. current through $R_L = R_{load}$), the smaller the steady-state error [1, 6, 7]; and the larger the capacitance C_i of the integrating component of the control circuit is, the smaller the transient oscillation of the control variable. However, because the phase margin [1, 6, 7] becomes smaller, the stability of the control is reduced.

8.11.6 PD Controller

The PD controller, see Figure 8.21a, consists of three OP amplifiers: the P controller in parallel to the D controller and a third one to invert the voltage signal. The first one provides the variable gain and the second one the differential property; both control networks rely on the inverting configuration such that the third OP amplifier is used to invert the output voltage of the two parallel-connected P and D controllers. A PD-controller with combined P- and D-networks is given in Figure 8.21b. The block diagram of the PD-controller of Figure 8.21a and b is represented in Figure 8.21c.

8.12 Summary

Ideal OP electronic amplifiers are important analog building components for designing controllers and active filters. The ideal OP amplifier has two inputs (v_+, v_-) and one output. The two inputs have an infinitely large input impedance so that the input currents $i_+ = i_- = 0$ and their input voltages (e.g. v_{in1}, v_{in2}) are

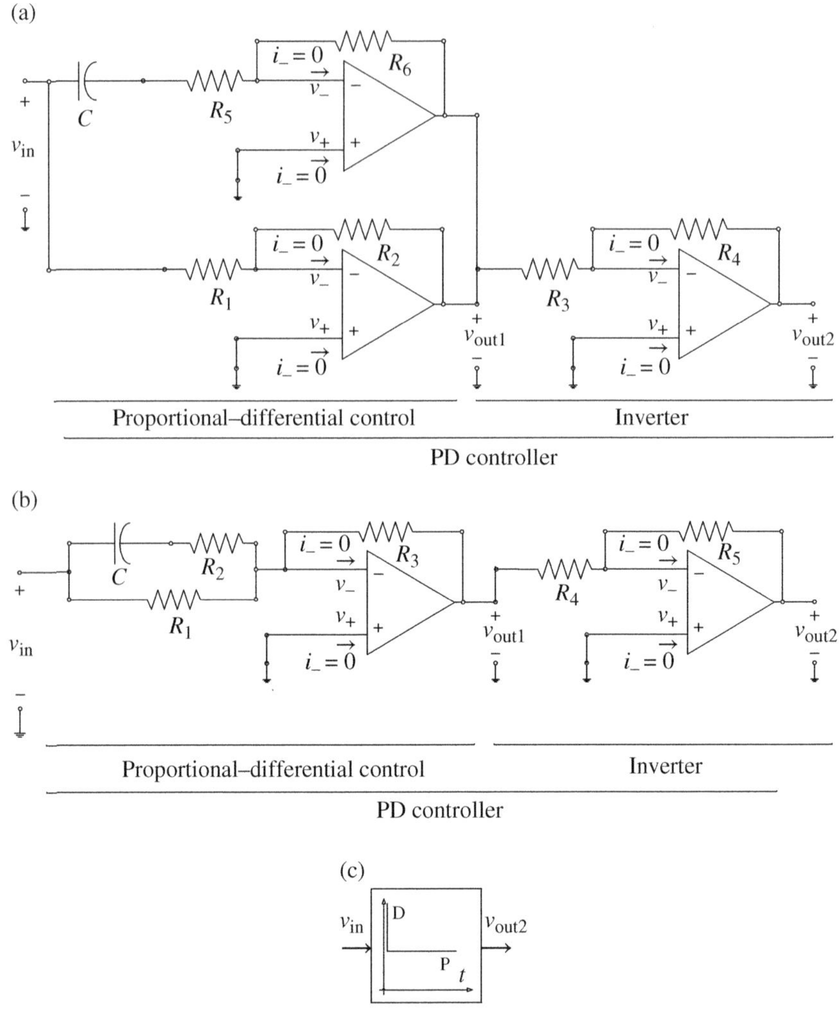

Figure 8.21 (a) PD controller with P and D circuits connected in parallel using two OP amplifiers. $R_5 = R_c \rightarrow 0$ represents the capacitor loss resistor. (b) PD controller with combined P and D circuits relying on one OP amplifier. $R_2 = R_c \rightarrow 0$ is the capacitor loss resistor. (c) Symbolic representation of PD-controller network of (a) and (b).

identical, that is, $v_+ = v_-$. The output impedance is zero; that is, an ideal OP amplifier can deliver any amount of output current to the following network. The open-loop gain $A_v \rightarrow \infty$, while the closed-loop gain G_v depends upon the feedback network connecting the output to one of the inputs and is in the range of 1–100 pu.

Power supply voltages $V_{DD} > 0$ and $V_{SS} < 0$ limit the output voltage excursions v_{out} and are in the range of ± 15 V or less.

First discussed is the noninverting OP amplifier and then the UG and inverting OP amplifiers. Next, summing, integrating, and differentiating networks, active filters, current-to-voltage converters, and various (e.g. P, PI, PID, PD) electronic controllers based on OP amplifiers are explained and applied. One control application is solved with Mathematica [32] software, and another is accomplished with PSPICE [34]. One concludes that proportional (P) control is the fastest regulation method, although the steady-state error $\varepsilon = 1/(1 + G)$ depends upon the overall gain [1, 6, 7] G: for a large gain G the steady-state error is small, but control oscillations might be large; for small G values the steady-state error is large, and the control response is slow. For these reasons P control is seldom employed for control applications.

The integrating control component makes the steady-state error ε zero, while the differential component anticipates the change of the control variable, and therefore it may increase the already existing transient fluctuation of the control variable. Note that Mathematica can model differential equations, while PSPICE mostly can model algebraic equations.

Ten application examples and 18 problems at the end of the chapter guide the reader to use the methods explained in this chapter.

Problems

8.1 Noninverting (NI) OP amplifier

The noninverting OP amplifier of Figure P8.1.1 is operated with $V_{DD} = +5$ V and $V_{SS} = -5$ V, and the input signal ranges from $v_{in} = 0.3$ to -0.6 V. What is the largest value of the closed-loop voltage (v) gain G_V^{NI} for which the circuit will perform as a linear amplifier? Assume ideal OP amplifier conditions ($v_- = v_+$, input currents to the OP amplifier $i_- = i_+ = 0$).

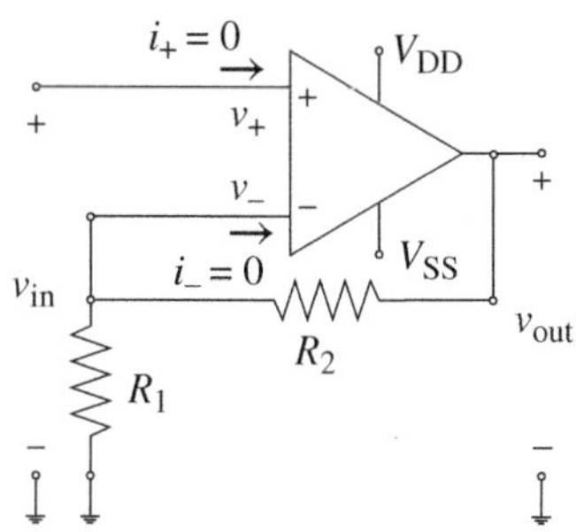

Figure P8.1.1 Noninverting (NI) OP amplifier.

8.2 Calculation of closed-loop voltage (v) gain G_v^{NI} for various noninverting (NI) configurations

Calculate the closed-loop voltage gain G_v^{NI} for the circuit of Figure P8.2.1a–c with resistor values for [3, 4]:

a Figure P8.2.1a $R_1 = 750\ \Omega$, $R_2 = 5.6\ \text{k}\Omega$;

b Figure P8.2.1b $R_1 = 2\ \text{k}\Omega$, $R_2 = 100\ \Omega$;

c Figure P8.2.1c $R_1 = 2\ \text{k}\Omega$, $R_2 = 133\ \Omega$.

Assume ideal OP amplifier conditions ($v_- = v_+$, input currents to the OP amplifier $i_- = i_+ = 0$).

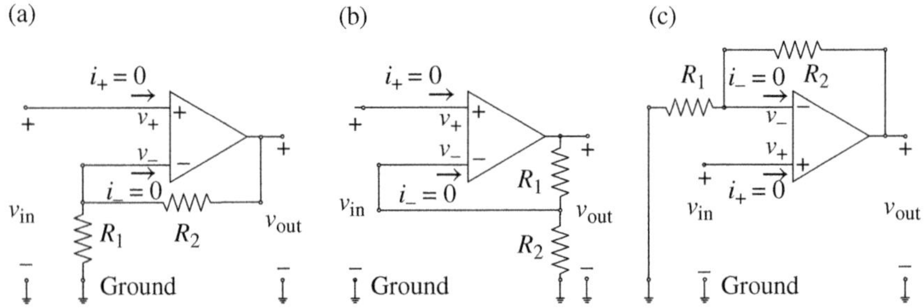

Figure P8.2.1 (a–c) Determination of closed-loop voltage gain G_v^{NI} for various noninverting OP amplifier configurations.

8.3 Calculation of closed-loop voltage gain G_v^{NI}

Calculate the closed-loop voltage gain G_v^{NI} for the circuit of Figure P8.3.1 with resistor values $R_1 = 237\ \Omega$, $R_2 = 3.6\ \text{k}\Omega$, $R_3 = 2\ \text{k}\Omega$ [3, 4].

Assume ideal OP amplifier conditions ($v_- = v_+$, input currents to the OP amplifier $i_- = i_+ = 0$).

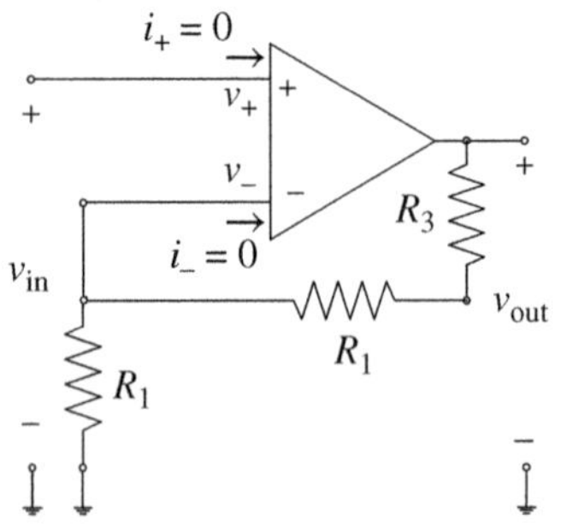

Figure P8.3.1 Determination of closed-loop voltage gain G_v^{NI} of noninverting (NI) OP amplifier configuration ($v_- = v_+$, input currents to the OP amplifier $i_- = i_+ = 0$).

8.4 Noninverting (NI) OP amplifier

The input voltage source (s) of Figure P8.4.1 is $v_s = 0.12\,\text{V}\cos(377t)$.

a Determine the output voltage v_{out} and find the maximum current through resistor R_L, with resistor values $R_1 = 237\,\Omega$, $R_2 = 3.6\,\text{k}\Omega$, $R_L = 2\,\text{k}\Omega$ [3, 4]. Assume ideal OP amplifier conditions.

b What is the maximum current $i_{L\,max}$ delivered to the load resistor R_L.

c What is the AC power delivered to R_L?

d What is the maximum current $i_{OP\,max}$ drawn from the OP amplifier?

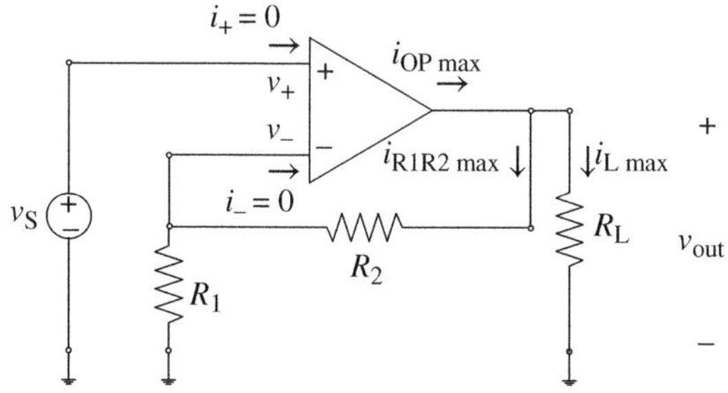

Figure P8.4.1 Noninverting (NI) OP amplifier supplying ohmic load R_L with AC power.

8.5 Noninverting (NI) OP amplifier

Determine the value of R_2 in order to have a closed-loop voltage gain of $G_V^{NI} = 50$ pu in Figure P8.5.1 for $R_1 = 4.7\,\text{k}\Omega$. Assume ideal noninverting OP amplifier conditions ($v_- = v_+$, input currents to the OP amplifier $i_- = i_+ = 0$).

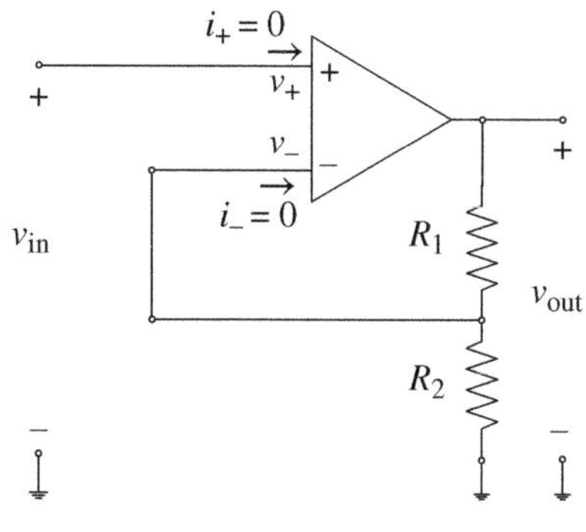

Figure P8.5.1 Calculation of load resistor R_2 of noninverting (NI) OP amplifier.

8.6 Unity-gain (UG) buffer network

Calculate the output DC voltage V_{out} as indicated in Figure P8.6.1 for the unity-gain buffer circuit with DC voltage $V_{DC} = 5$ V and $R_L = 3.6$ kΩ. Assume ideal OP amplifier conditions ($v_- = v_+$, input currents to the OP amplifier $i_- = i_+ = 0$).

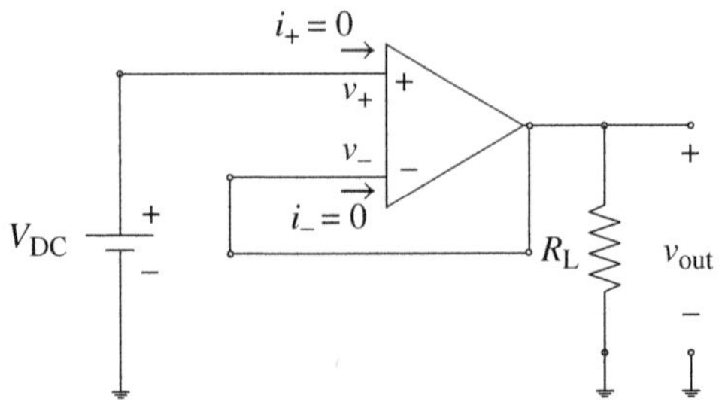

Figure P8.6.1 Unity-gain (UG) buffer network.

8.7 Inverting (IN) OP amplifier circuit

Derive an expression for the output voltage v_{out} for the input source (s) voltage $v_s = 0.1$ V sin ωt with resistor values $R_1 = 133$ Ω, $R_2 = 5.6$ kΩ, $R_L = 3.6$ kΩ [3, 4] (Figure P8.7.1). Assume ideal OP amplifier conditions ($v_- = v_+$, input currents to the OP amplifier $i_- = i_+ = 0$).

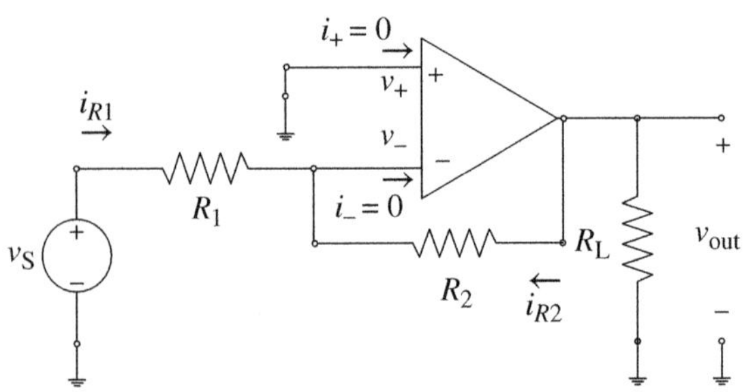

Figure P8.7.1 Inverting (IN) OP amplifier circuit.

8.8 Inverting (IN) OP amplifier circuit

Calculate the output DC voltage V_{out} across R_L in Figure P8.8.1 with resistor values $R_1 = 1000$ Ω, $R_2 = 4700$ Ω, $R_L = 2000$ Ω [3, 4], and the DC voltages $V_{DC1} = 10$ V and $V_{DC2} = 8$ V. Assume ideal OP amplifier conditions ($v_- = v_+$, input currents to the OP amplifier $i_- = i_+ = 0$).

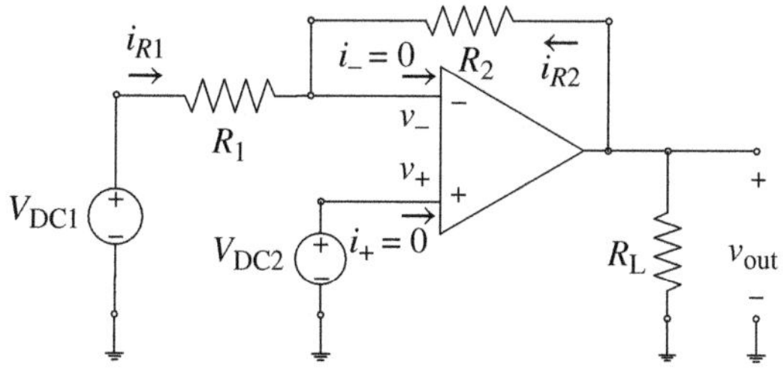

Figure P8.8.1 Inverting (IN) OP amplifier with DC voltages $V_{DC1} = 10\ \text{V}$ and $V_{DC2} = 8\ \text{V}$ at the v_- and v_+ inputs, respectively, of OP amplifier.

8.9 Differential amplifier

Determine the output voltage v_{out} in Figure P8.9.1 for $v_{in1} = 20\ \text{mV}$, $v_{in2} = 24\ \text{mV}$, $R_1 = 0.237\ \text{k}\Omega$, $R_2 = 9.1\ \text{k}\Omega$. Assume ideal OP amplifier conditions ($v_- = v_+$, input currents to the OP amplifier $i_- = i_+ = 0$).

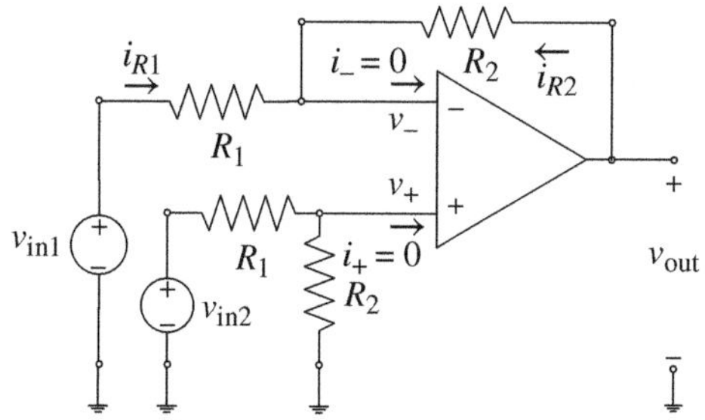

Figure P8.9.1 Differential amplifier with equal input resistances R_1.

8.10 Differential amplifier

For the differential amplifier circuit of Figure P8.10.1, show that if $(R_1/R_2) = (R_3/R_4)$, the output voltage is $v_{out} = (R_2/R_1)(v_{in1} - v_{in2})$. Assume ideal OP amplifier conditions ($v_- = v_+$, input currents to the OP amplifier $i_- = i_+ = 0$).

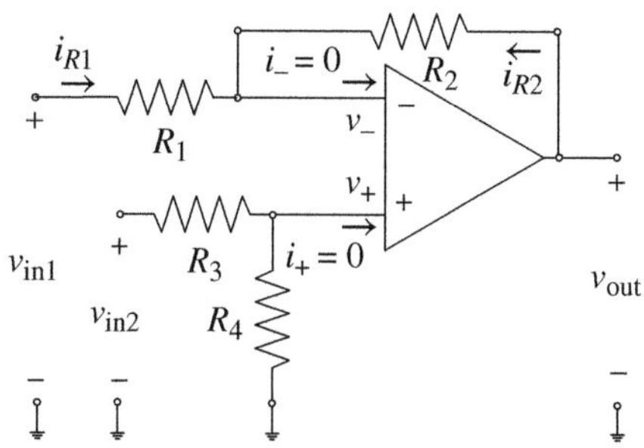

Figure P8.10.1 Differential amplifier with unequal input resistances R_1 and R_3.

8.11 OP amplifier with positive and negative feedback

Determine the closed-loop voltage gain G_v for the circuit of Figure P8.11.1, where $R_1 = 100\,\Omega$, $R_2 = 4.7\,\text{k}\Omega$, $R_3 = 9.1\,\Omega$, $R_4 = 4.7\,\text{k}\Omega$. Assume ideal OP amplifier conditions ($v_- = v_+$, input currents to the OP amplifier $i_- = i_+ = 0$). Are there values of R that should be avoided?

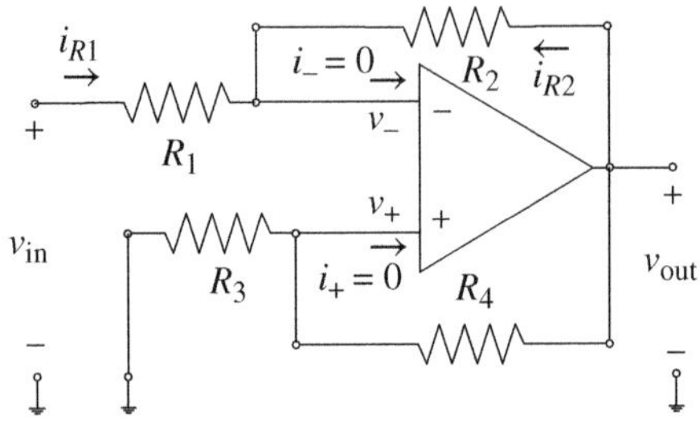

Figure P8.11.1 OP amplifier with positive and negative feedback paths.

8.12 Summing circuit

For the circuit of Figure P8.12.1, determine the output voltage v_{out} in terms of the input voltages v_{in1}, v_{in2}, and v_{in3}, for the resistor values $R_1 = 18\,\Omega$, $R_2 = 18\,\Omega$, $R_3 = 56\,\Omega$, $R_4 = 2\,\text{k}\Omega$ [3, 4]. Assume ideal OP amplifier conditions ($v_- = v_+$, input currents to the OP amplifier $i_- = i_+ = 0$).

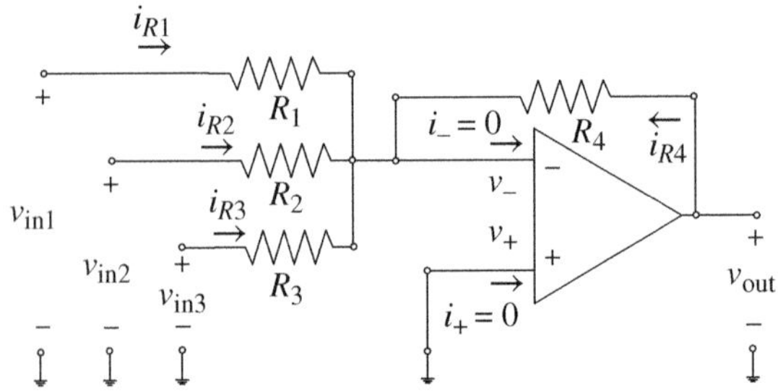

Figure P8.12.1 Summing circuit.

8.13 Temperature sensing network with Wheatstone bridge

The Wheatstone bridge of Figure P8.13.1 is part of a temperature sensing system where R_A is nominally/rated at $100\,\Omega$ and increases $1\,\Omega$ for every $1\,°\text{C}$ temperature increase. Select the value of R_2 that will result in a 1 V change at the output, v_{out}, for every degree temperature change at the sensor provided, with $R_1 = 4.7\,\text{k}\Omega$, $R_A(T) = [100 + \Delta T]\,\Omega$, $R_B = R_C = R_D = 100\,\Omega$, $V_{DC} = 12\,\text{V}$.

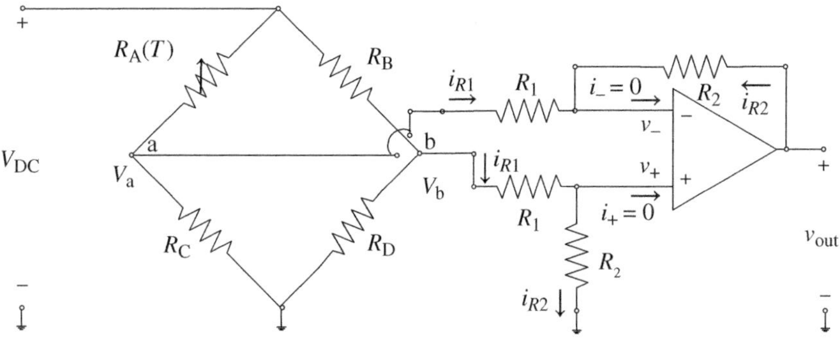

Figure P8.13.1 Temperature sensing network consisting of Wheatstone bridge and inverting OP amplifier.

8.14 Integrating network

a For the circuit of Figure P8.14.1a with the capacitor value $C = 150\,\mu F$ [4] and the resistor value $R = 4.7\,k\Omega$ [3, 4], calculate and plot the waveform v_{in} versus time t for the output voltage v_{out} as specified by Figure P8.14.1b.

b Show that for a constant input DC voltage V_{in}, the output voltage is the solution of the equation $C\frac{dv_{out}(t)}{dt} + \frac{V_{in}}{R} = 0$. Assume ideal OP amplifier conditions ($v_- = v_+$, input currents to the OP amplifier $i_- = i_+ = 0$).

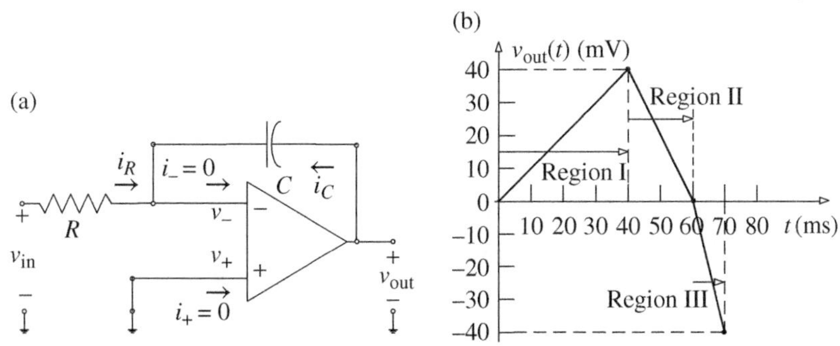

Figure P8.14.1 (a) Integrating network. (b) Given output voltage.

8.15 Differentiating network

a Show that the circuit of Figure P8.15.1a acts as a differentiator governed by the relation $v_{out}(t) = -RC\dfrac{dv_{in}(t)}{dt}$.

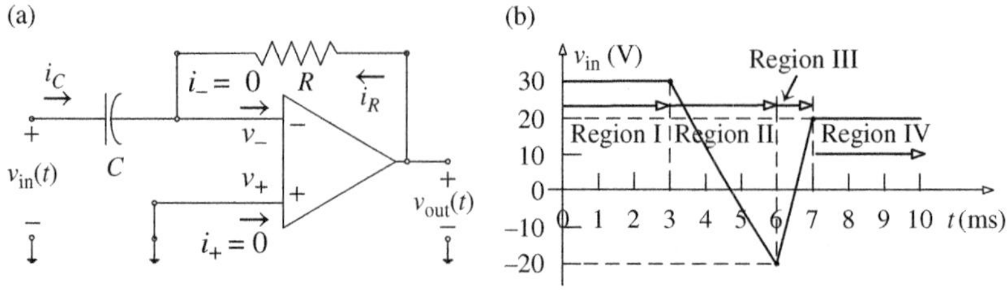

Figure P8.15.1 (a) Differentiating network. (b) Given input voltage.

b Plot the output voltage of Figure P8.15.1a as a function of time if the input voltage is given by Figure P8.15.1b, assuming resistor value $R = 9.1\,\text{k}\Omega$ [3, 4] and capacitor value $C = 0.22\,\mu\text{F}$ [4]. Assume ideal OP amplifier conditions ($v_- = v_+$, input currents to the OP amplifier $i_- = i_+ = 0$).

8.16 Low-pass filter circuit

A low-pass filter circuit is depicted in Figure P8.16.1.

a Plot the magnitude $M(\omega)$ of the closed-loop voltage gain $G(j\omega) = M(\omega)\angle\Phi(\omega)$ of this filter as a function of the angular frequency ω for the resistor values $R_1 = 0.988\,\text{k}\Omega$, $R_2 = 5.6\,\text{k}\Omega$ [3, 4], and the capacitor value $C = 22\,\mu\text{F}$ [4]. Assume ideal OP amplifier conditions ($v_- = v_+$, input currents to the OP amplifier $i_- = i_+ = 0$). It is recommended to use for the abscissa ω a logarithmic scale and for the ordinate $M(\omega)$ a linear scale. Compute and indicate on the $M(\omega)$ versus ω plot the half-power frequency ω_{HI} where $M(\omega)/\sqrt{2}$.

b Alternatively, the gain magnitude $M(\omega)$ can be represented in a logarithmic manner where the gain in dB (decibel) = 20 log $M(\omega)$. Show that the reduction of the gain at ω_{HI} is −3 dB. Note that the circuit of Figure P8.16.1 can be employed as a proportional–integral (PI) controller for control applications.

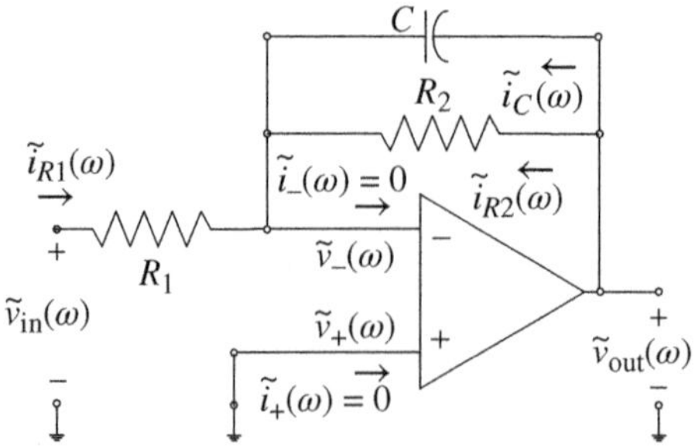

Figure P8.16.1 Low-pass filter circuit or proportional–integral (PI) controller.

8.17 Band-pass filter circuit

A band-pass filter circuit is depicted in Figure P8.17.1.

a Plot the magnitude $M(\omega)$ of the closed-loop voltage gain $G(j\omega) = M(\omega)\angle\Phi(\omega)$ of this filter as a function of angular frequency ω for the resistor values $R_1 = 2\ \text{k}\Omega$, $R_2 = 3.6\ \text{k}\Omega$ [3, 4], and the capacitor values $C_1 = 22\ \mu\text{F}$, $C_2 = 2.2\ \mu\text{F}$ [4]. Assume ideal OP amplifier conditions ($v_- = v_+$, input currents to the OP amplifier $i_- = i_+ = 0$). It is recommended to use for the abscissa ω a logarithmic scale and for the ordinate $M(\omega)$ a linear scale. Compute and indicate the half-power frequencies on the $M(\omega)$ versus ω plot ω_{HI} and ω_{LO} where $M(\omega)/\sqrt{2}$ is.

b Alternatively, the gain magnitude $M(\omega)$ can be represented in a logarithmic manner, where the gain$_{dB}$ in dB (decibel) = $20 \log M(\omega)$. Show that the reduction of the gain at ω_{HI} and ω_{LO} is -3 dB. Note that the circuit of Figure P8.17.1 can be employed as a proportional–integral–differential (PID) controller for control applications.

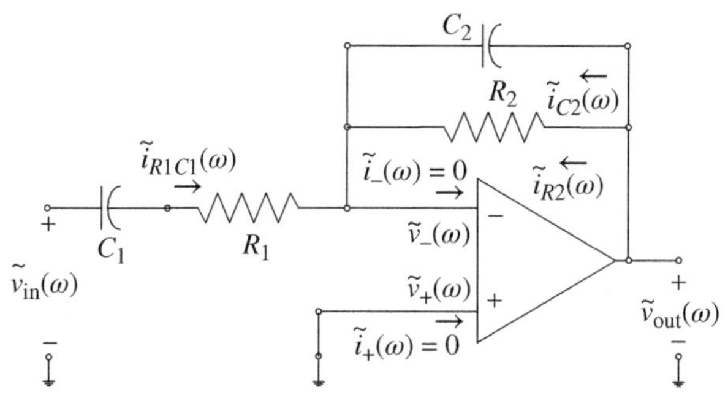

Figure P8.17.1 Band-pass filter circuit or proportional–integral–differential (PID) controller.

8.18 Photovoltaic (PV) panel supplying load

A small photovoltaic (PV) panel delivers a DC current of $I_{PV} = I_{in}$ at $V_{PV} = 6\ \text{V}$ to two ideal inverting OP amplifiers connected in series as depicted in Figure P8.18.1.

a Compute the load voltage V_{load} for $R_1 = 5\ \text{k}\Omega$, $R_2 = 10\ \text{k}\Omega$, $R_3 = 500\ \Omega$, $R_4 = 1\ \text{k}\Omega$, and $R_{load} = 5\ \Omega$.

b Compute the input power $P_{PV} = I_{PV}V_{PV}$ and the power P_{load} delivered to the load resistor R_{load}.

c Determine the power amplification $G_{DC\_power} = P_{load}/P_{PV}$.

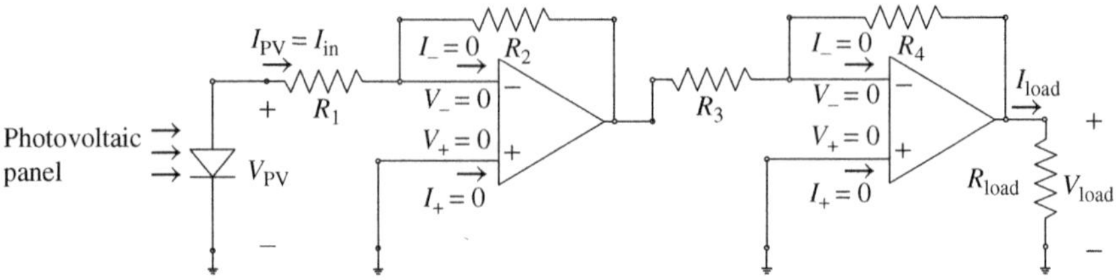

Figure P8.18.1 Small photovoltaic power plant with DC transmission network supplying R_{load}.

References

1 Fuchs, E.F. and Masoum, M.A.S. (2012). *Power Conversion of Renewable Energy Systems*, 1e, 2nd Printing. New York: Springer, 692 pp.

2 Sedra, A.S. and Smith, K.C. (1998). *Microelectronic Circuits*, 4e. Oxford: Oxford University Press.

3 Standard IEC 60063 (2019). Resistor values. EETech Media, LLC. http://www.resistorguide.com/resistor-values (accessed 15 May 2020).

4 Yang, W.Y., Kim, J., Park, K.W. et al. (2020). *Electronic Circuits with MATLAB®, PSpice®, and Smith Chart*, 1e. Wiley https://onlinelibrary.wiley.com/doi/pdf/10.1002/9781119598961.app7 (accessed 15 May 2020).

5 Kerns, D.V. Jr. and Irwin, J.D. (2004). *Essentials of Electrical and Computer Engineering*. Pearson Prentice Hall.

6 Leonhard, W. (1985). *Control of Electrical Drives*. Berlin: Springer-Verlag.

7 Kuo, B.C. (1967). *Automatic Control Systems*, 2e. Englewood Cliffs, NJ: Prentice-Hall, Inc.

8 Huang, H. and Fuchs, E.F. (1990). Successive nonlinear quadratic programming for high-power density and high-efficiency, permanent-magnet motors. *Proceedings of the International Conference on Electrical Machines*, Cambridge, MA (12–15 August 1990).

9 Huang, H., Anderson, W.M., and Fuchs, E.F. (1990). High-power density and high-efficiency motors for electric vehicle applications. *Proceedings of the International Conference on Electrical Machines*, Cambridge, MA (12–15 August 1990).

10 Fuchs, E.F. (2012). Alternating current machine with increased torque above and below rated speed for hybrid/electric propulsion systems. US Patent No. 8,183,814 B2, 22 May 2012.

11 Fuchs, E.F. (2013). Alternating current machine with increased torque above and below rated speed for hybrid/electric propulsion systems. U.S. Patent No. 8,436,569 B2, 7 May 2013.

12 Fuchs, E.F., Schraud, J., and Fuchs, F.S. (2008). Analysis of critical-speed increase of induction machines via winding reconfiguration with solid-state switches. *IEEE Transactions on Energy Conversion* 23 (3): 774–780.

13 Schraud, J., Fuchs, E.F., and Fuchs, H.A. (2008). Experimental verification of critical-speed increase of induction machines via winding reconfiguration with solid-state switches. *IEEE Transactions on Energy Conversion* 23 (2): 460–465.

14 Fuchs, E.F. and Hanna, W.J. (2002). Measured efficiency improvements of induction motors with thyristor/triac controllers. *IEEE Transactions on Energy Conversion* 17 (4): 437–444.

15 Fuchs, E.F. and Hanna, W.J. (2004). Closure of "measured efficiency improvements of induction motors with thyristor/triac controllers". *IEEE Transactions on Energy Conversion* 19 (3): 647–648.

16 Fuchs, E.F., Poloujadoff, M., and Neal, G.W. (1988). Starting performance of saturable three-phase induction motors. *IEEE Transactions on Energy Conversion* EC-3 (3): 624–635.

17 Appelbaum, J., Fuchs, E.F., and White, J.C. (1987). Optimization of three-phase induction motor design, part I: formulation of the optimization technique. *IEEE Transactions on Energy Conversion* EC-2 (3): 407–414.

18 Appelbaum, J., Khan, I.A., Fuchs, E.F., and White, J.C. (1987). Optimization of three-phase induction motor design, part II: the efficiency and cost of an optimal design. *IEEE Transactions on Energy Conversion* EC-2 (3): 415–422.

19 Fuchs, E.F., Chang, L.H., and Appelbaum, J. (1984). Magnetizing current, iron losses and forces of three-phase induction machines at sinusoidal and nonsinusoidal terminal voltages, part I: analysis. *IEEE Transactions on Power Apparatus and Systems* PAS-103 (11): 3303–3312.

20 Fuchs, E.F., Roesler, D.J., and Chang, L.H. (1984). Magnetizing current, iron losses and forces of three-phase induction machines at sinusoidal and nonsinusoidal terminal voltages, part II: results. *IEEE Transactions on Power Apparatus and Systems* PAS-103 (11): 3313–3326.

21 Geysen, W., DeJong, H., Fuchs, E.F., and Van Ranst, A. (1973). Flux distribution in a three-phase squirrel cage induction motor. *Revue Electricite, Belgium* 7 (8): 201–208.

22 Fardoun, A.A. and Fuchs, E.F. (1991). Generator operation of electronically commutated permanent-magnet machine drives. *Proceedings of the International Conference on the Evolution and Modern Aspects of Synchronous Machines*, Zürich, Switzerland (27–29 August 1991), pp. 407–412. Switzerland: Department of Electrical Machines und Antriebe, Eidgenoessische Technische Hochschule (ETH) Zuerich.

23 Fardoun, A.A., Fuchs, E.F., and Huang, H. (1992). Modeling and simulation of an electronically commutated permanent-magnet machine drive system using SPICE. *Conference Record of the 1992 IEEE Industry Applications Society Annual Meeting*, Houston, TX (4–9 October 1992).

24 Fuchs, E.F., Fardoun, A.A., Carlin, P.W., and Erickson, R.W. (1992). Permanent-magnet machines for operation with large speed variations. *Proceedings of Windpower 1992*, Seattle, WA (19–23 October 1992).

25 Yildirim, D., Fuchs, E.F., and Batan, T. (1998). Test results of a 20 kW, direct-drive, variable-speed wind power plant. *Proceedings of the International Conference on Electrical Machines*, Istanbul, Turkey (2–4 September 1998), pp. 2039–2044. Ankara, Turkey: Middle East Technical University (METU); Istanbul, Turkey: Istanbul Technical University (ITU).

26 Fuchs, E.F. and Masoum, M.A.S. (2008). *Power Quality in Power Systems and Electrical Machines*, 1e, 638. Elsevier/Academic Press.

27 Masoum, M.A.S. and Fuchs, E.F. (2015). *Power Quality in Power Systems and Electrical Machines*, 2e, 1123. Elsevier/Academic Press.

28 Nilsson, J.W. (1983). *Electric Circuits*. Addison-Wesley Publishing Company.

29 Erdelyi, E.A. and Fuchs, E.F. (1970). Nonlinear magnetic field analysis of DC machines, part I: theoretical fundamentals. *IEEE Transactions on Power Apparatus and Systems* PAS-89 (7): 1546–1554.

30 Fuchs, E.F. and Erdelyi, E.A. (1970). Nonlinear magnetic field analysis of DC machines, part II: application of the improved treatment. *IEEE Transactions on Power Apparatus and Systems* PAS-89 (7): 1555–1564.

31 Erdelyi, E.A., Fuchs, E.F., and Binkley, D.H. (1970). Nonlinear magnetic field analysis of DC machines, part III: equipotential plots drawn by computer. *IEEE Transactions on Power Apparatus and Systems* PAS-89 (7): 1565–1583.

32 Wolfram. What's new in Mathematica 11. http://www.wolfram.com/mathematica/new-in-11 (accessed 15 March 2020).

33 Mathworks. MATLAB. https://www.mathworks.com/products/matlab.html (accessed 15 March 2020).

34 PSpice http://www.pspice.com (accessed 15 March 2020).

9

Semiconductor Diodes and Switches

Without semiconductor switches, the design of renewable plants and automotive applications would be virtually impossible. For this reason, in an introductory text like this one for renewable energy, it is important to put the emphasis on electronic switches rather than on amplifiers. Starting from the elementary building block of semiconductor devices – the pn (abbreviation for positive and negative types of semiconductor materials) junction – circuit symbols, terminal characteristics, and circuit models are introduced and used for PSPICE (PC-based Simulation Program with Integrated Circuit Emphasis) analysis in Chapters 10 and 12. One can definitively state that without the inventions [1] of the diode in 1904, the bipolar junction transistor (BJT) in 1947, and subsequently field-effect transistors (FETs) and other semiconductor switches, the transition from fossil fuel power generation to renewable (e.g. wind and solar) sources would be far more difficult to achieve.

To facilitate the reading of this chapter, each abbreviation will be defined when used for the first time.

9.1 Introduction

Electrical devices and apparatus are listed [1] in alphabetical order, including their inventors and patent numbers. As can be seen, we heavily rely on the ingenuity of our predecessors. Efficient development of renewable energy systems is not possible without the deployment of semiconductor devices such as diodes (d), as well as line/load voltage-commutated and self-commutated switches. Mercury arc-type valves and rectifiers were invented in 1902 by Peter Cooper Hewitt [2, 3], as well as mercury arc inverters [4, 5]. The invention of the pn semiconductor material junction [6] and the subsequent invention of the bipolar transistor [7], including that of the metal–oxide–semiconductor field-effect transistor (MOSFET) [8], and the

Introduction to Energy, Renewable Energy, and Electrical Engineering: Essentials for Engineering Science (STEM) Professionals and Students, First Edition. Ewald F. Fuchs and Heidi A. Fuchs.

Companion website: www.wiley.com/go/fuchs/renewableenergyandelectricalengineering

invention of gate turn-off thyristors (GTOs) [9] and insulated-gate bipolar transistor (IGBTs) [10] and many others were instrumental in the rapid development of renewable energy sources [11, 12]. In parallel with semiconductor development, software such as Simulation Program with Integrated Circuit Emphasis (SPICE) and PC-based Simulation Program with Integrated Circuit Emphasis (PSPICE) [13–16] were made available to engineering professionals, permitting them to solve linear and nonlinear ordinary differential equations. In addition, software related to finite-difference (FD) [17] and finite-element (FE) [18] methods permit the solution of partial differential equations (Chapter 11), which greatly assist in the development of electric/hybrid drives for transportation and wind power applications. In this introductory text, we rely on either simplified or readily available models for semiconductor/power electronic components. Detailed information about power electronic diodes and switches can be gained from the excellent book by Sedra and Smith [19].

9.2 The pn Junction: Elementary Building Block of Semiconductor Diodes and Switches

The pn (positive–negative) junction is the basic building block [19] for all types of diodes, including semiconductor switches. It consists of a combination of 3-, 4-, and 5-valent lattice material such as boron (B), silicon (Si), and phosphorus (P), respectively: 4-valent material (Figure 9.1a) is doped with 3- (Figure 9.1b) and 5-valent (Figure 9.1c) materials. By implanting the latter two in the 4-valent lattice structure seen in Figure 9.2a and b, one obtains p-type and n-type semiconductor material, respectively. The p-type semiconductor consists of a silicon lattice/crystal structure in which boron atoms are introduced, generating one hole for each boron atom (see Figure 9.2a). As clarification, an absent electron is called a hole and represents a

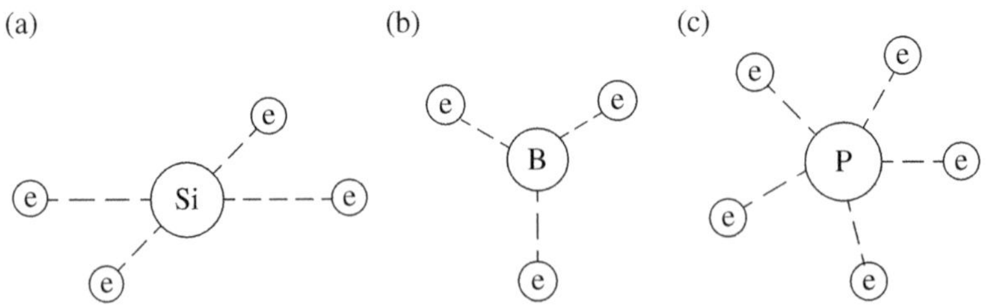

Figure 9.1 Quasi-three-dimensional representation of materials with (a) 4 (e.g. silicon [Si] is quadrivalent), (b) 3 (e.g. boron [B] is trivalent), and (c) 5 (e.g. phosphorus [P] is pentavalent) valence electrons.

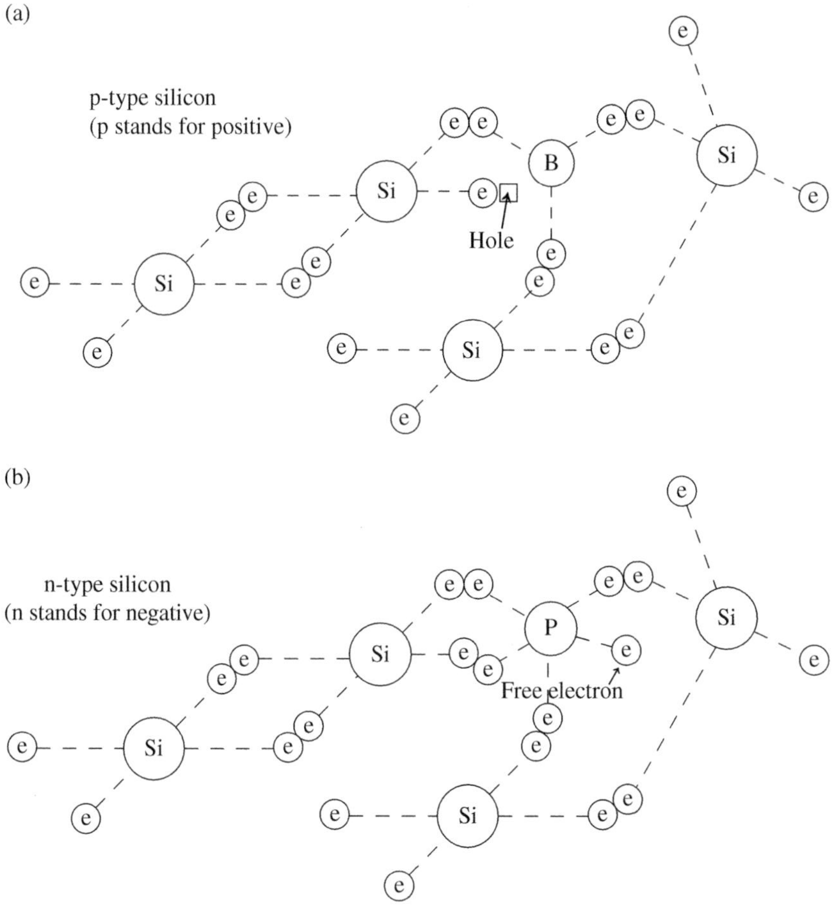

Figure 9.2 (a) Quasi-three-dimensional representation of material with 4 valence electrons implanted/doped with materials with trivalent (p-type silicon) and (b) pentavalent (n-type silicon) electrons, where Si stands for silicon, P for phosphorus, and e for valence electron.

positive charge. These can travel through diffusion and recombination [19] in a manner similar to free electrons. The n-type semiconductor consists of a silicon lattice/crystal structure in which phosphorus atoms are implanted, generating one free electron, or negative charge, for each phosphorus atom (see Figure 9.2b).

As a result of the doping illustrated in Figure 9.3a, one obtains a barrier voltage drop across the pn junction boundary, as shown in Figure 9.3b. The flow of electrons – in the direction opposite to the defined current – is detailed in Sedra and Smith [19].

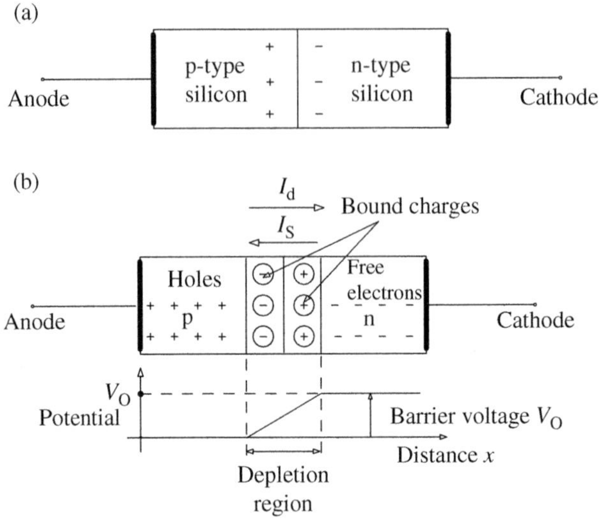

Figure 9.3 (a) Silicon-based pn junction [19] doped/implanted into the silicon crystal structure with p (positive) and n (negative) doping material causing p-type and n-type silicon such as boron and phosphorous, respectively. As a result one obtains a voltage drop V_o across the pn junction boundary as shown in (b). (b) The pn junction with no externally applied voltage (open circuited). The barrier voltage V_o is caused by the potential distribution within the depletion region causing the diffusion current I_d and the drift current I_s.

In the following treatment, we use only circuit symbols and PSPICE models for the analysis of circuits with diodes and switches [13–16].

As illustrated in Figure 9.4, the pn junction diode (see region I, where the insolation is $Q = 0$ for low voltages V_d) has the same characteristic as the light absorption photovoltaic (PV) cell (see region II with $Q \neq 0$ for high and low voltages V_c) and light-emitting diode (LED) (see region I for high voltages V_d and $Q \neq 0$). The highest efficiency for PV cells reaches 46% under laboratory conditions [22], while in practice for PV plants, efficiencies between 13 and 15%, depending upon nominal operating cell temperature (NOCT) and standard test conditions (STC), respectively, are reached (see Chapter 1, Appendix 1.A). Note the insolation on Earth has a maximum value of $Q_{max} = 1000\ \text{W/m}^2$.

In region I the device works as a diode for $V_d < 0.8$ V for the insolation $Q = 0$ and for $V_d = (2–3)$ V where the insolation $Q \neq 0$ light is emitted as a LED. In region II the device works as a PV cell absorbing the insolation $Q \neq 0$.

The exponential relation for a diode [19] is

$$I_d = I_S\left(e^{V_d/nV_T} - 1\right) \approx I_S e^{V_d/(nV_T)}, \tag{9.1}$$

where I_S is the saturation current, $n = 1$, and the thermal (T) voltage is $V_T = 25.2$ mV at a temperature of 20 °C.

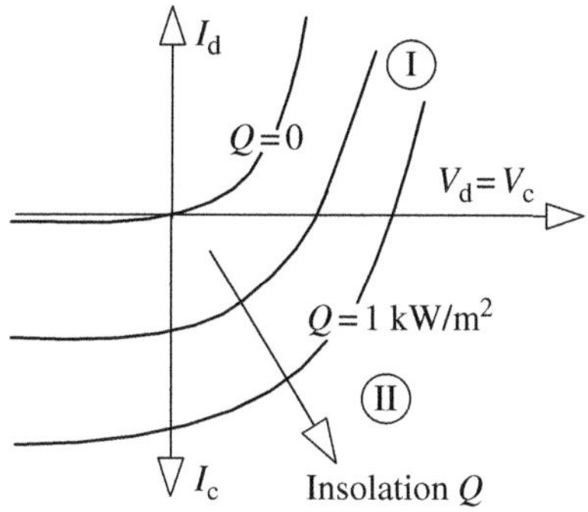

Figure 9.4 Characteristic of pn junction diode (d), LED [20, 21], photodiode, and photovoltaic cell (c).

Application Example 9.1 Calculation of diode saturation current I_S For $n = 1$ at a temperature of 20 °C, one obtains for $V_T = kT/q$ with the Boltzmann constant k, T the absolute temperature measured in kelvin, and q the electron charge produces the thermal voltage

$$V_T = 1.38 \cdot 10^{-23}(273 + 20)/1.6 \cdot 10^{-19} = 25.2\,\text{mV}. \tag{E9.1.1}$$

From Eq. (9.1) one gets

$$I_S = I_d e^{-V_d/(nV_T)}. \tag{E9.1.2}$$

For $I_d = 40$ A, $V_d = 0.7$ V with Eqs. (E9.1.1) and (E9.1.2), the saturation current is

$$I_S = 40\,\text{A}e^{-0.7\,\text{V}/25.2\,\text{mV}} = 34.54 \cdot 10^{-12}\,\text{A}.$$

The circuits for a diode and LED as used in high-efficiency lighting are depicted in Figure 9.5a and b, respectively.

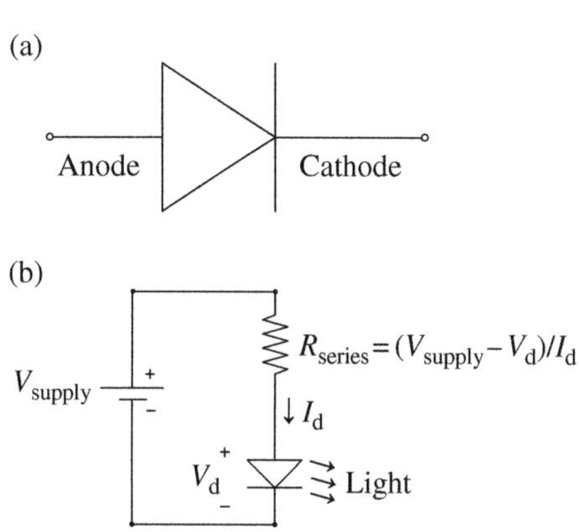

Figure 9.5 (a) Circuit symbol of a pn junction diode. (b) Circuit for a light-emitting diode (LED).

9.3 Zener Diode

Like a typical semiconductor diode, a Zener [23] diode lets current flow from the anode to the cathode but also allows current to flow in the opposite direction when its "Zener voltage" is reached. The symbol of a Zener diode is depicted in Figure 9.6a and its characteristic in Figure 9.6b. Details of how to use the Zener voltage and current values ($-V_Z$, $-I_z$) are explained in Sedra and Smith [19].

9.4 Varistor

A varistor's [24] electrical resistance $R_{\text{varistor}} = v_{\text{varistor}}/i_{\text{varistor}}$ varies with the applied voltage v, with its name coming from "variable resistor." Figure 9.7a shows the symbol for a varistor and Figure 9.7b its nonlinear current–voltage characteristic ($i_{\text{varistor}} - v_{\text{varistor}}$).

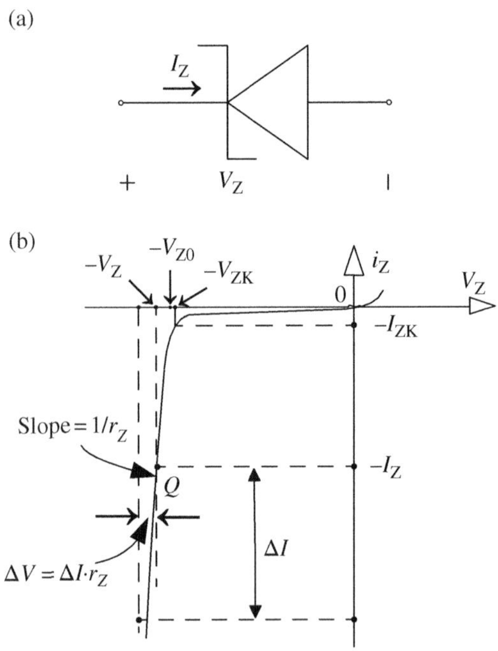

Figure 9.6 (a) Circuit symbol of a Zener (Z) diode; note that V_Z is positive. (b) Current–voltage ($i_Z - v_Z$) characteristic [19] of a Zener (Z) diode with a knee (K) or breakdown voltage of ($-V_{ZK}$), operating point voltage/current ($-V_Z$, $-I_Z$), and the intercept of the Zener voltage with the abscissa ($-V_{ZO}$).

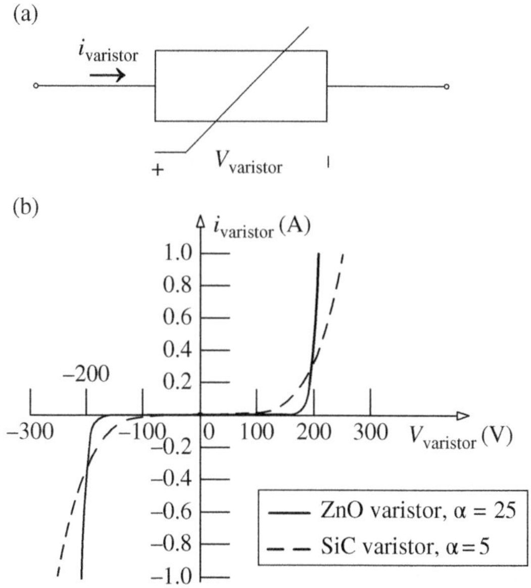

Figure 9.7 (a) Varistor symbol. (b) Varistor characteristic ($i_{varistor} - v_{varistor}$) where $I_{varistor} = kV^{\alpha}_{varistor}$. The term α describes the degree of nonlinearity. The dashed function shows the characteristic of a zinc oxide (ZnO), with high $\alpha = 25$ varistor, and the full-line function represents a silicon carbide (SIC) with low $\alpha = 5$ varistor.

9.5 Bipolar Junction Transistor (BJT)

The first bipolar junction transistor (BJT) [19] was invented at Bell Laboratories (Bell Labs) by William Shockley, Walter Brattain, and John Bardeen so late in 1947 that they did not publish their discovery until 1948. The BJT – either of the npn or pnp type (see Figure 9.8a) – is a self-commutated switch that can be turned on and off by the base (B) current $i_B = (I_B + i_b)$; that is, it is a current-controlled device that can amplify and switch the collector (C) current i_C. Its symbol within an amplifying/switching circuit is given in Figure 9.8b with base, collector, and emitter (E) terminals. The $i_C = (I_C + i_c)$ versus $v_{CE} = (V_{CE} + v_{ce})$ and $i_B = (I_B + i_b)$ versus $v_{BE} = (V_{BE} + v_{be})$ characteristics are shown in Figure 9.8c and d with the load line defining the graphical solutions for the small-signal amplifying and switch regions. The capital V and I symbols pertain to DC biasing (establishing operating point) values, and the lower base v and i symbols indicate instantaneous values.

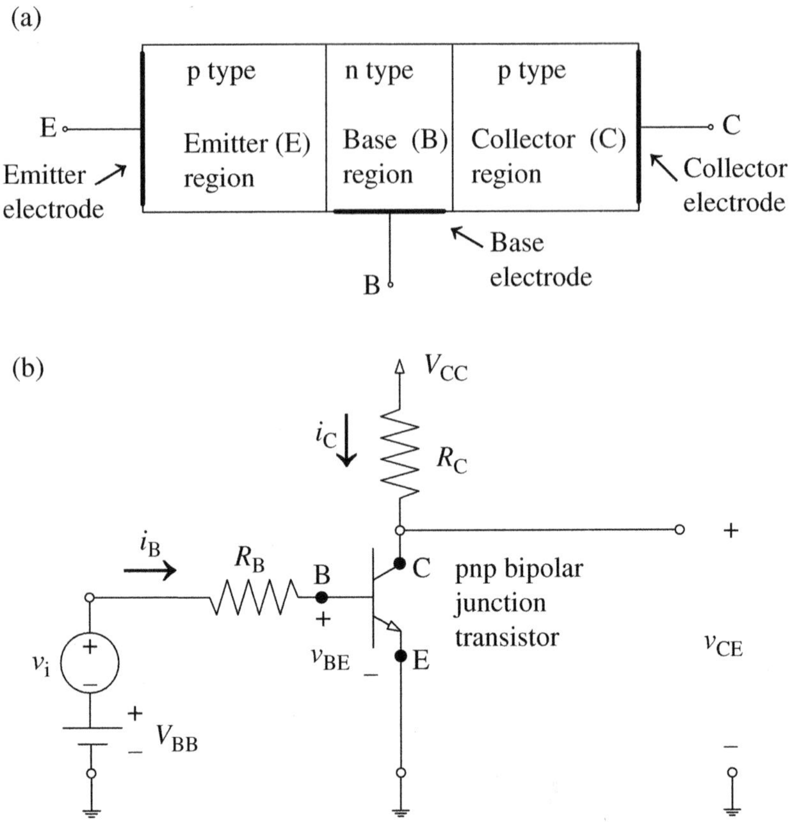

Figure 9.8 (a) pnp bipolar junction transistor (BJT). (b) Amplifier/switch circuit of pnp BJT consisting of input voltage ($V_{BB} + v_i$), base (B) resistor R_B, collector (C) resistor R_C, emitter (E), and power supply DC voltage V_{CC}. (c) ($i_C - v_{CE}$) characteristic of pnp BJT with load line for small-signal and switch operations indicated by the operating points operation as an "on" switch and operation as an "off" switch. The slope of the load line is ($-1/R_C$), (d) ($i_B - v_{BE}$) characteristic of pnp BJT with instantaneous load lines for small-signal operation. For DC analysis, $i_c = I_C$ and $i_B = I_B$.

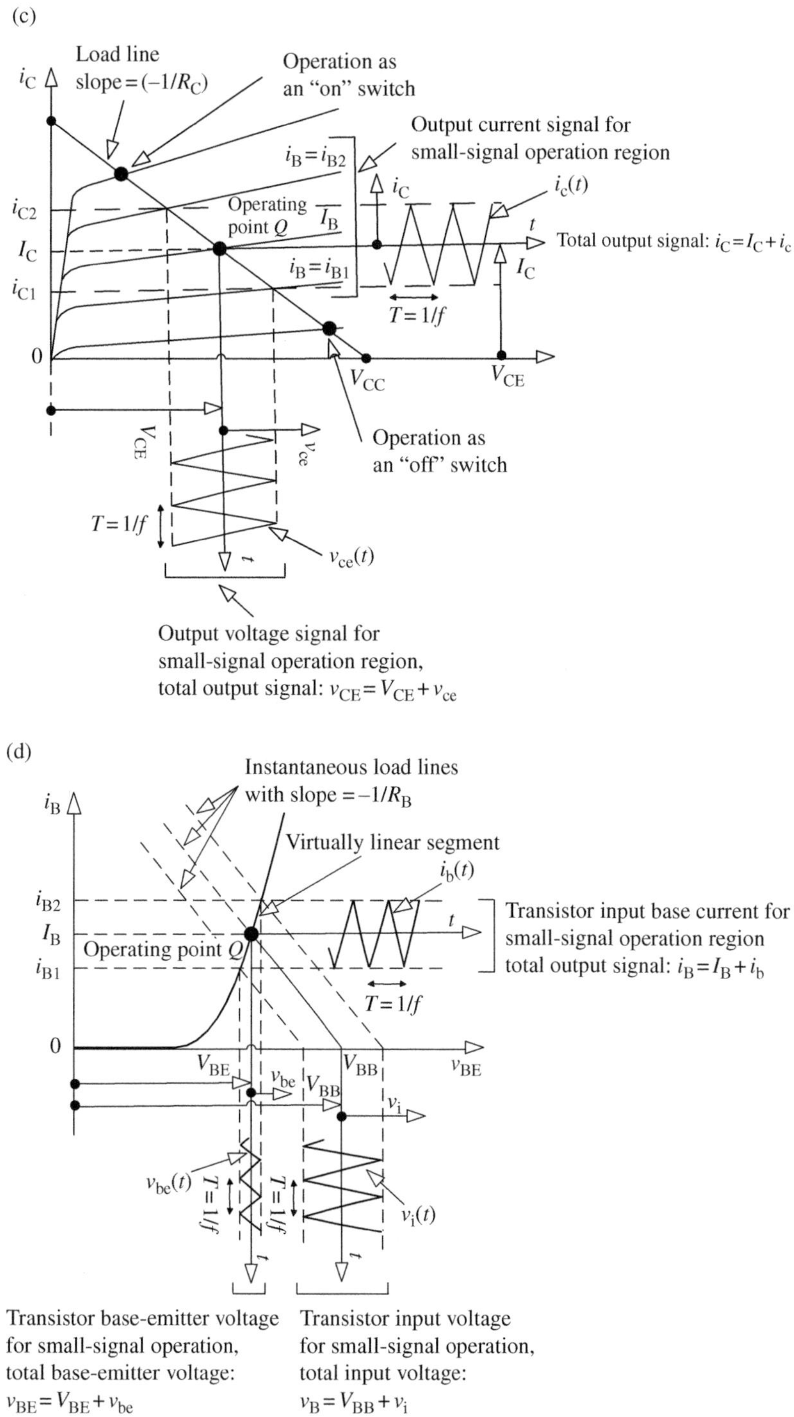

Figure 9.8 (Continued)

9.6 Metal–Oxide–Semiconductor Field-Effect Transistor (MOSFET)

In 1959, Mohamed Atalla and Dawon Kahng at Bell Labs achieved the first successful insulated-gate field-effect transistor (FET), which had been long anticipated by Lilienfeld, Heil, Shockley, and others by overcoming the "surface states" that blocked electric fields from penetrating into the semiconductor material. The MOSFET [8] of the n-type metal–oxide–semiconductor (NMOS) type (Figure 9.9a) is a self-commutated switch that can be turned on and off at will, depending upon the magnitude of the gate (G) to source (S) voltage v_{GS} indicated in Figure 9.9b, representing the amplifying/switching circuit of an enhancement-type MOSFET, and in Figure 9.9c and d its $(i_D - v_{DS})$ and $(i_D - v_{GS})$ characteristics, respectively, illustrating small-signal and switch operating regimes. While a BJT is a current-controlled device via the base current i_B, the NMOS is a voltage-controlled device, where v_{GS} can be in the range of +3 V or less and the gate current is virtually zero ($i_G \approx 0$). A FET is therefore very suitable for the design of integrated circuits (IC), compared with BJTs, where the controlling quantity is the base current. Note that the circuits and characteristics of BJT and n-type metal–oxide–semiconductor field-effect transistor (NMOSFET) (see Figures 9.8a–c and 9.9a–c, respectively) are very similar. The only differences are that in the NMOSFET, the gate current is about zero ($i_G \approx 0$), the control mechanism of the base current i_B is replaced by the gate voltage v_{GS}, and a MOSFET conducts in reverse direction (see diode in Figure 9.9b). The application of IC MOS amplifiers with nonlinear loads, resulting in complementary MOS (CMOS) logic circuits as employed in computers, is beyond the scope of this introductory text.

9.7 Thyristor (Current Gate) or Silicon-Controlled Rectifier (SCR)

The name "thyristor" stems from Greek and means "current gate" [25, 26]. A thyristor is also known under the name silicon-controlled rectifier (SCR). It is either a line-voltage- or load-voltage-commutated switch. That is, it can be turned on at will, but turn-off depends upon the zero crossing of its current, which must be accomplished by the thyristor's surrounding circuit. Figure 9.10a depicts its circuit symbol and Figure 9.10b the device current–voltage characteristic. It can be modeled by a self-commutated switch (e.g. MOSFET) in series [12] with a diode, as illustrated in Figure 9.10c.

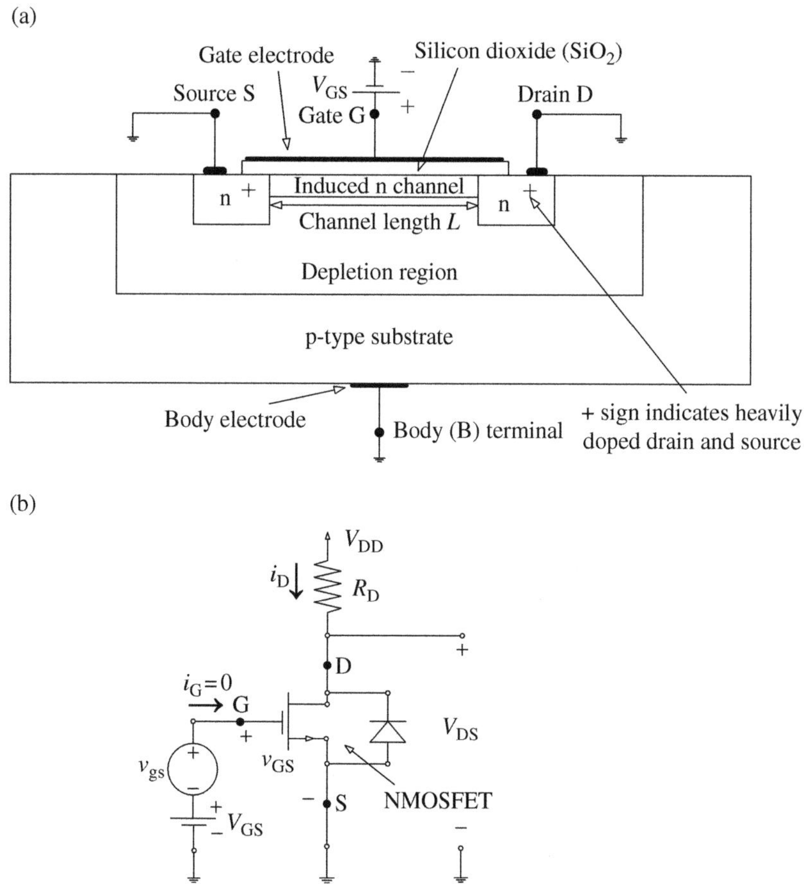

Figure 9.9 (a) Enhancement-type NMOSFET with a positive voltage applied to the gate. An n-channel is induced below the gate [19]. (b) Amplifier/switch circuit of enhancement MOSFET consisting of positive input voltage v_{GS} = (V_{GS} + v_{gs}), drain resistor R_D, and power supply DC voltage V_{DD}. Note drain (D), gate (G), and source (S). (c) (i_D − v_{DS}) characteristic of enhancement NMOSFET amplifier with load line illustrating small-signal and switch operation. The load line has the slope (−1/R_D), (d) (i_D − v_{GS}) characteristic of NMOSFET for small-signal operation [19]. For DC analysis i_D = I_D, and v_{GS} = V_{GS}.

(c)

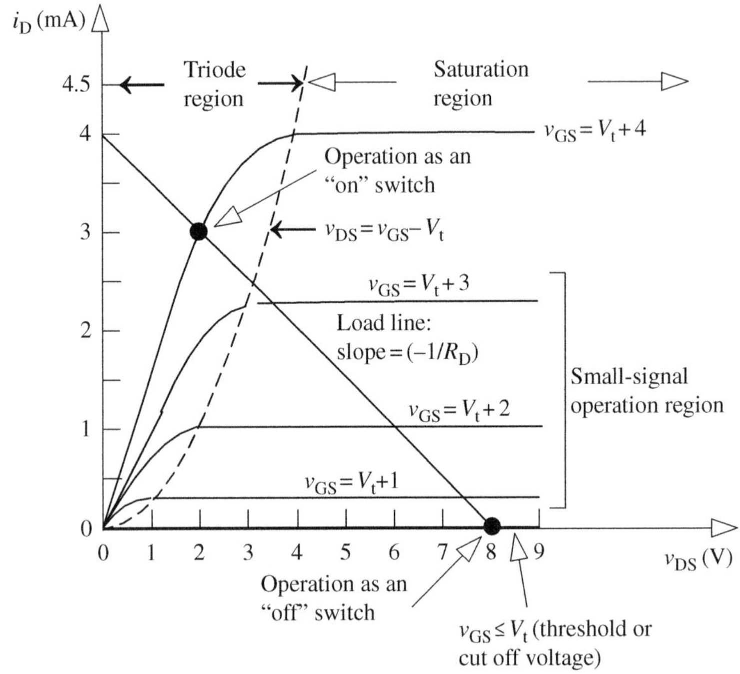

(d)

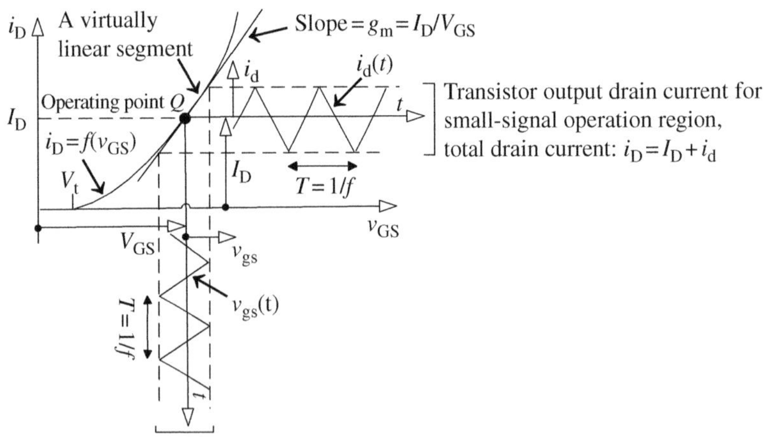

Transistor input gate-to-source voltage for small-signal operation region, total gate -to-source voltage : $v_{GS} = V_{GS} + v_{gs}$

Figure 9.9 (Continued)

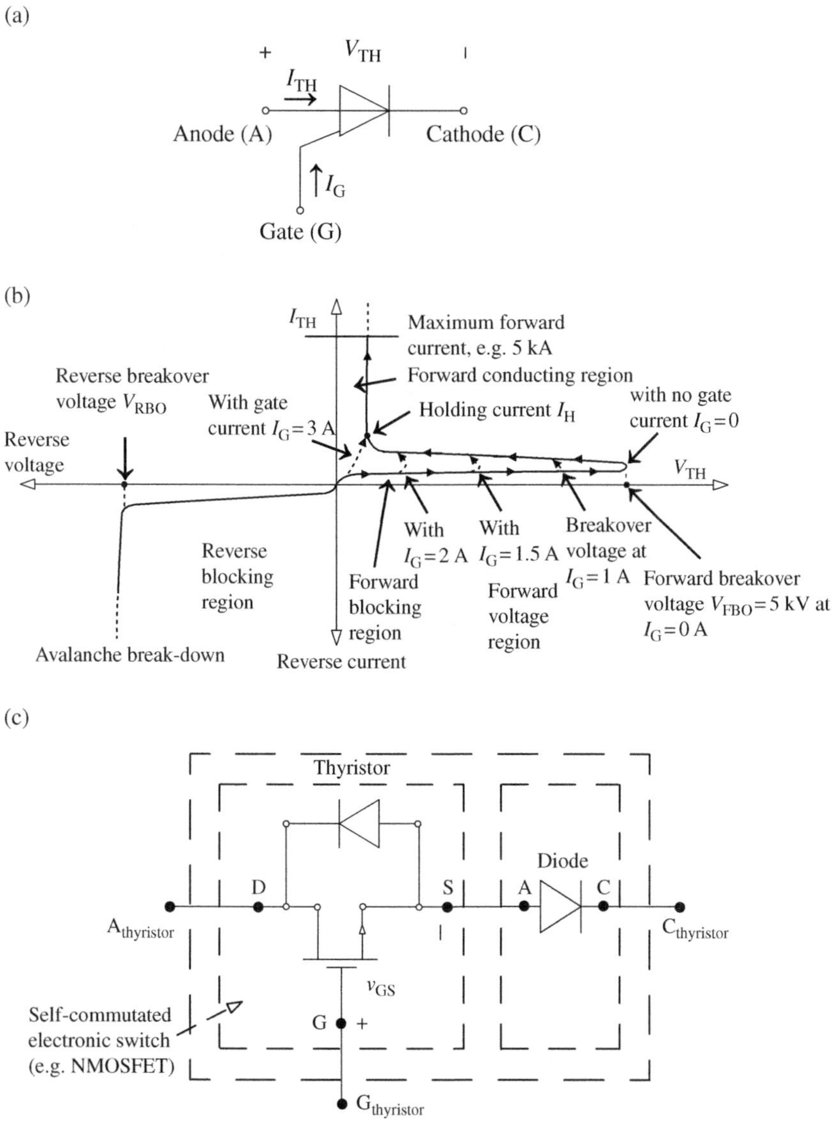

Figure 9.10 (a) Circuit symbol of a thyristor or SCR. (b) Current–voltage ($I_{TH} - V_{TH}$) characteristic of a thyristor or SCR where the gating current is positive $I_G > 0$. (c) Circuit model [12] of a thyristor or SCR, consisting of a self-commutated switch in series with a diode.

9.8 Triac

A triac [27] can be interpreted as two thyristors or SCRs connected in opposite direction (antiparallel), as illustrated in Figure 9.11a. The circuit symbol is shown in Figure 9.11b, and Figure 9.11c shows the current–voltage ($V_{triac} - I_{triac}$) characteristic of a triac.

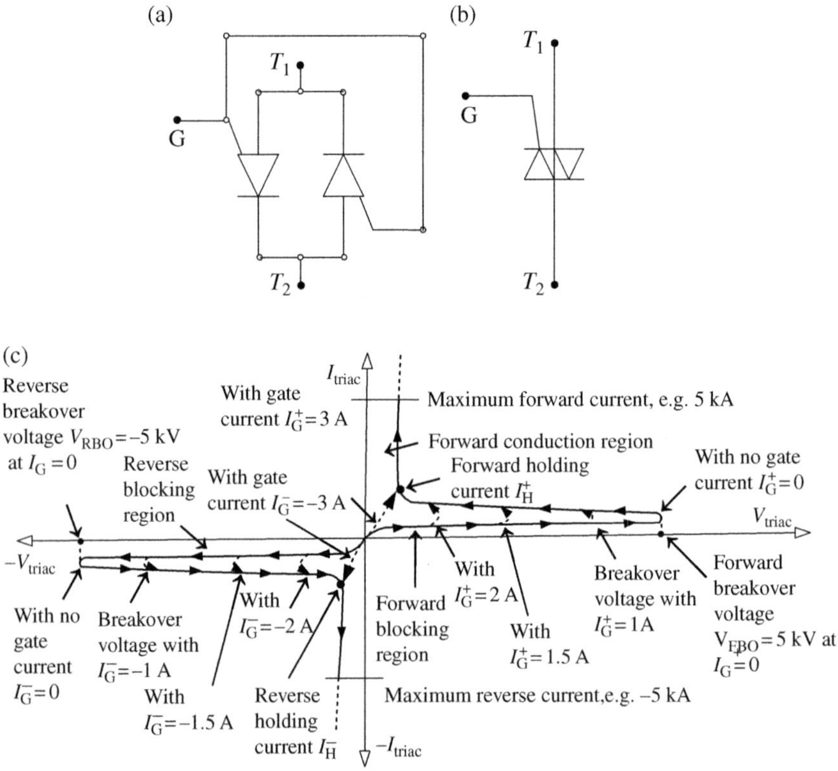

Figure 9.11 (a) Two anti-parallel-connected thyristors or SCRs, which are two building blocks for a triac. (b) Circuit symbol of a triac where I_{triac} flows in terminals T_1 and T_2 neglecting the current in terminal G (gate). (c) Current–voltage ($V_{triac} - I_{triac}$) characteristic of a triac, where the gating current I_G can be either positive I_G^+ or negative I_G^-.

9.9 Insulated-Gate Bipolar Transistor (IGBT)

The IGBT [10] is a self-commutated switch and is a cross between a BJT and a MOSFET, making it ideal as a semiconductor switching device. The IGBT incorporates the best parts of these two types of transistors – the high input impedance and high switching speeds of a MOSFET with the low saturation voltage of a BJT – and combines them to produce a switching device capable of handling large collector–emitter currents with virtually zero gate current ($i_G = 0$). IGBTs are mainly used in power electronics applications such as inverters, converters, and power supplies, where the demands of the solid-state switching device are not fully met by power BJTs and MOSFETs. High-current and high-voltage BJTs are available, but their switching speeds are slow, while MOSFETs may have higher switching speeds; however, high-voltage and high-current devices are expensive and difficult to achieve. The advantage of the IGBT over a BJT or MOSFET is that it offers greater power gain than the BJT combined with the higher voltage operation and lower input losses of the MOSFET. In effect, it is an FET integrated with a Darlington-type [19] configured BJT as shown in Figure 9.12a and b.

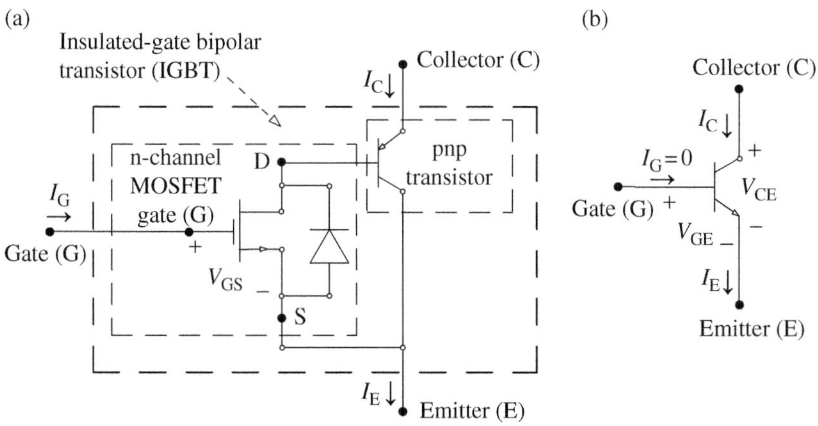

Figure 9.12 (a) Simplified equivalent circuit of IGBT consisting of Darlington [19] connection with $I_G = 0$. (b) Circuit symbol of an IGBT with $I_G = 0$, that is, $I_E = I_C$.

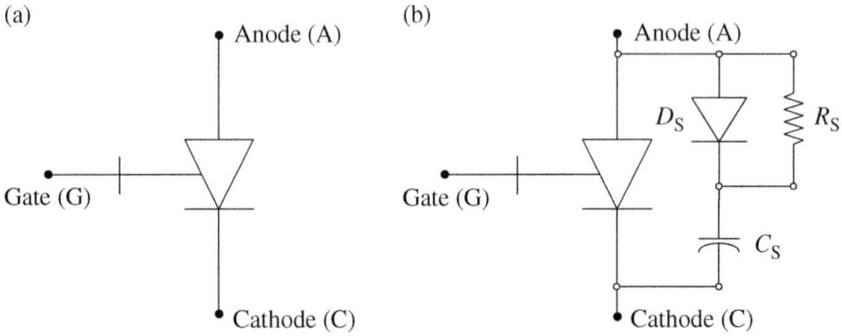

Figure 9.13 (a) Circuit symbol of GTO with anode (A), cathode (C), and gate (G). (b) GTO with snubbing (s) circuit.

9.10 Gate Turn-Off Thyristor (GTO)

The GTO [28] is a thyristor that can be turned on and off at will; that is, it is a self-commutated device. Its circuit symbol is depicted in Figure 9.13a. All electronic switches (e.g. diode, BJT, FET, BJT) require snubbing circuits to "snub out," or suppress, current and voltage spikes. If these snubbing circuits are not relied on, the rated current and voltage values are reduced, thus making the switches more expensive for a given application. These snubbing circuits are not shown in Figure 9.12, but are indicated in Figure 9.13b for the GTO.

9.11 Summary

Circuit symbols and characteristics of semiconductor switches with their applications to switches and amplifiers are presented. We mention auxiliary networks such as snubbing circuits, called snubbers (s), to reduce current and voltage spikes. Snubbers were introduced for mechanical relays a long time ago (e.g. patent of John K. Penrod Power Management Corporation [29], 1975) to increase the current and voltage rating of mechanical switches, and they permit us to increase these ratings of electronic switches as well. For power ratings of larger than 100 kW, rectifiers and inverters – based on diodes, (MOS)FETs, thyristors, IGBTs, and GTOs – snubbers are typically relied on. For large ratings, BJT snubbers play a less prominent role because they are controlled by the (base) current, which makes them less efficient compared with those controlled by the (gate) voltage, such as FETs and IGBTs. Chapters 10 and 12 will present rectifier and inverter applications suitable for renewable energy plants and automotive applications, such as hybrid and electric cars and trains. We make an effort to explain them without complicated analytical equations. Subsequently, PSPICE solutions will be presented.

References

1 The Editors of Encyclopaedia Britannica. Conductors, insulators, and semiconductors. https://www.britannica.com/technology/semiconductor-diode (accessed 17 May 2020).
2 Edison Tech Center. Mercury arc rectifiers. https://edisontechcenter.org/MercArcRectifiers.html (accessed 15 May 2020).
3 Hewitt, P.C., inventor. (1901). Mercury-vapor lamp. U.S. Patent 682,692 on September 17, 1901.
4 Rissik, H. (1941). *Mercury-Arc Current Converters*. Pitman.
5 Abram, L., McBreen, J.P., and Sherlock, J. (1958). DC winder drives using mercury-arc rectifier/inverters. *Proc. IEE – Part A: Power Eng.* 105 (20): 77–89.
6 Electronics Tutorials. PN junction diode. http://www.electronics-tutorials.ws/diode/diode_3.html (accessed 15 May 2020).
7 All About Circuits. Bipolar junction transistors. http://www.allaboutcircuits.com/textbook/semiconductors/chpt-2/bipolar-junction-transistors (accessed 15 May 2020).
8 Computer History Museum. 1960: Metal oxide semiconductor (MOS) transistor demonstrated. http://www.computerhistory.org/siliconengine/metal-oxide-semiconductor-mos-transistor-demonstrated (accessed 15 May 2020).
9 Electronics Notes. Gate turn-off thyristor, GTO. http://www.radio-electronics.com/info/data/semicond/thyristor/gate-turn-off-thyristor.php (accessed 15 May 2020).
10 Electronics Tutorials. Insulated-gate-bipolar-transistor. http://www.electronics-tutorials.ws/power/insulated-gate-bipolar-transistor.html (accessed 15 May 2020).
11 Masoum, M.A.S. and Fuchs, E.F. (2015). *Power Quality in Power Systems and Electrical Machines*, 2e. Elsevier/Academic Press, 1123 pp.
12 Fuchs, E.F. and Masoum, M.A.S. (2012). *Power Conversion of Renewable Energy Systems*, 1e, 2nd Printing. USA: Springer, 692 pp.
13 Roberts, G.W. and Sedra, A.S. (1996). *SPICE*, 2e. New York: Oxford University Press.
14 Tuinenga, P.W. (1992). *SPICE, A Guide to Circuit Simulation and Analysis Using PSpice*, 2e. Englewood Cliffs, NJ: Prentice Hall.
15 Banzhaf, W. (1992). *Computer-Aided Circuit Analysis Using PSpice*, 2e. Englewood Cliffs, NY: Regents/Prentice Hall.

16 Rashid, M.H. (1992). *Spice for Power Electronics and Electric Power*. Englewood Cliffs, NJ: Prentice Hall.

17 Fuchs, E.F. and McNaughton, G.A. (1982). Comparison of first-order finite difference and finite element algorithms for the analysis of magnetic fields, part I: theoretical analysis. *IEEE Trans. Power Appar. Syst.* 101 (5): 1170–1180; part II: numerical results. *IEEE Trans. Power Appar. Syst.* 101(5): 1181–1201 (1982).

18 Chari, M. V. K. (1970). Finite element analysis of nonlinear magnetic fields in electric machines (March 1970). PhD thesis. McGill University, Canada.

19 Sedra, A.S. and Smith, K.C. (1998). *Microelectronic Circuits*, 4e. New York/Oxford: Oxford University Press.

20 Circuits Today. Invention history of light emitting diode (LED). http://www.circuitstoday.com/invention-history-of-light-emitting-diode-led (accessed 15 May 2020).

21 Van Zeghbroeck, B. (2011). Principles of Electronic Devices, (self-published). Dept. of Electrical, Computer and Energy Engineering, University of Colorado at Boulder, Campus Box 425, Boulder, CO 80309. https://scholar.google.de/scholar?q=Van+Zeghbroeck,+B.+(2011).+Principles+of+Electronic+Devices&hl=en&as_sdt=0&as_vis=1&oi=scholart.

22 The National Renewable Energy Laboratory, 15013 Denver West Parkway, Golden, CO 80401. https://www.nrel.gov/pv/assets/pdfs/best-research-cell-efficiencies.20200406.pdf.

23 Krenz, H.J. (1987). *An Introduction to Electrical Circuits and Electronic Devices, A Laboratory Approach*. Englewood Cliffs, NJ: Prentice-Hall, Inc.

24 Resistor Guide. Varistor. http://www.resistorguide.com/varistor (accessed 15 May 2020).

25 All About Circuits The silicon-controlled-rectifier (scr). http://www.allaboutcircuits.com/textbook/semiconductors/chpt-7/silicon-controlled-rectifier-scr (accessed 15 May 2020).

26 Arsov, G.L. and Mirčevski, S. (2010). The sixth decade of the thyristor. *Electronics* 14 (1):3. http://electronics.etfbl.net/journal/Vol14No1/xPaper_01.pdf (accessed 15 May 2020).

27 American Microsemiconductor, Inc. http://www.americanmicrosemi.com/tutorials/triac.htm (accessed 15 May 2020).

28 Mitsubishi Electric. Mitsubishi high power semiconductors: feature and application of gate turn-off thyristors. http://www.mitsubishielectric.com/semiconductors/files/manuals/gtothyristors.pdf (accessed 15 May 2020).

29 Penrod, J.K. Arc suppressor circuit. US3982137A. https://patents.google.com/patent/US3982137A/en (accessed 15 May 2020).

10

Applications of Semiconductor Switches Using PSPICE Uncontrolled and Controlled AC–DC Converters (Rectifiers), AC Voltage and Current Regulators and Controllers, and DC–AC Converters (Inverters)

Rectifiers, inverters, and alternating current (AC) voltage and current controllers are indispensable for renewable energy systems that generate and consume electric energy at variable speed and torque. While in Chapter 9 the basic components of semiconductor devices [1] are addressed, in this chapter we highlight their deployment via learning by progressively more complicated application examples, without going into the details of power electronic switches. An analytical approach for simple circuits is confirmed by PSPICE [2–5] (Personal Simulation Program with Integrated Circuit Emphasis), and solutions for more complicated circuit designs are achieved with computer-aided analysis based on PSPICE. Commercially available standard resistor and capacitor values [6] are used throughout. Any resistor and capacitor values can be approximately obtained through series and parallel connections. In order to stress the application of this text to renewable electric energy systems [7], in the application examples we mostly use voltages larger than 100 V and currents larger than 5 A. Most texts concerned with (micro-) electronics deal with voltages and currents in the few volt and milliampere ranges, respectively.

To make reading this chapter easier, each abbreviation will be defined when used for the first time, where the first characters are repeated in parentheses, for example, "alternating current (AC)."

The objective of rectification from alternating current (AC) to direct current (DC) via rectifiers is to obtain a unipolar quantity (e.g. current, voltage, power, energy) with a minimum of harmonic variation determined from Fourier analysis. Measures for rectification are the average values of abovementioned quantities, such as average current I_{avg}, average voltage V_{avg}, average power P_{avg}, and average energy E_{avg}.

Introduction to Energy, Renewable Energy, and Electrical Engineering: Essentials for Engineering Science (STEM) Professionals and Students, First Edition. Ewald F. Fuchs and Heidi A. Fuchs.

Companion website: www.wiley.com/go/fuchs/renewableenergyandelectricalengineering

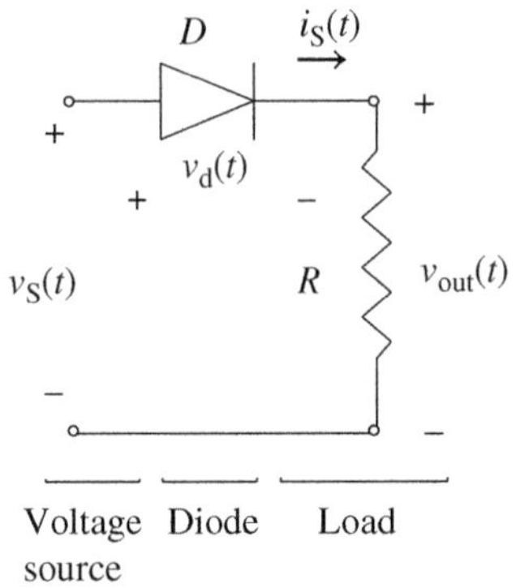

Figure 10.1 Rectifier supplying pulsating DC current from single-phase AC voltage source (s) $v_s(t) = v_{out}(t) = \sqrt{2} \cdot 120\,\text{V} \sin \omega t$ through diode D to resistor $R = 3.0\,\Omega$.

The objective of inversion from DC to AC is to obtain a sinusoidal/cosinusoidal quantity (e.g. current, voltage) with a minimum of harmonic distortion determined from Fourier analysis. Measures for the quality of inversion via inverters are the total harmonic distortion [8] of current i (THD_i) and total harmonic distortion of voltage v (THD_v).

10.1 Half-Wave, Single-Phase Rectification

Although half-wave rectification results in poor power quality [8] due to the generation of large even current harmonics and – through their voltage drops across the system impedance – also voltage harmonics, it is desirable to first explain the rectification based on simple rectifier configurations. Thereafter, we will discuss full-wave, single-phase, and three-phase rectification, which are very important for renewable energy systems, one reason being that three-phase systems require the least amount of conduction material (e.g. copper, aluminum) to process and transmit electric energy and have fewer power quality problems than single-phase systems.

The following half-wave, single-phase diode rectifier with resistor without capacitor is analyzed. Figure 10.1 depicts the simplest rectifier consisting of semiconductor diode (see Figures 9.3–9.5a in Chapter 9) in series with resistor R, and Figure 10.2a–c illustrate common approximations used for diode characteristics.

Application Example 10.1 Ideal diode characteristic
The application of Figure 10.2a to the circuit of Figure 10.1 results with $i_s(t) = v_s(t)/R$ in the pulsating load/source (s) current $i_s(t) = \sqrt{2} \cdot 120\,\text{A} \sin \omega t/3.0 = \sqrt{2} \cdot 40\,\text{A} \sin \omega t$. The source voltage $v_s(t)$, load output voltage, and load current $i_s(t)$ are shown in Figure E10.1.1a. The amplitude of the voltage is $a_v = 169.68\,\text{V}$, and

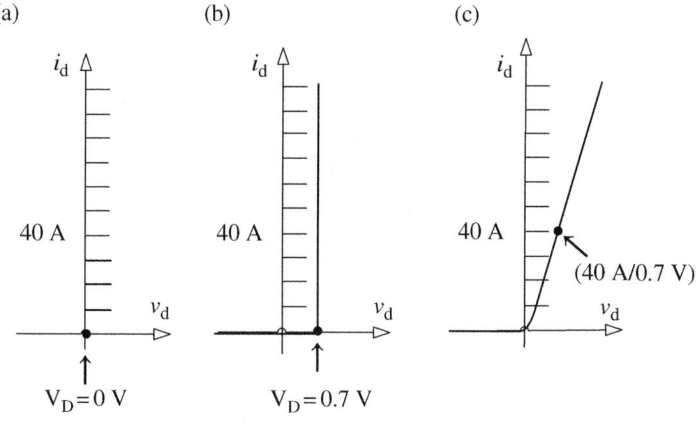

Figure 10.2 Common approximations [1] for diode characteristics: (a) ideal characteristic, (b) constant voltage characteristic with $V_D = 0.7$ V, and (c) exponential characteristic [1] $I_D = I_s \cdot (e^{V_D/nV_T} - 1) \approx I_s \cdot e^{V_D/nV_T}$, with $I_s = 34.54 \cdot 10^{-12}$ A and $V_T = 25.2$ mV; one obtains the operating point ($I_D = 40$ A/$V_D = 0.7$ V) as explained in Chapter 9.

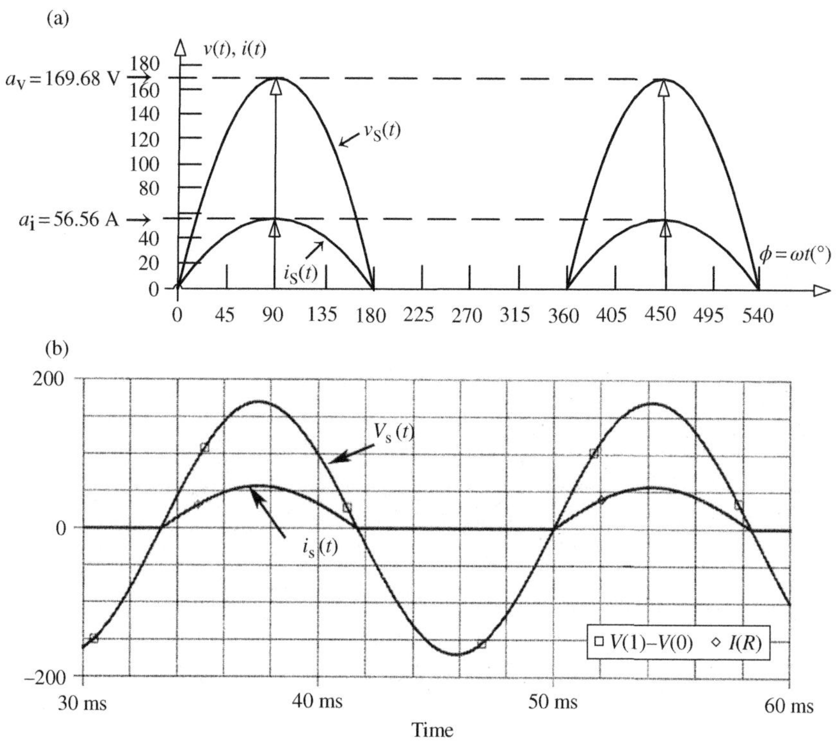

Figure E10.1.1 (a) Source voltage $v_s(t)$, load output voltage $v_{out}(t) = v_s(t)$, and load output current $i_s(t)$ of rectifier with resistor $R = 3\ \Omega$ load based on ideal ($i_d - v_d$) characteristic of diode. (b) Source voltage $v_s(t) = V(1) - V(0)$ and source current $i_s(t) = I(R)$ of rectifier with resistor $R = 3\ \Omega$ based on ideal characteristic of diode.

Table E10.1.1a Voltage harmonics of $v_s(t) = v_{out}(t)$ of Figure E10.1.1a in terms of volts and in percent referred to voltage amplitude $a_v = 169.68$ V based on closed-form solution.

DC component	1	2	4	6
54.01 V/31.83%	84.32 V/50%	35.99 V/21.22%	7.19 V/4.24%	3.08 V/1.82%

Table E10.1.1b PSPICE program [2–5] for ideal ($i_d - v_d$) characteristic with R = 3 Ω load.

```
Application Example 10.1
*AC source, ideal diode, R=3 ohms
Vs 1 0 sin(0V 169.68V 60Hz)
*ideal diode D
D 1 2 IDEAL
*Load
R 2 0 3
*Model
.Model IDEAL D(IS=0.01E-12)
*Analysis request
.Tran 10us 60ms 30ms 10us
*Output request
.Probe
.FOUR 60 10 V(R)
.end
```

the amplitude of the current is $a_i = 56.56$ A. A Fourier analysis of the pulsating quantities as shown in Figure E10.1.1a is

$$y = \frac{1}{\pi} + \frac{1}{2}\sin x - \frac{2}{\pi}\left(\frac{\cos 2x}{1\cdot 3} + \frac{\cos 4x}{3\cdot 5} + \frac{\cos 6x}{5\cdot 7} + \dots\right). \quad \text{(E10.1.1)}$$

From Eq. (E10.1.1) one obtains the average value for the voltage $v_s(t) = v_{out}(t)$, $V_{avg} = a_v/\pi = 54.012$ V, and the average value for the current $i_s(t)$ $I_{avg} = a_i/\pi = 56.56/\pi = 18.00$ A. Clearly this is not a good rectification, because the average values are very low compared to the amplitudes of $v_s(t)$ and $i_s(t)$, a_v and a_i, respectively. In addition the odd and even harmonics are large, as can be noted from Table E10.1.1a. Due to the resistive load, the percentage current harmonic spectrum is the same as that of Table E10.1.1a.

With PSPICE program of Table E10.1.1b, one obtains Figure E10.1.1b.

Conclusions

Both sets of data, the one obtained from Eq. (E10.1.1), as shown in Table E10.1.1a, and the PSPICE values in Table E10.1.1c, are comparable; that is, Figure E10.1.1a is confirmed by PSPICE simulation, resulting in Figure E10.1.1b. This enables us to obtain the results of all subsequent application examples based on PSPICE.

Table E10.1.1c Voltage harmonics of $v_s(t) = v_{out}(t)$ of Figure E10.1.1b in terms of volts and in percent referred to voltage amplitude $a_v = 169.68$ V based on PSPICE numerical solution. The PSPICE simulation yields the data: DC COMPONENT = 5.361 093E + 01 V.

Harmonic order number	Frequency (Hz)	Fourier component (V)	Normalized Fourier component (PU)	Phase (DEG)	Normalized phase (DEG)
1	60	84.32	1.000E + 00	−1.440E + 02	0.000E + 00
2	120	35.99	4.268E − 01	−1.800E + 01	2.700E + 02
3	180	0.1647	1.953E − 03	1.080E + 02	5.400E + 02
4	240	7.194	8.531E − 02	5.400E + 01	6.300E + 02
5	300	0.097	1.150E − 03	−1.800E + 02	5.400E + 02
6	360	3.081	3.654E − 02	1.260E + 02	9.900E + 02
7	420	0.068 44	8.116E − 04	−1.080E + 02	9.000E + 02
8	480	1.710	2.028E − 02	−1.620E + 02	9.900E + 02
9	540	0.0527	6.254E − 04	−3.597E + 01	1.260E + 03
10	600	1.087	1.290E − 02	−9.000E + 01	1.350E + 03

Total harmonic voltage distortion $THD_v = 4.374\,829E + 01\%$.

Application Example 10.2 Diode characteristic approximated by $V_D = 0.7$ V
If we assume a constant voltage approximation of Figure 10.2b with $V_D = 0.7$ V, then Figure E10.2.1 follows and is obtained via the PSPICE program of Table E10.2.1. The voltage harmonics of $v_{out}(t)$ are listed in Table E10.2.2.

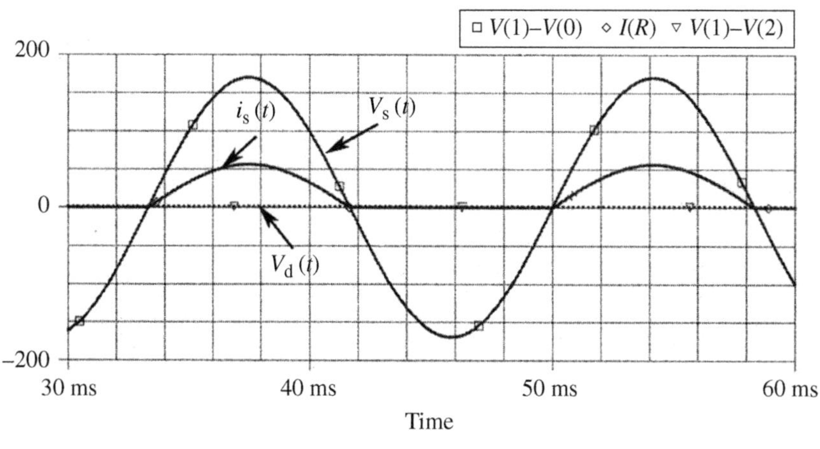

Figure E10.2.1 Source voltage $v_s(t) = V(1) - V(0)$, diode voltage $v_d(t) = V(2) - V(3) = V_D$, and source current $i_s(t) = I(R)$ of rectifier with resistor $R = 3\ \Omega$ based on $(i_d - v_d)$ characteristic of diode with $V_D = 0.7$ V.

Table E10.2.1 PSPICE program [2–5] for (i_d – v_d) characteristic with V_D = 0.7 V with R = 3 Ω load.

```
Application Example 10.2          *Model
*AC source                        .Model IDEAL D(IS=0.01E-12)
Vs 1 0 sin(0V 169.68V 60Hz)       *Analysis request
*Ideal diode D                    .Tran 10us 60ms 30ms 10us
D 1 2 ideal                       *Output request
VD 2 3 0.7V                       .Probe
*Load                             .FOUR 60 10 V(R)
R 3 0 3                           .end
```

Table E10.2.2 Voltage harmonics of $v_{out}(t)$ of Figure E10.2.1 in terms of volts and in percent referred to voltage amplitude a_v = 169.68 V.

DC component	1	2	4	6
53.26 V/31.39%	83.9 V/49.45%	36.00 V/21.22%	7.20 V/4.24%	3.08 V/1.82%

Application Example 10.3 Diode characteristic approximated by exponential function

If the exponential current diode relationship is used (Figure 10.2c: 40 A at 0.7 V), then Figure E10.3.1 results. The PSPICE program is listed in Table E10.3.1, and the voltage harmonics of $v_{out}(t)$ are given in Table E10.3.2.

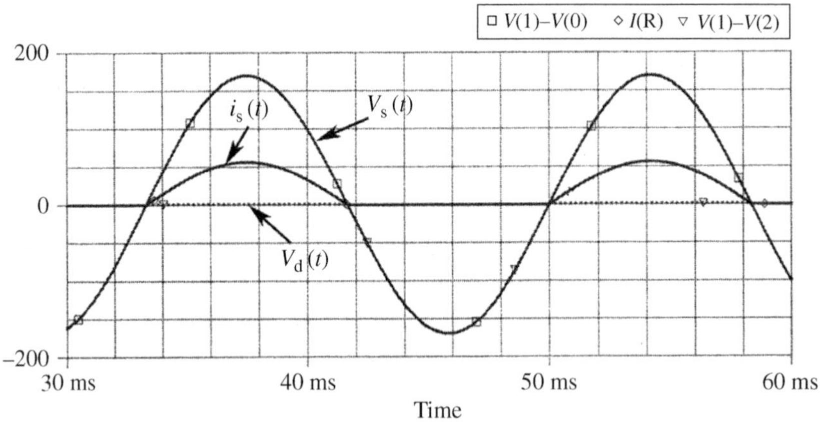

Figure E10.3.1 Source voltage $v_s(t)$ = $V(1)$ – $V(0)$, diode voltage $v_d(t)$ = $V(1)$ – $V(2)$ and source current $i_s(t)$ = $I(R)$ of rectifier with resistor R = 3 Ω based on exponential (i_d – v_d) characteristic of diode.

Table E10.3.1 PSPICE program for exponential ($i_d - v_d$) characteristic with $R = 3\ \Omega$ load.

```
Application Example 10.3
*AC source
Vs 1 0 sin(0V 169.68V 60Hz)
*Real diode
D 1 2 real
*Load
R 2 0 3
*Model
.Model real D(IS=34.54E-12)
*Analysis request
.Tran 10us 60ms 30ms 10us
*Output request
.Probe
.FOUR 60 10 V(R)
.end
```

Table E10.3.2 Voltage harmonics of $v_{out}(t)$ of Figure E10.3.1 in terms of volts and in percent referred to voltage amplitude $a_v = 169.68$ V.

DC component	1	2	4	6
53.60 V/31.59%	84.30 V/49.68%	36.00 V/21.22%	7.20 V/4.24%	3.08 V/1.82%

Conclusions

The type of diode approximation does not significantly influence the DC component and harmonics generated by the rectification of an AC voltage. One concludes that the ideal diode approximation [1] $i_d = I_s \cdot (e^{v_d/nV_T} - 1) \approx I_s \cdot e^{v_d/nV_T}$ with $I_s = 0.01\text{E}{-12}$ A and $V_T = 25.2$ mV at 20 °C is adequate for the above application examples.

Application Example 10.4 Rectifier with the three different diode approximations of Figure 10.2a–c and inductive load ($R = 3\ \Omega$, $L = 5$ mH).

Most resistors have an inductivity L due to the coiled resistance wire (see Figure E10.4.1), resulting in Figures E10.4.2–E10.4.4 and the Tables E10.4.1–E10.4.6 for the three ($i_d - v_d$) approximations of a diode.

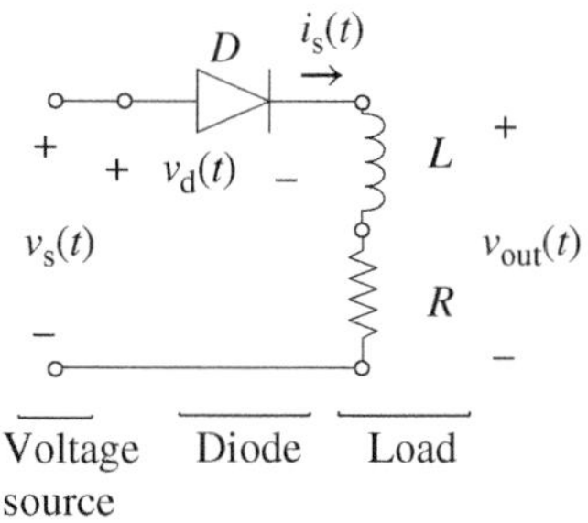

Figure E10.4.1 Rectifier supplying pulsating DC current from single-phase voltage source (s) $v_s(t) = \sqrt{2} \cdot 120$ V sin ωt through diode D to resistor $R = 3\ \Omega$ including inductivity $L = 5$ mH.

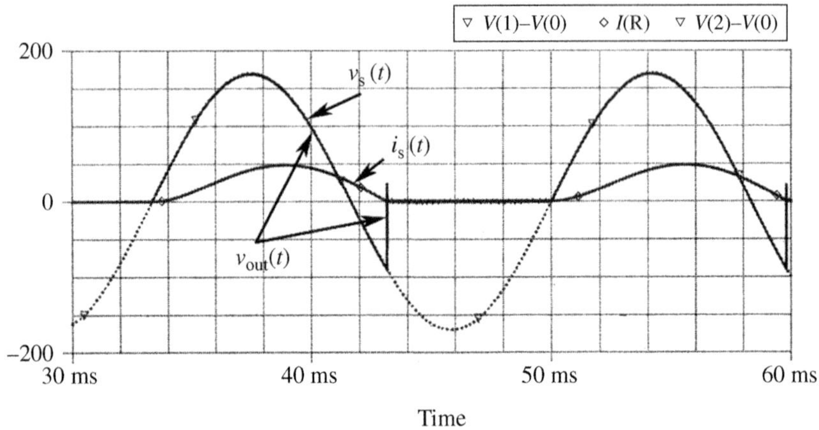

Figure E10.4.2 Source voltage $v_s(t) = V(1) - V(0)$, output load voltage $v_{out}(t) = V(2) - V(0)$, and source current $i_s(t) = I(R)$ of rectifier with resistor $R = 3\ \Omega$ and inductivity $L = 5$ mH based on ideal ($i_d - v_d$) characteristic of diode. Note the lagging phase shift of $i_s(t)$ due to inductance L and the spike of $v_{out}(t)$ when current $i_s(t)$ reaches zero value due to $v_L = L(di/dt)$.

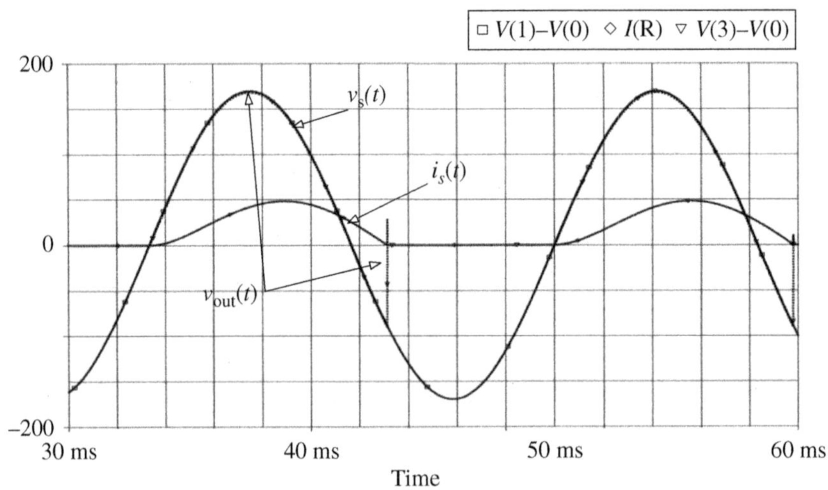

Figure E10.4.3 Source voltage $v_s(t) = V(1) - V(0)$, output voltage $v_{out}(t) = V(3) - V(0)$, and current $i_s(t) = I(R)$ of rectifier with resistor $R = 3\ \Omega$ and inductivity $L = 5$ mH based on constant voltage drop $V_D = 0.7$ V ($i_d - v_d$) characteristic of diode. Note the lagging phase shift of $i_s(t)$ due to inductance L and the spike of $v_{out}(t)$ when current $i_s(t)$ reaches zero value due to $v_L = L$ (di/dt).

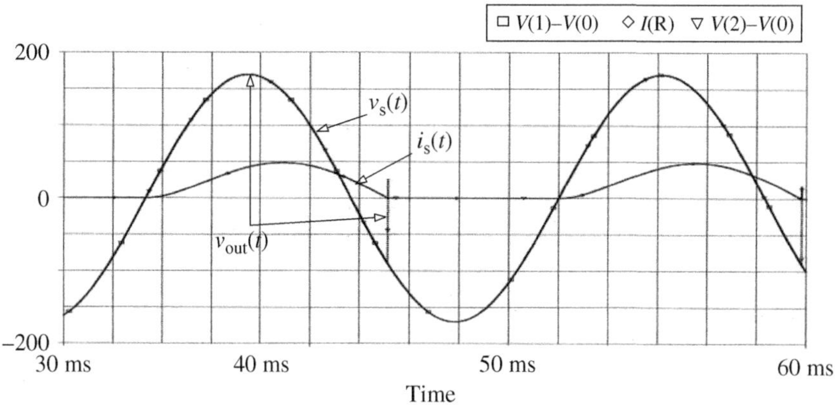

Figure E10.4.4 Source voltage $v_s(t) = V(1) - V(0)$, output load voltage $v_{out}(t) = V(2) - V(0)$, and source current $i_s(t) = I(R)$ of rectifier with resistor $R = 3\ \Omega$ and inductivity $L = 5$ mH based on exponential ($i_d - v_d$) characteristic of diode. Note the lagging phase shift of $i_s(t)$ due to inductance L and the spike of $v_{out}(t)$ when current $i_s(t)$ reaches zero value due to $v_L = L(di/dt)$.

Table E10.4.1 PSPICE program for ideal diode ($i_d - v_d$) characteristic with $R = 3\ \Omega$ and $L = 5$ mH load.

```
Application Example 10.4a          *Model
*AC source                         .Model IDEAL D(IS=0.01E-12)
Vs 1 0 sin(0V 169.68V 60Hz)        *Analysis request
*Ideal diode D                     .Tran 10us 60ms 30ms 10us
D 1 2 IDEAL                        *Output request
*Load                              .Probe
L 2 3 5m                           .FOUR 60 10 V(R)
R 3 0 3                            .end
```

Table E10.4.2 Voltage harmonics of $v_{out}(t)$ of Figure E10.4.2 in terms of volts and in percent referred to voltage amplitude $a_v = 169.68$ V.

DC component	1	2	4	6
49.67 V/29.27%	74.20 V/43.73%	26.32 V/15.51%	3.92 V/2.31%	1.20 V/0.71%

Table E10.4.3 PSPICE program for diode ($i_d - v_d$) characteristic modeled with constant voltage drop V_D = 0.7 V with R = 3 Ω and L = 5 mH load.

```
Application Example 10.4b
*AC source, 0.7V diode drop, load R=3 ohms
Vs 1 0 sin(0V 169.68V 60Hz)
*Ideal diode D
D 1 2 ideal
VD 2 3 0.7V
*Load
L 3 4 5m
R 4 0 3
*Model
.Model IDEAL D(IS=0.01E-12)
*Analysis request
.Tran 10us 60ms 30ms 10us
*Output request
.Probe
.FOUR 60 10 V(R)
.end
```

Table E10.4.4 Voltage harmonics of $v_{out}(t)$ of Figure E10.4.3 in terms of volts and in percent referred to voltage amplitude a_v = 169.68 V.

DC component	1	2	4	6
49.12 V/28.95%	73.79 V/43.49%	26.34 V/15.52%	3.95 V/2.33%	1.22 V/0.72%

Table E10.4.5 PSPICE program for exponential diode ($i_d - v_d$) characteristic with R = 3 Ω and L = 5 mH load.

```
Application Example 10.4c
*AC source, exponential diode
*characteristic, load R=3 ohms, L=5mH
Vs 1 0 sin(0V 169.68V 60Hz)
*Ideal and real diodes in series
D2 1 2 real
*Load
L 2 3 5mH
R 3 0 3
*Model
.Model real D(IS=34.54E-12)
*Analysis request
.Tran 10us 60ms 30ms 10us
*Output request
.Probe
.FOUR 60 10 V(R)
.end
```

Table E10.4.6 Voltage harmonics of $v_{out}(t)$ of Figure E10.4.4 in terms of volts and in percent referred to voltage amplitude a_v = 169.68 V.

DC component	1	2	4	6
49.45 V/29.14%	74.18 V/43.72%	26.32 V/15.51%	3.92 V/2.31%	1.20 V/0.71%

Conclusions

The inductance of a resistor decreases the average value of the DC component of the output voltage of a rectifier and its harmonic components. The inductance L is responsible for the large negative voltage spike – due to $v_L = L(di/dt)$ – when the current reaches a zero value, at turn-off caused by the diode, and L causes the output current to lag the output voltage, that is, a lagging phase shift between current and voltage. As noted in previous examples, the ideal 0.7 V and exponential diode approximations generate about the same results. For this reason, in all subsequent examples, ideal diode characteristics will be assumed.

Application Example 10.5 Half-wave, single-phase diode rectifier with resistor $R = 3\ \Omega$, inductor $L = 1$ mH, and capacitor $C = 500\ \mu F$

In order to obtain unipolar current $i_R(t)$ and voltage $v_{out}(t)$ with low harmonic content from an AC source, the rectifier must be equipped with a capacitor C to smooth the AC pulsations/ripples. Figure E10.5.1 depicts the rectifier consisting of semiconductor diode D (see Figures 9.3 and 9.4, Chapter 9) in series with a resistor $R = 3\ \Omega$ with a paralleled capacitor $C = 500\ \mu F$, resulting in a capacitive time constant $\tau = RC = 3 \cdot 500 \cdot 10^{-6}$ s $= 1.5$ ms corresponding to a fraction of one 60 Hz period.

The application of Figure 10.2a to the circuit of Figure E10.5.1 results with $R = 3\ \Omega$ and $C = 500\ \mu F$ in the output voltage, source voltage, and various currents as shown in Figure E10.5.2. The PSPICE program is displayed in Table E10.5.1, and the voltage harmonics of $v_{out}(t)$ are itemized in Table E10.5.2.

Figure E10.5.3 depicts the rectifier consisting of semiconductor diode D (see Figures 9.3 and 9.4 in Chapter 9) in series with a resistor $R = 3\ \Omega$ and inductor $L = 1$ mH with a paralleled capacitor $C = 500\ \mu F$.

The application of Figure 10.2a to the circuit of Figure E10.5.3 results with $R = 3\ \Omega$, $L = 1$ mH, $C = 500\ \mu F$ in Figure E10.5.4, where the source voltage $v_s(t)$, output load voltage $v_{out}(t)$, source current $i_s(t)$, capacitor current $i_C(t)$, and resistor current

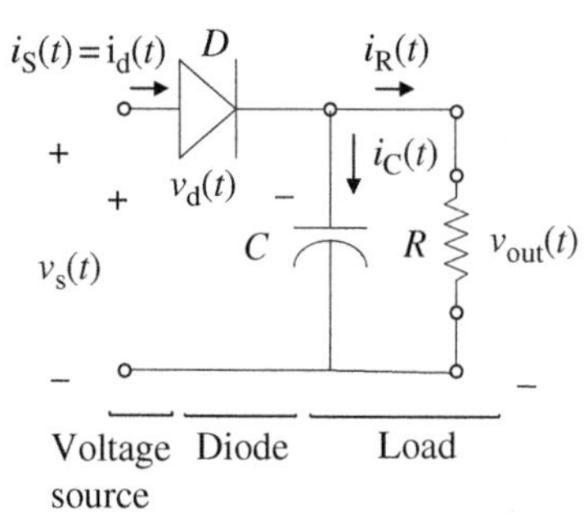

Figure E10.5.1 Rectifier supplying pulsating DC current from single-phase voltage source (s) $v_s(t) = \sqrt{2} \cdot 120$ V $\sin \omega t$ through diode D to resistor $R = 3\ \Omega$ and capacitor [6] $C = 500\ \mu F$.

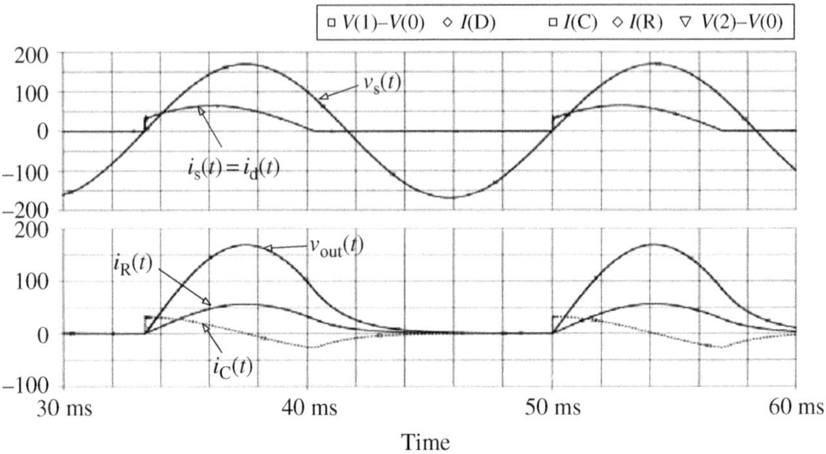

Figure E10.5.2 Top plot: source voltage $v_s(t) = V(1) - V(0)$; source current $i_s(t)$ is equal to diode current $i_s(t) = i_d(t) = I(D)$. Bottom plot: output load voltage $v_{out}(t) = V(2) - V(0)$, resistor current $i_R(t) = I(R)$, and capacitor current $i_C(t) = I(C)$ of rectifier based on ideal $(i_d - v_d)$ characteristic of diode with $R = 3\ \Omega$ and $C = 500\ \mu F$. C causes the source current $i_s(t)$ to lead the source voltage $v_s(t)$, that is, a leading phase shift between current and voltage. See Table E10.5.1 for PSPICE program.

Table E10.5.1 PSPICE program for ideal diode $(i_d - v_d)$ characteristic, $R = 3\ \Omega$, and $C = 500\ \mu F$.

```
Application Example 10.5a
*AC source, ideal diode, R=3 ohms, C=500uF
Vs 1 0 sin(0V 169.68V 60Hz)
*Ideal diode D
D 1 2 IDEAL
*Load
R 2 0 3
C 2 0 500u
*Model
.Model IDEAL D(IS=0.01E-12)
*Analysis request
.Tran 10us 60ms 30ms 10us
*Output request
.Probe
.FOUR 60 10 V(R)
.end
```

Table E10.5.2 Voltage harmonics of $v_{out}(t)$ of Figure E10.5.2 in terms of volts and in percent referred to voltage amplitude $a_v = 169.68$ V.

DC component	1	2	4	6
57.62 V/33.96%	82.28 V/48.49%	31.54 V/18.59%	4.78 V/2.82%	1.48 V/0.87%

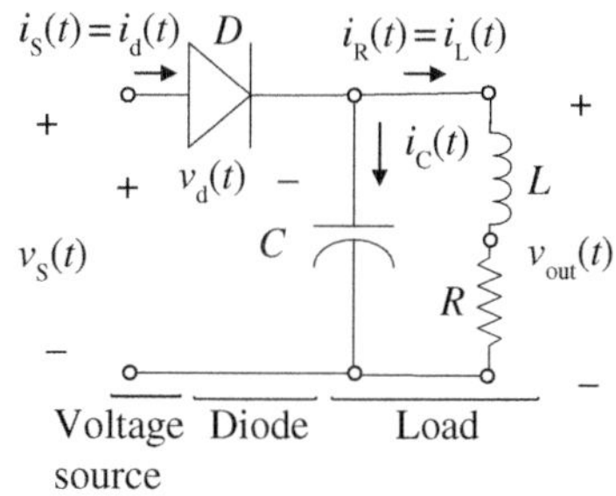

Figure E10.5.3 Rectifier supplying pulsating DC current from single-phase voltage source (s) $v_s(t) = \sqrt{2} \cdot 120\,V \sin \omega t$ through diode D to resistor $R = 3\,\Omega$ including inductivity $L = 1$ mH and capacitor $C = 500\,\mu F$.

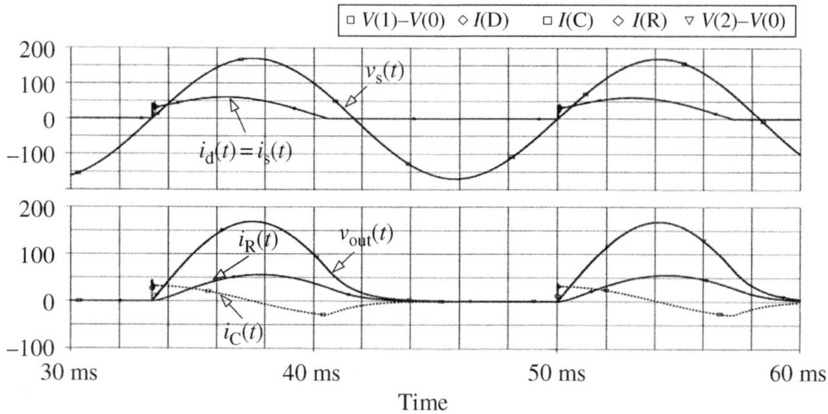

Figure E10.5.4 Top plot: source voltage $v_s(t) = V(1) - V(0)$ and source current equal diode current $i_s(t) = i_d(t) = I(D)$. Bottom plot: output load voltage $v_{out}(t) = V(2) - V(0)$, resistor current $i_R(t) = I(R)$, and capacitor current $i_C(t) = I(C)$ of rectifier based on ideal $(i_d - v_d)$ characteristic of diode with $R = 3\,\Omega$, $L = 1$ mH, and $C = 500\,\mu F$.

$i_R(t)$ are shown. See Table E10.5.3 for PSPICE program and Table E10.5.4 for the DC value and harmonic components.

Conclusions

The inductance L in series with the resistor decreases the DC (unipolar, average) value of the output voltage; that is, for any rectifier inductances should be minimized. In addition, any inductance increases current oscillations. Depending upon the size of the capacitance C and the inductance L, the input current $i_s(t)$ is either leading or lagging the input source voltage $v_s(t)$. Application Examples 10.1–10.5 demonstrate the operating principle of a rectifier but fail to enable the reader to construct a suitable rectifier with low harmonic voltages and currents. In the following we will analyze a so-called half-wave, single-phase diode peak rectifier.

Table E10.5.3 PSPICE program for ideal diode ($i_d - v_d$) characteristic, $R = 3\ \Omega$, $L = 1$ mH, and $C = 500\ \mu$F.

```
Application Example 10.5b
*AC source, ideal diode, R=3 ohms, C=500uF
Vs 1 0 sin(0V 169.68V 60Hz)
*Ideal diode D
D 1 2 IDEAL
*Load
R 2 3 3
L 3 0 1m
C 2 0 500u
*Model
.Model IDEAL D(IS=0.01E-12)
*Analysis request
.Tran 10us 60ms 30ms 10us
*Output request
.Probe
.FOUR 60 10 V(R)
.end
```

Table E10.5.4 Voltage harmonics of $v_{out}(t)$ of Figure E10.5.4 in terms of volts and in percent referred to voltage amplitude $a_v = 169.68$ V.

DC component	1	2	4	6
55.75 V/32.86%	82.88 V/48.85%	31.92 V/18.81%	4.75 V/2.80%	1.44 V/0.85%

Application Example 10.6 Half-wave, single-phase diode peak rectifier
The circuit of Figure E10.5.1 will be relied on to design a peak rectifier where the DC component is about 98% of the amplitude a_v of the source voltage of $v_s(t)$. With $R = 30\ \Omega$ and $C = 20$ mF a capacitive time constant of $\tau = RC = 30 \cdot 20 \cdot 10^{-3}$ s = 600 ms results, corresponding to 36 periods of 60 Hz. Figure E10.6.1 shows the source

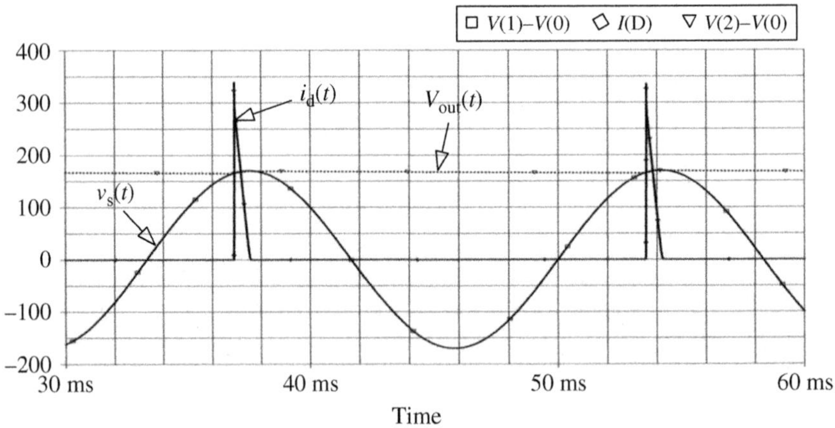

Figure E10.6.1 Source voltage $v_s(t) = V(1) - V(0)$ as well as the output voltage $v_{out}(t) = V(2) - V(0)$ and the charging current through the diode $i_d(t) = I(D)$ on a large scale from −200 to 400 V and 0 to 340 A, respectively.

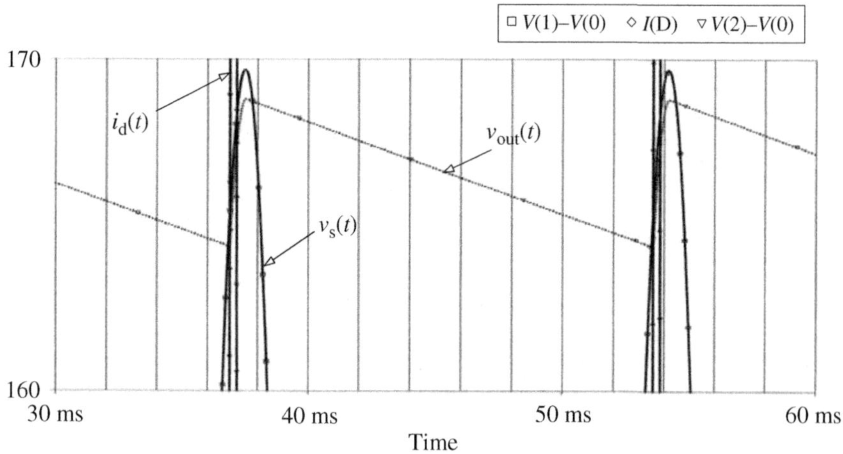

Figure E10.6.2 Source voltage $v_s(t) = V(1) - V(0)$ as well as the output voltage $v_{out}(t) = V(2) - V(0)$ and the charging current through the diode $i_d(t) = I(D)$ on a small scale from 160 to 170 V and 160 to 170 A, respectively.

Table E10.6.1 PSPICE program for ideal diode ($i_d - v_d$) characteristic.

```
Application Example 10.6
*half-wave, single-phase peak rectifier
*with 2.5 percent deviation from a_v
*AC source, ideal diode, R=30 ohms, C=20mF
Vs 1 0 sin(0V 169.68V 60Hz)
*ideal diode D
D 1 2 IDEAL
*load
R 2 0 30
C 2 0 20m
*Model
.Model IDEAL D(IS=0.01E-12)
*Analysis request
.Tran 10us 60ms 30ms 10us
*output request
.Probe
.FOUR 60 10 V(R)
.end
```

voltage $v_s(t)$ and the output voltage $v_{out}(t)$ as well as the charging current through the diode $i_d(t)$ on a large scale from −170 to 170 V and 0 to 340 A, respectively. Figure E10.6.2 illustrates the output voltage within the range from 160 to 170 V. The perturbation from a constant output voltage of 166.7 V is about ±2 V, corresponding to ±1.2%, which is acceptable for peak rectifiers. The PSPICE program is detailed in Table E10.6.1, and the voltage harmonics of $v_{out}(t)$ are shown in Table E10.6.2.

Table E10.6.2 Voltage harmonics of $v_{out}(t)$ of Figures E10.6.1 and E10.6.2 in terms of volts and in percent referred to voltage amplitude a_v = 169.68 V.

DC component	1	2	4	6
166.70 V/98.24%	1.47 V/0.87%	0.73 V/0.43%	0.24 V/0.14%	0.23 V/0.14%

Conclusions

The design of a peak rectifier based on half-wave operation is not advisable because it requires a large capacitance C. It is advantageous to employ a full-wave rectification – as discussed later in this chapter – which reduces the capacitance required of the time constant τ = RC for the same output voltage variation ($\pm$1.2%) by about a factor of two. Half-wave rectifiers generate even current harmonics in the power system and are therefore not recommended.

Circuits with Zener diodes [1] are very popular for voltage regulators and limiters. The breakdown of the pn diodes at negative voltages (see Figure 9.6a and b) is relied on.

Application Example 10.7 Voltage regulator with time delay

Figure E10.7.1a depicts the circuit of a voltage regulator with time delay for the linearly increasing input voltage $v_{in}(t)$, as given in Figure E10.7.1b. If $v_{in}(t) \leq V_z$ prior to time t_0, then the output voltage $v_{out}(t) = 0$. For $v_{in}(t) > V_z$ after the time t_0 has been reached, the output voltage $v_{out}(t)$ is linearly increasing, for which at any time the relation $v_{out}(t) = v_{in}(t) - V_z$ is valid.

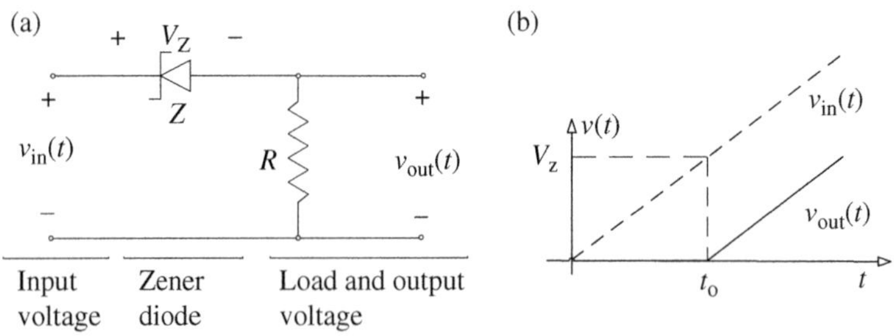

Figure E10.7.1 (a, b) A Zener diode with the voltage V_z can be applied to a circuit to delay the input voltage $v_{in}(t)$ applied to a load resistance R, resulting in $v_{out}(t)$.

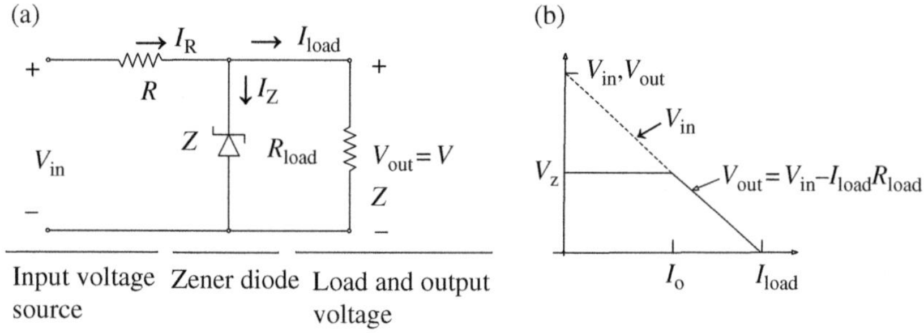

Figure E10.8.1 (a, b) Single-phase Zener diode applied to a circuit to limit the voltage V_{out} applied to a load R_{load}.

Conclusions

Zener diodes can be employed to introduce a delay of a time-dependent voltage.

For safety applications voltage limiters [1] are designed, as demonstrated in Application Example 10.8.

Application Example 10.8 Single-phase regulator with voltage limiter
Figure E10.8.1a shows a voltage limiter based on a Zener diode. Such a circuit is mainly relied on to prevent high voltages – for a given input voltage V_{in} for $I_{load} < I_o$ – from entering sensitive circuits.

Conclusions

Zener diodes can be used to limit voltages so that following circuits are not exposed to excessive voltages.

In order to control the output load voltage $v_{out}(t)$ of a rectifier, the diode is replaced by a device whose turn-on time can be changed such as a thyristor [7, 8] that can be turned on via the gating current $i_G(t)$ signal and must rely for turn-off on the naturally occurring zero crossing of its current $i_{TH}(t)$, as demonstrated in Application Examples 10.9 and 10.10.

Application Example 10.9 Half-wave, single-phase thyristor (TH) rectifier with resistor *R*
Figure E10.9.1 represents the half-wave, single-phase thyristor circuit supplying pulsating current to a resistor *R*. The corresponding PSPICE program is listed in Table E10.9.1. Note that the thyristor is modeled [7] in the PSPICE program based on Figure 9.10c in Chapter 9. As discussed there, thyristors are either load

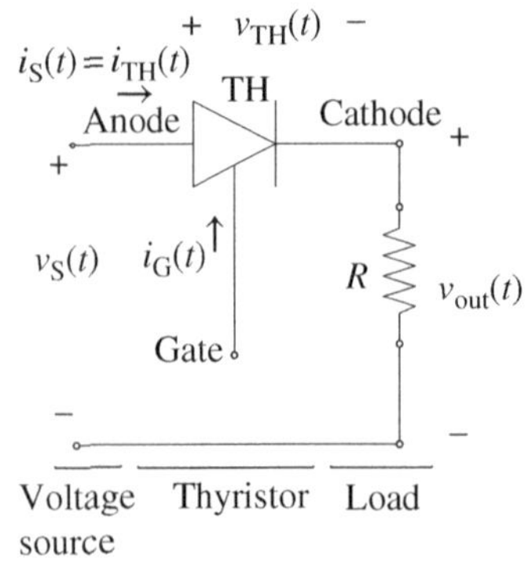

Figure E10.9.1 Rectifier supplying pulsating DC current from single-phase voltage source (s) $v_s(t) = \sqrt{2} \cdot 120\,\text{V} \sin \omega t$ through thyristor TH to resistor $R = 3\,\Omega$.

or line voltage commutated switches; that is, the device cannot be turned off at will, but relies on the zero crossing of its current, which may depend either on the resistor (load) R or the source/line voltage $v_s(t)$. A great disadvantage of a thyristor is that it cannot be turned off at will, that is, at any time.

The thyristor is modeled [7] by a series connection of a self-commutated device (e.g. metal–oxide–semiconductor field-effect-transistor [MOSFET]) and a diode. The conduction duration of the MOSFET must at frequency $f = 60\,\text{Hz}$ exceed half a period length $T/2 = 8.333\,\text{ms}$ so that the diode is responsible for turn-off and not the MOSFET. In most cases the pulse time duration between 8.333 and 10 ms is chosen. The thyristor gating angle of $\alpha = 60°$ corresponds to the MOSFET switching delay time $t_d = 2.7778\,\text{ms}$; the rise (t_r) and fall (t_f) times are assumed to be zero. Due to the selection of the MOSFET model, the gating voltage is very large, e.g. $v_{GS} = 100\,\text{V}$. The voltage harmonics of $v_{out}(t)$ are shown in Table E10.9.2.

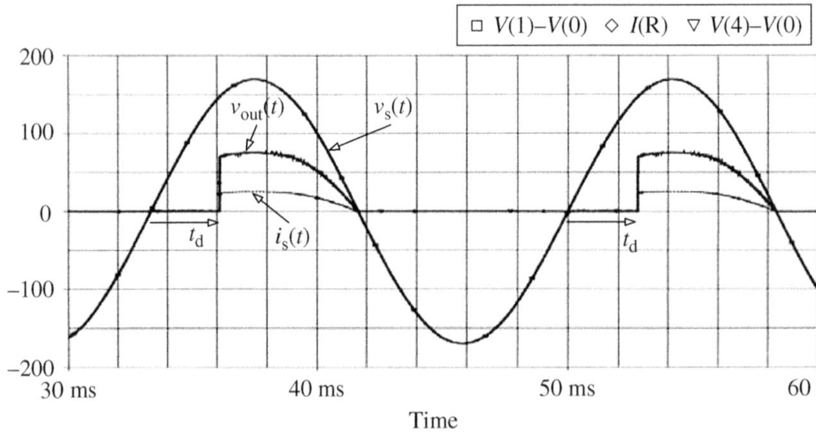

Figure E10.9.2 Source voltage $v_s(t) = V(1) - V(0)$, output load voltage $v_{out}(t) = V(4) - V(0)$, and source current $i_s(t) = i_{TH}(t) = i_R(t) = I(R)$ of thyristor (TH) loaded with $R = 3\,\Omega$.

Table E10.9.1 PSPICE program for the signals of Figure E10.9.2.

```
Application Example 10.9
*AC source, ideal diode, MOSFET, R=3
ohms
Vs 1 0 sin(0V 169.68V 60Hz)
*ideal diode model D in series with
*MOSFET model, 16.67% of 16.66ms no
conduction,
*frequency f=60Hz, thyristor firing angle
alpha=60 *degrees,
*rise and fall times are zero, pulse duration
must be *larger than half a period, e.g.
8.3333ms to about 10ms.
vmosfet 2 3 pulse(0 100 2.7778m 0 0
8.3333m +16.6667m)
mosfet 1 2 3 3 qfet
D 3 4 IDEAL
*Load
R 4 0 3
*Model
.model IDEAL D(IS=0.01E-12)
.model qfet NMOS(level 3 gamma=0
delta=0 eta=0 +theta=0 kappa=0.2
vmax=0 xj=0 tox=100n uo=600 +phi=0.6
rs=42.69m kp=20.87u l=2u w=500u
vto=0.3
+rd=0.19 cbd=200n pb=0.8 mj=0.5
cgso=3.5n +cgdo=3.5n rg=0.2 is=10f)
*Analysis request
.Tran 1us 60ms 30ms 1us
*Output request
.Probe
.FOUR 60 10 V(R)
.options abstol=10m chgtol=10m
reltol=0.1 vntol=100m
.end
```

Note: All comment lines must be preceded with an asterisk (*) while continued calculation statements must be preceded with a plus sign (+).

Table E10.9.2 Voltage harmonics of $v_{out}(t)$ of Figure E10.9.2 in terms of volts and in percent referred to voltage amplitude $a_v = 169.68$ V.

DC component	1	2	4	6
19.217 V/11.33%	33.56 V/19.78%	21.44 V/12.64%	3.937 V/2.32%	3.995 V/2.35%

Conclusions

A thyristor is a phase-controlled switch where the firing angle α ranges from 5° to 175°. Because of the parasitic capacitances of the thyristor, a control near 0° and 180° is not advisable.

Application Example 10.10 Half-wave, single-phase thyristor TH rectifier with resistor *R* and capacitor *C*

In order to reduce the harmonic content of the output voltage $v_{out}(t)$, one adds a capacitor C in parallel to the resistor R. The capacitor acts as a short circuit (filter for higher harmonics), and the goal is to obtain a DC output voltage across the resistor R. Figure E10.10.1 shows the circuit and Figure E10.10.2 depicts source voltage $v_s(t)$, output load current $i_R(t)$, and source current $i_s(t) = i_{TH}(t)$ of rectifier based on thyristor loaded with $R = 3\ \Omega$ and $C = 500\ \mu$F. Note the load voltage $v_{out}(t)$ is proportional to the load current $i_R(t)$. Table E10.10.1 lists its PSPICE program.

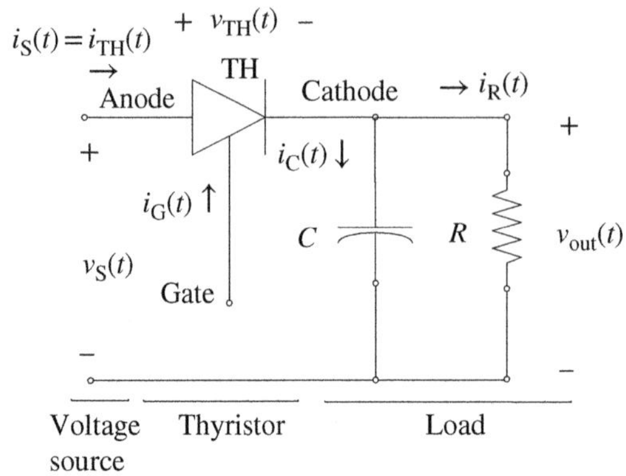

Figure E10.10.1 Rectifier supplying pulsating DC current from single-phase voltage source (s) $v_s(t) = \sqrt{2} \cdot 120\,V \sin \omega t$ through thyristor TH to resistor $R = 3\,\Omega$ and capacitor $C = 500\,\mu F$.

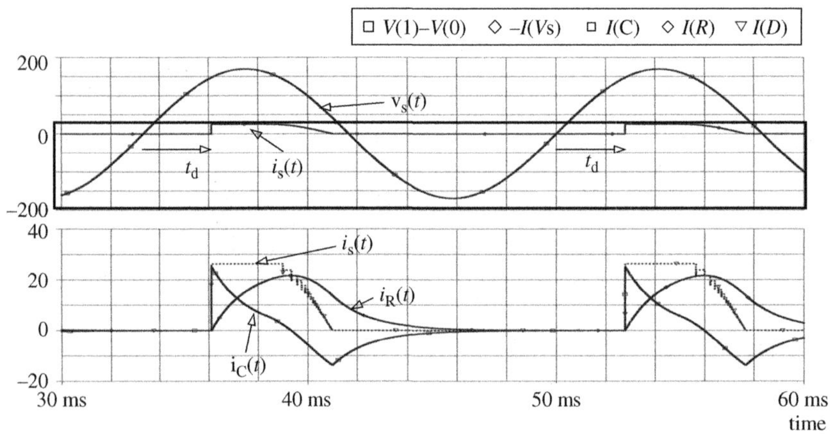

Figure E10.10.2 Top plot: source voltage $v_s(t) = V(1) - V(0)$ and input source current $i_s(t) = -I(V_s)$ where the current leaves the source $v_s(t)$ and therefore the minus (−) sign. Bottom plot: load current $i_R(t) = I(R)$, thyristor or source current $i_s(t) = i_{TH}(t) = I(D) = -I(V_s)$, and capacitor current $i_C(t) = I(C)$ of rectifier based on thyristor loaded with $R = 3\,\Omega$ and $C = 500\,\mu F$.

Similar timing considerations apply as for Application Example 10.9. The voltage harmonics of $v_{out}(t)$ are listed in Table E10.10.2.

Conclusions

In order to improve the controllability of rectification, the load or line voltage commutated thyristor switch can be replaced by a self-commutated device such as a field-effect transistor, in particular a MOSFET as discussed in Chapter 9, Section 9.6. Such a self-commutated switch can be turned on or off at any time, and thus the

Table E10.10.1 PSPICE program for the signals of Figure E10.10.2.

```
Application Example 10.10
*AC source, ideal diode, MOSFET, R=3
ohms
Vs 1 0 sin(0V 169.68V 60Hz)
*ideal diode model D in series with
*MOSFET model, no conduction or
thyristor angle is
*alpha=60 degrees of 360 degrees
(16.66ms) at f=60Hz
vmosfet 2 3 pulse(0 100 2.7778m 0 0 10m
16.6667m)
mosfet 1 2 3 3 qfet
D 3 4 IDEAL
*Load
R 4 0 3
C 4 0 500u
*Model
.model IDEAL D (IS=0.01E-12)
.model qfet NMOS(level 3 gamma=0
delta=0 eta=0
+theta=0 kappa=0.2 vmax=0 xj=0
tox=100n uo=600
+phi=0.6 rs=42.69m kp=20.87u l=2u
w=500u vto=0.3
+rd=0.19 cbd=200n pb=0.8 mj=0.5
cgso=3.5n
+cgdo=3.5n rg=0.2 is=10f)
*Analysis request
.Tran 1us 60ms 30ms 1us
*Output request
.Probe
.FOUR 60 10 V(R)
.options abstol=1m chgtol=1m reltol=0.1
vntol=10m
.end
```

Note: All comment lines must be preceded with an asterisk (*) while continued calculation statements must be preceded with a plus sign (+).

Table E10.10.2 Voltage harmonics of $v_{out}(t) = R \cdot i_R(t)$ of Figure E10.10.2 in terms of volts and in percent referred to voltage amplitude $a_v = 169.68$ V.

DC component	1	2	4	6
18.05 V/10.64%	28.21 V/16.63%	15.13 V/8.92%	1.03 V/0.61%	1.52 V/0.896%

rectification of an AC voltage can be refined and better controlled/adjusted, as compared to the thyristor circuits of Application Examples 10.9 and 10.10.

Application Example 10.11 Half-wave, single-phase rectifier supplying power to a resistor *R* where a MOSFET is in series with a diode *D*

Figure E10.11.1 shows the circuit of a diode in series with a MOSFET switched at a low frequency of $f_{switch} = 600$ Hz and supplying power to a resistive load R. As can be seen, the pulsating waveshape of the output load current (Figure E10.11.2) is not acceptable from a power quality point of view because of the pulsating input current with the $f_s = 600$ Hz switching frequency component. The PSPICE program is displayed in Table E10.11.1, and the voltage harmonics of $v_{out}(t)$ are shown in Table E10.11.2.

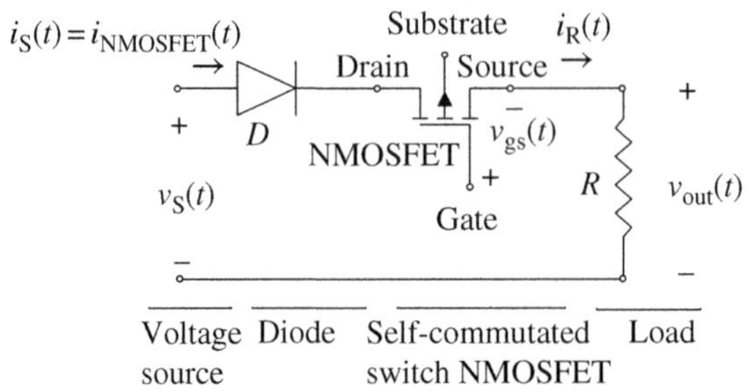

Figure E10.11.1 Rectifier supplying pulsating DC current from single-phase voltage source (s) $v_s(t) = \sqrt{2} \cdot 120\ \text{V} \sin \omega t$ through MOSFET in series with diode to resistor $R = 3\ \Omega$.

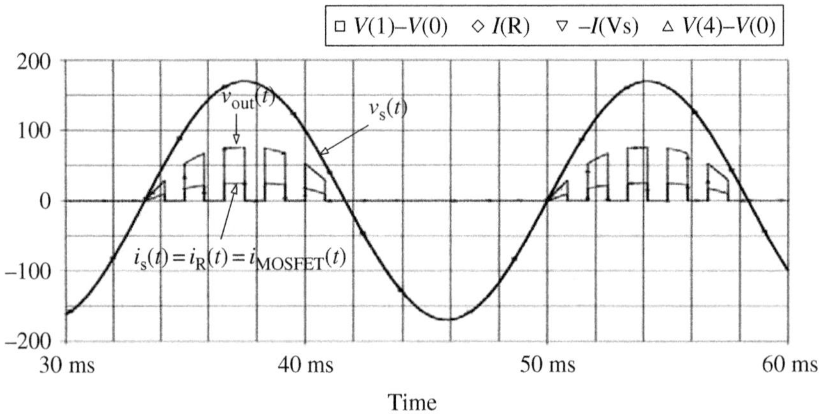

Figure E10.11.2 Source voltage $v_s(t) = V(1) - V(0)$, output load voltage $v_{out}(t) = V(4) - V(0)$, and source current $i_s(t) = i_{MOSFET}(t) = i_R(t) = I(R)$ of rectifier based on MOSFET switch loaded with $R = 3\ \Omega$.

Table E10.11.1 PSPICE program for the signals of Figure E10.11.2.

Application Example 10.11	.model qfet NMOS(level 3 gamma=0 delta=0
*AC source, ideal diode,	+eta=0 theta=0 kappa=0.2 vmax=0
MOSFET, R=3 ohms	+xj=0 tox=100n uo=600 phi=0.6 rs=42.69m
Vs 1 0 sin(0V 169.68V 60Hz)	+kp=20.87u l=2u w=500u vto=0.3
*ideal diode model D in series	+rd=0.19 cbd=200n pb=0.8 mj=0.5 cgso=3.5n
with	+cgdo=3.5n rg=0.2 is=10f)
*MOSFET model, fs=600Hz	*Analysis request
vmosfet 2 3 pulse(0 100 0 0 0	.Tran 1us 60ms 30ms 1us
0.83333m 1.6667m)	*Output request
mosfet 1 2 3 3 qfet	.Probe
D 3 4 IDEAL	.FOUR 60 10 V(R)
*Load	.options abstol=10m chgtol=10m reltol=0.1
R 4 0 3	vntol=100m
*Model	.end
.model IDEAL D(IS=0.01E-12)	

Note: All comment lines must be preceded with an asterisk (*) while continued calculation statements must be preceded with a plus sign (+).

Table E10.11.2 Voltage harmonics of $v_{out}(t)$ of Figure E10.11.2 in terms of volts and in percent referred to voltage amplitude a_v = 169.68 V.

DC component	1	2	4	6	8	10
13.10 V/ 7.72%	20.18 V/ 11.89%	7.64 V/ 4.50%	2.48 V/ 1.46%	0.23 V/ 0.136%	4.83 V/ 2.85%	16.70 V/ 9.84%

Conclusions

A self-commutated switch (e.g. MOSFET) permits more precise current control than a load- or line-commutated switch (e.g. diode, triac, thyristor) by changing the duty ratio or duty cycle δ = (pulse width)/(period length) = (pw/per) from 0.05 to 0.95. Because of the reverse-recovery currents due to parasitic switch capacitances, the duty ratio cannot be near 0 or 1. The pulse magnitude of the gating voltage (e.g. 100 V) and the switch model (e.g. w = 500 u) must be adjusted depending upon the magnitude of the current.

Application Example 10.12 Half-wave, single-phase rectifier supplying power to a resistor *R* and capacitor *C* where a MOSFET is in series with a diode *D*

The advantage of the circuit of Figure E10.12.1 as compared with that of Figure E10.11.1 is that the f_s = 600 Hz harmonic in the load voltage $v_{out}(t)$ is largely

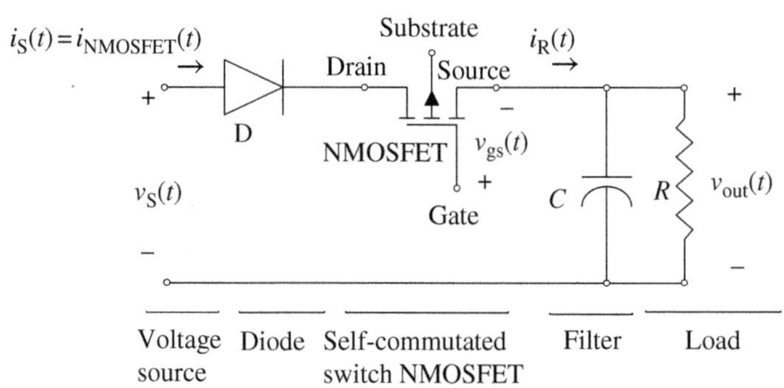

Figure E10.12.1 Rectifier supplying pulsating DC current from single-phase voltage source (s) $v_s(t) = \sqrt{2} \cdot 120$ V sin ωt through MOSFET to resistor R = 3 Ω and capacitor C = 500 μF. The self-commutated switch is modeled by a MOSFET in series with a diode D.

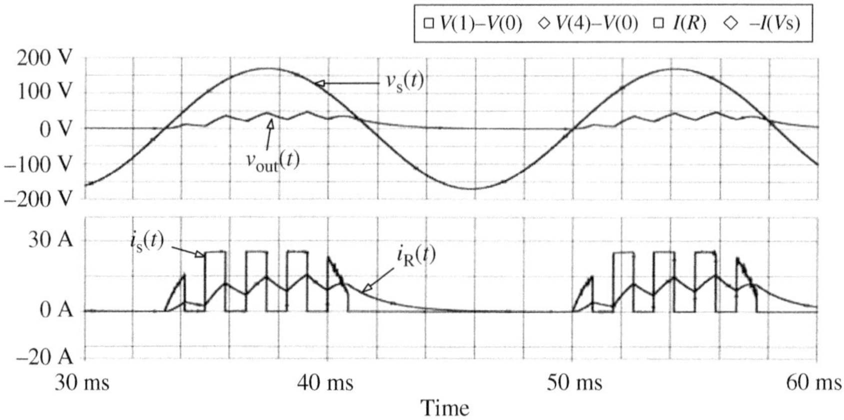

Figure E10.12.2 Top plot: source voltage $v_s(t) = V(1) - V(0)$ and output load voltage $v_{out}(t) = V(4) - V(0)$. Bottom plot: source current $i_s(t) = i_{MOSFET}(t) = -I(V_s)$, and load current $i_R(t) = I(R)$ of rectifier based on MOSFET switch loaded with $R = 3\ \Omega$ and filter capacitance $C = 500\ \mu F$.

Table E10.12.1 PSPICE program for the signals of Figure E10.12.2.

```
Application Example 10.12
*AC source, ideal diode,
MOSFET, R=3 ohms
Vs 1 0 sin(0V 169.68V 60Hz)
*ideal diode model D in series
with
*MOSFET model, fs=600Hz
with C=500uF
vmosfet 2 3 pulse(0 100 0 0 0
0.83333m 1.6667m)
mosfet 1 2 3 3 qfet
D 3 4 IDEAL
*Load
R 4 0 3
C 4 0 500u
*Model
.model IDEAL D(IS=0.01E-12)
.model qfet NMOS(level 3 gamma=0 delta=0 eta=0
theta=0 +kappa=0.2 vmax=0 xj=0 tox=100n
uo=600 phi=0.6 +rs=42.69m kp=20.87u
+l=2u w=500u vto=0.3 rd=0.19
+cbd=200n pb=0.8 mj=0.5 cgso=3.5n cgdo=3.5n
rg=0.2 +is=10f)
*Analysis request
.Tran 1us 60ms 30ms 1us
*Output request
.Probe
.FOUR 60 10 V(R)
.options abstol=10m chgtol=10m reltol=0.1
vntol=100m
.end
```

Note: All comment lines must be preceded with an asterisk (*) while continued calculation statements must be preceded with a plus sign (+).

eliminated (Figure E10.12.2) through the capacitor C filter. The PSPICE program is displayed in Table E10.12.1, and the voltage harmonics of $v_{out}(t)$ are shown in Table E10.12.2.

Table E10.12.2 Voltage harmonics of $v_{out}(t)$ of Figure E10.12.2 in terms of volts and in percent referred to voltage amplitude $a_v = 169.68$ V.

DC component	1	2	4	6	8	10
15.03 V/ 8.86%	19.46 V/ 11.47%	4.43 V/ 2.61%	1.46 V/ 0.86%	0.51 V/ 0.30%	0.84 V/ 0.50%	3.34 V/ 1.97%

Conclusions

A self-commutated switch (e.g. MOSFET) permits more precise current control than a load- or line-commutated switch (e.g. diode, triac, thyristor) by changing the duty ratio δ = (pulse width)/(period length) = pw/per from 0.05 to 0.95. The addition of a parallel capacitor C to the load R acts like a filter reducing the switching harmonic amplitudes. The disadvantage of half-controlled rectifier circuits is the generation of even harmonics, which – if of significant magnitude – must not enter the main grid. The pulse magnitude of the gating voltage (e.g. 100 V) and the switch model (e.g. w = 500 u) must be adjusted depending upon the magnitude of the current. The switching frequency f_s should be larger than 600 Hz in order to reduce current ripple.

10.2 Full-Wave, Single-Phase Rectification

The disadvantage of half-wave rectification is the generation of even current and voltage harmonics detrimental for power system operation due to poor power quality [8]. To eliminate even harmonics, full-wave rectification is preferred and used throughout electronic devices such as radios, television sets, computers, printers, etc.

Application Example 10.13 Full-wave, single-phase diode bridge rectifier with resistor *R*

The simplest full-wave rectifier is presented in Figure E10.13.1, where a diode bridge supplies current/power to a resistor. During the positive half-wave of $v_s(t)$ diodes D_1 and D_2 conduct current (behave like a short circuit), while for the negative half-wave diodes D_3 and D_4 are short circuits. The great disadvantage of this circuit is the pulsating nature of the output load current (Figure E10.13.2), which is proportional to the output voltage $v_{out}(t)$. The PSPICE program is listed

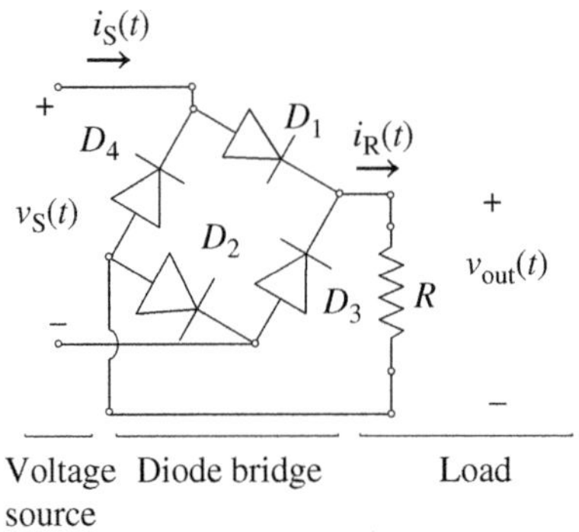

Figure E10.13.1 Rectifier supplying pulsating DC current from single-phase voltage source (s) $v_s(t) = \sqrt{2} \cdot 120\text{ V} \sin \omega t$ through diode bridge to resistor $R = 3\ \Omega$.

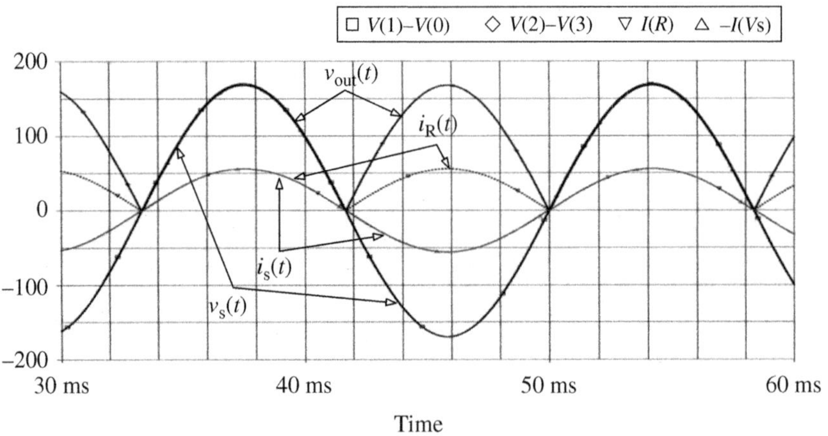

Figure E10.13.2 Source voltage $v_s(t) = V(1) - V(0)$, output load voltage $v_{out}(t) = V(2) - V(3)$, load current $i_R(t) = I(R)$, and source current $i_s(t) = -I(V_s)$ of rectifier based on diode bridge loaded with $R = 3\ \Omega$.

in Table E10.13.1, and the voltage harmonics of $v_{out}(t)$ are displayed in Table E10.13.2.

Conclusions

A diode bridge rectifies the input voltage resulting in large even harmonics that are detrimental to power system operation [8]. To mitigate these harmonics a filter capacitor is installed in Application Example 10.14. The identity $i_s(t) = -I(V_s)$ exists because PSPICE assumes that the current $I(V_s)$ flows into the source, while in Figure E10.13.1 the current $i_s(t)$ flows out of the source.

Table E10.13.1 PSPICE program for the signals of Figure E10.13.2.

```
Application Example 10.13
*AC source, 4 ideal diodes, diode *bridge R=3 ohms
Vs 1 0 sin(0V 169.68V 60Hz)
*ideal diodes D1, D2, D3, D4
D1 1 2 IDEAL
D2 3 0 IDEAL
D3 0 2 IDEAL
D4 3 1 IDEAL
*Load
R 2 3 3
*Model
.Model IDEAL D(IS=0.01E-12)
*Analysis request
.Tran 10us 60ms 30ms 10us
*Output request
.Probe
.FOUR 60 10 V(R)
.end
```

Table E10.13.2 Voltage harmonics of $v_{out}(t)$ of Figure E10.13.2 in terms of volts and in percent referred to voltage amplitude $a_v = 169.68$ V.

DC component	1	2	4	6
106.42 V/62.72%	0.003 V/0.002%	72 V/42.4%	14.37 V/8.47%	6.15 V/3.63%

Application Example 10.14 Full-wave, single-phase diode bridge rectifier with resistor *R* and parallel capacitor *C*

The circuit of Figure E10.14.1 represents a diode bridge rectifier, which finds some use in electronic devices where a variable output voltage is not required. The capacitor C acts as a filter and provides a pulsating output voltage $v_{out}(t)$ with reduced even harmonics (Figure E10.14.2). The PSPICE program is itemized in Table E10.14.1, and the voltage harmonics of $v_{out}(t)$ are detailed in Table E10.14.2.

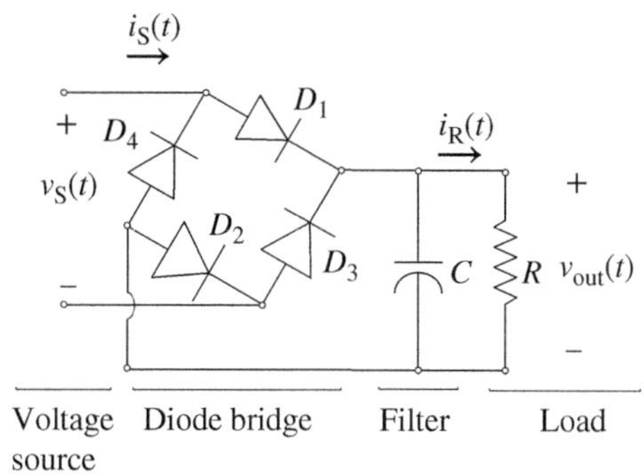

Figure E10.14.1 Rectifier supplying pulsating DC current from single-phase voltage source (s) $v_S(t) = \sqrt{2} \cdot 120$ V $\sin \omega t$ through diode bridge to resistor $R = 3\ \Omega$ and capacitor $C = 500\ \mu$F.

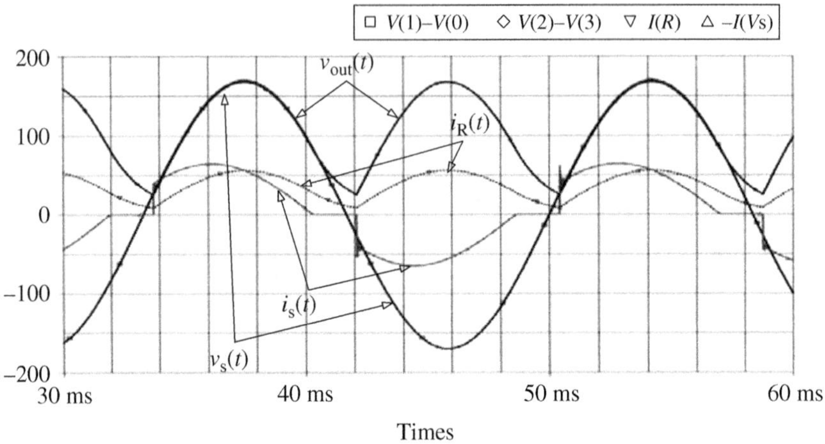

Figure E10.14.2 Source voltage $v_s(t) = V(1) - V(0)$, output load voltage $v_{out}(t) = V(2) - V(3)$, load current $i_R(t) = I(R)$, and source current $i_s(t) = -I(V_s)$ of rectifier based on diode bridge loaded with $R = 3\ \Omega$ and capacitor $C = 500\ \mu F$.

Table E10.14.1 PSPICE program for the signals of Figure E10.14.2.

```
Application Example 10.14
*AC source, 4 ideal diodes, diode bridge, R=3
*ohms, filter capacitor C=500u
Vs 1 0 sin(0V 169.68V 60Hz)
*ideal diodes D1, D2, D3, D4
D1 1 2 IDEAL
D2 3 0 IDEAL
D3 0 2 IDEAL
D4 3 1 IDEAL
*Load
R 2 3 3
C 2 3 500u
*Model
.Model IDEAL D(IS=0.01E-12)
*Analysis request
.Tran 10us 60ms 30ms 10us
*Output request
.Probe
.FOUR 60 10 V(R)
.options abstol=0.1m chgtol=0.1m
reltol=0.1 +vntol=0.1m
.end
```

Note: All comment lines must be preceded with an asterisk (*) while continued calculation statements must be preceded with a plus sign (+).

Table E10.14.2 Voltage harmonics of $v_{out}(t)$ of Figure E10.14.2 in terms of volts and in percent referred to voltage amplitude $a_v = 169.68$ V.

DC component	1	2	4	6
109.15 V/64.33%	0.001 V/0.000%	66.30 V/39.07%	9.57 V/5.64%	2.90 V/1.71%

Conclusions

A full-wave, single-phase diode bridge rectifies the input voltage, resulting in even harmonics that are reduced by a filter capacitor C, making this a useful circuit. It is advisable to increase the capacitance (e.g. $C = 1$ mF) and the load resistor (e.g. $R = 30\ \Omega$), resulting in a time constant for the peak rectification of $\tau_p = RC = 30$ ms, which represents about two 60 Hz periods.

Application Example 10.15 Full-wave, single-phase half-controlled thyristor bridge rectifier with resistor *R* and capacitor *C*

This circuit of Figure E10.15.1 consists of a half-controlled bridge with two thyristors and two diodes. The thyristors are modeled by a metal–oxide–semiconductor field-effect transistor (MOSFET) connected in series with a diode. The MOSFET turns on the thyristor and the diode turns off the thyristor, if a zero current crossing occurs.

One thyristor is modeled by a series connection of a self-commutated device (e.g. MOSFET) and a diode (D). The conduction duration of the MOSFET must at frequency $f = 60$ Hz exceed half a period length $(T/2) = 8.333$ ms so that the diode is responsible for turn-off and not the MOSFET. In most cases the pulse length between 8.333 and 10 ms is chosen. The thyristor gating angle of $\alpha = 30°$ corresponds to switching delay time of $t_d = 1.39$ ms at $f = 60$ Hz; the rise (t_r) and fall (t_f) times are assumed to be zero (see Figure E10.15.2). Due to the

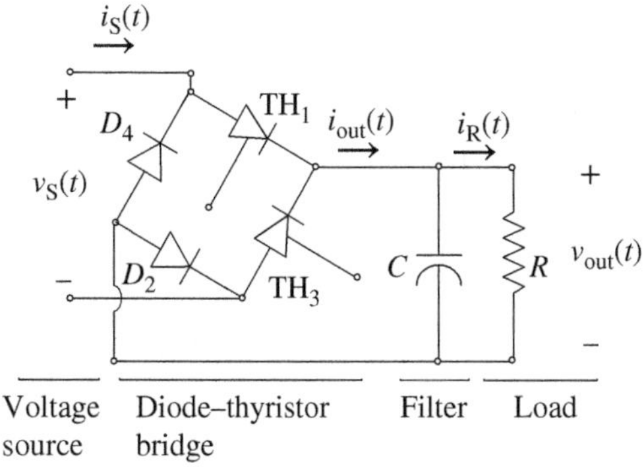

Figure E10.15.1 Full-wave, single-phase half-controlled diode/thyristor bridge rectifier with resistor $R = 3\ \Omega$ and capacitor $C = 500\ \mu$F.

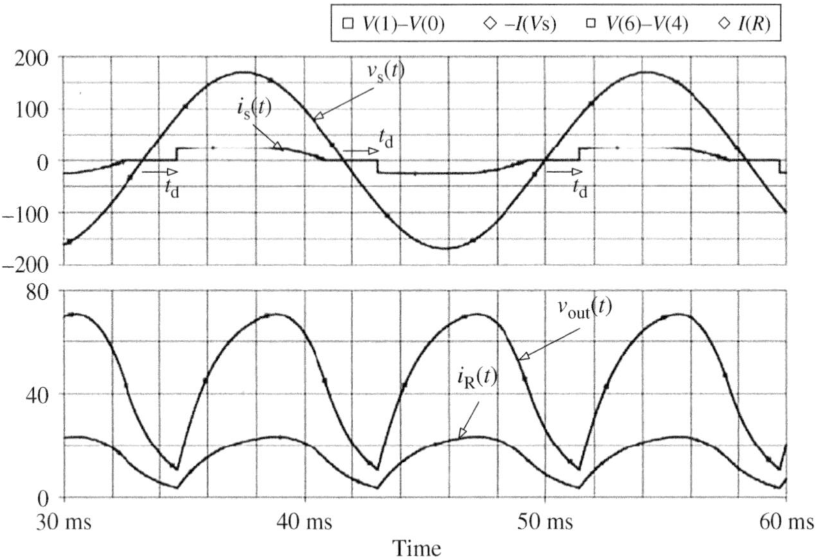

Figure E10.15.2 Top plot: source voltage $v_S(t) = V(1) - V(0)$ and source current $i_S(t) = -I(V_s)$. Bottom plot: output load voltage $v_{out}(t) = V(6) - V(4)$ and load current $i_R(t) = I(R)$ of rectifier based on half-controlled diode/thyristor bridge loaded with $R = 3\,\Omega$ and filter capacitor $C = 500\,\mu F$.

selection of the MOSFET model, the magnitude of the gating voltage is very large, e.g. $v_{GS} = 100$ V. The relation between the two gating angles α_1 and α_3 of thyristors TH_1 and TH_3 is $\alpha_3 - \alpha_1 \equiv (T/2) = (16.6667/2)$ ms $= 8.33$ ms for 30° at 60 Hz $\alpha_1 \equiv t_d = 1.39$ ms and $\alpha_2 \equiv 9.72$ ms ≥ 8.333 ms. The PSPICE program is shown in Table E10.15.1, and the voltage harmonics of $v_{out}(t)$ are presented in Table E10.15.2.

Conclusions

A diode/thyristor half-controlled bridge controls the output voltage by changing the firing angle α in the range 5°–175°.The pulse magnitude of the gating voltage (e.g. 100 V) and the switch model (e.g. w = 500 u) must be adjusted depending upon the magnitude of the current.

A similar circuit to that of Figure E10.15.1 is shown in Figure E10.15.3 with an input transformer and with small inductors serving as current snubbers suppressing reverse-recovery currents [7, 8] due to parasitic (unwanted) capacitances of the two thyristors and the two diodes.

One thyristor is modeled by a series connection of a self-commutated device (e.g. MOSFET) and a diode (D). The conduction duration of the MOSFET must at

Table E10.15.1 PSPICE program for the signals of Figure E10.15.2.

```
Application Example 10.15a
*AC source
Vs 1 0 sin(0V 169.68V 60Hz)
*half-controlled thyristor bridge, thyristor
modelled by
*ideal diode and mosfet in series, R=3ohms
and C=500uF for alpha=30 degrees
(=1.39ms)
D1 1 3 IDEAL
vmosfet1 5 6 pulse(0 100 1.39m 0 0 10m
16.667m)
mosfet1 3 5 6 6 qfet
D3 0 8 IDEAL
vmosfet3 7 6 pulse(0 100 9.72m 0 0 10m
16.667m)
mosfet3 8 7 6 6 qfet
D2 4 0 IDEAL
D4 4 1 IDEAL
*Load
R 6 4 3
C 6 4 500u
*Model
.model IDEAL D(IS=0.01E-12)
.model qfet NMOS(level 3 gamma=0
delta=0 eta=0
+theta=0 kappa=0.2 vmax=0 xj=0
tox=100n uo=600
+phi=0.6 rs=42.69m kp=20.87u l=2u
w=500u vto=0.3
+rd=0.19 cbd=200n pb=0.8 mj=0.5
cgso=3.5n
+cgdo=3.5n rg=0.2 is=10f)
*Analysis request
.Tran 1us 60ms 30ms 1us
*Output request
.Probe
.FOUR 60 10 V(R)
.options abstol=10m chgtol=10m
reltol=0.1 vntol=100m
.end
```

Note: All comment lines must be preceded with an asterisk (*) while continued calculation statements must be preceded with a plus sign (+).

Table E10.15.2 Voltage harmonics of $v_{out}(t)$ of Figure E10.15.2 in terms of volts and in percent referred to voltage amplitude $a_v = 169.68$ V.

DC component	1	2	4	6
47.40 V/27.94%	0.02 V/0.01%	27.44 V/16.17%	5.81 V/3.42%	1.42 V/0.84%

frequency $f = 60$ Hz exceed half a period length $(T/2) = 8.333$ ms so that the diode is responsible for turn-off and not the MOSFET. In most cases the pulse length between 8.333 and 10 ms is chosen. The thyristor gating angle of $\alpha = 30°$ or MOSFET switching delay time $t_d \equiv 1.39$ ms at 60 Hz is chosen; the rise (t_r) and fall (t_f) times are assumed to be zero (see Figure E10.15.4). Due to the selection of the MOSFET model, the magnitude of the gating voltage is very large (e.g. $v_{GS} = 200$ V). The relation between the two gating angles α_1 and α_3 of thyristors TH_1 and TH_3 is $\alpha_3 - \alpha_1 \equiv (T/2) = (16.6667/2)$ ms $= 8.33$ ms for $\alpha = 30°$ at $f = 60$ Hz, $\alpha_1 \equiv t_d = 1.39$ ms, and $\alpha_2 \equiv 9.72$ ms. The PSPICE program is listed in Table E10.15.3, and the voltage harmonics of $v_{out}(t)$ are displayed in Table E10.15.4.

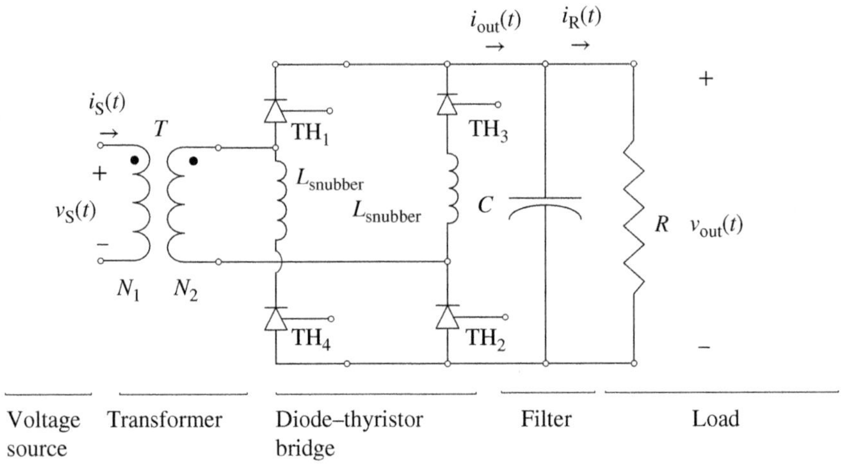

Figure E10.15.3 Full-wave, single-phase half-controlled diode/thyristor bridge rectifier with step-up transformer T, resistor $R = 3\ \Omega$, capacitor $C = 500\ \mu F$, and current-spike limiting (snubber) inductors $L_{snubber}$.

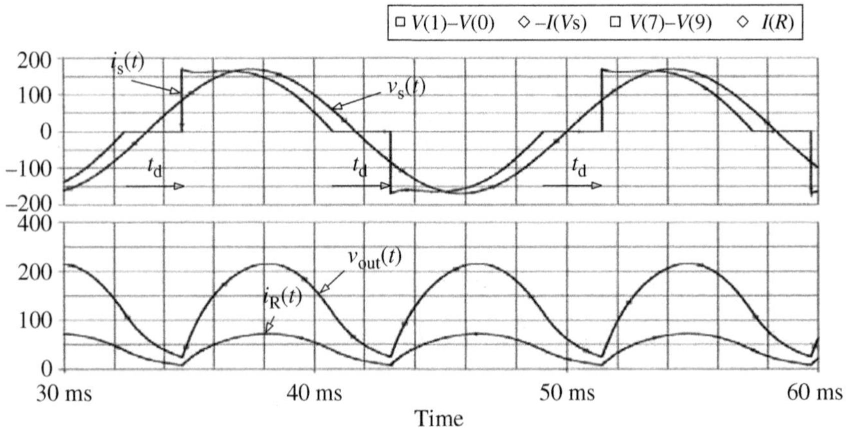

Figure E10.15.4 Top plot: source voltage $v_s(t) = V(1) - V(0)$ and source current $i_s(t) = -I(V_s)$. Bottom plot: output load voltage $v_{out}(t) = V(7) - V(9)$ and load current $i_R(t) = I(R)$ of rectifier based on half-controlled diode/thyristor bridge fed by transformer T loaded with $R = 3\ \Omega$ and capacitor $C = 500\ \mu F$.

Table E10.15.3 PSPICE program for the signals of Figure E10.15.4.

```
Application Example 10.15b
* AC source
Vs 1 0 sin(0V 169.68V 60Hz)
*half-controlled thyristor bridge, thyristor
modelled by *ideal diode and mosfet in
series, R=3ohms and *C=500uF including
transformer
*transformer
L1 100 0 50
L2 300 4 200
K12 L1 L2 0.999999999
R1 100 1 0.01
R2 300 3 0.01
Risolation 0 4 1000000
D1 3 5 IDEAL
vmosfet1 6 7 pulse(0 200 1.39m 0 0 8.335m
16.667ms)
mosfet1 5 6 7 7 qfet
D3 11 12 IDEAL
vmosfet3 13 7 pulse(0 200 9.72m 0 0 8.335m
16.667ms)
mosfet3 12 13 7 7 qfet
D2 9 4 IDEAL
D4 9 8 IDEAL
*Load
Ls1 8 3 10p
Ls2 4 11 10p
R 7 9 3
C 7 9 500u
*Model
.model IDEAL D(IS=0.01E-12)
.model qfet NMOS(level 3 gamma=0
delta=0 eta=0 +theta=0 kappa=0.2
vmax=0 xj=0 tox=100n uo=600 +phi=0.6
rs=42.69m kp=20.87u l=2u w=500u
vto=0.3 +rd=0.19 cbd=200n pb=0.8
mj=0.5 cgso=3.5n +cgdo=3.5n rg=0.2
is=10f)
*Analysis request
.Tran 1us 60ms 30ms 1us
*Output request
.Probe
.FOUR 60 10 V(R)
.options abstol=10m chgtol=10m
reltol=0.1 vntol=100m
.end
```

Note: All comment lines must be preceded with an asterisk (*) while continued calculation statements must be preceded with a plus sign (+).

Table E10.15.4 Voltage harmonics of $v_{out}(t)$ of Figure E10.15.4 in terms of volts and in percent referred to voltage amplitude a_v = 169.68 V.

DC component	1	2	4	6
137.44 V/81.00%	0.05 V/0.03%	90.12 V/53.11%	14.23 V/8.38%	5.91 V/3.48%

Conclusions

Transformers are used to adjust the voltage and current to suitable levels for a given electric circuit. Thyristors control the output voltage by changing the firing angle α in the range 5°–175°. The pulse magnitude of the gating voltage (e.g. 200 V) and the switch model (e.g. $w = 500$ u) must be adjusted depending upon the magnitude of the current.

Application Example 10.16 Full-wave, single-phase MOSFET rectifier with resistor R and filter capacitors C_{f1} and C_{f2}

Until now, most discussed diode and rectifier circuits have poor power quality because the source current is not sinusoidal. To improve the source current waveshape, self-commutated, high-frequency (e.g. 6 kHz and larger) switches are employed, which can be turned on and off at will. Figure E10.16.1 shows a MOSFET-controlled rectifier with a diode bridge, filter capacitors C_{f1}, C_{f2}, and a load R at a switching frequency of $f_{switch} = 6$ kHz (see Figure E10.16.2). The PSPICE program is listed in Table E10.16.1.

Conclusions

Table E10.16.2 lists the low-frequency harmonics of the output voltage $v_{out}(t) \approx V_{DC} = 13.56$ V, but it does not show the switching harmonic of frequency $f_{switch} = 6$ kHz. To prevent these current harmonics from entering the voltage source $v_s(t)$, a

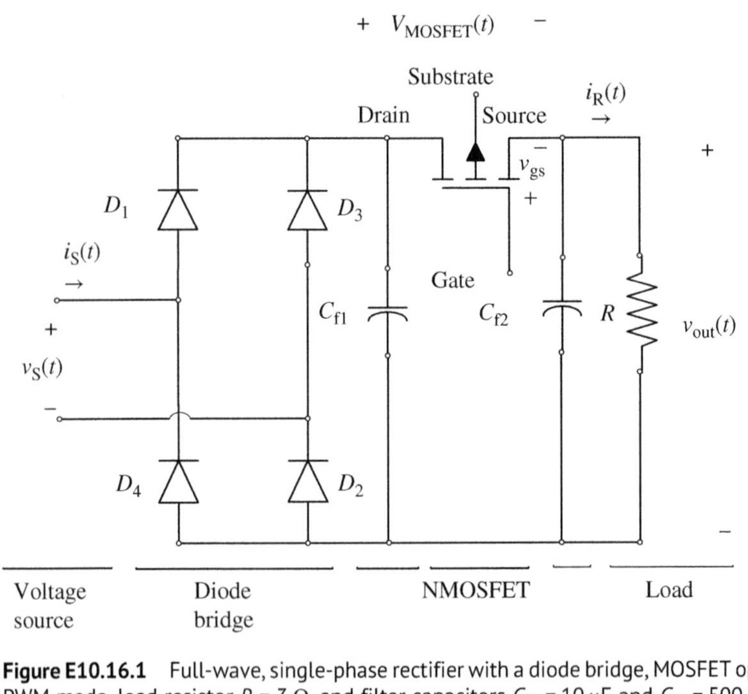

Figure E10.16.1 Full-wave, single-phase rectifier with a diode bridge, MOSFET operated in PWM mode, load resistor $R = 3\ \Omega$, and filter capacitors $C_{f1} = 10\ \mu F$ and $C_{f2} = 500$ mF.

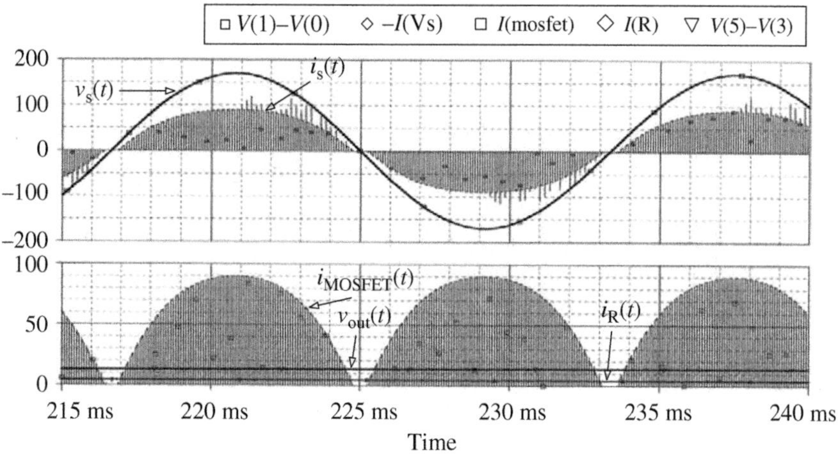

Figure E10.16.2 Top plot: source voltage $v_s(t) = V(1) - V(0)$ and input or source current $i_s(t) = -I(V_s)$. Bottom plot: output load voltage $v_{out}(t) = V(5) - V(3)$, MOSFET drain current $i_{MOSFET}(t)$ = ID(mosfet), and load current $i_R(t) = I(R)$ of rectifier based on self-commutated switch (e.g. MOSFET) loaded with $R = 3\ \Omega$ with filter capacitors $C_{f1} = 10\ \mu F$ and $C_{f2} = 500$ mF.

Table E10.16.1 PSPICE program for the signals of Figure E10.16.2.

```
Application Example 10.16
* AC source
Vs 1 0 sin(0V 169.68V 60Hz)
*Diode bridge with MOSFET
voltage control through
*pulse width (pw/per)=50%
(PWM) R=3ohms and
*Cf1=10uF, Cf2=500uF,
fswitch=6kHz, VDC=12V
D1 1 2 IDEAL
D3 0 2 IDEAL
D2 3 0 IDEAL
D4 3 1 IDEAL
vmosfet 4 5 pulse(0 200 0 0 0
0.0833ms 0.16667ms)
mosfet 2 4 5 5 qfet
*Filter
Cf1 2 3 10u
Cf2 5 3 500m
*Load
R 5 3 3
*Model
.model IDEAL D(IS=0.01E-12)
.model qfet NMOS(level 3 gamma=0 delta=0 eta=0
+theta=0 kappa=0.2 vmax=0 xj=0 tox=100n uo=600
+phi=0.6 rs=42.69m kp=20.87u l=2u w=500u
vto=0.3
+rd=0.19 cbd=200n pb=0.8 mj=0.5 cgso=3.5n
+cgdo=3.5n rg=0.2 is=10f )
*Analysis request
.Tran 1us 240ms 215ms 1us
*Output request
.Probe
.FOUR 60 10 V(R)
.options abstol=10m chgtol=10m reltol=0.1
+vntol=100m
.end
```

Note: All comment lines must be preceded with an asterisk (*) while continued calculation statements must be preceded with a plus sign (+).

Table E10.16.2 Voltage harmonics of $v_{out}(t)$ of Figure E10.16.2 in terms of volts and in percent referred to voltage amplitude $a_v = 169.68$ V.

DC component	1	2	3	4
13.56 V/8.00%	0.27 V/0.16%	0.16 V/%	0.09 V/0.05%	0.06 V/0.04%

tuned LC filter at 6 kHz must be employed; such a filter is not used in Figures E10.16.1 and E10.16.2. The combination of a diode bridge with a self-commutated switch (e.g. MOSFET) is a popular and inexpensive rectifying circuit applied to circuits with low voltages and currents (e.g. 10–1000 V and 0.1–100 A, respectively). In this case, the only drawback is that one obtains for large input and MOSFET instantaneous currents (e.g. about 100 A), only a small DC output voltage (e.g. 13.56 V). The pulse magnitude (e.g. 200 V) and the switch model (e.g. $w = 500$ u) must be adjusted depending upon the magnitude of the current.

10.3 AC Current Controllers

Application Example 10.17 Full-wave, single-phase triac AC controller with resistor *R*

The employment of triacs at low-power applications is very popular because the output current and voltage can be controlled. The disadvantage of the circuit in Figure E10.17.1 is that the output current and voltage are pulsating (see Figure E10.17.2), which results in poor rectifier performance from a power quality point of view. The PSPICE program is shown in Table E10.17.1, and the voltage harmonics of $v_{out}(t)$ are presented in Table E10.17.2.

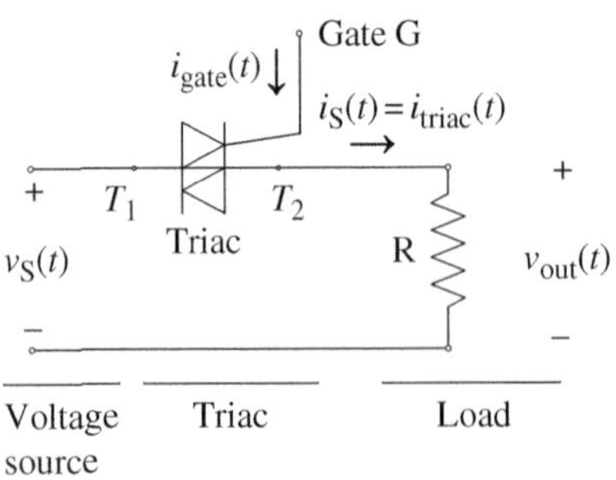

Figure E10.17.1 Controller supplying AC current with variable magnitude from single-phase voltage source (s) $v_s(t) = \sqrt{2} \cdot 120$ V sin ωt through triac to resistor $R = 3\ \Omega$. The relation between the two gating angles α_1 and α_3 of thyristors TH_1 and TH_2 is $\alpha_2 - \alpha_1 \equiv (T/2) = (16.6667/2)$ ms = 8.33 ms for $\alpha = 60°$ at $f = 60$ Hz, $\alpha_1 \equiv 2.778$ ms = t_d, and $\alpha_2 \equiv 11.111$ ms.

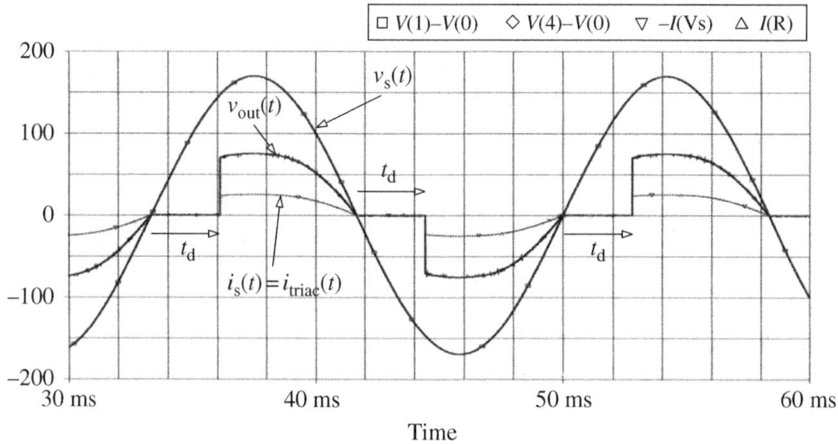

Figure E10.17.2 Source voltage $v_s(t) = V(1) - V(0)$, output load voltage $v_{out}(t) = V(4) - V(0)$, and input or source and load current $i_s(t) = -I(V_s) = i_{triac}(t) = I(R)$ of AC controller based on triac loaded with $R = 3\ \Omega$.

Table E10.17.1 PSPICE program for the signals of Figure E10.17.2.

```
Application Example 10.17
*AC source, ideal diode,
MOSFET, R=3 ohms
Vs 1 0 sin(0V 169.68V 60Hz)
*ideal diode model D in series
with
*MOSFET model, 16.67% of
16.66ms no conduction
vmosfet1 2 3 pulse(0 100
2.7778m 0 0 10m 16.6667ms)
mosfet1 1 2 3 3 qfet
D1 3 4 IDEAL
vmosfet2 6 7 pulse(0 100
11.111m 0 0 10m 16.6667ms)
mosfet2 4 6 7 7 qfet
D2 7 1 IDEAL
*Load
R 4 0 3
*Model
.model IDEAL D(IS=0.01E-12)
.model qfet NMOS(level 3 gamma=0 delta=0 eta=0
+theta=0 kappa=0.2 vmax=0 xj=0 tox=100n
uo=600 +phi=0.6 rs=42.69m kp=20.87u l=2u
w=500u vto=0.3
+rd=0.19 cbd=200n pb=0.8 mj=0.5 cgso=3.5n
+cgdo=3.5n rg=0.2 is=10f )
*Analysis request
.Tran 1us 60ms 30ms 1us
*Output request
.Probe
.FOUR 60 10 I(R)
.options abstol=10m chgtol=10m reltol=0.1
+vntol=100m
.end
```

Note: All comment lines must be preceded with an asterisk (*) while continued calculation statements must be preceded with a plus sign (+).

Table E10.17.2 Current harmonics of $i_R(t)$ of Figure E10.17.2 in terms of amperes and in percent referred to current amplitude $a_i = 169.68 \cdot V/3\ \Omega = 56.56$ A.

DC component	1	3	5	7
0.00/0.00%	22.36 A/39.53%	5.50 A/9.72%	3.87 A/6.84%	1.81 A/3.20%

Conclusions

Triacs are phase-controlled, inexpensive switches applied to circuits with medium voltages and currents (e.g. 20–3000 V and 1–100 A, respectively). This is an example for phase control of the current. The pulse magnitude (e.g. 100 V) and the switch model (e.g. $w = 500$ u) must be adjusted depending upon the magnitude of the desired output current.

Application Example 10.18 Full-wave, single-phase triac AC controller with resistor *R* and capacitor *C* as filter

The circuit of Figure E10.18.1 provides a controllable AC output voltage and current and is frequently used for higher voltages and currents. The triac is modeled by two antiparallel-connected thyristors modeled by two self-commutated switches such as a MOSFET and a diode in series. The MOSFETs turn on the triac and the diodes turn it off, if a current zero crossing occurs (see Figure E10.18.2). The relation between the two gating angles α_1 and α_3 of thyristors TH_1 and TH_2 is $\alpha_2 - \alpha_1 \equiv (T/2) = (16.6667/2)$ ms $= 8.33$ ms for 60° at $f = 60$ Hz $\alpha_1 \equiv 2.778$ ms $= t_d$ and $\alpha_2 \equiv 11.111$ ms. The PSPICE program is detailed in Table E10.18.1, and the current harmonics of $i_R(t)$ are shown in Table E10.18.2.

Conclusions

Triacs are phase-controlled, inexpensive switches applied to circuits with medium voltages and currents (e.g. [20–3000] V and [1–100] A, respectively). The filter capacitor *C* reduces the voltage/current harmonics at the output. The pulse magnitude of the gating voltage (e.g. 100 V) and the switch model (e.g. $w = 500$ u) must be adjusted depending upon the desired magnitude of the output current.

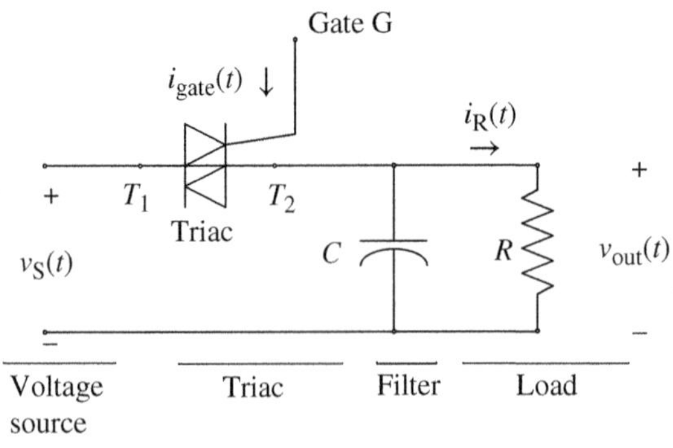

Figure E10.18.1 Controller supplying AC current from single-phase voltage source (s) $v_s(t) = \sqrt{2} \cdot 120$ V sin ωt through triac to resistor $R = 3\ \Omega$ and filter capacitor $C = 500\ \mu$F.

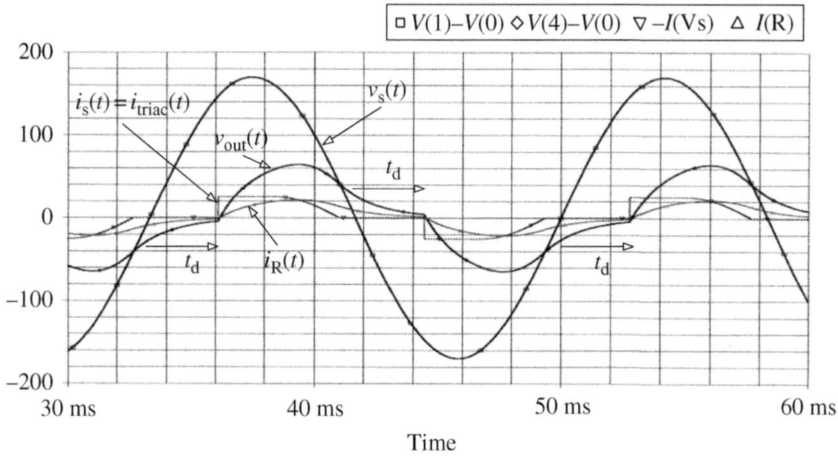

Figure E10.18.2 Source voltage $v_s(t) = V(1) - V(0)$, source current $i_s(t) = i_{triac}(t) = -I(V_s)$, output load voltage $v_{out}(t) = V(4) - V(0)$, and load current $i_R(t) = I(R)$ of rectifier based on triac loaded with $R = 3\ \Omega$ and filter capacitor $C = 500\ \mu F$.

Table E10.18.1 PSPICE program for the signals of Figure E10.18.2.

```
Application Example 10.18
*AC source, ideal diode, MOSFET, R=3
ohms
Vs 1 0 sin(0V 169.68V 60Hz)
*ideal diode model D in series with
*MOSFET model, 16.67% of 16.66ms no
conduction
vmosfet1 2 3 pulse(0 100 2.7778m 0 0
10m 16.6667ms)
mosfet1 1 2 3 3 qfet
D1 3 4 IDEAL
vmosfet2 6 7 pulse(0 100 11.111m 0 0
10m 16.6667ms)
mosfet2 4 6 7 7 qfet
D2 7 1 IDEAL
*Load
R 4 0 3
C 4 0 500u
*Model
.model IDEAL D(IS=0.01E-12)
.model qfet NMOS(level 3 gamma=0 delta=0
eta=0 +theta=0 kappa=0.2 vmax=0 xj=0
tox=100n uo=600 +phi=0.6 rs=42.69m
kp=20.87u l=2u w=500u vto=0.3
+rd=0.19 cbd=200n pb=0.8 mj=0.5
cgso=3.5n +cgdo=3.5n rg=0.2 is=10f)
*Analysis request
.Tran 1us 60ms 30ms 1us
*Output request
.Probe
.FOUR 60 10 I(R)
.options abstol=10m chgtol=10m reltol=0.1
+vntol=100m
.end
```

Note: All comment lines must be preceded with an asterisk (*) while continued calculation statements must be preceded with a plus sign (+).

Table E10.18.2 Current harmonics of $i_R(t)$ of Figure E10.18.2 in terms of amperes and in percent referred to current amplitude a_i=169.68 V/3 Ω = 56.56 A.

DC component	1	3	5	7
0.00/0.00%	18.83 A/33.29%	3.77 A/6.67%	1.09 A/1.94%	0.56 A/0.98%

Application Example 10.19 Single-phase pulse-width-modulated (PWM) AC current controller using two antiparallel-connected MOSFETs in series with two diodes

The control of the AC current with line-commutated switches such as thyristors is inexpensive; however, the phase-control approach generates current harmonics and in turn introduces voltage harmonics in the grid [7, 8]. These unwanted signals can be mitigated by filters [8]. A better but more expensive approach is the reliance on self-commutated electronic switches such as MOSFETs whereby the pulse width (pw) is variable. Note that the duty cycle or duty ratio is defined as δ/T = (pulse width)/(period duration) = (pw/per). Figure E10.19.1 shows the circuit diagram with two antiparallel-connected MOSFETs each in series with a diode. The PSPICE program is displayed in Table E10.19.1, and the current harmonics of $i_R(t)$ are itemized in Table E10.19.2.

One MOSFET is connected in series with a diode, and both MOSFETs are connected in antiparallel. Note the diode of a MOSFET conducting in reverse direction (Chapter 9, Figure 9.9b) is not shown. Each of the two switches is modeled by one MOSFET with a diode connected in series, whereby the diode turns the current off at zero crossing (see Figures E10.19.2 and E10.19.3). The PSPICE program is detailed in Table E10.19.3, and the current harmonics of $i_R(t)$ are shown in Table E10.19.4.

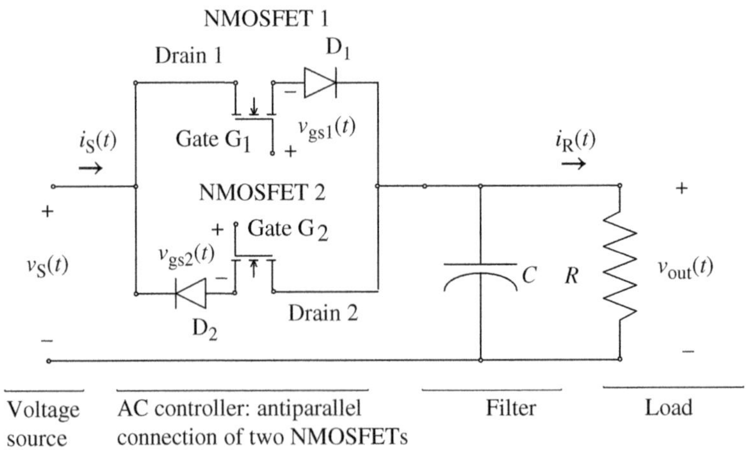

Figure E10.19.1 Controller supplying variable AC current from single-phase voltage source (s) $v_s(t) = \sqrt{2} \cdot 120\ \text{V} \sin \omega t$ through two antiparallel MOSFETs controlled with pulse width modulation to resistor $R = 3\ \Omega$ and filter capacitor $C = 500\ \mu\text{F}$.

Table E10.19.1 PSPICE program for the signals of Figure E10.19.2.

```
Application Example 10.19a
*AC source, ideal MOSFET switched at
fs=6kHz, R=3 *ohms,
*C=500uF, variable pulse width pw=
(0.16667-*0.08333)ms, (pw/per)
=0.5pu=50%
Vs 1 0 sin(0V 169.68V 60Hz)
*MOSFET model, fs=6000Hz with
C=500uF
vmosfet1 2 3 pulse(0 200 0 0 0 0.08333m
0.16667m)
mosfet1 1 2 3 3 qfet
D1 3 4 IDEAL
vmosfet2 6 7 pulse(0 200 0 0 0 0.08333m
0.16667m)
mosfet2 4 6 7 7 qfet
D2 7 1 IDEAL
*Load
R 4 0 3
C 4 0 500u
 *Model
.model IDEAL D(IS=0.01E-12)
.model qfet NMOS(level 3 gamma=0
delta=0 eta=0 +theta=0 kappa=0.2 vmax=0
xj=0 tox=100n uo=600 +phi=0.6 rs=42.69m
kp=20.87u l=2u w=500u vto=0.3
+rd=0.19 cbd=200n pb=0.8 mj=0.5
cgso=3.5n +cgdo=3.5n rg=0.2 is=10f)
*Analysis request
.Tran 1us 60ms 30ms 1us
*Output request
.Probe
.FOUR 60 10 I(R)
.options abstol=10m chgtol=10m reltol=0.1
+vntol=100m
.end
```

Note: All comment lines must be preceded with an asterisk (*) while continued calculation statements must be preceded with a plus sign (+).

Table E10.19.2 Current harmonics of $i_R(t)$ of Figure E10.19.2 in terms of amperes and in percent referred to current amplitude $a_i = 169.68\ \text{V}/3\ \Omega = 56.56\ \text{A}$.

DC component	1	3	5	7
0.00/0.00%	28.26 A/50%	0.38 A/0.67%	0.01 A/0.02%	0.00 A/0.00%

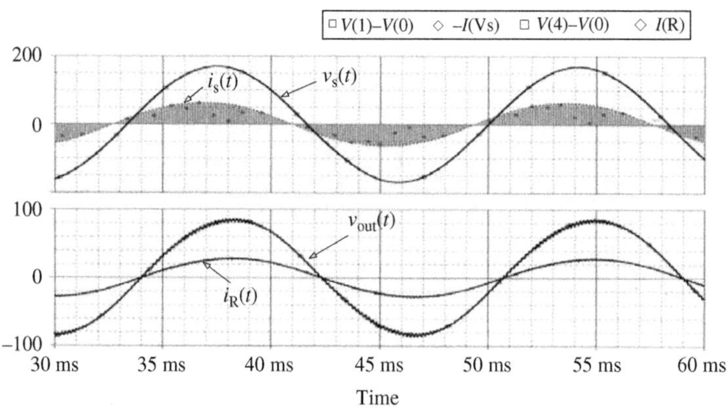

Figure E10.19.2 Top plot: source voltage $v_s(t) = V(1) - V(0)$ and input or source current $i_s(t) = -I(V_s)$. Bottom plot: output load voltage $v_{out}(t) = V(4) - V(0)$ and load current $i_R(t) = I(R)$ of AC current controller with pulse width pw = 0.083 33 ms per period or a duty ratio of (pw/per) = 0.50 pu or 50% with load $R = 3\ \Omega$ and filter capacitor $C = 500\ \mu\text{F}$.

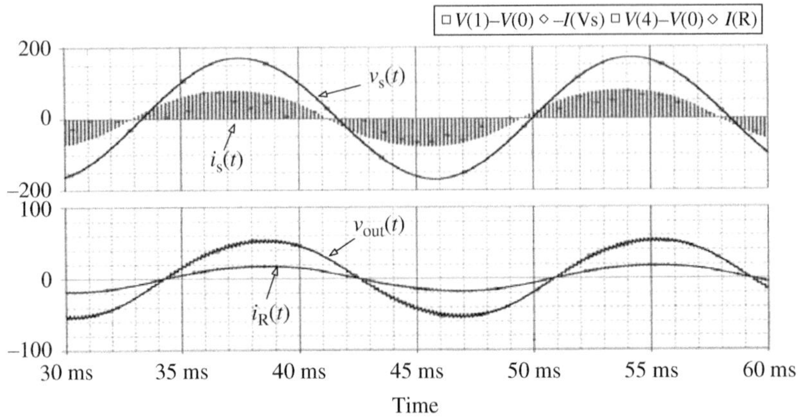

Figure E10.19.3 Top plot: source voltage $v_s(t) = V(1) - V(0)$ and input source current $i_s(t) = -I(V_s)$. Bottom plot: output load voltage $v_{out}(t) = V(4) - V(0)$ and load current $i_R(t) = I(R)$ of AC current controller with pulse width pw = 0.041 65 ms per period or a duty ratio of (pw/per) = 0.25 pu = 25% with load $R = 3\,\Omega$ and filter capacitor $C = 500\,\mu F$.

Table E10.19.3 PSPICE program for the signals of Figure E10.19.3.

```
Application Example 10.19b
*AC source, ideal MOSFET switched at
fs=6kHz, R=3 *ohms, C=500uF, variable
pulse width pw=(0.16667-*0.125)ms, (pw/
per)=0.25pu=25%
Vs 1 0 sin(0V 169.68V 60Hz)
*MOSFET model, fs=6000Hz with
C=500uF
vmosfet1 2 3 pulse(0 200 0 0 0 0.04165ms
0.16667ms)
mosfet1 1 2 3 3 qfet
D1 3 4 IDEAL
vmosfet2 6 7 pulse(0 200 0 0 0 0.04165ms
0.16667ms)
mosfet2 4 6 7 7 qfet
D2 7 1 IDEAL
*Load
R 4 0 3
C 4 0 500u
*Model
.model IDEAL D(IS=0.01E-12)
.model qfet NMOS(level 3 gamma=0
delta=0 eta=0 +theta=0 kappa=0.2
vmax=0 xj=0 tox=100n uo=600 +phi=0.6
rs=42.69m kp=20.87u l=2u w=500u
vto=0.3
+rd=0.19 cbd=200n pb=0.8 mj=0.5
cgso=3.5n +cgdo=3.5n rg=0.2 is=10f )
*Analysis request
.Tran 1us 60ms 30ms 1us
*Output request
.Probe
.FOUR 60 10 I(R)
.options abstol=10m chgtol=10m
reltol=0.1
+vntol=100m
.end
```

Note: All comment lines must be preceded with an asterisk (*) while continued calculation statements must be preceded with a plus sign (+).

Table E10.19.4 Current harmonics of $i_R(t)$ of Figure E10.19.3 in terms of amperes and in percent referred to current amplitude $a_i = 169.68\,V/3\,\Omega = 56.56\,A$.

DC component	1	3	5	7
0.00/0.00%	17.92 A/31.68%	0.45 A/0.80%	0.023 A/0.04%	0.00 A/0.00%

Conclusions

The AC current through a load R can be controlled in a sinusoidal manner through the duty ratio $\delta = $ (pulse width)/(period duration) = (pw/per), that is, the pw of the MOSFET gating voltage $v_{gs}(t)$. By employing PWM control, the harmonic content of the load current based on the duty ratio can be made very small via elimination of the switching harmonics through a filter (e.g. $C = 500\ \mu F$). The duty ratio δ varies between the range $0.01 \leq \delta \geq 0.99$. The pulse magnitude of the gating voltage (e.g. 200 V) and the switch model (e.g. $w = 500$ u) must be adjusted depending upon the magnitude of the current.

10.4 Clippers and Clampers

Circuits that limit the output voltage waveforms are called clippers, and circuits that shift the input voltage either up or down are called positive or negative clampers, respectively.

Application Example 10.20 Clipper and clamper

Simple clippers consist of input voltage, resistances, diodes, and batteries as illustrated in Figure E10.20.1. Figure E10.20.2 shows the input and output voltages of the clipper of Figure E10.20.1, where the amplitude of the sinusoidal input voltage of $v_{in}(t) = 141.4\ V \sin \omega t$ is limited at the output to the voltage $v_{out}(t)$ with a positive excursion of 75 V and a negative excursion of −50 V. Table E10.20.1 lists the PSPICE program for the input and output voltage of Figure E10.20.2.

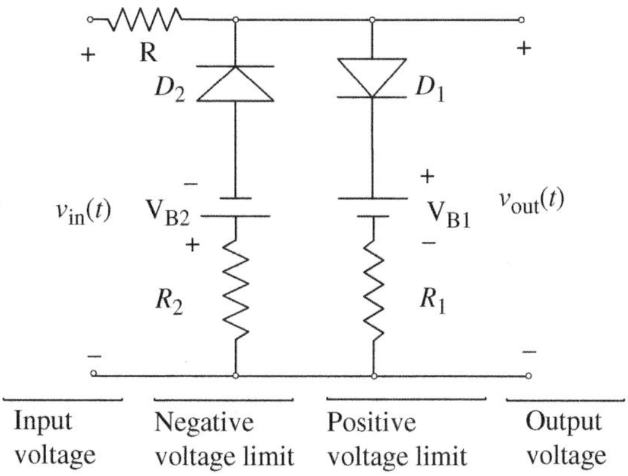

Figure E10.20.1 Clipping circuit limiting the positive and negative excursions of a sinusoidal voltage.

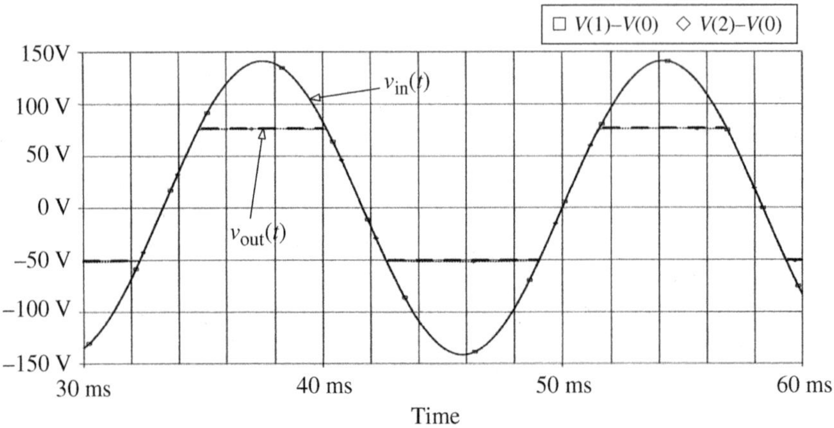

Figure E10.20.2 Input $v_{in}(t) = V(1) - V(0)$ and output $v_{out}(t) = V(2) - V(0)$ voltages of the clipping circuit of Figure E10.20.1, where the input voltage with amplitude 141.4 V is limited to positive (75 V) and negative (−50 V) excursions.

Table E10.20.1 PSPICE program for the signals of Figure E10.20.2.

```
Application Example 10.20a
* Asymmetric clipper, AC
source, ideal diode characteristic
Vin 1 0 sin(0V 141.4V 60Hz)
R 1 2 100
* Ideal diodes
D2 3 2 ideal
D1 2 5 ideal
VB2 4 3 50
VB1 5 6 75
R2 4 0 1
R1 0 6 1
*Model diode
.Model ideal D(IS=0.01E-12)
*Analysis request
.Tran 10us 60ms 30ms 10us
*Output request
.Probe
.end
```

Note: All comment lines must be preceded with an asterisk (*) while continued calculation statements must be preceded with a plus sign (+).

Conclusions

A clipper is employed to limit the most positive and most negative excursions of an input signal or voltage to v_{max} and v_{min}, respectively, in order to protect the following circuit from excessive voltages.

A positive clamping circuit is relied on to have well-defined positive voltages, where the most negative value of the waveform resides at about 0 V. Figure E10.20.3 depicts a positive clamping circuit consisting of capacitor C, resistor R, and diode D (see Figure E10.20.4). The PSPICE program is presented in Table E10.20.2.

Figure E10.20.3 Positive clamping circuit maintaining at the output voltage the waveform of the input voltage, but shifting the input voltage upward so that practically only positive output voltage values occur.

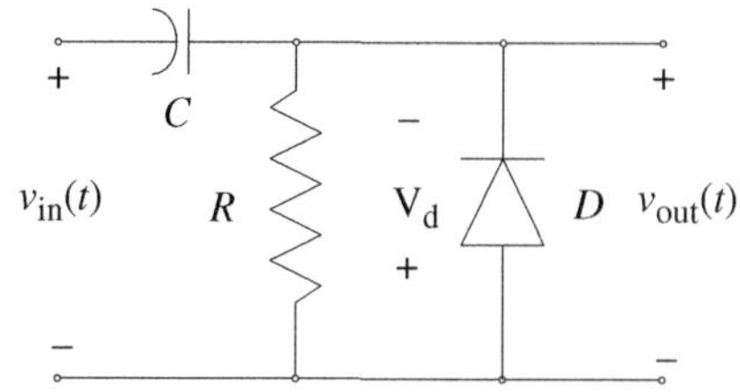

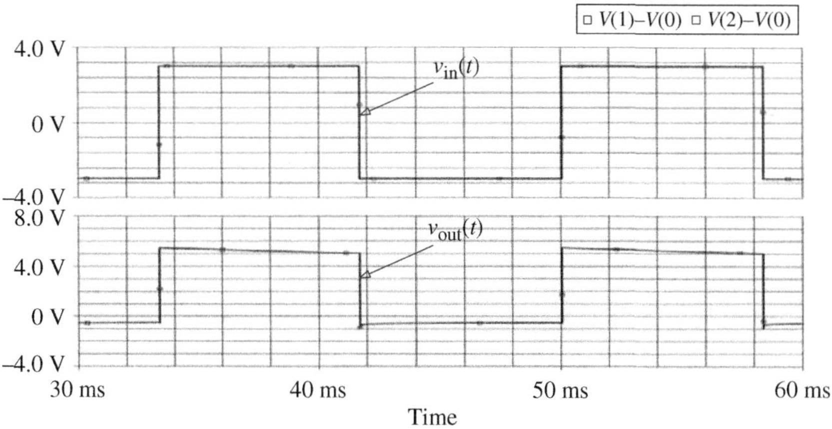

Figure E10.20.4 Top plot: rectangular input $v_{in}(t) = V(1) - V(0)$. Bottom plot: output $v_{out}(t) = V(2) - V(0)$ voltages of the clamping circuit of Figure E10.20.3, where the input voltage with maximum positive and negative excursions of 3.0 V is limited to positive (6.0 V $- V_d$) and negative ($-V_d$) excursions, whereby V_d correspond to the rated diode voltage drop of the clamping diode and is about $V_d = 0.7$ V.

Table E10.20.2 PSPICE program for the signals of Figure E10.20.4.

```
Application Example 10.20b
* Positive clamping circuit, rectangular AC input
*voltage source, real diode characteristic, 60 Hz voltage
Vin 1 0 pulse(-3V 3V 50us 0.1us 0.1us 8.333ms 16.666ms)
C 1 2 1m
* Real diode
D 0 2 real
R 2 0 100
*Model diode
.Model real D(IS=1E-12)
*Analysis request
.Tran 10us 60ms 30ms 10us
*Output request
.Probe
.end
```

Note: All comment lines must be preceded with an asterisk (*) while continued calculation statements must be preceded with a plus sign (+).

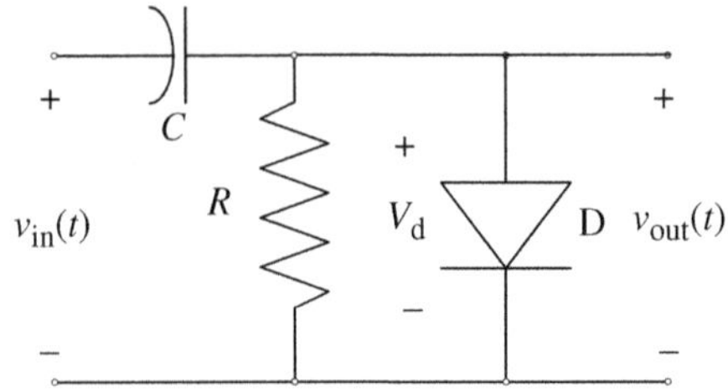

Figure E10.20.5 Negative clamping circuit maintaining at the output voltage the waveform of the input voltage, but shifting the input voltage downward so that practically only negative output voltage values occur.

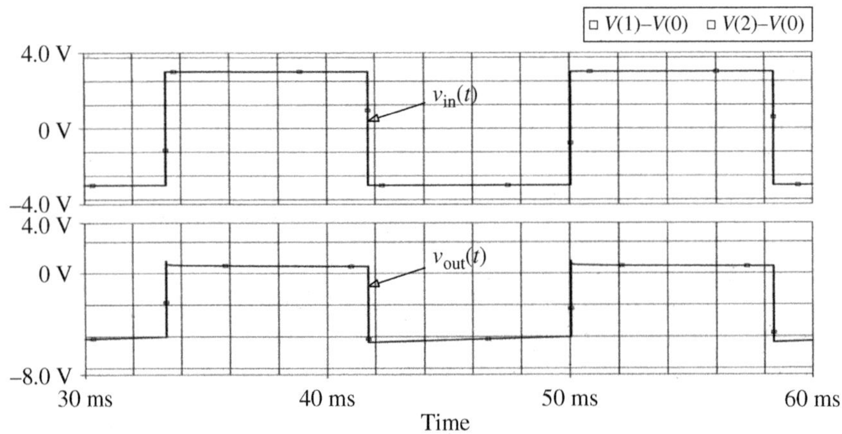

Figure E10.20.6 Top plot: input $v_{in}(t) = V(1) - V(0)$. Bottom plot: rectangular input and output $v_{out}(t) = V(2) - V(0)$ voltages of the clamping circuit of Figure E10.20.5, where voltage with maximum positive and negative excursions of 3.0 V is limited to negative ($-6.0\,V + V_d$) and positive (V_d) excursions, whereby V_d correspond to the rated diode voltage drop of the clamping diode and is about $V_d = 0.7\,V$.

Conclusions

A positive clamping circuit is used to have the most negative excursion clamped to a diode voltage drop, e.g. $V_d = -0.7\,V$.

A negative clamping circuit is relied on to have well-defined negative voltages, where the most positive value of the waveform resides at about 0 V. Figure E10.20.5 depicts a negative clamping circuit consisting of capacitor C, resistor R, and diode D (see Figure E10.20.6; Table E10.20.3). In Figure E10.20.6 the positive excursion is clamped to a diode voltage drop of $V_d = 0.7\,V$.

Table E10.20.3 PSPICE program for the signals of Figure E10.20.6.

```
Application Example 10.20c
* Negative clamping circuit, rectangular AC input
*voltage source, real diode characteristic, 60 Hz voltage
Vin 1 0 pulse(-3V 3V 5us 0.1us 0.1us 8.333ms 16.666ms)
C 1 2 1m
* Real diode
D 2 0 real
R 2 0 1000
*Model diode
.Model real D(IS=1E-12)
*Analysis request
.Tran 10us 60ms 30ms 10us
*Output request
.Probe
.end
```

10.5 Three-Phase Rectifiers

Power systems in the majority of countries are three phase, because with three-phase systems the maximum amount of electric energy can be transmitted with the minimum amount of conducting material (e.g. copper, aluminum) [7, 8]. Single-phase systems cannot generate a rotating field as utilized in rotating machines; two-phase systems require four wires; balanced three-phase systems require three wires because the current of the three phases adds up to zero, and the so-called neutral does not conduct any current at balanced operation; and four-phase systems require eight conductors, six-phase systems require 12 conductors, and so on. For renewable energy generation three-phase converters – such as rectifiers and inverters – are essential.

Application Example 10.21 Full-wave, three-phase diode bridge rectifier
A simple uncontrolled three-phase rectifier is depicted in Figure E10.21.1, where a diode bridge is relied on to transform AC to DC voltage and current and supplying

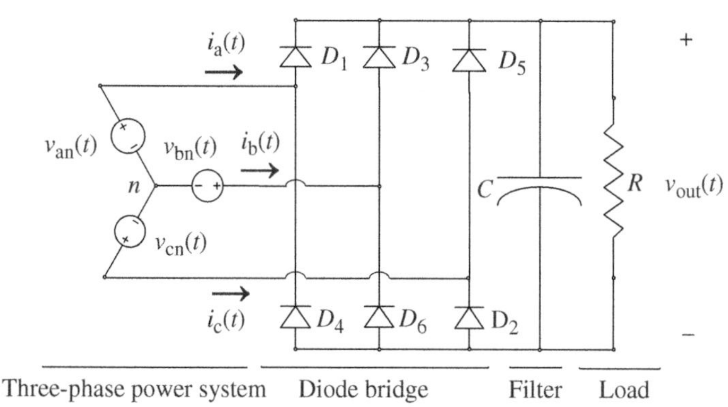

Figure E10.21.1 Full-wave, three-phase rectifier employing a diode bridge feeding an R load with filter capacitance C; a, b, and c stand for the three phases.

power to an R load with filter capacitance C (see Figure E10.21.2). The PSPICE program is displayed in Table E10.21.1, and the fundamental and harmonic currents of $i_{an}(t)$ are itemized in Table E10.21.2.

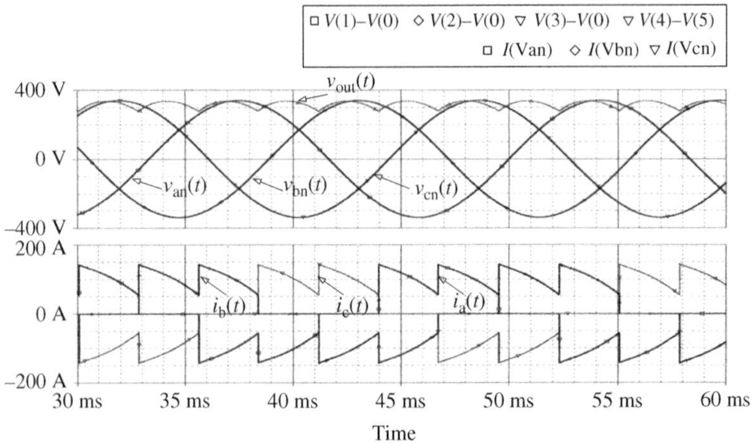

Figure E10.21.2 Top plot: source voltages $v_{an}(t) = V(1) - V(0)$, $v_{bn}(t) = V(2) - V(0)$, $v_{cn}(t) = V(3) - V(0)$ and output load voltage $v_{out}(t) = V(4) - V(5)$. Bottom plot: source currents $i_a(t) = I$ (Van), $i_b(t) = I$(Vbn), and $i_c(t) = I$(Vcn) of rectifier based on three-phase diode rectifier loaded with $R = 3\ \Omega$ and capacitor $C = 500\ \mu F$.

Table E10.21.1 PSPICE program for the signals of Figure E10.21.2.

```
Application Example 10.21
*Full-wave, three-phase diode bridge
rectifier
Van 1 0 SIN(0 339.36V 60Hz 0 0 0)
Vbn 2 0 SIN(0 339.36V 60Hz 0 0 120)
Vcn 3 0 SIN(0 339.36V 60Hz 0 0 240)
Rshunt 0 5 10Meg
D1 1 4 D1N4001
D3 2 4 D1N4001
D5 3 4 D1N4001
D4 5 1 D1N4001
D6 5 2 D1N4001
D2 5 3 D1N4001
R 4 5 3
C 4 5 500u
.model D1N4001 D(IS=1E-12)
.tran .05u 60ms 30ms iu uic
.options abstol=1m chgtol=1m reltol=1m
vntol=1m
.four 60 I(Van)
.probe
.end
```

Note: All comment lines must be preceded with an asterisk (*) while continued calculation statements must be preceded with a plus sign (+).

Table E10.21.2 Fundamental current (1) and harmonics of $i_a(t)$ of Figure E10.21.2.

DC component	1	2	5	7	9
−4.16 A	204.2 A	8.75 A	47.09 A	20.87 A	2.48 A

Conclusions

The dominant current harmonics at the input of a three-phase system are the 5th, 7th, 11th, etc. As compared with half-wave rectification, full-wave rectification requires a relatively small filter capacitance *C*.

To control the DC current and voltage of a rectifier, a line- or load-commutated switch such as a thyristor can be used, as demonstrated in Application Example 10.22. These types of rectifiers, including gate turn-off thyristors (GTOs), can be employed at very high voltages and currents as used for high-voltage DC transmission lines.

Application Example 10.22 Full-wave, three-phase thyristor bridge rectifier

A controlled three-phase rectifier is depicted in Figure E10.22.1 where thyristors are relied on to transform AC to DC voltage and current. Thyristor firing angles will be investigated for $\alpha = 0°$ and $\alpha = 50°$. Because of parasitic capacitances generating reverse-recovery current, which can be mitigated by snubbing inductances L_{s1}, L_{s3}, and L_{s5} in practical applications, the $\alpha = 0°$ condition should be avoided, and it is acceptable to operate the firing angle under the condition $0.05° \leq \alpha \leq 175°$. To easily explain the gating sequence in this text, the firing angle is assumed to be zero, as shown in Figure E10.22.2. At the outset it is important to note that the applied voltage phase sequence must be the same as the sequence of the gating voltages of the six thyristors. For this reason Van 1 0 SIN(0 339.36V 60Hz 0 0 0), Vbn 2 0 SIN(0 339.36V 60Hz 0 0 -120), Vcn 3 0 SIN(0 339.36V 60Hz 0 0 -240) were

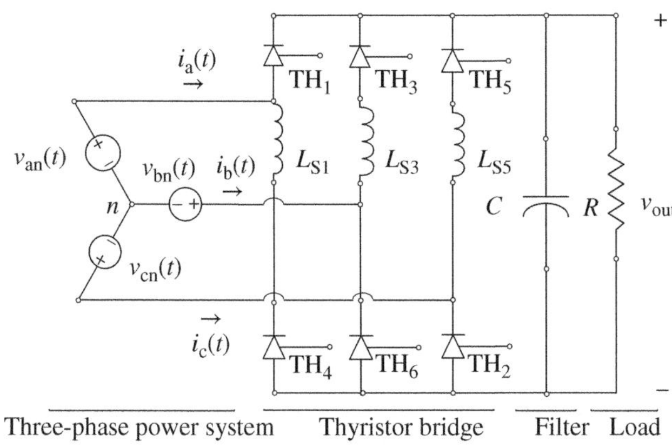

Figure E10.22.1 Full-wave, three-phase thyristor rectifier supplying power to *R* load with filter capacitance *C*; a, b, and c stand for the three phases.

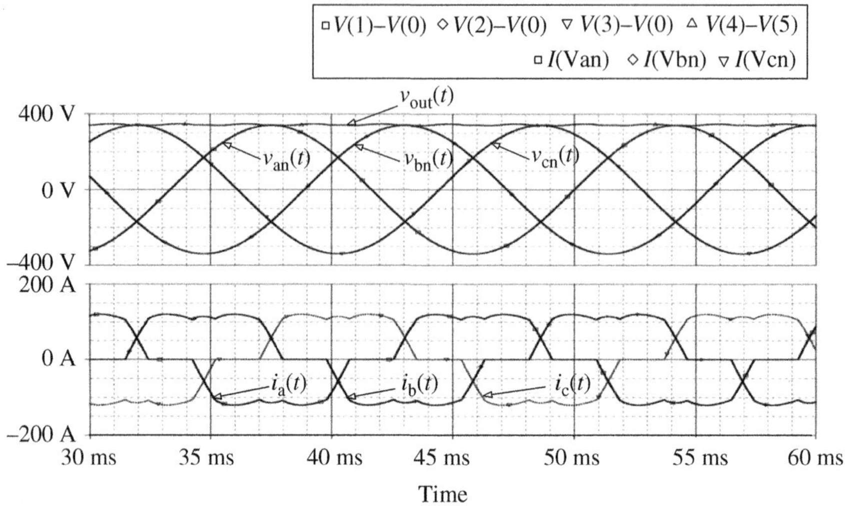

Figure E10.22.2 Top plot: source voltages $v_{an}(t) = V(1) - V(0)$, $v_{bn}(t) = V(2) - V(0)$, $v_{cn}(t) = V(3) - V(0) = V(3)$ (note by definition $V(0) = 0$) and output load voltage $v_{out}(t) = V(4) - V(5)$. Bottom plot: source currents $i_a(t) = I(\text{Van})$, $i_b(t) = I(\text{Vbn})$, $i_c(t) = I(\text{Vcn})$ of rectifier based on three-phase thyristor rectifier with firing angle $\alpha = 0°$ loaded with $R = 3\ \Omega$ and capacitor $C = 500\ \mu\text{F}$.

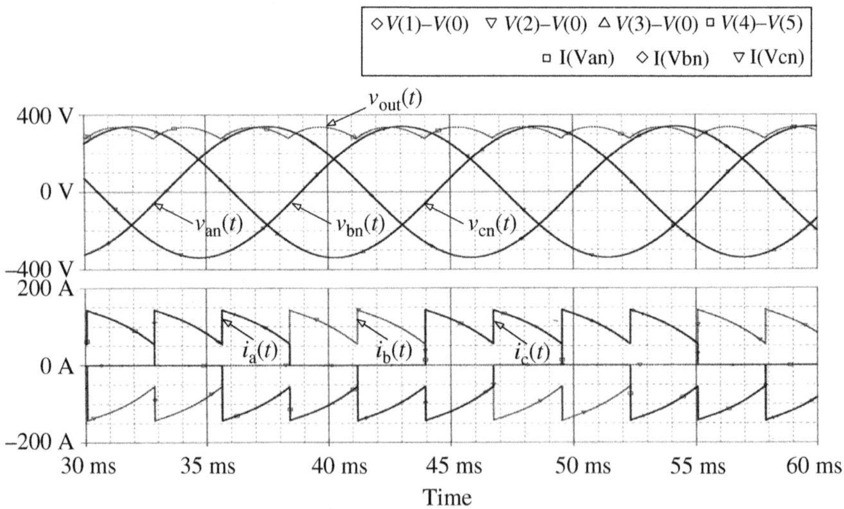

Figure E10.22.3 Top plot: source voltages $v_{an}(t) = V(1) - V(0)$, $v_{bn}(t) = V(2) - V(0)$, $v_{cn}(t) = V(3) - V(0)$, and output load voltage $v_{out}(t) = V(4) - V(5)$. Bottom plot: source currents $i_a(t) = I(\text{Van})$, $i_b(t) = I(\text{Vbn})$, $i_c(t) = I(\text{Vcn})$ of rectifier based on three-phase thyristor rectifier with firing angle $\alpha = 50°$ loaded with $R = 3\ \Omega$ and capacitor $C = 500\ \mu\text{F}$.

Table E10.22.1 PSPICE program for the signals of Figure E10.22.2, firing angle $\alpha = 0°$.

```
Application Example 10.22a, alpha=0 degrees
*Full-wave, three-phase thyristor bridge
rectifier
Van 1 0 SIN(0 339.36V 60Hz 0 0 0)
Vbn 2 0 SIN(0 339.36V 60Hz 0 0 -120)
Vcn 3 0 SIN(0 339.36V 60Hz 0 0 -240)
Rshunt 0 5 1000000
vmosfet1 1a 1b pulse(0 300 0m 1u 1u 8.33ms
+16.667m)
mosfet1 1 1a 1b 1b qfet
D1 1b 4 D1N4001
vmosfet3 2a 2b pulse(0 300 5.555m 1u 1u
8.33m +16.667m)
mosfet3 2 2a 2b 2b qfet
D3 2b 4 D1N4001
vmosfet5 3a 3b pulse(0 300 11.11m 1u 1u
8.33m +16.667m)
mosfet5 3 3a 3b 3b qfet
D5 3b 4 D1N4001
vmosfet4 5a 5b pulse(0 300 8.333m 1u 1u
8.33m +16.667m)
mosfet4 5 5a 5b 5b qfet
D4 5b 5c D1N4001
Ls4 5c 1 1p
vmosfet6 6a 6b pulse(0 300 13.888m 1u 1u
8.33m
+16.667m)
mosfet6 5 6a 6b 6b qfet
D6 6b 6c D1N4001
Ls6 6c 2 1p
vmosfet2 7a 7b pulse(0 300 19.443m 1u
1u 8.33m +16.667m)
mosfet2 5 7a 7b 7b qfet
D2 7b 7c D1N4001
Ls2 7c 3 1p
R 4 5 3
C 4 5 500u
.model D1N4001 D(IS=1E-12)
.model qfet NMOS(level 3 gamma=0
delta=0 eta=0 +theta=0 kappa=0.2
vmax=0 xj=0 tox=100n uo=600
+phi=0.6 rs=42.69m kp=20.87u l=2u
w=500u vto=0.3
+rd=0.19 cbd=200n pb=0.8 mj=0.5
cgso=3.5n
+cgdo=3.5n rg=0.2 is=10f)
.tran .05u 60ms 30ms 1u uic
.options abstol=100m chgtol=100m
reltol=100m vntol=100m
.four 60 I(Van)
.probe
.end
```

Note: All comment lines must be preceded with an asterisk (*) while continued calculation statements must be preceded with a plus sign (+).

chosen, and the corresponding gating signals (see Table E10.22.1) occur for mosfet1 at time 0 ms, for mosfet3 at 5.555 ms, for mosfet5 at 11.11 ms, for mosfet4 at 8.333 ms, for mosfet6 at 13.888 ms, and for mosfet2 at 19.443 ms. The definition of the sequence of thyristors in Figure E10.22.1 is universally accepted in the engineering community. As a result the gating voltages of TH_1 and TH_4 are 180° apart, that of TH_3 and TH_6 are180° apart as well, and the same applies to TH_5 and TH_2. All gating signals last 120°. If the firing angle α is nonzero, then all gating signals/voltages are shifted in time by α degrees. PSPICE program in Table E10.22.2 determines the signals of Figure E10.22.3 for firing angle $\alpha = 50°$. The fundamental and harmonic currents of $i_a(t)$ are shown in Tables E10.22.3 and E10.22.4 for $\alpha = 0$ and 50°, respectively.

Conclusions

The advantage of either line- or load-commutated thyristors is that the output voltage can be controlled; however, thyristor rectifiers generate nonsinusoidal

Table E10.22.2 PSPICE program for the signals of Figure E10.22.3, firing angle α = 50°.

```
Application Example 10.22b,
alpha=50 degrees
*Full-wave, three-phase thyristor
bridge rectifier
Van 1 0 SIN(0 339.36V 60Hz 0 0 0)
Vbn 2 0 SIN(0 339.36V 60Hz 0 0 -120)
Vcn 3 0 SIN(0 339.36V 60Hz 0 0 -240)
Rshunt 0 5 1000000
vmosfet1 1a 1b pulse(0 300 2.31m 1u
1u 8.33ms +16.667m)
mosfet1 1 1a 1b 1b qfet
D1 1b 4 D1N4001
vmosfet3 2a 2b pulse(0 300 7.869m 1u
1u 8.33m +16.667m)
mosfet3 2 2a 2b 2b qfet
D3 2b 4 D1N4001
vmosfet5 3a 3b pulse(0 300 13.399m
1u 1u 8.33m +16.667m)
mosfet5 3 3a 3b 3b qfet
D5 3b 4 D1N4001
vmosfet4 5a 5b pulse(0 300 10.639m
1u 1u 8.33m +16.667m)
mosfet4 5 5a 5b 5b qfet
D4 5b 5c D1N4001
Ls4 5c 1 1p
vmosfet6 6a 6b pulse(0 300 16.199m
1u 1u 8.33m
+16.667m)
mosfet6 5 6a 6b 6b qfet
D6 6b 6c D1N4001
Ls6 6c 2 1p
vmosfet2 7a 7b pulse(0 300 21.732m 1u 1u 8.33m
+16.667m)
mosfet2 5 7a 7b 7b qfet
D2 7b 7c D1N4001
Ls2 7c 3 1p
R 4 5 3
C 4 5 500u
.model D1N4001 D(IS=1E-12)
.model qfet NMOS(level 3 gamma=0 delta=0
eta=0
+theta=0 kappa=0.2 vmax=0 xj=0 tox=100n
uo=600 +phi=0.6 rs=42.69m kp=20.87u l=2u
w=500u vto=0.3
+rd=0.19 cbd=200n pb=0.8 mj=0.5 cgso=3.5n
+cgdo=3.5n rg=0.2 is=10f )
.tran .05u 60ms 30ms 1u uic
.options abstol=100m chgtol=100m
reltol=100m
+vntol=100m
.four 60 I(Van)
.probe
.end
```

Note: All comment lines must be preceded with an asterisk (*) while continued calculation statements must be preceded with a plus sign (+).

Table E10.22.3 Fundamental current (1) and harmonics of $i_a(t) = I(\text{Van})$ of Figure E10.22.3, firing angle α = 0°.

DC component	1	2	5	7	9
0.00	124.2 A	0.02 A	14.24 A	2.36 A	0.0 A

Table E10.22.4 Fundamental current (1) and harmonics of $i_a(t) = I(\text{van})$ of Figure E10.22.3, firing angle α = 50°.

DC component	1	2	5	7	9
−0.04 A	115.7 A	0.15 A	32.26 A	17.85 A	0.46 A

input AC currents and voltages – the latter caused by harmonic voltage drops across the power system impedance due to the nonsinusoidal AC current – requiring filters at the input and output of the rectifiers. For this reason the tendency is to employ self-commutated switches such as MOSFETs and IGBTs, which can be switched or turned-on and turned-off at will at high frequency, such as 6 kHz and higher. The pulse magnitude (e.g. 300 V) and the switch model (e.g. w = 500 u) must be adjusted depending upon the magnitude of the current.

Application Example 10.23 Full-wave, three-phase MOSFET rectifiers
A controlled three-phase rectifier is depicted in Figure E10.23.1 where one MOSFET is relied on to transform AC voltage and current to DC voltage $V_{out}(t)$ as shown in Figure E10.23.2. The DC current is not shown. The PSPICE program for the voltages and currents of Figure E10.23.2 is presented in Table E10.23.1, and the fundamental and harmonic currents of Figure E10.23.2 are listed in Table E10.23.2.

Conclusions

The introduction of the filter inductance $L_f = 1$ mH not only improves the unipolar output voltage but also reduces its magnitude. The pulse magnitude (e.g. 200 V) and the switch model (e.g. w = 500 u) must be adjusted depending upon the magnitude of the current. The current amplitude of the switching harmonic is large and cannot be accepted without any additional filter across the three-phase AC source.

An alternative MOSFET three-phase bridge rectifier is depicted in Figure E10.23.3, where six MOSFETs and six diodes are used. The diodes prevent the

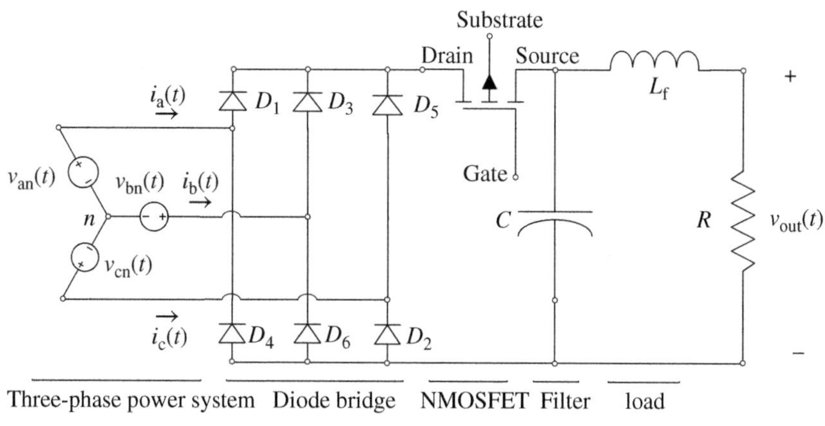

Figure E10.23.1 Full-wave, three-phase rectifier with diode bridge employing one self-commutated MOSFET switch operates in PWM mode supplying power to R load with C_f = 500 μF filter via the inductor L_f = 1 mH.

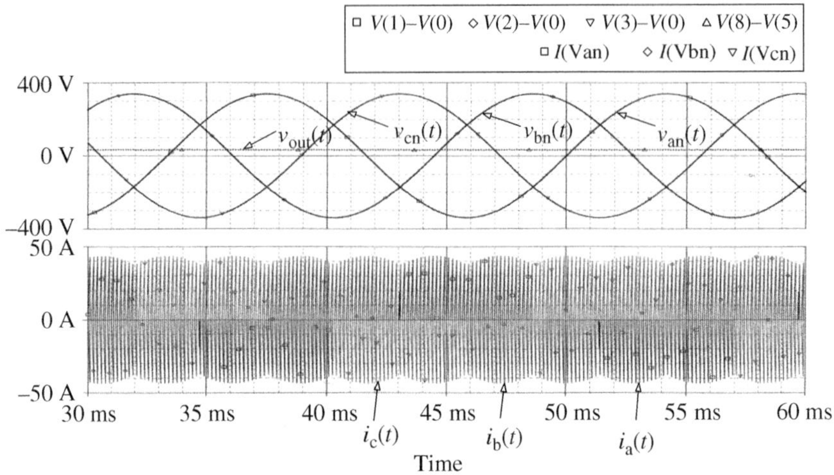

Figure E10.23.2 Top plot: source voltages $v_{an}(t) = V(1) - V(0)$, $v_{bn}(t) = V(2) - V(0)$, $v_{cn}(t) = V(3) - V(0)$, and output load voltage $v_{out}(t) = V(8) - V(5)$. Bottom plot: source current $i_a(t) = I(\text{Van})$, $i_b(t) = I(\text{Vbn})$, $i_c(t) = I(\text{Vcn})$ of rectifier based on rectifier with one self-commutated switch (MOSFET) loaded with $R = 3\ \Omega$ and capacitor $C_f = 500\ \mu F$ via the inductor $L_f = 1$ mH.

Table E10.23.1 PSPICE program for the signals of Figure E10.23.2.

```
Application Example 10.23a
*Full-wave, three-phase diode bridge rectifier
*with one mosfet
Van 1 0 SIN(0 339.36V 60Hz 0 0 0)
Vbn 2 0 SIN(0 339.36V 60Hz 0 0 120)
Vcn 3 0 SIN(0 339.36V 60Hz 0 0 240)
Rshunt 0 5 10Meg
D1 1 4 D1N4001
D3 2 4 D1N4001
D5 3 4 D1N4001
D4 5 1 D1N4001
D6 5 2 D1N4001
D2 5 3 D1N4001
vmosfet 6 7 pulse(0 200 0m 1u 1u 0.0833ms
+ 0.16667m)
mosfet 4 6 7 7 qfet
Lf 7 8 1m
R 8 5 3
Cf 8 5 500u
.model D1N4001 D(IS=1E-12)
.model qfet NMOS(level 3 gamma=0
delta=0 eta=0 theta=0 kappa=0.2
vmax=0 xj=0 tox=100n uo=600
+phi=0.6 rs=42.69m kp=20.87u l=2u
w=500u vto=0.3
+rd=0.19 cbd=200n pb=0.8 mj=0.5
cgso=3.5n
+cgdo=3.5n rg=0.2 is=10f )
.tran .0005u 60ms 30ms 1u uic
.options abstol=100m chgtol=100m
reltol=100m
+vntol=100m
.four 60 I(Van)
.probe
.end
```

Note: All comment lines must be preceded with an asterisk (*) while continued calculation statements must be preceded with a plus sign (+).

Table E10.23.2 Fundamental current (1) and harmonics of $i_a(t) = I(Van)$ of Figure E10.23.2.

DC component	1	2	5	7	9
−0.01 A	12.35 A	0.01 A	2.76 A	1.41 A	0.08 A

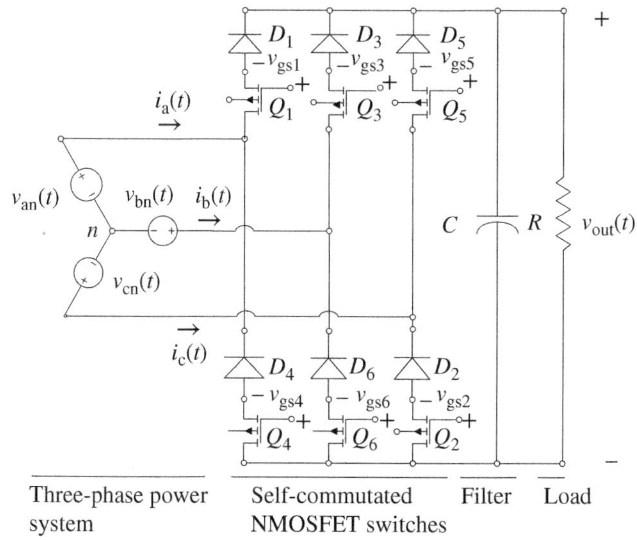

Figure E10.23.3 Full-wave, three-phase MOSFET rectifier employing six self-commutated switches supplying power to R load with $C = C_f$ filter; a, b, and c stand for the three phases.

MOSFETs conducting in reverse direction. The absence of these diodes changes the control of this rectifier, and a negative unipolar output voltage results. The PSPICE program for the voltages and currents of Figure E10.23.4 is displayed in Table E10.23.3.

Conclusions

Low-frequency harmonic current components have small magnitudes. It is advisable to rely on a tuned L_f C_f filter to mitigate the large 6 kHz current switching harmonics not indicated in Table E10.23.4. Deletion of the six diodes causes reverse conduction of MOSFETs, resulting in a negative output voltage $v_{out}(t)$. The pulse magnitude of the gating voltage (e.g. 300 V) and the switch model (e.g. $w = 500$ u) must be adjusted depending upon the magnitude of the current.

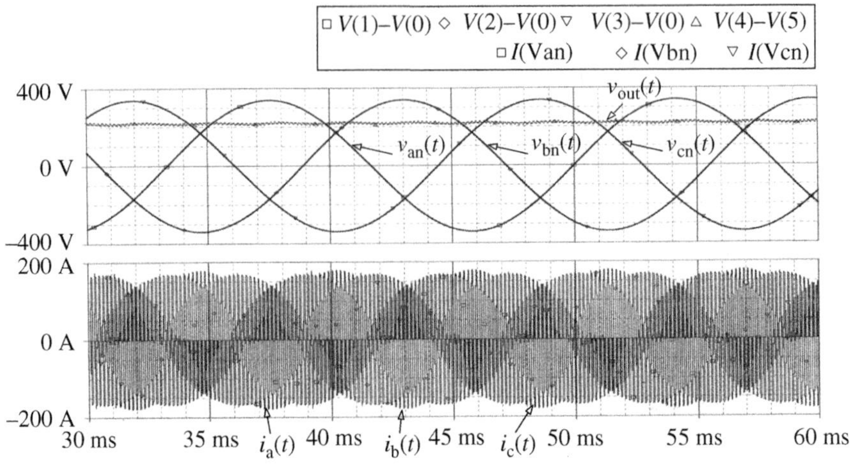

Figure E10.23.4 Top plot: source voltages $v_{an}(t) = V(1) - V(0)$, $v_{bn}(t) = V(2) - V(0)$, $v_{cn}(t) = V(3) - V(0)$, and output load voltage $v_{out}(t) = V(4) - V(5)$. Bottom plot: source currents $i_a(t) = I(\text{Van})$, $i_b(t) = I(\text{Vbn})$, $i_c(t) = I(\text{Vcn})$ of rectifier based on rectifier with six self-commutated switches (MOSFET) operated in PWM mode and loaded with $R = 3\ \Omega$ and capacitor $C_f = 500\ \mu\text{F}$.

Table E10.23.3 PSPICE program for the signals of Figure E10.23.4.

```
Application Example 10.23b
*Full-wave, three-phase mosfet
bridge rectifier
*with 6 mosfets and 6 diodes to block
*negative current flow
*switching frequency fs=6kHz
Van 1 0 SIN(0 339.36V 60Hz 0 0 0)
Vbn 2 0 SIN(0 339.36V 60Hz 0 0
-120)
Vcn 3 0 SIN(0 339.36V 60Hz 0 0 -240)
Rshunt 0 5 1000000
vmosfet1 1a 1b pulse(0 300 0m 1u 1u
0.0833ms +0.16667m)
mosfet1 1 1a 1b 1b qfet
D1 1b 4 D1N4001
vmosfet3 2a 2b pulse(0 300 0.05555m
1u 1u 0.0833m +0.16667m)
mosfet3 2 2a 2b 2b qfet
D3 2b 4 D1N4001
vmosfet5 3a 3b pulse(0 300 0.1111m
1u 1u 0.0833m +0.16667m)
mosfet5 3 3a 3b 3b qfet
D5 3b 4 D1N4001
vmosfet4 5a 5b pulse(0 300 0.08333m
1u 1u 0.0833m +0.16667m)
mosfet4 5 5a 5b 5b qfet
D4 5b 1 D1N4001
vmosfet6 6a 6b pulse(0 300 0.13888m 1u 1u
0.0833m
+0.16667m)
mosfet6 5 6a 6b 6b qfet
D6 6b 2 D1N4001
vmosfet2 7a 7b pulse(0 300 0.19443m 1u 1u
0.0833m +0.16667m)
mosfet2 5 7a 7b 7b qfet
D2 7b 3 D1N4001
R 4 5 3
Cf 4 5 500u
.model D1N4001 D(IS=1E-12)
.model qfet NMOS(level 3 gamma=0 delta=0
eta=0 +theta=0 kappa=0.2 vmax=0 xj=0
tox=100n uo=600 +phi=0.6 rs=42.69m
kp=20.87u l=2u w=500u
+vto=0.3 rd=0.19 cbd=200n pb=0.8 mj=0.5
+cgso=3.5n cgdo=3.5n rg=0.2 is=10f)
.tran .05u 60ms 30ms 1u uic
.options abstol=100m chgtol=100m reltol=100m
+vntol=100m
.four 60 I(Van)
.probe
.end
```

Note: All comment lines must be preceded with an asterisk (*) while continued calculation statements must be preceded with a plus sign (+).

Table E10.23.4 Fundamental current (1) and harmonics of $i_a(t) = I(\text{Van})$ of Figure E10.23.4.

DC component	1	2	5	7	9
0.04 A	74.87 A	0.03 A	1.28 A	0.87 A	0.02 A

Application Example 10.24 Full-wave, three-phase diode bridge rectifier with step-down delta/wye input transformer supplied by power system with finite impedances

Perform a PSPICE analysis for the circuit of Figure E10.24.1 where a three-phase diode bridge rectifier with filter $C_f = 500\ \mu\text{F}$ serves a load $R_{load} = 10\ \Omega$. An ideal Δ/(ungrounded Y) step-down transformer with a turns ratio of $(N_1/N_2) = 12$ may be assumed. The power system impedance per phase is $R_{syst} = 0.10\ \Omega$, $X_{syst} = 0.25\ \Omega$ at $f = 60$ Hz, the line-to-line voltages of the grid are $v_{AB}(t) = \sqrt{2} \cdot 12\,472\ \text{V} \cos \omega t$, $v_{BC}(t) = \sqrt{2} \cdot 12\,472\ \text{V} \cos(\omega t - 120°)$, $v_{CA}(t) = \sqrt{2} \cdot 12\,472\ \text{V} \cos(\omega t - 240°)$, and assume ideal diodes D_1–D_6.

Plot at steady-state operation:

1) line-to-line voltages $v_{AB}(t)$, $v_{BC}(t)$, $v_{CA}(t)$
2) primary phase (ph) voltages $v_{Aph}(t)$, $v_{Bph}(t)$, $v_{Cph}(t)$
3) secondary transformer line-to-line voltages $v_{ab}(t)$, $v_{bc}(t)$, $v_{ca}(t)$
4) secondary transformer phase (ph) currents $i_{aph}(t) = I(L_4)$, $i_{bph}(t) = I(L_5)$, $i_{cph}(t) = I(L_6)$
5) output voltage $v_{out}(t)$
6) source or line (L) currents $i_{AL}(t)$, $i_{BL}(t)$, $i_{CL}(t)$.

Subject $v_{AB}(t) = V(1,2)$, $v_{ab}(t) = V(11,12)$, $i_{AL}(t) = I(\text{RsysA})$, $i_{aph}(t) = I(L_4)$, and $v_{out}(t) = V(15,14)$ to Fourier analyses (up to the 12th harmonic); are they different?

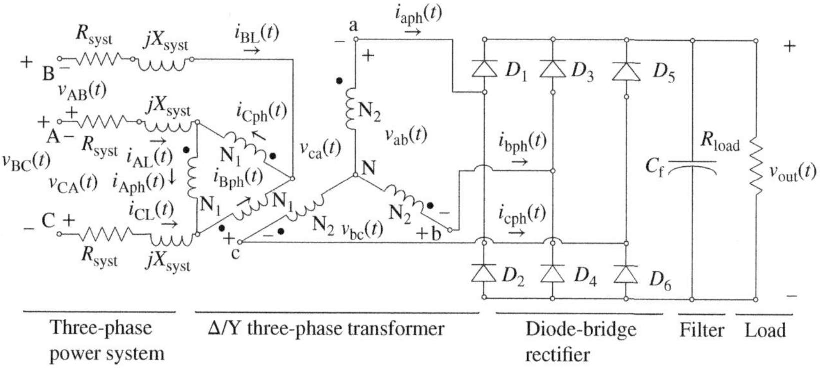

Figure E10.24.1 Three-phase Δ/(ungrounded Y) step-down transformer supplying the load $R_{load} = 10\ \Omega$ via a diode bridge rectifier with filter $C_f = 500\ \mu\text{F}$.

Solution

Figures E10.24.2 and E10.24.3 illustrate the above listed voltages and currents. The PSPICE program for Figures E10.24.2 and E10.24.3 is shown in Table E10.24.1, and the fundamental and harmonic voltages for $v_{AB}(t)$, $V_{ab}(t)$

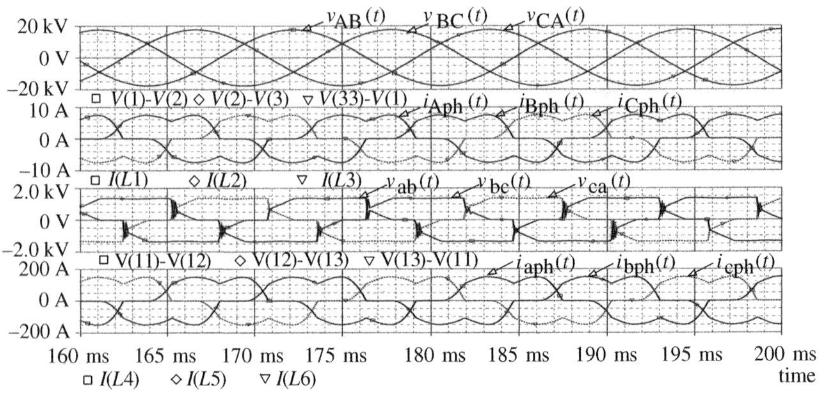

Figure E10.24.2 Top plot: line-to-line voltages $v_{AB}(t) = V(1) - V(2)$, $v_{BC}(t) = V(2) - V(3)$, $v_{CA}(t) = V(33) - V(1)$. Second plot from top: primary transformer phase currents $i_{Aph}(t) = I(L1)$, $i_{Bph}(t) = I(L2)$, $i_{Cph}(t) = I(L3)$. Third plot from top: secondary line-to-line transformer voltages $V_{ab}(t) = V(11) - V(12)$, $v_{bc}(t) = V(12) - V(13)$, $v_{ca}(t) = V(13) - V(11)$. Bottom plot: secondary transformer phase currents $i_{aph}(t) = I(L4)$, $i_{bph}(t) = I(L5)$, $i_{cph}(t) = I(L6)$.

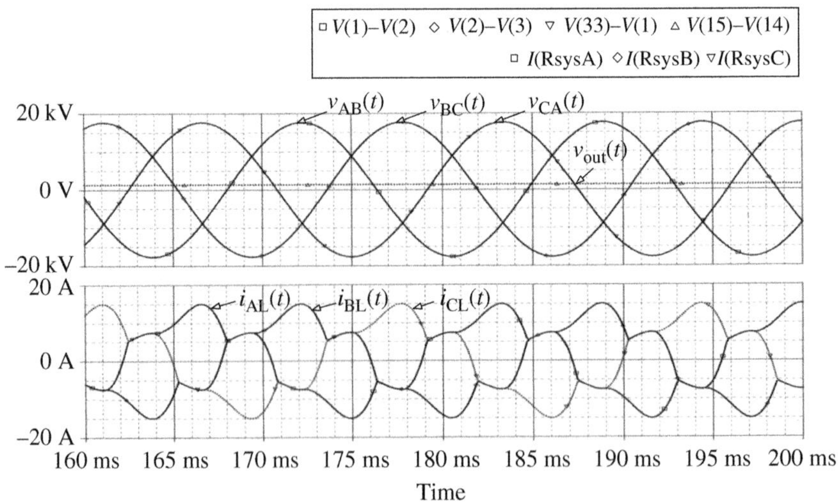

Figure E10.24.3 Top plot: line-to-line voltages $v_{AB}(t) = V(1) - V(2)$, $v_{BC}(t) = (V2) - (V3)$, $v_{CA}(t) = V(33) - V(1)$, and output voltage $v_{out}(t) = v_{load}(t) = V(15) - V(14)$. Bottom plot: source or line currents $i_{AL}(t) = I(RsysA)$, $i_{BL}(t) = I(RsysB)$, $i_{CL}(t) = I(RsysC)$.

Table E10.24.1 PSPICE input program for Figures E10.24.2 and E10.24.3.

```
Application Example 10.24
* Three-phase delta/(ungrounded Y) step-down
*transformer supplying the load Rload via a diode-*bridge
rectifier with filter Cf
*Primary line-to-line voltages
VAB 1 2 sin(0 17635.24 60 0 0 -30)
VBC 2 3 sin(0 17635.24 60 0 0 -150)
VCA 33 1 sin(0 17635.24 60 0 0 -270)
*Grid system impedances
RsysA 6 1 100m
RsysB 4 2 100m
RsysC 5 3 100m
LsysA 7 6 1326.3u
LsysB 9 4 1326.3u
LsysC 8 5 1326.3u
*Step-down transformer ratio VAB/Vab=12.0
*=ratio of primary to seconday line-to-line voltages
L1 7 8 4000
L2 9 7 4000
L3 8 40 4000
L4 11 10 10
L5 12 10 10
L6 13 10 10
K14 L1 L4 0.9999
K25 L2 L5 0.9999
K36 L3 L6 0.9999
*R8 8 0 10meg
*Diode bridge
D1 11 15 Dideal
D4 14 11 Dideal
D3 13 15 Dideal
D6 14 13 Dideal
D5 12 15 Dideal
D2 14 12 Dideal
*Filter
Cf 15 14 500u
*Load
Rload 15 14 10
*Auxiliary parameters
R 33 3 1u
Rn 10 0 10meg
R3 40 9 1u
R100 1 0 1meg
*Model of diodes
.model Dideal d(is=1m)
.tran 100u 200m 160m 100u
uic
.four 60 12 V(1,2) V(11,12)
I(RsysA) I(L4)
+V(15,14)
.options abstol=1m
chgtol=0.1m reltol=0.1
+vntol=1m
.options itl4=200 itl5=0
.probe
.end
```

Note: All comment lines must be preceded with an asterisk (*) while continued calculation statements must be preceded with a plus sign (+).

Table E10.24.2 Fundamental voltage and harmonics of $v_{AB}(t) = V(1) - V(2)$ of Figure E10.24.2.

DC component	1	2	5	7	11
0.55 V	17 635 V	0.53 V	0.14 V	0.14 V	0.16 V

are displayed in Tables E10.24.2 and E10.24.3, respectively. The fundamental and harmonic currents for $i_{AL}(t)$, $i_{aph}(t)$ are listed in Tables E10.24.4 and E10.24.5, and the fundamental and harmonic voltages for $v_{load}(t)$ are detailed in Table E10.24.6.

Table E10.24.3 Fundamental voltage and harmonics of $v_{ab}(t) = V(11) - V(12)$ of Figure E10.24.2.

DC component	1	2	5	7	11
−0.94 V	1466 V	3.10 V	208.7 V	99.52 V	59.53 V

Table E10.24.4 Fundamental current and harmonics of $i_{AL}(t) = I(\text{RsysA})$ of Figure E10.24.3.

DC component	1	2	5	7	11
−0.01 A	12.83 A	0.004 A	2.74 A	0.94 A	0.37 A

Table E10.24.5 Fundamental current (1) and harmonics of $i_{aph}(t) = I(L4)$ of Figure E10.24.2.

DC component	1	2	5	7	11
−0.03 A	148.10 A	0.05 A	31.60 A	10.88 A	4.33 A

Table E10.24.6 Fundamental voltage and harmonics of $v_{load}(t) = V(15) - V(14) = v_{out}(t)$ of Figure E10.24.3.

DC component	1	2	5	6	7	11
1349.2 V	0.11 V	0.04 V	0.05 V	15.33 V	0.02 V	0.02 V

Conclusions

A delta/wye three-phase transformer blocks triplen harmonics (e.g. 3rd, 6th, 9th, ...) and the phase shift between the line-to-line and phase voltages is (−30°). The rectifier generates a significant amount of harmonics within secondary voltages and currents, as well as within the primary currents flowing into the grid system. To prevent such an injection into the power system, filters across the terminals A, B, and C are required.

10.6 Three-Phase Inverters

Inverters converting DC voltages and currents to AC quantities are relied on for brushless DC machine drives [9–17], photovoltaic (PV) power plants [7, 8], and wind power plants [18–23] where the AC power system voltage is given and the inverter current is controlled, resulting in so-called current-controlled voltage source inverters. The relation between the DC input voltage and the AC output voltage of the inverter is for standard inverter cases given by $V_1 = mV_{DC}/2\sqrt{2}$ where V_1 is the phase rms voltage value of the power system and V_{DC} is the total applied DC voltage at the input of the inverter. The parameter m is the modulation index with the range $0 < m < 1.0$. The waveshaping inductance is L_{wi} where ($i = a$, b, c) must be sufficiently large so that an acceptable PWM waveshape of the AC current with a low amount of switching harmonic amplitudes can be achieved. Lastly it is important to know that here only self-commutated switches (e.g. MOSFET, GTO) are used for the design of inverters. Single-phase inverters are relied on for low-power applications (e.g. output power of 10 kW), while three-phase inverters are employed for large ratings – up to 10 MW output power for wind power plants. Hybrid/electric automobiles [9–17] have output power ratings up to 300 kW.

Application Example 10.25 Brushless DC machine drive [9–17]

This type of motor drive consists of an inverter operating in either full-on or pulse-width-modulation (PWM) mode feeding a permanent-magnet or synchronous machine (e.g. train à grande vitesse [TGV] in France), which can be either operated as a motor or a generator in order to recover some of the braking energy [7]. Such permanent-magnet drives can have a high ($\eta = 96\%$) efficiency, the speed can be controlled by changing the inverter frequency, and the torque can be altered by changing the current phase with respect to the induced voltage of the permanent-magnet machine. The only complication with this drive is the need to sense the rotor speed by employing a position encoder. Some designs do not rely on a position sensor/encoder, but compute the speed as a function of time.

A three-phase brushless motor/generator drive operating in full-on mode is depicted in Figure E10.25.1, where six self-commutated switches (e.g. MOSFETs) are used. Brushless DC machines (more precisely, called electronically commutated permanent-magnet machines [ECPMs]) are desirable drives for electric and hybrid automobiles due to their low weight for a given output power. The commutation principle discussed in Chapter 11 for DC machines is applied to ECPM, where the DC voltage of the battery is converted/transformed into AC voltage for motoring and vice versa from AC to DC for regeneration during braking action [7]. The position control of ECPM is performed with either a Hall sensor with rotary encoder or a computational technique predicting the rotor position. The control of the torque is based on either flux weakening/addition, where the rotor and the stator fluxes are shifted in space with respect to one another, or by relying on PWM for the control of the stator current [10, 11]. The speed control is solely determined by the full-on frequency of the machine imposed by the inverter.

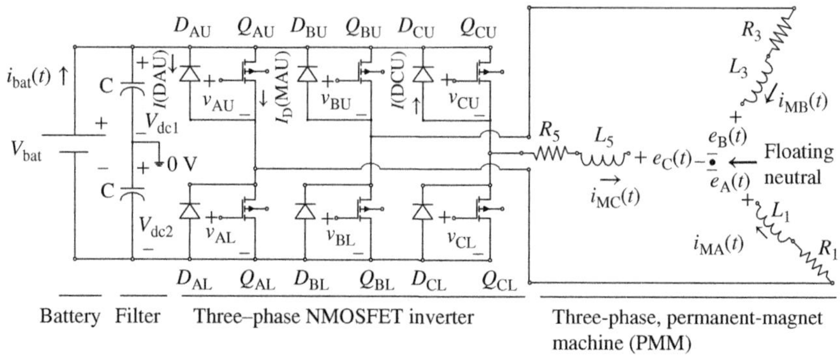

Figure E10.25.1 Brushless DC machine consisting of battery, inverter, and three-phase permanent-magnet machine. A, B, and C stand for the three phases; L_1, L_3, L_5 are leakage inductances; U stands for upper and L for lower MOSFET.

The objective of this example is to calculate the battery voltage required for a brushless DC motor so that the maximum phase current excursion is i_{MAmax} = 100 A during starting with a fundamental switching frequency of f_1 = 6 Hz. The inverter is a six-pulse, six-step, or full-on inverter consisting of six MOSFET switches. The electric machine is a three-phase permanent-magnet motor represented by induced voltages $e_A(t)$, $e_B(t)$, $e_C(t)$, resistances, and leakage inductances (of stator phase-to-phase windings, because there is no rotor winding) for all three phases. The induced voltage of the stator winding (phase A) of the permanent-magnet motor is

$$e_A(t) = V_A = 1.28\text{ V}\sin(\omega t + \theta), \tag{E10.25.1}$$

where $\omega = 2\pi f_1$. For low-speed operation f_1 = 6 Hz, and θ is the phase shift between induced voltage and stator currents. Correspondingly,

$$e_B(t) = V_B = 1.28\text{ V}\sin(\omega t + 120° + \theta), \tag{E10.25.2}$$

$$e_C(t) = V_C = 1.28\text{ V}\sin(\omega t + 240° + \theta). \tag{E10.25.3}$$

The stator resistance $R_1 = R_3 = R_5$ and the stator leakage inductance $L_1 = L_3 = L_5$ of one of the stator phases are 0.5 Ω and 50 μH, respectively. Note that the leakage inductances are defined between the three stator phases and not between stator and rotor phases. Also, PSPICE cannot accept "e" as induced voltage and the identifications R_s and L_s.

The magnitude of the gating voltages of the six MOSFETs is V_{Gmax} = 15 V. The gating signals with their phase sequence are shown in Figure E10.25.2. Note that the phase sequence of the induced voltages $e_A(t)$, $e_B(t)$, and $e_C(t)$ (as defined in Eqs. (E10.25.1)–(E10.25.3)) and that of the gating signals (see Figure E10.25.2)

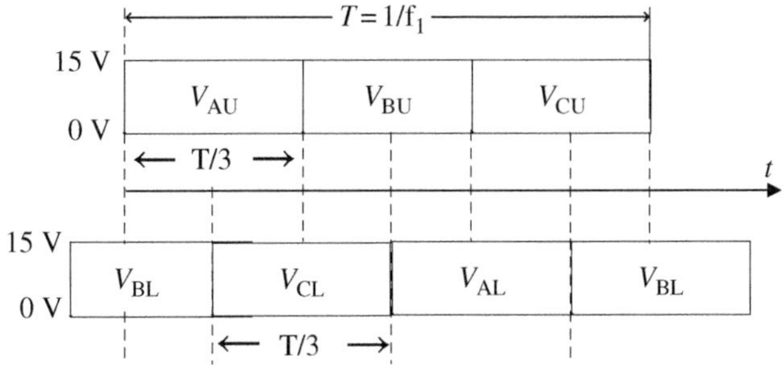

Figure E10.25.2 Sequence of gating signals for brushless DC motor/machine in six-step (six-pulse or full-on mode) operation. See gating signals in Table E10.25.1, where $V_{AU} = V_{fet1}$, $V_{BU} = V_{fet2}$, $V_{CU} = V_{fet3}$, $V_{AL} = V_{fet4}$, $V_{BL} = V_{fet5}$, $V_{CL} = V_{fet6}$, and $f_1 = 6$ Hz.

Table E10.25.1 PSPICE program for the signals of Figure E10.25.3.

```
Application Example 10.25
*brushless DC motor drive at low
speed, f1=6Hz
Va 1a st SIN(0 1.28V 6Hz 0 0 0)
Vb 3a st SIN(0 1.28V 6Hz 0 0 120)
Vc 5a st SIN(0 1.28V 6Hz 0 0 240)
Vdc1 7 0 DC 75.0V
Vdc2 8 0 DC -75.0
Rshunt st 0 10Meg
*5 deg delay at 60Hz = 0.231 ms
*50 deg delay at 60 Hz = 2.31 ms
*150 deg delay at 60 Hz= 6.94 ms
Vfet1 sp1 2 pulse(0 15 0 1u 1u
55.55ms 166.66ms)
Vfet2 sp2 6 pulse(0 15 55.556ms 1u
1u 55.556ms +166.67ms)
Vfet3 sp3 4 pulse(0 15 111.11ms 1u
1u 55.556ms +166.67ms)
Vfet4 sp4 8 pulse(0 15 83.333ms 1u
1u 55.556ms +166.67ms)
Vfet5 sp5 8 pulse(0 15 138.89ms 1u
1u 55.556ms +166.67ms)
Vfet6 sp6 8 pulse(0 15 27.778ms 1u
1u 55.556ms +166.67ms)
Mau 7 sp1 2 2 SMM
Mbu 7 sp2 6 6 SMM
Mcu 7 sp3 4 4 SMM
Mal 2 sp4 8 8 SMM
Mbl 6 sp5 8 8 SMM
Mcl 4 sp6 8 8 SMM
L1 1a 1 50u
R1 1 2 0.5
L3 3a 3 50u
R3 3 4 0.5
L5 5a 5 50u
R5 5 6 0.5
C1 7 0 100u
C2 0 8 100u
Dau 2 7 D1N4001
Dbu 6 7 D1N4001
Dal 8 2 D1N4001
Dbl 8 6 D1N4001
Dcu 4 7 D1N4001
Dcl 8 4 D1N4001
.model D1N4001 D(IS=1E-12)
.model SMM nmos(level=3 gamma=0 delta=0
eta=0
+theta=0 kappa=0 vmax=0 xj=0 tox=100n
uo=600
+phi=0.6 rs=42.69m kp=20.87u l=2u w=2.9
vto=3.487 +rd=0.19 cbd=200n pb=0.8 mj=0.5
cgso=3.5n
+cgdo=100p rg=1.2 is=10f)
.tran 5u 500ms 150ms 1u uic
.options abstol=1m chgtol=1m reltol=10m
vntol=10m
.four 6 12 I(R1)
.probe
.end
```

Note: All comment lines must be preceded with an asterisk (*) while continued calculation statements must be preceded with a plus sign (+).

must be the same. If these phase sequences are not the same, then no periodic solution for the machine currents $i_{MA}(t)$, $i_{MB}(t)$, $i_{MC}(t)$ can be obtained, where M stands for machine and A, B, C for the phase identification.

The models of the enhancement MOSFETs and those of the (external) freewheeling diodes are as follows:

Model for self-commutated electronic switch (MOSFET):
.MODEL SMM NMOS(LEVEL = 3 GAMMA = 0 DELTA = 0 ETA
= 0 THETA = 0
+ KAPPA = 0 VMAX = 0 XJ = 0 TOX = 100 n UO = 600 PHI = 0.6 RS = 42.69
+ M KP = 20.87 u
+ L = 2 u W = 2.9 VTO = 3.487 RD = 0.19 CBD = 200 n PB = 0.8 MJ = 0.5 CGSO
+ = 3.5 n
+ CGDO = 100 p RG = 1.2 IS = 10 f)

Model for diode:
.MODEL D1N4001 D(IS = 1p)

Using PSPICE, compute the required battery voltage V_{bat} and plot the drain current of MOSFET Q_{AU} (e.g. $i_{QAU}(t)$), the reverse current of MOSFET Q_{CU} (e.g. $i_{DCU}(t)$), and the motor current of phases A, B, and C (e.g. $i_{MA}(t)$, $i_{MB}(t)$, $i_{MC}(t)$) for the phase angle of the induced voltages $\theta = 0°$. Note that the gating signal frequency of the MOSFETs corresponds at low speed (e.g. during starting) to the frequency $f = f_1 = 6$ Hz; that is, full-on mode operation exists.

The step size for the numerical solution should be in the neighborhood of $\Delta t = 0.05$ μs; and to eliminate computational transients due to inconsistent initial conditions, compute at least three periods of all quantities, and plot the last (third) period of $i_{MA}(t)$.

Solution

The chosen amplitudes of the induced voltages V_A, V_B, and V_C are 1.28 V; the battery voltage selected is $V_{bat} = V_{dc1} + V_{dc2} = 150$ V with $V_{dc1} = 75$ V and $V_{dc2} = 75$ V; and the maximum phase current excursion is assumed to be $i_{MAmax} = 100$ A. Note V_{dc1} and V_{dc2} are the DC voltages across the two capacitors C of Figure E10.25.1.

Figure E10.25.3 shows the motor stator currents for all three phases A, B, C, while Figure E10.25.4 represents the current time functions of the reverse conducting upper diode of phase A, and the drain current of the upper MOSFET of phase A.

Conclusions

The phase currents turn out to have about a rectangular waveform with i_{MAmax} = 100 A, and therefore the fundamental phase current amplitude is

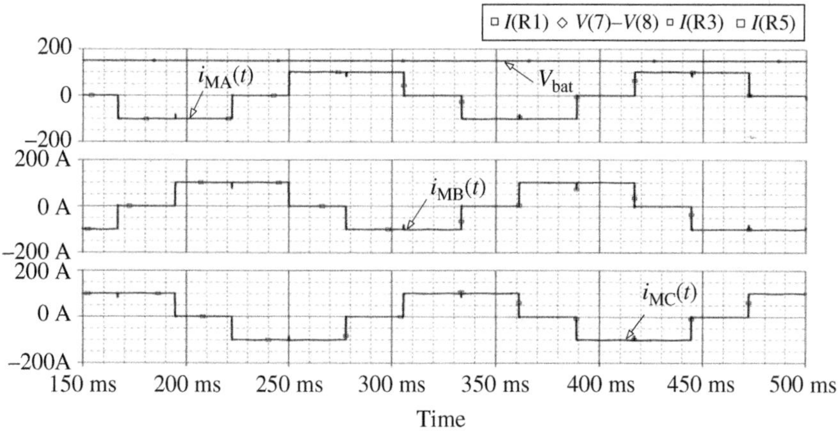

Figure E10.25.3 Top plot: motor (stator) current $i_{MA}(t) = I(R1)$ and applied battery voltage $V_{bat} = 150$ V. Middle plot: motor (stator) current $i_{MB}(t) = I(R3)$. Bottom plot: motor (stator) current $i_{MC}(t) = I(R5)$, based on full-on motoring mode [11] during start-up with low frequency $f_1 = 6$ Hz at $\theta = 0°$; maximum phase A current excursion is $i_{MAmax} = 100$ A.

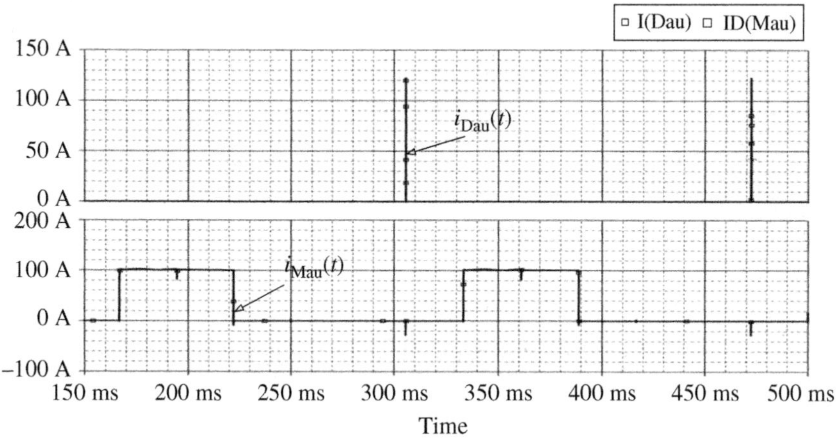

Figure E10.25.4 Top plot: reverse conducting current of diode $i_{Dau}(t) = I(\text{Dau})$ of phase A. Bottom plot: drain (D) current of upper MOSFET $i_{Mau}(t) = \text{ID(Mau)}$, based on full-on motoring mode [11] during start-up with low frequency $f_1 = 6$ Hz at $\theta = 0°$; maximum phase A current excursion is $i_{MAmax} = 100$ A.

somewhat larger (e.g. 110.7 A). The third harmonic in Table E10.25.2 is zero because three-phase drive configurations cannot have triplen (3, 6, 9, 12, ...) harmonic current/voltage components. The switch model (e.g. $w = 2.9$) must be adjusted depending upon the magnitude of the current. The power at starting is

Table E10.25.2 Current harmonics of $i_{MA}(t) = I(R1)$ of Figure E10.25.3 in terms of amperes and in percent referred to maximum excursion of $I(R1)$, $i_{MAmax} = 100$ A.

DC component	1	3	5	7
0.00/0.00%	110.7 A/110.7%	0.00 A/0.00%	22.19 A/22.18%	15.80 A/15.80%

$P_{out} = 3 \cdot 1.28\ \text{V}/\sqrt{2} \cdot 110.7/\sqrt{2}\ \text{A} = 212.6$ W, which is very small; however, the torque is large due to the AC rms current $I_{rms} = 110.7/\sqrt{2}\ \text{A} = 78.29$ A. This indicates that it is advisable to equip each electric car with several (e.g. 2 or 4) electric brushless DC machine drives. This is also supported by the calculations performed in Problems 10.1–10.4 where each brushless DC machine has a maximum output power – as a function of the switching frequency f_1 – in the range from 50 to 100 hp. Also, because of safety considerations, the battery voltage should not exceed 400 V_{DC}. Presently 300 V DC batteries are used in most electric cars, which are available and have an energy storage ranging from 17.6 kWh (the Smart Fortwo Ed [2017]) to 75 kWh [the Tesla Model S 75D], which permit travel ranges from 160 to 490 km, respectively.

Application Example 10.26 Battery supplies $P_{supply} = 56.62$ kW into the rms 380 $V_{L\text{-}L}$ power system via a three-phase current-controlled voltage source inverter at a leading displacement power factor of cos (36.87°) = 0.80 (consumer notation of power factor)

The storage power plant consists of a battery and a full-wave, three-phase bridge MOSFET current-controlled voltage source inverter [17]. The rms per-phase voltage is $V_{ph} = 380\ \text{V}/1.732 = 219.4$ V, and the per-phase system impedance of the power system (sys) is $Z_{sys} = (R_{sys} + j\omega L_{sys})$ with $R_{sys} = 0.59\ \Omega = R_M$ and $L_{sys} = 2.838\ \text{mH} = L_M$, where M stands for main power system and m for milli. The battery and the full-wave, three-phase current-controlled voltage source inverter supplying energy into the power system is presented in Figure E10.26.1. The displacement fundamental power factor is $\cos\theta = 0.80$ leading, where the phase current leads the phase voltage by $\theta = 36.87°$ (that is, the consumer definition of current; reactive power Q is supplied to grid). Such inverters are commonly used for wind power and large photovoltaic plants when the voltage is imposed by the grid. The current fed into the grid is controlled, but the voltage of the grid cannot be changed by such current-controlled voltage source inverters. Note that there must be a phase-lock loop [20, 22] (PLL) for the control/gating of the inverters, where the frequency of the power system is sensed and applied to the inverter gating circuit. The waveshaping (w) inductors have the impedance $Z_w = (R_w + j\omega L_w)$, where $R_w = 10\ \text{m}\Omega$ and $L_w = 0.10$ mH. The battery (bat) voltage is $V_{bat} = 900$ V.

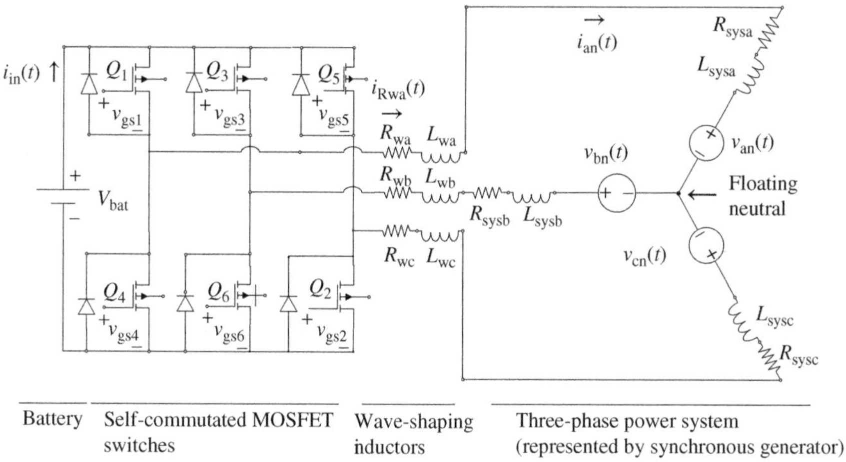

Figure E10.26.1 Full-wave, three-phase current-controlled voltage source inverter feeding power into the grid by employing six self-commutated MOSFET switches operated in PWM mode feeding power into a three-phase grid; a, b, and c stand for the three phases.

Perform a PSPICE analysis to compute and plot battery current = input current to inverter = $i_{in}(t)$ and power system currents $i_{an}(t)$, $i_{bn}(t)$, $i_{cn}(t)$ at leading displacement power factor. Also plot power system voltages $v_{an}(t)$, $v_{bn}(t)$, $v_{cn}(t)$. The modulation index m for this case is defined by $V_{ph} = mV_{bat}/(2\sqrt{2})$ or $m = V_{ph} \cdot 2\sqrt{2}/V_{bat} = 219.4\ \text{V} \cdot 2 \cdot \sqrt{2}/(900\ \text{V}) = 0.689$, which is smaller than 1.0 and is therefore acceptable for leading power factor operation, where V_{bat} must be relatively large as compared to unity or lagging power factors. The resulting current waves should be therefore sinusoidal/cosinusoidal.

Plot the voltages and currents V_{bat}, $i_{in}(t)$, $v_{ref1}(t)$, $v_{ref2}(t)$, $v_{ref3}(t)$, $i_{an}(t)$, $i_{bn}(t)$, $i_{cn}(t)$, $v_{an}(t)$, $v_{bn}(t)$, $v_{cn}(t)$.

Solution

The PSPICE program is listed in Table E10.26.1, and the plots of the requested voltages and currents are shown in Figures E10.26.2–E10.26.4.

Note that leading displacement power factors require a relatively large V_{bat} value as compared to the AC phase voltage amplitudes.

Conclusions

The fundamental phase currents lead the fundamental phase voltages by 36.87° – supplying reactive power Q to the power system – and its waveshapes are sinusoidal/cosinusoidal, as can be noted from Table E10.26.2. This is due to the rather

Table E10.26.1 PSPICE program for the signals of Figures E10.26.2–E10.26.4.

```
Application Example 10.26
*Current-controlled pulse-width-
modulated
*(PWM) voltage-source inverter supplying
from
* a storage battery energy into the power
system
Vbat 2 0 900
*Switches
msw1 2 11 10 10 qfet
dsw1 10 2 diode
msw5 2 21 20 20 qfet
dsw5 20 2 diode
msw3 2 31 30 30 qfet
dsw3 30 2 diode
msw4 10 41 0 0 qfet
dsw4 0 10 diode
msw2 20 51 0 0 qfet
dsw2 0 20 diode
msw6 30 61 0 0 qfet
dsw6 0 30 diode
*Waveshaping inductors
lwa 10 15 0.1m
lwb 20 25 0.1m
lwc 30 35 0.1m
Rwa 15 16 10m
Rwb 25 26 10m
Rwc 35 36 10m
*Current references in terms of voltages
vref1 12 0 sin(0 152.1 60 0 0 36.87)
vref2 22 0 sin(0 152.1 60 0 0 -83.13)
vref3 32 0 sin(0 152.1 60 0 0 -203.13)
eout1 13 0 15 16 100
eout2 23 0 25 26 100
eout3 33 0 35 36 100
*Errorsignals produced as difference
between
*vref and eout
rdiff1 12 13a 1k
rdiff2 22 23a 1k
rdiff3 32 33a 1k
cdiff1 12 13a 1u
cdiff2 22 23a 1u
cdiff3 32 33a 1u
rdiff4 13a 13 1k
rdiff5 23a 23 1k
rdiff6 33a 33 1k
ecin1 14 0 12 13a 2
ecin2 24 0 22 23a 2
*Gating for lower switches
egs4 41 0 poly(1) (11,10) 50 -1
egs5 51 0 poly(1) (21,20) 50 -1
egs6 61 0 poly(1) (31,30) 50 -1
*Filter is deleted because PSpice is limited
to
*64 nodes
*Parameters of power system with line-to-
*neutral induced (phase) voltages ea, eb, ec
RM1 16 18 0.59
LM1 18 19 2.838m
Vout1 19 123 sin(0 310.23 60 0 0 0)
RM2 26 28 0.59
LM2 28 29 2.838m
Vout2 29 123 sin(0 310.23 60 0 0 -120)
RM3 36 38 0.59
LM3 38 39 2.838m
Vout3 39 123 sin(0 310.23 60 0 0 -240)
*Subcircuit for comparator
.subckt comp 1 2 9 10
rin 1 3 2.8k
r1 3 2 20meg
e2 4 2 3 2 50
r2 4 5 1k
d1 5 6 zenerdiode1
d2 2 6 zenerdiode2
e3 7 2 5 2 1
r3 7 8 10
c3 8 2 10n
r4 3 8 100k
e4 9 10 8 2 1
*Models for zener diodes
.model zenerdiode1 D (Is=1p BV=0.1)
.model zenerdiode2 D (Is=1p BV=50)
.ends comp
*Model for switch
.model qfet nmos(level=3 gamma=0
kappa=0
+tox=100n rs=42.69m kp=20.87u l=2u
+ w=2.9 delta=0 eta=0 theta=0 vmax=0
xj=0
+uo=600 phi=0.6 vto=3.487 rd=0.19
cbd=200n
+pb=0.8 mj=0.5 cgso=3.5n cgdo=100p
rg=1.2
+is=10f )
*Model for diodes
.model diode d(is=1p)
*Options to aid convergence
```

Table E10.26.1 (Continued)

ecin3 34 0 32 33a 2	.options abstol=0.01m chgtol=0.01m
vtrial 5 0 pulse(-10 10 0 86.5u 86.5u 0.6u	+reltol=50m vntol=1m itl5=0 itl4=400
+173.6u)	*Transient analysis
*Gating signals for upper switches	.tran 5u 350m 330m 5u
*as result of comparison between	*Plotting of traces
*triangular waveform and error signal	.probe
xgs1 14 5 11 10 comp	*Fourier analysis
xgs2 24 5 21 20 comp	.four 60 12 I(Rwa)
xgs3 34 5 31 30 comp	.end

Note: All comment lines must be preceded with an asterisk (*) while continued calculation statements must be preceded with a plus sign (+).

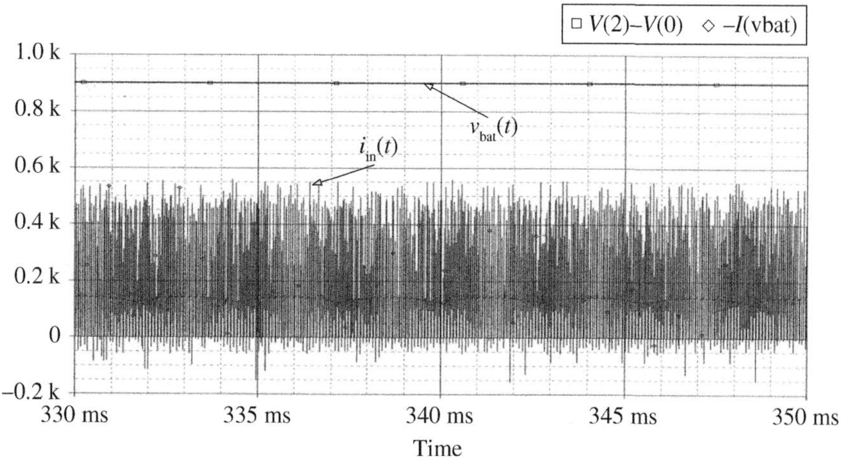

Figure E10.26.2 Battery voltage $V_{bat} = V(2) - V(0)$ and battery current or input current to inverter $i_{in}(t) = -I(\text{vbat})$.

low chosen modulation (m) index $m = 0.689 \leq 1.0$. If a larger m were chosen, say, $m = 1.0$, then a sinusoidal/cosinusoidal current waveshape would not be obtained.

The real power delivered to the power system is $P_{supply} = 3 \cdot (152.1/\sqrt{2}) \cdot (310.23/\sqrt{2}) \cdot 0.80 = 56.62\,\text{kW}$, the reactive power is $Q_{supply} = 3 \cdot (152.1/\sqrt{2}) \cdot (310.23/\sqrt{2}) \cdot 0.60 = 42.47\,\text{kVAr}$, and the apparent power $S = (P^2 + Q^2)^{1/2} = (56.620^2 + 42.470^2)^{1/2} = 70.78\,\text{kVA}$. This checks with $S_{supply} = 3 \cdot (152.1/\sqrt{2}) \cdot (310.23/\sqrt{2}) = 70.78\,\text{kVA}$. The switch model (e.g. $w = 2.9$) must be adjusted depending upon the magnitude of the current.

It is well known that in conventional and smart power systems, very large (e.g. 1000 MW) generator units (e.g. nuclear, hydro, steam) serve as frequency leaders,

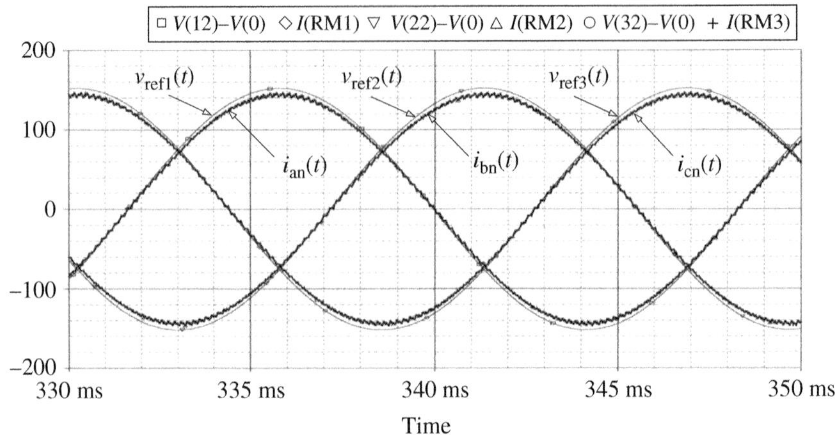

Figure E10.26.3 Reference currents of inverter $v_{ref1}(t) = V(12) - V(0)$, $v_{ref2}(t) = V(22) - V(0)$, $v_{ref3}(t) = V(32) - V(0)$ (note that in the PSPICE program the reference currents are defined as voltages) and actually computed inverter currents $i_{an}(t) = I(\text{RM1})$, $i_{bn}(t) = I(\text{IRM2})$, $i_{cn}(t) = I(\text{RM3})$.

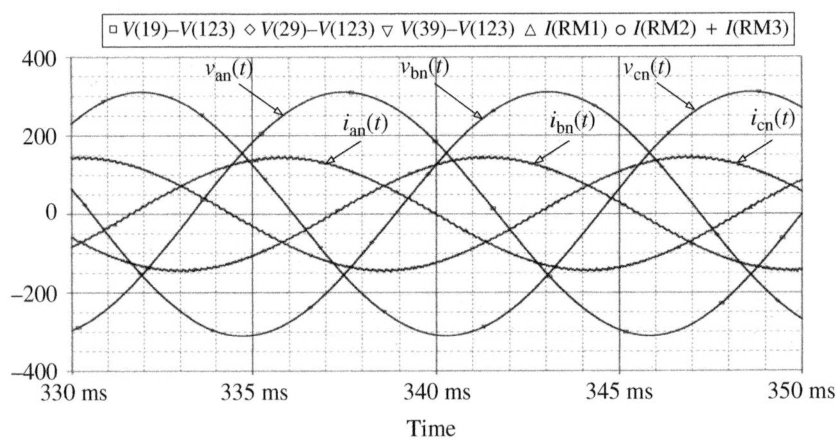

Figure E10.26.4 Power system phase voltages $v_{an}(t) = V(19) - V(123)$, $v_{bn}(t) = V(29) - V(123)$, $v_{cn}(t) = V(39) - V(123)$ and phase currents $i_{an}(t) = I(\text{RM1})$, $i_{bn}(t) = I(\text{RM2})$, $i_{cn}(t) = I(\text{IRM3})$ leading the phase voltages by 36.87° (consumer notation of currents; reactive power is supplied to power system by inverter).

Table E10.26.2 Current harmonics of $i_{an}(t) = I(\text{Rsysa}) = I(\text{Rwa})$ of Figures E10.26.3 and E10.26.4 in terms of amperes in percent of reference current amplitude of 152.1 A.

DC component	1	3	5	7
0.09 A/0.00%	145.7 A/95.79%	2.57 A/1.69%	0.036 A/0.024%	0.756 A/0.497%

where the nominal frequency of 60 Hz may change – due to load/generation imbalance – within 59–61 Hz. This means that inverters of relatively small renewable power plants (e.g. 100 MW) must sense the actual frequency and phase of the power system through phase-lock loop [20, 22] (PLL) and adjust their gating frequencies accordingly so that active filtering equipment works as intended. Passive filters do not have such a timing problem because they have sufficient filter bandwidth. A PLL circuit is required to sense the power system frequency and adjust the gating voltages/currents of inverters. This can be accomplished with a toolbox available in MATLAB/Simulink wherein the gating frequency can be changed as a function of the (main) power system frequency. If such a PLL circuit is not employed, then one cannot correctly operate an inverter of a renewable power plant and compensate harmonic generation of power systems with inverters due to the changing nominal fundamental power system frequency. For example, a 295 Hz (corresponding to 59 Hz) signal in the power system cannot be compensated by a 300 Hz (corresponding to 60 Hz) component as generated by an inverter gated with a constant frequency of 60 Hz.

10.7 Design of a Photovoltaic (PV) Power Plant

Although PV power plants [7, 8] operate at intermittent times, they represent an important albeit small (e.g. 8.7% in 2018 of total generated energy of Germany [24]) component in the design of renewable energy systems if the insolation is in the range of 400–800 W/m^2.

Application Example 10.27 Design of a maximum power point (MPP) $P_{AC\_supply} = 35$ kW photovoltaic (PV) power plant

A PV plant supplies power into the grid at maximum insolation $Q_s = 1000$ W/m^2 via a three-phase current-controlled inverter as discussed in Application Example 10.26. P_{MPP} is the output power of the photovoltaic array at maximum insolation of 1000 W/m^2.

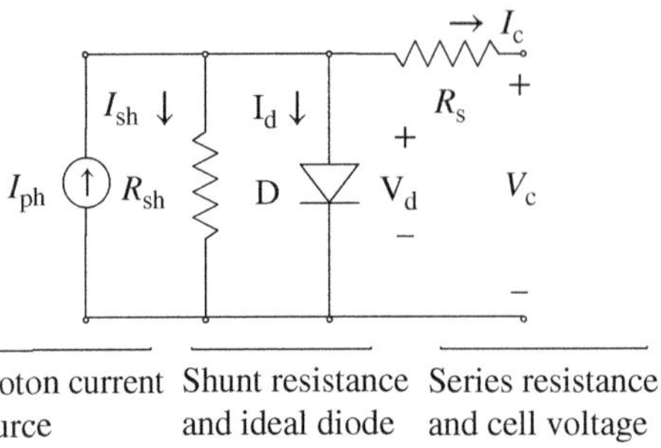

Figure E10.27.1 Equivalent circuit of a solar cell.

a) For a solar cell, see Figure E10.27.1, with the data $R_s = 0.05\ \Omega$, $R_{sh} = 1\ \text{k}\Omega$, $I_r = 0.0005$ A, $T_c = 313.2\ \text{K} \equiv T_{rat} = 40°\text{C}$, maximum insolation $Q_s = 1000\ \text{W/m}^2$, $I_{ph} = 90$ mA. Derive the nonlinear convex $(V_c - I_c)$ characteristic of a solar cell (c), that is, the cell voltage V_c as a function of the cell current I_c, and find the maximum output power $P_c = P_{max} = P_{MPP}$ of this solar cell at the maximum power point.

Solution

The $(V_c - I_c)$ relation of a solar cell is

$$V_c = -R_s I_c + \frac{1}{\Lambda}\ln\left[\frac{I_{ph} - \left(1 + \frac{R_s}{R_{sh}}\right)I_c - \frac{V_c}{R_{sh}}}{I_r} + 1\right], \tag{E10.27.1a}$$

where

$$A_0 = 1.95 + 0.02 \cdot 1000 = 21.95 \tag{E10.27.1b}$$

$$\Lambda = \frac{q}{A_0 k T_c} = \frac{1.6 \cdot 10^{-19}}{21.95 \cdot 1.37 \cdot 10^{-23} \cdot 313.2} = 1.699 \approx 1.7\left[\frac{1}{\text{V}}\right] \tag{E10.27.1c}$$

$$I_{ph} = I_{sh} + I_d + I_c, \tag{E10.27.1d}$$

$$V_d = R_s I_c + V_c. \tag{E10.27.1e}$$

Calculation of the open-circuit (oc) voltage $V_{c(oc)}$ for $R_{sh} \rightarrow \infty$, $R_s = 0$, $I_c = 0$, $V_{c(oc)} = V_d$, and $I_{ph} = 0.09$ A. With Eq. (E10.27.1a) follows

$$V_{c(oc)} = \frac{1}{\Lambda}\ln\left[\frac{I_{ph}}{I_r} + 1\right], \tag{E10.27.2a}$$

or

$$V_{c(oc)} = \frac{1}{1.7}\ln\left[\frac{0.09}{0.0005} + 1\right] = 3.058\ \text{V} \tag{E10.27.2b}$$

Calculation of the short-circuit (sc) current $I_{c(sc)}$ for $R_{sh} \to \infty$, $R_s = 0.05\ \Omega$, $I_c = I_{c(sc)}$, $V_c = 0$, and $I_{ph} = 0.09$ A. With Eq. (E10.27.1a) follows

$$R_s I_{c(sc)} = \frac{1}{\Lambda} \ln \left[\frac{I_{ph} - \left(1 + \frac{R_s}{R_{sh}}\right) I_{c(sc)}}{I_r} + 1 \right], \qquad \text{(E10.27.3a)}$$

$$\begin{aligned} 0.05 \cdot I_{c(sc)} &= \frac{1}{1.7} \ln \left[\frac{0.09 - \left(1 + \frac{0.05}{1000}\right) I_{c(sc)}}{0.0005} + 1 \right] \\ &= \frac{1}{1.7} \ln \left[180 - 2000 I_{c(sc)} + 1 \right]; \end{aligned} \qquad \text{(E10.27.3b)}$$

this nonlinear algebraic equation can be solved by iterations as follows:

assume $I_{c(sc)}^{(1)} = 0.085$ A, resulting with Eq. (E10.27.3b) in the inequality 0.004 25 V $\neq$ 1.4105 V; $I_{c(sc)}^{(2)} = 0.0899$ A results with Eq. (E10.27.3b) in the inequality 0.004 495 V $\neq$ 0.107 25 V; $I_{c(sc)}^{(3)} = 0.089\,995$ A results with Eq. (E10.27.3b) in the approximate equality 0.004 499 8 V $\approx$ 0.005 85 V, which has a sufficiently small error.

Calculation of functional values near the knee of the $(V_c - I_c)$ function yields the power $P_{max\_cell} = P_{MPP\_cell} = V_{cMPP} \cdot I_{cMPP}$ at the maximum power point (MPP) of the solar cell characteristic:

For $I_{c(0.08\ A)} = 0.08$ A and $R_{sh} \to \infty$, one obtains the voltage from Eq. (E10.27.1a)

$$V_{c(0.08\ A)} = -0.05 \cdot 0.08 + \frac{1}{1.7} \ln \left[\frac{0.09 - 0.08}{0.0005} \right] + 1 = 1.787\ \text{V}.$$

For $I_{c(0.07\ A)} = 0.07$ A and $R_{sh} \to \infty$, one obtains the voltage from Eq. (E10.27.1a).

$$V_{c(0.07\ A)} = -0.05 \cdot 0.07 + \frac{1}{1.7} \ln \left[\frac{0.09 - 0.07}{0.0005} \right] + 1 = 2.181\ \text{V}.$$

For $I_{c(0.06\ A)} = 0.06$ A and $R_{sh} \to \infty$, one obtains the voltage from Eq. (E10.27.1a).

$$V_{c(0.06\ A)} = -0.05 \cdot 0.06 + \frac{1}{1.7} \ln \left[\frac{0.09 - 0.06}{0.0005} \right] + 1 = 2.415\ \text{V}.$$

The above calculated data are shown in Figure E10.27.2. One concludes that at the knee of the $(V_c - I_c)$ function, the voltage and current pertaining to the maximum power value $P_{max\_cell} = P_{MPP\_cell}$ are $V_{cMPP\_cell} = 2.181$ V and $I_{cMPP\_cell} = 70$ mA, respectively, resulting in the maximum power $P_{MPP\_cell} = 0.153$ W.

b) Based on the solar characteristic of Figure E10.27.2, design a PV power plant with an AC output of $P_{AC\text{-}supply} = 35$ kW when the plant operates at an

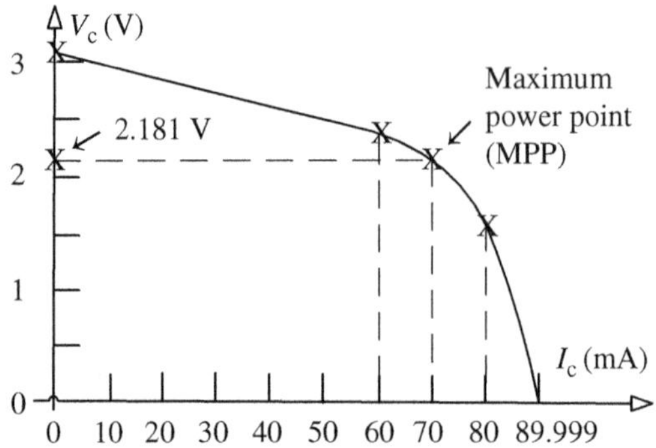

Figure E10.27.2 Nonlinear convex ($V_c - I_c$) characteristic of a solar cell.

insolation of $Q_s = 1000\ \text{W/m}^2$ at the maximum power point with a modulation index of $m \approx 1.0$.

For N_s solar cells in series per string and N_p strings in parallel, one can write $V_{DC} = N_s V_c$ and $I_{DC} = N_p I_c$ resulting with Eq. (E10.27.1a) and setting $R_{sh} \to \infty$ in the modified $V_{DC} - I_{DC}$ relation.

$$V_{DC} = -R_s \frac{N_s}{N_p} I_{DC} + \frac{N_s}{\Lambda} \ln \left[\frac{I_{ph} - \frac{I_{DC}}{N_p}}{I_r} + 1 \right]. \tag{E10.27.4}$$

In order to mitigate the effect of shading, snow, and dust, it is common practice to use bypass diodes for set/string of 18 in series connected solar cells, as depicted in Figure E10.27.3.

Solution

Provided one chooses a $V_{DC} = 620\ \text{V}$, one obtains for $V_{MPP} = 2.181\ \text{V}$ of Figure E10.27.2 $N_s = 620\ \text{V}/(2.181\ \text{V}) \approx 284$ solar cells in series and for $I_{MPP} = 0.07\ \text{A}$ of Figure E10.27.2 $N_p = 56\ \text{A}/(0.07\ \text{A}) \approx 807$ solar cells in parallel. The losses in the inverter are neglected.

The power plant consists of PV array and a full-wave, three-phase bridge MOSFET current-controlled voltage source inverter [17]. The per-phase voltage is $V_{ph} = 380\ \text{V}/1.732 = 219.4\ \text{V}$ and the per-phase system impedance is $Z_{sys} = (R_{sys} + j\omega L_{sys})$ with $R_{sys} = 0.59\ \Omega = R_M$ and $L_{sys} = 2.838\ \text{mH} = L_M$, where M stands for main power system and m for milli. The battery representing the solar array and the full-wave, three-phase current-controlled voltage source inverter supplying energy into the power system is presented in Figure E10.26.1. The displacement power factor is assumed to be $\cos\theta = 1.0$, where the line-to neutral voltage and line current are in phase. The current fed into the grid is controlled, but the voltage of the grid cannot be changed by such current-controlled voltage source inverters. The

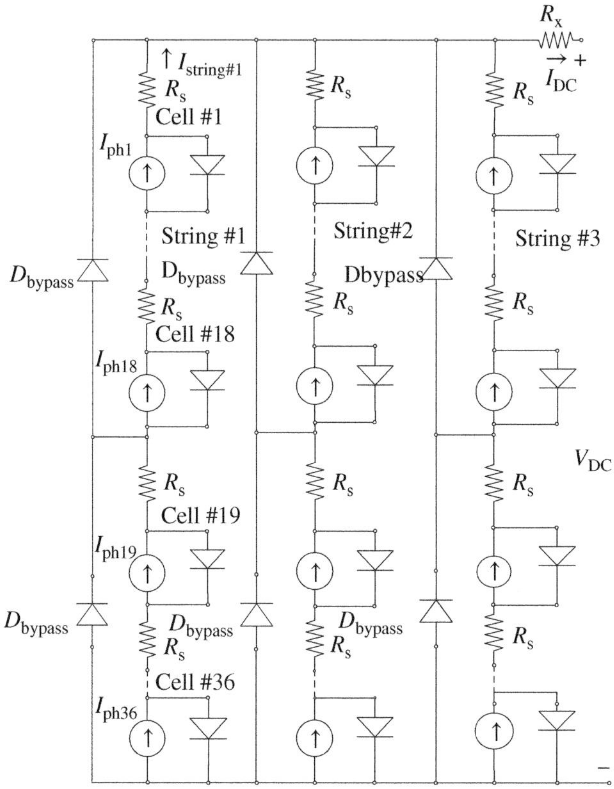

Figure E10.27.3 Equivalent circuit of a solar PV array or panel with bypass diodes D_{bypass} bridging 18 solar cells by one D_{bypass}. One string consists of 36 cells, and there are three strings shown.

waveshaping inductors have the impedance $Z_w = (R_w + j\omega L_w)$, where $R_w = 8\ \text{m}\Omega$ and $L_w = 0.08$ mH. The battery voltage is $V_{bat} = 620$ V. Perform a PSPICE analysis to compute and plot DC current = input current to inverter = $i_n(t)$ and power system currents $i_{an}(t)$, $i_{bn}(t)$, $i_{cn}(t)$ at unity power factor. Also plot power system voltages $v_{an}(t)$, $v_{bn}(t)$, $v_{cn}(t)$. The modulation index m for this case is defined by $V_{ph} = mV_{bat}/(2\sqrt{2})$ or $m = V_{ph} \cdot 2\sqrt{2}/V_{bat} = 219.4\ \text{V} \cdot 2 \cdot \sqrt{2}/(620\ \text{V}) \approx 1.0$, which is acceptable: the resulting current waves should be therefore approximately sinusoidal/cosinusoidal.

The PSPICE program is listed in Table E10.27.1, and the plots of the requested voltages and currents are shown in Figures E10.27.4–E10.27.6, and Table E10.27.2 depicts the harmonic components of the inverter line current.

Table E10.27.1 PSPICE program for the signals of Figure E10.27.4–E10.27.6.

```
Application Example 10.27
*Current-controlled pulse-width-
modulated *(PWM) voltage-source
inverter feeding *energy into the power
system
Vbat 2 0 620
*Switches
msw1 2 11 10 10 qfet
dsw1 10 2 diode
msw5 2 21 20 20 qfet
dsw5 20 2 diode
msw3 2 31 30 30 qfet
dsw3 30 2 diode
msw4 10 41 0 0 qfet
dsw4 0 10 diode
msw2 20 51 0 0 qfet
dsw2 0 20 diode
msw6 30 61 0 0 qfet
dsw6 0 30 diode
*Waveshaping inductors
lwa 10 15 0.08m
lwb 20 25 0.08m
lwc 30 35 0.08m
Rwa 15 16 8m
Rwb 25 26 8m
Rwc 35 36 8m
*current references in terms of voltages
vref1 12 0 sin(0 107.46 60 0 0 0)
vref2 22 0 sin(0 107.47 60 0 0 -120)
vref3 32 0 sin(0 107.46 60 0 0 -240)
eout1 13 0 15 16 100
eout2 23 0 25 26 100
eout3 33 0 35 36 100
*Errorsignals produced as difference
*between vref and eout
rdiff1 12 13a 1k
rdiff2 22 23a 1k
rdiff3 32 33a 1k
cdiff1 12 13a 1u
cdiff2 22 23a 1u
cdiff3 32 33a 1u
rdiff4 13a 13 1k
rdiff5 23a 23 1k
rdiff6 33a 33 1k
ecin1 14 0 12 13a 2
ecin2 24 0 22 23a 2
ecin3 34 0 32 33a 2
vtrial 5 0 pulse(-10 10 0 86.5u 86.5u
xgs3 34 5 31 30 comp
*Gating for lower switches
egs4 41 0 poly(1) (11,10) 50 -1
egs5 51 0 poly(1) (21,20) 50 -1
egs6 61 0 poly(1) (31,30) 50 -1
*Filter is deleted because PSpice is limited to
*64 nodes
*Parameters of power system with line-to-
*neutral induced (phase) voltages ea, eb, ec
RM1 16 18 0.59
LM1 18 19 2.838m
Vout1 19 123 sin(0 219.4 60 0 0 0)
RM2 26 28 0.59
LM2 28 29 2.838m
Vout2 29 123 sin(0 219.4 60 0 0 -120)
RM3 36 38 0.59
LM3 38 39 2.838m
Vout3 39 123 sin(0 219.4 60 0 0 -240)
*Subcircuit for comparator
.subckt comp 1 2 9 10
rin 1 3 2.8k
r1 3 2 20meg
e2 4 2 3 2 50
r2 4 5 1k
d1 5 6 zenerdiode1
d2 2 6 zenerdiode2
e3 7 2 5 2 1
r3 7 8 10
c3 8 2 10n
r4 3 8 100k
e4 9 10 8 2 1
*Models for zener diodes
.model zenerdiode1 D (Is=1p BV=0.1)
.model zenerdiode2 D (Is=1p BV=50)
.ends comp
*Model for switch
.model qfet nmos(level=3 gamma=0 kappa=0
+tox=100n rs=42.69m kp=20.87u l=2u
+ w=2.9 delta=0 eta=0 theta=0 vmax=0 xj=0
+uo=600 phi=0.6 vto=3.487 rd=0.19
cbd=200n
+pb=0.8 mj=0.5 cgso=3.5n cgdo=100p rg=1.2
+is=10f )
*Model for diodes
.model diode d(is=1p)
*Options to aid convergence
.options abstol=0.01m chgtol=0.01m
+reltol=50m vntol=1m itl5=0 itl4=400
```

Table E10.27.1 (Continued)

+0.6u 173.6u) *Gating signals for upper switches *as result of comparison between xgs1 14 5 11 10 comp xgs2 24 5 21 20 comp	*Transient analysis .tran 5u 350m 330m 5u *Plotting of traces .probe *Fourier analysis .four 60 12 I(Rwa) .end

Note: All comment lines must be preceded with an asterisk (*) while continued calculation statements must be preceded with a plus sign (+).

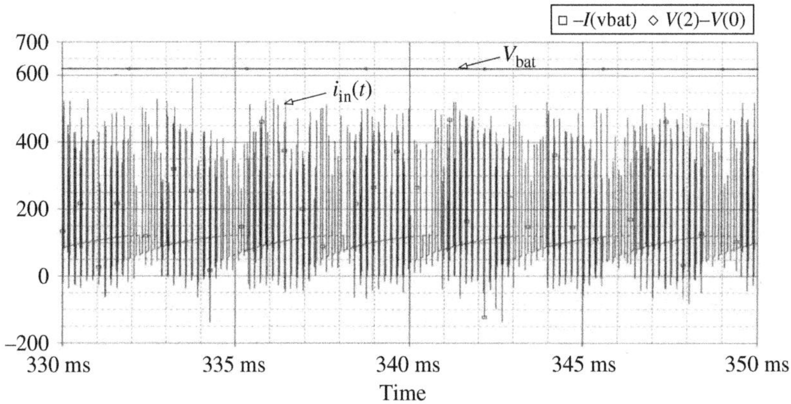

Figure E10.27.4 Battery voltage $V_{bat} = V(2) - V(0)$ and battery current or input current to inverter $i_{in}(t) = -I(\text{vbat})$.

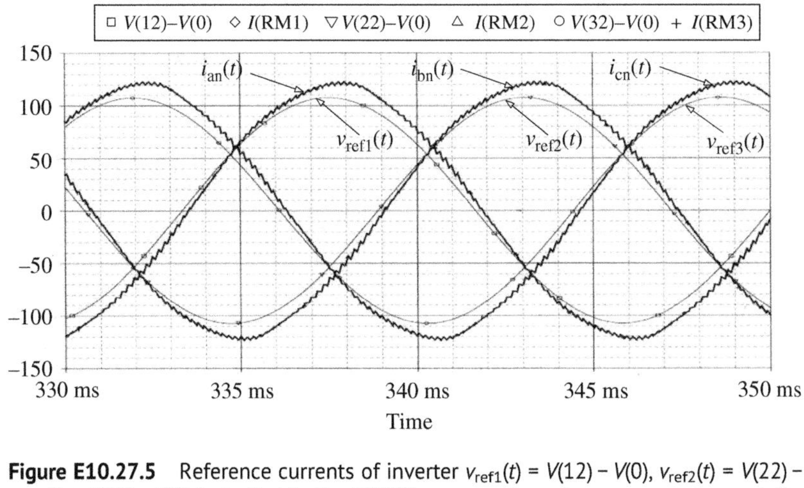

Figure E10.27.5 Reference currents of inverter $v_{ref1}(t) = V(12) - V(0)$, $v_{ref2}(t) = V(22) - V(0)$, $v_{ref3}(t) = V(32) - V(0)$ (note that in the PSPICE program the reference currents are defined as voltages) and actually computed inverter currents $i_{an}(t) = I(\text{RM1})$, $i_{bn}(t) = I(\text{IRM2})$, $i_{cn}(t) = I(\text{RM3})$.

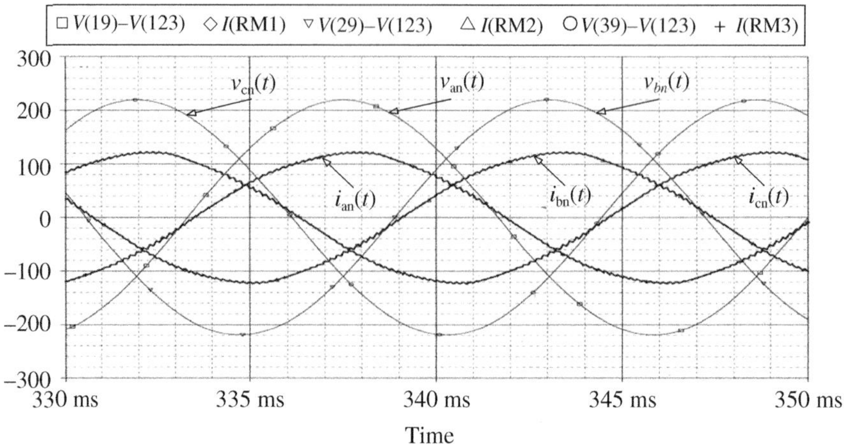

Figure E10.27.6 Power system phase voltages $v_{an}(t) = V(19) - V(123)$, $v_{bn}(t) = V(29) - V(123)$, $v_{cn}(t) = V(39) - V(123)$ and phase currents $i_{an}(t) = I(RM1)$, $i_{bn}(t) = I(RM2)$, $i_{cn}(t) = I(IRM3)$ lagging the phase voltages by 3.5° (consumer notation of currents; reactive power is absorbed from power system by inverter).

Table E10.27.2 Current harmonics of $i_{an}(t) = I(Rsysa) = I(Rwa) = I(RM1)$ of Figures E10.27.5 and E10.27.6 in terms of amperes in percent of reference current amplitude of 107.46 A.

DC component	1	3	5	7
0.00/0%	120.4 A/112%	0.01 A/0.00%	1.32 A/1.20%	0.74 A/0.69%

The phase shift between $i_{anref}(t) = V(12) - V(0)$ and $i_{an}(t) = I(RM1)$ is $-3.5°$; that is, the lagging displacement power factor is $\cos\Phi_{inverter} = \cos(-3.5°) = 0.998$ based on consumer system definition due to proportional (P) P control as defined in the subcircuit of PSPICE program in Table E10.27.1 and due to the high (e.g. $m = 1.0$) modulation index. For the same reason, the amplitude of $i_{an}(t) = I(RM1)$ is about 16% larger than that of the reference current $v_{ref1}(t) = V(12) - V(0)$. If the modulation index m were chosen to be less than 1.0, then the actual currents $i_{an}(t)$, $i_{bn}(t)$, $i_{cn}(t)$ would more closely follow the reference currents $v_{ref1}(t)$, $v_{ref2}(t)$, $v_{ref3}(t)$.

Conclusions

The line currents are about in phase with the line-to-neutral voltages, and its waveshapes are sinusoidal/cosinusoidal, as can be noted from Table E10.27.2. This is due to the chosen modulation index $m \approx 1.0$. If a larger m is chosen, say, $m = 1.1$, then a sinusoidal/cosinusoidal current waveshape will not be obtained. The real power delivered to the power system is $P_{AC\_supply} = 3 \cdot (107.46\,A/\sqrt{2}) \cdot (219.4\,V/\sqrt{2}) \approx 35\,kW$, and the apparent power $S_{supply} \approx P_{AC\_supply}$. The switch model (e.g. $w = 2.9$) must be adjusted depending upon the magnitude of the current.

Please refer back to Application Example 10.26 about the necessity of a phase-lock loop (PLL).

10.8 Design of a Wind Power Plant

Variable-speed wind power plants [18–23] are commonly designed and available based on nonsalient pole synchronous and permanent-magnet generators. In Application Example 10.28, the latter design is chosen, which was analyzed, designed, and tested as part of master of science [23] and doctoral [22] dissertations.

Application Example 10.28 Wind power plant with power rating of $P_{out\_inverter} = 219\,kW$

Figure E10.28.1 shows the components of a variable-speed, direct-drive wind power plant: a three-phase step-up delta/wye input transformer with diode bridge rectifier fed by permanent-magnet generator (PMG) with finite impedances. An inverter supplies at unity displacement (power) factor $P_{out\_inverter} = 219\,kW$ into the three-phase power system with finite impedances. Figures E10.28.2–E10.28.4 show details of the block diagram of Figure E10.28.1.

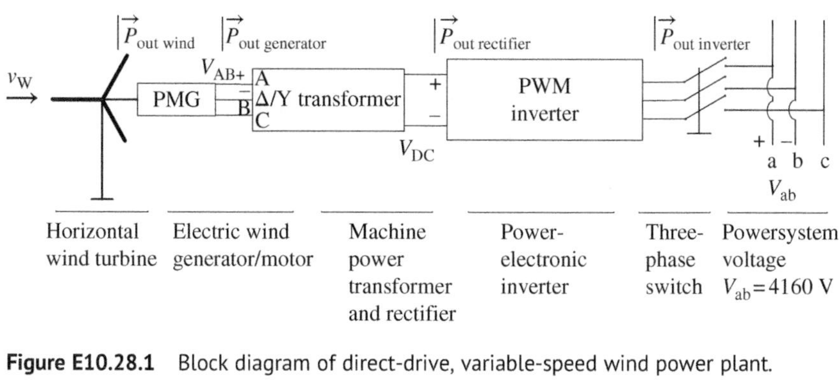

Figure E10.28.1 Block diagram of direct-drive, variable-speed wind power plant.

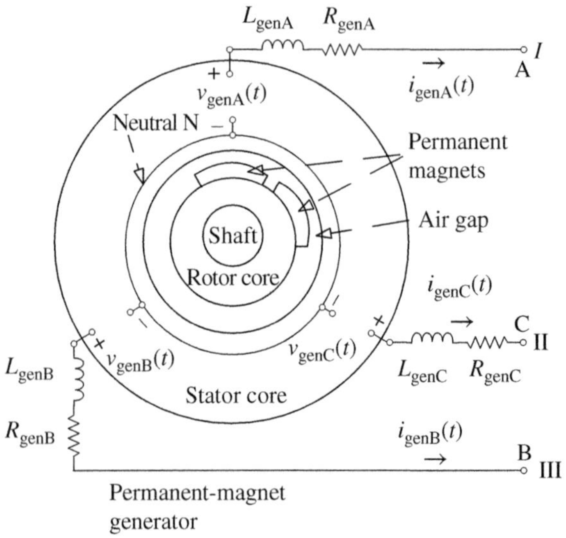

Figure E10.28.2 Direct-drive (without mechanical gear), variable-speed three-phase permanent-magnet generator.

The design data of the wind power plant components are:

a) wind speed $v_w = 5$ m/s and wind-turbine efficiency coefficient $c_p = 0.45$.
b) three-phase permanent-magnet generator (PMG) with speed $n_m = 60$ rpm, frequency $f = 6$ Hz, and generator (gen) rms voltage $V_{L\text{-}N} = 2799.5$ V. $R_{gen} = 0.1\ \Omega$, and $L_{gen} = 100\ \mu$H.
c) three-phase Δ/(ungrounded Y) ideal step-up transformer with turns ratio $(N_2/N_1) = 1.7$.
d) diode bridge rectifier with filter $L_f = 100$ mH and $C_f = 10$ mF.
e) current-controlled voltage source inverter supplying energy to the power system with a line-to-line rms voltage of $V_{ab} = 4160$ V or line-to-neutral voltage $v_{an}(t) = \sqrt{2}(4160/\sqrt{3})$ V cos ωt = 3396.2 V cos ωt, v_{bn}(t) = $\sqrt{2}(4160/\sqrt{3})V$ cos ωt = 3396.2 V cos(ωt − 120°), $v_{cn}(t) = \sqrt{2}(4160/\sqrt{3})$V cos ωt = 3396.2 V cos(ωt − 240°) at about unity power factor cos θ = 1, $R_{sys} = 50$ mΩ, and $L_{sys} = 0.265$ mH.

The design of this wind power plant proceeds from the power system and the inverter (Figure E10.28.4), delta/wye transformer, and rectifier (Figure E10.28.3) via permanent-magnet generator to the wind turbine (Figure E10.28.2).

1) Perform a PSPICE analysis for the circuit of Figure E10.28.4 where a three-phase MOSFET current-controlled voltage source inverter supplies power $P_{out\_inverter} = 219$ kW into the three-phase power system. The power system

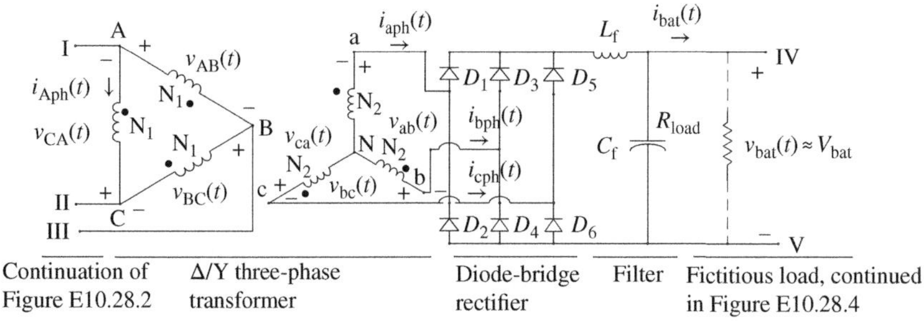

Figure E10.28.3 Three-phase Δ/(ungrounded Y) step-up transformer supplying the equivalent load R_{load} = 216.27 Ω via a diode bridge rectifier with L_fC_f filter.

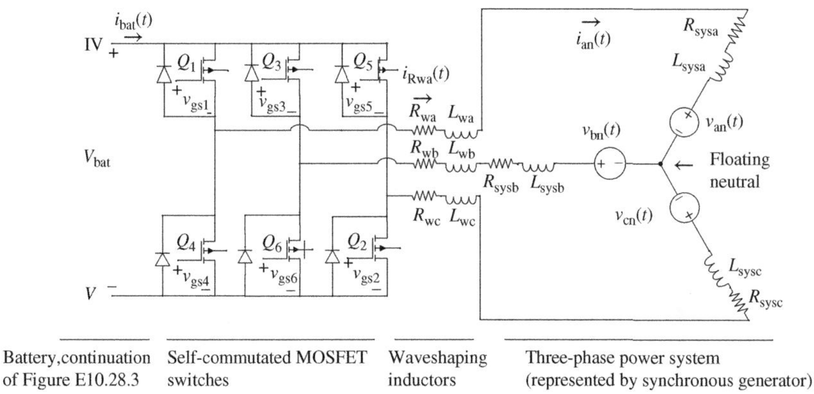

Figure E10.28.4 Three-phase current-controlled voltage source inverter supplying $P_{out\_inverter}$ = 219 kW to three-phase power system.

impedance per phase is R_{sys} = 50 mΩ, L_{sys} = 0.265 mH at f = 60 Hz. The line-to-neutral voltages of the power system are $v_{an}(t) = \sqrt{2} \cdot (V_{an})\cos \omega t = \sqrt{2} \cdot (4160/\sqrt{3})$ V $\cos \omega t$ = 3396.2 V $\cos \omega t$, $v_{bn}(t) = \sqrt{2} \cdot (4160/\sqrt{3})$ V $\cos(\omega t - 120°)$ = 3396.2 V $\cos(\omega t - 120°)$, $v_{cn}(t) = \sqrt{2} \cdot (4160/\sqrt{3})$ V $\cos(\omega t - 240°)$ = 3396.2 V $\cos(\omega t - 240°)$, and the rms phase current delivered to the power system is $I_{ph\_rms} = P_{out\_inverter}/(3 \cdot V_{rms} \cdot \cos \theta) = 219\,000/(3 \cdot 2401.8 \cdot 1.0) = 30.394$ A or $i_{an}(t) = 42.977$ A $\cos \omega t$, $i_{bn}(t) = 42.977$ A $\cos(\omega t - 120°)$, $i_{cn}(t) = 42.977$ A $\cos(\omega t - 240°)$. The modulation index is chosen to be $m = 0.8$. With $V_{an} = m \cdot V_{DC}/(2\sqrt{2})$ follows for the battery voltage $V_{bat} = V_{DC} = (2\sqrt{2}) \cdot V_{an}/m = (2\sqrt{2})\cdot(4160\text{ V}/\sqrt{3})/0.8 = 8490.5$

V. Waveshaping (w) inductances and resistances are $L_{wa} = L_{wb} = L_{wc} = 50$ mH and $R_{wa} = R_{wb} = R_{wc} = 10$ mΩ, respectively.

2) Plot and subject at steady-state operation line-to-neutral voltages $v_{an}(t)$, $v_{bn}(t)$, $v_{cn}(t)$, line currents $i_{an}(t)$, $i_{bn}(t)$, $i_{cn}(t)$ and required rectifier output unipolar voltage $v_{bat}(t)$ and $i_{bat}(t)$ to Fourier analyses (up to the 12th harmonic).
3) Plot the reference line currents $v_{ref1} = i_{anref}(t)$, $v_{ref2} = i_{bnref}(t)$, $v_{ref3} = i_{cnref}(t)$ and the actual line currents $i_{an}(t)$, $i_{bn}(t)$, $i_{cn}(t)$.
4) Perform a PSPICE analysis for the circuit of Figures E10.28.2 and E10.28.3, where a three-phase delta/wye transformer and a three-phase diode bridge rectifier with filter $C_f = 10$ mF, $L_f = 100$ mH serves the equivalent load R_{load} so that $V_{bat} = 8490.5$ V. The generator impedance per phase is $R_{gen} = 0.10\ \Omega$, $L_{gen} = 0.100$ mH, and assume ideal diodes D_1–D_6.
5) Plot and subject at steady-state operation the transformer secondary line-to-line voltages $v_{ab}(t)$, $v_{bc}(t)$, $v_{ca}(t)$, transformer secondary line currents $i_{aph}(t)$, $i_{bph}(t)$, $i_{cph}(t)$, to Fourier analyses (up to the 12th harmonic).
6) Plot and subject at steady-state operation the transformer primary line-to line voltages $v_{AB}(t)$, $v_{BC}(t)$, $v_{CA}(t)$, transformer primary input line currents $i_{genA}(t)$, $i_{genB}(t)$, $i_{genC}(t)$ to Fourier analyses (up to the 12th harmonic).

Solution

Inverter and power system analyses: The PSPICE solution of the program listed in Table E10.28.1 yields Figures E10.28.5–E10.28.7. Note that the PSPICE solution required more than 30 minutes of computation time on a laptop computer.

The phase shift between $v_{nref1}(t) = V(12) - V(0)$ and $i_{an}(t) = I(\text{Rwa})$ is $-7.2°$; that is, lagging displacement power factor $\cos\theta_{inverter} = \cos(-7.2°) = 0.994$ based on consumer system definition due to P control as defined in the subcircuit of PSPICE program in Table E10.28.1. All harmonics are negligibly small except the switching harmonics related to the switching frequency of $f_{switching} = 2876.9$ Hz. With the computed lagging displacement power factor angle of $\theta = -7.2°$, one obtains for the computed inverter output currents $i_{an}(t) = 42.977\text{ A}\cos(\omega t - 7.2°)$, $i_{bn}(t) = 42.977\text{ A}\cos(\omega t - 127.2°)$, $i_{cn}(t) = 42.977\text{ A}\cos(\omega t - 247.2°)$, resulting in apparent output inverter power $S_{out\_inverter} = 3 \cdot (2404.5) \cdot 30.39$ VA = 219.25 kVA or $P_{out_inverter} = S_{out_inverter}\cos\theta_{inverter} = 219.25\cos(-7.2°)$kW = 217.85 kW, which is close to 219 kW. The DC input power neglecting all harmonic power contributions is $P_{bat} = V_{bat} \cdot I_{bat} = 8490.5 \cdot (-29.98) = 254.55$ kW. Note the minus sign means that the voltage V_{bat} generates a negative current $I_{bat} = -29.98$ A entering the battery and not the inverter. The inverter efficiency, including the losses in the waveshaping inductors and the power system, is therefore

Table E10.28.1 PSPICE input program for Figure E10.28.4.

```
Application Example 10.28a
*Wind power plant
*Current-controlled pulse-width-
modulated
*(PWM) voltage-source inverter feeding
*energy into the power system
Vbat 2 0 8490.5
*switches
msw1 2 11 10 10 qfet
dsw1 10 2 diode
msw5 2 21 20 20 qfet
dsw5 20 2 diode
msw3 2 31 30 30 qfet
dsw3 30 2 diode
msw4 10 41 0 0 qfet
dsw4 0 10 diode
msw2 20 51 0 0 qfet
dsw2 0 20 diode
msw6 30 61 0 0 qfet
dsw6 0 30 diode
*waveshaping inductors
lwa 10 15 50m
lwb 20 25 50m
lwc 30 35 50m
Rwa 15 16 10m
Rwb 25 26 10m
Rwc 35 36 10m
*current references in terms of voltages
vref1 12 0 sin(0 42.977 60 0 0 0)
vref2 22 0 sin(0 42.977 60 0 0 -120)
vref3 32 0 sin(0 42.977 60 0 0 -240)
eout1 13 0 15 16 100
eout2 23 0 25 26 100
eout3 33 0 35 36 100
*errorsignals produced as difference
between
*vref and eout
rdiff1 12 13a 1k
rdiff2 22 23a 1k
rdiff3 32 33a 1k
cdiff1 12 13a 1u
cdiff2 22 23a 1u
cdiff3 32 33a 1u
rdiff4 13a 13 1k
rdiff5 23a 23 1k
rdiff6 33a 33 1k
*gating for lower switches
egs4 41 0 poly(1) (11,10) 50 -1
egs5 51 0 poly(1) (21,20) 50 -1
egs6 61 0 poly(1) (31,30) 50 -1
*Filter is deleted because PSpice is limited to
*64 nodes
*parameters of power system with line-to-
*neutral induced (phase) voltages ea, eb, ec
RM1 16 18 50m
LM1 18 19 265u
Vout1 19 123 sin(0 3396.2 60 0 0 0)
RM2 26 28 50m
LM2 28 29 265u
Vout2 29 123 sin(0 3396.2 60 0 0 -120)
RM3 36 38 50m
LM3 38 39 265u
Vout3 39 123 sin(0 3396.2 60 0 0 -240)
*subcircuit for comparator
.subckt comp 1 2 9 10
rin 1 3 2.8k
r1 3 2 20meg
e2 4 2 3 2 50
r2 4 5 1k
d1 5 6 zenerdiode1
d2 2 6 zenerdiode2
e3 7 2 5 2 1
r3 7 8 10
c3 8 2 10n
r4 3 8 100k
e4 9 10 8 2 1
*models for zener diodes
.model zenerdiode1 D (Is=1p BV=0.1)
.model zenerdiode2 D (Is=1p BV=50)
.ends comp
*model for switch
.model qfet nmos(level=3 gamma=0 kappa=0
+tox=100n rs=42.69m kp=20.87u l=2u
+ w=2.9 delta=0 eta=0 theta=0 vmax=0 xj=0
+uo=600 phi=0.6 vto=3.487 rd=0.19
cbd=200n
+pb=0.8 mj=0.5 cgso=3.5n cgdo=100p rg=1.2
+is=10f )
*model for diodes
.model diode d(is=1p)
*options to aid convergence
.options abstol=0.01m chgtol=0.01m
```

(Continued)

Table E10.28.1 (Continued)

```
ecin1 14 0 12 13a 2
ecin2 24 0 22 23a 2
ecin3 34 0 32 33a 2
vtrial 5 0 pulse(-10 10 0 86.5u 86.5u
0.6u
+173.6u)
*gating signals for upper switches
*as result of comparison between
*triangular waveform and error signal
xgs1 14 5 11 10 comp
xgs2 24 5 21 20 comp
xgs3 34 5 31 30 comp
+reltol=50m vntol=1m itl5=0 itl4=400
*transient analysis
.tran 10u 200m 180m 10u
*plotting of traces
.probe
*Fourier analysis
.four 60 7 V(123,19) V(12,0) I(Rwa), V(2,0), +I
(Vbat)
.end
```

Note: All comment lines must be preceded with an asterisk (*) while continued calculation statements must be preceded with a plus sign (+).

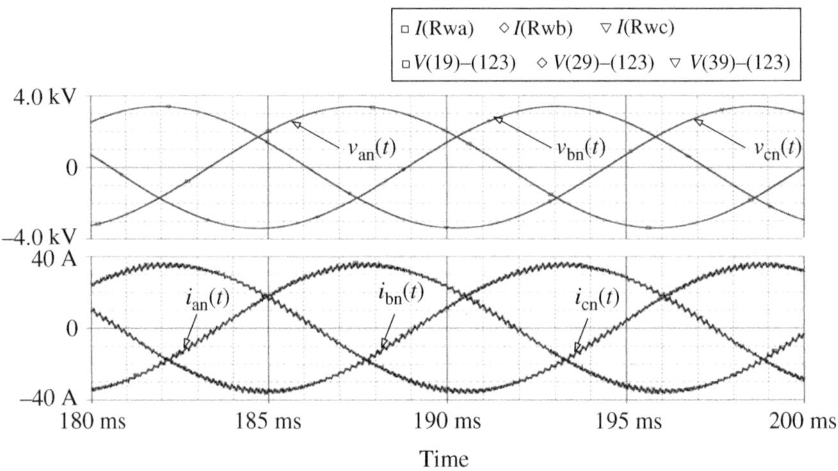

Figure E10.28.5 Top plot: steady-state line-to-neutral voltages $v_{an}(t) = V(19) - V(123)$, $v_{bn}(t) = V(19) - V(123)$, $v_{cn}(t) = V(39) - V(123)$. Bottom plot: actual line currents $i_{an}(t) = I(\text{Rwa})$, $i_{bn}(t) = I(\text{Rwb})$, $i_{cn}(t) = I(\text{Rwc})$. The phase shift between $v_{an}(t)$ and $i_{an}(t)$ is −7.2°; that is, lagging displacement power factor $\cos\theta = \cos(-7.2°) = 0.994$ based on consumer displacement power factor definition.

$\eta_{inverter} = 217.85/254.55 = 0.86$ or 86%. The voltage $V_{bat} = 8490.5$ V and the current $I_{bat} = (-29.98$ A) correspond to a load resistance of $R_{load} = V_{bat}/I_{bat} = 8490.5$ V/ 29.98 A = 283.2 Ω. Figure E10.28.7 shows that the actual line currents $i_{an}(t)$, $i_{bn}(t)$, $i_{cn}(t)$ do not closely follow the reference currents $i_{anref}(t)$, $i_{bnref}(t)$, $i_{cnref}(t)$: the reason is that the PSPICE program is based on a proportional (P) control algorithm in

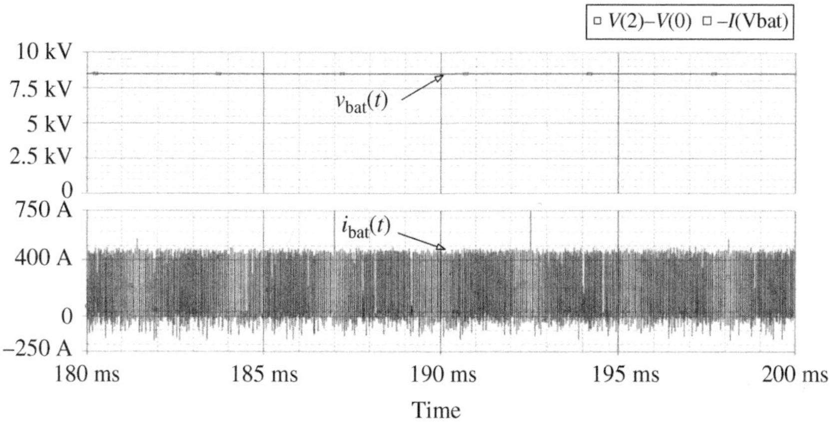

Figure E10.28.6 Top plot: required (given) rectifier output unipolar voltage $v_{bat}(t) = V(2) - V(0)$. Bottom plot: current $i_{bat}(t) = -I(\text{Vbat})$ supplied by unipolar voltage V_{bat}.

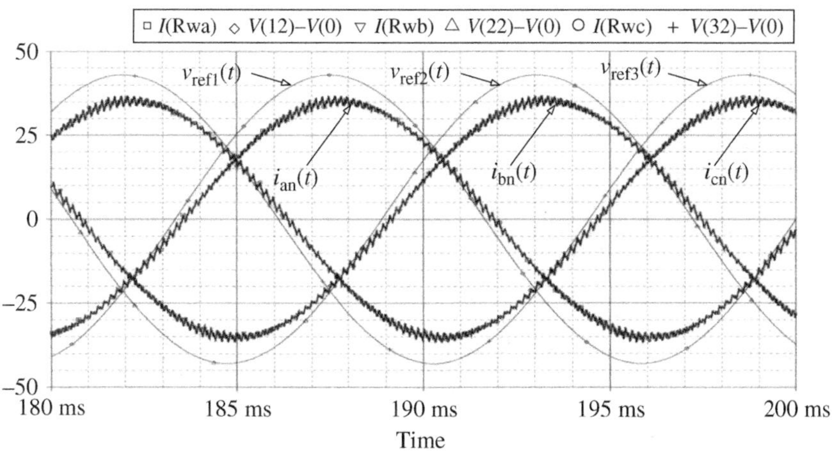

Figure E10.28.7 Reference line currents $v_{ref1}(t) = V(12) - V(0)$, $v_{ref2}(t) = V(22) - V(0)$, $v_{ref3}(t) = V(32) - V(0)$; note that in the PSPICE program the reference currents are defined as voltages and actual line currents $i_{an}(t) = I(\text{Rwa})$, $i_{bn}(t) = I(\text{Rwb})$, $i_{cn}(t) = I(\text{Rwc})$.

the subcircuit for comparator. An integral (I) control would assure that the controlled variables would closely follow its references.

Evaluation of PSPICE results with respect to power system and inverter: At the outset of the above analysis, all losses occurring in the power system, the waveshaping impedance, and gating losses of the inverter were neglected. In addition, it was assumed that proportional (P) control exists for the generation of the AC currents supplied into the power system. As a result, the actually computed phase currents I(Rwa), I(Rwb), I(Rwc) supplied to the power system will be less than the reference currents vref1, vref2, vref3 of Table E10.28.1. Thus, due to these imperfections, the amplitude of the actually computed three-phase currents is 35.24 A instead of 42.977 A for the battery voltage $V_{bat} = 8490.5$ V. Because of these losses in the power system, the efficiency of the power system including waveshaping losses is 86%, as computed above.

Rectifier and generator analyses: The PSPICE program for the rectifier and generator equivalent circuits of Figures E10.28.3 and E10.28.2 is listed in Table E10.28.2, and the pertinent voltages and currents of the rectifier circuit are shown in Figures E10.28.8–E10.28.10. The fundamental including harmonics of currents and voltages are listed in Tables E10.28.3–E10.28.5.

Table E10.28.2 PSPICE input program for Figures E10.28.2 and E10.28.3.

```
Application Example 10.28b
*wind power plant
* three-phase delta/(ungrounded Y) step-up
*transformer, supplying from a permanent-*magnet
generator operating at 6 Hz, the load *Rload via a
diode-bridge rectifier with filter
*Lf, Cf
*primary line-to-neutral voltages
VAB 1 2 sin(0 3958.5 6 0 0 0)
VBC 2 3 sin(0 3958.5 6 0 0 -120)
VCA 33 1 sin(0 3958.5 6 0 0 -240)
*generator resistances and inductances
RgenA 6 1 0.1
RgenB 4 2 0.1
RgenC 5 3 0.1
LgenA 7 6 100u
LgenB 9 4 100u
LgenC 8 5 100u
*step-up transformer ratio *Vab/VAB=17/10=ratio
of secondary to *primary line-to-line voltages
L1 7 8 10
L2 9 7 10
L3 8 40 10
L4 11 10 17
L5 12 10 17
L6 13 10 17
K41 L4 L1 0.9999
K52 L5 L2 0.9999
K63 L6 L3 0.9999
*R8 8 0 10meg
*diode bridge
D1 11 15 Dideal
D4 14 11 Dideal
D3 13 15 Dideal
D6 14 13 Dideal
D5 12 15 Dideal
D2 14 12 Dideal
*filter
Lf 15 16 100m
Cf 16 14 10m
*load
Rload 16 14 240
*auxiliary parameters
Rm 33 3 1u
Rn 10 0 10meg
R3 40 9 1u
R100 1 0 1meg
*model of diodes
.model Dideal d(is=1m)
.tran 100u 16 15.8 100u uic
.four 6 12 V(1,2) V(11,12)
I(RgenA) I(L1) I(L4) V(16,14)
.options abstol=1m chgtol=0.1m
reltol=0.1 vntol=1m
.options itl4=200 itl5=0
.probe
.end
```

Note: All comment lines must be preceded with an asterisk (*) while continued calculation statements must be preceded with a plus sign (+).

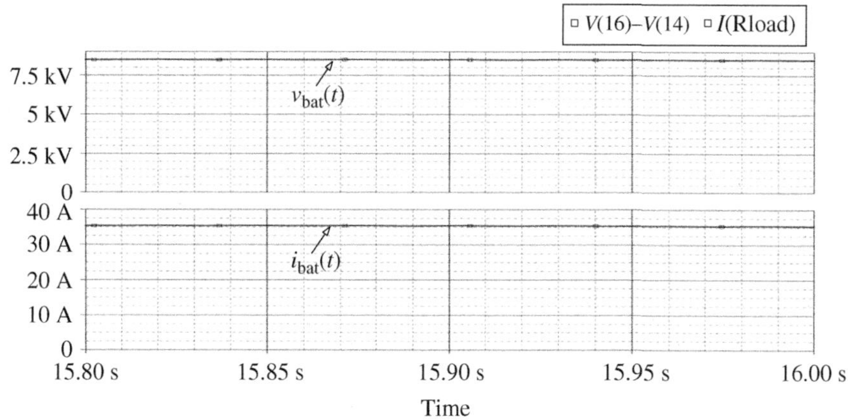

Figure E10.28.8 Top plot: required (given) rectifier output unipolar voltage $v_{bat}(t) = V(16) - V(14)$. Bottom plot: required rectifier output current $i_{bat}(t) = I(\text{Rload})$.

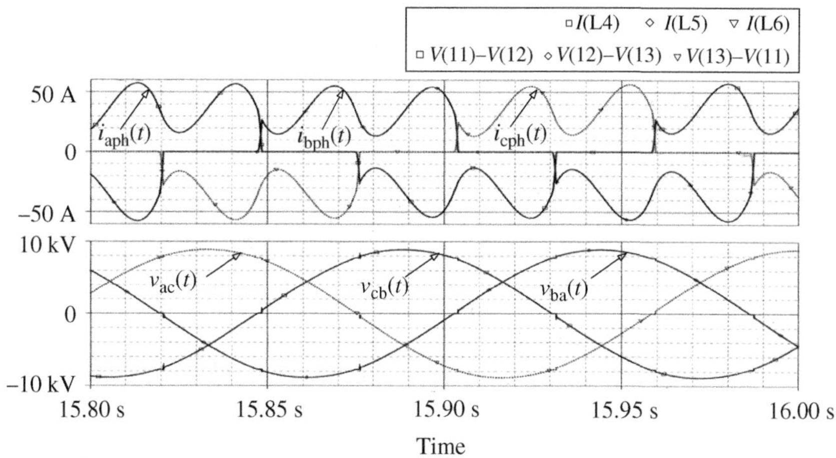

Figure E10.28.9 Top plot: secondary phase transformer currents $i_{aph}(t) = I(L4)$, $i_{cph}(t) = I(L5)$, $i_{bph}(t) = I(L6)$. Bottom plot: secondary line-to-line transformer voltages $v_{ac}(t) = V(11) - V(12)$, $v_{cb}(t) = V(12) - V(13)$, $v_{ba}(t) = V(13) - V(11)$.

The design of the wind turbine is based on the Lanchester–Betz–Joukowsky limit [7] $W_{max} = (8Av^3\rho)/27$ including an estimated permanent-magnet generator efficiency of $\eta_{generator} = 80\%$, rectifier efficiency of 90%, and inverter and power system efficiency of 86% and results in the required output power of the wind turbine $P_{out\_wind} = P_{out\_inverter}/(\eta_{generator} \cdot \eta_{rectifier} \cdot \eta_{inverter}) = 217.85\ \text{kW}/$

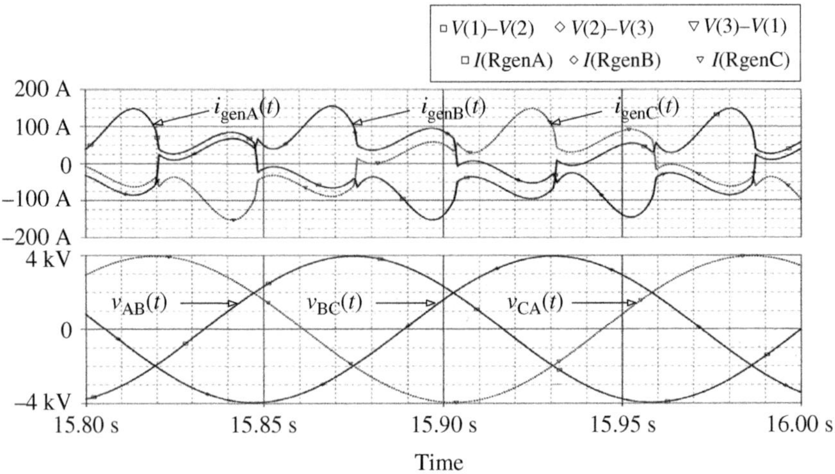

Figure E10.28.10 Top plot: primary line transformer currents $i_{genA}(t)$ = I(RgenA), $i_{genB}(t)$ = I(RgenB), $i_{genC}(t)$ = I(RgenC). Bottom plot: primary line-to-line transformer voltages $v_{AB}(t)$ = V(1) − V(2), $v_{BC}(t)$ = V(2) − V(3), $v_{CA}(t)$ = V(3) − V(1).

Table E10.28.3 DC voltage and harmonics of $V_{bat}(t)$ = V(16) − V(14) of Figure E10.28.8.

DC component	1	5	6	12
8494.4 V	4.09 V	0.06 V	9.13 V	0.58 V

Table E10.28.4 Fundamental current (1) and harmonics of $i_{aph}(t)$ = I(L4) of Figure E10.28.9.

DC component	1	5	7	11
0.84 A	39.41 A	16.11 A	10.26 A	3.40 A

Table E10.28.5 Fundamental current (1) and harmonics of $i_{genA}(t)$ = I(RgenA) of Figure E10.28.10.

DC component	1	2	5	7	11
−4.65 A	93.18 A	1.82 A	36.36 A	23.20 A	7.70 A

$(0.8 \cdot 0.9 \cdot 0.86) = 351.82$ kW. For a wind velocity of $v_w = 5$ m/s and a maximum wind-turbine efficiency coefficient of $c_{p\_max} = 16/27 = 0.592\,59$ at sea level – where $\rho = 1.2255$ kg/m^3 is the air density – one obtains the blade length/radius R_{blade} from the maximum power relation $W_{max} = (8 A v^3 \rho)/27 = 351.820$ kW, where $A = \pi R_{blade}^2$ is the rotor-swept area, as

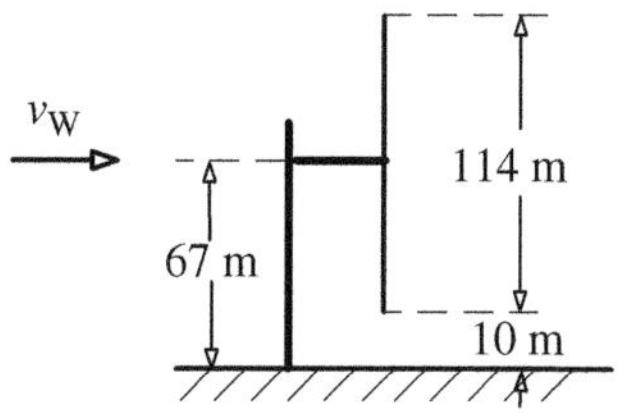

Figure E10.28.11 Design dimensions of tower of wind power plant.

$$R_{blade}^{cp=0.592\,59} = \sqrt{\frac{27 \cdot 351\,820}{8 \cdot 5^3 \cdot 1.2255 \cdot \pi}} = 49.67 \text{ m}.$$

For the nonideal wind-turbine efficiency coefficient of $c_p = 0.45$, one obtains $P_{max} = W_{max}(0.59259/0.45) = 463\,300$ W or

$$R_{blade}^{cp=0.45} = \sqrt{\frac{27 \cdot 463\,300}{8 \cdot 5^3 \cdot 1.2255 \cdot \pi}} = 57 \text{ m}.$$

The design of the tower is illustrated in Figure E10.28.11.

Summary of wind power plant operation:

- Inverter output power: $P_{out\_inverter} = 217.85$ kW
- Inverter efficiency $\eta_{inverter} = 86\%$
- DC input power to inverter: $P_{bat} = V_{bat} \cdot I_{bat} = 8490.5 \cdot (-29.98) = 254.55$ kW, neglecting harmonics
- $P_{out\_rectifier} = 8500.6 \cdot 30.016 = 255.19$ kW, neglecting harmonics
- rectifier efficiency: $\eta_{rectifier} = 90\%$
- $P_{rectifier\_in} = 255.19 \text{ kW}/0.9 = 283.54$ kW
- Permanent-magnet generator efficiency: $\eta_{generator} = 80\%$
- $P_{generator\_in} = 283.54 \text{ kW}/0.8 = 354.43$ kW
- The overall efficiency of the wind power plant is $\eta_{wind\_power\_plant} = \eta_{inverter} \cdot \eta_{rectifier+transformer} \cdot \eta_{generator} = 0.86 \cdot 0.9 \cdot 0.8 = 62\%$ or $P_{out\_inverter}/P_{generator\_in} = (217.85 \text{ kW}/354.43 \text{ kW}) \cdot 100\% = 62\%$

The overall efficiency of a wind power plant is small due to the many required components (e.g. generator, transformer, rectifier, and inverter). Because of the nonsinusoidal generator currents [20, 22] at low frequency (fundamental of 6 Hz), it is advisable to install at the output of the permanent-magnet generator a filter. Finally, the efficiencies of the wind power plant (Figure E10.28.11), DC or AC transmission lines, and power system must be taken into account as well, further decreasing the overall efficiency.

Note that an inverter supplying power to the power system at leading (consumer notation) displacement power factor requires a larger input DC voltage than lagging (consumer notation) displacement power factors.

Further work: A delta/wye (ungrounded) three-phase transformer blocks triplen harmonics (e.g. 3rd, 6th, 9th, ...) and the phase shift between the line-to-line and phase voltages is (−30°). The rectifier generates a significant amount of harmonics within secondary currents as well as within the primary currents flowing into the permanent-magnet generator. To prevent such an injection into the generator, passive harmonic filters across the terminals A, B, C, and a, b, c are required. In addition, it will be advisable to employ a delta/zigzag [7] transformer and the PWM rectifier of Application Example 10.23, both reducing harmonics (e.g. 5th, 7th, 11th, ...) significantly. The pulse magnitude of the gating voltage (e.g. 20 V) and the switch model (e.g. $w = 2.9$) must be adjusted depending upon the magnitude of the current. Please refer back to Application Example 10.26 about the necessity of phase-lock loops (PLL).

10.9 Efficiency Increase of Induction Motors Based on Semiconductor Controllers and Influence of Harmonics on Power System Components

In 1977 Frank Nola [25] published the displacement power factor increase (e.g. $\cos\theta$ from 0.3 to 0.55) and efficiency increase (e.g. η by 5–10%) of induction motors by employing semiconductor (e.g. thyristor/triac) controllers; while the displacement power factor increase occurs at all loading conditions, the efficiency increase occurs at light and medium loads only. At rated load the efficiency is decreased due to the losses of the semiconductor controller [26–30].

Unfortunately semiconductor controllers introduce harmonics, which can be prevented from entering the motor by passive or active filters: by doing so losses occur in the filter, which reduces the abovementioned efficiency increase somewhat. To minimize the use of filters, the aging of appliances is investigated [31–33], and the influence of harmonics on power system protection is measured [34–37]. Renewable energy systems have increased [38–40] harmonic voltage/current problems, and it is important to reduce harmonics in distribution systems through optimization [41–45].

10.10 Power Quality and the Use of Input and Output Filters for Rectifiers and Inverters

Power quality [7, 8] is a measure for the deviation of sinus/cosinusoidal voltages and currents at steady-state operation of electric apparatuses. Such power quality issues are of increased concern with the deployment of converters such as rectifiers and inverters. The term converter relates to power electronic equipment converting electricity to voltage/current of different frequency. For example, an AC–DC converter is called a rectifier, which changes alternating voltage/current to unipolar voltage/current; the DC–AC converter is called an inverter because it changes unipolar voltage/current to alternating voltage/current. DC–DC converters and AC–AC converters – the latter also called cycloconverters – are employed for special applications only and are not discussed in this text. All types of converters generate harmonics that cause power quality problems within conducting materials, and therefore filters [8] must be used to prevent their harmful effects. If three-phase power systems and full-wave switching actions are relied on, mostly the 5th, 7th, 11th, ... harmonics occur, while for single-phase systems with half-wave switching all (2nd, 3rd, 4th, 5th, ...) harmonics occur. In all drive trains, either motoring or regeneration can be applied. This is done in electric trains and electric and hybrid cars to convert braking energy through regeneration to electric energy, which can be either stored in the supply AC system or in the onboard battery. The energy savings due to regeneration for automobiles is in the range of 20%, depending upon road and braking conditions [7]. In the following application example, the drive train of a railroad locomotive or streetcar with a single-phase overhead line/catenary will be simulated with PSPICE, whereby the single-phase AC voltage V_{system} exists between the overhead/catenary wire and the grounded rail. Figure 10.3 shows the voltage–current ($v - i$) plane where motoring and regeneration regions are identified. Motoring occurs at constant and increasing torque (corresponding to i) and speed (corresponding to v), and regeneration occurs when the torque is negative at positive speed, thus converting braking energy into electric form. Figures 10.4 and 10.5 illustrate block diagrams of motoring and regeneration modes.

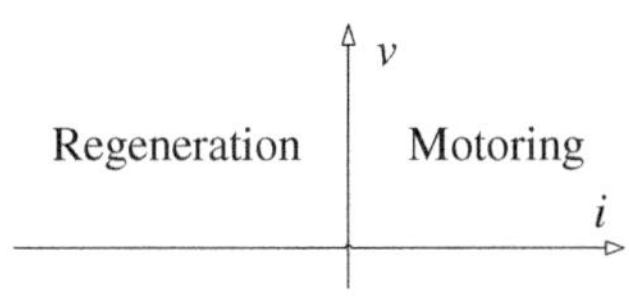

Figure 10.3 Voltage–current (v-i) diagram of motoring and regeneration modes of an electric drive.

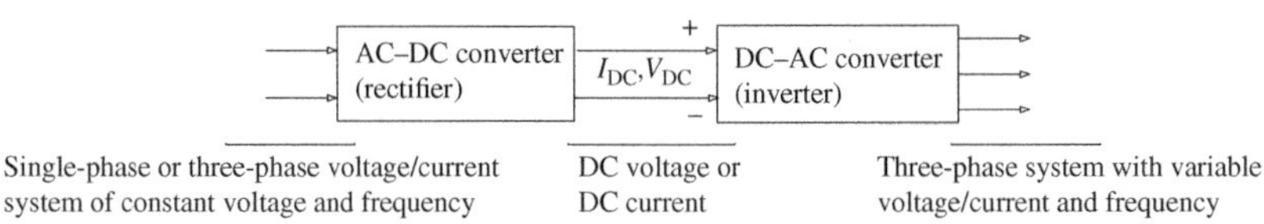

Figure 10.4 Block diagram of motoring mode converting electric energy of constant voltage and frequency, via a rectifier and an inverter, to variable voltage/current of drive train with variable electric frequency. Note that a rectifier can be operated as an inverter and vice versa.

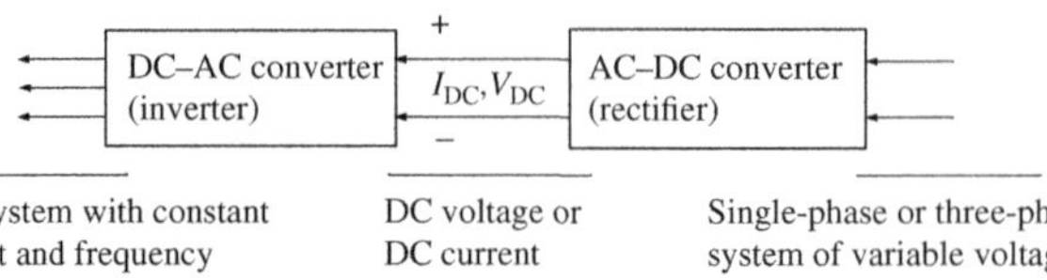

Figure 10.5 Block diagram of regeneration mode converting mechanical braking energy of drive train with variable voltage/current and electric frequency, via a rectifier and an inverter, to constant voltage and frequency. Note that an inverter can be operated as a rectifier and vice versa.

Application Example 10.29 Electric drive-train circuit of a locomotive

Consider an electric drive-train circuit with four drive units each with a rated output power of 1.28 MW or for four drive units 5.14 MW; this small output power rating is for a train with one engine/locomotive and a few railroad cars only. Power quality [46] in railroad power systems is a severe problem because of the single-phase supply. Nicola Tesla's invention [47] and IEEE reference books [48–51] are sources of important information for power system analysis and protection with respect to machines, transformers, and relays. Railroad power systems are single-phase power supplies with, for example, an overhead line voltage of $V_{syst\_rms} = 15$ kV and a frequency of $f = 16.667$ Hz in Germany, $f = 50$ Hz in France, and $f = 60$ Hz in the United States. Employing an input single-phase transformer with rectifier as shown in Figure E10.29.1, and a series-connected inverter as depicted in Figure E10.29.2, permits locomotives with either induction motors in Germany or synchronous motors in France to operate in a single-phase power

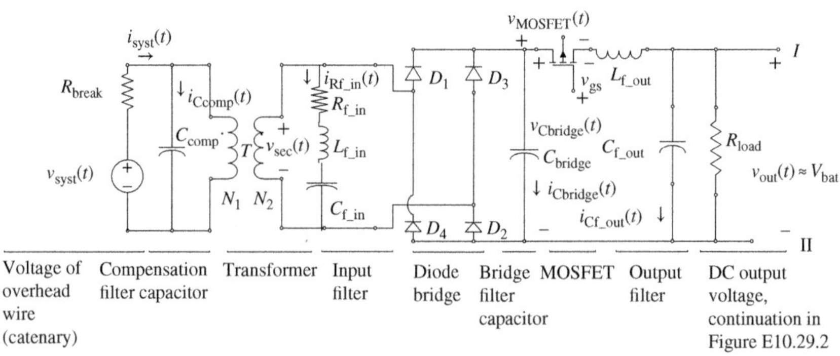

Figure E10.29.1 Input transformer ($N_1 : N_2$) with diode bridge PWM MOSFET switch rectifier with input (R_{f_in}, L_{f_in}, C_{f_in}) and output (L_{f_out}, C_{f_out}) filters and compensation capacitors C_{comp} and C_{bridge}. R_{load} represents an equivalent load resistor providing the necessary output DC power for the inverter of Figure E10.29.2.

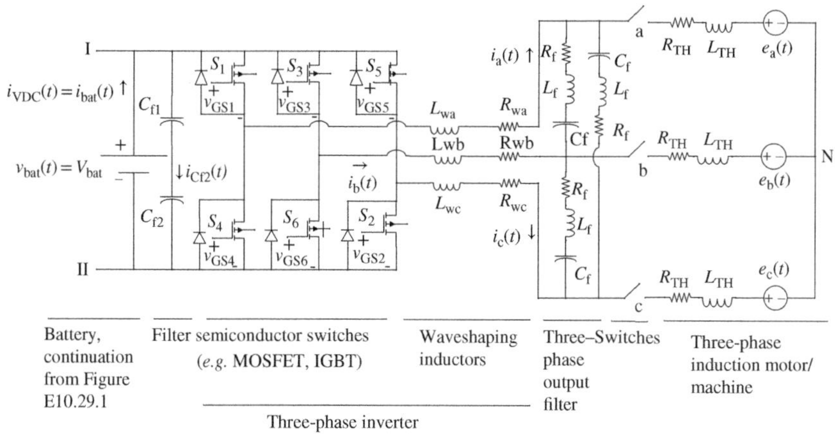

Figure E10.29.2 Current-controlled PWM voltage source inverter [7, 8, 17] feeding a three-phase squirrel-cage induction motor. Continuation of the circuit depicted in Figure E10.29.1.

system with different voltages and frequencies. This means a locomotive need not be exchanged if the train crosses a border between two countries.

The equivalent per-phase circuit parameters of a high efficiency η, high displacement power factor $\cos\theta$, deep-rotor bar, three-phase induction machine are $V_{ph\_rms} = 4000\,\text{V}$, $p = 4$ poles, $R_s = 1.2\,\Omega$, $X_s = 4\,\Omega$, $X_m = 150\,\Omega$, $R_r' = 0.4\,\Omega$, $X_r' = 5\,\Omega$, deep-rotor bars, wye-connected stator winding $\cos\theta = 0.86$ lagging. The parameters of the waveshaping inductors are the inductance $L_w = 10\,\text{mH}$ and the resistance $R_W = 10\,\text{m}\Omega$. All other parameters can be gained from the PSPICE program of Table E10.29.1.

Rated operation of three-phase induction motor drive train at $f_1 = 60$ Hz, PSPICE analysis based on Figure E10.29.2: From the formulae of Section 12.4 and Application Example 12.11 in Chapter 12, one obtains for the above circuit parameters the Thévenin per-phase resistance $R_{TH} = 1.1386\,\Omega$ and for the Thévenin per-phase (linear, unsaturated) inductance $L_{TH} = X_{TH}/(2\pi \cdot 60) = 3.9054/(2\pi \cdot 60) = 10.359\,\text{mH}$. For rated slip $s_{rat} = 0.02$ pu, $f_1 = 60$ Hz, and the Thévenin (TH) per phase voltage $\tilde{V}_{TH} = 3896\angle 0.446°\text{V}$ the rotor per phase current is $\tilde{I}_r' = 169.85\angle -22.399°\text{A}$. From Figure E12.11.1a and b follows through the application of Kirchhoff's voltage law for the induced phase voltage

$$\tilde{E} = \tilde{V}_{TH} - \tilde{I}_r'(R_{TH} + jX_{TH}) = 3896\angle 0.446°\text{V} - 169.85\angle -22.399°\text{A} \cdot (1.1386 + j3.9054)\Omega = 3464.3\angle -8.3627°\text{V}.$$

The instantaneous induced phase voltages of the induction machine (see Figure 12.16) are now with the amplitude $\sqrt{2} \cdot 3464.3\,\text{V} = 4898.5\,\text{V}$: $e_a(t) = 4898.5$

Table E10.29.1 PSPICE program for the signals of Figures E10.29.3–E10.29.6.

```
Application Example E10.29a
*Current-controlled pulse-width-
modulated
*(PWM) voltage-source inverter feeding
an
*induction motor, srated=0.02, f=60Hz
VDC 2 0 14140
Cf1 2 200 1mF
Cf2 200 0 1mF
Rf 200 0 1meg
*Switches
msw1 2 11 10 10 qfet
dsw1 10 2 diode
msw5 2 21 20 20 qfet
dsw5 20 2 diode
msw3 2 31 30 30 qfet
dsw3 30 2 diode
msw4 10 41 0 0 qfet
dsw4 0 10 diode
msw2 20 51 0 0 qfet
dsw2 0 20 diode
msw6 30 61 0 0 qfet
dsw6 0 30 diode
*Waveshaping inductors
l_w1 10 15 50m
l_w2 20 25 50m
l_w3 30 35 50m
R_w1 15 16 10m
R_w2 25 26 10m
R_w3 35 36 10m
*Current references in terms of voltages
vref1 12 0 sin(0 240.17 60 0 0 -22.399)
vref2 22 0 sin(0 240.17 60 0 0 -142.399)
vref3 32 0 sin(0 240.17 60 0 0 -262.399)
eout1 13 0 15 16 100
eout2 23 0 25 26 100
eout3 33 0 35 36 100
*Errorsignals produced as difference
between
*vref and eout
rdiff1 12 13a 1k
rdiff2 22 23a 1k
rdiff3 32 33a 1k
cdiff1 12 13a 1u
cdiff2 22 23a 1u
cdiff3 32 33a 1u
rdiff4 13a 13 1k
rdiff5 23a 23 1k
rdiff6 33a 33 1k
*Filter is deleted because PSpice is limited to
*64 nodes
*lfi1 16 15b 45u
*lfi2 26 25b 45u
*lfi3 36 35b 45u
*rfi1 15b 15c 0.01
*rfi2 25b 25c 0.01
*rfi3 35b 35c 0.01
*cfi1 15c 26 10.3u
*cfi2 25c 36 10.3u
*cfi3 35c 16 10.3u
*Parameters of power system with line-to-
*neutral induced voltages ea, eb, ec
RTH1 16 18 1.1386
LTH1 18 19 10.359m
Vout1 19 123 sin(0 4898.5 60 0 0 -8.3627)
RTH2 26 28 1.1386
LTH2 28 29 10.359m
Vout2 29 123 sin(0 4898.5 60 0 0 -128.3627)
RTH3 36 38 1.1386
LTH3 38 39 10.359m
Vout3 39 123 sin(0 4898.5 60 0 0 -248.3627)
*Subcircuit for comparator
.subckt comp 1 2 9 10
rin 1 3 2.8k
r1 3 2 20meg
e2 4 2 3 2 50
r2 4 5 1k
d1 5 6 zenerdiode1
d2 2 6 zenerdiode2
e3 7 2 5 2 1
r3 7 8 10
c3 8 2 10n
r4 3 8 100k
e4 9 10 8 2 1
*Models for zener diodes
.model zenerdiode1 D (Is=1p BV=0.1)
.model zenerdiode2 D (Is=1p BV=50)
.ends comp
*Model for switch
.model qfet nmos(level=3 gamma=0
kappa=0
+tox=100n rs=42.69m kp=20.87u l=2u
+ w=2.9 delta=0 eta=0 theta=0 vmax=0
xj=0
+uo=600 phi=0.6 vto=3.487 rd=0.19
cbd=200n
+pb=0.8 mj=0.5 cgso=3.5n cgdo=100p
rg=1.2
```

Table E10.29.1 (Continued)

```
ecin1 14 0 12 13a 2
ecin2 24 0 22 23a 2
ecin3 34 0 32 33a 2
vtria1 5 0 pulse(-50 50 0 173.3u 173.3u
0.6u
+347.2u)
*Gating signals for upper switches
xgs1 14 5 11 10 comp
xgs2 24 5 21 20 comp
xgs3 34 5 31 30 comp
*Gating signals for lower switches
egs4 41 0 poly(1) (11,10) 50 -1
egs5 51 0 poly(1) (21,20) 50 -1
egs6 61 0 poly(1) (31,30) 50 -1
+is=10f )
*Model for diodes
.model diode d(is=1p)
*Options to aid convergence
.options abstol=100m chgtol=100m
+reltol=100m vntol=100m itl5=0
itl4=20000
+itl1=200 itl2=200
*Transient analysis
.tran 10u 350m 300m 10u
*Plotting of traces
.probe
*Fourier analysis
.four 60 12 I(R_w1)
.end
```

Note: All comment lines must be preceded with an asterisk (*) while continued calculation statements must be preceded with a plus sign (+).

V $\cos(\omega t - 8.3627°)$, $e_b(t) = 4898.5$ V $\cos(\omega t - 128.3627°)$, $e_c(t) = 4898.5$ V $\cos(\omega t - 248.3627°)$.

The required DC voltage of the inverter is approximately governed by the relation [7] $V_{DC} = 2\sqrt{2}V_1/m$, where $V_1 = E = 4000$ V and the chosen modulation index $m = 0.8$ results in $V_{DC} = V_{bat} = 2\sqrt{2} \cdot 4000\,\text{V}/0.8 = 14\,140$ V.

The instantaneous Thévenin-adjusted currents are with the amplitude $\sqrt{2} \cdot 169.85 = 240.17$ A: $i_a(t) = 240.17\,\text{A}\cos(\omega t - 22.399°)$, $i_b(t) = 240.17\,\text{A}\cos(\omega t - 142.399°)$, $i_c(t) = 240.17\,\text{A}\cos(\omega t - 262.399°)$.

The output power and torque equations are given in Application Example 12.11: they are for this machine $P_{in} = 3 \cdot 169.85\,\text{A}\angle -22.399° \cdot 3896\,\text{V}\angle 0.446° = 1985.2\cos(-21.953°) = 1841.3$ kW and developed torque $\tau_d = 9183.2$ Nm. The ohmic losses are approximately (neglecting the losses in the magnetizing branch) $P_{loss\Omega} = 3(R_{TH} + R'_r) \cdot \left|\vec{I}'_r\right|^2 = 133.16$ kW. Estimating the iron core loss as 33% of the ohmic losses $P_{loss\_core} = (1/3) \cdot P_{loss\Omega} = 44.387$ kW, resulting in the total losses of $P_{loss\_total} = 177.55$ kW. The output power of the induction motor including losses is then $P_{out\_based\ on\ loss} = 1663.8$ kW, and the efficiency of the induction machine $\eta_{induction\_machine} = (P_{in} - P_{loss\_total})/P_{in} = 0.904$ pu. The efficiency of the inverter including the waveshaping inductor losses is estimated to be $\eta_{inverter} = 0.90$ pu. The overall efficiency of the circuit of Figure E10.29.2 is therefore $\eta_{inverter\,+\,motor} = 0.904 \cdot 0.90 = 0.814$ pu.

Figures E10.29.3–E10.29.6 depict the simulation results. The corresponding PSPICE program is listed in Table E10.29.1.

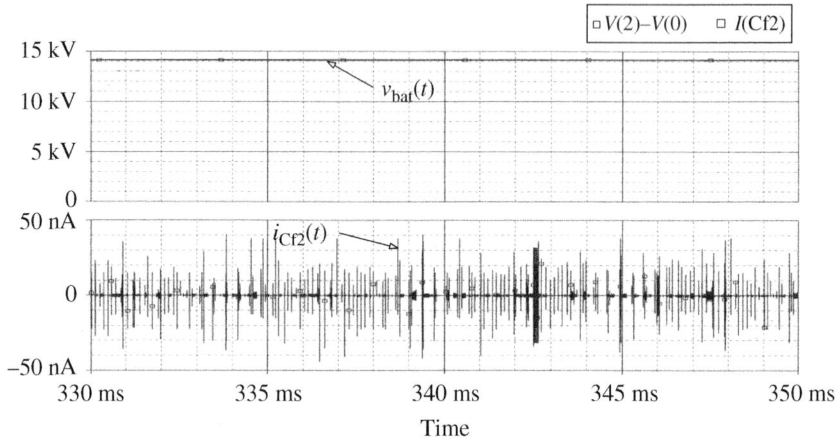

Figure E10.29.3 Top plot: input DC voltage to inverter $v_{out}(t) = V_{bat}(t) \approx V_{bat} = V(2) - V(0)$. Bottom plot: harmonic current through Cf2 input capacitor $i_{Cf2}(t) = I(Cf2)$.

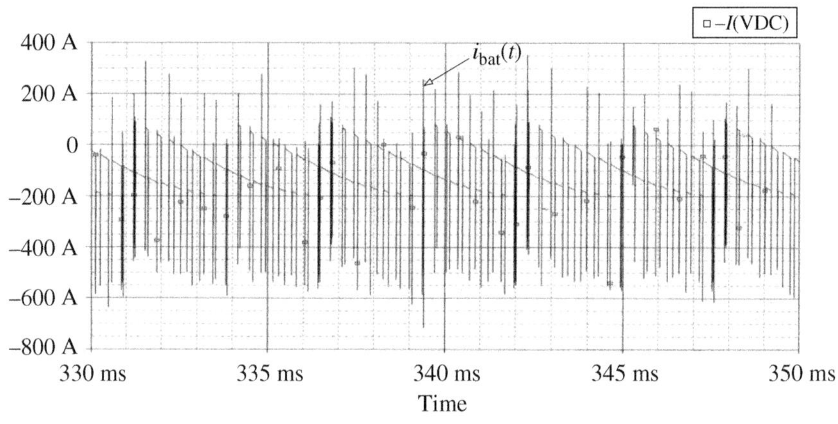

Figure E10.29.4 Harmonic current through DC input source of inverter $i_{bat}(t) = I(VDC)$.

Summary of inverter operation: The output power of the induction motor based on losses is $P_{out\_based\ on\ loss} = 1663.8$ kW. From Figure E10.29.6, one obtains a phase shift between reference current $i_{a\_reference}(t) = v_{ref1}(t)$ and calculated current $i_a(t) = I(RTH1)$ of 17.65° at a fundamental amplitude (see Table E10.29.2) of 197.7 A. Because the reference current $v_{ref1}(t)$ lags 14.036° (see Table E10.29.1), the output voltage $v_{out1}(t)$, and the actual current $i_a(t)$ lag the reference current $v_{ref1}(t)$

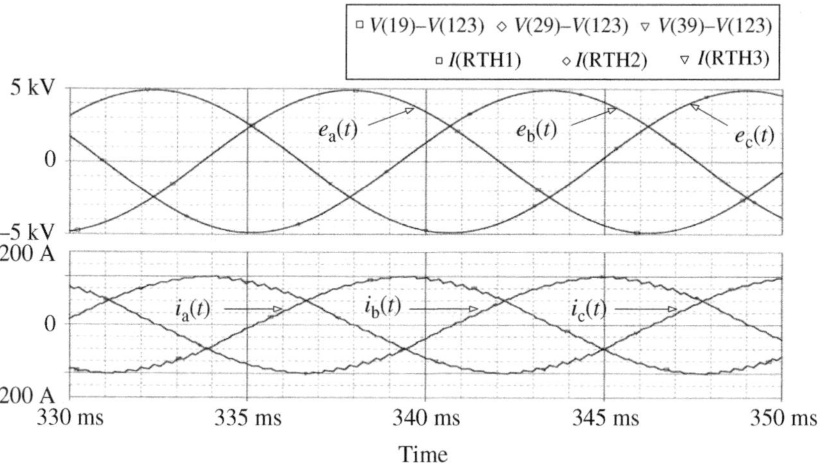

Figure E10.29.5 Top plot: induced voltages of induction machine $e_a(t) = v_{out1}(t) = V(19) - V(123)$, $e_b(t) = v_{out2}(t) = V(29) - V(123)$, $e_c(t) = v_{out3}(t) = V(39) - V(123)$. Bottom plot: calculated induction motor currents $i_a(t) = I(RTH1)$, $i_b(t) = I(RTH2)$, $i_c(t) = I(RTH3)$.

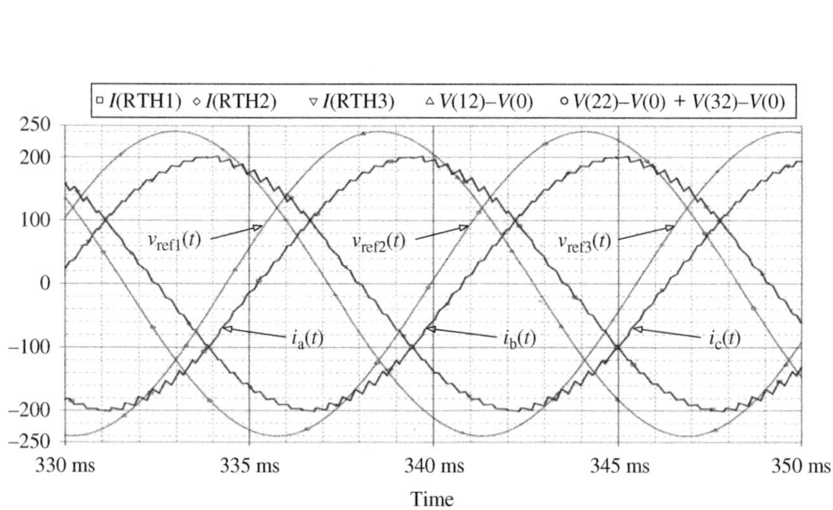

Figure E10.29.6 Comparison of reference currents $v_{ref1}(t) = V(12) - V(0)$, $v_{ref2}(t) = V(22) - V(0)$, $v_{ref3}(t) = V(32) - V(0)$ (note that in the PSPICE program the reference currents are defined as voltages) and calculated currents $i_a(t) = I(RTH1)$, $i_b(t) = I(RTH2)$, $i_c(t) = I(RTH3)$.

Table E10.29.2 Fundamental component (1) and harmonics of $i_a(t)$ = I(RTH1) = I(R_w1).

DC component	1	5	7	11
0.11 A	197.7 A	0.90 A	1.83 A	0.34 A

by 18.72° (see Figure E10.29.6), the output power based on PSPICE-calculated currents is $P_{out\_based\ on\ PSPICE} = 3 \cdot (197.7\ A/\sqrt{2})4000\ V \cdot \cos(-14.036° - 18.72°) = 3 \cdot (197.7\ A/\sqrt{2}) \cdot 4000\ V \cdot 0.841 = 1411.0\ kW$. The difference between $P_{out\_based\ on\ loss}$ and $P_{out\_based\ on\ PSPICE}$ is due to the proportional (P) control of the current introducing a phase shift of 18.72°; an integral (I) control would make at steady state the calculated current identical with the reference one. Note that the induction motor currents of Figure E10.29.6 are practically sinusoidal and contain a small amount of switching harmonic of $f_{switching} = 2.88$ kHz. Also, the input filter capacitors C_{f1} and C_{f2} absorb hardly (about 50 nanoampere; see Figure E10.29.3) any switching current harmonics, because the DC voltage source absorbs (see Figure E10.29.4) more than 500 A at 2.88 kHz current. The reason for this harmonic current distribution is that a DC source represents and behaves like a very large capacitor; thus the input filter (C_{f1}, C_{f2}) and the three output filters (R_F, L_F, C_F) are not needed, unless the filter design is changed. The pulse magnitude of the gating circuit (e.g. 50 V) and the switch model (e.g. $w = 2.9$) must be adjusted depending upon the magnitude of the current.

Rated operation of three-phase induction motor drive train at $f_1 = 16.667$ Hz, PSPICE analysis based on Figure E10.29.1: The high overhead line voltage of $V_{syst\_rms} = 15$ kV at 16.667 Hz of the German railroad system causes power quality problems that are severe unless proper mitigation techniques such as power factor correction and harmonic cancellations through filters are employed. To be flexible from an input voltage point of view, the input single-phase transformer must have variable turns ratio N_1/N_2, and the PWM switch (MOSFET) must be operated with a duty ratio $\delta = 50\%$ at a switching frequency of 600 Hz. If a larger unipolar (DC) voltage is required, then $\delta > 50\%$, and if a smaller one is necessary, then $\delta < 50\%$ can be selected. This enables locomotives to be supplied with variable AC voltage with frequencies ranging from 16.667 to 60 Hz or larger. It appears that the 16.667 Hz operation causes more severe power quality issues than 50 or 60 Hz. The PSPICE program for Figures E10.29.7–E10.29.11 is listed in Table E10.29.3, and the fundamental and harmonic currents of $i_{syst}(t)$ are detailed in Table E10.29.4. Table E10.29.5 presents the DC component and harmonics for $V_{out}(t)$.

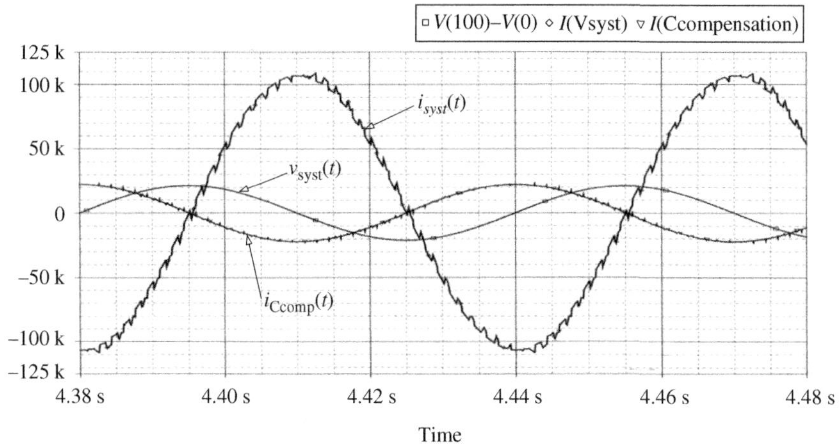

Figure E10.29.7 Input AC voltage to primary of step-up transformer $v_{syst}(t) = V(100) - V(0)$, input AC current $i_{syst}(t) = I(\text{Vsyst})$, and $i_{Ccomp}(t) = I(\text{Ccompensation})$.

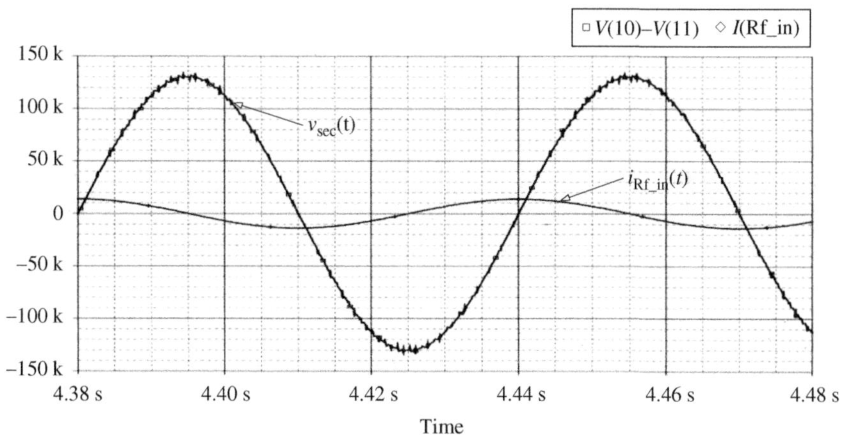

Figure E10.29.8 Secondary transformer voltage $v_{sec}(t) = V(10) - V(11)$ and current through input (R_{f_in} L_{f_in} C_{f_in}) filter $i_{Rf\_in}(t) = I(\text{Rf\_in})$.

Summary of rectifier operation: The DC output voltage is $V_{out} = 15.0918$ kV at a load resistance of $R_{load} = 150\ \Omega$, resulting in a DC current of $I_{DC} = V_{DC}/R_{load} = 15.0918\ \text{kV}/150\ \Omega = 100.61$ A or an output power of $P_{DC} = 15091.8\ \text{V} \cdot 100.61\ \text{A} = 1.52$ MW, which is larger than the required inverter output power of 1.284 MW; this will take into account the inverter losses.

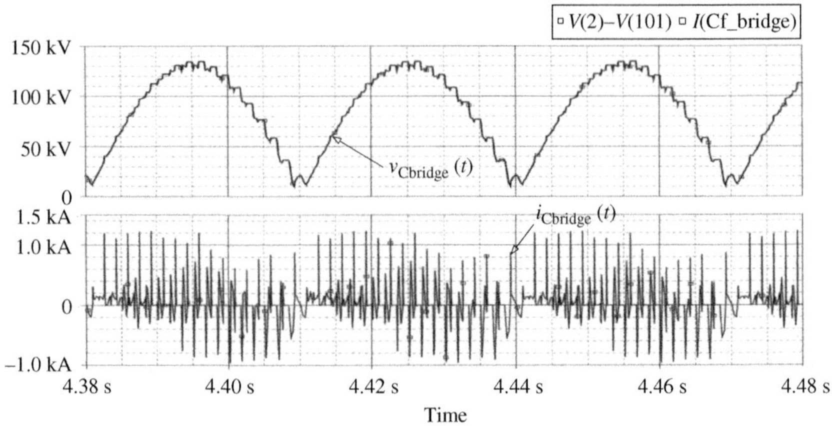

Figure E10.29.9 Top plot: unipolar output voltage of diode bridge $v_{Cbridge}(t) = V(2) - V(101)$. Bottom plot: harmonic current through bridge capacitor $i_{Cbridge}(t) = I(Cf\_bridge)$.

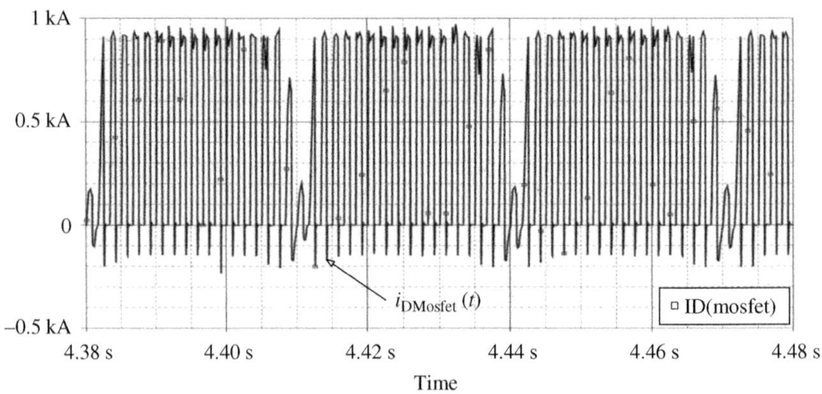

Figure E10.29.10 Current through MOSFET switch $i_{DMosfet}(t)$ = ID(mosfet) operating at a duty ratio or duty cycle of $\delta = 50\%$.

Conclusions

The single-phase power system approach, required by railroad applications due to the single overhead wire, results in large harmonics that must be mitigated by various filters. The power factor is very low if only a few trains are served at any time. This is due to the 15 kV line voltage and the relatively low current, which is in this case 102.95 A for one drive train. If a large number (e.g. 800) of trains are served at

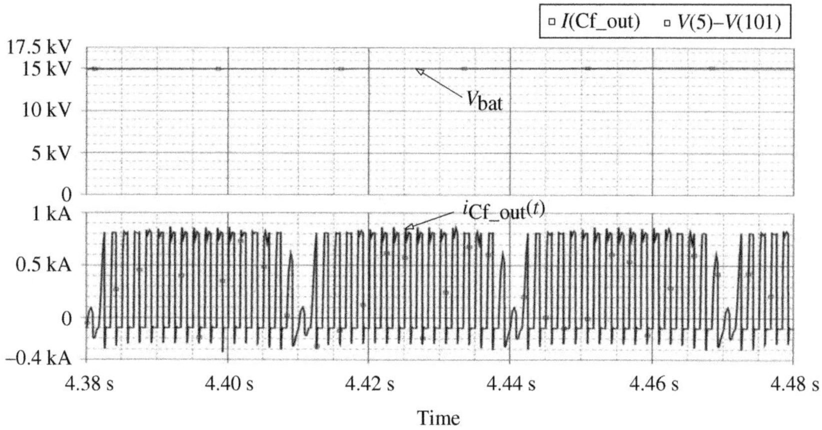

Figure E10.29.11 Top plot: unipolar (or DC) output voltage $v_{out}(t) \approx V_{bat} = V(5) - V(101)$. Bottom plot: output filter current $i_{Cf\_out}(t) = I(Cf\_out)$.

Table E10.29.3 PSPICE program for the signals of Figures E10.29.7–E10.29.11.

```
Application Example 10.29b
* AC source
Vsyst 100 0 sin(0V 21210V 16.6667Hz)
*Transformer
Ccompensation 1 0 10m
L1 1 0 20
L2 10 11 750
K12 L1 L2 0.9999999
Rbreak 1 100 1m
Rmegohm1 1 10 10000
*Diode bridge with MOSFET (PWM) voltage
*control via duty cycle delta=50%
*Rload=150ohms, Ccompensation=10mF,
*Cf_in=1mF, Rf_in=1m ohms, Lf_in=1mH,
*Lf_out=10mH, Cf_out=100mF,
*fswitch=600Hz, Vout=15.442kV,
Pout=1.59MW
D1 10 2 IDEAL
D3 11 2 IDEAL
D2 101 11 IDEAL
D4 101 10 IDEAL
vmosfet 3 4 pulse(0 50 0 0.9u 0.9u 0.8333ms
+1.6667ms)
mosfet 2 3 4 4 qfet
*Filters
Rf_in 10 12 1m
Lf_in 12 13 1m
Cf_in 13 11 1m
Cf_bridge 2 101 10u
Lf_out 4 5 10m
Cf_out 5 101 100m
*Equivalent load resistance
Rload 5 101 150
 *Model
.model IDEAL D(IS=1.0E-12)
.model qfet NMOS(level 3 gamma=0
delta=0
+eta=0 theta=0 kappa=0.2 vmax=0
xj=0
+tox=100n uo=600 phi=0.6 rs=42.69m
+kp=20.87u l=2u w=2.9
vto=0.3rd=0.19
+cbd=200n pb=0.8 mj=0.5 cgso=3.5n
+cgdo=3.5n rg=0.2 is=10f)
*Analysis request
.Tran 1ms 4.48s 4.38s 1ms
*Output request
.Probe
.FOUR 16.667 12 V(Rload), I(Vsyst)
.options abstol=100m chgtol=100m
reltol=100m
+vntol=100m itl5=0 itl4=20000,
itl1=200 itl2=200
.end
```

Note: All comment lines must be preceded with an asterisk (*) while continued calculation statements must be preceded with a plus sign (+).

Table E10.29.4 Fundamental component (1) and harmonics of $i_{syst}(t) = I(\text{Vsyst})$.

DC component	1	3	4	7	8	10	11	12
18.25 A	107 000 A	818.3 A	229.9 A	247.6 A	541.8 A	192.3 A	106.9 A	108.2 A

Table E10.29.5 DC component (1) and harmonics of $V_{out}(t) = V(\text{Rload}) = V(5) - V(101)$.

DC component	1	2	3	4	5	6	7	8
15 091.8 V	52.47 V	25.08 V	17.51 V	11.38 V	10.54 V	9.56 V	7.7 V	6.79 V

any one time from the same supply system, then the power factor improves due to the 800 drive trains involved requiring 82 360 A. The magnitude of the voltage gating pulse of the switch (e.g. 50 V) and the switch model (e.g. $w = 500$ u) must be adjusted depending upon the magnitude of the current.

10.11 Summary

In Chapter 9 the circuits of most important semiconductor switches (e.g. diode, phase-controlled thyristor and triac, MOSFET, IGBT) are presented without justification. Understanding the equivalent circuits of such switches is beyond the scope of an introductory course, and thus they are not discussed. However, to use such devices within electric circuits that are important for renewable energy systems, many application examples are solved with the software PSPICE. The student/engineer learns by example, starting out with very simple diode circuits, progressing to thyristor and triac circuits – which can be turned on at will, but must be turned off by the surrounding network – to MOSFETs, which can be turned on and off at will.

Very few semiconductor circuits applied to renewable energy are amenable to closed-form solutions at transient and even at steady-state operation. The reason is that semiconductor devices represent nonlinear networks and have parasitic capacitances that significantly influence the performance of electronic circuits (e.g. the reverse-recovery currents of diodes and other switches). Instead of solving a system of ordinary nonlinear differential equations based on closed-form solution techniques, it was decided to solve (for transient and steady-state conditions) these systems of differential equations using PSPICE, a computer-aided design approach. Starting with very simple diode rectifiers, voltage regulators and limiters, and clippers and clamps, application examples illustrate the use of thyristors, triacs, and MOSFETs as they are applied to converter (e.g. rectifiers and inverters) circuits including brushless DC machines, PV and wind power plants, and drive trains for electric/hybrid cars and rail applications. The increase of the efficiency of induction motors with thyristor/triac controllers and important topics of power quality in railway electrification, as well as IEEE power system requirements, conclude this chapter.

Problems

10.1 Design of a small (30 hp) brushless DC machine drive train for a hybrid automobile at low output power and low speed based on Application Example 10.25.

For a battery voltage of $V_{bat} = 400$ V a fundamental switching frequency of $f_1 = 60$ Hz resulting in $\omega = 2\pi f_1 = 367.98$ rad/s, the induced stator voltages are $e_A = 190\,\text{V}\sin(\omega t + \theta)$, $e_B = 190\,\text{V}\sin(\omega t + 120° + \theta)$, $e_C = 190\,\text{V}\sin(\omega t + 240° + \theta)$, the stator phase resistance $R_1 = 0.5\ \Omega$, the stator leakage inductance $L_1 = L_{1\ell} = 60\ \mu\text{H}$, gating voltage $V_{Gmax} = 15$ V, phase angle $\theta = 30°$, $C_1 = C = 100\ \mu\text{F}$, $C_2 = C = 100\ \mu\text{F}$. Using PSPICE, compute and plot the current within MOSFET Q_{AU}, e.g. $i_{QAU}(t)$ and the motor current of A phase $i_{MA}(t)$ as well as the stator voltage $e_A(t)$ of Figure E10.25.1. What is the output power $P_{out} = 3(190/\sqrt{2}) \cdot i_{MAmax\_avg}$? Note that 1 horsepower ≈ 746 watts.

10.2 Design of a (100 hp) brushless DC machine drive train for an electric automobile at rated output power and rated speed based on Application Example 10.25.

For a battery voltage of $V_{bat} = 600$ V and a fundamental switching frequency of $f_1 = 500$ Hz resulting in $\omega = 2\pi f_1 = 3141.5$ rad/s, the induced stator voltages are $e_A = 190\,\text{V}\sin(\omega t + \theta)$, $e_B = 190\,\text{V}\sin(\omega t + 120° + \theta)$, $e_C = 190\,\text{V}\sin(\omega t + 240° + \theta)$, the stator phase resistance $R_1 = 0.3\ \Omega$, the stator leakage inductance $L_1 = L_{1\ell} = 30\ \mu\text{H}$, gating voltage $V_{Gmax} = 15$ V, phase angle $\theta = 0°$. $C_1 = C = 200\ \mu\text{F}$, $C_2 = C = 200\ \mu\text{F}$. Using PSPICE, compute and plot the current within MOSFET Q_{AU}, e.g. $i_{QAU}(t)$ and the motor current of phase A $i_{MA}(t)$, as well as the stator voltage $e_A(t)$ of Figure E10.25.1.

What is the output power $P_{out} = 3(190/\sqrt{2}) \cdot i_{MAmax\_avg}$? Note that 1 horsepower ≈ 746 watts.

10.3 Design of a large (100 hp) brushless DC machine drive train for an electric automobile at rated power and high speed based on Application Example 10.25 with flux weakening.

For a battery voltage of $V_{bat} = 600$ V and a fundamental switching frequency of $f_1 = 1500$ Hz resulting in $\omega = 2\pi f_1 = 9424.5$ rad/s, the induced stator voltages are $e_A = 190\text{ V}\sin(\omega t + \theta)$, $e_B = 190\text{ V}\sin(\omega t + 120^\circ + \theta)$, $e_C = 190\text{ V}\sin(\omega t + 240^\circ + \theta)$, the stator phase resistance $R_1 = 0.3\ \Omega$, the stator leakage inductance $L_1 = L_{1\ell} = 30\ \mu$H, gating voltage $V_{Gmax} = 15$ V, phase angle $\theta = 0^\circ$. $C_1 = C = 200\ \mu$F, $C_2 = C = 200\ \mu$F. Using PSPICE, compute and plot the current within MOSFET Q_{AU}, e.g. $i_{QAU}(t)$ and the motor current of phase A or $i_{MA}(t)$ of Figure E10.25.1. What is the output power $P_{out} = 3(190/\sqrt{2}) \cdot i_{MAmax\_avg}$? Note that 1 horsepower ≈ 746 watts.

10.4 Design of a large (100 hp) brushless DC machine drive train for an electric automobile at rated power and very high speed based on Application Example 10.25 with flux weakening compensation [14, 15].

For a battery voltage of $V_{bat} = 700$ V and a fundamental frequency of $f_1 = 3000$ Hz resulting in $\omega = 2\pi f_1 = 18\,849$ rad/s, the induced stator voltages are $e_A = 190\text{ V}\sin(\omega t + \theta)$, $e_B = 190\text{ V}\sin(\omega t + 120^\circ + \theta)$, $e_C = 190\text{ V}\sin(\omega t + 240^\circ + \theta)$, the stator phase resistance $R_1 = 0.15\ \Omega$, the stator leakage inductance $L_1 = L_{1\ell} = 15\ \mu$H, gating voltage $V_{Gmax} = 15$ V, phase angle $\theta = 0^\circ$. $C_1 = C = 200\ \mu$F, $C_2 = C = 200\ \mu$F. Using PSPICE, compute and plot the current within MOSFET Q_{AU}, e.g. $i_{QAU}(t)$ and the motor current of phase A, e.g. $i_{MA}(t)$ of Figure E10.25.1. What is the output power $P_{out} = 3(190/\sqrt{2}) \cdot i_{MAmax\_avg}$? Note that 1 horsepower ≈ 746 watts.

10.5 Design of a storage battery supplying an AC peak (maximum) power $P_{peak} = 7$ kW into the three-phase power system with line-to-line rms voltage of $V_{L\text{-}L} = 230$ V based on Application Example 10.26.

The storage power plant consists of a battery and a full-wave, three-phase bridge MOSFET current-controlled voltage source inverter [17]. The per-phase rms voltage is $V_{ph} = 230\text{ V}/1.732 = 132.8$ V at $f = 60$ Hz with an amplitude of $132.8\sqrt{2}$ V $= 187.78$ V. The per-phase system impedance of the power system (sys) is $Z_{sys} = (R_{sys} + j\omega L_{sys})$ with $R_{sys} = 0.30\ \Omega = R_M$ and $L_{sys} = 2$ mH $= L_M$, where M stands for main power system and m for milli. The battery and the full-wave, three-phase current-controlled voltage source inverter supplying energy into the power system is presented in Figure E10.26.1. The displacement fundamental power factor is cos θ = 0.90 lagging, where the phase current lags the phase voltage by $\theta = -28.713°$ (that is, consumer definition of current; reactive power Q is absorbed from the grid). The waveshaping (w) inductors have the impedance $\tilde{Z}_w = (R_w + j\omega L_w)$, where $R_w = 10\ \text{m}\Omega$ and $L_w = 0.10$ mH. The battery (bat) voltage is $V_{bat} = 600$ V. Perform a PSPICE analysis to compute and plot battery current = input current to inverter = $i_{in}(t)$ and power system currents $i_{an}(t)$, $i_{bn}(t)$, $i_{cn}(t)$ at leading displacement power factor. Also plot power system voltages $v_{an}(t)$, $v_{bn}(t)$, $v_{cn}(t)$. The modulation index m for this case is defined by $V_{ph} = mV_{bat}/(2\sqrt{2})$ or $m = V_{ph} \cdot 2\sqrt{2}/V_{bat} = 132.8\ \text{V} \cdot 2 \cdot \sqrt{2}/(600\ \text{V}) = 0.626$, which is smaller than 1.0, therefore acceptable for lagging power factor operation, and the resulting current waves should be sinusoidal/cosinusoidal. Plot the voltages and currents V_{bat}, $i_{in}(t)$, $v_{ref1}(t)$, $v_{ref2}(t)$, $v_{ref3}(t)$, $i_{an}(t)$, $i_{bn}(t)$, $i_{cn}(t)$, $v_{an}(t)$, $v_{bn}(t)$, $v_{cn}(t)$.

10.6 Design of a storage battery supplying an AC peak power $P_{peak} = 1080$ kW at $f = 60$ Hz into the three-phase power system with line-to-line rms voltage of $V_{L\text{-}L} = 2400$ V based on Application Example 10.26.

The storage power plant consists of a battery and a full-wave, three-phase bridge MOSFET current-controlled voltage source inverter [17]. The per-phase rms voltage is $V_{ph} = 2400\ \text{V}/1.732 = 1385.7$ V at $f = 60$ Hz with an amplitude of $1385.7\sqrt{2}\ \text{V} = 1959.4$ V. The per-phase system impedance of the power system (sys) is $\tilde{Z}_{sys} = (R_{sys} + j\omega L_{sys})$ with $R_{sys} = 0.50\ \Omega = R_M$ and $L_{sys} = 4\ \text{mH} = L_M$, where M stands for main power system and m for milli. The battery and the full-wave, three-phase current-controlled voltage source inverter supplying energy into the power system is presented in Figure E10.26.1. The displacement fundamental power factor is cos θ = 0.80 leading, where the phase current leads the phase voltage by $\theta = 36.87°$ (that is, consumer definition of current; reactive power Q is supplied to grid). The waveshaping (w) inductors have the impedance $\tilde{Z}_w = (R_w + j\omega L_w)$, where $R_w = 10\ \text{m}\Omega$ and $L_w = 0.10$ mH. The battery (bat) voltage is $V_{bat} = 6000$ V. Perform a PSPICE analysis to compute and plot battery current = input current to inverter = $i_{in}(t)$ and power system currents $i_{an}(t)$, $i_{bn}(t)$, $i_{cn}(t)$ at leading displacement power factor. Also plot power system voltages $v_{an}(t)$, $v_{bn}(t)$, $v_{cn}(t)$. The modulation index m for this case is defined by $V_{ph} = mV_{bat}/(2\sqrt{2})$ or $m = V_{ph} \cdot 2\sqrt{2}/V_{bat} = 1385.7\ \text{V} \cdot 2 \cdot \sqrt{2}/(6000\ \text{V}) = 0.653$, which is smaller than 1.0 and therefore acceptable for leading power factor operation, where V_{bat} must be relatively large as compared to either unity or lagging power factors: the resulting current waves should therefore be sinusoidal/cosinusoidal. Plot the voltages and currents V_{bat}, $i_{in}(t)$, $v_{ref1}(t)$, $v_{ref2}(t)$, $v_{ref3}(t)$, $i_{an}(t)$, $i_{bn}(t)$, $i_{cn}(t)$, $v_{an}(t)$, $v_{bn}(t)$, $v_{cn}(t)$.

10.7 Design of a photovoltaic power plant supplying $P_{peak} = 900$ kW at $f = 60$ Hz AC power into the three-phase power system based on Application Example 10.27 with small waveshaping impedance $\tilde{Z}_w = (R_w + j\omega L_w)$ and inductances $L_w = L_{wa} = L_{wb} = L_{wc} = 0.01$ mH.

Assuming a storage battery voltage of $V_{bat} = 4500$ V, then one obtains for $V_{MPP} = 2.181$ V of Figure E10.27.2 $N_s = 4500\text{ V}/(2.181\text{ V}) \approx 2063$ solar cells in series and for $I_{MPP} = 0.07$ A of Figure E10.27.2 $N_p = 200\text{ A}/(0.07\text{ A}) \approx 286$ solar cells in parallel. The losses in the inverter are neglected. The power plant consists of the PV array and a full-wave, three-phase bridge MOSFET current-controlled voltage source inverter [17]. The per-phase rms voltage is $V_{ph} = 2400\text{ V}/1.732 = 1386$ V, and the amplitude of the phase voltage is $\sqrt{2} \cdot V_{ph} = \sqrt{2} \cdot 1386\text{ V} = 1959.8$ V. The per-phase system impedance is $\tilde{Z}_{sys} = (R_{sys} + j\omega L_{sys})$ with $R_{sys} = 0.60\ \Omega = R_M$ and L_{sys}=1 mH $= L_M$, where M stands for main power system and m for milli. The battery representing the solar array and the full-wave, three-phase current-controlled voltage source inverter supplying energy into the power system is presented in Figure E10.26.1. The displacement power factor is assumed to be unity, that is, $\cos\theta = 1.0$ or $\theta = 0^\circ$, where the line-to neutral voltage and line current are in phase. The current fed into the grid is controlled, but the voltage of the grid cannot be changed by such current-controlled voltage source inverters. The waveshaping inductors have the impedance $\tilde{Z}_w = (R_w + j\omega L_w)$, where $R_w = 10$ mΩ and $L_w = 0.01$ mH.

Perform a PSPICE analysis to compute and plot DC current = input current to inverter = $i_{in}(t)$ and power system currents $i_{an}(t)$, $i_{bn}(t)$, $i_{cn}(t)$ at unity power factor. Also plot power system voltages $v_{an}(t)$, $v_{bn}(t)$, $v_{cn}(t)$. The modulation index m for this case is defined by $V_{ph} = mV_{bat}/(2\sqrt{2})$ or $m = V_{ph} \cdot 2\sqrt{2}/V_{bat} = 1386\text{ V} \cdot 2 \cdot \sqrt{2}/(4500\text{ V}) \approx 0.87$, which is acceptable: the resulting current waves should be therefore approximately sinusoidal/cosinusoidal with a large harmonic content of the AC currents.

10.8 Design of a PV power plant supplying $P_{peak} = 900$ kW at $f = 60$ Hz AC power into the three-phase power system based on Application Example 10.27, with large waveshaping impedance $\tilde{Z}_w = (R_w + j\omega L_w)$ and inductances $L_w = L_{wa} = L_{wb} = L_{wc} = 2$ mH.

Assuming a storage battery voltage of $V_{bat} = 4500$ V, then one obtains for $V_{MPP} = 2.181$ V of Figure E10.27.2 $N_s = 4500$ V/(2.181 V) ≈ 2063 solar cells in series and for $I_{MPP} = 0.07$ A of Figure E10.27.2 $N_p = 200$ A/(0.07 A) ≈ 286 solar cells in parallel. The losses in the inverter are neglected. The power plant consists of a PV array and a full-wave, three-phase bridge MOSFET current-controlled voltage source inverter [17]. The per-phase rms voltage is $V_{ph} = 2400\,\text{V}/1.732 = 1386$ V, and the per-phase system impedance is $\tilde{Z}_{sys} = (R_{sys} + j\omega L_{sys})$ with $R_{sys} = 0.60\,\Omega = R_M$ and $L_{sys} = 1\,\text{mH} = L_M$, where M stands for main power system and m for milli. The battery representing the solar array and the full-wave, three-phase current-controlled voltage source inverter supplying energy into the power system is presented in Figure E10.26.1. The displacement power factor is assumed to be unity, that is, $\cos\theta = 1.0$ or $\theta = 0°$, where the line-to-neutral voltage and line current are in phase. The current fed into the grid is controlled, but the voltage of the grid cannot be changed by such current-controlled voltage source inverters. The waveshaping inductors have the impedance $\tilde{Z}_w = (R_w + j\omega L_w)$, where $R_w = 10\,\text{m}\Omega$ and $L_w = 2$ mH.

Perform a PSPICE analysis to compute and plot DC current = input current to inverter = $i_{in}(t)$ and power system currents $i_{an}(t)$, $i_{bn}(t)$, $i_{cn}(t)$ at unity power factor. Also plot power system voltages $v_{an}(t)$, $v_{bn}(t)$, $v_{cn}(t)$. The modulation index m for this case is defined by $V_{ph} = mV_{bat}/(2\sqrt{2})$ or $m = V_{ph} \cdot 2\sqrt{2}/V_{bat} = 1386\,\text{V} \cdot 2 \cdot \sqrt{2}/(4500\,\text{V}) \approx 0.87$, which is acceptable: the resulting current waves should be therefore approximately sinusoidal/cosinusoidal with a small harmonic content of the AC currents.

10.9 Design of a wind power plant supplying $P_{out\_inverter} = 547$ kW AC power into the three-phase power system based on Application Example 10.28.

The design of this wind power plant proceeds from the power system and the inverter (Figure E10.28.4), delta/wye transformer, and rectifier (Figure E10.28.3) via permanent-magnet generator to the wind turbine (Figure E10.28.2).

The design data of the components of the wind power plant are:

a wind speed $v_w = 7.5$ m/s and wind-turbine efficiency coefficient $c_p = 0.44$

b three-phase permanent-magnet generator (PMG) with speed $n_m = 8$ rpm, frequency $f = 8$ Hz requiring [23] a $f = (8\,\text{rpm}/60) \cdot 120$ poles/2 = 8 Hz pole generator, and generator (gen) rms voltage $V_{L\text{-}N} = 3223.5$ V. $R_{gen} = 0.1\,\Omega$, and $L_{gen} = 100\,\mu\text{H}$

c three-phase Δ/(ungrounded Y) ideal step-up transformer with turns ratio $(N_2/N_1) = 1.7$

d diode bridge rectifier with filter $L_f = 100$ mH, and $C_f = 1$ mF

e current-controlled voltage source inverter supplying energy to the power system with a line-to-line rms voltage of $V_{ab} = 8320$ V or line-to-neutral voltage $v_{an}(t) = \sqrt{2}(8320/\sqrt{3})$V cos ωt = 6792.4 V cos ωt, R_{sys}=50 mΩ, and L_{sys}=0.265 mH at a DC input voltage of $V_{DC} = 16\,981$ V.

1 Perform a PSPICE analysis for the circuit of Figure E10.28.4 where a three-phase MOSFET current-controlled voltage source inverter supplies power $P_{out\_inverter} = 547$ kW into the three-phase power system. The power system impedance per phase is $R_{sys} = 50$ mΩ, $L_{sys} = 265$ μH at f = 60 Hz. The line-to-neutral voltages of the power system are $v_{an}(t) = \sqrt{2} \cdot (V_{an})$ cos $\omega t = \sqrt{2} \cdot (8320/\sqrt{3})$V cos $\omega t = \sqrt{2} \cdot 4803.7$ V cos ωt = 6792.4 V cos ωt, $v_{bn}(t) = \sqrt{2} \cdot (8320/\sqrt{3})$V cos($\omega t$ – 120°) = 6792.4 V cos(ωt – 120°), $v_{cn}(t) = \sqrt{2} \cdot (8320/\sqrt{3})$V cos($\omega t$ – 240°) = 6792.4 V cos(ωt – 240°), and the modulation index is $m = 0.8$. The displacement power factor (consumer notation) is cos $\theta = 0.8$ lagging or $\theta = -36.87°$. With $V_{an} = m \cdot V_{DC}/(2\sqrt{2})$ follows for the battery voltage $V_{bat} = V_{DC} = (2\sqrt{2}) \cdot V_{an}/m = (2\sqrt{2}) \cdot (8320\text{ V}/\sqrt{3})/0.8 = 16\,981$ V, where the modulation index is $m = 0.8$. Waveshaping (w) inductances and resistances $L_w = L_{wa} = L_{wb} = L_{wc} = 50$ mH and $R_{wa} = R_{wb} = R_{wc} = 10$ mΩ, respectively.

2 Plot and subject at steady-state operation line-to-neutral voltages $v_{an}(t)$, $v_{bn}(t)$, $v_{cn}(t)$, line currents $i_{an}(t)$, $i_{bn}(t)$, $i_{cn}(t)$, and required rectifier output unipolar voltage $v_{bat}(t)$ and $i_{bat}(t)$ to Fourier analyses (up to the 12th harmonic).

3 Plot the reference line currents $i_{anref}(t)$, $i_{bnref}(t)$, $i_{cnref}(t)$ and the actual line currents $i_{an}(t)$, $i_{bn}(t)$, $i_{cn}(t)$.

4 Perform a PSPICE analysis for the circuit of Figures E10.28.3 and E10.28.2, where a three-phase delta/wye transformer and a three-phase diode bridge rectifier with filter $C_f = 1$ mF, $L_f = 100$ mH serves the equivalent load $R_{load} = V_{bat}/I_{bat} = 16\,981/35.4556 = 478.94$ Ω so that $V_{bat} = 16\,981$ V, where $I_{bat} = (-)35.4556$ A is obtained as a result of evaluating Table P10.9.1. The generator impedance per phase is $R_{gen} = 0.10$ Ω, $L_{gen} = 0.100$ mH, and assume ideal diodes D_1–D_6.

5 Plot and subject at steady-state operation the transformer secondary line-to-line voltages $v_{ab}(t)$, $v_{bc}(t)$, $v_{ca}(t)$, transformer secondary line currents $i_{aph}(t)$, $i_{bph}(t)$, $i_{cph}(t)$, to Fourier analyses (up to the 12th harmonic).

6 Plot and subject at steady-state operation the transformer primary line-to line voltages $v_{AB}(t)$, $v_{BC}(t)$, $v_{CA}(t)$, transformer primary input line currents $i_{genA}(t)$, $i_{genB}(t)$, $i_{genC}(t)$ to Fourier analyses (up to the 12th harmonic).

References

1 Sedra, A.S. and Smith, K.C. (1998). *Microelectronic Circuits*, 4e. New York: Oxford University Press.

2 Roberts, G.W. and Sedra, A.S. (1996). *SPICE*, 2e. New York: Oxford University Press.

3 Tuinenga, P.W. (1992). *SPICE, A Guide to Circuit Simulation & Analysis Using PSPICE*, 2e. Englewood Cliffs, NJ: Prentice Hall.

4 Schivon, M. (2016). OrCAD Lite 17.2-2016 Capture+PSpice, FlowCAD EDA-Software Vertriebs GmbH, Feldkirchen, Germany.

5 Rashid, M.H. (1992). *SPICE for Power Electronics and Electric Power*. Englewood Cliffs, NJ: Prentice Hall.

6 Yang, W.Y., Kim, J., Park, K.W. et al. (2020). Appendix G: standard values of resistors, capacitors, and inductors. In: *Electronic Circuits with MATLAB®, PSpice®, and Smith Chart*. Wiley https://onlinelibrary.wiley.com/doi/pdf/10.1002/9781119598961.app7 (accessed 20 May 2020).

7 Fuchs, E.F. and Masoum, M.A.S. (2012). *Power Conversion of Renewable Energy Systems*, 1e, 2nd Printing. USA: Springer, 692 pp.

8 Masoum, M.A.S. and Fuchs, E.F. (2015). *Power Quality in Power Systems and Electrical Machines*, 2e. London: Elsevier/Academic Press, 1123 pp.

9 Huang, H. and Fuchs, E.F. (1990). Successive nonlinear quadratic programming for high-power density and high-efficiency, permanent-magnet motors. *Proceedings of the International Conference on Electrical Machines*, Cambridge, MA (12–15 August 1990).

10 Huang, H., Anderson, W.M., and Fuchs, E.F. (1990). High-power density and high-efficiency motors for electric vehicle applications. *Proceedings of the International Conference on Electrical Machines*, Cambridge, MA (12–15 August 1990).

11 Fardoun, A.A., Fuchs, E.F., and Huang, H. (1994). Modeling and simulation of an electronically commutated permanent-magnet machine drive system using SPICE. *IEEE Transactions on Industry Applications* **30** (4): 927–937.

12 Fuchs, E.F., Schraud, J., and Fuchs, F.S. (2008). Analysis of critical-speed increase of induction machines via winding reconfiguration with solid-state switches. *IEEE Transactions on Energy Conversion* **23** (3): 774–780.

13 Schraud, J., Fuchs, E.F., and Fuchs, H.A. (2008). Experimental verification of critical-speed increase of induction machines via winding reconfiguration with solid-state switches. *IEEE Transactions on Energy Conversion* **23** (2): 460–465.

14 Fuchs, E.F. (2012). Alternating current machine with increased torque above and below rated speed for hybrid/electric propulsion systems. US Patent No. 8,183,814 B2, 22 May 2012.

15 Fuchs, E.F. (2013). Alternating current machine with increased torque above and below rated speed for hybrid/electric propulsion systems. US Patent No. 8,436,569 B2, 7 May 2013.

16 Fuchs, E.F. and Rosenberg, L.T. (1974). Analysis of an alternator with two displaced stator windings. *IEEE Transactions on Power Apparatus and Systems* **PAS-93** (6): 1776–1786.

17 Teltsch, M. (1997). Adjustable-speed drive for an electric car with large rated torque at low rated speed. MS thesis. University of Colorado at Boulder.

18 Fuchs, E.F. (1992). Electronically commutated variable-speed, permanent-magnet drive train for wind power plants. *Power Engineering Review* **12** (12): 7–8.

19 Carlin, P.W., Fingersh, L.J., and Fuchs E.F. (1996). Preliminary results from the variable-speed, direct-drive generating system. *Presentation at the 15th ASME ETCE Wind Energy Symposium*, Houston, TX (29 January–1 February 1996) (Abstract).

20 Yildirim, D., Fuchs, E.F., and Batan, T. (1998). Test results of a 20 kW, direct-drive, variable-speed wind power plant. *Proceedings of the International Conference on Electrical Machines*, Istanbul, Turkey (2–4 September 1998), pp. 2039–2044.

21 Fardoun, A.A. (1994). A novel wind-power train: permanent-magnet with flux weakening feeding a controlled real/reactive power converter. PhD thesis. Department of Electrical and Computer Engineering, University of Colorado at Boulder.

22 Yildirim, D. (1999). Commissioning of 30 kVA variable-speed, direct–drive wind power plant. PhD thesis. Department of Electrical and Computer Engineering, University of Colorado at Boulder.

23 Fingersh, L.J. (1995). The testing of a low-speed, permanent-magnet generator for wind power applications. MS thesis. Department of Electrical Engineering, University of Colorado, Boulder, CO.

24 Harry Wirth, compiler (2020). Recent facts about photovoltaics in Germany. Karin Schneider, Press and Public Relations. Last update: 26 March 2020. https://www.ise.fraunhofer.de/content/dam/ise/en/documents/publications/studies/recent-facts-about-photovoltaics-in-germany.pdf (accessed 20 May 2020).

25 Nola, F. (1977). NASA Marshall Space Flight Center, Huntsville, AL. US-Patent-4,052,648, NASA Tech Brief MFS-23 280.

26 Fuller, J.F. and Hanna, W.J. (1982). *An Investigation of Motor Characteristics When Connected to a Nola Controller*. Final report. University of Colorado, Boulder, CO. DAAK 70-80-C-0136.

27 *Micro Energy Smart, Single-Phase Power Planner 115/230 V/104 042-20/10 A; Three-Phase Power Planner 230 V Line-to-Line/12 A Line Current* (1996). Manufactured and Marketed, Coast Energy Management, Inc., Tempe, AZ.

28 Fuchs, E.F. and Lin, D. (1996). *Testing of Power Planners at the University of Colorado*. Final report. University of Colorado, Boulder, CO.

29 Fuchs, E.F. and Hanna, W.J. (2002). Measured efficiency improvements of induction motors with thyristor/triac controllers. *IEEE Transactions on Energy Conversion* **17** (4): 437–444.

30 Fuchs, E.F. and Hanna, W.J. (2004). Closure of "measured efficiency improvements of induction motors with thyristor/triac controllers". *IEEE Transactions on Energy Conversion* **19** (3): 647–648.

31 Fuchs, E.F., Roesler, D.J., and Alashhab, F.S. (1987). Sensitivity of electrical appliances to harmonics and fractional harmonics of the power system's voltage, part I. *IEEE Transactions on Power Delivery* **TPWRD-2** (2): 437–444.

32 Fuchs, E.F., Roesler, D.J., and Kovacs, K.P. (1987). Sensitivity of electrical appliances to harmonics and fractional harmonics of the power system's voltage, part II. *IEEE Transactions on Power Delivery* **TPWRD-2** (2): 445–453.

33 Fuchs, E.F., Roesler, D.J., and Kovacs, K.P. (1986). Aging of electrical appliances due to harmonics of the power system's voltage. *IEEE Transactions on Power Delivery* **TPWRD-1** (3): 301–307.

34 Fuller, J.F., Fuchs, E.F., and Roesler, D.J. (1988). Influence of harmonics on power system distribution protection. *IEEE Transactions on Power Delivery* **TPWRD-3** (2): 546–554.

35 Fuller, J.F., Fuchs, E.F., and Roesler, D.J. (1992). Influence of harmonics on power distribution system protection. In: *Protective Relaying for Power Systems II* (ed. S.H. Horowitz), 29–37. Piscataway, NJ: IEEE Press.

36 Fuchs, E.F., Roesler, D.J., and Masoum, M.A.S. (2004). Are harmonic recommendations according to IEEE and IEC too restrictive? *IEEE Transactions on Power Delivery* **19** (4): 1775–1786.

37 Fuchs, E.F., Roesler, D.J., and Masoum, M.A.S. (2007). Closure of "are harmonic recommendations according to IEEE and IEC too restrictive?". *IEEE Transactions on Power Delivery* **22** (2): 1263–1264.

38 Fuchs, E.F. and Fuchs, W.L. (2011). Complementary control of intermittently operating renewable sources with short-and long-term storage plants. In: *Energy Storage* (ed. R. Carbone). London: InTech Open Access Publisher.

39 Fuchs, W.L. and Fuchs, E.F. (2012). Frequency variations of power system due to switching of renewable energy sources. *International Conference on Renewable Energies and Power Quality (ICREPQ'12)*, Santiago de Compostela (Spain) (28–30 March 2012), 6 pages.

40 Fuchs, W.L. and Fuchs, E.F. (2013). Frequency variations of power system due to switching of renewable energy sources. *Journal of Energy and Power Engineering* **7** (1): 1–9.

41 Masoum, M.A.S., Jafarian, A., Ladjevardi, M. et al. (2004). Fuzzy approach for optimal placement and sizing of capacitor banks in the presence of harmonics. *IEEE Transactions on Power Delivery* **19** (2): 822–829.

42 Masoum, M.A.S., Ladjevardi, M., Jafarian, A., and Fuchs, E.F. (2004). Optimal placement, replacement and sizing of capacitor banks in distorted distribution networks by genetic algorithms. *IEEE Transactions on Power Delivery* **19** (4): 1794–1801.

43 Masoum, M.A.S., Ladjevardi, M., Fuchs, E.F., and Grady, W.M. (2004). Application of local variations and maximum sensitivities selection for optimal placement of shunt capacitor banks under nonsinusoidal operating conditions. *International Journal of Electrical Power and Energy Systems* **26** (10): 761–769.

44 Ladjevardi, M., Masoum, M.A.S., and Fuchs, E.F. (2004). Optimal placement and sizing of fixed and switched capacitor banks under nonsinusoidal operating conditions. *IUST-International Journal of Engineering Science* **15** (4): 9–18.

45 Masoum, M.A.S., Jafarian, A., Ladjevardi, M. et al. (2005). Closure of "Fuzzy approach for optimal placement and sizing of capacitor banks in the presence of harmonics". *IEEE Transactions on Power Delivery* **20** (2): 1214–1216.

46 Mousavi, G.S.M., Adel Tabakhpour, L., Fuchs, E.F., and Al-Haddad, K. (2015). Power quality issues in railway electrification: a comprehensive perspective. *IEEE Transactions on Industrial Electronics* **62** (5): 3081–3090.

47 Tesla, N. (1888). *A New System of Alternate Current Motors and Transformers*. New York: American Institute of Electrical Engineers.

48 The IEEE Red Book (1976). *IEEE Recommended Practice for Electric Power Distribution for Industrial Plants*, Standard 141-1976. The Institute of Electrical and Electronic Engineers, Inc.

49 The IEEE Gray Book (1974). *IEEE Recommended Practice for Electric Power Systems in Commercial Buildings*, Standard 241-1974. The Institute of Electrical and Electronic Engineers, Inc.

50 The IEEE Brown Book (1980). *IEEE Recommended Practice for Power System Analysis*, Standard 399-1980. The Institute of Electrical and Electronic Engineers, Inc.

51 The IEEE Buff Book (1975). *IEEE Recommended Practice for Protection and Coordination of Industrial and Commercial Power Systems*, Standard 242-1975. The Institute of Electrical and Electronic Engineers, Inc.

11

DC Machines Serving as Role Models for AC Rotating Machine Operation and Electronic Converters

11.1 Introduction

Although direct-current (DC) machines (either motors or generators) are rarely used any longer, they have superior properties: conversion of DC to alternating current (AC) through mechanical commutators in motor operation and AC-to-DC conversion in generator operation with excellent control of electric (e) torque T_e and mechanical (m) angular velocity ω_m. In other words, DC machines are a precursor of electronic inverters converting DC to AC and electronic rectifiers changing AC to DC. The commutation process converting DC to AC and AC to DC and control of torque and speed can be applied to AC machines (Chapter 12) combined with electronic converters (Chapters 9 and 10) by converting DC to AC and vice versa. Figure 11.1 depicts a DC machine that can be used either as motor or as a generator. Intentionally a very old design [1] is illustrated, which has an open structure revealing many design details, as compared with modern DC machines.

To make the reading of this chapter – containing many abbreviations – easier, each abbreviation will be defined when used for the first time (e.g. "direct current [DC]").

Figure 11.2 shows the disassembled machine [1] of Figure 11.1 where most components are easily identifiable, such as rotor with ventilator, stator with terminal box, left- and right-hand-side end bells, brush holder, and other components.

A rotor lamination (iron) sheet is depicted in Figure 11.3 with rotor slots, which are in most cases rectangular in order to accommodate the rotor winding turns, as will be discussed next. The magnetic rotor teeth are then in most cases trapezoidal. Figure 11.4 shows a rectangular slot (a), which carries a coil consisting of winding turns covered by a nonmagnetic wedge to protect the winding within the slot. There are single-layer armature windings (b) and double-layer armature windings (c). An armature winding consists of a group of either concentrated or distributed coils, and a coil contains one or several turns. The term "armature winding"

Introduction to Energy, Renewable Energy, and Electrical Engineering: Essentials for Engineering Science (STEM) Professionals and Students, First Edition. Ewald F. Fuchs and Heidi A. Fuchs.

Companion website: www.wiley.com/go/fuchs/renewableenergyandelectricalengineering

Figure 11.1 Assembled DC machine [1].

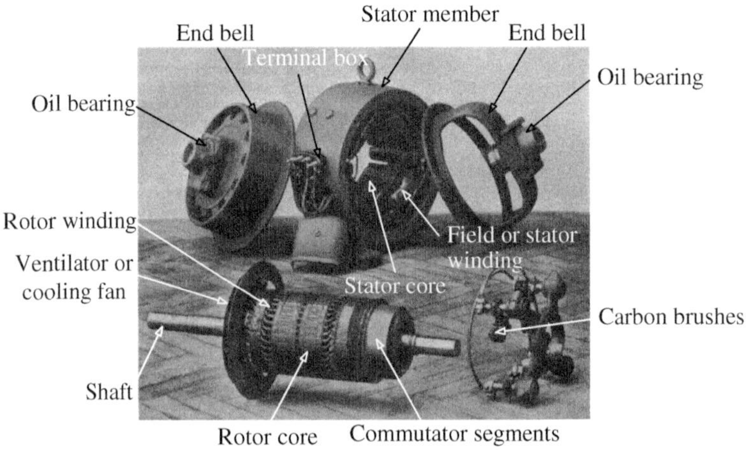

Figure 11.2 Disassembled DC machine [1] of Figure 11.1.

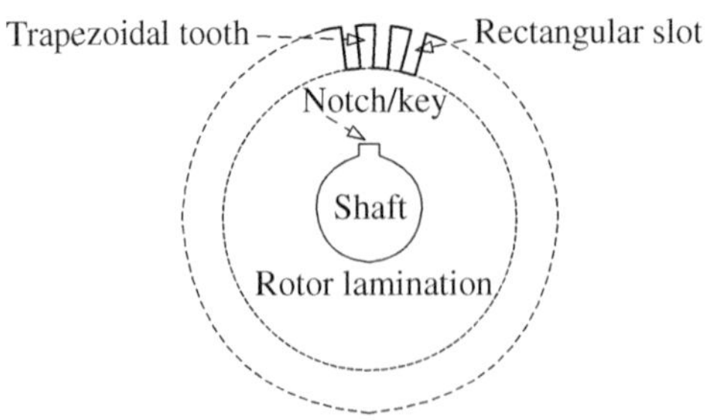

Figure 11.3 Rotor lamination with rectangular slots to house the rotor winding.

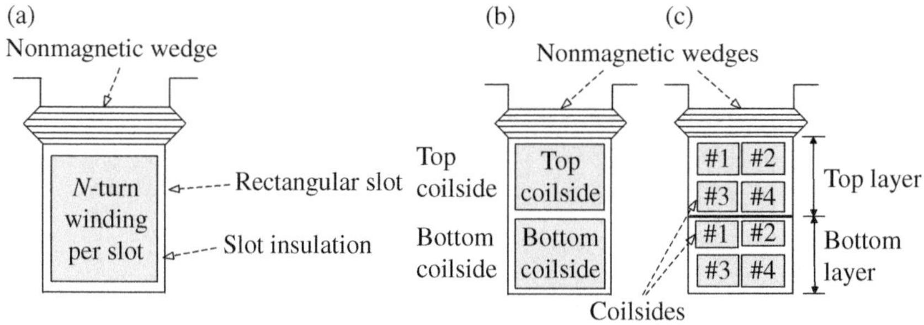

Figure 11.4 Types of rotor slots housing different winding coils: (a) single turn of winding, (b) single-layer armature winding with top and bottom coil sides, and (c) double-layer armature winding.

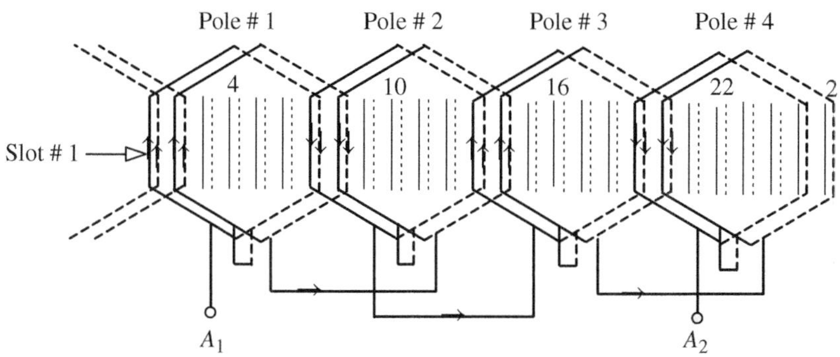

Figure 11.5 Double-layer lap winding with $p = 4$ poles and $N = 24$ full-pitch coils residing in $N = 24$ slots. One full pitch corresponds to $N/p = 24/4 = 6$ slots.

denotes a winding in which the induction process occurs. Usually for DC machines the armature winding resides on the rotor and is also called rotor winding.

A coil consists of one or several turns and resides in two rotor slots, as illustrated in Figure 11.5. This figure shows the location of N coils within the rectangular rotor slots in a double-layer arrangement.

The iron cores of stator and rotor consist of thin (e.g. less than 1 mm) magnetic iron laminations or sheets in which rectangular slots are punched, as depicted in Figure 11.3. Figure 11.6 presents an example of a complete four-pole rotor/armature winding – with two north (N) and two south (S) poles – connected to commutator segments as described below, on which the carbon brushes slide and make connection between the stationary parts of the machine and the rotating reference frame of the moving rotor. The rotor winding is usually called the armature

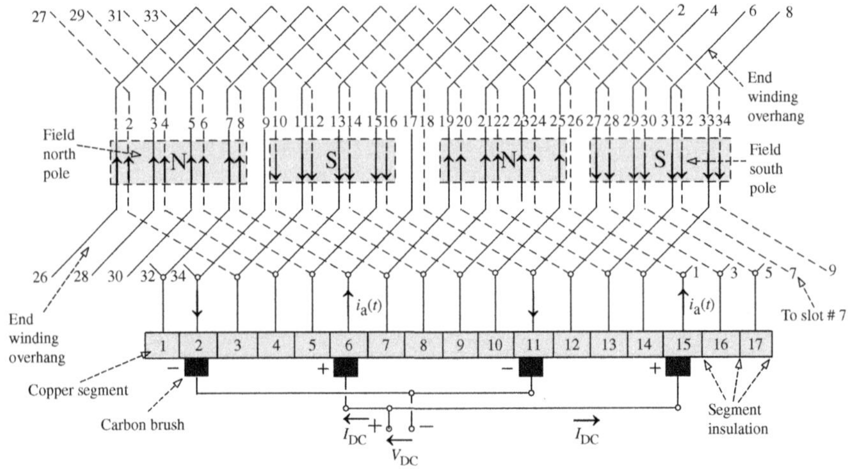

Figure 11.6 Developed (straightened out) four-pole rotor/armature winding of a DC machine – with two north (N) and two south (S) poles – with commutator segments attached to the terminals of the winding coils on which the four (two positive and two negative) carbon brushes slide. There are 34 rotor slots indicated by the numbers 1–34 and 17 commutator segments that are insulated from each other.

winding, the stator winding is named field winding, and the windings around the commutating poles or interpoles are called commutating field windings assisting (reducing sparks) the commutation of the rotor current from a negative value to a positive value and vice versa. As indicated in Figure 11.9a, there are compensating field and series windings located on the main-pole shoes. The commutating, series, and compensating field windings are only used for high-performance machines where precise speed and torque control performances are required to counteract flux weakening due to the superposition of the positive field/stator and the counteracting armature winding/rotor magnetic fields. Flux weakening will be discussed later in this and other chapters.

There are many types of DC machines depending upon the arrangement of field/stator, armature/rotor, compensation, series, and commutating windings as shown in Figure 11.9a. The most important ones are discussed below. After explaining the mechanical commutation process, we describe torque production based on the Lorentz force law relation [2], which strictly applies only to the interaction of an air-gap field with currents flowing in conductors residing in the air gap.

In Appendix 11.A we apply conformal mapping [3, 4] to the magnetic field calculation of a rectangular slot as they occur in all rotating and linear electric machines. Conformal mapping is used for determining solutions to the Laplace

[5] partial differential equation on planar domains that appear in many physical problems, such as fluid mechanics, aerodynamics, thermomechanics, electrostatics, elasticity, and magnetics. Electromagnetic machines such as motors and generators have windings housed in slots distributed in a uniform manner around the circumference of the rotor to provide mechanical stability and to protect the winding from excessive magnetic field exposure, because the largest amount of the magnetic field is conducted by the magnetic rotor and stator teeth, as shown in Figure 11.9b. The currents of the armature (rotor) of an electromagnetic machine generate magnetic fields, which can be compensated by windings located on the stationary member of the machine (stator) as depicted in Figure 11.9a and b. To make the determination of magnetic field calculations possible, the finite-difference (FD) and finite-element (FE) algorithms (Appendix 11.A) for various node/mesh/grid structures are introduced and applied to DC machine two-dimensional field configurations. These magnetic field distributions show that an insignificant amount of flux is passed through slots, and for this reason the Lorentz force [2] calculation cannot be applied to a winding residing in slots; instead, the forces acting on the iron-core teeth and windings must be determined from the Maxwell stress [6–9]. The residing of windings within slots protects the winding conductors from mechanical vibrations due to changing magnetic fields.

We introduce the calculation of linear and operating-point-dependent inductances from flux plots in Appendix 11.B. Application examples relating to steady-state and transient torque and speed control problems are solved using Mathematica [10] software program. We show that compensation/reinforcement of field/stator flux produces stable transient current conditions, while flux weakening – where the field/stator flux is partially reduced by the counteracting armature/rotor flux – reduces the transient stability of electric machines. Real-world applications of DC machines are electric drives where the torque and the speed can be independently controlled. For example, such drives were applied in electric cars, locomotives, wind power plants, rolling mills in steel plants, and other variable-speed applications up to the time when semiconductor switches were invented about 70 years ago. The application of DC machines is now largely replaced by AC machines (Chapter 12) fed by inverters (Chapters 9 and 10).

11.2 Mechanical Commutation Concept

Electromechanical (e.g. DC machines, AC induction machines, AC synchronous machines, permanent-magnet machines) and, in general, electric/magnetic energy conversion processes (e.g. transformers) are based on the corollary

For energy conversion to take place, the interacting magnetic fields of the primary (e.g. stator) and secondary (e.g. rotor) inside the conversion device must be stationary with respect to one another.

There are at least five types of DC machines:

1) Separately excited, employed for speed/angular velocity and torque control, where two variable DC voltage sources must be available.
2) Separately excited with armature reaction compensation and commutating/interpole windings to reduce commutation sparks, where the air-gap field between two adjacent poles can be altered through the superposition of the fields induced by the armature and commutating windings (designs for very large machines used, for example, in steel mills).
3) Separately and cumulatively connected DC generator: if cumulatively connected, then generator represents a voltage source; if differentially (fluxes of field and series windings are subtractive) connected, then generator represents a current source.
4) Self-excited DC generator, useful for applications when only one voltage--current source is available.
5) Series-connected DC motor, where large torque at low speed/angular velocity is desirable.

Figure 11.7a shows the configuration of a two-pole ($p = 2$) separately excited machine, where the primary (stator) DC machine field is stationary with angular velocity $\omega_s = 0$ and the secondary (rotor/armature) field is stationary with respect to the rotating rotor and rotates with the angular velocity $\omega_r \neq 0$ with respect to the stator due to the rotor terminals connected to two (copper) slip rings residing side by side on the shaft. Because $\omega_r \neq \omega_s$, neither an electromagnetic force nor torque develops. Figure 11.7b illustrates the stationary stator magnetomotive force (mmf) $\vec{F}_s$ with the stator flux density $\vec{B}_s$ and the nonstationary rotor mmf $\vec{F}_r$ with the rotor flux density $\vec{B}_r$ as indicated in Figure 11.7a. The mmfs are not mechanical forces but are defined by

$$\vec{F}_k = i_k \cdot N_k, [\text{At}], \tag{11.1}$$

that is, current (i_k) times number of turns (N_k) of a winding, measured in ampere-turns or [At], where k is either s (stator) or r (rotor). Note that the arrow (→) on the variables represents vectors and tilde (~) on the variables defines phasors, as discussed in Chapter 4.

The mmfs $\vec{F}_s$ and $\vec{F}_r$ are depicted in Figure 11.7b together with the magnetic flux densities $\vec{B}_s$ and $\vec{B}_r$, which are defined via the magnetic field strength:

$$\vec{H}_k = \text{mmf}/(\text{radial air-gap length}) = \vec{F}_k/g\ [\text{At/m}], \tag{11.2}$$

where k represents either s or r and g (gap) is the one-sided radial air-gap length.

(a)

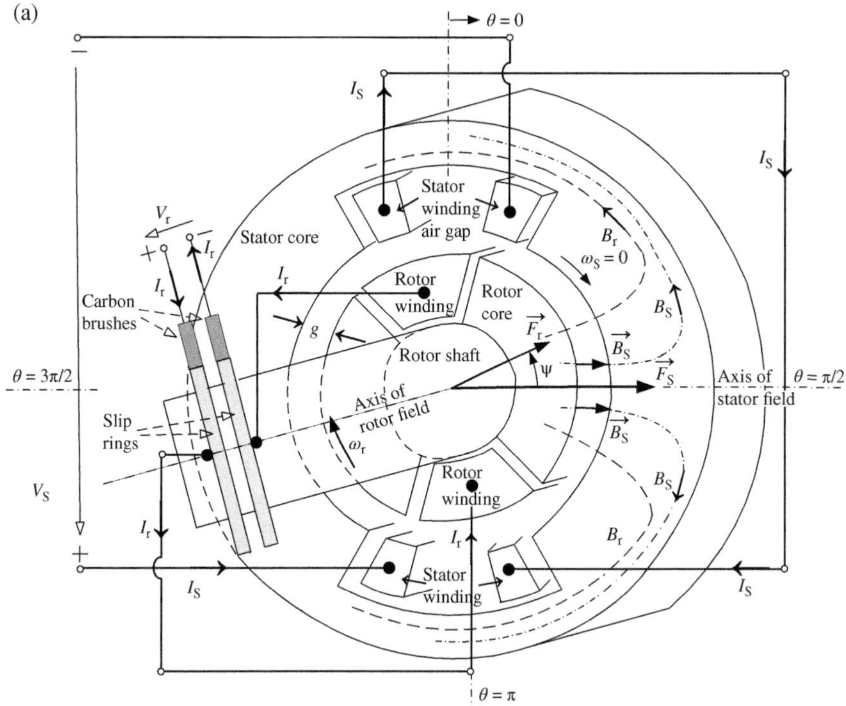

Figure 11.7a Two-pole direct-current (DC) machine with slip rings consists of a stationary stator (s) and a rotating rotor (r). The stator is excited by the mmf $\vec{F}_s$ rotating with angular velocity $\omega_s = 0$ with respect to the stator, and the rotor is excited by the mmf $\vec{F}_r$ rotating with angular velocity $\omega_r \neq 0$ with respect to the stator. The angle ψ has been chosen arbitrarily. The field lines go from north (left-hand-side) to south (right-hand-side) pole. Note that the stator winding is also called field winding and the rotor winding is usually called armature winding.

The magnetic flux density is defined as

$$\vec{B}_K = \mu_0 \cdot \vec{H}_K \text{ [tesla or T]}, \tag{11.3}$$

where $\mu_0 = 4\pi \cdot 10^{-7}$ [H/m] is the (universal constant) permeability of free space, or vacuum permeability.

The magnetic flux is given by the integral/summation

$$\phi = \int_S^{\text{pole-pitch}} B(\theta)\mathrm{d}S, \text{ [Vs]} \tag{11.4}$$

where dS represents the differential surface of the (pole-pitch) surface S and $B(\theta)$ the flux density distribution over one pole pitch or half a period $0 \le \theta \le \pi$, as illustrated in Figure 11.7b.

(b)

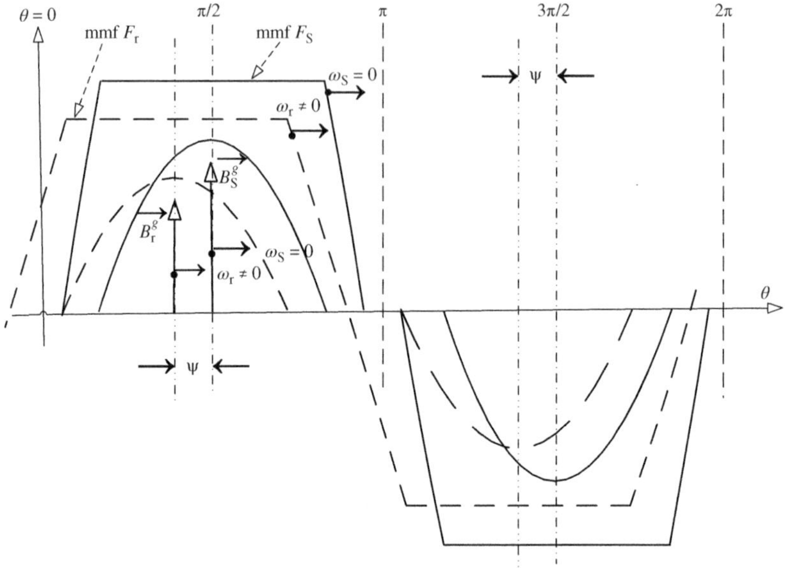

Figure 11.7b Instantaneous magnetomotive forces (mmfs) F_s and F_r; flux density air-gap vectors (phasors; see Chapter 4) $\vec{B}_s^g$ and $\vec{B}_r^g$ on the surfaces of the stator and rotor, respectively.

In order to produce torque in an electric machine, the interacting magnetic fields (e.g. flux density) must be stationary with respect to another per the above corollary. Both above discussed fields $\vec{B}_k$ are not stationary with respect to one another, and therefore no electric torque is produced in the configuration of Figure 11.7a: $T_e = 0$.

If in Figure 11.7a the rotor field and the stator windings are replaced by a permanent magnet and a three-phase winding, permanent-magnet machines may be used for direct-drive wind power plants [11, 12] and electric/hybrid cars [13, 14], as discussed in Chapter 12. Provided one replaces the rotor winding with two brushes by a three-phase winding with three brushes and the stator winding by another three-phase winding, then a doubly fed induction generator (DFIG) [15] is obtained. DFIGs have been recently used for hydropower plants in order to reduce power delivery response time to a few minutes by quickly adjusting angular velocity and torque to meet changing load demands. This stresses the importance of DC machines as a role model for speed and torque control for the development of new electric drives (e.g. motors and generators) – important to renewable power generation and utilization.

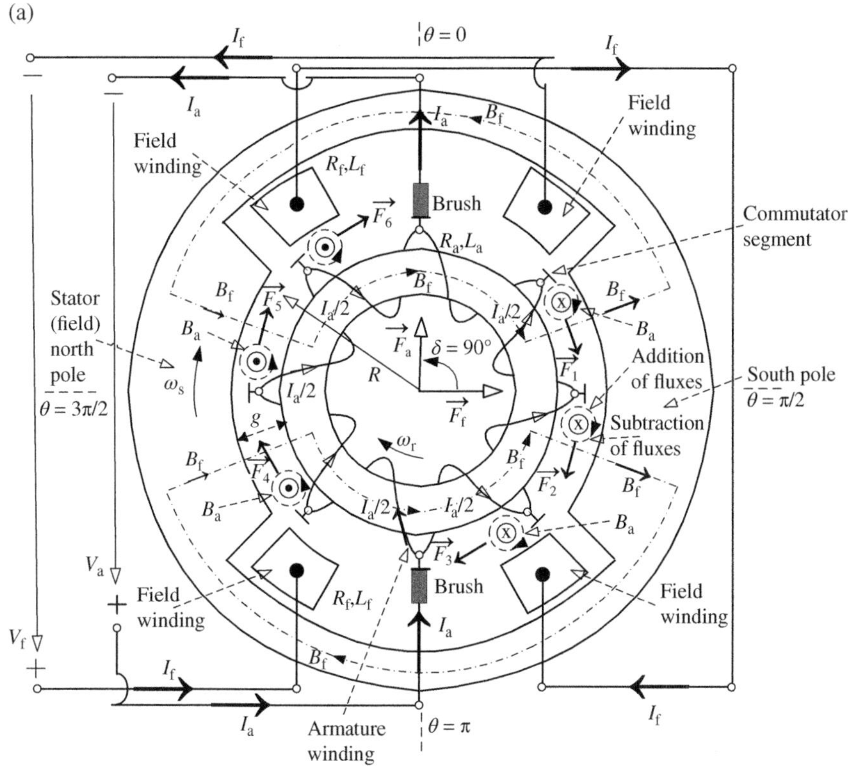

Figure 11.8a Two-pole DC machine consisting of stationary stator (s) with angular velocity $\omega_s = 0$ and rotating armature or rotor (r) with angular velocity $\omega_r \neq 0$ is excited in stator by DC current but in rotor by AC current due to the commutator that represents the copper segments connected to the turns of the rotor winding. Field lines go from north (left-hand-side) to south (right-hand-side) pole. Note that the field winding is also called stator winding and the armature winding is usually called rotor winding.

In order to achieve a stationary relationship between the two (stator and rotor) interacting magnetic fields, a mechanical commutator was invented [16, 17] as described in an early textbook [18]. The commutator consists of copper segments attached to one another through insulation material; each segment is connected to at least one turn of the rotor/armature winding. Figure 11.8a depicts a simple separately excited configuration, where the two side-by-side slip rings of Figure 11.7a are replaced by a great number of so-called commutator (copper) segments held together by nonconducting material and forming a ring – a so-called Gramme [19] ring – on which carbon brushes slide to make the connection to the DC source $V_s = V_a$. As a consequence of the conversion of DC to AC current due to the commutator, the rotor magnetic field flux density B_a or mmf $\vec{F}_a$ is stationary with respect to the stator mmf $\vec{F}_f$, although the mechanical rotor rotates with ω_r.

In Figure 11.8a the stator field (f) $\vec{F}_f$ is stationary with respect to the stator, and the rotor or also called armature (a) field $\vec{F}_a$ is stationary with respect to the stator as well. As a result, both fields are stationary with respect to one another – as required by corollary – and electric torque is produced due to the six magnetically generated forces $\vec{F}_1$–$\vec{F}_6$. Figure 11.8b demonstrates that the two interacting mmfs $\vec{F}_f$ and $\vec{F}_a$ are stationary with respect to one another and have the angular frequency $\omega_s = 0$ with respect to the stator. The magnetic flux density $\vec{B}_f^g$ of the field winding and the magnetic flux density $\vec{B}_a^g$ of the armature winding – both measured in the radial center of the air gap – are plotted as well.

Mechanical Force and Torque Production Base on Vector Product $\vec{C} = \vec{A} \times \vec{B}$ [20, 21]: The magnetically generated mechanical force at the nth armature winding residing in the air gap (and not in slots) of Figure 11.8a is defined based on the Lorentz force law as

$$\vec{F}_n = \ell \cdot \left(\vec{I}_a \times \vec{B}_{fn}\right), \tag{11.5}$$

(b)

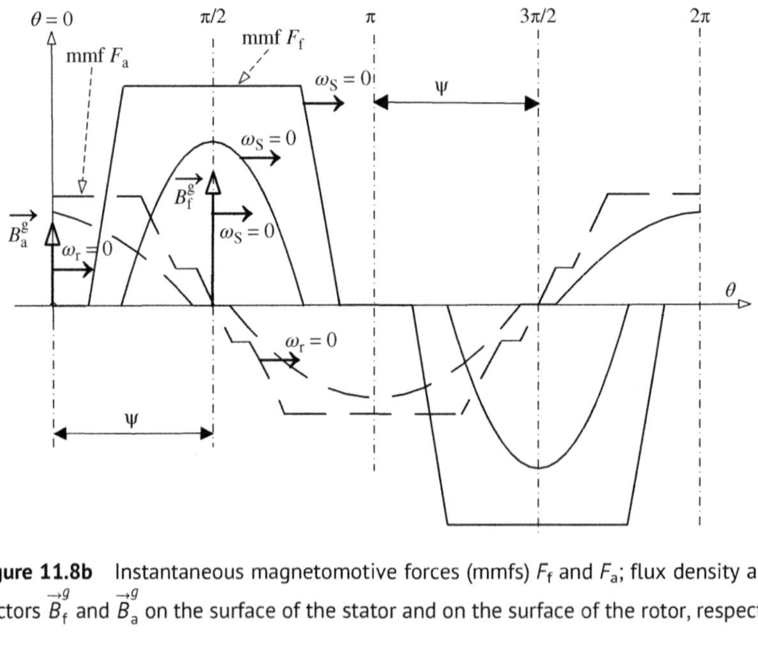

Figure 11.8b Instantaneous magnetomotive forces (mmfs) F_f and F_a; flux density air-gap vectors $\vec{B}_f^g$ and $\vec{B}_a^g$ on the surface of the stator and on the surface of the rotor, respectively.

where ℓ is the ideal length of the north and south pole iron cores and $\vec{B}_{fn}$ is the air-gap flux density vector of the nth armature winding turn carrying the current $\vec{I}_a$. The associated torque is

$$\vec{T}_{en} = R \cdot \vec{F}_n = R \cdot \ell \cdot \left(\vec{I}_a \times \vec{B}_{fn}\right), \tag{11.6}$$

where radius R is the radial length from the center of the machine rotor to the middle of the air gap. Considering only magnitudes, one obtains for the nth armature winding turn

$$\left|\vec{T}_{en}\right| = T_{en} = R \cdot \ell \cdot (I_a \cdot B_{fn}) \cdot \sin\delta. \tag{11.7}$$

For the torque angle $\delta = 90°$, the maximum torque becomes

$$T_{en\_max} = R \cdot \ell \cdot B_{fn} \cdot I_a. \tag{11.8}$$

The total torque is the summation of the $N = 6$ individually developed torque contributions within the air gap under the north and south poles. Provided the air-gap circumference below the north and south poles relates to the entire circumference (see Figure 11.8a) as (13/15), then one gets for the total torque developed by the machine:

$$T_e = \sum_{n=1}^{N=6} T_{en\_max} = \sum_{n=1}^{N=6} R \cdot \ell \cdot B_{fn} \cdot I_a = \sum_{n=1}^{N=6} \frac{A_n}{2\pi\left(\frac{13}{15}\right)} B_{fn} \cdot I_a = k_e \cdot \phi \cdot I_a. \tag{11.9}$$

Assuming that the associated area with any armature winding turn is A_n and the corresponding flux density is B_{fn}, the magnetic flux is

$$\phi = \sum_{n=1}^{N=6} A_n \cdot B_{fn} \tag{11.10}$$

or

$$k_e \cdot \phi = \sum_{n=1}^{N=6} \frac{A_n B_{fn}}{2\pi\left(\frac{13}{15}\right)} \ [\text{Vs/rad}]. \tag{11.11}$$

The total developed torque by the DC machine is

$$T_e = k_e \cdot \phi \cdot I_a \ [\text{VAs/rad}], \tag{11.12}$$

and the total developed induced armature voltage of the DC machine, derived in a similar manner, is based on Eq. (3.12) where the induced voltage is $v = L(\mathrm{d}i/\mathrm{d}t)$, if the inductance L is constant. Provided the inductance $L(x)$ is a function of the spatial coordinate x, then one can write for the induced voltage

$$v \equiv e = d\{L(x,t)i(t)\}/\mathrm{d}t = \frac{d}{\mathrm{d}t}\{L(x)i(t)\} = L(x)\frac{\mathrm{d}i(t)}{\mathrm{d}t} + i(t)\frac{\mathrm{d}L(x)}{\mathrm{d}x}\frac{\mathrm{d}x}{\mathrm{d}t}, \tag{11.13a}$$

or for rotary motion occurring in rotating electric machines, the inductance is a function of the angular displacement θ and represents the angular displacement

$$v \equiv e = d\{L(\theta,t)i(t)\}/\mathrm{d}t = \frac{d}{\mathrm{d}t}\{L(\theta)i(t)\} = L(\theta)\frac{\mathrm{d}i(t)}{\mathrm{d}t} + i(t)\frac{\mathrm{d}L(\theta)}{\mathrm{d}\theta}\frac{\mathrm{d}\theta}{\mathrm{d}t}. \tag{11.13b}$$

The first term of the induced voltage Eqs. (11.13a) and (11.13b) is called the "transformer voltage" occurring predominantly in transformers as discussed in Chapter 6. The second term is denoted the "speed voltage," occurring mainly in linear and rotating electric machines. Both voltages occur in homopolar machines [22, 23]. With $\mathrm{d}\theta/\mathrm{d}t = \omega_\mathrm{m}$, and with $i(t) = I$ resulting in $I \cdot \mathrm{d}L(\theta)/\mathrm{d}\theta = k_\mathrm{e}\phi$, one can write

$$e_\mathrm{a} = I \cdot \frac{\mathrm{d}L(\theta)}{\mathrm{d}\theta}\frac{\mathrm{d}\theta}{\mathrm{d}t} = k_\mathrm{e}\phi \cdot \omega_\mathrm{m}\ [\mathrm{V}]. \tag{11.14}$$

The Right-Hand-Side Rule for Torque Production: The application of the vector product relation $\vec{C} = \vec{A} \times \vec{B}$ leads to the right-hand-side rule depicted in Figure 11.8c for the armature winding turns in the air gap below the north and south poles.

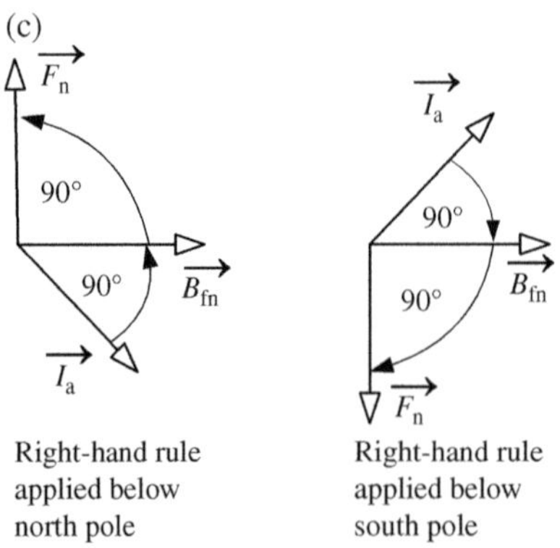

Figure 11.8c Application of the right-hand-side rule producing the mechanical force $\vec{F}_\mathrm{n} = \ell \cdot \left(\vec{I}_\mathrm{a} \times \vec{B}_\mathrm{fn}\right)$, where ℓ is the axial length of the machine.

Addition and Subtraction of Interacting Stator and Rotor Fluxes: The direction of the forces $\vec{F}_n$ developed as indicated in Figure 11.8a also can be found by observing the addition and subtraction of interacting stator and rotor fluxes. The side of the winding where the cancelling of the stator and rotor fluxes occurs is identical to the direction of the developed force.

Minimization of Forces Acting on Windings Through the Deployment of Slots: To minimize the mechanical forces acting on windings, they are embedded in stator/field and rotor/armature slots as evident from Figure 11.9a and b where the magnetic fields practically do not pass through slots; see also Figures 11.A.1 and 11.B.2. Unfortunately, due to the commutator consisting of hundreds or thousands of segments, the transition of the carbon brushes moving from segment to segment, which are insulated, induces sparks [24] because of the short circuiting of the individual coils connected to adjacent copper segment pairs. Magnetic

(a)

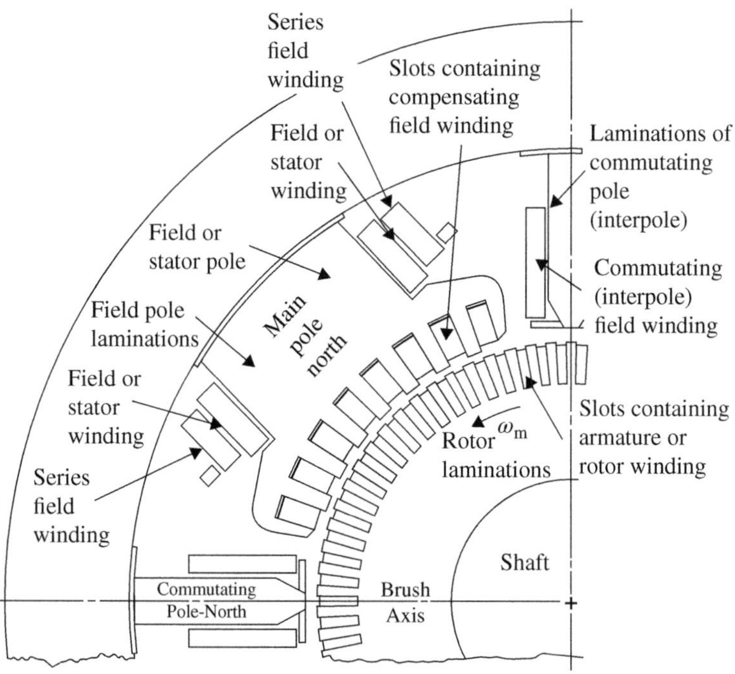

Figure 11.9a Schematic cross section of one pole pitch (half a period) of a large separately excited four-pole DC machine with commutating poles, also called commutating/interpoles, and field and series windings residing on stator field poles that also contain a compensating winding [25] in order to cancel the effect of the rotor/armature flux on the stator/field flux, thus reducing/eliminating the so-called armature reaction, which is responsible for flux weakening.

(b)

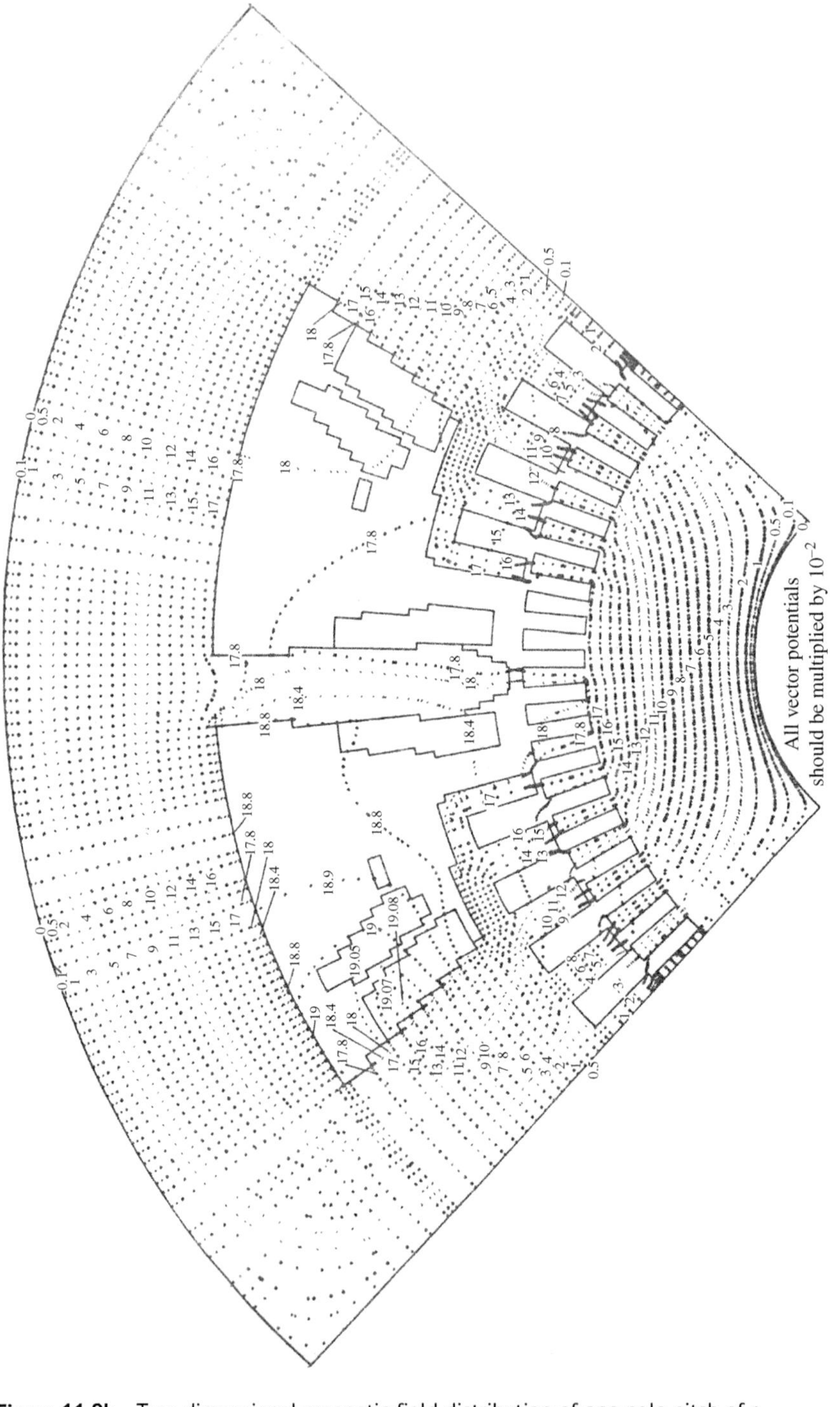

Figure 11.9b Two-dimensional magnetic field distribution of one pole pitch of a separately excited four-pole DC machine with commutating/interpole windings and compensation windings residing within slots of the field-pole shoes, reducing field distortion due to armature reaction [25].

(c)

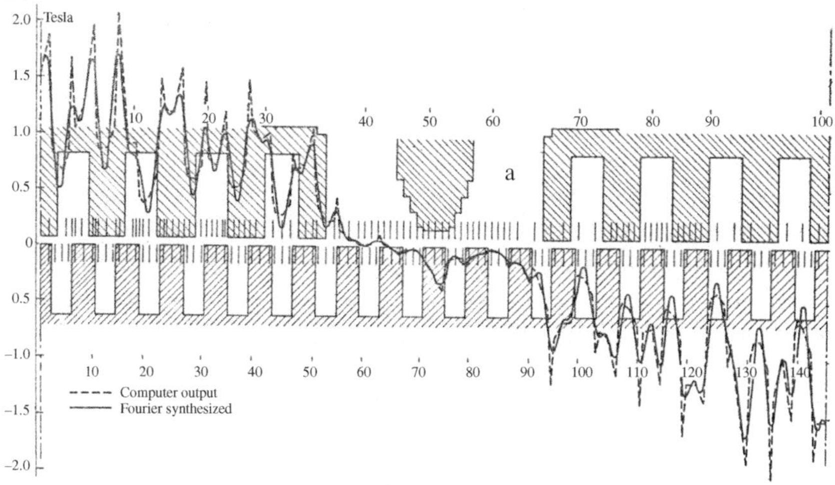

Figure 11.9c Air-gap flux density of Figure 11.9, expanded rotor position 1 indicated by "a." Note that the flux density contains a great deal of spatial harmonics due to stator and rotor slotting [25].

commutating/interpoles [26–28] with appropriate windings, as indicated in Figure 11.9a–c, can reduce sparking to an acceptable level. Figure 11.9b illustrates the two-dimensional magnetic vector potential (with unit Wb/m = Vs/m) distribution within one pole pitch (half a magnetic period). Figure 11.9c illustrates the associated flux density within the radial air gap. Note that this flux density has spatial harmonics due to stator slotting [29]. As can be seen from the flux plots, the magnetic flux passing through the slots is very small, and therefore the force/torque computation based on the Lorentz force law is not valid; instead, the Maxwell stress must be relied on for accurate force/torque calculation in electric machines [6–9, 29].

The idea of mechanical commutation as described above can be extended to permanent-magnet machines and inverters [9], where electronic commutation is used to change DC to AC for driving mode and AC to DC for regeneration [9] operation in automotive electric drives [13, 14], and for direct-drive [11, 12] wind power plants commutating AC of variable frequency to nominal frequency of either 60 or 50 Hz, as described in Chapters 10 and 12. The disadvantages of DC machines and electronic converters are the sparks due to commutation, which introduce power quality problems for telecommunication circuits [29]. In addition, the energy dissipated by the sparking deteriorates copper segments of DC machines and converter switches. However, the great advantage of DC machines and converters is their ease of voltage/speed and current/torque control, which is indispensable for renewable power generation and utilization (e.g. heat pumps, railroad drives).

11.3 Equivalent Circuits and Voltage–Current Diagrams of Separately, Cumulatively, Differentially, Self-Excited, and Series-Connected DC Machines

There are several types of DC machines: separately excited; separately excited with armature reaction compensation; separately excited with armature reaction winding compensation and commutating/interpole windings to reduce commutation sparks whereby the field near the interpole region is altered; separately excited with cumulatively connected winding; separately excited with differentially connected winding; self-excited; and series connected. Equivalent circuits and some associated voltage/speed/current diagrams of these different connections are depicted below (see Figures 11.10a–d, 11.11a, b, 11.12a–c, 11.13, and 11.14a, b).

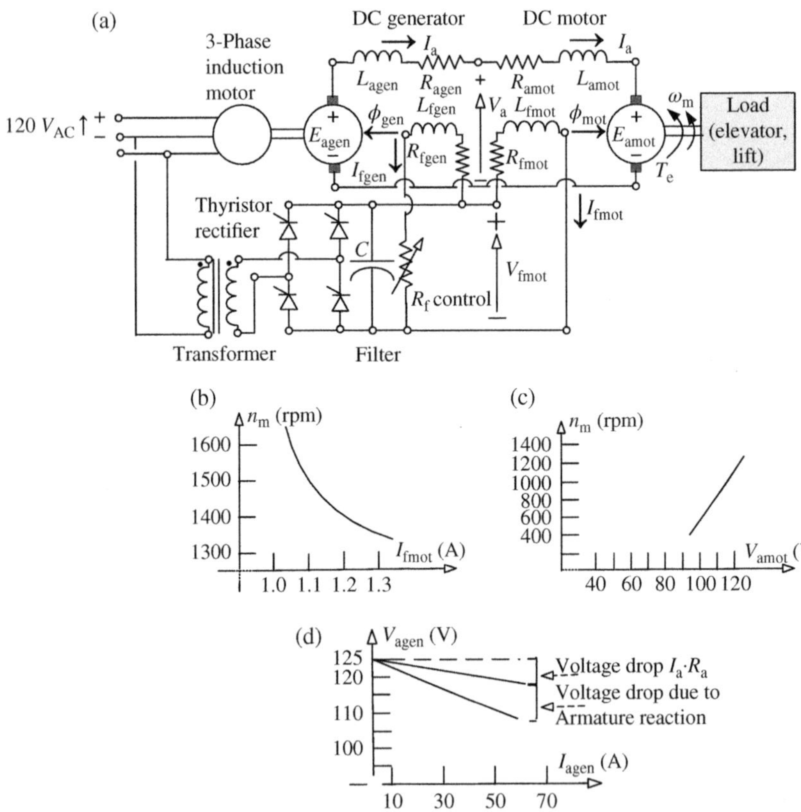

Figure 11.10 (a) Ward Leonard system [30] or generator/motor set applied to elevator and lift variable-speed drives, (b) motor speed as a function of motor field current, (c) motor speed as a function of motor armature voltage, and (d) generator armature voltage as a function of generator armature current.

The most prominent control application for separately excited DC machines is the Ward Leonard control application [30] for elevators and lifts, demonstrated in Figure 11.10a–d. The various characteristics of the separately excited DC motor pertain to a rated output power $P_{out} = 15$ hp $\equiv 11.185$ kW at a speed range of $n_m =$ (1150–2600) rpm and an excitation/field current of $I_f = (0.95–1.3)$ A. The various characteristics of the separately excited DC generator pertain to a rated output power $P_{out} = 12.06$ hp $\equiv 9.0$ kW at a speed of $n_m = 1150$ rpm and an excitation/field current of $I_f = (0.88–2.63)$ A. The relation between hp (horsepower) and kW (kilowatt) is

$$1\,\text{hp} = 0.745\,699\,\text{kW}. \tag{11.15}$$

Recall that any DC machine can be used either as a motor or a generator. The relation between the speed in revolutions per second (rps) or revolutions per minute (rpm) and the angular velocity ω_m (rad/s) is

$$\omega_m = 2\pi \cdot n_{rps} = 2\pi \cdot \text{n}_{rpm}/60, \tag{11.16}$$

and usually n_{rpm} is denoted as n_m. The mechanical power output of a motor is defined as $P_{out} = \omega_m T_e$ (see Figures 11.10a–d, 11.11a, b, and 11.14a, b), and the mechanical power input to a generator is defined as $P_{in} = \omega_m T_e$ (see Figures 11.12a–c and 11.13).

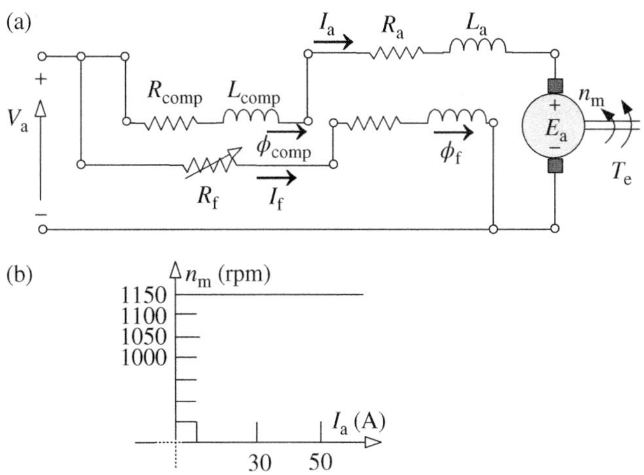

Figure 11.11 (a) Separately excited DC motor with armature reaction compensation to maintain constant speed independent of load, (b) speed n_m load current I_a diagram. The compensating flux ϕ_{comp} cancels the armature reaction that causes flux weakening and maintains the flux at ϕ_f = constant.

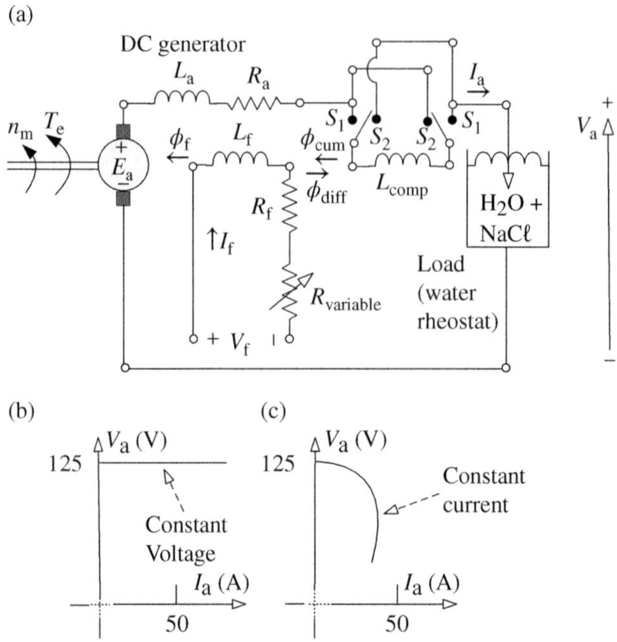

Figure 11.12 (a) Separately excited generator with cumulatively/differentially connected winding resulting in (b) voltage source and (c) current source.

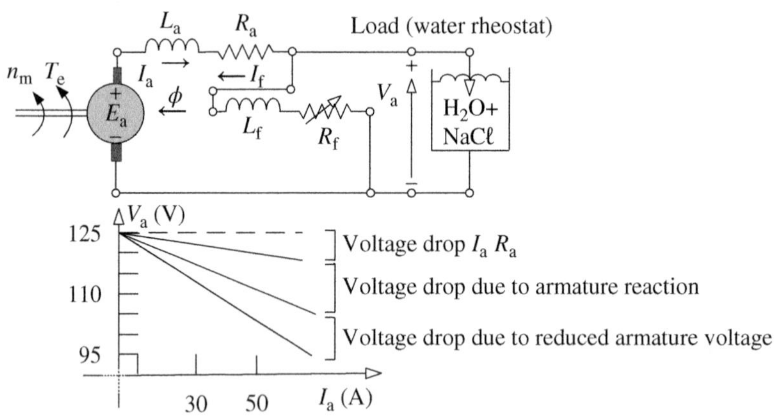

Figure 11.13 Self-excited generator used when no independent voltage source is available.

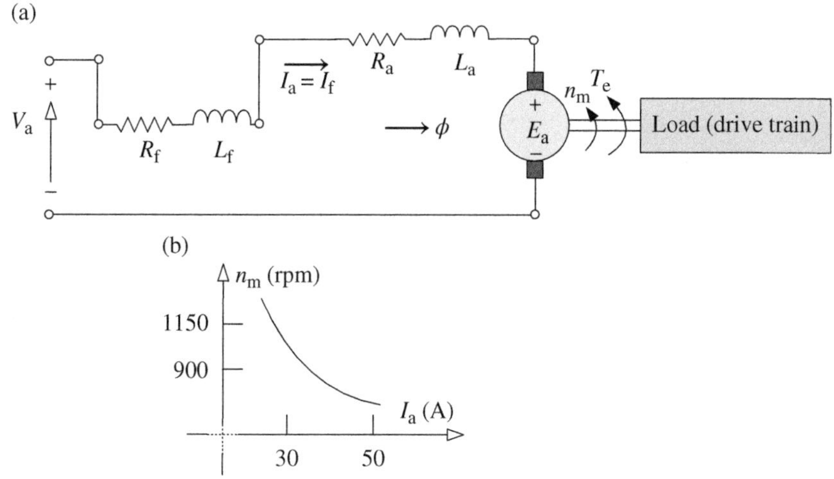

Figure 11.14 (a) Series excited motor with one independent voltage source V_a for drives requiring high torque at low speed and low torque at high speeds, such as rail/drive applications, and (b) speed (n_m) versus load (I_a) characteristic.

The no-load field of a separately excited machine is shown in Appendix 11.B, Figure 11.B.2. From such a field the self-, mutual, and leakage inductances of the required field winding with the associated flux ϕ_τ^{ps} are determined from flux plots for one pole pitch.

Application Example 11.1 Separately excited DC motor

From the equivalent circuit parameters of Appendix 11.B, one obtains armature (a) resistance of $R_a = 0.3\ \Omega$ and the flux coefficient $(k_e\ \phi) = 4.32$ Vs/rad. In addition, given parameters are steady-state terminal armature voltage $V_a = 240$ V, armature current $I_a = 100$ A, field (f) current $I_f = 10$ A, field resistance $R_f = 0.42\ \Omega$, and field control resistance $R_{fcontrol} = R_{ftotal} - R_f = 23.58\ \Omega$.

Solution

armature input power $P_a = V_a \cdot I_a = 240 \cdot 100\ \text{W} = 24\ \text{kW}$,
total field resistance $R_{ftotal} = V_a/I_f = 240/10 = 24\ \Omega$,
input (in) power to field circuit $P_f = V_a \cdot I_f = 240 \cdot 10 = 2.4\ \text{kW}$,
total input power $P_{in} = P_a + P_f = 26.4\ \text{kW}$.
With the relation $E_a = k_e\phi\omega_m$ follows the induced voltage $E_a = V_a - R_a \cdot I_a = 240\ \text{V} - 30\ \text{V} = 210\ \text{V}$,
angular velocity $\omega_m = E_a/k_e\phi = 210/4.32 = 48.61$ rad/s,
electrical torque $T_e = k_e\phi I_a = 4.32 \cdot 100 = 432$ Nm,

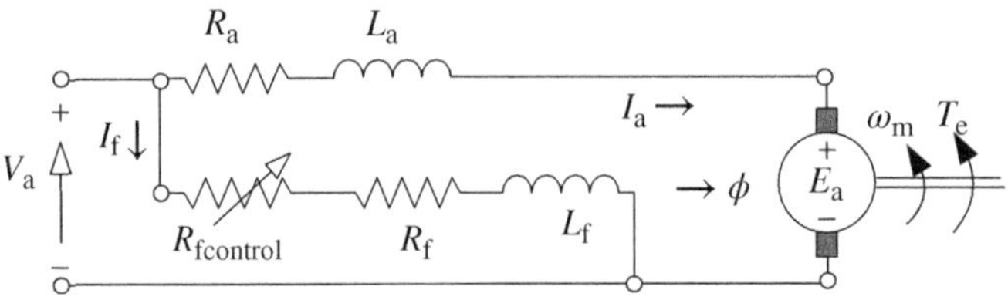

Figure E11.1.1 Equivalent circuit of a separately excited DC motor. For steady-state analysis the parameters armature leakage inductance L_a = 5 mH and field winding self-inductance L_f = 14.4 mH are not used; they are important for transient operation only.

and output (out) power $P_{out} = T_e\omega_m = 432 \cdot 48.61 = 21$ kW resulting in the efficiency $\eta = P_{out}/P_{in} = 21/26.4 = 0.795$ neglecting iron-core losses.

11.4 Speed and Torque Control

Application Example 11.2 Transient analysis of a separately excited DC machine (motor/generator) with armature compensation (addition or weakening of flux) winding in stator based on Mathematica [10, 31–36] and equivalent circuit Figure E11.2.1.

A $P = 60$ kW, $V_a = 240$ V, $n_m = 1150$ rpm, $k_e\phi = 2.0$ Vs/rad separately excited DC machine has the armature resistance $R_a = 0.2\,\Omega$, armature inductance $L_a = 8$ mH, polar moment of inertia $J = 0.4$ kgm^2, damping coefficient of the viscous torque $B = 0.8$ Nm/(rad/s), torque at standstill $T_s = 0$, Coulomb frictional torque T_c = Tic = 5 Nm, and the coefficient of windage torque $c_w = 0.002$ Nm/(rads)^2. The terminal armature voltage $V_a(t)$ and the mechanical load torque doing the work $T_m(t)$ are

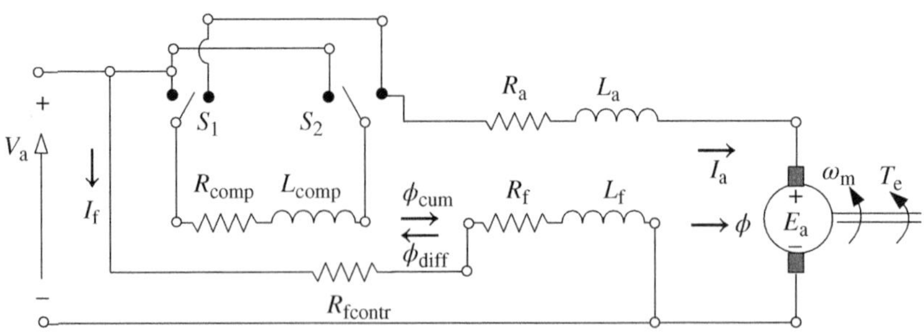

Figure E11.2.1 Equivalent circuit of DC machine with (cumulative) flux addition (ϕ_{cum}), (differential) flux weakening (ϕ_{diff}), and constant flux. The torque components T_c, T_m, T_s and coefficients B and c_w are not shown.

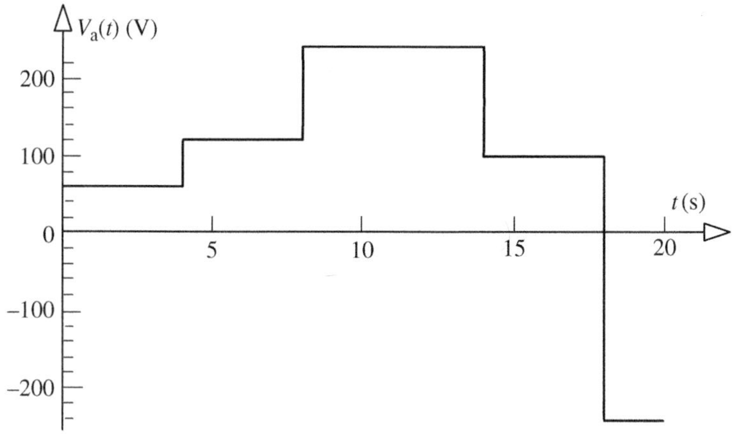

Figure E11.2.2 Applied terminal armature voltage $V_a(t)$ as a function of time.

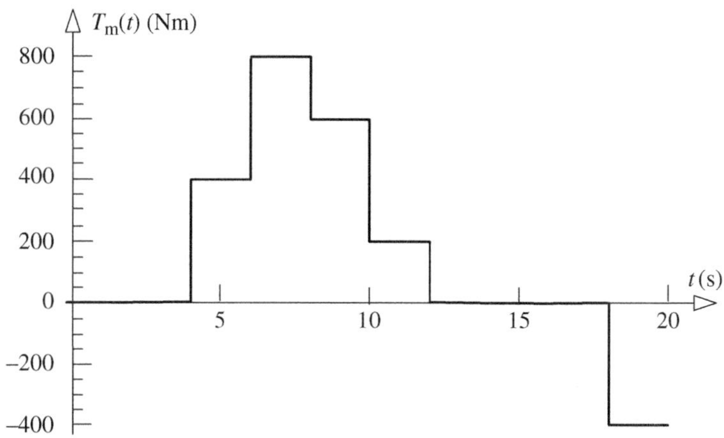

Figure E11.2.3 Applied mechanical shaft torque $T_m(t)$ as a function of time.

variable, depicted in Figures E11.2.2 and E11.2.3, respectively. Investigate the behavior of the machine as a motor under variable load/speed conditions and as a generator with reverse angular velocity due to negative load torque for the flux conditions:

a) Constant, rated flux,
b) Increased flux kphi[t_] = (1 + 0.0004 * Ia[t]) * kphi[t],
c) Decreased flux kphi[t_] = (1 − 0.00055 * Ia[t]) * kphi[t].

Solution

The pertinent differential and algebraic equations for this application example are

$$\frac{di_a}{dt} = \frac{1}{L_a}[V_a - R_a i_a - (k_e\phi)\omega_m],$$

$$\frac{d\omega_m}{dt} = \frac{1}{J}\left[(k_e\phi)i_a - T_m - B\omega_m - T_c - c(\omega_m)^2 - T_s\right],$$

electric shaft torque $T_e = (k_e\phi)i_a$, and the output power at shaft $P = T_e\omega_m$.

a) Perform a Mathematica analysis for a constant or rated/nominal DC flux Φ excitation of the machine, where $k_e\phi$

   kphi[t_]:=If[t < 8., 2., 2.];

   at a variable input armature voltage $V_a(t)$, and the required mechanical torque $T_m(t)$ as depicted in Figures E11.2.2 and E11.2.3, respectively. The armature current, the angular velocity, the electrical torque, and the output power are shown in Figures E11.2.4–E11.2.7, respectively. Table E11.2.1 contains the list of the Mathematica statements for cases a, b, and c.

b) Perform a Mathematica analysis for an armature current $I_a(t)$-dependent flux excitation of the machine, where

   kphi[t_] = (1 + 0.0004*Ia[t])*kphi[t];

   i.e. when a field flux increase occurs as a function of $I_a(t)$ through overcompensation of armature reaction proposed in Figure 11.11a and b and a variable input voltage $V_a(t)$ and the mechanical torque $T_m(t)$ depicted in Figures E11.2.2

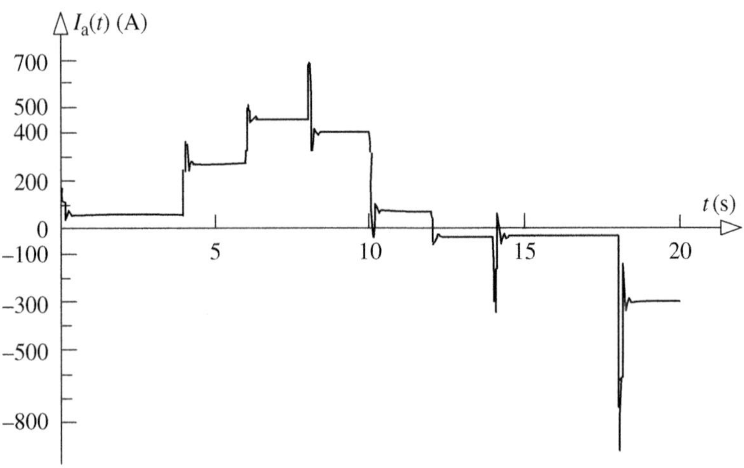

Figure E11.2.4 Calculated armature current $I_a(t)$ as a function of time.

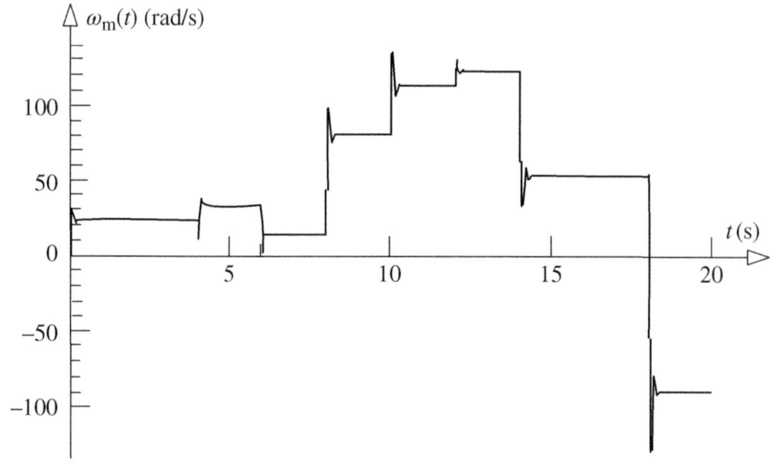

Figure E11.2.5 Calculated angular velocity $\omega_m(t)$ as a function of time.

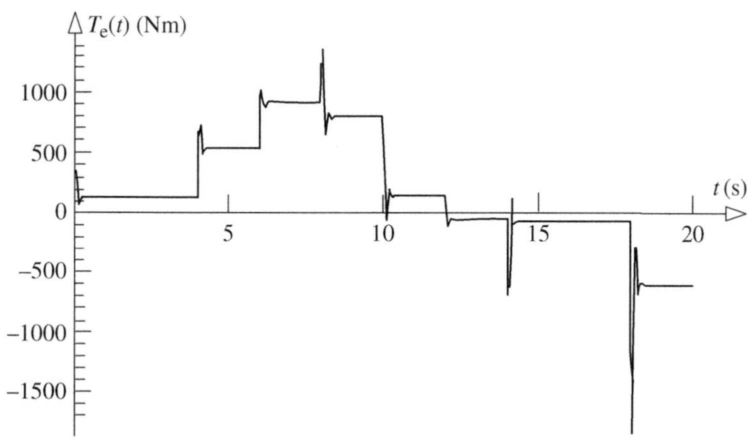

Figure E11.2.6 Calculated electrical torque $T_e(t)$ as a function of time.

and E11.2.3, respectively. The armature current, the angular velocity, the electrical torque, and the output power are presented in Figures E11.2.8–E11.2.11, respectively.

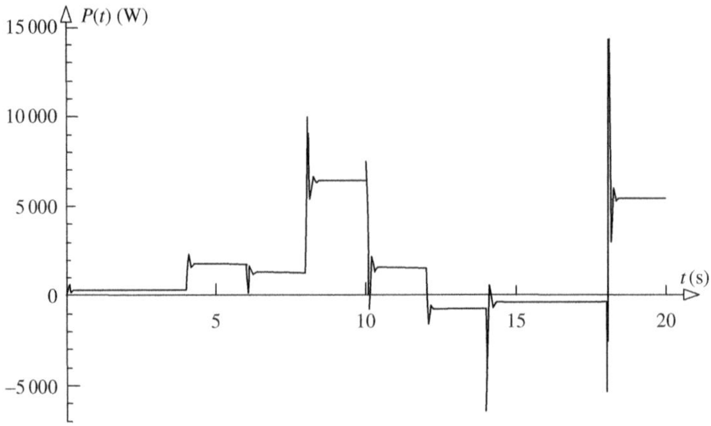

Figure E11.2.7 Calculated output power $P(t)$ as a function of time.

Table E11.2.1 Mathematica statements for cases a, b, and c.

```
Remove[T, TL, Va, Wm, sol, eqn1, eqn2, ic1, ic2, kphi, Tm, Tic, cw, P, Ia]

La=0.008;
Ra=0.2;
J=0.4;
B=0.8;
Ts=0
Va[t_]:=If[t
kphi[t_]:=If[t
Tm[t_]:=If[t
Tic[t_]:=If[t
cw[t_]:=If[t
*for flux weakening*
kphi[t_]=(1-0.00055*Ia[t])*kphi[t];
(*for flux addition*)
(*kphi[t_]=(1+0.0004*Ia[t])*kphi[t];*)
ic1=Ia[0]==0;
ic2=Wm[0]==0;
Te[t_]:=kphi[t]*Ia[t];
P[t_]:=Te[t]*Wm[t];
eqn1=Ia'[t]==(1/La)*(Va[t]-Ra*Ia[t]-kphi[t]*Wm[t]);
eqn2=Wm'[t]==(1/J)*(kphi[t]*Ia[t]-Tm[t]-B*Wm[t]-Tic[t]-cw[t]*(Wm[t])^2-Ts);
sol=NDSolve[{eqn1,eqn2,ic1,ic2},{Ia[t],Wm[t]},{t,0,20},MaxSteps→10000,Method→
{"ExplicitRungeKutta","StiffnessTest"→False}];
Plot[Evaluate[Va[t]], {t, 0, 20}, PlotRange → All, AxesLabel → {"t (s)", "Va (V)"}]
Plot[Evaluate[Tm[t]], {t, 0, 20}, PlotRange → All, AxesLabel → {t (s), Tm (Nm)}]
Plot[Ia[t]/.sol, {t, 0, 20}, PlotRange → All, AxesLabel → {"t (s)", "Ia (A)"}]
Plot[Wm[t]/.sol, {t, 0, 20}, PlotRange → All, AxesLabel → {"t (s)", "Wm (rad/s)"}]
Plot[Te[t]/.sol, {t, 0, 20}, PlotRange → All, AxesLabel → {"t (s)", "Te (Nm)"}]
Plot[P[t]/.sol, {t, 0, 20}, PlotRange → All, AxesLabel → {"t (s)", "P (W)"}]
```

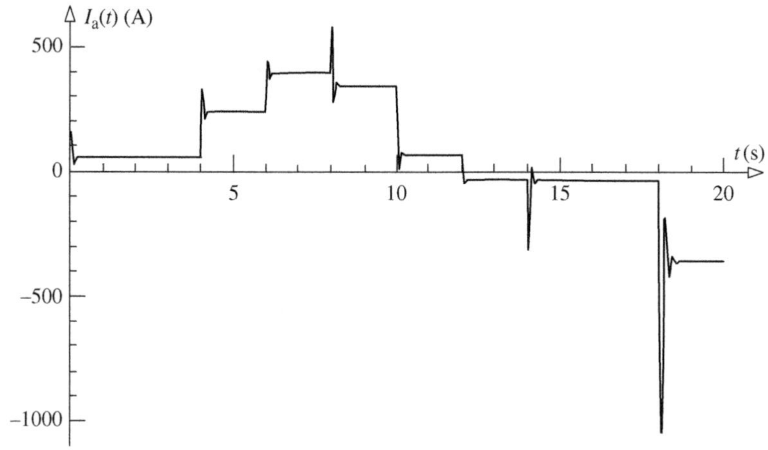

Figure E11.2.8 Calculated armature current $I_a(t)$ as a function of time.

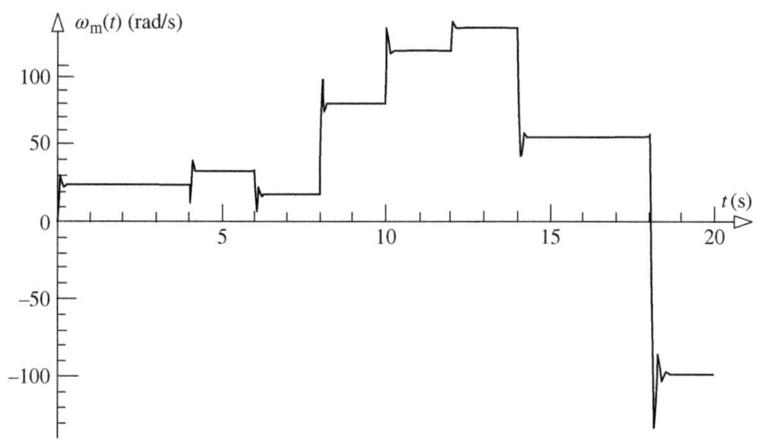

Figure E11.2.9 Calculated angular velocity $\omega_m(t)$ as a function of time.

Conclusions

An increase/addition of the flux makes the transient response more stable resulting in reduced armature current $I_a(t)$ as shown in Figures E11.2.4 and E11.2.8 and is therefore desirable [37–39].

c) Perform a Mathematica analysis for an armature current $I_a(t)$-dependent flux excitation of the machine, where

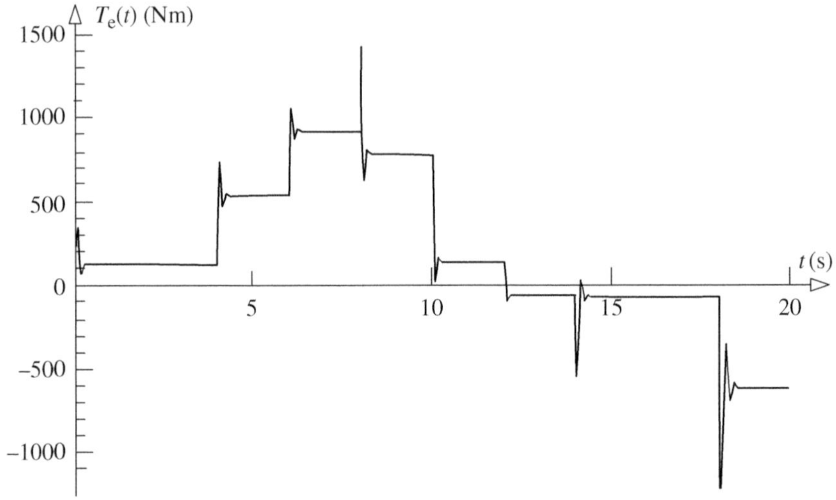

Figure E11.2.10 Calculated electrical torque $T_e(t)$ as a function of time.

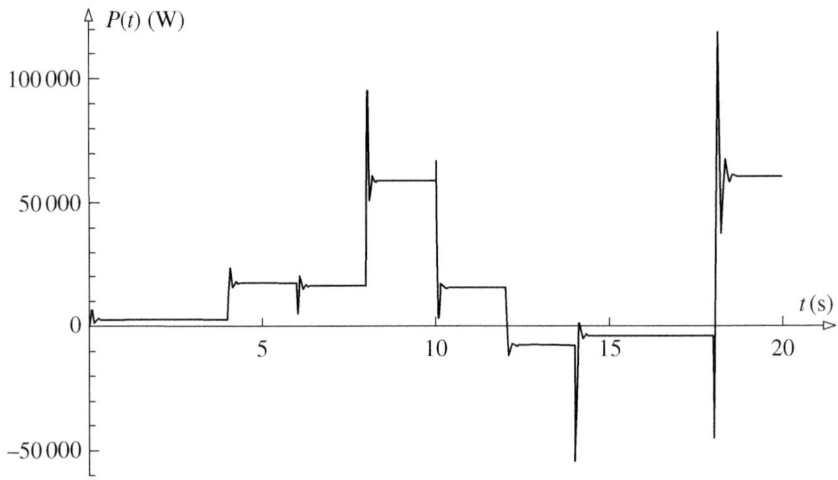

Figure E11.2.11 Calculated output power $P(t)$ as a function of time.

$$\text{kphi}[\text{t\_}] = (1 - 0.00055 * \text{Ia}[t]) * \text{kphi}[t];$$

i.e. when field flux weakening occurs as a function of $I_a(t)$ (e.g. due to armature reaction) as it occurs in all uncompensated machines and a variable input voltage $V_a(t)$ and the mechanical torque $T_m(t)$ as depicted in Figures E11.2.2 and E11.2.3, respectively. The armature current, the angular velocity, the electrical torque, and the output power are displayed in Figures E11.2.12–E11.2.15, respectively.

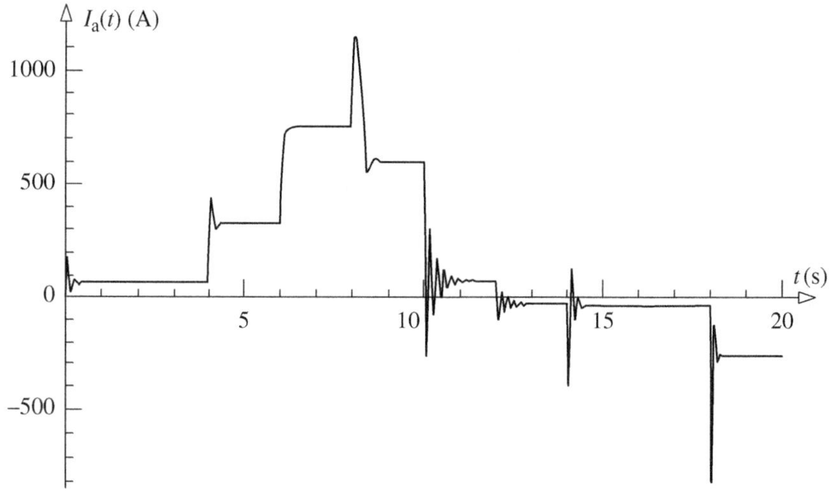

Figure E11.2.12 Calculated armature current $I_a(t)$ as a function of time.

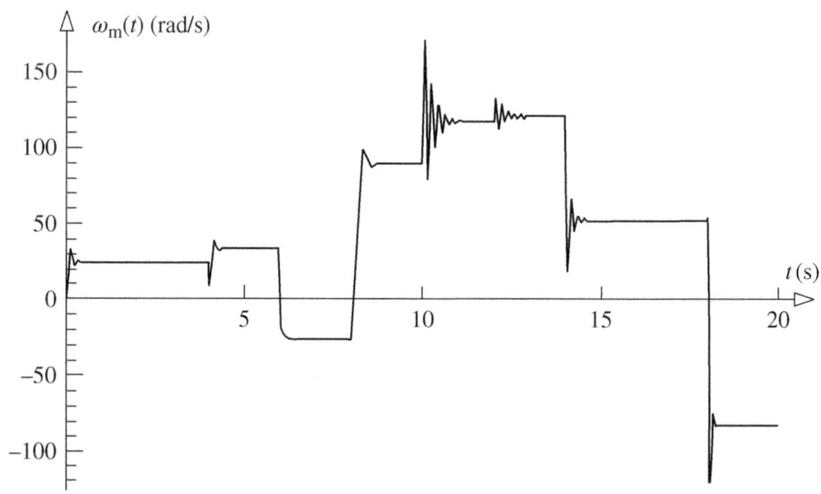

Figure E11.2.13 Calculated angular velocity $\omega_m(t)$ as a function of time.

Conclusions

A decrease/weakening of the field flux (e.g. due to armature reaction) makes the transient response less stable resulting in an increased armature current $I_a(t)$ as shown in Figures E11.2.4 and E11.2.12 and is therefore undesirable [37–39].

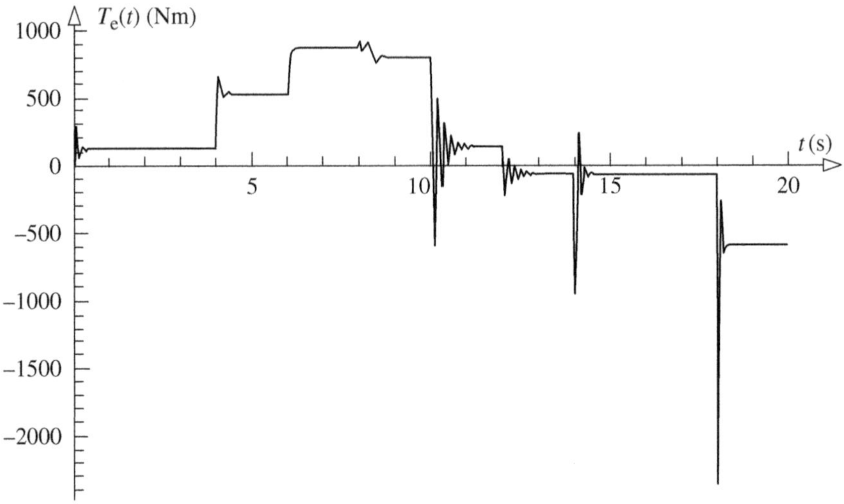

Figure E11.2.14 Calculated electrical torque $T_e(t)$ as a function of time.

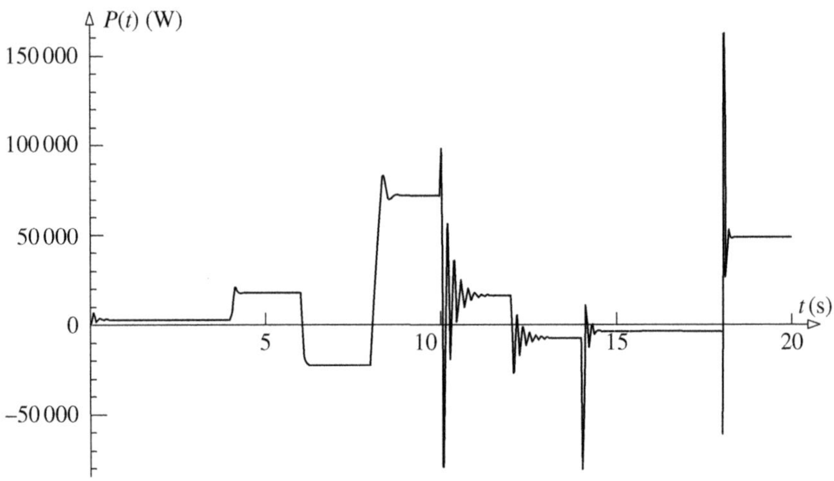

Figure E11.2.15 Calculated output power $P(t)$ as a function of time.

11.5 Summary

The commutation process and control of electric torque $T_e(t)$ and angular velocity $\omega_m(t)$ characteristics of DC machines – either motors or generators – can also be applied to AC machines combined with electronic rectifiers converting AC to DC and vice versa with inverters. The addition of (cumulative) flux makes the operation more stable, while the weakening of (differential) flux induces transient oscillations and results in less stable operation. The compensation of flux weakening is important because it results in precise torque/speed applications critical to electric automotive drives [37, 38]. Machines presently used for renewable energy generation and utilization, such as direct-drive permanent-magnet machines and DFIGs, are by-products of DC machine development where torque and angular velocity can be independently controlled [9, 15].

In Appendix 11.A, conformal mapping is applied to the magnetic field calculation within a rectangular slot as they occur in all rotating and linear electric machines. To make the determination of magnetic field calculations possible, the first-order, FD, and FE algorithms for various node/mesh/grid structures are introduced and applied to two-dimensional field configurations. These magnetic field distributions show that an insignificant amount of flux is passed through slots. For this reason, the Lorentz force [6–9, 20, 21] calculation cannot be applied; instead, the forces acting on the iron-core teeth must be determined from the Maxwell stress [6–9]. The calculation of the magnetic forces is difficult with the Maxwell stress, and therefore most textbooks rely on the Lorentz force approach, which is not correct.

Appendix 11.B introduces the calculation of linear and operating-point-dependent nonlinear inductances from flux plots. Application examples relating to steady-state and transient torque and speed control problems are solved using Mathematica.

In summary, one can state that for electromechanical (e.g. DC machines, AC induction machines, AC synchronous machines, permanent-magnet machines) and electric (e.g. transformers) energy conversion processes to occur, the corollary, as stated in Section 11.2, must be satisfied.

Problems

11.1 Starting of separately excited DC motor and its associated operating characteristics

a The measurement of the armature resistance R_a is accomplished with a water rheostat (equipment that controls DC current by varying its resistance $R_{rheostat}$) to limit the armature current I_a as illustrated in

Figure P11.1.1. For the measurement of the armature resistance, according to Figure P11.1.1, the following voltage and current values of Table P11.1.1 are recorded. Calculate and plot the armature resistance R_a as a function of I_a from the values listed in Table P11.1.1. Explain how the plasma between carbon brushes and copper segments of the commutator influences the value of $R_a = R_{a\_armature\ winding} + R_{a\_carbon\ brushes} + R_{a\_plasma}$. Estimate the value of $(R_{a\_armature\ winding} + R_{a\_carbon\ brushes})$.

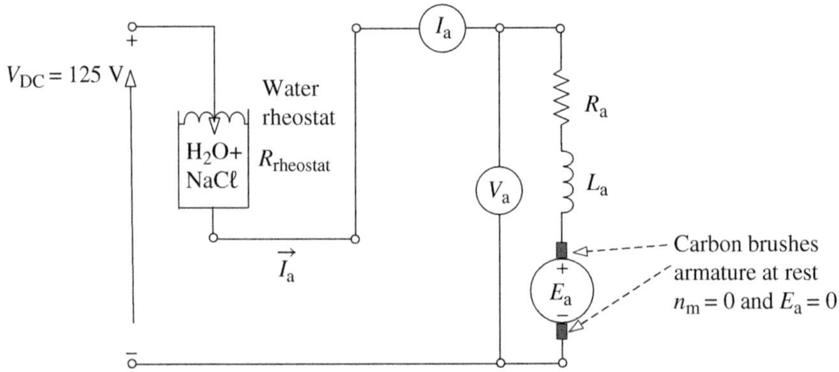

Figure P11.1.1 Circuit [39] for measuring the armature resistance R_a.

Table P11.1.1 Voltage–current values for voltage-drop method of measuring the armature resistance R_a [Ω] of separately excited DC motor, recorded within a few minutes.

V_a (V)	0.6	2.1	4.9
I_a (A)	5	20	70

b The starting of separately excited DC motors must be accomplished with a starting box as shown in Figure P11.1.2. Such starting boxes, where at start the induced voltage of the motor E_a is zero, must be used for most electric rotating machines. The starting current I_a is recorded with the field fully excited $I_{f\_rated} = 1.75$ A, $V_a = 125$ V, and with the starter in the first position ($R_{start} = 0$), one obtains $I_{a\_start\_max} = 179$ A. If the starting is completed, with the field current fully excited, and the starter is fully on (when attracted by the magnet as indicated in Figure P11.1.2, $R_{start} = R_{start1} + R_{start2} = 0.64\ \Omega$ short circuited by relay$_1$ and relay$_2$), then the armature current is $I_{a\_starting\ completed} = 8$ A at the terminal voltage $V_a = 125$ V. Explain the behavior of the armature current and estimate the starting current I_{astart_max}.

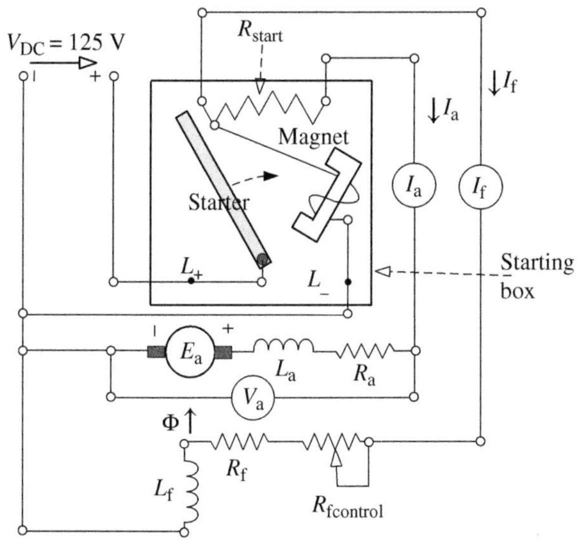

Figure P11.1.2 Starting circuit [39] of separately excited DC motor including manually operated starting box.

Figure P11.1.3 illustrates an automatic starting box consisting of two starting resistors R_{start1} and R_{start2}, which are bypassed by relays relay$_1$ and relay$_2$ after successful starting.

From Table P11.1.1 one obtains for the armature resistance $R_a = 4.9$ V/70 A $= 0.07\ \Omega$, resulting in an initial maximum starting current of $I_{a\_max} = 125$ V/0.07 $\Omega = 179$ A, as depicted in Figure P11.1.4. The starting of this motor lasts about 16 s, as indicated in Figure P11.1.4. Note that the maximum speed with a small overshoot is reached within 1 s.

c Steady-state speed characteristics of separately excited DC motor at no load ($T_e \approx 0$) are given by V_a, I_a, n_m, and I_f of Table P11.1.2. Plot $V_a = f(I_a)$ as given in Table P11.1.2, and $R_a \approx 0.06\ \Omega$ as obtained in Part a for I_a values much larger than 70 A, say 179 A. Calculate/estimate based on the data of Table P11.1.2 the $(k_e\phi)$-values as a function of the listed operating points given by n_m, and I_a from $(k_e\phi) = E_a/\omega_m$, where $\omega_m = 2\pi \cdot n_{rps} = (2\pi \cdot n_{rpm})/60$ with the speed measured in revolutions per second (rps) or revolutions per minute (rpm).

d Plot $(k_e\phi)$ as a function of I_f and explain this nonlinear relationship in terms of saturation and armature reaction.

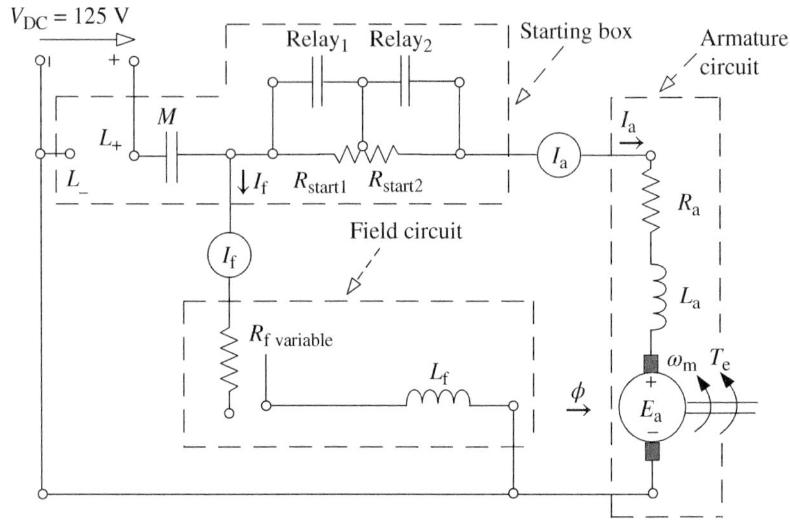

Figure P11.1.3 Separately excited DC motor with automatic starting box consisting of two starting box resistances R_{start1}, R_{start2}. M is the main relay, and relay$_1$, relay$_2$ bypass/short-circuit the starting box resistances R_{start1}, R_{start2}, respectively.

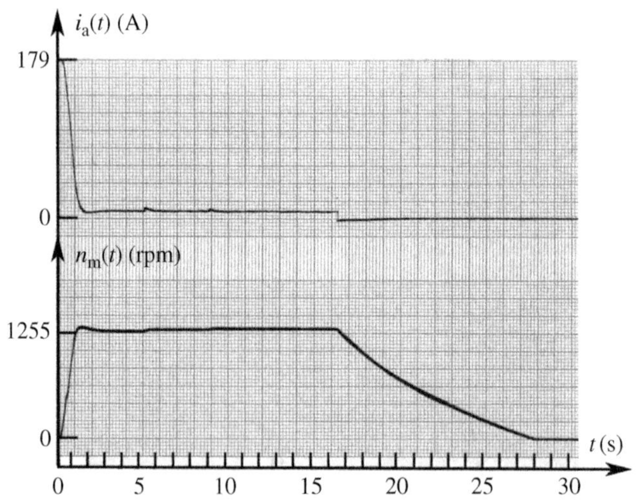

Figure P11.1.4 Transient armature current $i_a(t)$ in ampere (A) and transient speed $n_m(t)$ in revolutions per minute (rpm) during start-up of a separately excited DC motor with automatic starting box consisting of two resistances R_{start1} and R_{start2}.

Table P11.1.2 Variation of speed n_m with excitation current I_f for no-load conditions, $T_e \approx 0$.

V_a (V)	125	125	125	124	124	124	124	124
I_a (A)	9.5	10.0	10.5	10.8	11	11.5	12.0	13.0
n_m (rpm)	995	1042	1104	1152	1204	1293	1321	1511
I_f (A)	2.5	2.25	2.0	1.88	1.75	1.63	1.5	1.25

11.2 Steady-state rated analysis of separately excited DC motor.

A $V_a = 125$ V, $n_m = 265$ rpm separately excited DC motor has armature resistance $R_a = 0.105\ \Omega$, field resistance of $(R_f + R_{f\,control}) = 36.77\ \Omega$, and flux coefficient $(k_e\phi) = 4.32$ Vs/rad. The rotational (friction, windage) loss is $P_{friction\_windage} = 455$ W, and the iron-core loss is $P_{core} = 250$ W.

Compute for rated conditions

a angular speed ω_m,
b induced voltage E_a,
c armature current I_a,
d armature input power P_a,
e field current I_f,
f losses in field winding P_f,
g total input power P_{in},
h electric torque T_e,
i output power P_{out}, and
j efficiency η.

11.3 Steady-state operating characteristics of separately excited DC generator.

Figure P11.3.1 depicts the equivalent circuit where a water rheostat is used as a load. The motor of Problem 11.2 is now employed as a generator with the data $V_a = 125$ V, $n_m = 265$ rpm, a separately excited DC motor with armature resistance $R_a = 0.105\ \Omega$, field resistance of $(R_f + R_{f\,control}) = 36.77\ \Omega$, and flux coefficient $(k_e\phi) = 4.32$ Vs/rad. Rotational (friction, windage) loss is $P_{friction\_windage} = 255$ W, and iron-core loss is $P_{core} = 250$ W.

In Table P11.3.1 the armature voltage V_a, armature current I_a, and the speed n_m are listed, while the excitation current I_f is maintained constant.

a Plot V_a as a function of I_a and apply the least squares line or regression line [40] for the $V_a = f(I_a)$ characteristic to obtain a linear equation for $V_a = f(I_a)$.

b Determine the least squares error [40] E between the regression line and given actual data points.

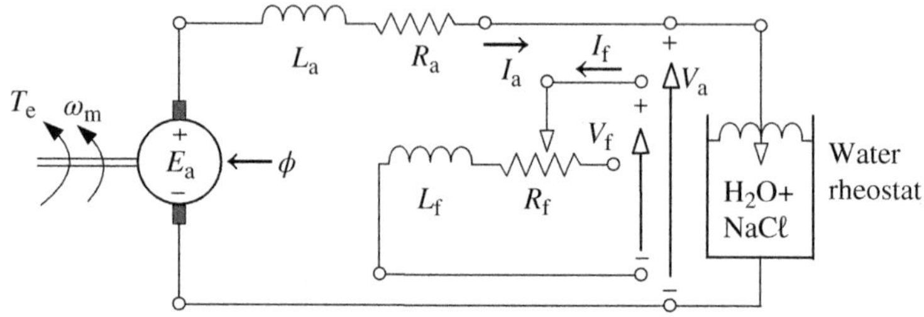

Figure P11.3.1 Equivalent circuit of separately excited DC generator supplying power to water rheostat.

Table P11.3.1 Armature voltage V_a, armature current I_a, and speed n_m.

V_a (V)	125	118	114	111	108	107
I_a (A)	10	15	30	45	60	72
n_m (rpm)	280	277	272	271	270	270

c Determine the voltage drop due to armature $V_{armature\_reaction}$ at I_a = 72 A.

d What is the reduction of flux ϕ due to armature reaction? That is, what is the influence of the armature (leakage) inductance L_a on the terminal voltage V_a? Compute for rated conditions

e angular speed ω_m,

f specify the armature current I_a,

g induced voltage E_a,

h armature output power P_a,

i field current I_f,

j losses in field winding P_f,

k electric input torque T_e,

l total input power P_{in}, and

m efficiency η.

11.4 Steady-state analysis of separately excited DC generator

A P_{out} = 9.7 kW, V_a = 107 V, n_m = 1100 rpm, $k_e\phi$ = 1.3 Vs/rad separately excited DC generator has armature resistance R_a = 0.3 Ω and field resistance of R_f = 36.77 Ω; see Figure P11.3.1. Rotational (friction, windage) loss is $P_{friction\_windage}$ = 890 W and iron-core loss is P_{core} = 500 W.

Compute for rated conditions

a angular speed ω_m,

b armature output power $P_a = P_{out}$ delivered to water rheostat,

c armature current I_a,
d induced voltage $E_{a,}$
e field current I_f,
f losses in field winding P_f,
g electric input torque T_e,
h total input power P_{in}, and
i efficiency η.

11.5 Transient analysis of separately excited DC machine as defined by Figure E11.1.1 using Mathematica [10, 31–36].

Hint: Before this problem is solved, it is advised to read and understand Application Example 11.2.

A $P = 10$ kW, $V_a = 125$ V, $n_m = 173$ rpm, $(k_e\phi) = 4.32$ Vs/rad separately excited DC machine has the armature resistance $R_a = 0.07\,\Omega$, armature inductance $L_a = 5$ mH, polar moment of inertia $J = 0.8$ kgm^2, damping coefficient of the viscous torque $B = 0.3$ Nm/(rad/s), torque at standstill $T_s = 0$, Coulomb frictional torque $T_c = \text{Tic} = 100$ Nm, and the coefficient of windage torque $c_w = 0.002$ Nm/(rads)^2. Investigate the behavior of the machine as a motor under variable load/speed conditions and as a generator with reverse angular velocity whereby the terminal armature voltage $V_a(t)$ is defined by

Va[t_]:=If[t < 55.0, If[t < 35, If[t < 8, If[t < 4, 8, 75], 125], 100], – 100]

(P11.5.1)

The mechanical load torque doing the work $T_m(t)$ is variable as defined below:

Tm[t_]:=If[t < 65.0, If[t < 60.0, If[t < 55.0, If[t < 40.0, If[t < 15.0, If[t < 10.0, If[t < 5.0, 0, 200], 300], 400], 200], 100.0], 100.0], – 400]

(P11.5.2)

a Perform a Mathematica analysis for a constant or rated/nominal DC excitation of the machine, where $(k_e\phi) = 4.32$ Vs/rad, that is,

kphi[t_]:=If[*t* < 8.0, 4.32, 4.32]

and a variable input voltage $V_a(t)$ as defined above Eq. (P11.5.1), whereby the mechanical torque is given by Eq. (P11.5.2).

b Perform a Mathematica analysis for an armature current $I_a(t)$-dependent DC excitation of the machine, where

kphi[t_] = (1 + 0.002∗Ia[*t*])∗kphi[*t*]

i.e. when a field increase, due to cumulative (cum) overcompensation, occurs (see Figure E11.2.1) as a function of I_a through overcompensation of armature reaction as is done in Application Example 11.2, Part b, and a variable input voltage $V_a(t)$ and the mechanical torque $T_m(t)$ are as defined above by Eqs. (P11.5.1) and (P11.5.2), respectively.

c Perform a Mathematica analysis for an armature current $I_a(t)$-dependent DC excitation of the machine, where

$$\text{kphi}[\text{t}\_] = (1 - 0.0005 * \text{Ia}[t]) * \text{kphi}[t]$$

i.e. when a field decrease due to differential (diff) weakening occurs (see Figure E11.2.1) as a function of I_a, due to armature reaction in all uncompensated machines, whereby a variable input voltage $V_a(t)$ and the mechanical torque $T_m(t)$ are as defined above by Eqs. (P11.5.1) and (P11.5.2), respectively.

References

1 Richter, R. (1924). *Elektrische Maschinen, Volume 1: Allgemeine Berechnungselemente, Die Gleichstrommaschinen*. Berlin: Verlag Julius Springer.

2 Nave, C.R. http://hyperphysics.phy-astr.gsu.edu/hbase/magnetic/magfor.html. (accessed 22 May 2020).

3 Olver, P.J. (2018). Complex analysis and conformal mapping, pp. 1–84. http://www.math.umn.edu/~olver/ln_/cml.pdf (accessed 22 May 2020).

4 Frey, K. (1921). Anwendungen der konformen Abbildung auf praktische Probleme des Elektromaschinenbaus. Arbeiten aus dem Elektrotechnischen Institut der Universiät Karlsruhe, Band 4, 9 February 1921.

5 Weisstein, E.W. Laplace's equation. Last updated 1 May 2020. http://mathworld.wolfram.com/LaplacesEquation.html (accessed 22 May 2020).

6 Stratton, J.A. (1941). *Electromagnetic Theory*, 153–155. New York: McGraw-Hill Book Company, Inc.

7 Carpenter, C.J. (1960). Surface integral methods of calculating forces on magnetized iron parts. *Proceedings of the IEE-Part C: Monographs* 106C: 19–28.

8 Reichert, K., Freundl, H., and Vogt, W. (1976). The calculation of forces and torques within numerical magnetic field calculation methods. *Proceedings of Compumag Conference*, London.

9 Fuchs, E.F. and Masoum, M.A.S. (2012). *Power Conversion of Renewable Energy Systems*, 1e, 2nd printing. USA: Springer, 692 pp.

10 Wolfram. Try Mathematica for free. https://www.wolfram.com/mathematica/trial (accessed 22 May 2020).

11 Yildirim, D. (1999). Commissioning of 30 kVA variable-speed, direct-drive wind power plant. Doctoral dissertation. University of Colorado at Boulder.

12 Yildirim, D., Fuchs, E.F., and Batan, T. (1998). Test results of a 20 kW, direct-drive, variable-speed wind power plant. *Proceedings of the International Conference on Electrical Machines*, Istanbul, Turkey (2–4 September 1998), pp. 2039–2044.

13 Huang, H. and Fuchs, E.F. (1990). Successive nonlinear quadratic programming for high-power density and high-efficiency, permanent-magnet motors. *Proceedings of the International Conference on Electrical Machines,* Cambridge, MA (12–15 August 1990).

14 Huang, H., Anderson, W.M., and Fuchs, E.F. (1990). High-power density and high-efficiency motors for electric vehicle applications. *Proceedings of the International Conference on Electrical Machines*, Cambridge, MA (12–15 August 1990).

15 Dubey, G.K. (1989). *Power Semiconductor Controlled Drives*. Englewood Cliffs, NJ 07632: Prentice Hall.

16 Thomson, E. (1881). Commutators for dynamo electric machines. US Patent 242,488, 7 June 1881.

17 Jacobs, H. (1881). Commutator for magneto electric machines. US Patent 246,612, 6 September 1881.

18 Arnold, E. (1907). *Die Gleichstrommaschine: Bd. Konstruktion, Berechnung und Arbeitsweise*. Berlin: Julius Springer Verlag.

19 Florida Center for Instructional Technology, College of Education, University of South Florida. Gramme ring armature. http://etc.usf.edu/clipart/20800/20802/gramringarm_20802.htm (accessed 22 May 2020).

20 Nave, C.R. http://hyperphysics.phy-astr.gsu.edu/hbase/vvec.html (accessed 22 May 2020).

21 DigiCert Global Root CA. https://cnx.org/resources/db92fac564a0ca02ce646330f1d43c178c509a86/Figure_23_07_02a.jpg (accessed 22 May 2020).

22 Fuchs, E.F. and Frank, U.V. (1983). High-speed motors with reduced windage and eddy current losses, part I: mechanical design. *ETZ – Archiv* 5 (H.1): 17–23.

23 Fuchs, E.F. and Frank, U.V. (1983). High-speed motors with reduced windage and eddy current losses, part II: magnetic design. *ETZ – Archiv* 5 (H.2): 55–62.

24 Taylor, P.L. (1970). Improving commutation in d.c. machines by the use of flux traps. *Proceedings of the Institution of Electrical Engineers* 117 (7): 1269–1276.

25 Erdelyi, E.A. and Fuchs, E.F. (1970). *Non-Linear Electrical Generator Study: Final Report on Phase "B", Part I Non-Linear D.C. Generators*. Contract No. N00014-67-A-405-003, Project No. 097-374/5-19-70 (Code 473). Prepared for the Office of Naval Research, Arlington, VA.

26 Erdelyi, E.A. and Fuchs, E.F. (1970). Nonlinear magnetic field analysis of DC machines, part I: theoretical fundamentals. *IEEE Transactions on Power Apparatus and Systems* PAS-89 (7): 1546–1554.

27 Fuchs, E.F. and Erdelyi, E.A. (1970). Nonlinear magnetic field analysis of DC machines, part II: application of the improved treatment. *IEEE Transactions on Power Apparatus and Systems* PAS-89 (7): 1555–1564.

28 Erdelyi, E.A., Fuchs, E.F., and Binkley, D.H. (1970). Nonlinear magnetic field analysis of DC machines, part III: equipotential plots drawn by computer. *IEEE Transactions on Power Apparatus and Systems* PAS-89 (7): 1565–1583.

29 Masoum, M.A.S. and Fuchs, E.F. (2015). *Power Quality in Power Systems and Electrical Machines*, 2e. Amsterdam: Elsevier/Academic Press, 1123 pp.

30 Chapman, S.J. (1991). *Electric Machinery Fundamentals*, 2e. New York: McGraw-Hill, Inc.

31 Hastings, C., Mischo, K., and Morrison, M. (2016). *Hands-on Start to Wolfram Mathematica and Programming with the Wolfram Language*, 2e. Wolfram Media, Inc., 494 pp.

32 Wolfram, S. (2017). *An Elementary Introduction to the Wolfram Language*, 2e. Wolfram Media, Inc.

33 Wolfram (2020). Mathematica quick revision history – Mathematica 12.1. http://www.wolfram.com/mathematica/quick-revision-history.html (accessed 22 May 2020).

34 Wolfram. https://account.wolfram.com/auth/sign-in#downloads (accessed 22 May 2020).

35 Wolfram Language & System Documentation Center. Stiffness detection. http://reference.wolfram.com/language/tutorial/NDSolveStiffnessTest.html#1559192564 (accessed 22 May 2020).

36 Wolfram Language & System Documentation Center. Advanced numerical differential equation solving in the Wolfram language. http://reference.wolfram.com/language/tutorial/NDSolveOverview.html (accessed 22 May 2020).

37 Fuchs, E.F. (2012). Alternating current machine with increased torque above and below rated speed for hybrid/electric propulsion systems. US Patent No. 8,183,814 B2, 22 May 2012.

38 Fuchs, E.F. (2013). "Alternating current machine with increased torque above and below rated speed for hybrid/electric propulsion systems. US Patent No. 8,436,569 B2, 7 May 2013.

39 Hanna, W.J. and Fuller, J.F. (1975). *EE 354 Laboratory Notes*. Electrical Engineering Department of the University of Colorado at Boulder.

40 Wolfram Language & System Documentation Center. LeastSquares. https://reference.wolfram.com/language/ref/LeastSquares.html (accessed 22 May 2020).

Appendix 11.A Magnetic Field Computation Based on Numerical Methods

Closed-Form Solutions of Laplace Equation Based on Conformal Mapping: The magnetic field within a slot pitch with given finite slot depth and width is found using the method of conformal transformation. Frey [1] in his doctoral dissertation investigated several magnetic field configurations. However, Frey did not solve the complex elliptic integral for the case of the slot with finite dimensions, and therefore the analysis [2] represents an extension of Frey [1]. Two different

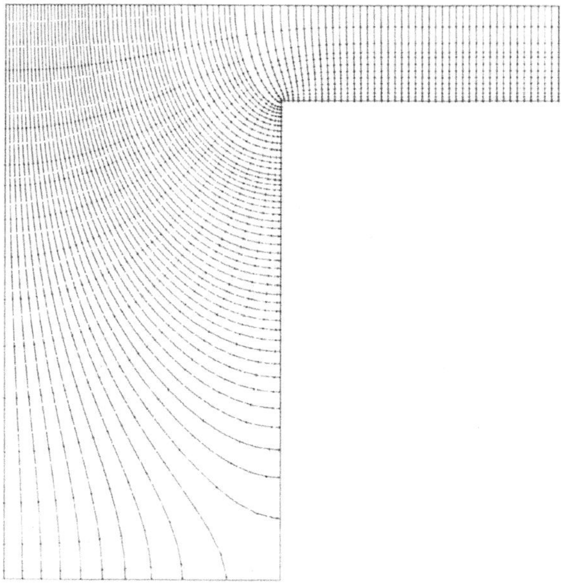

Figure 11.A.1 Vector equipotential lines (specified by potential ϕ_i) for a rectangular slot [2]. The solid lines enclose unit flux tubes, and the dashed lines divide some of the unit flux tubes into tubes of equal flux.

transformations are used to solve the complex elliptic integral, and both transformations yield identical results as shown in Figure 11.A.1 up to the eighth position after the decimal point of the coordinate values (x_i, y_i) of the potential ϕ_i.

Unfortunately, the magnetic field calculation for electric machines is very difficult to perform based on conformal transformation, and nonlinear effects such as saturation cannot be taken into account. For this reason, engineers have searched during the past 50 years for alternate ways of computing magnetic fields; two possibilities are numerical approaches based on the FD [3–12] or the finite-element (FE) [5–9, 13] methods.

Finite-Difference (FD) Method: A general first-order triangular algorithm in rectangular (x, y) coordinates was derived by Winslow [2, 3]. The vectors used for the derivation are defined in Figure 11.A.2b. The vector potential A_o at node o is a function of the potentials A_ℓ of the surrounding $N = 6$ vertices and the current density $J_{\ell+1/2}$ usually measured in (A/mm^2) within the N triangles with the area $(3a_{\ell+1/2})$ and the reluctivity $\nu_{\ell+1/2}$ measured in (ampere-turns/Vs), where Vs is called weber or in short (Wb). The following relation for Figure 11.A.2a with $N = 6$ triangular meshes is [2]

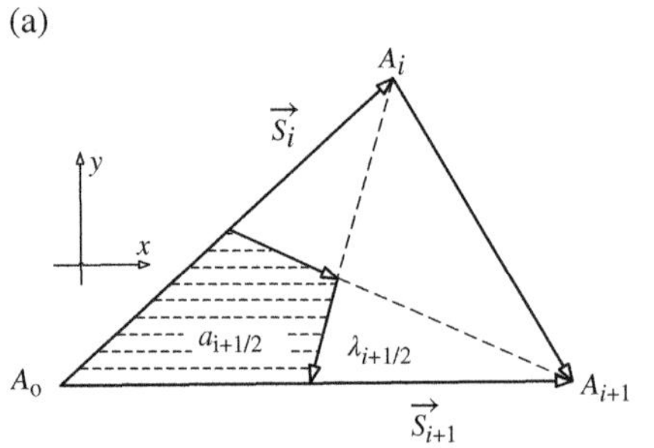

Figure 11.A.2a Vectors used for flux calculation [3]. The potential A_o is a function of the potentials A_ℓ of the surrounding nodes ranging from $\ell = (1\ldots i\ldots N)$ vertices of the current density $J_{\ell+1/2}$ within the N triangles with the area $a_{\ell+1/2}$ and the reluctivity $\nu_{\ell+1/2}$.

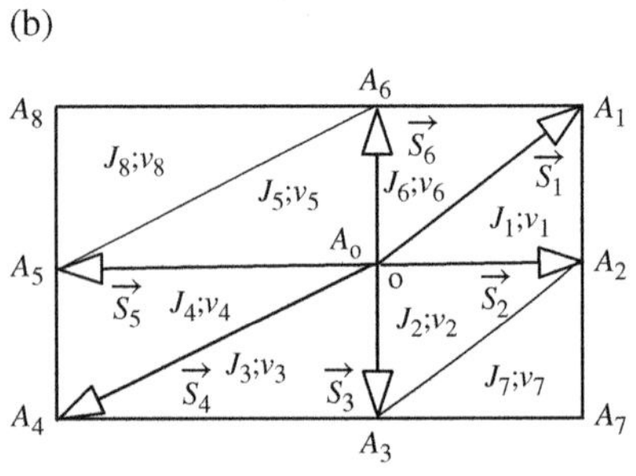

Figure 11.A.2b Vectors used for flux calculation based on triangular first-order finite-difference algorithm [2] as applied to rectangular six-element mesh block with one axis (horizontal) of symmetry.

$$\frac{1}{2}\sum_{\ell=1}^{N=6}\left\{\frac{\nu_{\ell+½}}{\left(\vec{S}_\ell\angle-\pi/2\right)\cdot\vec{S}_{\ell+1}}\vec{S}_{\ell+1}\cdot\left(\vec{S}_{\ell+1}-\vec{S}_\ell\right)+\frac{\nu_{\ell-½}}{\left(\vec{S}_\ell\angle-\pi/2\right)\cdot\vec{S}_\ell}\vec{S}_{\ell-1}\cdot\left(\vec{S}_{\ell-1}-\vec{S}_\ell\right)\right\}$$

$$(A_\ell - A_o) = -\sum_{\ell=1}^{N=6} J_{\ell+½}\cdot a_{\ell+1/2} \tag{11.A.1a}$$

where the area defined as

$$a_{\ell+1/2} = a_{\ell+1/2}^{\text{total}}/3 \tag{11.A.1b}$$

is the area of the shaded quadrilateral associated with a given vertex as shown in Figure 11.A.2a and the vector $\left(\vec{S}_\ell\angle-\pi/2\right)$ represents the vector $\vec{S}_\ell$ (see Eq. (11.A.1a)) rotated clockwise by an angle $\pi/2$.

A_ℓ are the unknown vector potentials, and $\nu_{\ell+½}$ are the material's reluctivities or magnetic resistances. $J_{\ell+½}$ are the current densities, $a_{\ell+1/2}$ are the areas of the triangular elements, and $\vec{S}_\ell$ are the vectors from the central point A_o to the ℓth point with the unknown vector potential A_ℓ, as seen in Figure 11.A.2b. Note that the numbers are cyclic; that is, $\ell - 1 = 6$ for $\ell = 1$, and $\ell + 1 = 1$ for $\ell = 6$. If the inner products, identified by $(\cdot)$ in Eq. (11.A.1a), are evaluated for the rectangular mesh block of Figure 11.A.2b, it becomes clear that the potentials A_1 and A_4 have no effect on the solution since their coefficients are zero for the rectangular mesh constraints $J_1 = J_6$; $J_3 = J_4$; $J_2 = J_7$; $J_5 = J_8$; $\nu_1 = \nu_6$; $\nu_3 = \nu_4$; $\nu_2 = \nu_7$; $\nu_5 = \nu_8$. Figure 11.A.2c illustrates the numerically calculated magnetic field distribution based on the vector potential A_ℓ by applying Eqs. (11.A.1a) and (11.A.1b) to a rectangular slot pitch configuration.

A general first-order algorithm for polar (r, φ) coordinates was derived by Jackson and Erdelyi [4] and subsequently applied by Fuchs et al. [5–9]. The derivation

(c)

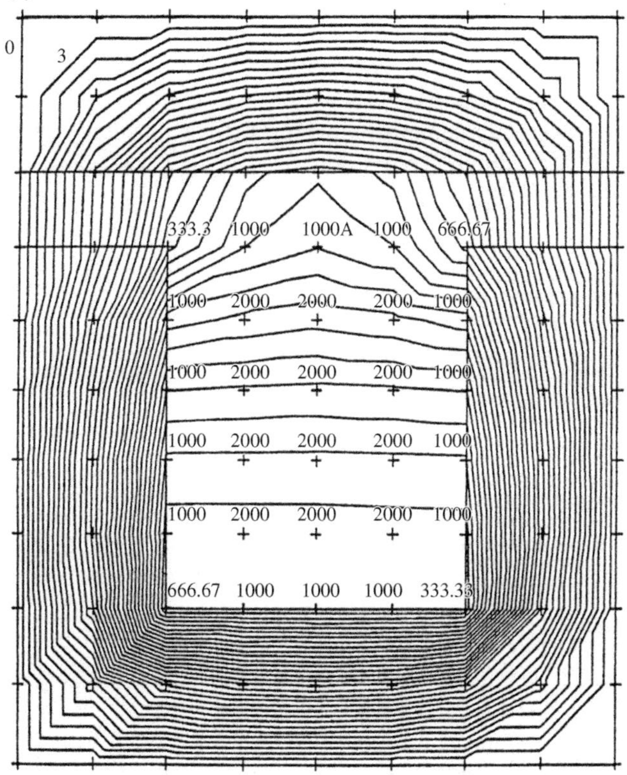

Figure 11.A.2c Equidistant vector potential (A_ℓ) solution [2] for a simple one-slot configuration based on the triangular first-order finite-difference algorithm of Winslow. All vector potentials must be multiplied by 10^{-3} Wb/m. Current distribution and resulting magnetic field are asymmetric.

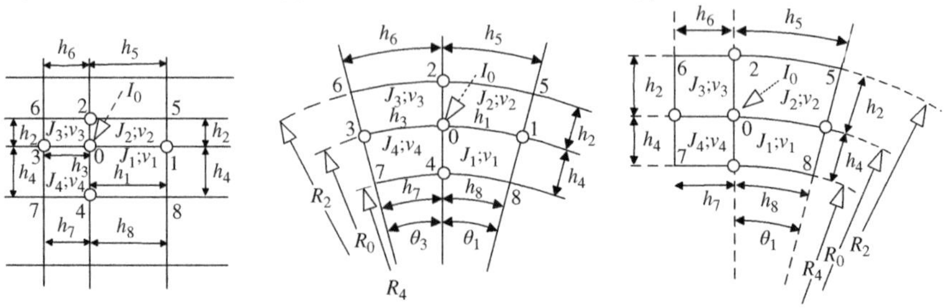

Figure 11.A.3 Relations used for flux calculation [2] based on rectangular (a), polar (b), or a combination of both also called polycentric (c) first-order finite-difference algorithm.

of the first-order orthogonal algorithm, based on Stokes' theorem [10], is given by Erdelyi and Fuchs [11]. The individual mesh blocks in the grid may be square, rectangular, polar, or a combination of these. The relations used for the first-order orthogonal algorithm are defined in Figure 11.A.3a–c.

In order to facilitate the logic of the computer program [12], it is advantageous to use an algorithm valid for all the different types of meshes. Using Stokes' theorem, it has been found that the vector potential at a grid point (J, K), denoted as A_o, can be expressed as

$$A_o = \left(I_o + \sum_{i=1}^{4} \alpha_i A_i \right) \Big/ \sum_{i=1}^{4} \alpha_i, \text{ for } i = 1, 2, 3, 4 \tag{11.A.2}$$

In this equation the α_i are coefficients that are functions of the mesh geometry and the reluctivities, also called magnetic resistances, as follows:

$$\alpha_1 = (\nu_1 h_2 + \nu_1 h_4)/2h_1; \alpha_2 = [\nu_2(h_1 + h_5) + \nu_3(h_3 + h_6)]/4h_2, \tag{11.A.3}$$

$$\alpha_3 = (\nu_3 h_2 + \nu_4 h_4)/2h_3; \alpha_4 = [\nu_4(h_3 + h_7) + \nu_1(h_1 + h_8)]/4h_4. \tag{11.A.4}$$

The current term I_o, a point function dependent on the current densities and mesh geometry, is

$$I_o = \frac{1}{8}[J_1(h_1 + h_8)h_4 + J_2(h_1 + h_5)h_2 + J_3(h_3 + h_6)h_2 + J_4(h_3 + h_7)h_4. \tag{11.A.5}$$

Figure 11.A.3d illustrates the numerically calculated magnetic field distribution based on the vector potential A_o by applying Eqs. (11.A.2)–(11.A.5) to a rectangular slot pitch configuration.

Finite-Element Method: A general algorithm for the first-order triangular elements was derived by Chari [13]. The definition of vertex coordinates for the algorithm is defined in Figure 11.A.4a.

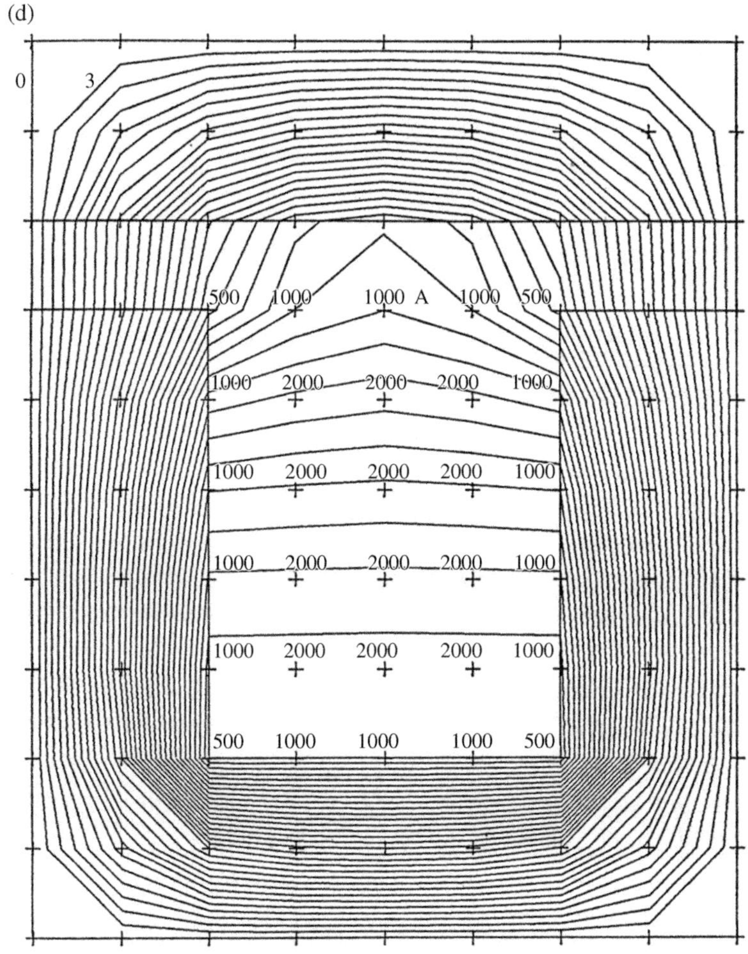

Figure 11.A.3d Equidistant vector potential (A_o) solution [2] for a simple one-slot configuration based on the square finite-difference algorithm of Jackson and Erdelyi [4]. All vector potentials must be multiplied by 10^{-3} Wb/m. Current distribution and resulting magnetic field are symmetric.

For only one triangle of a triangular grid/mesh system as shown in Figure 11.A.4a, a suitable discrete first-order expression of the energy functional [13]

$$F = \sum\left(\iint\left[\int_0^B \nu b \, \mathrm{d}b\mathrm{d}x\mathrm{d}y\right] - \iint J \cdot A\mathrm{d}x\mathrm{d}y\right) \tag{11.A.6}$$

is given by the matrix equation

(a)

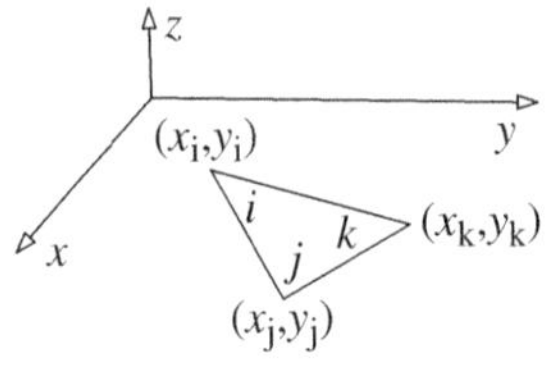

Figure 11.A.4a Definition of vertex coordinates used for flux calculation based on first-order triangular finite-element algorithm [2].

(b)

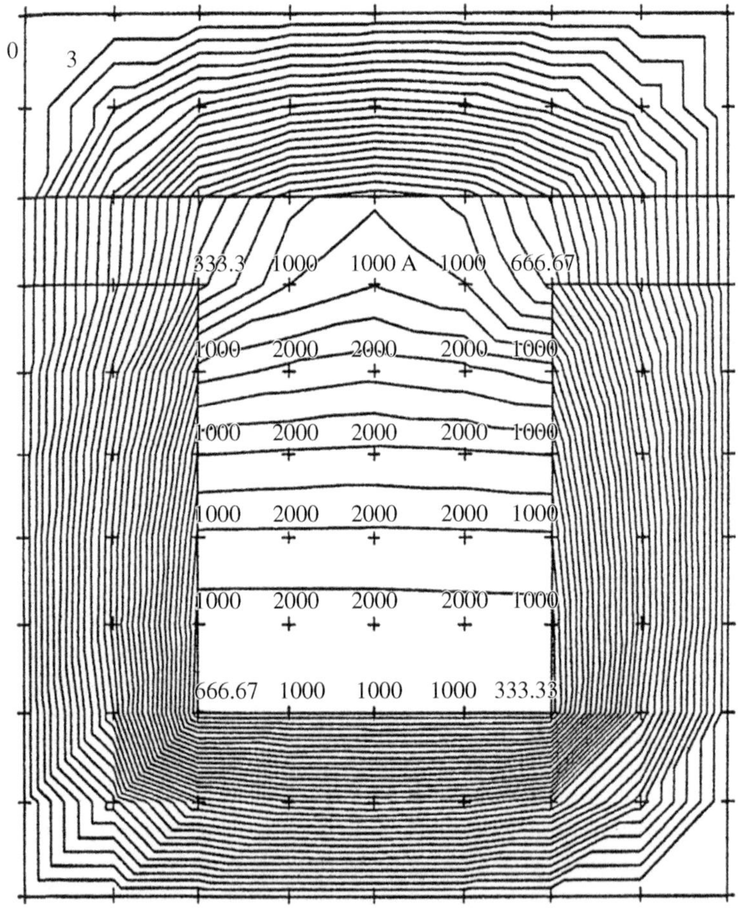

Figure 11.A.4b Equidistant vector potential (A_ℓ) solution [2] for a simple one-slot configuration based on the triangular first-order finite-element algorithm of Fuchs and McNaughton [2] and Chari [13]. All vector potentials must be multiplied by 10^{-3} Wb/m. Current distributions and resulting magnetic field are asymmetric.

$$\frac{\upsilon}{4\Delta}\begin{vmatrix} (b_i^2 + c_i^2) & (b_i b_j + c_i c_j) & (b_i b_k + c_i c_k) \\ (b_j b_i + c_j c_i) & \left(b_j^2 + c_j^2\right) & (b_j b_k + c_j c_k) \\ (b_k b_i + c_k c_i) & (b_k b_j + c_k c_j) & (b_k^2 + c_k^2) \end{vmatrix} \cdot \begin{vmatrix} A_i \\ A_j \\ A_k \end{vmatrix} = J\Delta \begin{vmatrix} 1/3 \\ 1/3 \\ 1/3 \end{vmatrix}, \tag{11.A.7}$$

where $b_i = Y_j - Y_k$; $c_i = X_k - X_j$; $b_j = Y_k - Y_i$; $c_j = X_i - X_k$; $b_k = Y_i - Y_j$; $c_k = X_j - X_i$ and the Δ is the area of the triangle under consideration. Figure 11.A.4b illustrates the numerically calculated magnetic field distribution based on the vector potential A_ℓ by applying Eq. (11.A.7) to a rectangular slot pitch configuration.

Conclusions

The algorithms of the finite-difference and finite-element methods are identical. That is, for the same mesh/element system, one obtains an identical system of simultaneous equations that can be solved for linear configurations (neglecting saturation) by closed-form matrix solvers or by iterative solution methods. For nonlinear equation systems taking saturation into account, iterative solution methods (e.g. Newton–Raphson, block iteration [14–16]) must be used.

References

1 Frey, K. (1921). Anwendungen der konformen Abbildung auf praktische Probleme des Elektromaschinenbaus. Arbeiten aus dem Elektrotechnischen Institut der Universiät Karlsruhe, Band 4, 9 February 1921.

2 Fuchs, E.F. and McNaughton, G.A. (1981). *Comparison of First-Order Finite Element with First-Order Partial Difference Methods as Applied to Magnetic Field Computations*. Report prepared for the General Electric Company by the Department of Electrical Engineering, University of Colorado, Boulder, CO, 31 October 1981.

3 Winslow, A.M. (1967). Numerical solution of the quasilinear Poisson equation in a nonuniform triangle mesh. *Journal of Computational Physics* 2: 149–172.

4 Jackson, R.F. and Erdelyi, E.A. (1968). Combination and separation of coordinates and modular programming for DC machine fields. *IEEE Transactions on Power Apparatus and Systems* 87: 659–664.

5 Fuchs, E.F. and McNaughton, G.A. (1982). Comparison of first-order finite difference and finite element algorithms for the analysis of magnetic fields, part I:

theoretical analysis. *IEEE Transactions on Power Apparatus and Systems* PAS-101 (5): 1170–1180.

6 McNaughton, G.A. and Fuchs, E.F. (1982). Comparison of first-order finite difference and finite element algorithms for the analysis of magnetic fields, part II: numerical results. *IEEE Transactions on Power Apparatus and Systems* PAS-101 (5): 1181–1201.

7 Fuchs, E.F. and McNaughton, G.A. (1982). Properties of orthogonal and triangular grids for the analysis of magnetic fields based on the finite difference and finite element methods, part I: theoretical analysis. *Acta Technica* 27 (2): 168–205.

8 McNaughton, G.A., Fuchs, E.F., and Siegl, M. (1982). Properties of orthogonal and triangular grids for the analysis of magnetic fields based on the finite difference and finite element methods, part II: numerical results based on Gaussian elimination. *Acta Technica* 27 (2): 206–238.

9 Fuchs, E.F. and Siegl, M. (1982). Properties of orthogonal and triangular grids for the analysis of magnetic fields based on the finite difference and finite element methods, part III: comparison of iterative solutions. *Acta Technica* 27 (3): 261–290.

10 Morse, P.M. and Feshbach, H. (1953). Stokes' theorem. In: *Methods of Theoretical Physics, Part I*, 43. New York: McGraw-Hill https://mathworld.wolfram.com/StokesTheorem.html (accessed 22 May 2020).

11 Erdelyi, E.A. and Fuchs, E.F. (1970). Nonlinear magnetic field analysis of DC machines, part I: theoretical fundamentals. *IEEE Transactions on Power Apparatus and Systems* PAS-89 (7): 1546–1554.

12 Younglove, B.L. and Johnson S.D. (1983). *User's Manual for Program Fields.* Final report. Department of Electrical Engineering, University of Colorado, Boulder, Colorado 30309.

13 Chari, M.V.K. (1970). Finite element analysis of nonlinear magnetic fields in electric machines. PhD thesis. McGill University, Canada.

14 Fuchs, E.F. (1976). Industrial experience with the relaxation solutions of the quasi-Poissonian differential equation. *Proceedings of International Conference on Electrical Machines*, Vienna, Austria (13–15 September 1976), pp. D20-1–D20-10.

15 Fuchs, E.F. and Pohl, G. (1981). Computer generated polycentric grid design and a novel dynamic acceleration of convergence for the iterative solution of magnetic fields based on the finite difference method. *IEEE Transactions on Power Apparatus and Systems* PAS-100 (8): 3911–3920.

16 Fuchs, E.F. and Senske, K. (1981). Comparison of iterative solutions of the finite difference method with measurements as applied to Poisson's and the diffusion equations. *IEEE Transactions on Power Apparatus and Systems* PAS-100 (8): 3983–3992.

Appendix 11.B Sample Calculation of Self- and Leakage Inductances and Flux of a DC Machine Field Winding from Flux Plots

Inductance and flux calculations based on flux plots are performed for a 16-pole DC machine [1] including iron-core saturation of stator and rotor with axial iron-core length of $\ell_{core} = 0.20$ m as shown in Figure 11.B.1 and 11.B.2. The (*B*-*H*) characteristics of the stator and rotor are listed in Tables 11.B.1 and 11.B.2 where *B* is the flux density and *H* is the magnetic field intensity.

Figure 11.B.1 illustrates the grid or node system employed for the numerical field calculation based on the FD method, which is iterative because of the (*B*-*H*) nonlinearities of Tables 11.B.1 and 11.B.2.

For the determination of the self-, mutual, and leakage inductances and flux ϕ, the following vector potential values must be defined [2]: $A^{bottom}_{max\_s}$ is the maximum vector potential in the stator bottom coil, $A^{top}_{max\_s}$ is the maximum vector potential

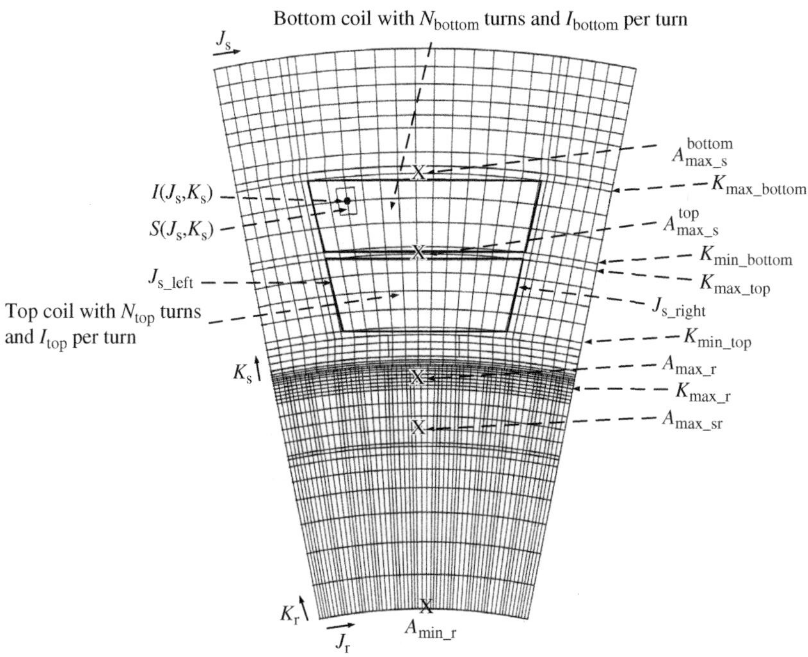

Figure 11.B.1 Grid or node system for the numerical analysis of a 16-pole DC machine [1] for the solution in terms of vector potential for one pole pitch or half a period. *X* indicates the location of the vector potential values A_{min_r}, A_{max_r}, A_{max_sr}, $A^{top}_{max\_s}$, and $A^{bottom}_{max\_s}$.

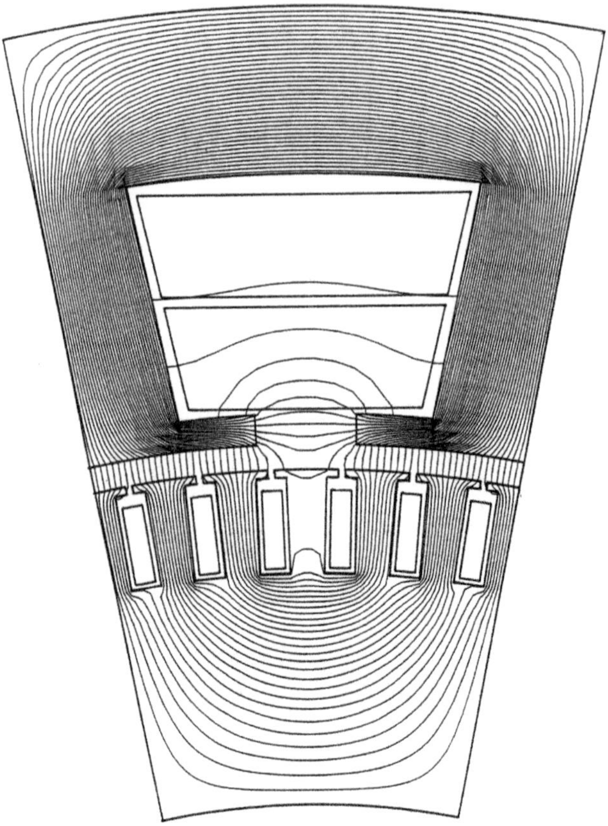

Figure 11.B.2 Magnetic field or vector potential distribution based on numerical analysis of a 16-pole DC machine [1] used for the determination of the self-, mutual, and leakage inductances and flux ϕ for pole pitch or half a period. The potential difference between two field lines is $\Delta A = 0.5 \cdot 10^{-3}$ Wb/m.

in the stator top coil, and A_{max_r} is the maximum vector potential in the rotor that does not contribute to the steady-state energy conversion. Starting with the flux linkage definition λ, one obtains the flux linkages per ampere of one filament $L = \lambda/I = (N \cdot \phi)/I$, which represents the inductance. The flux linkages λ are defined as the product of number of turns N times the flux ϕ. I is the current flowing in the N-turn coil producing the total flux ϕ and the $\lambda = (N \cdot \phi)$ flux linkages. The summation of all flux linkages per ampere over the cross-sectional area S_C of one (e.g. bottom or top) field coil yields the field coil leakage inductance per pole pitch:

Table 11.B.1 (B-H) characteristic for stator iron core, where T stands for tesla ≡ (Vs)/m^2 ≡ Wb/m^2 and k stands for 1000.

B (T)	0	0.05	0.1	0.15	0.2	0.25	0.3	0.35	0.4	0.45	0.5	0.55	0.6
H (A/m)	0	57.5	115	172.5	230	287.5	345	402.5	460	517.5	575	632.5	690
B (T)	0.65	0.7	0.75	0.8	0.85	0.9	0.95	1.0	1.05	1.1	1.15	1.2	1.25
H (A/m)	747.5	805	862.5	920	977.5	1035	1092.5	1150	1250	1380	1500	1700	1900
B (T)	1.3	1.35	1.4	1.45	1.5	1.55	1.6	1.65	1.7				
H (A/m)	2159	2500	3000	3500	4100	5050	6300	8k	10k				

Table 11.B.2 (B-H) characteristic for rotor iron core, where T stands for tesla ≡ (Vs)/m^2 ≡ Wb/m^2 and k stands for 1000.

B (T)	0	0.05	0.1	0.15	0.2	0.25	0.3	0.35	0.4	0.45	0.5	0.55	0.6
H (A/m)	0	30	60	90	120	150	180	210	240	270	300	330	360
B (T)	0.65	0.7	0.75	0.8	0.85	0.9	0.95	1.0	1.05	1.1	1.15	1.2	1.25
H (A/m)	390	420	450	480	510	540	570	600	660	730	820	910	1050
B (T)	1.3	1.35	1.4	1.45	1.5	1.55	1.6	1.65	1.7	1.75	1.8	1.85	1.9
H (A/m)	1 300	1 730	3 000	4 500	6 400	8 400	10 800	13 600	17k	22k	34k	50k	80k
B (T)	1.95	2.00	2.05	2.1	2.15								
H (A/m)	116k	160k	196k	235k	320k								

$$L_{\ell\tau}^{\mathrm{ps}} = \ell_{\mathrm{core}} \cdot \left(\frac{1}{S_{\mathrm{C}}}\right) \sum_{J_{\mathrm{s}}} \sum_{K_{\mathrm{s}}} \frac{\phi_{\ell}(J_{\mathrm{s}}, K_{\mathrm{s}}) \cdot s(J_{\mathrm{s}}, K_{\mathrm{s}})}{I(J_{\mathrm{s}}, K_{\mathrm{s}})} \cdot N, \tag{11.B.1}$$

where ℓ_{core} is the axial core length of a pole; p stands for pole, s for stator, ℓ for leakage, and τ for pole pitch; $s(J_{\mathrm{s}}, K_{\mathrm{s}})$ is the area element of a field (e.g. bottom or top) coil filament at the grid point $(J_{\mathrm{s}}, K_{\mathrm{s}})$; and $I(J_{\mathrm{s}}, K_{\mathrm{s}})$ is the current element in the mesh block around $(J_{\mathrm{s}}, K_{\mathrm{s}})$. It is advantageous to define the weighted number of turns N of a filament as

$$N = \left(\frac{s(J_{\mathrm{s}}, K_{\mathrm{s}})}{S_{\mathrm{t}}}\right) W_{\mathrm{f}}, \tag{11.B.2}$$

where $W_{\mathrm{f}} = (N_{\mathrm{bottom}} + N_{\mathrm{top}})$ is the actual number of one field winding per pole and $S_{\mathrm{t}} = S_{\mathrm{C}}/W_{\mathrm{f}}$ is the cross-sectional area of one turn. The above inductance expression now reads

$$L_{\ell\tau}^{\mathrm{ps}} = \ell_{\mathrm{core}} \sum_{J_{\mathrm{s}}} \sum_{K_{\mathrm{s}}} \frac{\phi_{\ell}(J_{\mathrm{s}}, K_{\mathrm{s}})}{I(J_{\mathrm{s}}, K_{\mathrm{s}})} \cdot \left(s(J_{\mathrm{s}}, K_{\mathrm{s}})/S_{\mathrm{C}}\right)^2 \cdot W_{\mathrm{f}}^2. \tag{11.B.3}$$

Within one pole pitch there are two field coil cross sections each of area S_{C}. The area of the bottom field coil is limited by the grid lines $K_{\mathrm{min\_bottom}}$, $K_{\mathrm{max\_bottom}}$ and by $J_{\mathrm{s\_left}}$, $J_{\mathrm{s\_right}}$. The area of the top field coil is limited by the grid lines $K_{\mathrm{min\_top}}$, $K_{\mathrm{max\_top}}$ and by $J_{\mathrm{s\_left}}$, $J_{\mathrm{s\_right}}$. The maximum potential (kernel) in the armature/rotor coil area $A_{\mathrm{max\_}r}$ occurs at the vertical coordinate $K_{\mathrm{max\_r}}$. The total field/stator leakage (ℓ stands for leakage) inductance, including the two end regions, is now

$$L_{\ell\tau}^{\mathrm{ps}} = \ell_{\mathrm{core}} \left[\left(L_{\ell\tau}^{\mathrm{ps}}\right)_{\mathrm{bottom}} + \left(L_{\ell\tau}^{\mathrm{ps}}\right)_{\mathrm{top}} + 2 \cdot \left(L_{\ell\tau}^{\mathrm{ps}}\right)_{\mathrm{end\ region}} \right]. \tag{11.B.4}$$

The quantity $\left(L_{\ell\tau}^{\mathrm{ps}}\right)_{\mathrm{bottom}}$ is the field leakage inductance per unit length of all the filaments between the vertical grid lines $K_{\mathrm{min\_bottom}}$, $K_{\mathrm{max\_bottom}}$ and the horizontal grid lines $J_{\mathrm{s\_left}}$, $J_{\mathrm{s\_right}}$ of the bottom field coil:

$$\left(L_{\ell\tau}^{\mathrm{ps}}\right)_{\mathrm{bottom}} = \sum_{J=J_{\mathrm{s\_left}}}^{J_{\mathrm{s\_right}}} \sum_{K=K_{\mathrm{min\_bottom}}}^{K_{\mathrm{max\_bottom}}} \frac{\left(A(J_{\mathrm{s}}, K_{\mathrm{s}}) - A_{\mathrm{max\_sr}}\right)}{I(J_{\mathrm{s}}, K_{\mathrm{s}})} \left(s(J_{\mathrm{s}}, K_{\mathrm{s}})/S_{\mathrm{C}}\right)^2 \cdot W_{\mathrm{f}}^2. \tag{11.B.5}$$

The quantity $\left(L_{\ell\tau}^{\mathrm{ps}}\right)_{\mathrm{top}}$ is the field leakage inductance per unit length of all the filaments between the vertical grid lines $K_{\mathrm{min\_top}}$, $K_{\mathrm{max\_top}}$ and the horizontal grid lines $J_{\mathrm{s\_left}}$, $J_{\mathrm{s\_right}}$ of the top field coil:

$$\left(L_{\ell\tau}^{\text{ps}}\right)_{\text{top}} = \sum_{J=J_{\text{s\_left}}}^{J_{\text{s\_right}}} \sum_{K=K_{\text{min\_top}}}^{K_{\text{max\_top}}} \frac{\left(A(J_{\text{s}},K_{\text{s}}) - A_{\text{max\_sr}}\right)}{I(J_{\text{s}},K_{\text{s}})} \cdot \left(s(J_{\text{s}},K_{\text{s}})/s_{\text{C}}\right)^2 \cdot W_{\text{f}}^2. \tag{11.B.6}$$

In Eqs. (11.B.5) and (11.B.6), either vector potential $A_{\text{max\_sr}}$ or $A_{\text{max\_r}}$ can be used because according to Figure 11.B.2, these two values are about the same; an average value $(A_{\text{max\_sr}} + A_{\text{max\_r}})/2$ could be chosen as well.

$\left(L_{\ell\tau}^{\text{ps}}\right)_{\text{end}-\text{region}}$ is the field leakage inductance of one end region, which is in most cases small as compared with $\left(L_{\ell\tau}^{\text{ps}}\right)_{\text{bottom}}$ and $\left(L_{\ell\tau}^{\text{ps}}\right)_{\text{top}}$.

For $A_{\text{min\_r}}$ one obtains the field self-inductance

$$L_{\text{ff}\tau}^{\text{ps}} = \ell_{\text{core}}\left[\left(L_{\text{ff}\tau}^{\text{ps}}\right)_{\text{bottom}} + \left(L_{\text{ff}\tau}^{\text{ps}}\right)_{\text{top}} + 2\cdot\left(L_{\ell\tau}^{\text{ps}}\right)_{\text{end region}}\right] \tag{11.B.7}$$

with

$$\left(L_{\text{ff}\tau}^{\text{ps}}\right)_{\text{bottom}} = \sum_{J=J_{\text{s\_left}}}^{J_{\text{s\_right}}} \sum_{K=K_{\text{min\_bottom}}}^{K_{\text{max\_bottom}}} \frac{\left(A(J_{\text{s}},K_{\text{s}}) - A_{\text{min\_r}}\right)}{I(J_{\text{s}},K_{\text{s}})} \cdot \left(s(J_{\text{s}},K_{\text{s}})/s_{\text{C}}\right)^2 \cdot W_{\text{f}}^2 \tag{11.B.8}$$

and

$$\left(L_{\text{ff}\tau}^{\text{ps}}\right)_{\text{top}} = \sum_{J=J_{\text{s\_left}}}^{J_{\text{s\_right}}} \sum_{K=K_{\text{min\_top}}}^{K_{\text{max\_top}}} \frac{\left(A(J_{\text{s}},K_{\text{s}}) - A_{\text{min\_r}}\right)}{I(J_{\text{s}},K_{\text{s}})} \cdot \left(s(J_{\text{s}},K_{\text{s}})/s_{\text{C}}\right)^2 \cdot W_{\text{f}}^2, \tag{11.B.9}$$

where $A_{\text{min\_r}}$ (=0) for Eqs. (11.B.8) and (11.B.9).

The total flux linked with the rotor for one pole pitch is

$$\phi_{\tau}^{\text{ps}} = \ell_{\text{core}}\left[A_{\text{max\_sr}} - A_{\text{min\_r}}\right] = \ell_{\text{core}} A_{\text{max\_sr}}. \tag{11.B.10}$$

The (self-) armature/rotor inductance can be obtained in a similar manner as the field/stator coil inductances by exciting the armature. Operating-point-dependent inductances for field (stator) and armature (rotor) circuits can be obtained based on analyses where the reluctivity distribution is operating point dependent [1–5].

The above cited relations can be used if the detailed vector potential distribution is available as a computer file. However, if only a flux plot or printed data are available, the following simplified approach can be relied on to calculate circuit inductances.

Simplified Equivalent Circuit Parameters of the Machine as Depicted in Figures 11.B.1 and 11.B.2: Operating-point-dependent inductances are obtained from the above listed relations. Here, approximate formulas for the field leakage and self-inductance values are obtained from the flux linkage (λ)–current (I) relation for a N-turn coil:

$$L = \lambda/I = (N \cdot \phi)/I; \tag{11.B.11}$$

for the bottom coil one obtains for the leakage inductance per pole pitch τ

$$L^{\text{ps}}_{\ell\tau\_\text{bottom}} = \ell_{\text{core}}\, N_{\text{bottom}} \frac{\phi_{\text{bottom}}}{I_{\text{bottom}}} = \ell_{\text{core}}\, N_{\text{bottom}} \left\{A^{\text{bottom}}_{\text{max\_s}} - A_{\text{max\_r}}\right\}/I_{\text{bottom}}, \tag{11.B.12}$$

and for the top coil correspondingly

$$L^{\text{ps}}_{\ell\tau\_\text{top}} = \ell_{\text{core}}\, N_{\text{top}} \frac{\phi_{\text{top}}}{I_{\text{top}}} = \ell_{\text{core}}\, N_{\text{top}} \left\{A^{\text{top}}_{\text{max\_s}} - A_{\text{max\_r}}\right\}/I_{\text{top}}. \tag{11.B.13a}$$

The field self-inductance is per pole pitch τ

$$L^{\text{ps}}_{\text{ff}\tau\_\text{bottom}} = \ell_{\text{core}}\, N_{\text{bottom}}\, A^{\text{bottom}}_{\text{max\_s}}/I_{\text{bottom}}, \tag{11.B.13b}$$

$$L^{\text{ps}}_{\text{ff}\tau\_\text{top}} = \ell_{\text{core}}\, N_{\text{top}}\, A^{\text{top}}_{\text{max\_s}}/I_{\text{top}}, \tag{11.B.14}$$

or

$$L^{\text{ps}}_{\text{ff}\tau} = \ell_{\text{core}} \left\{N_{\text{bottom}} A^{\text{bottom}}_{\text{max\_s}}/I_{\text{bottom}} + N_{\text{top}} A^{\text{top}}_{\text{max\_s}}/I_{\text{top}}\right\}. \tag{11.B.15}$$

The core flux is per pole pitch τ

$$\phi^{\text{ps}}_{\tau} = \ell_{\text{core}}\{A_{\text{max\_sr}} - A_{\text{min\_r}}\} = \ell_{\text{core}} A_{\text{max\_sr}}. \tag{11.B.16}$$

With the definition $E_{\text{a}} = k_{\text{e}}\phi\omega_{\text{m}}$, where E_{a} is the induced voltage and ω_{m} the angular velocity, one obtains for the flux coefficient $k_{\text{e}}\phi = p \cdot \phi^{\text{ps}}_{\tau}$.

Numerical Values for Inductances [1–5]: For an axial iron-core length $\ell_{\text{core}} = 0.20$ m, $\text{area}_{\text{bottom\_coil}} = 1576.3\ \text{mm}^2$, $\text{area}_{\text{top\_coil}} = 1296.3\ \text{mm}^2$. $I_{\text{bottom}} = I_{\text{top}} = I_{\text{f}} = 10$ A, $N_{\text{bottom}} = N_{\text{top}} = N = 200$ turns, $I_{\text{top}} N_{\text{top}} = I_{\text{bottom}} N_{\text{bottom}} = 2000$ A turns. Note that the bottom coil has a larger cross section conducting the same ampere-turns as the top coil because of reduced cooling properties due to reduced ventilation. $\text{area}^{\text{top}}_{\text{wire}} = \dfrac{1296.3\ \text{mm}^2}{200\ \text{turns}} = 6.4815\ \text{mm}^2/\text{turn}$ or a current density of $j^{\text{top}}_{\text{wire}} = \dfrac{10\ \text{A}}{\frac{6.4815\ \text{A}}{\text{mm}^2}} = 1.5429\ \text{A}/\text{mm}^2$.

Field/Stator Leakage Inductance

With $A_{\text{max\_r}} = 1.35 \cdot 10^{-3}$ Wb/m, $A^{\text{bottom}}_{\text{max\_s}} = 1.85 \cdot 10^{-3}$ Wb/m, and $A^{\text{top}}_{\text{max\_s}} = 1.75 \cdot 10^{-3}$ Wb/m follow the field leakage inductance values

$$L^{\text{ps}}_{\ell\tau_{\text{bottom}}} = 200 \cdot 0.2 \frac{(1.85 - 1.35)}{10} \cdot 10^{-3} = 2\,\text{mH},$$

$$L^{\text{ps}}_{\ell\tau_{\text{top}}} = 200 \cdot 0.2 \frac{(1.75 - 1.35)}{10} \cdot 10^{-3} = 1.6\,\text{mH},$$

or

$$L^{\text{ps}}_{\ell\tau} = L^{\text{ps}}_{\ell\tau\_\text{bottom}} + L^{\text{ps}}_{\ell\tau\_\text{top}} = (2 + 1.6)\text{mH} = 3.6\,\text{mH},$$

neglecting the end-winding leakage inductance.

Field/Stator Self-Inductance
The field self-inductance is

$$L^{\text{ps}}_{f\tau\_\text{bottom}} = 200 \cdot 0.2 \frac{(1.85 - 0)}{10} \cdot 10^{-3} = 7.4\,\text{mH},$$

$$L^{\text{ps}}_{f\tau\_\text{top}} = 200 \cdot 0.2 \frac{(1.75 - 0)}{10} \cdot 10^{-3} = 7.0\,\text{mH},$$

or the total self-inductance per pole pitch is

$$L^{\text{ps}}_{f\tau} = L^{\text{ps}}_{f\tau\_\text{bottom}} + L^{\text{ps}}_{f\tau\_\text{top}} = (7.4 + 7.0)\text{mH} = 14.4\,\text{mH},$$

neglecting the end-winding leakage inductance.

Rotor Flux
The rotor flux per pole pitch is

$\phi^{\text{ps}}_{\tau} = \ell_{\text{core}}(A_{\max\_\text{sr}} - A_{\min\_\text{r}}) = 0.2(1.35) = 0.27\,\text{Wb}$. The relation between induced voltage and angular velocity is $E_{\text{a}} = k_{\text{e}}\phi\omega_{\text{m}}$ where the flux coefficient is

$$k_{\text{e}}\phi = p \cdot \phi^{\text{ps}}_{\tau} = 16 \cdot 0.27 = 4.32\,\text{Wb/rad}.$$

The field winding resistance is $R_{\text{f}} = \rho \frac{\text{length}}{q} N$, where $q = \text{area}^{\text{top}}_{\text{wire}} = 6.4815\,\text{mm}^2/\text{turn}$, length $= 4 \cdot \ell_{\text{core}} = 0.8$ m taking the end-winding portions into account; with $N = 200$ turns, $\rho = 1.73 \cdot 10^{-8}\,\Omega\text{m}$ one obtains

$R_{\text{f}} = 1.73 \cdot 10^{-8}\Omega\text{m} \frac{0.800\,\text{m}}{6.4815 \cdot 10^{-6}\,\text{m}^2} 200\,\text{turns} = 427 \cdot 10^{-3}\,\Omega = 0.427\,\Omega$. Note that "turns" is not a unit.

The armature/rotor leakage inductance L_{a} can be obtained in a similar manner by exciting the armature winding only and only taking into account the flux, which is not linked with the field winding: $L_{\text{a}} = 5\,\text{mH}$. Operating-point-dependent inductances are obtained as described for reactances [2–5].

References

1 Younglove, B.L. and Johnson S.D. (1983). *User's Manual for Program Fields*. Final report. Department of Electrical Engineering, University of Colorado, Boulder, CO.

2 Fuchs, E.F. (1970). Numerical determination of synchronous, transient, and subtransient reactances of a synchronous machine. PhD thesis. University of Colorado, Boulder, CO.

3 Fuchs, E.F. and Erdelyi, E.A. (1972). Determination of waterwheel alternator steady-state reactances from flux plots. *IEEE Transactions on Power Apparatus and Systems* PAS-91 (6): 2510–2527.

4 Fuchs, E.F. and Erdelyi, E.A. (1972). Determination of waterwheel alternator transient reactances from flux plots. *IEEE Transactions on Power Apparatus and Systems* PAS-91 (5): 1795–1802.

5 Fuchs, E.F. and Erdelyi, E.A. (1974). Nonlinear salient pole alternator subtransient reactances and damper winding currents. *IEEE Transactions on Power Apparatus and Systems* PAS-93 (6): 1871–1892.

12

Permanent-Magnet, Induction, and Synchronous Machines

Their Performance at Variable Speed and Torque

All types of polyphase machines (e.g. permanent-magnet machines [PMMs], induction or asynchronous machines [IMs], and synchronous machines [SMs]) have at least on one member – either stator or rotor – a polyphase winding generating a rotating magnetic field with an angular synchronous (s) velocity measured in radians per second $\omega_s = 2\pi f/(p/2)$, where f is the current/voltage frequency measured in hertz (Hz) and p is the number of poles, which must be an even number. After explaining simple polyphase windings, as invented by Nikola Tesla [1], we explain several types of polyphase AC machines based on linear (unsaturated) equivalent circuits, resulting in phasor diagrams for steady-state operation. Flux patterns pertaining to nonlinear (saturated) load conditions (e.g. no-load, full-load, and starting operation) are obtained that can be employed for the calculation of linearized leakage and mutual inductances/reactance, which are instrumental in understanding linearized equivalent circuits and phasor diagrams. To date, no other textbook has presented such detailed nonlinear flux patterns calculated from finite-difference (FD) and finite-element (FE) methods that visualize and assist in the understanding of rotary energy conversion apparatuses.

PMMs excited by an inverter are usually called "brushless DC machines," the design of which can be based on either the intersection point of the load line with the magnet characteristic or on magnetic field plots. Both approaches will be taught. We will explain the definition of the machine impedances, in particular the definition of the leakage inductances with respect to two phases of the polyphase stator winding. PSPICE (Personal Simulation Program with Integrated Circuit Emphasis) software [2–4] solutions will highlight speed/torque control. Additionally, we will present applications of PMMs to automotive and wind power.

IMs are modified from PMMs by replacing the rotor's permanent magnets with a polyphase (wound or squirrel-cage type) circuit, whereby two rotating fields

Introduction to Energy, Renewable Energy, and Electrical Engineering: Essentials for Engineering Science (STEM) Professionals and Students, First Edition. Ewald F. Fuchs and Heidi A. Fuchs.

Companion website: www.wiley.com/go/fuchs/renewableenergyandelectricalengineering

interact at the same synchronous angular velocity $\omega_s = 2\pi f/(p/2)$ or synchronous speed $n_s^{(\text{rpm})}$ measured in rpm, producing torque at any given mechanical (m) speed measured either in revolutions per minute (rpm) $n_m^{(\text{rpm})}$ or in revolutions per second (rps) $n_m^{(\text{rps})} = n_m^{(\text{rpm})}/60$. If the rotor rotates at a mechanical angular velocity ω_m, then the rotating magnetic field of the rotor rotates with respect to the rotor with the angular velocity $\omega_r = 2\pi f_r/\left(\frac{P}{2}\right)$, where f_r is the frequency of the rotor three-phase currents and voltages. Note that the number of poles in stator and rotor is always the same and is an even number: that is, the rotating stator field of IM induces the rotor field with the proper number of poles in an automatic manner. After discussing equivalent circuits, we will explain the various types of IM and the speed/torque control highlighted through PSPICE examples with respect to automotive and wind power applications.

Non-salient and salient pole SMs are obtained similarly to PMM by replacing the rotor's permanent magnets with a field coil, which is supplied by DC current via two slip rings. Their equivalent circuits as well as speed and torque control will be outlined for drive applications, such as those for the French railroad "train à grande vitesse" (TGV) and the recently developed doubly fed induction generator (DFIG) for hydropower plants equipped with a three-phase winding on the stator and one on the rotor, where a short response time for load changes is required.

12.1 Revolving Magnetic Field

Tesla [1] obtained 111 American patents from 1886 to 1926, including one describing the two-phase revolving magnetic field of an electric rotating machine, which subsequently was extended by other engineers to three-phase configurations and those with an even higher number of phases. It turns out that the three-phase circuit requires the smallest amount of conduction material (e.g. copper, aluminum) for conversion/transmitting of electrical to mechanical energy and is therefore mostly employed in practice [5, 6].

Figure 12.1 illustrates phases a–a′ and b–b′ of Tesla's two-phase (phase number $m = 2$), two-pole ($p = 2$) stator winding arranged in a counterclockwise (ccw) manner, representing the mathematical positive sense, where θ is the circumferential angular coordinate. In Figures 12.1–12.5 concentrated phase windings are used in order to simplify the analysis. Later on Figures 12.8 and 12.9 employ distributed windings as used in real machines.

The magnetomotive force (mmf) as defined in Chapters 6 and 11 of coil a–a′ of Figure 12.1 is illustrated in Figure 12.2, where for the two-pole configuration of Figure 12.1, one north (N) and one south (S) pole can be identified. The magnetic flux pattern runs from the north to the south pole. From experiments [7] it is well

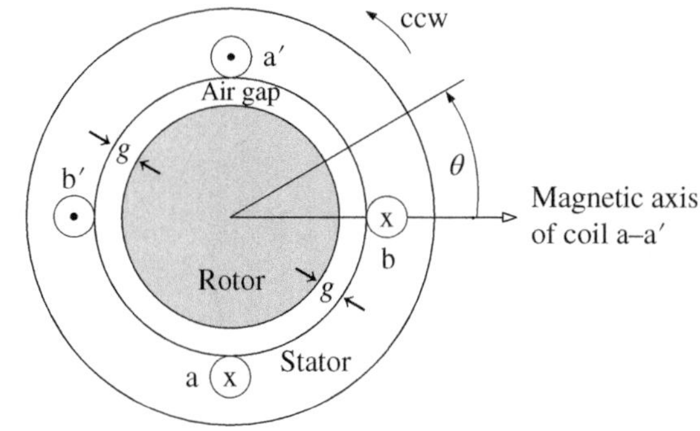

Figure 12.1 Two-phase, two-pole stator winding consisting of concentrated windings a–a′ and b–b′ where the magnetic axis of coil a–a′ is indicated.

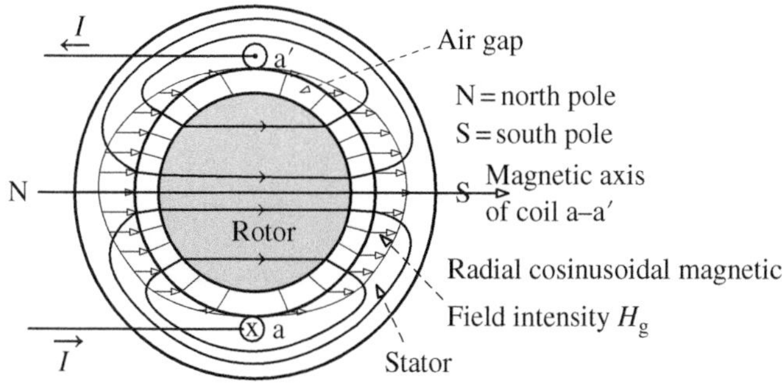

Figure 12.2 Radial fundamental cosinusoidal magnetic field intensity H_g of concentrated phase belt a–a′.

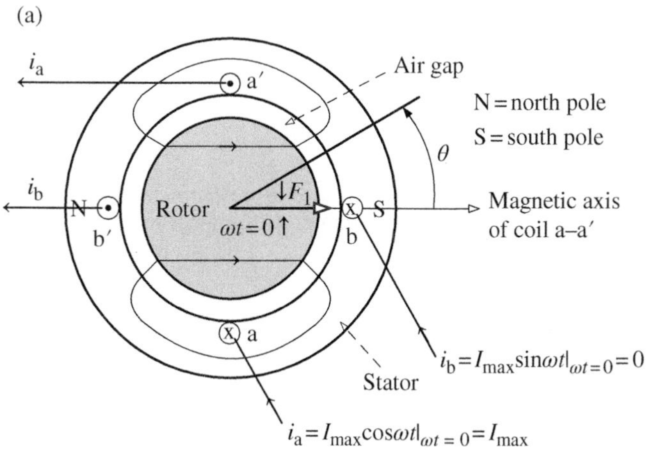

Figure 12.3a Instantaneous magnetic field intensity $H_g(\omega t = 0)$ or $F_1(\omega t = 0)$ of concentrated phase belts a–a′ and b–b′ for $\omega t = 0$ or $t = 0$; rotor rotates with angular velocity ω in ccw direction.

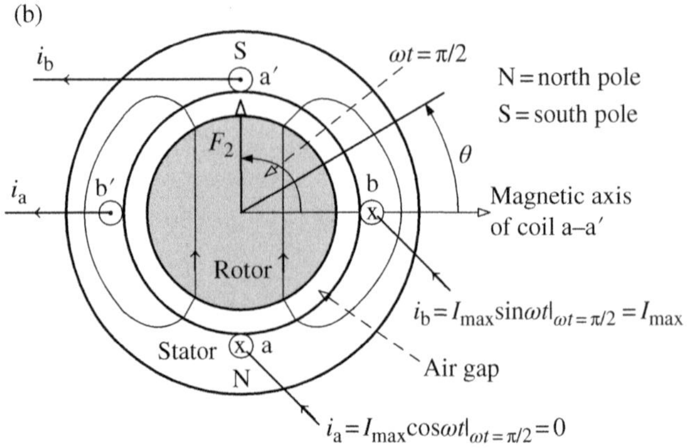

Figure 12.3b Instantaneous magnetic field intensity $H_g(\omega t = \pi/2)$ or $F_2(\omega t = \pi/2)$ of concentrated phase belts a–a′ and b–b′ for $\omega t = \pi/2$ or $t = T/4$; rotor rotates with angular velocity ω in ccw direction.

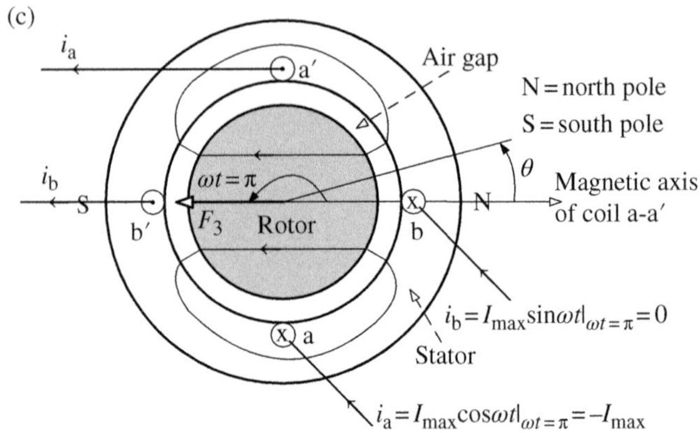

Figure 12.3c Instantaneous magnetic field intensity $H_g(\omega t = \pi)$ or $F_3(\omega t = \pi)$ of concentrated phase belts a–a′ and b–b′ for $\omega t = \pi$ or $t = T/2$; rotor rotates with angular velocity ω in ccw direction.

known that the radial magnetic field intensity H_g is approximately a square wave, which has a fundamental cosinusoidal component for the given magnetic axis of coil/phase belt a–a′, as illustrated in Figure 12.2.

The mmf of the fundamental cosinusoidal magnetic field intensity H_g is

$$F = H_g \cdot g, \tag{12.1}$$

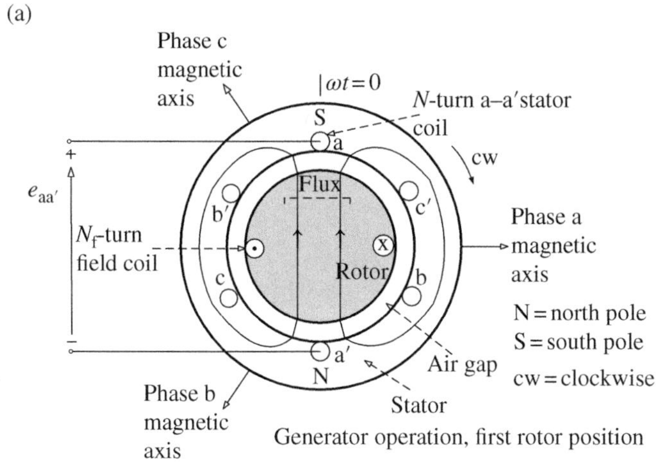

Figure 12.4a Three-phase stator winding and rotating DC winding on rotor. The transformation of free electrons to endowed electrons (see Chapter 1, Appendix 1.B) at rotor position 1 corresponding to ωt = 0° results in a voltage or potential across the three concentrated stator windings or phases due to Faraday's law. The rotor rotates with angular velocity ω in cw direction.

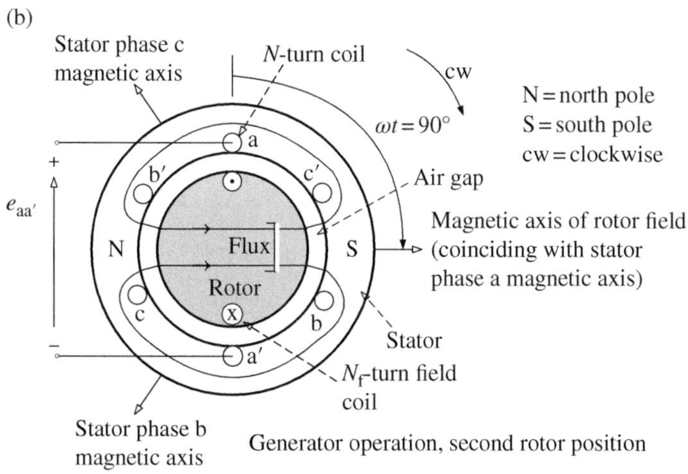

Figure 12.4b Three-phase stator winding and rotating DC winding on rotor. The transformation of free electrons to endowed electrons (see Chapter 1, Appendix 1.B) at rotor position 2, corresponding to ωt = 90°, results in a voltage or potential across the three concentrated stator windings or phases due to Faraday's law. The rotor rotates with angular velocity ω in cw direction.

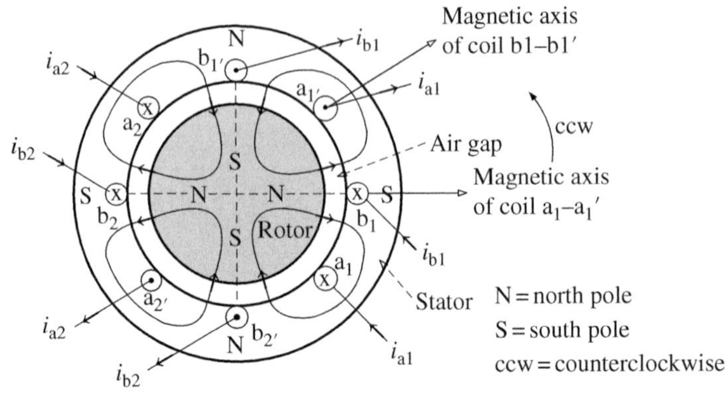

Figure 12.5 Four-pole configuration with concentrated two-phase windings. The rotor rotates with angular velocity ω in ccw direction.

where g is the constant one-sided air-gap length. Introducing in Figure 12.2 the second coil b–b′, one obtains Figure 12.1 representing the two-phase, two-pole stator winding as proposed by Tesla.

For coil/phase belt a–a′ of Figures 12.1 and 12.2 with N turns, the mmf is

$$F_a = N \cdot I \cdot \cos\theta, \tag{12.2}$$

and for coil b–b′ of Figure 12.1, one obtains correspondingly when the displacement of 90° is taken into account:

$$F_b = N \cdot I \cdot \sin\theta. \tag{12.3}$$

If the DC current I in Eqs. (12.2) and (12.3) is replaced by AC excitation currents

$$i_a = I_{max} \cdot \cos\omega t, \tag{12.4a}$$

$$i_b = I_{max} \cdot \sin\omega t, \tag{12.4b}$$

where $\omega = 2\pi f$, max stands for maximum/amplitude value of $i_a(t)$, $i_b(t)$, and electric frequency $f_e = f = 60$ Hz resulting in the electric time period $T_e = T = 1/f = 16.667$ ms, then one gets the rotating magnetic fields generated by the two coils at different instants of time, as depicted in Figure 12.3a–c representing the time-dependent magnetic field distributions of a two-phase, two-pole machine.

Proof that a two-phase current system results in a revolving (rotating) wave:

The mmf of phase belt a–a′ with N winding turns is

$$F_a(\theta, t) = N \cdot i_a \cdot \cos\theta = N \cdot I_{max} \cdot \cos\omega t \cdot \cos\theta, \tag{12.5a}$$

and the mmf of phase belt b–b′ is

$$F_b(\theta, t) = N \cdot i_b \cdot \sin\theta = N \cdot I_{\max} \cdot \sin\omega t \cdot \sin\theta. \tag{12.5b}$$

The addition of both (Eqs. 12.5a and 12.5b) mmf results in the total mmf of

$$F_{2\text{-phase}} = F_a(\theta, t) + F_b(\theta, t) \tag{12.5c}$$

or, using trigonometric relations,

$$F_{2\text{-phase}} = N \cdot I_{\max} \cdot \cos(\omega t - \theta). \tag{12.5d}$$

This relation is the analytical expression for a rotating (revolving) wave as discovered by Tesla. One can observe the maximum of this rotating wave when

$$(\omega t - \theta) = 0, \tag{12.5e}$$

or the angular displacement is

$$\omega t = \theta. \tag{12.5f}$$

The angular velocity of the rotating wave is then

$$\frac{d\theta}{dt} = \omega. \tag{12.5g}$$

For an m-phase current system

$$F_{m\text{-phase}} = (m/2) \cdot N \cdot I_{\max} \cdot \cos(\omega t - \theta), \tag{12.5h}$$

and in particular for an $m = 3$ phase system

$$F_{3\text{-phase}} = (3/2) \cdot N \cdot I_{\max} \cdot \cos(\omega t - \theta). \tag{12.5i}$$

For a $p = 2$ pole configuration, the following relation holds:

$$\omega_m = d\theta/dt = \omega_s = \omega, \tag{12.6}$$

where $\omega_m = d\theta/dt$ is the mechanical (m) synchronous (s) angular velocity of the (mechanical) rotor, ω_s is the mechanical synchronous angular velocity of rotating mmf or magnetic field with respect to the stator, and ω is the electrical angular velocity $\omega = 2\pi f$ where f is the frequency of applied currents (e.g. $i_a(t)$, $i_b(t)$, $i_c(t)$).

A rotating electric machine can have $p = 2, 4, 6, 8, 10, \ldots$ (even) poles. Now we are going to derive the angular velocity relations for a $p = 4$ pole machine with two phases. In Figure 12.5 we assume

$$i_a = i_{a1} = i_{a2} = I_{\max} \cos\omega t|_{\omega t = 0} = I_{\max}, \tag{12.7}$$

$$i_b = i_{b1} = i_{b2} = I_{\max} \sin\omega t|_{\omega t = 0} = 0. \tag{12.8}$$

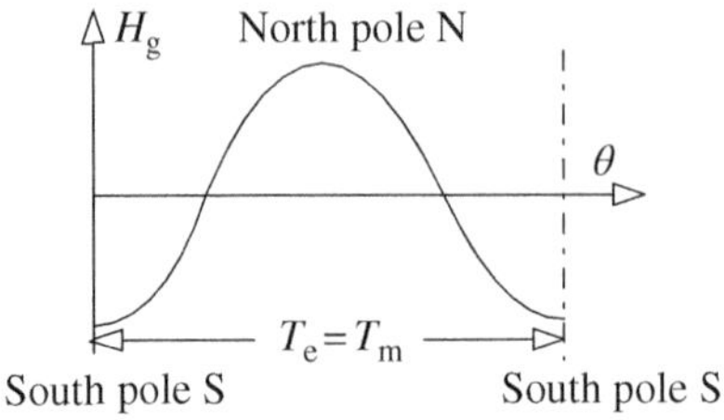

Figure 12.6 Magnetic field intensity H_g in the air gap around the circumference of Figure 12.3a–c (two-pole configuration).

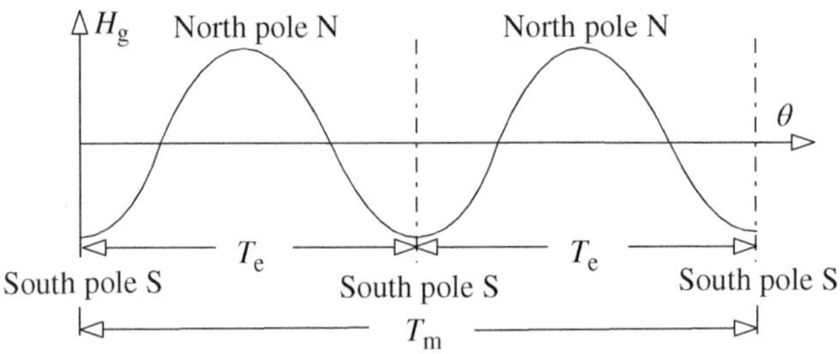

Figure 12.7 Magnetic field intensity H_g in the air gap around the circumference of Figure 12.5 (four-pole configuration).

Moving around the air-gap circumference of Figure 12.3a–c, one can sense one north pole and one south pole. That is, for $p = 2$ poles, one can plot the magnetic field intensity H_g as illustrated in Figure 12.6, where the mechanical (m) period T_m is identical to the electrical (e) period T_e.

For the four-pole configuration of Figure 12.5, the magnetic field intensity H_g is shown in Figure 12.7, where the mechanical period T_m corresponds to two electrical periods T_e.

Figures 12.8 and 12.9 show three-phase, two- and four-pole windings, respectively, as employed in practice.

Conclusions

1) For a $p = 2$ pole machine as illustrated in Figure 12.3a–c

$$T_e = T_m, \tag{12.9}$$

with $f = 1/T$, it follows $1/f_e = 1/f_m$, and one obtains the identity

$$f_e = f_m, \tag{12.10}$$

with $\omega = \omega_e = 2\pi f$ follows

$$\omega_e = \omega_m = 2\pi f_e = 2\pi f. \tag{12.11}$$

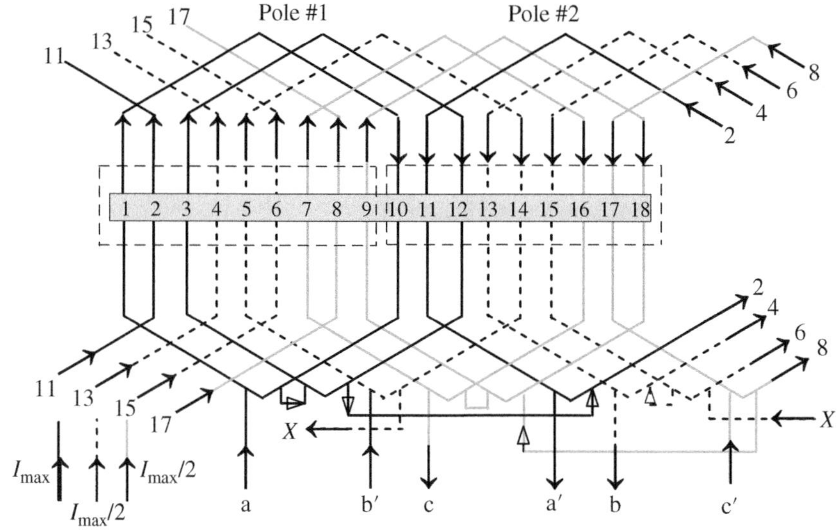

Figure 12.8 Three-phase, two-pole distributed winding located in 18 stator slots where in terminal a the current $i_a = I_{max}\cos\,\omega t|_{\omega t=0} = I_{max}$ is injected, in terminal b the current $i_b = I_{max}\cos(\omega t - 120°)|_{\omega t=0} = -I_{max}/2$, and in terminal c the current $i_c = I_{max}\cos(\omega t - 240°)|_{\omega t=0} = -I_{max}/2$. The terminals a, b, c are called "line" terminals, and when a′, b′, c′ are connected together the connecting point is called "neutral" n, which is not shown. The thickness of the arrows indicates the amount and polarity of the currents injected.

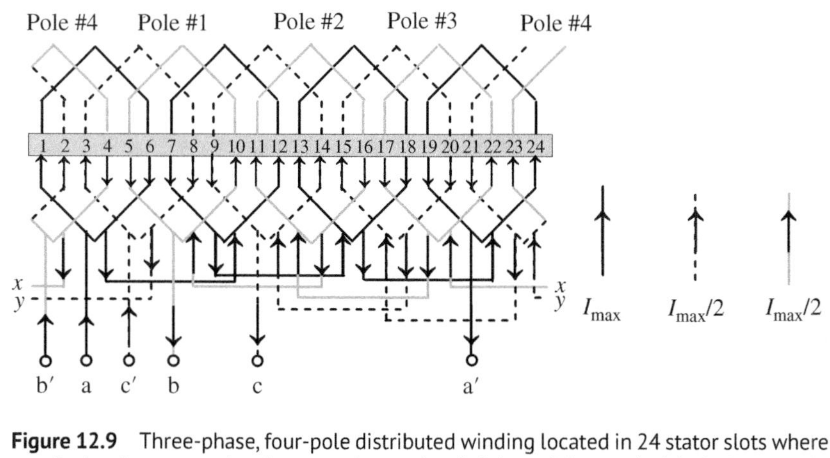

Figure 12.9 Three-phase, four-pole distributed winding located in 24 stator slots where in terminal a the current $i_a = I_{max}\cos\,\omega t|_{\omega t=0} = I_{max}$ is injected, in terminal b the current $i_b = I_{max}\cos(\omega t - 120°)|_{\omega t=0} = -I_{max}/2$, and in terminal c the current $i_c = I_{max}\cos(\omega t - 240°)|_{\omega t=0} = -I_{max}/2$.

The electric frequency f is the power system frequency, either $f = 60$ Hz or 50 Hz as used in different countries.

2) For a $p = 4$ pole machine as shown in Figure 12.6

$$2T_e = T_m, \tag{12.12}$$

with the electric frequency (f_e) $f_e = 1/T_e$, and mechanical frequency (f_m) follows $2/f_e = 1/f_m$, then one obtains the identity

$$f_e = 2f_m, \tag{12.13}$$

with $\omega = \omega_e = 2\pi f$, follows

$$\omega_e = 2\omega_m = 2\pi f_e = 2\pi f. \tag{12.14}$$

3) For a p-pole machine one can generalize Eq. (12.14):

$$\omega_e = (p/2)\,\omega_m = (p/2)\,\omega_s = 2\pi f_e = 2\pi f. \tag{12.15}$$

12.2 Permanent-Magnet Materials

PMMs of rare earth elements such as dysprosium (Dy), neodymium–iron–boron (NdFeB) where Fe stands for iron, and samarium–cobalt (SmCo) – and others such as ferrites [8, 9] as well as Ceramic 7 and AlNiCo 5 and 8 – gain importance in the application of drive motors for electric/hybrid automobiles [10–12] and generators for direct-drive wind power plants [13–18]. Figure 12.10 shows the PMMs as published by IBS Magnet [8]. Figure 12.11 illustrates the temperature dependency of NdFeB material as published by Shin-Etsu [9]. Shin-Etsu's *N* Series NdFeB magnets have one of the best performances of all rare earth magnets; moreover, NdFeB costs less than SmCo magnets. Applications of NdFeB are expanding rapidly in the electric car and wind power plant businesses as their cost and performance advantages gain recognition. Corrosion resistance has been improved through various types of surface processing, including nickel (Ni) plating. However, the thermal characteristics of Shin-Etsu *R* Series [9] SmCo magnets are superior to those of NdFeB magnets, maintaining high magnetism and reliability even in challenging high-temperature and opposed magnetic field environments. SmCo magnets also offer high Curie temperatures with less coercivity change due to temperature – properties unmatched by other magnets. Other advantages include high performance and energy output, thanks to the application of Sm_2Co_{17} compound rather than conventional crystalline $SmCo_5$, a result of detailed metallurgical analysis.

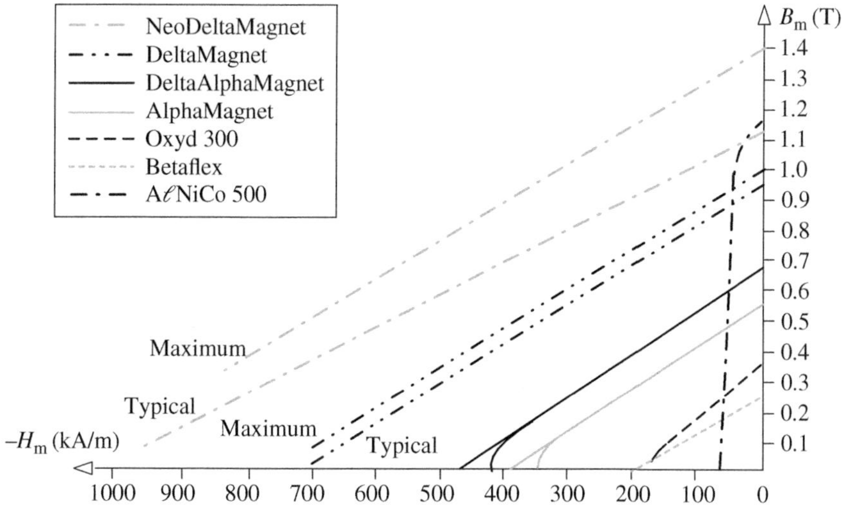

Figure 12.10 Various permanent-magnet material characteristics [8], where NeoDeltaMagnet corresponds to the neodymium–iron–boron (NdFeB) magnet material. Maximum indicates the maximal reachable characteristic, and Typical represents a typical characteristic. Characteristics are represented in a simplified manner. In terms of units, note 1 Oersted = 1 Oe = $(1000/4\pi)$ A/m ≈ 79.557 kA/m or 1 kA/m · 12.57 ≈ 1 Oe.

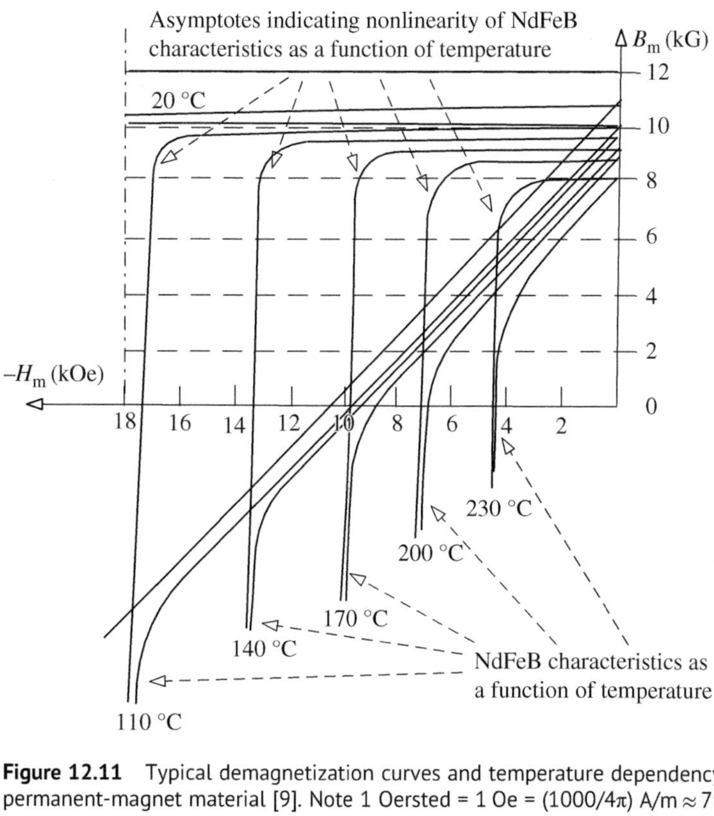

Figure 12.11 Typical demagnetization curves and temperature dependency of NdFeB permanent-magnet material [9]. Note 1 Oersted = 1 Oe = $(1000/4\pi)$ A/m ≈ 79.557 kA/m.

Figure 12.12 shows the choice of the flux density B_m and the required magnetic (m) field intensity H_m based on the selection of the load line for NdFeB magnets. Load line 2 is chosen for the application of the minimum PMM, resulting in the maximum energy product $(-B_m \cdot H_m)_{max}$. The election of the maximum energy product is not advantageous for electric machines because the resulting air-gap flux density B_g is low. Instead, the selection of load line 1 yields a desirable air-gap flux density of about $B_m = 0.93$ T.

For electric machines with NdFeB material requiring high air-gap flux density B_g, a lower energy product than $(-B_m \cdot H_m)_{max}$ occurs. Note kAt = kA, because t (turn) is not a unit, and 1 Oersted = 1 Oe = $(1000/4\pi)$ A/m ≈ 79.557 kA/m. The remanent (r) or residual flux density is $B_r = 1.07$ T, and the coercivity (c) of the magnetic field intensity is $H_c = -835.35$ kA/m. The operating point is determined by the intersection of the permanent-magnet characteristic with the load line (e.g. load line 1), resulting in $B_{mo} = 0.93$ T and $(-H_{mo}) = 99.5$ kA/m.

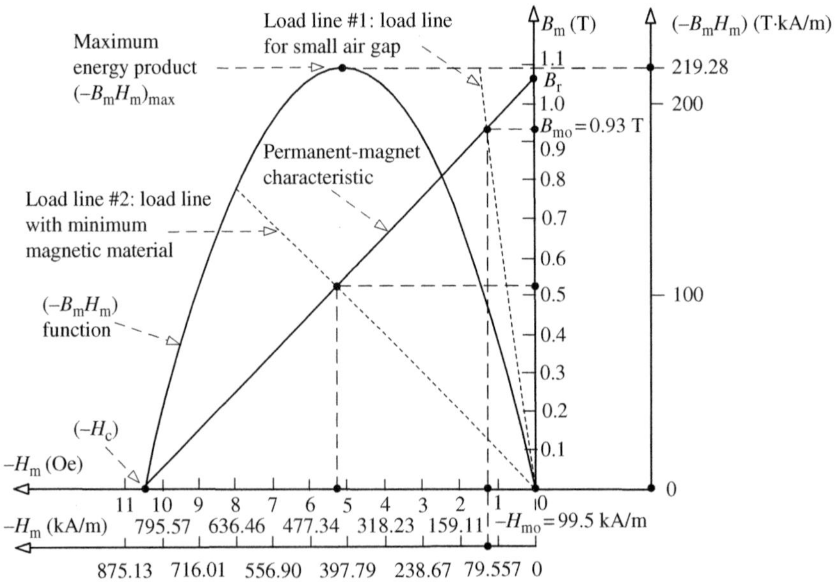

Figure 12.12 Load line 1: load line for small air-gap and large flux density as required for PMM drives but requires more than minimum permanent-magnet material. Load line 2: load line for maximum energy product $(-B_m \cdot H_m)_{max}$ requiring minimum permanent-magnet material.

Application Example 12.1 Maximum energy product $(-B_m H_m)_{max}$
According to Figure 12.12, a PMM with a small air gap yields for $B_{mo} = 0.93$ T and $(-H_{mo}) = 99.5$ kAt/m an energy product of $(-B_m \cdot H_m) = 92.54$ TkA/m (load line 1), while load line 2 with minimum magnetic material results for $B_{mo} = 0.525$ T and $(-H_{mo}) = 417.67$ kAt/m in the maximum energy product of $(-B_m \cdot H_m)_{max} =$ 219.28 TkAt/m.

Figure 12.13a depicts the partial cross section of a 12-pole wind power generator [13–18] with a small air gap.

Ampere's law [5, 6], known as Stokes' theorem in terms of physics/mathematics applied to magnetic circuits, relates the current density $\vec{J}$, which is enclosed by the integration path C, defining the integration surface S, to the magnetic field intensity $\vec{H}$ as given by Eq. (12.16), where $\overrightarrow{dS}$ and $\overrightarrow{dl}$ are the incremental (differential) integration surface and length, respectively. This equation relates a surface integral to a line integral:

$$\int_S^{\text{surface}} \vec{J} \cdot \overrightarrow{dS} = \oint_C^{\text{contour}} \vec{H} \cdot \overrightarrow{d\ell}. \tag{12.16}$$

(a)

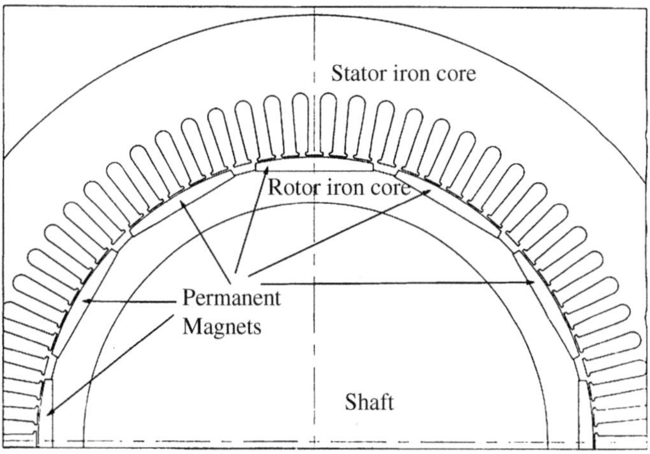

Figure 12.13a Partial cross section of a 12-pole wind power generator [13–18].

Application Example 12.2 Application of Ampere's law in three and two dimensions in free space

Figures E12.2.1 and E12.2.2 illustrate the application of Ampere's law to three and two dimensions, respectively. Figure E12.2.1 demonstrates the application of Ampere's law or Stokes' theorem to a three-dimensional configuration consisting of a surface S penetrated by the current density field $\vec{J}$, resulting in the magnetic field intensity $\vec{H}$ surrounding the current density field along path C.

The application of Ampere's law to the permanent magnets in Figure 12.13a is illustrated [19–21] in Figure E12.2.3, where two current sheets next to PMM are employed to impose the field intensity H_{m} within the magnet.

By applying Ampere's law (Eq. 12.16) to Figure E12.2.3, one obtains the relation for no-load condition when stator currents are zero:

$$I = J \cdot S = H_{\text{m}} \cdot \ell_{\text{m}} + H_{\text{g}} \cdot \ell_{\text{g}} + H_{\text{ts1}} \cdot \ell_{\text{ts1}} + H_{\text{bs}} \cdot \ell_{\text{bs}} + H_{\text{ts2}} \cdot \ell_{\text{ts2}} + H_{\text{slot}} \cdot \ell_{\text{slot}} + H_{\text{br3}} \cdot \ell_{\text{br3}} + H_{\text{br2}} \cdot \ell_{\text{br2}} + H_{\text{br1}} \cdot \ell_{\text{br1}}, \quad \text{(E12.2.1)}$$

where I represents the total current within a current sheet measured in A or At where t stands for turns (which is not a unit), S is the area covered by the current

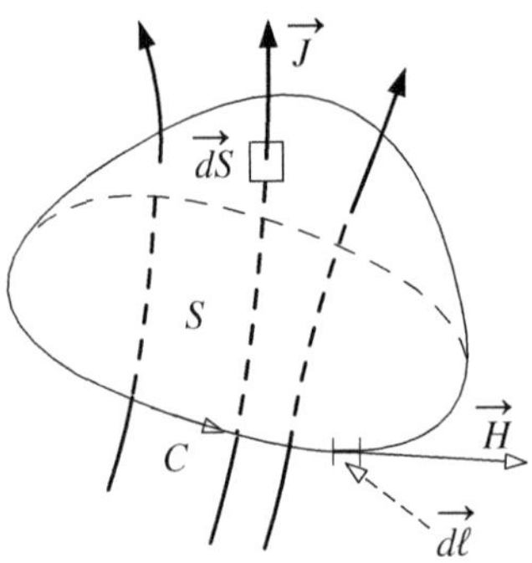

Figure E12.2.1 Graphical interpretation of Ampere's law in three-dimensional free space.

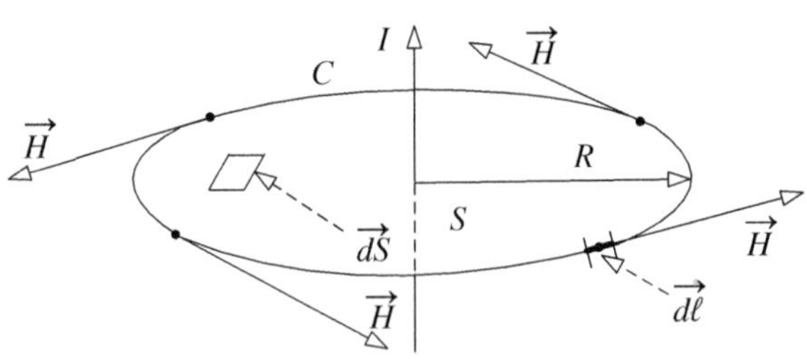

Figure E12.2.2 Application of Ampere's law to a single wire in two-dimensional free space with current I.

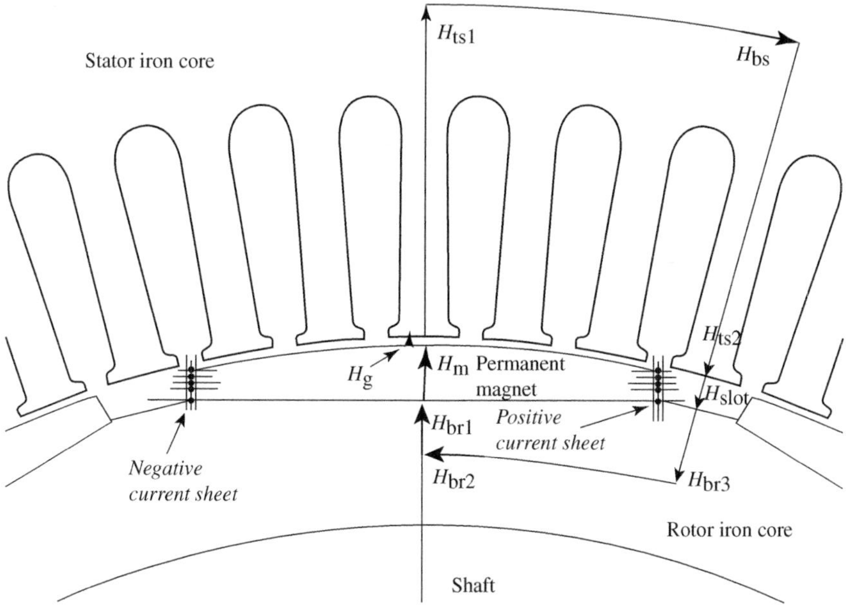

Figure E12.2.3 Application of Ampere's law in two dimensions to permanent-magnet material based on a polycentric grid/mesh system in a wind generator (see Chapter 11, Appendix 11.A, Figure 11.A.3a–c). The positive current sheet is defined by the current (*I*) distributed over the surface *S* in the air gap on the right-hand side of a permanent magnet. The same applies to the negative current sheet with negative current (−*I*).

density J, and m, g, t, s, b, and r stand for magnet, gap, tooth, stator, back iron, and rotor, respectively. The length ℓ_i of the magnetic circuit corresponds to the magnetic field intensities H_i, where i stands for m, g, ts1, ts2, bs, slot, br3, br2, and br1. Because of

$$|H_{\mathrm{m}} \cdot \ell_{\mathrm{m}}| \gg |H_{\mathrm{g}} \cdot \ell_{\mathrm{g}}| + |H_{\mathrm{ts1}} \cdot \ell_{\mathrm{ts1}}| + |H_{\mathrm{bs}} \cdot \ell_{\mathrm{bs}}| + |H_{\mathrm{ts2}} \cdot \ell_{\mathrm{ts2}}| + |H_{\mathrm{slot}} \cdot \ell_{\mathrm{slot}}| + |H_{\mathrm{br3}} \cdot \ell_{\mathrm{br3}}| + |H_{\mathrm{br2}} \cdot \ell_{\mathrm{br2}}| + |H_{\mathrm{br1}} \cdot \ell_{\mathrm{br1}}|, \qquad \text{(E12.2.2)}$$

one can write as a good approximation

$$I \approx |H_{\mathrm{m}}| \cdot \ell_{\mathrm{m}}, \qquad \text{(E12.2.3)}$$

where I is the total current associated with one current sheet measured in either A or At. The maximum height ℓ_{m} of the permanent magnet associated with H_{m} in Figure E12.2.3 is measured in meter (m); this results in $|H_{\mathrm{m}}|$ with the unit A/m. The above approximate relation corroborates equation (8) of Brauer et al. [21].

(b)

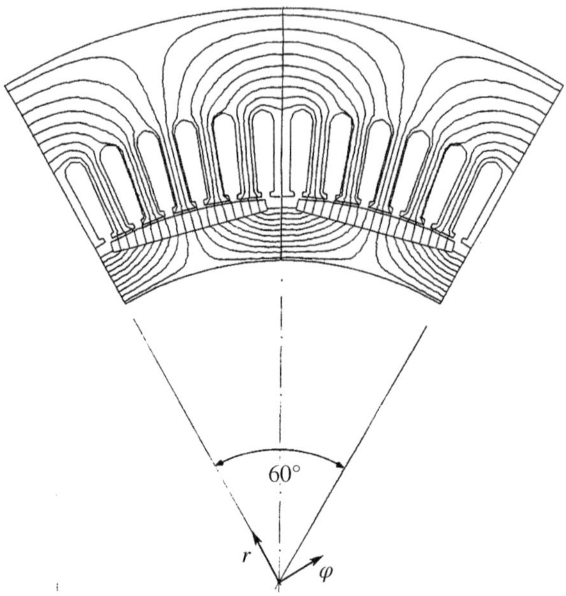

Figure 12.13b No-load magnetic field $B(r, \varphi)$ of the 12-pole generator of Figure 12.13a based on a polycentric grid/mesh system with the two-dimensional coordinates (r, φ) containing 12 stator slots and two permanent magnets attached to the rotor iron (see Chapter 11, Appendix 11.A, Figure 11.A.3a–c).

Figure 12.13b illustrates within the cross section of Figure 12.13a the no-load magnetic field $B(r, \varphi)$, whereby stator currents are zero.

12.3 Designs of Permanent-Magnet Machines (PMMs)

PMMs can be designed by relying on the load line (Figure 12.12) or by computing the field distribution (Figures 12.13b and E12.3.1). In Application Example 12.3 we use the load-line approach, while in Application Example 12.4 we employ the FD method for the calculation of the two-dimensional magnetic field. PMMs are prime candidates for electric drives of electric or hybrid cars because of their high power density per weight, measured in kW/(kg-force), as addressed in Application Example 12.5. Similarly, these machines are also suited for direct-drive, variable-speed wind power generators with a high number of poles. The flux weakening/addition, discussed in Chapter 11 for DC machines, makes the application of PMMs to variable speed possible – as reported in Application Example 12.6.

Because of the strong permanent magnet, the flux generated by the additional field winding as shown in Figure E12.6.1 is not very effective.

Application Example 12.3 Design of a permanent-magnet generator for a direct-drive, variable speed 20 kW or 30 kVA wind power plant [17, 18]

A 20 kW permanent-magnet generator is to be designed for the speed range of 40–150 revolutions per minute (rpm) and a corresponding no-load line-to-line voltage range of 200–760 V. Design and measured data of this generator are presented by Fingersh [18]. The design and measurements of the entire wind power plant (generator, resonant rectifier, and inverter, including control circuits) are detailed in [13–17]. The permanent-magnet generator has the cross section of Figures 12.13a, E12.2.3, and 12.13b, with 72 stator slots. The number of poles is $p = 12$, and its rated synchronous (s) speed is $n_s = n_{rated} = 60$ rpm, corresponding to a stator voltage/current frequency of $f = (60\text{ rpm}/60) \cdot (12\text{ poles}/2) = 6$ Hz.

a) Apply Ampere's law to Figure E12.3.1, assuming that the stator and rotor iron cores have a large relative permeability ($\mu_r \to \infty$).
b) Apply the continuity of flux condition for the areas A_m and A_{gap} where $\ell_m = 0.007$ m is the effective height of the magnet, $\ell_{width} = 0.0635$ m,

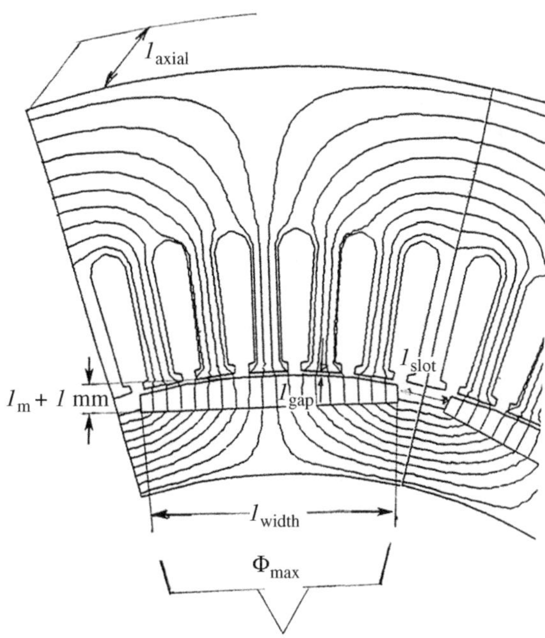

Figure E12.3.1 Application of Ampere's law and specifications of geometrical dimensions (lengths ℓ_m, ℓ_{width}, ℓ_{gap}, ℓ_{axial}, ℓ_{slot}) and maximum flux Φ_{max}. A_m is the magnet surface area within one pole pitch, and A_{gap} is the air-gap surface area of one pole pitch on stator surface.

$\ell_{slot} = 0.00747$ m, $\ell_{gap} = 0.001$ m, and the axial length of the stator core is $\ell_{axial} = 0.334$ m, resulting in the magnet cross section per pole of $A_m = \ell_{width} \cdot \ell_{axial} = 0.0212\ \text{m}^2$, and an air-gap cross section per pole $A_{gap} = (\ell_{width} + \ell_{slot})\ \ell_{axial} = 0.0237\ \text{m}^2$. See Figure E12.3.1 for the definition of these parameters, where A_m is the surface area of one permanent magnet and A_{gap} is the air-gap area of one pole pitch on the stator surface.

c) Provided the relative permeability of the stator and rotor cores is $\mu_r \to \infty$, compute the load line of this machine. Plot the load line. Is the NdFeB magnet operated at the maximum energy product?

d) What is the recoil permeability of the NdFeB magnet material?

e) Compute the flux Φ_{max} provided there are $N_{ph} = 360$ turns per stator phase, the (pitch and distribution) winding factor is $k_w = k_p \cdot k_b = 0.9$ (full-pitch winding: $k_p = 1.0$, breadth factor: $k_b = 0.9$), and the iron-core stacking factor is $k_{fe} = 0.96$. Calculate the induced no-load phase and line-to-line voltages.

f) Two 14 AWG wires are connected in parallel at a stator cross section of $A_{st\_slot} = 243.06\ \text{mm}^2$: determine the copper fill factor k_{cu}.

Solution

a) Ampere's law:

$$H_m \ell_m + H_{gap} \ell_{gap} = 0. \tag{E12.3.1}$$

b) Continuity of flux:

$$\Phi = B_m A_m = B_{gap} A_{gap}. \tag{E12.3.2}$$

c) Constituent relation:

$$B_{gap} = \mu_0 H_{gap}. \tag{E12.3.3}$$

From Eq. (E12.3.1)

$$H_m \ell_m = -H_{gap} \ell_{gap}, \tag{E12.3.4}$$

with Eq. (E12.3.3)

$$H_m \ell_m = -B_{gap} \ell_{gap} / \mu_0, \tag{E12.3.5}$$

and with Eq. (E12.3.2)

$$B_{gap} = B_m \left(A_m / A_{gap}\right), \tag{E12.3.6}$$

the load line becomes

$$H_{\mathrm{m}} = -\left(\frac{A_{\mathrm{m}}}{A_{\mathrm{gap}}}\right)\left(\frac{\ell_{\mathrm{gap}}}{\ell_{\mathrm{m}}}\right)\frac{B_{\mathrm{m}}}{\mu_0}. \tag{E12.3.7}$$

Introducing the given numerical values into load-line equation (E12.3.7), with the permeability of free space $\mu_0 = 4\pi \cdot 10^{-7}$ H/m, gives

$$H_{\mathrm{m}} = -\left(\frac{0.0212}{0.0237}\right)\left(\frac{0.001}{0.007}\right)\frac{1}{4\pi \cdot 10^{-7}}B_{\mathrm{m}}\left(\frac{\mathrm{kA}}{\mathrm{m}}\right) = -101.7B_{\mathrm{m}}\left(\frac{\mathrm{kA}}{\mathrm{m}}\right)$$

$$= -1.28\,B_{\mathrm{m}}\,(\mathrm{Oe}). \tag{E12.3.8}$$

This equation is plotted in Figure 12.12 (load line 1). Note that the permanent-magnet material is not operated at the point of maximum energy product: $(-B_{\mathrm{mo}} \cdot H_{\mathrm{mo}}) = -0.93\,\mathrm{T} \cdot 99.5\,\mathrm{kA/m} = -92.535\,\mathrm{TkA/m}$.

d) Recoil permeability of NdFeB (slope of the permanent-magnet characteristic) is obtained from Figure 12.12 with $B_{\mathrm{r}} = 1.07$ T and $H_{\mathrm{c}} = -835.35$ kA/m

$$\mu_{\mathrm{R}} = \frac{(y_2 - y_1)}{(x_2 - x_1)} = \frac{(B_{\mathrm{r}} - 0)}{(0 - H_{\mathrm{c}})} = \frac{1.07}{835.35 \cdot 10^3} = 1.28 \cdot 10^{-6}\,\frac{\mathrm{H}}{\mathrm{m}} \tag{E12.3.9}$$

or

$$\frac{\mu_{\mathrm{R}}}{\mu_0} = 1.28 \cdot \frac{10^{-6}}{4\pi \cdot 10^{-7}} = 1.02. \tag{E12.3.10}$$

The recoil permeability μ_{R} cannot be lower than the permeability of free space μ_0; however, the lower the recoil permeability, the better and more effective the permanent-magnet material.

e) Compute the flux Φ_{max} and calculate the induced no-load phase and line-to-line voltages.

The maximum flux is, according to Figure E12.3.1,

$$\Phi_{\mathrm{max}} = B_{\mathrm{m}} \cdot A_{\mathrm{m}}; \tag{E12.3.11}$$

the flux is defined as

$$\Phi(t) = \Phi_{\mathrm{max}} \cdot \cos \omega t. \tag{E12.3.12}$$

Therefore, the maximum phase-induced voltage is

$$E_{\mathrm{max}} = N(\mathrm{d}\Phi/\mathrm{d}t) = \omega N_{\mathrm{ph}} k_{\mathrm{w}} k_{\mathrm{fe}} \Phi_{\mathrm{max}}. \tag{E12.3.13}$$

From the intersection of the load line with the magnet characteristic as depicted in Figure 12.12, one obtains graphically $B_{\mathrm{mo}} = 0.93$ T and $(-H_{\mathrm{mo}}) = 99.5$ kAt/m. Thus,

$$\Phi_{max} = B_{mo} \cdot A_m = 0.93 \cdot 0.0212 = 0.01972\,\text{Wb}, \quad \text{(E12.3.14)}$$

and the induced no-load phase (ph) or line (L)-to-neutral (N) voltage is now with $N_{ph} = 360$ turns and Eq. (E12.3.13):

$$e_{ph}(t) = 2\pi \cdot 6 \cdot 360 \cdot 0.9 \cdot 0.96 \cdot 0.01972 \sin \omega t = 231 \sin \omega t = E_{maxL\text{-}N} \cdot \sin \omega t. \quad \text{(E12.3.15)}$$

The maximum (max) or amplitude of the no-load induced voltage is

$$E_{maxL\text{-}N} = 231\,\text{V}, \quad \text{(E12.3.16a)}$$

$$E_{rmsL\text{-}N} = E_{maxL\text{-}N}/\sqrt{2} = 163\,\text{V}, \quad \text{(E12.3.16b)}$$

and the no-load line-to-line voltage of the fundamental voltage, neglecting harmonics and the voltage induced in the two stator end windings, is

$$V_{L\text{-}L\_fundamental} = \sqrt{3}E_{rmsL\text{-}N} = 283\,\text{V}. \quad \text{(E12.3.17)}$$

Measured [18] rms voltage is $\sqrt{(283)^2 + (8.49)^2 + (16.98)^2} = 283.63$ V based on the waveshape of Figure E12.3.2, which includes 7th and 11th harmonic voltages of 3 and 6%, respectively.

f) A 14 AWG wire has the nominal diameter [5] of 0.0641 inches = 1.628 mm; with two wires in parallel the equivalent wire cross section is $A_{equiv\_wire} = \pi(1.628/2)^2(2)\,\text{mm}^2 = 4.16\,\text{mm}^2$. The number of stator slots occupied for

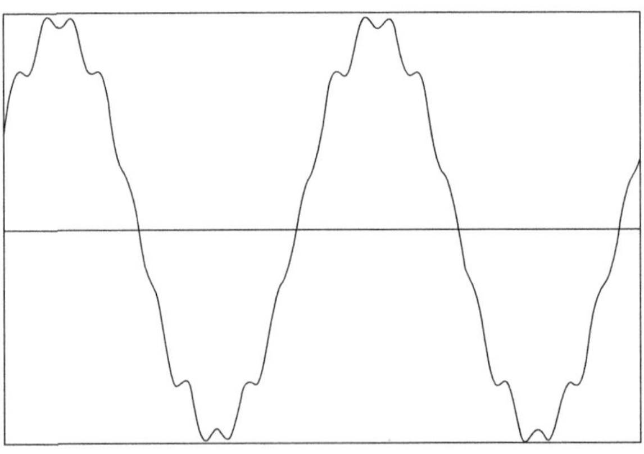

Figure E12.3.2 Measured (see figure 9 of Fingersh [18]) nonsinusoidal no-load line-to-line voltage waveshape at full load with a 7th harmonic of 3% and an 11th harmonic of 6% due to the pitch factor of $k_p = 1.0$, where harmonics are not entirely eliminated. The root-mean-square (rms) value of this nonsinusoidal no-load voltage is 300 V (see figure 3 of Fingersh [18]). This harmonic distortion is acceptable because the generator supplies voltage and current to a three-phase rectifier, where the input voltage need not be sinusoidal. A design error of {(300 − 283)/300} · 100% = 5.7% is acceptable.

one phase winding of the three-phase winding is 72/3 = 24 slots. There are $N_{ph} = 360$ equivalent (two wires in parallel) turns. Since each winding phase belt resides within two stator slots, the 360 turns of one phase occupy 12 slots per one phase belt side. The total copper cross section per stator slot is then $A_{cu\_slot} = (360/12) \cdot 4.16\ mm^2 = 124.8\ mm^2$, and the cross section of one stator slot is given as $A_{st\_slot} = 243.06\ mm^2$; therefore, the copper fill factor k_{cu} is then

$$k_{cu} = A_{cu\_slot}/A_{st\_slot} = 124.8/243.06 = 0.513. \tag{E12.3.18}$$

This value takes into account all insulation components and is acceptable for a ventilated machine; see chapter 4, section 4.2.13.4 of Masoum and Fuchs [6].

Application Example 12.4 High power density and high-efficiency permanent-magnet motors [6, 11–13] for electric vehicle applications with field distributions based on a polycentric grid/mesh system (see Appendix 11.A, Figure 11.A.3a–c)

In 1990, for their client Bayerische Motoren Werke (BMW AG) of Munich, Germany, Unique Mobility, Inc. [10–12] of Englewood, Colorado, researched permanent-magnet motors as they can be used for automobiles by optimization to obtain the highest output power for a given weight without sacrificing performance. They obtained a power density per weight of 50 kW/(13.513 kg-force) = 3.7 kW/(kg-force). For the 50 kW machine Figure E12.4.1a and b show the no-load and full-load fields, respectively. The efficiency of this machine is 96.33% at a

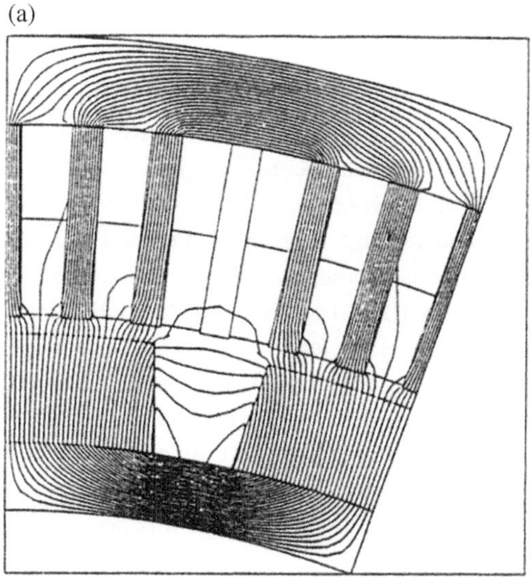

Figure E12.4.1a Symmetric no-load field of 50 kW machine [10].

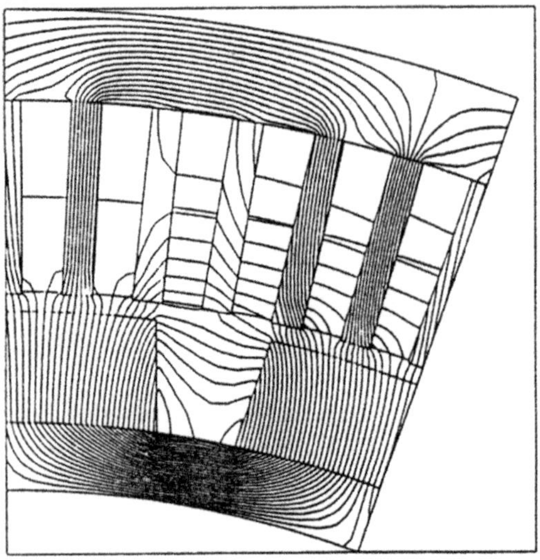

Figure E12.4.1b Asymmetric rated full-load field of 50 kW machine [10].

temperature rise of $\Delta T = 100\,°C$, speed 3523.7 rpm, and voltage $E = 212$ V at a slot-fill factor of $k_{cu} = 66.90\%$ for a forced cooled machine. Design details are available in the paper by Huang and Fuchs [10]. Note that these machines can run either as a motor or as a generator, charging the battery through regeneration during braking; about 20–30% of the kinetic energy of the moving car can be recovered [5] by recharging the battery, depending upon the drive mode (e.g. frequent braking).

Application Example 12.5 High power density and high-efficiency permanent-magnet motors [5, 6, 11, 12] for electric vehicle applications with field distributions based on a polycentric grid/mesh system (see Chapter 11, Appendix 11.A, Figure 11.A.3a–c)

A 20 kW PMM can be modeled based on the no-load (Figure E12.5.1a) and rated-load (Figure E12.5.1b) fields as well as three reactances: direct-axis, quadrature-axis, and leakage reactances. Figure E12.5.1c depicts the magnetic field when the stator is excited only for the calculation of the direct (d)-axis reactance X_d, and Figure E12.5.1d presents the field distribution for the calculation of the quadrature (q)-axis reactance X_q. While for a conventional synchronous machine the X_d is always larger than the X_q (see Section 12.5.1), for PMM this is not true and $X_q > X_d$. Leakage reactances are usually defined between stator and rotor windings (see Chapter 11, Appendix 11.B). As evident from Figure E12.5.1a and b, there is no rotor winding for PMM, and therefore the stator (s) leakage (ℓ) reactance $X_{s\ell}$ of PMM is defined by the leakage flux between two stator phases.

Figure E12.5.1a Symmetric no-load field of 20 kW PMM [16].

(a)

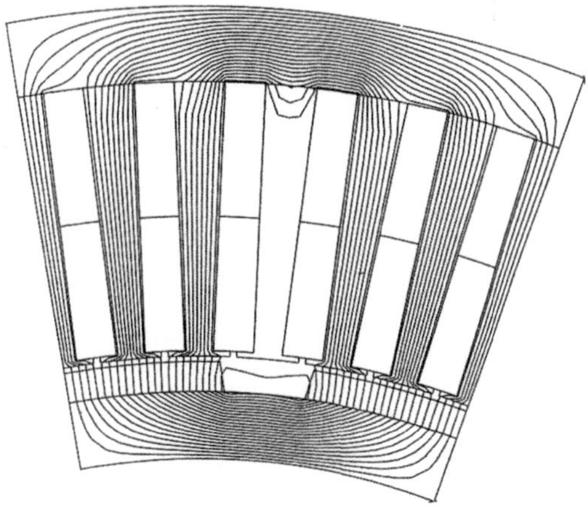

Figure E12.5.1b Asymmetric rated-load field of 20 kW PMM [16].

(b)

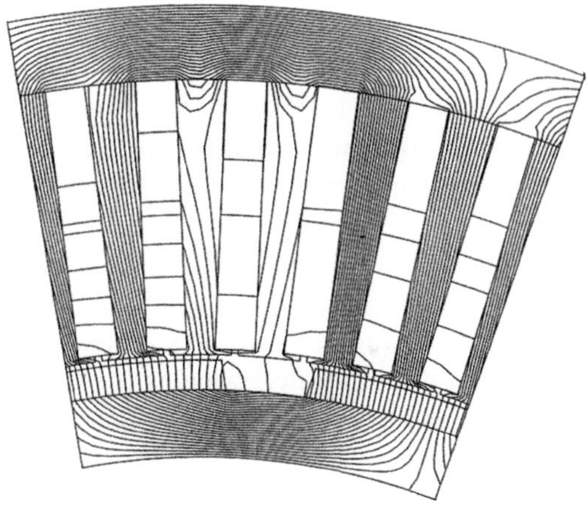

Figure E12.5.1c Calculation of direct-axis X_d and stator leakage $X_{s\ell}$ reactances from flux distribution [12]. The flux plot is not symmetric because the winding factor $k_w = k_p \cdot k_b < 1.0$ and the reluctivity distribution corresponds to rated load [16].

(c)

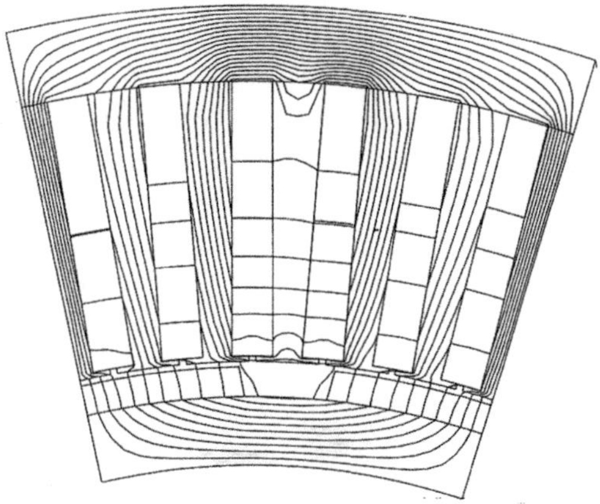

(d)

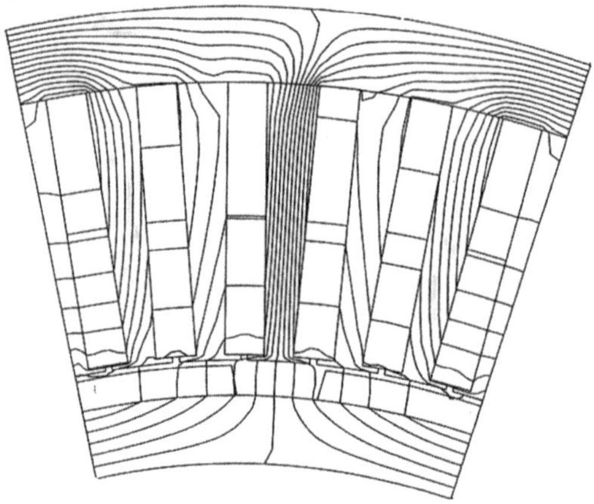

Figure E12.5.1d Calculation of quadrature-axis X_q and stator leakage $X_{s\ell}$ reactances from flux distribution [12]. The flux plot is not symmetric because the winding factor $k_w = k_p \cdot k_b < 1.0$ and the reluctivity distribution corresponds to rated load [16].

Application Example 12.6 Field distributions for PMM with flux weakening/addition [12–16] based on a polycentric grid/mesh system (see Chapter 11, Appendix 11.A, Figure 11.A.3a–c)

The principle of flux weakening/addition as discussed in Chapter 11 for DC machines can be applied to PMM as well, thus changing their output characteristics. Figure E12.6.1 depicts a PMM where the magnetic flux generated by a field winding is superposed (added/subtracted) with that of the permanent magnet (PM). The only problem with this arrangement is that the relatively weak flux of the additional rotor field winding has to modify the strong flux generated by the PM. For such a superposition of fluxes, it is therefore advisable to employ a relatively weak PM.

12.3.1 Speed and Torque Control of PMM

Brushless DC machines or electronically commutated permanent-magnet machines (ECPMMs) are desirable drives for electric and hybrid automobiles due to their low weight for a given output power. Figure 12.14a illustrates the brushless DC circuit, consisting of a battery, inverter, and PMM. Such electronically commutated machines are commonly operated in two modes: the full-on mode, as in Figure 12.14b, and the pulse-width-modulation (PWM) mode, as depicted in Figure 12.14c. The commutation principle as discussed in Chapter 11 for DC machines is applied to ECPMM where the DC voltage of the battery is converted/transformed into AC voltage for motoring and vice versa from AC to DC for regeneration during braking action. The position control of an ECPMM rotor is performed with either a Hall sensor, rotary encoder, or

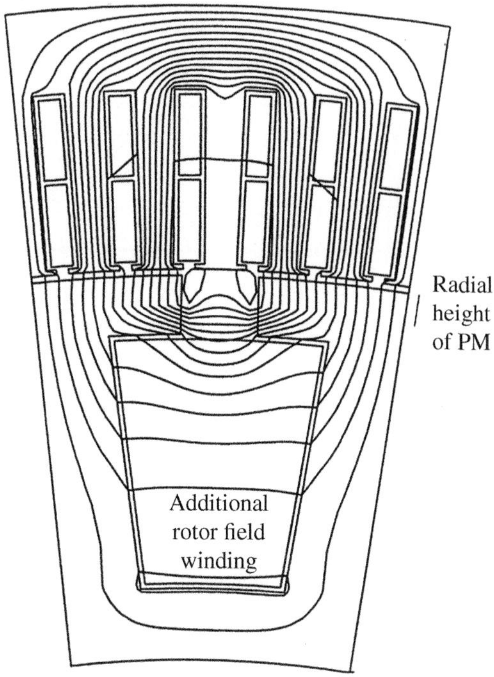

Figure E12.6.1 PMM with flux distribution at no load with flux-weakening (FW) [16] ampere-turns of $(AT)_{FW} = -10\,500$ At. The radial height of the permanent magnet (PM) is indicated by a line on the right-hand side of the pole pitch.

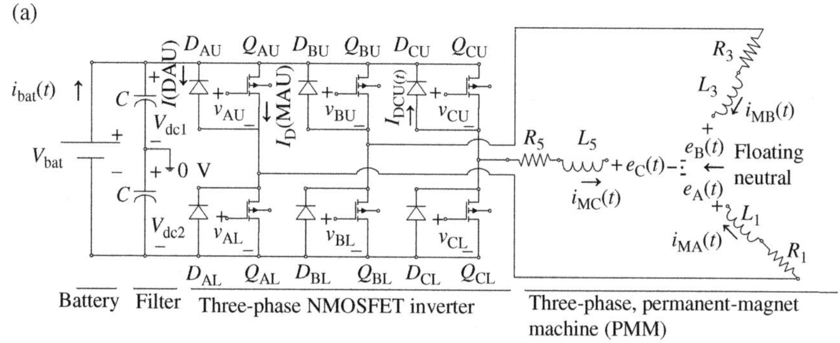

Figure 12.14a Brushless DC machine consisting of battery, inverter, and three-phase permanent-magnet machine. A, B, C stand for the three phases; L_1, L_3, and L_5 for leakage inductances; U for upper, and L for lower.

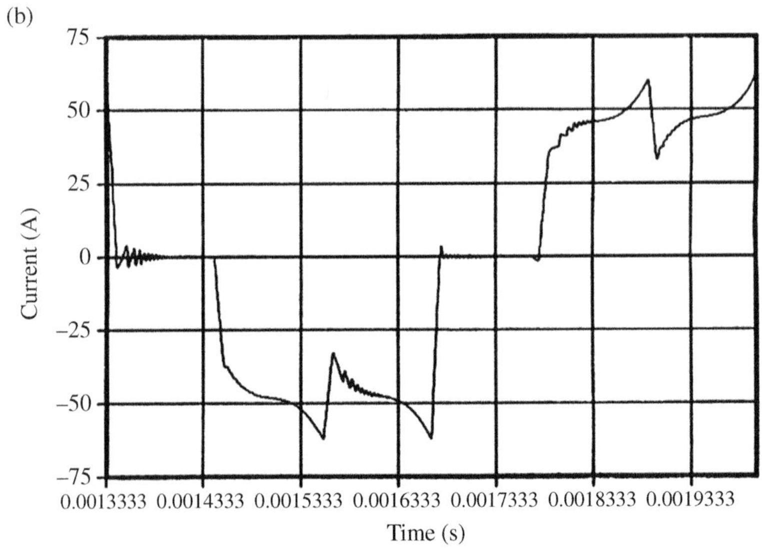

Figure 12.14b Motor (stator) current waveshape based on full-on motoring mode [11].

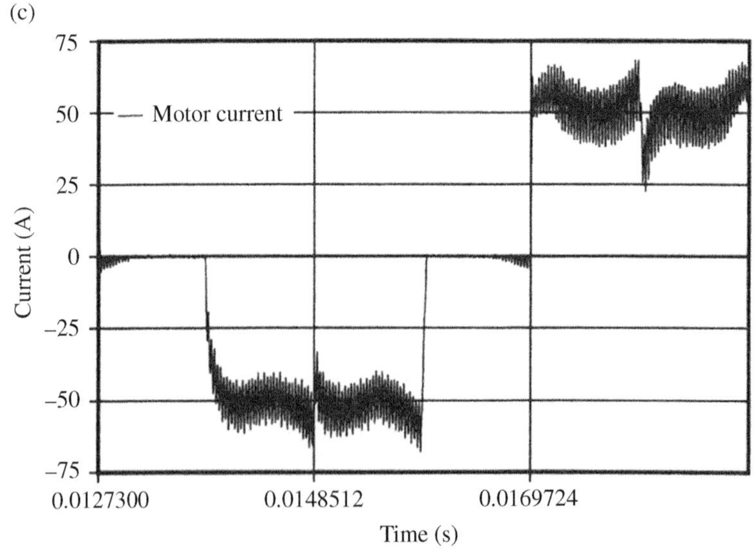

Figure 12.14c Motor (stator) current waveshape based on PWM motoring mode [11, 12].

(d)

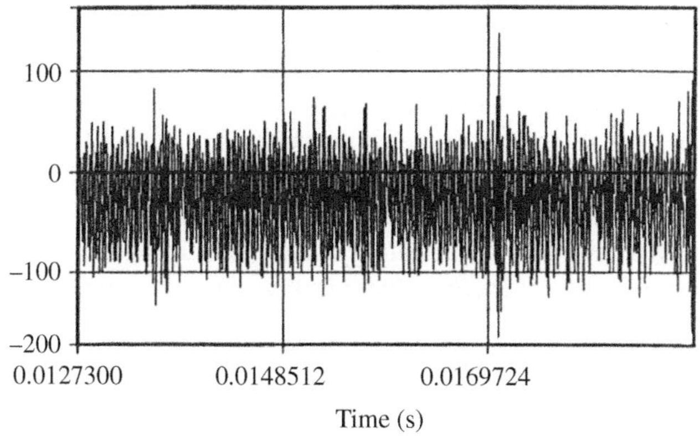

Figure 12.14d Current supplied during regeneration to battery, whereby the average current is negative [11, 12] due to PSPICE [3] battery/source definition.

computational technique predicting the rotor position. The control of the torque is based on either flux weakening/addition, where the rotor and the stator fluxes are shifted in space with respect to one another, or by relying on PWM for the control of the stator current magnitudes. The speed control is solely determined by the full-on frequency of the machine imposed by the inverter.

12.3.2 Applications of PMM to Automobiles and Wind Power Plants

Application Example 12.7 Brushless DC motor/machine

The objective of this example is to calculate the battery voltage required for a brushless DC motor [10–18] and to introduce the compensation of flux weakening (CFW) [22, 23], resulting in high power and torque per weight even at high speeds, as well as adjustable-speed (proportional to switching frequency f_1) and torque (proportional to input current) control of such inverter-fed machines in full-on mode (analyzed with PSPICE [3]). Electric or hybrid cars commonly use batteries with a voltage of $V_{bat} = 300\ V_{DC}$. The inverter is a six-pulse or six-step or full-on inverter consisting of six self-commutated metal–oxide–semiconductor field-effect transistor (MOSFET) switches. The electric machine is a three-phase PMM represented by induced voltages (e_A, e_B, e_C), resistances, and leakage inductances (with respect to two stator phase windings) for all three phases. The induced voltage of the stator winding (phase A) of the PMM is

$$e_A = 160\ \text{V} \sin(\omega t + \theta), \tag{E12.7.1a}$$

where $\omega = 2\pi f_1$ and $f_1 = 1500$ Hz. Correspondingly,

$$e_B = 160\,\text{V}\sin(\omega t + 120^\circ + \theta), \quad \text{(E12.7.1b)}$$

$$e_C = 160\,\text{V}\sin(\omega t + 240^\circ + \theta). \quad \text{(E12.7.1c)}$$

The resistance R_s and the leakage inductance $L_{s\ell}$ of one of the phases are 0.5 Ω and 50 μH, respectively. The magnitude of the gating voltages of the six MOSFETs is $V_{Gmax} = 15\,\text{V}$. Figure E12.7.1a shows the gating signals with their phase sequence. Note that the phase sequence of the induced voltages (e_A, e_B, e_C), as defined in Eqs. E12.7.1a–E12.7.1c, and that of the gating signals (see Figure E12.7.1a) must be the same. If these phase sequences are not identical, then no periodic solution for the machine currents (i_{MA}, i_{MB}, i_{MC}) can be obtained.

The models of the enhancement MOSFETs and those of the (external) freewheeling diodes are as follows:

Model for self-commutated electronic switch (MOSFET):
.MODEL SMM NMOS(LEVEL = 3 GAMMA = 0 DELTA = 0 ETA = 0 THETA = 0
+ KAPPA = 0 VMAX = 0 XJ = 0 TOX = 100 n UO = 600 PHI = 0.6 RS = 42.69
+M KP = 20.87 u L = 2 u W = 2.9 VTO = 3.487 RD = 0.19 CBD = 200 n PB = 0.8
+ MJ = 0.5 CGSO = 3.5 n CGDO = 100 p RG = 1.2 IS = 10 f)

Model for diode:
.MODEL D1N4001 D(IS=10^{-12})

a) Using PSPICE [3], compute and plot the current of MOSFET Q_{AU} (e.g. i_{QAU}) and the motor current of phase A (e.g. i_{MA}) for the phase angles of the induced voltages $\theta = 0°$, $\theta = +30°$, $\theta = +60°$, $\theta = -30°$, and $\theta = -60°$, $L_{s\ell} = 50\,\mu\text{H}$, amplitudes of 160 V for e_A, e_B, e_C. The gating signal frequency of the MOSFET

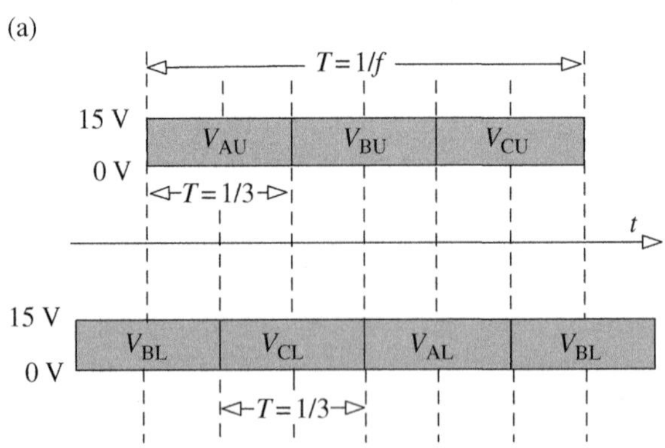

Figure E12.7.1a Sequence of gating signals for brushless DC motor in six-step (six-pulse) operation.

corresponds to the frequency $f_1 = 1500$ Hz; that is, full-on mode operation exists. For switching sequence, see Figure E12.7.1a.

b) Compute and plot the current of MOSFET Q_{AU} (e.g. i_{QAU}) and the motor current of phase A (e.g. i_{MA}) for the phase angles of the induced voltages $\theta = 0°$, $L_{s\ell} = 50$ μH, amplitudes of 80 V for e_A, e_B, e_C. The gating signal frequency of the MOSFET corresponds to the frequency $f_2 = 3000$ Hz; that is, full-on mode operation with CFW exists. For switching sequence, see Figure E12.7.1a.
c) Compute and plot the current of MOSFET Q_{Au} (e.g. i_{QAU}) and the motor current of phase A (e.g. i_{MA}) for the phase angles of the induced voltages $\theta = 0°$, $L_{s\ell} = 50$ μH, amplitudes of 80 V for e_A, e_B, e_C. The gating signal frequency of the MOSFET corresponds to the frequency $f_3 = 750$ Hz; that is, full-on mode operation exists. For switching sequence, see Figure E12.7.1a.
d) If the full-on switching sequence is not identical with the phase sequence of the induced voltages e_a, e_b, e_c, then the phase currents are asymmetric, and no rotating circular (neglecting harmonics) field will be obtained. The full-on frequency is $f_1 = 1500$ Hz.

Note the following: The step size for the numerical solution should be in the neighborhood of $\Delta t = 0.05$ μs. To eliminate computational transients due to inconsistent initial conditions, compute at least three periods of all quantities and plot the last (third) period of i_{QAU} and i_{MA} for all five cases, where θ assumes the values given above.

Solution

For any renewable energy application such as electric/hybrid cars, wind power plants, heat pumps, etc., the speed and the torque of the motor must be controllable:

1) *Speed control:* The speed of brushless DC motors/generators can be controlled with the full-on switching frequency $f = f_1$.
2) *Torque control:* The torque can be adjusted in the six-step mode by the phase-shift angle θ or by compensation of field weakening [22, 23] by switching the stator number of turns via an application-specific integrated circuit (ASIC) as well as by changing the current through pulse width modulation (PWM).
   a) Performance of gating with full-on mode with $f_1 = 1500$ Hz is depicted in Figure E12.7.1b–f, and the PSPICE code is given in Table E12.7.1.
   b) The field weakening at high speeds can be compensated [22, 23] by replacing the phase winding with N_{rat} turns by a reconnected phase winding with $N_{rat}/2$ turns if a series connection of N_{rat} turns is replaced by a parallel connection with $N_{rat}/2$ turns. This reduces the induced voltage (proportional to N) to 80 V, the leakage inductance per phase ($L_1 = L_3 = L_5$ proportional to N^2) to 12.5 μH, and the resistance $R_1 = R_3 = R_5$ (from series to

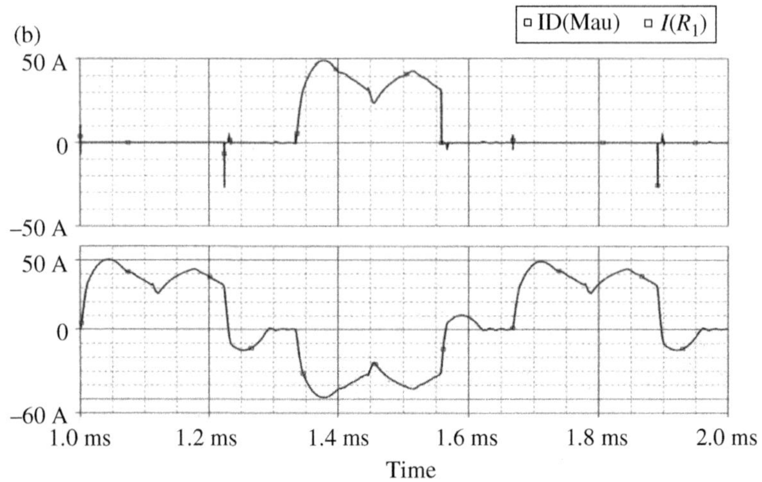

Figure E12.7.1b Upper graph, drain (D) current of upper MOSFET of phase a ID(Mau) $\approx i_{QAU}$; lower graph, motor current through phase a $I(R_1)$ at $\theta = 0°$, and $f = f_1 = 1500$ Hz. The fundamental motor current is 41.7 A.

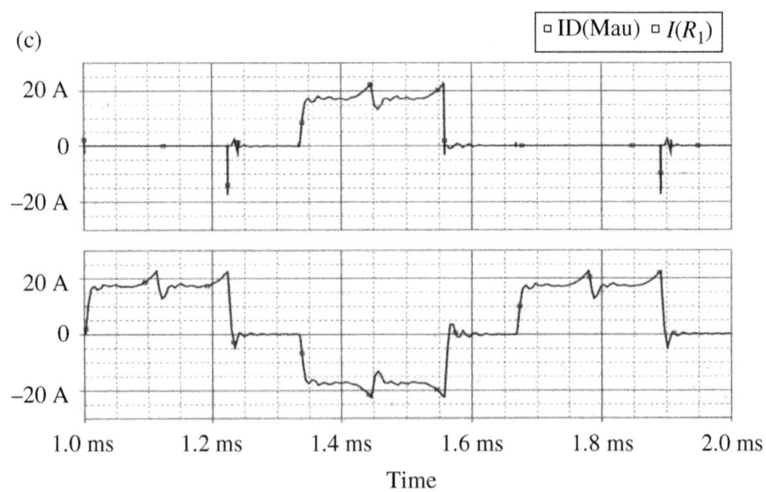

Figure E12.7.1c Upper graph, drain current of upper MOSFET of phase a ID(Mau) $\approx i_{QAU}$; lower graph, motor current through phase a at $I(R_1)$ $\theta = 30°$, $f = f_1 = 1500$ Hz. The fundamental motor current is 19.5 A.

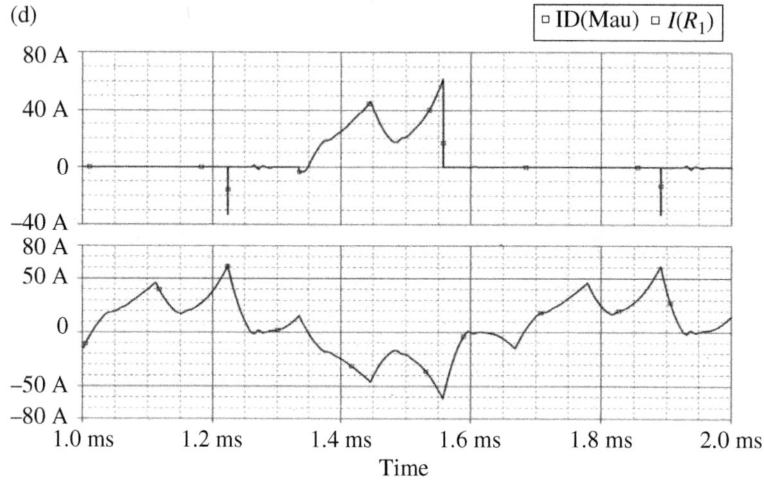

Figure E12.7.1d Upper graph, drain current of upper MOSFET of phase a ID(Mau) ≈ i_{QAU}; lower graph, motor current through phase a $I(R_1)$ at $\theta = 60°$, $f = f_1 = 1500$ Hz. The fundamental motor current is 33.7 A.

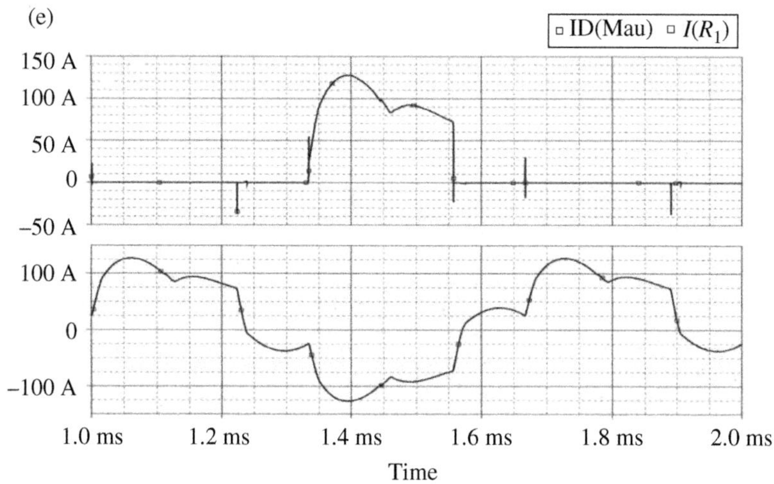

Figure E12.7.1e Upper graph, drain current of upper MOSFET of phase a ID(Mau) ≈ i_{QAU}; lower graph, motor current through phase a $I(R_1)$ at $\theta = -30°$, $f = f_1 = 1500$ Hz. The fundamental motor current is 115.0 A.

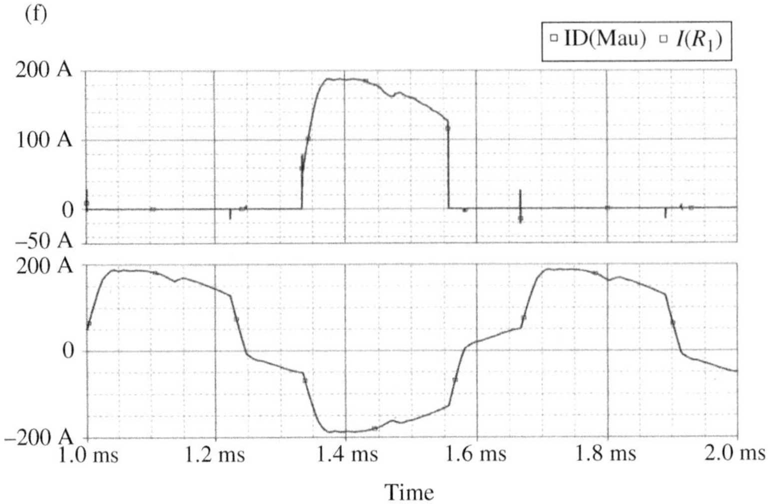

Figure E12.7.1f Upper graph, drain current of upper MOSFET of phase a ID(Mau) ≈ i_{QAU}; lower graph, motor current through phase a $I(R_1)$ at $\theta = -60°$, $f = f_1 = 1500$ Hz. The fundamental motor current is 191.1 A.

Table E12.7.1 PSPICE program codes for Application Example 12.7, part a, Figure E12.7.1b–f.

```
Application Example 12.7a
*Brushless DC motor drive for theta=0°, ea, eb, ec with 160V
*amplitude, f1=1500 Hz, L1=L3=L5=50 uH, rated operation
Va 1a st SIN(0 160V 1500Hz 0 0 0)
Vb 3a st SIN(0 160V 1500Hz 0 0 120)
Vc 5a st SIN(0 160V 1500Hz 0 0 240)
Vdc1 7 0 DC 150V
Vdc2 8 0 DC -150
Rshunt st 0 10Meg
*5 deg delay = 0.231 ms
*50 deg delay = 2.31 ms
*150 deg delay = 6.94 ms
Vfet1 sp1 2 pulse(0 15 0 1u 1u 222.22us 666.66us)
Vfet2 sp2 6 pulse(0 15 222.22us 1u 1u 222.22us 666.66us)
Vfet3 sp3 4 pulse(0 15 444.44us 1u 1u 222.22us 666.66us)
Vfet4 sp4 8 pulse(0 15 333.33us 1u 1u 222.22us 666.66us)
Vfet5 sp5 8 pulse(0 15 555.51us 1u 1u 222.22us 666.66us)
Vfet6 sp6 8 pulse(0 15 111.11us 1u 1u 222.22us 666.66us)
```

Table E12.7.1 (Continued)

```
Mau 7 sp1 2 2 SMM
Mbu 7 sp2 6 6 SMM
Mcu 7 sp3 4 4 SMM
Mal 2 sp4 8 8 SMM
Mbl 6 sp5 8 8 SMM
Mcl 4 sp6 8 8 SMM
L1 1a 1 50u
R1 1 2 0.5
L3 3a 3 50u
R3 3 4 0.5
L5 5a 5 50u
R5 5 6 0.5
C1 7 0 100u
C2 0 8 100u
Dau 2 7 D1N4001
Dbu 6 7 D1N4001
Dal 8 2 D1N4001
Dbl 8 6 D1N4001
Dcu 4 7 D1N4001
Dcl 8 4 D1N4001
.model D1N4001 D(IS=1E-12)
.model SMM nmos(level=3 gamma=0 delta=0 eta=0 theta=0
+kappa=0 vmax=0 xj=0 tox=100n uo=600 phi=0.6 rs=42.69m kp=20.87u
+l=2u w=2.9 vto=3.487 rd=0.19 cbd=200n pb=0.8 mj=0.5 cgso=3.5n
+cgdo=100p rg=1.2 is=10f )
.tran 5u 2m 1ms 5u uic
.options abstol=1m chgtol=1m reltol=10m vntol=10m
.four 1500 I(L1)
.probe
.end
```

parallel connection) to 0.25 Ω. The frequency is $f_2 = 3000$ Hz, which somewhat increases the winding resistance due to the skin effect. This increase is neglected here. See Figure E12.7.1g and Table E12.7.2.

c) The frequency is $f_3 = 750$ Hz, which somewhat decreases the winding resistance due to the skin effect. This decrease is neglected here. Influence of compensation of field weakening at low speeds is shown in Figure E12.7.1h, and the PSPICE program code is listed in Table E12.7.3.

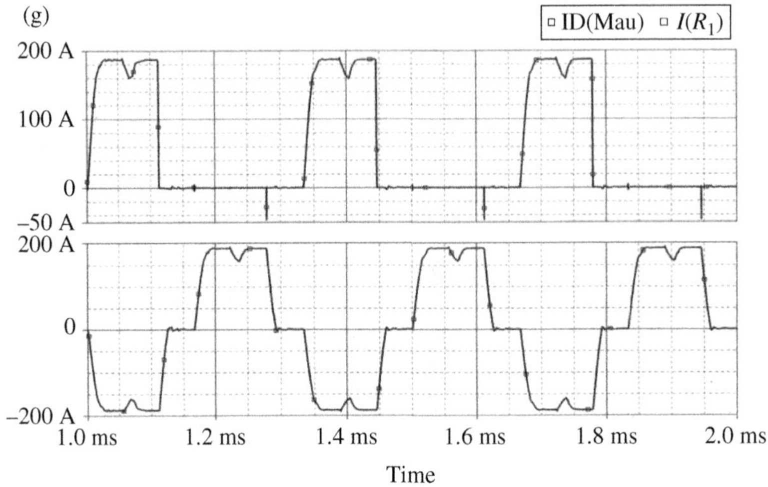

Figure E12.7.1g Upper graph, drain current of upper MOSFET of phase a ID(Mau) ≈ i_{QAU}; lower graph, motor current through phase a $I(R_1)$ at $\theta = 0°, f = f_2 = 3000$ Hz with compensation of field weakening [22, 23]. The fundamental motor current is 199.2 A.

Table E12.7.2 PSPICE program codes for Application Example 12.7, part b (see Figure E12.7.1g).

```
Application Example 12.7b
*Brushless DC motor drive for theta=0 degrees, ea, eb, ec with 80 V amplitude,
*f2=3000 Hz, L1=L3=L5=12.5 uH, R1=R3=R5=0.125 ohms,
*compensation of field weakening
Va 1a st SIN(0 80V 3000Hz 0 0 0)
Vb 3a st SIN(0 80V 3000Hz 0 0 120)
Vc 5a st SIN(0 80V 3000Hz 0 0 240)
Vdc1 7 0 DC 150V
Vdc2 8 0 DC -150
Rshunt st 0 10Meg
Vfet1 sp1 2 pulse(0 15 0 1u 1u 111.11us 333.33us)
Vfet2 sp2 6 pulse(0 15 111.11us 1u 1u 111.11us 333.33us)
Vfet3 sp3 4 pulse(0 15 222.22us 1u 1u 111.11us 333.33us)
Vfet4 sp4 8 pulse(0 15 166.66us 1u 1u 111.11us 333.33us)
Vfet5 sp5 8 pulse(0 15 277.77us 1u 1u 111.11us 333.33us)
Vfet6 sp6 8 pulse(0 15 55.55us 1u 1u 111.11us 333.33us)
```

Table E12.7.2 (Continued)

```
Mau 7 sp1 2 2 SMM
Mbu 7 sp2 6 6 SMM
Mcu 7 sp3 4 4 SMM
Mal 2 sp4 8 8 SMM
Mbl 6 sp5 8 8 SMM
Mcl 4 sp6 8 8 SMM
L1 1a 1 12.5u
R1 1 2 0.25
L3 3a 3 12.5u
R3 3 4 0.25
L5 5a 5 12.5u
R5 5 6 0.25
C1 7 0 100u
C2 0 8 100u
Dau 2 7 D1N4001
Dbu 6 7 D1N4001
Dal 8 2 D1N4001
Dbl 8 6 D1N4001
Dcu 4 7 D1N4001
Dcl 8 4 D1N4001
.model D1N4001 D(IS=1E-12)
.model SMM nmos(level=3 gamma=0 delta=0 eta=0 theta=0
+kappa=0 vmax=0 xj=0 tox=100n uo=600 phi=0.6 rs=42.69m kp=20.87u
+l=2u w=2.9 vto=3.487 rd=0.19 cbd=200n pb=0.8 mj=0.5 cgso=3.5n
+cgdo=100p rg=1.2 is=10f)
.tran 5u 2m 1m 5u uic
.options abstol=1m chgtol=1m reltol=10m vntol=10m
.four 3000 I(L1)
.probe
.end
```

d) The change of the sequence of e_a, e_b, e_c to e_b, e_a, e_c in Table E12.7.1 results in Figure E12.7.1i, where the phase currents are neither identical in magnitude nor are shifted by 120° from one another. This happens because the switching sequence of the full-on gating signals is not the same as that of the induced voltages e_b, e_a, e_c. The full-on frequency is $f_1 = 1500$ Hz.

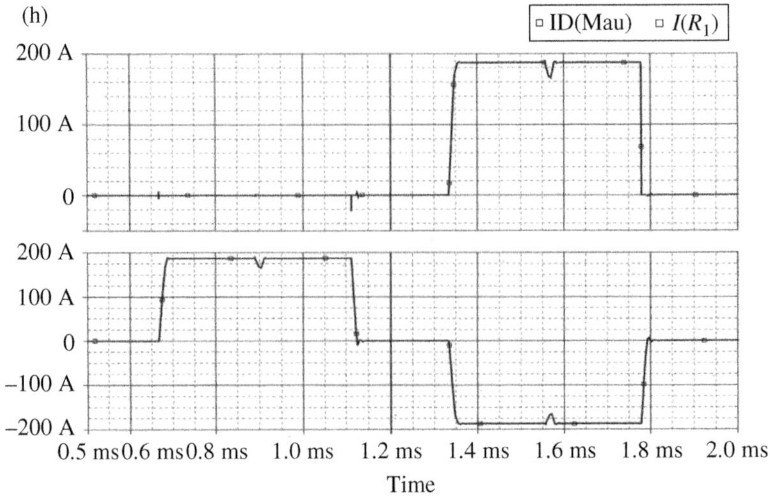

Figure E12.7.1h Upper graph, drain current of upper MOSFET of phase a ID(Mau) $\approx i_{QAU}$; lower graph, motor current through phase a $I(R_1)$ at $\theta = 0°$ and $f = f_3 = 750$ Hz with compensation of field weakening [22, 23]. The fundamental motor current is 199 A.

Table E12.7.3 PSPICE program codes for Application Example 12.7, part c (see Figure E12.7.1h).

```
Application Example 12.7c
*Brushless DC motor drive for theta=0°, ea, eb, ec with 80 V amplitude,
*f3=750 Hz, L1=L3=L5=12.5µH, compensation of field weakening
Va 1a st SIN(0 80V 750Hz 0 0 0)
Vb 3a st SIN(0 80V 750Hz 0 0 120)
Vc 5a st SIN(0 80V 750Hz 0 0 240)
Vdc1 7 0 DC 150V
Vdc2 8 0 DC -150
Rshunt st 0 10Meg
Vfet1 sp1 2 pulse(0 15 0 1u 1u 444.44us 1333.33us)
Vfet2 sp2 6 pulse(0 15 444.44us 1u 1u 444.44us 1333.33us)
Vfet3 sp3 4 pulse(0 15 888.88us 1u 1u 444.44us 1333.33us)
Vfet4 sp4 8 pulse(0 15 666.66us 1u 1u 444.44us 1333.33us)
Vfet5 sp5 8 pulse(0 15 1111.11us 1u 1u 444.44us 1333.33us)
Vfet6 sp6 8 pulse(0 15 222.22us 1u 1u 444.44us 1333.33us)
Mau 7 sp1 2 2 SMM
Mbu 7 sp2 6 6 SMM
Mcu 7 sp3 4 4 SMM
```

Table E12.7.3 (Continued)

```
Mal 2 sp4 8 8 SMM
Mbl 6 sp5 8 8 SMM
Mcl 4 sp6 8 8 SMM
L1 1a 1 12.5u
R1 1 2 0.25
L3 3a 3 12.5u
R3 3 4 0.25
L5 5a 5 12.5u
R5 5 6 0.25
C1 7 0 100u
C2 0 8 100u
Dau 2 7 D1N4001
Dbu 6 7 D1N4001
Dal 8 2 D1N4001
Dbl 8 6 D1N4001
Dcu 4 7 D1N4001
Dcl 8 4 D1N4001
.model D1N4001 D(IS=1E-12)
.model SMM nmos(level=3 gamma=0 delta=0 eta=0 theta=0
+kappa=0 vmax=0 xj=0 tox=100n uo=600 phi=0.6 rs=42.69m kp=20.87u
+l=2u w=2.9 vto=3.487 rd=0.19 cbd=200n pb=0.8 mj=0.5 cgso=3.5n
+cgdo=100p rg=1.2 is=10f )
.tran 5u 2m 1m 5u uic
.options abstol=1m chgtol=1m reltol=10m vntol=10m
.four 750 I(L1)
.probe
.end
```

Conclusions

Compensation of flux weakening (CFW) increases the output torque and the power per unit of weight compared with flux-weakening operation. The advantage of the PSPICE approach is that solutions for nonsinusoidal induced voltages can be obtained as well. The brushless DC machine is well suited for hybrid/electric cars because the battery voltage can be in the range of 300–600 V.

Electric/hybrid cars: PMMs have a high output power for a given weight [10], such as 3.6 kW/kg-force. Drawbacks such as brittleness, temperature sensitivity, and the relatively high cost of $50–100/kg-force may eventually limit their future application to electric vehicle and wind power applications. The efficiencies [12] of

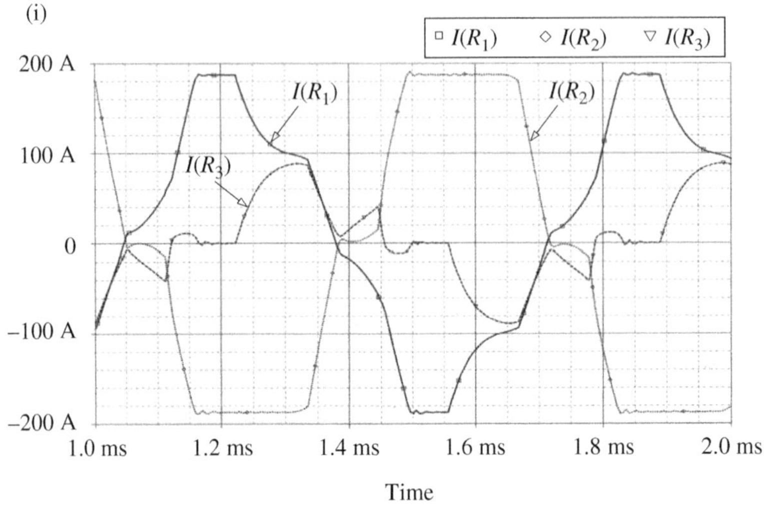

Figure E12.7.1i Phase sequence of full-on gating signals is not the same as that of the induced voltages e_b, e_a, e_c, and for this reason the amplitudes and phase shifts of the three phases are different, so no rotating magnetic field is established. Note phase b_{motor} connected to phase $a_{inverter}$, phase a_{motor} connected to phase $b_{inverter}$, and phase c_{motor} connected to phase $c_{inverter}$ as depicted in Figure 12.14a.

PMM at full-on mode reach 98%, while for PWM mode they are about 86%. The inverter/controller of the brushless DC machine (see Figure 12.14a) has efficiencies of 98 and 85% for full-on mode and PWM, respectively. In the regeneration mode the efficiencies are 93 and 94% for the PMM and the controller/inverter, respectively. Although these efficiencies are very high, the recovered braking energy efficiency through regeneration [5, 6] ranges from 20 to 30%. While the regenerated energy and that stored in the vehicle battery is greatest in stop-and-go traffic, it is zero in steady traffic when no braking is required. Figure 12.15 illustrates an air-cooled PMM with fins [11]; alternatively, water cooling can be used, reducing the size of the machine.

Application Example 12.8 Direct-drive, variable-speed wind power plant
The block diagram of a direct-drive, variable-speed 20 kW/30 kVA wind power plant [10–18] is shown in Figure E12.8.1. Current-controlled voltage source inverters are mostly operated at unity power factor in order to limit the required DC input voltage [5, 6]. A lagging power factor (generator reference system), see Figure E12.8.2, requires a larger DC input voltage than at unity power factor. Conversely, a leading power factor (generator reference system) permits the

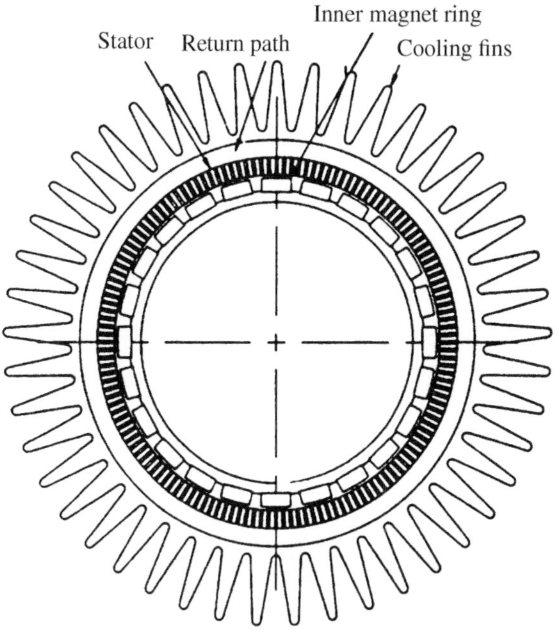

Figure 12.15 Single inner magnet external fin-cooled motor [11].

application of a lower DC voltage than at unity power factor. This is the reason why the 20 kW can only be operated at 30 kVA with a leading power factor. One problem with large (20 kW and larger) PMM is that the rotor with the permanent magnets requires a mounting gear because of the large magnetic forces attracting the stator core to the permanent-magnet poles if the mounting does not occur in a concentric manner. In addition, the frequency of the inverter feeding power into the grid must have exactly the same (slowly changing) frequency as the power system; this requires a phase-lock loop circuit as indicated in Figure E12.8.1. The synchronization of the inverter with the power grid must be performed so that current transients are minimized [15, 17]. Figure E12.8.3 depicts the transient current spike during synchronization of the 20 kW wind power plant with the power grid [17]. This transient current is due to the instantaneously charged capacitances of the inverter. The two three-phase transformers, 208 : 480 V and 480 : 240 V connected in series, can be replaced by one 208 : 240 V transformer. Figure E12.8.2 shows the phasor diagram of the permanent-magnet generator employing the previously discussed direct-axis, quadrature-axis, and stator leakage reactances X_d, X_q, and $X_{s\ell}$, respectively.

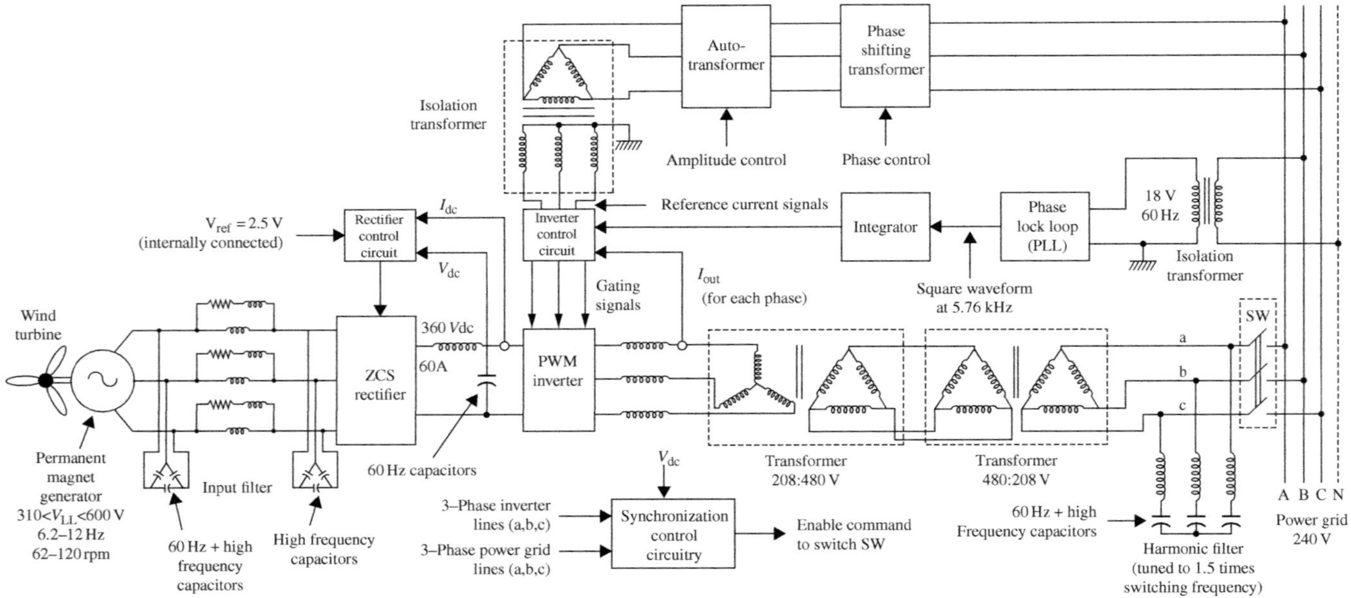

Figure E12.8.1 Block diagram of 20 kW/30 kVA direct-drive, variable-speed wind power plant [15, 17] where the generator supplies power through a zero-current-switch (ZCS) resonant rectifier and a PWM current-controlled voltage source inverter (VSI).

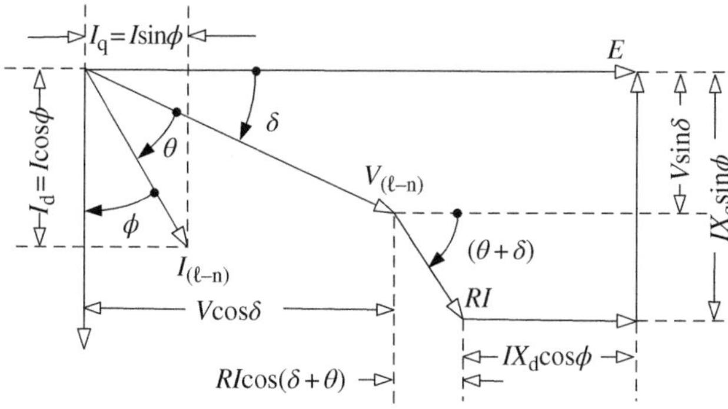

Figure E12.8.2 Phasor diagram of permanent-magnet generator [18].

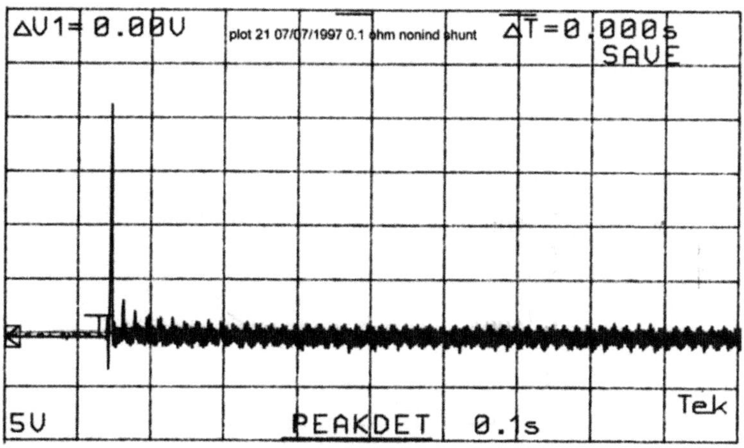

Figure E12.8.3 Transient current during synchronization of inverter with power grid [15, 17]. The inverter voltage is leading the power system voltage by $\theta = 24°$. Scale for abscissa: 0.1 s/div, scale for ordinate: 50 A/div.

Wind power plants with large output powers (e.g. 2–8.8 MW) are operated with 1/8 revolutions per second (rps), that is, 60/8 = 7.5 revolutions per minute (rpm). Smaller plants (e.g. kW range) can be operated at higher rpm as is illustrated in Figure E12.8.1.

Conclusions

A so-called smart grid consists of renewable energy sources (e.g. microgrids) and the main power grid, which serves as a frequency leader with frequency f_{power_grid}. Most renewable energy sources (e.g. photovoltaic, wind) are operating in an intermittent manner and must be connected and disconnected from the main power grid if there is no energy production. While the disconnection is not difficult, the connection of a microgrid to the main power grid requires the frequency of the microgrid to be $f_{micro\_grid} = f_{power\_grid}$, and the phases of the microgrid and power grid voltages must be about the same $\Phi_{micro\_grid\_i} - \Phi_{power\_grid\_i} \approx 0$, where i indicates the three phases A, B, C: both conditions are achieved by using a so-called phase-lock loop. Of course, the terminal instantaneous voltage amplitudes of both grids must be about the same as well. Otherwise, problems occur; the supply of electric energy of Guangdong in southern China via high-voltage DC lines (HVDC) led to a breakup (as discussed in the Preface) of the AC grid due to the lack of a frequency-leading AC power plant and through insufficient reactive power to control voltage and frequency.

IMs represent an alternative to PMM for electric/hybrid cars and wind power plant generators, having no brittleness, no severe temperature sensitivity, and a low cost. However, the efficiencies reached with IM are lower than those of PMM.

12.4 Three-Phase (Polyphase) IMs: Balanced Operation

It can be shown that three-phase systems require the smallest amount of conduction material (e.g. copper, aluminum) for the conversion/transmission of electric power from point 1 to point 2. This applies to IMs as well and is why Tesla's two-phase approach is generally not used. At balanced operation, only three wires are needed – even for three-phase configurations – and the sum of the three currents adds to zero, so no return path for the currents from point 1 to point 2 is normally required:

$$I_{max} \cos \omega t + I_{max} \cos (\omega t - 120°) + I_{max} \cos (\omega t - 240°) = 0. \qquad (12.17)$$

Any IM can be either operated as a motor or as a generator.

12.4.1 Basic Principle of Operation

A basic law of energy conversion is (see Chapter 11, Section 11.2) as follows:

> For steady-state torque production, the interacting stator and rotor magnetic fields must be stationary with respect to one another.

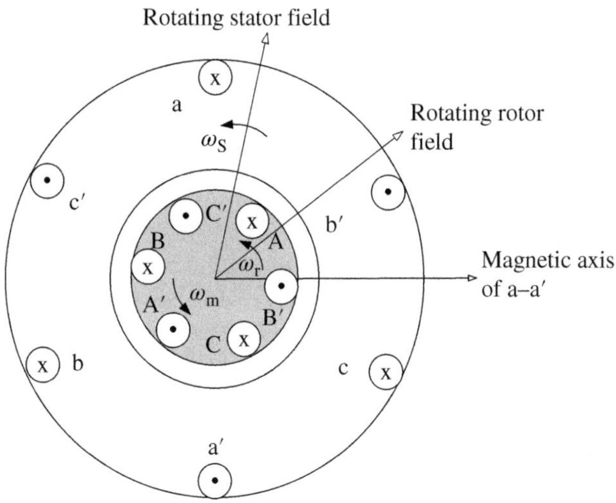

Figure 12.16 Cross section of three-phase IM with three stator windings and three rotor windings.

The three-phase winding configuration of an IM is depicted in Figure 12.16, where the stator carries the three windings a–a′, b–b′, and c–c′; the three windings A–A′, B–B′, and C–C′ are installed on the rotor. Also indicated is the associated angular velocity of stator rotating field ω_s with respect to stator, angular mechanical velocity of rotor ω_m with respect to stator, and angular velocity of rotating field of rotor with respect to rotating rotor ω_r.

ω_s, ω_m, and ω_r are rotating in the same angular direction and therefore pertain to motor operation.

Satisfying the above law for energy conversion, one can state

$$\omega_s = \omega_m + \omega_r. \tag{12.18}$$

1) Mechanical (subscript m) rotor member rotates with respect to the stationary stator with ω_m.
2) Stator (subscript s) field rotates with respect to the stationary stator with ω_s.
3) Rotor (subscript r) field rotates with respect to the rotating rotor with ω_r.

In order to better understand Eq. (12.18), we perform two experiments:

Experiment 1: Rotor is at rest (locked or blocked) and does not rotate (stationary with respect to stator), which means $\omega_m = 0$. The angular velocity of the rotor field is the same as that of the stator field

$$\omega_r = \omega_s = (2/p)2\pi f; \tag{12.19}$$

as a result, the IM behaves like a three-phase transformer discussed in Chapter 6. Note p is the number of poles.

Experiment 2: Rotor rotates at the same angular velocity as the stator field, that is,

$$\omega_m = \omega_s. \tag{12.20}$$

Because the rotor does not rotate relative to the stator field, there are – according to Faraday's law (see Chapter 6) – no rotor voltages/currents induced, and their frequency does not exist; $f_r = 0$ or $\omega_r = 2\pi f_r = 0$, resulting in

$$\omega_s = \omega_m + \omega_r = \omega_m = (2/p)2\pi f. \tag{12.21}$$

Generalizing the above results, we can state that $\omega_s = \omega_m + \omega_r$ is satisfied.

Definition of slip s: In IM, the rotating stator field (angular velocity ω_s with respect to the stator) induces in the rotating rotor winding with angular velocity ω_m the rotating rotor field with angular velocity ω_r with respect to the rotor; no external excitation of the rotor is required. Because of this induced rotor field, such a machine is called induction machine. As a consequence, the rotor and the stator field cannot rotate in synchronism ($\omega_m \neq \omega_s$), and the rotor must have a slip s with respect to the stator field: this is called the induction motor principle.

Induction motor principle: One defines the slip s as

$$s = (\omega_s - \omega_m)/\omega_s, \tag{12.22}$$

or

$$s \cdot \omega_s = \omega_s - \omega_m = \omega_r, \tag{12.23}$$

where $s \cdot \omega_s = \omega_r$ is the mechanical angular velocity of the rotating rotor field with respect to the rotating rotor. To check the validity of Eq. (12.23), we perform two more experiments:

Experiment 1: Block or lock the rotor so that it cannot rotate: $\omega_m = 0$. The IM works now like a three-phase transformer, and the rotating stator field induces in the stationary rotor winding a rotating rotor field with $\omega_r = \omega_s$; thus the slip $s = (\omega_s - \omega_m)/\omega_s = \omega_s/\omega_s = 1$.

Experiment 2: Rotate the rotor in synchronism with the rotating stator field; that is, $\omega_m = \omega_s$. There is no relative motion between rotating rotor winding and rotating stator field; thus no rotor field will be induced in the rotating rotor winding and $\omega_r = 0$, with $s = (\omega_s - \omega_m)/\omega_s = (\omega_s - \omega_s)/\omega_s = 0$. In general, for any ω_m relation, Eq. (12.22) is satisfied. From this follows

$$\omega_m = \omega_s(1 - s) = \omega_s - s \cdot \omega_s = \omega_s - \omega_r \tag{12.24}$$

where

$$\omega_s = (2/p) \cdot 2\pi f, \tag{12.25}$$

p is the pole number of either the stator or rotor windings (in IM the pole number is the same in stator and rotor, which applies to any rotating electric machine), and f is the electric frequency of stator voltages/currents.

Correspondingly,

$$\omega_r = (2/p) \cdot 2\pi f_r, \tag{12.26}$$

where f_r is the electrical frequency of the rotor voltages/currents.

The relation between stator and rotor frequencies f and f_r, respectively, is obtained from

$$\omega_m = \omega_s - s \cdot \omega_s = \omega_s - \omega_r = (2/p) \cdot 2\pi f - s \cdot (2/p) \cdot 2\pi f; \tag{12.27}$$

with ω_s proportional to f and ω_r proportional to $(s \cdot f)$, one gets $\omega_s/\omega_r = f/f_r = f/(sf)$ where s is the slip or

$$f_r = sf. \tag{12.28}$$

Relation between speed measured in revolutions per minute (n^{rpm}) and frequency f measured in hertz or 1/s, where s stands for the unit of seconds, is discussed next: The synchronous (denoted by subscript s) speed measured in revolutions per second (rps) is

$$n_s^{rps} = \frac{\omega_s}{2\pi} = \frac{\left(\frac{2}{p}\right) 2\pi f}{2\pi} = \frac{2}{p} f, \tag{12.29}$$

where the speed in revolutions per minute (rpm) is

$$n_s^{rpm} = 60 \cdot n_s^{rps}. \tag{12.30}$$

For $f = 60$ Hz one obtains

$$p = 2 \rightarrow n_s^{rpm} = \frac{60 \cdot 120}{2} = 3600 \text{ rpm}, \tag{12.31a}$$

$$p = 4 \rightarrow n_s^{rpm} = \frac{60 \cdot 120}{4} = 1800 \text{ rpm}, \tag{12.31b}$$

etc.

For $f = 50$ Hz the speed values are

$$p = 2 \rightarrow n_s^{rpm} = \frac{50 \cdot 120}{2} = 3000 \text{ rpm}, \tag{12.32a}$$

$$p = 4 \rightarrow n_s^{rpm} = \frac{50 \cdot 120}{4} = 1500 \text{ rpm}, \tag{12.32b}$$

etc.

Motor operation: The slip s is positive and must be in the range $0 \le s \le 1$, at steady state $0 \le s_{rated} \le 0.1$; from $s = (\omega_s - \omega_m)/\omega_s$ follows $\omega_m < \omega_s$.

Generator operation: The slip s is negative and for steady state must be in the range $-0.1 \leq s_{rated} \leq 0$; from $s = (\omega_s - \omega_m)/\omega_s$ follows $\omega_m > \omega_s$.

12.4.2 Equivalent Circuits

Balanced constant-speed operation of IM (either motor or generator) may be modeled by the per-phase wye (Y) equivalent circuit shown below.

In the above equivalent circuit, the following quantities are defined:

- $\tilde{V}_s$ = line-to-neutral stator phase voltage applied at the terminals (a, n), usually used as phase reference. a stands for phase a and n for neutral
- $\tilde{I}_s$ = stator (s) phase current
- R_s = phase stator winding resistance
- X_s = phase leakage reactance associated with the stator flux that does not link the rotor winding (see definition of leakage inductance of Appendix 11.B in Chapter 11)
- X_m = per-phase magnetizing (m) reactance due to mutual flux between stator and rotor windings
- $\tilde{I}_r'$ = rotor (r) phase current referred (reflected) to the stator
- $\tilde{X}_r'$ = phase leakage reactance associated with the rotor flux referred – indicated by a prime (′) – to stator that does not link the stator winding (see definition of leakage inductance of Appendix 11.B in Chapter 11)
- $\tilde{R}_r'$ = phase rotor winding resistance referred (′) to stator
- s = slip of the mechanical rotor, which is a measure of rotor speed $n_m^{rpm} = n_s^{rpm}(1 - s)$.

An alternative equivalent circuit to Figure 12.17 is depicted in Figure 12.18a.

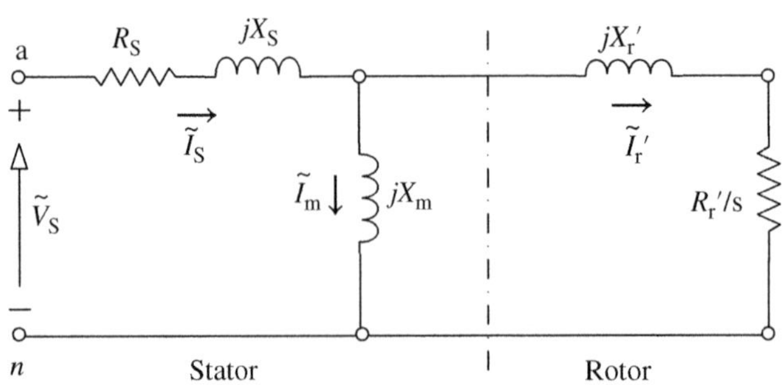

Figure 12.17 Wye equivalent circuit of IM.

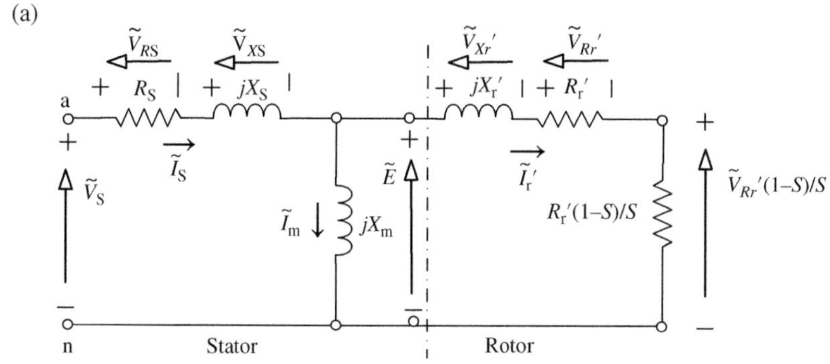

Figure 12.18a Alternative wye equivalent circuit of IM.

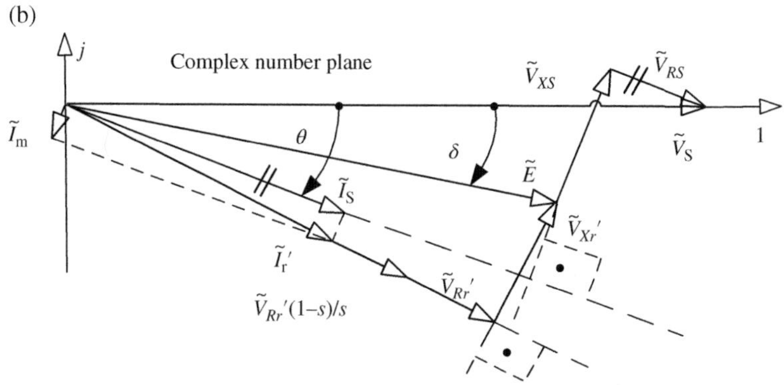

Figure 12.18b Phasor diagram (not to scale) at rated-load operation based on circuit of Figure 12.18a. The two phasors $\tilde{I}_s$ and $\tilde{V}_{Rs}$ are collinear or parallel as indicated by the symbol //. Θ is the power factor angle between phasors $\tilde{V}_s$ and $\tilde{I}_s$; δ is the load angle between phasors $\tilde{V}_s$ and $\tilde{E}$.

Note

$$R_r' + R_r'\frac{(1-s)}{s} = R_r' + \frac{R_r'}{s} - R_r' = \frac{R_r'}{s}. \tag{12.33}$$

Total output power (P) developed (d) is

$$P_d = 3\left|\tilde{I}_r'\right|^2 R_r'\frac{(1-s)}{s}. \tag{12.34}$$

Total output torque (τ) developed is

$$\tau_d = \frac{P_d}{\omega_m} = \frac{3\left|\tilde{I}_r'\right|^2 R_r'/s}{\omega_s}. \tag{12.35}$$

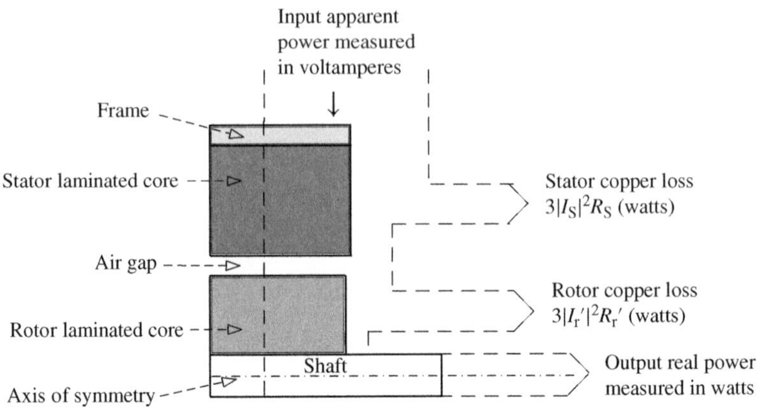

Figure 12.19 Power and loss distribution within a three-phase IM, whereby the stator and rotor iron-core losses as well as windage/friction losses are ignored.

The relation for horsepower in terms of the meter–kilogram–second–ampere (MKSA) system is 1 hp ≈ 746 W. Figure 12.19 illustrates the power/loss distribution in a three-phase IM, where the stator and rotor iron-core losses as well as the frictional/windage losses are disregarded. The mechanical design of an IM is presented in Fuchs and Masoum [5]

Application Example 12.9 No-load operation of IM (slip $s = 0$)

A three-phase Y-connected wound-rotor (r) IM [24] has the parameters $V_{\text{L-L}} = 460$ V, $P_{\text{out}} = 20$ hp, $f = 60$ Hz, four poles $p = 4$, rated (rat) slip $s_{\text{rat}} = 3\%$, and the following impedances (resistances, reactances) per phase referred to the stator (s) circuit: $R_s = 0.64\,\Omega$, $X_s = 1.1\,\Omega$, $R_r' = 0.33\,\Omega$, $X_r' = 0.46\,\Omega$, and $X_m = 26\,\Omega$. The iron (ferrum)-core loss ($R_{fe} \to \infty$) and the windage/friction losses are small and can be neglected: $P_{\text{windage/friction}} = 0$. IM always operates at lagging (consumer system) displacement power factor due to the magnetizing (mutual) reactance X_m between stator and rotor and the stator and rotor leakage reactances X_s and X_r', respectively.

a) Compute the line (L)-to-neutral (N) voltage $\tilde{V}_s = \tilde{V}_{L-N}$ and the stator current $\tilde{I}_s$ at no load, that is, $\tilde{I}_r' = 0$.
b) Draw the phasor diagram for voltage $\tilde{V}_{L-N}$ and current $\tilde{I}_s$.

Solution

The line-to-neutral voltage is defined as the reference phasor resulting in $\tilde{V}_s = \tilde{V}_{L-N} = (V_{L-L}/\sqrt{3})\angle 0° = (460/\sqrt{3})\angle 0°\text{V} = 265.6\angle 0°\text{V}$. From Figure 12.17 one obtains for no-load condition, where $\tilde{I}_r' = 0$ the voltage/current relation

$$\tilde{V}_s = \tilde{V}_{L-N} = \tilde{I}_s(R_s + jX_s + jX_m). \tag{E12.9.1}$$

Introducing in Eq. (E12.9.1) the above defined parameter values, one obtains for the stator current

$$\tilde{I}_s = \frac{\tilde{V}_s}{R_s + jX_s + jX_m} = \frac{265.6\angle 0°\text{V}}{(0.64\,\Omega + j1.1\,\Omega + j26\,\Omega)}$$

$$= \frac{265.6\angle 0°\text{V}}{\sqrt{(0.64)^2 + (27.1)^2}\angle 88.65°\Omega} = \left|\tilde{I}_s\right|\angle\theta = 9.8\angle(0° - 88.65°)\,\text{A} = 9.8\angle -88.65°\text{A}. \tag{E12.9.2}$$

Figure E12.9.1 defines the phasor voltages and currents, and Figure E12.9.2 depicts the not-to-scale phasor diagram at no load whereby the line-to-neutral voltage $\tilde{V}_s = \tilde{V}_{L-N} = 265.6\angle 0°\text{V}$ is chosen as reference direction within the complex (j, 1) number Gaussian plane.

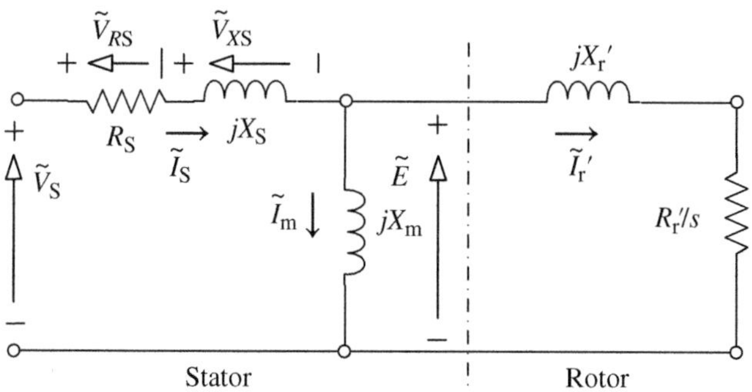

Figure E12.9.1 Definition of phasor voltages and currents of stator for slip $s = 0$ resulting in zero rotor current $\tilde{I}_r' = 0$.

Figure E12.9.2 Phasor diagram in the (j, 1) complex number Gaussian plane (not to scale) at no-load operation, where the rotor current $\tilde{I}_r' = 0$. The two phasors $\tilde{I}_s$ and $\tilde{V}_{Rs}$ are collinear or parallel, as indicated by the symbol //.

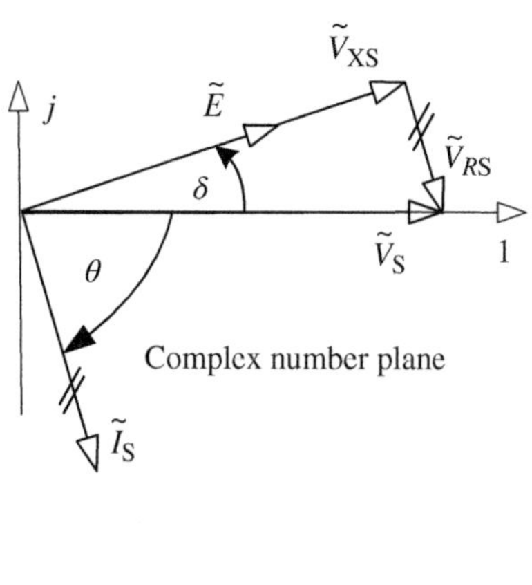

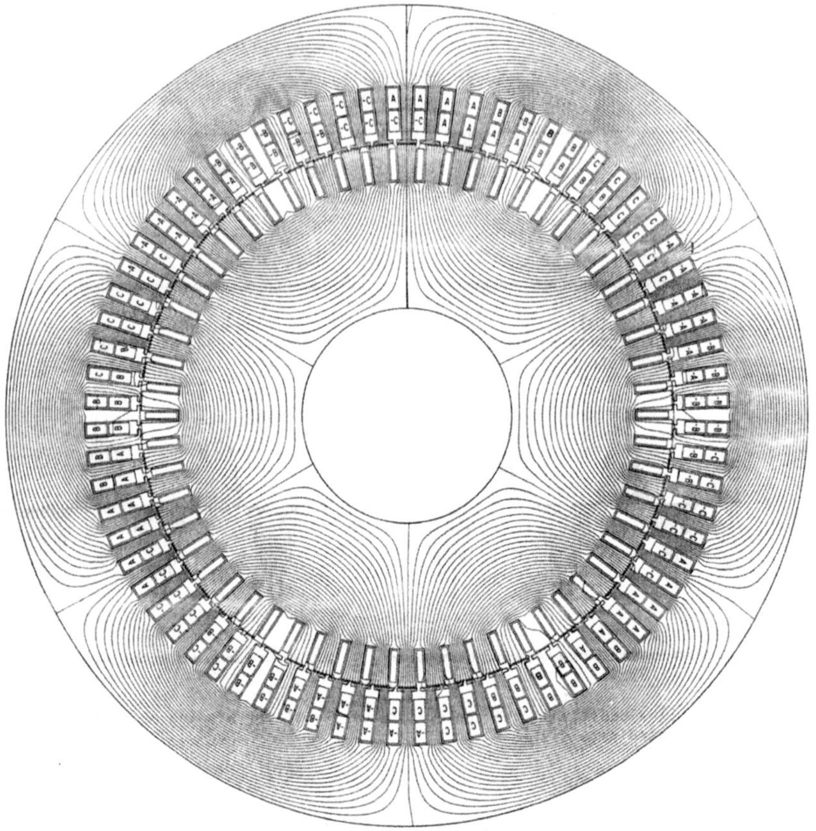

Figure E12.9.3 Saturated no-load magnetic field of three-phase squirrel-cage IM.

The FD/FE methods [25–27] outlined in Chapter 11, Appendix 11.A, Figure 11. A.3a–c are employed to compute the magnetic fields for induction motors. Figure E12.9.3 shows the no-load field of a three-phase squirrel-cage IM.

Application Example 12.10 Stator iron-core forces at no-load rated voltage operation of three-phase IM

The stator iron-core forces can be computed from surface integrals [28–31] and are the origin of audible noise. Figure E12.10.1 shows such radially acting forces at no-load operation of a three-phase IM.

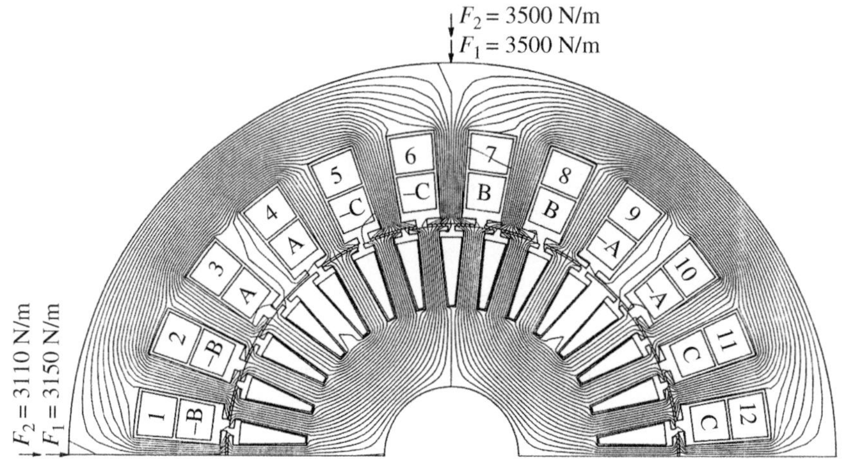

Figure E12.10.1 Saturated no-load magnetic field [30–35] of a three-phase IM based on a polycentric grid/mesh system (see Chapter 11, Appendix 11.A, Figure 11.A.3a–c) with calculated stator iron-core forces generating audible noise.

Application Example 12.11 Rated-load and starting operations of IM with rated slip $s = s_{rated} = 0.03$ per unit (pu) and slip at starting (start) $s_{start} = 1.0$ pu, respectively

Compute for the motor data given in Application Example 12.9 the following machine parameters and data:

a) Thévenin (TH) voltage $\widetilde{V}_{TH}$ and Thévenin impedance $\widetilde{Z}_{TH} = R_{TH} + jX_{TH}$ for the stator equivalent circuit Figure 12.17.
b) Compute the rotor current $\widetilde{I}_r'$ for rated slip $s_{rat} = 0.03$ pu $= 3\%$.
c) Determine the developed (d) output power P_d in kW and horsepower hp (note 1 hp = 0.746 kW).
d) Calculate the developed torque τ_{d_rat} at rated operation.
e) Calculate the starting torque $\tau_{d\text{-start}}$.
f) Draw the not-to-scale phasor diagram for rated operation.

Solution

a) Calculation of the Thévenin voltage $\widetilde{V}_{TH}$ and impedance $\widetilde{Z}_{TH} = R_{TH} + jX_{TH} = Z_{TH}$, where the impedance $\widetilde{Z}_{TH}$ is a complex number sometimes displayed without a tilde (~). All other complex numbers are characterized with a tilde. In Figure E12.11.1a the original stator circuit is replaced by the Thévenin equivalent stator circuit, resulting in Figure E12.11.1b.

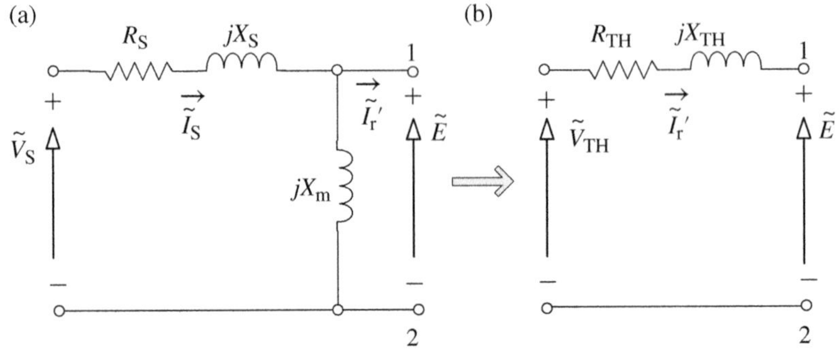

Figure E12.11.1 (a) The original stator circuit is replaced by (b) Thévenin equivalent stator circuit.

The Thévenin voltage $\widetilde{V}_{TH}$ is obtained by applying the voltage divider rule:

$$\widetilde{V}_{TH} = \widetilde{V}_s \cdot \frac{jX_m}{R_s + jX_s + jX_m} = 265.6\angle 0° \cdot \frac{j26}{0.64 + j1.1 + j26}$$

$$= \frac{265.6\angle 0° \cdot 26\angle 90°}{\sqrt{(0.64)^2 + (27.1)^2}\ \angle 88.65°} = 254.75\angle 1.35°\text{V}, \quad \text{(E12.11.1)}$$

and the Thévenin impedance $\widetilde{Z}_{TH}$ is found by the parallel connection of the two impedances jX_m and $(R_s + jX_s)$:

$$\widetilde{Z}_{TH} = jX_m \cdot \frac{R_s + jX_s}{R_s + jX_s + jX_m} = j26 \cdot \frac{0.64 + j1.1}{0.64 + j27.1}$$

$$= \left[26\sqrt{(0.64)^2 + (1.1)^2}\angle(59.81 + 90)°\right] \Big/ \left[\sqrt{(0.64)^2 + (27.1)^2}\angle 88.65°\right]$$

$$= 1.221\angle 61.17°\Omega = R_{TH} + jX_{TH} = 1.221\cos(61.17°) + j1.221\sin(61.17°)$$

$$= (0.59 + j1.07)\Omega, \quad \text{(E12.11.2)}$$

$$R_{TH} = 0.59\ \Omega, \text{and } X_{TH} = 1.07\ \Omega. \quad \text{(E12.11.3)}$$

As a result one obtains the equivalent circuit, which is simpler because no cross-impedance branch is present, as indicated in Figure E12.11.1c.

b) The rotor current is now at $s_{rat} = 0.03$ pu

$$\widetilde{I}_r = \widetilde{V}_{TH} \Big/ \left[\left(R_{TH} + \frac{R_r'}{s_{rat}}\right) + j(X_{TH} + X_r')\right], \quad \text{(E12.11.4)}$$

(c)

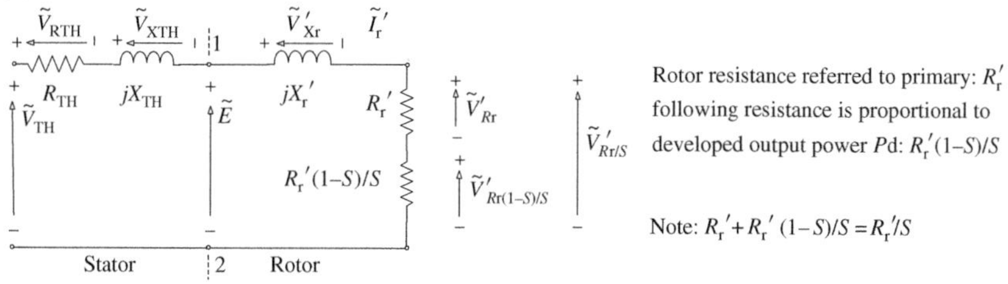

Figure E12.11.1c Simplified equivalent circuit employing Thévenin equivalent circuit for the stator circuit.

$$\tilde{I}_r' = \frac{254.75\angle 1.35°\text{V}}{\left[\left(0.59 + \frac{0.33}{0.03}\right) + j(1.07 + 0.46)\right]} = \frac{254.75\angle 1.351.35°\text{V}}{11.59 + j1.53}$$

$$= (254.75\angle 1.35°\text{V}) \Big/ \left[\sqrt{134.33 + 2.34}\angle 7.5°\Omega = 21.79\text{ A}\angle\theta = -6.17°.\right.$$

c) Total (three-phase) developed rated output power at the shaft is based on Eq. (12.34)

$$P_d = 3 \cdot (21.79)^2 \cdot 0.33(1 - 0.03)/0.03 = 15.2\text{ kW} \equiv 20.4\text{ hp}.$$

d) From Eq. (12.35) one finds the total rated torque for $p = 4$ and $n_s = 1800$ rpm, resulting in angular velocity (or frequency) $\omega_s = (1800/60) \cdot 2\pi = 188.48$ rad/s:

$$\tau_d = \frac{\left[3 \cdot |\tilde{I}_r'|^2 \cdot \left(\frac{R_r'}{s}\right)\right]}{\omega_s} = \left[3 \cdot (21.79)^2 \cdot \frac{\frac{0.33}{0.03}}{188.49}\right] = 83.127\text{ Nm}. \tag{E12.11.5}$$

e) Correspondingly, the starting torque at $s = 1.0$ pu is

$$\tilde{I}_r' = \frac{254.75\angle 1.35°\text{V}}{\left[\left(0.59 + \frac{0.33}{1.0}\right) + j(1.07 + 0.46)\right]} = \frac{254.75\angle 1.35°\text{V}}{0.92 + j1.53}$$

$$= \frac{254.75\angle 1.35°\text{V}}{\left[\sqrt{0.8464 + 2.34}\angle 58.98°\right]\Omega} = 142.69\text{ A}\angle\theta = -57.631°,$$

$$\tau_d = \frac{\left[3 \cdot |\tilde{I}_r'|^2 \cdot \left(\frac{R_r'}{s}\right)\right]}{\omega_s} = \left[\frac{3 \cdot (142.69)^2 \cdot \left(\frac{0.33}{1.0}\right)}{188.49}\right] = 106.94\text{ Nm}.$$

f) Figure E12.11.1d shows the not-to-scale phasor diagram at rated operation, where the voltage/current phasors are defined in Figure E12.11.1c.

(d)

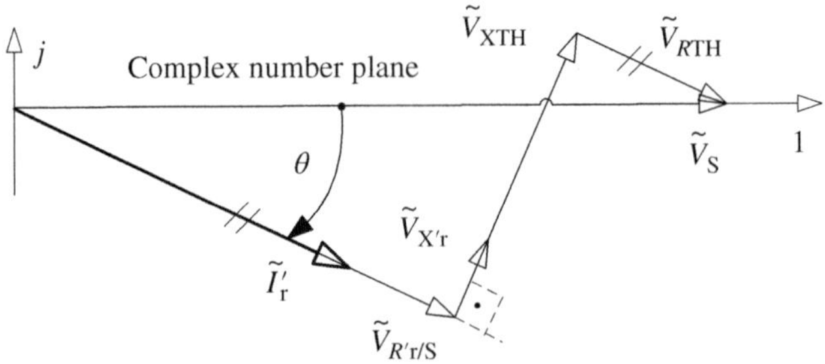

Figure E12.11.1d Phasor diagram (not to scale) for rated operation based on Thévenin equivalent circuit, Figure E12.11.1c. The three phasors $\tilde{I}'_r$, $\tilde{V}_{RTH}$, and $\tilde{V}_{Rr'}/s$ are collinear or parallel, as indicated by the symbol //.

Application Example 12.12 Starting operation [32–35] of squirrel-cage-type IM with $s = 1.0$

The starting of induction motors depends upon the saturation of the stator and rotor teeth as depicted by the field distribution under rated, saturated (nonlinear) conditions [5]. The following field plots for unsaturated (linear) conditions represent a first step for the calculation of the parameters and under saturated conditions [35] at rated voltage.

(a)

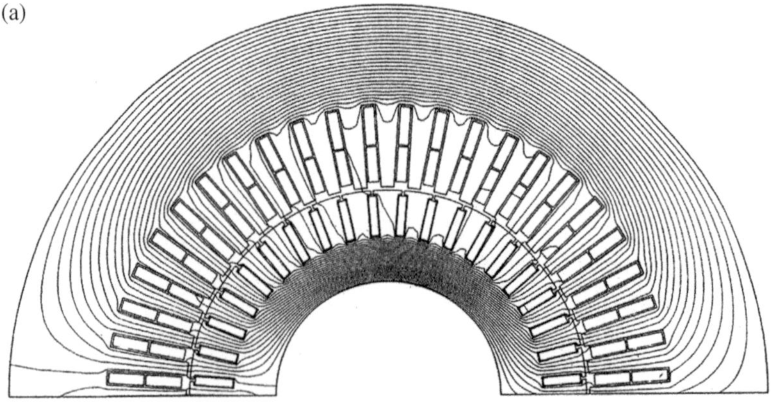

Figure E12.12.1a Magnetic field for the calculation of unsaturated (linear) stator leakage reactance [35] based on a polycentric grid/mesh system (see Chapter 11, Appendix 11.A, Figure 11.A.3a–c). The asymmetry is due to the winding pitch factor $k_w < 1$.

(b)

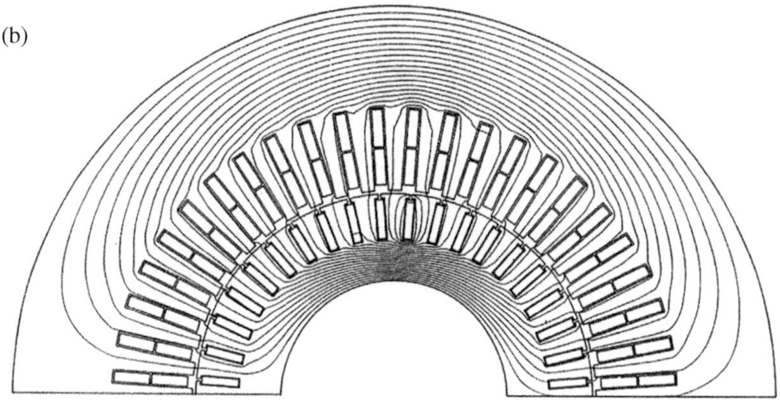

Figure E12.12.1b Magnetic field for the calculation of unsaturated (linear) rotor leakage reactance [35] based on a polycentric grid/mesh system (see Chapter 11, Appendix 11.A, Figure 11.A.3a–c). The asymmetry is due to the even number of rotor slots within one pole pitch.

(c)

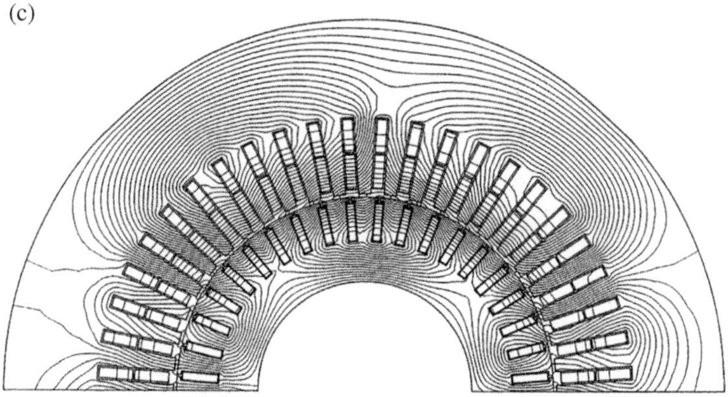

Figure E12.12.1c Magnetic field for the calculation of starting torque at rated stator voltage and unsaturated (linear) rotor iron cores [35] based on a polycentric grid/mesh system (see Chapter 11, Appendix 11.A, Figure 11.A.3a–c).

Figure E12.12.1d illustrates the influence of saturation on the performance of induction motors at starting: saturation [5] increases the starting torque and starting current significantly compared with unsaturated conditions. The influence of saturation is very beneficial for induction motor starting.

(d)

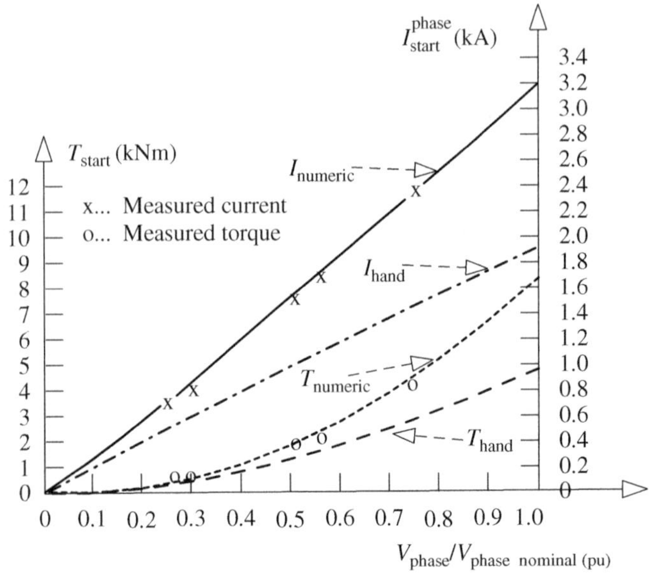

Figure E12.12.1d Starting current per phase I_{start}^{phase} and total (three-phase) torque [35] T_{start} of squirrel-cage IM at unsaturated (I_{hand} and T_{hand} are based on linear equivalent circuit) and saturated ($I_{numeric}$ and $T_{numeric}$ are based on magnetic field calculation) conditions as a function of terminal voltage.

12.4.3 Types of Induction Machines

There are different types of three-phase IM: squirrel-cage, wound-rotor, and doubly fed induction [5, 6] machines. The two latter types require three slip rings on which carbon brushes slide for the three-phase rotor excitation. The last one is frequently used in hydro and wind power plants, where the speed must be variable to improve the power delivery response, and is known as doubly fed induction generators.

12.4.4 Speed and Torque Control with Semiconductor Converters and Controllers of IM as Applied to Heat Pumps, Automobiles, Trains, and Wind Power Plants

The objective of this section is threefold: (i) the calculation of the battery voltage required for an IM as used in wind power plants and transportation drives [5, 6, 22, 23, 36–41]; (ii) the introduction of the compensation of flux weakening [22, 23] (CFW) resulting in high power and torque outputs even at high speeds; and

(iii) adjustable-speed and torque control of such inverter-fed machines analyzed with PSPICE [3].

Renewable energy drives such as heat pumps, electric/hybrid vehicles, trains with energy recovery during braking, and variable-speed wind power plants are equipped with power electronic converters/controllers such as rectifiers and inverters, which permit operation at variable speed without the employment of mechanical gears [22, 23]. Figure 12.20a depicts a current-controlled PWM inverter [41] feeding a three-phase induction motor at variable speed and load as they occur in electric/hybrid cars. In the following example, the Thévenin (TH) equivalent IM circuit of Figure E12.11.1c is relied on to provide the input data R_{TH}, L_{TH}, $e_a(t)$, $e_b(t)$, $e_c(t)$, $i_a(t)$, $i_b(t)$, and $i_c(t)$ at $f_1 = f_{rat} = 60$ Hz for the PSPICE program as listed in Table 12.1.

a) Compute the IM performance for rated operation at $s_{rat} = 0.03$ pu.
b) Compute the IM performance for starting at $s = 1.0$ pu.
c) Operation at increased frequency $f_2 = 360$ Hz with flux weakening.
d) An example for an increased frequency of $f_{high} = 360$ Hz will be investigated by simultaneously reducing the stator (Thévenin) leakage inductance and induced voltages $e_a(t)$, $e_b(t)$, $e_c(t)$ as proposed by patents [22, 23] advocating compensation of field weakening.

Wind power plants can be equipped either with squirrel-cage IM or DFIGs as well as PM generators. Variable-speed IM are also employed in trains [40], where recovery of braking energy is possible.

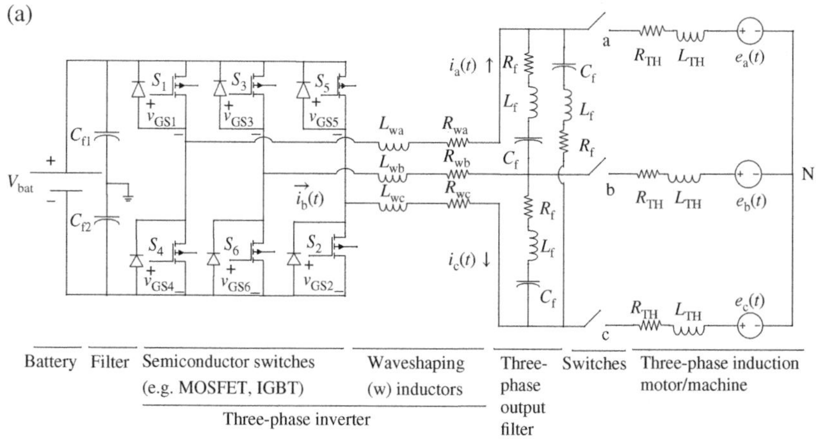

Figure 12.20a Current-controlled PWM voltage source inverter [41] feeding a three-phase squirrel-cage IM.

Table 12.1 PSPICE input program for PWM inverter operation at a gating signal voltage (vtria1) frequency of 5.76 kHz at lagging displacement (power) factor (consumer notation, underexcited, delivering real power to three-phase IM and absorbing reactive power) at rated slip s_{rat} = 0.03 pu at f_1 = 60 Hz as well as s = 1.0 at f_1 = 60 Hz indicated by (**).

Three-phase inverter-IM drive	*lfi2 26 25b 45u
*Current-controlled pulse-width-	*lfi3 36 35b 45u
modulated	*rfi1 15b 15c 0.01
*(PWM) voltage-source inverter	*rfi2 25b 25c 0.01
feeding an	*rfi3 35b 35c 0.01
*induction motor at rated	*cfi1 15c 26 10.3u
operation with	*cfi2 25c 36 10.3u
*srated = 0.03, f1=60Hz,	*cfi3 35c 16 10.3u
*(**) refers to starting at s=1.0 at	*parameters of power system with
f1=60 Hz	line-to-
vsuppl 2 0 939.3	*neutral induced voltages ea, eb, ec
*switches	RM1 16 18 0.59
msw1 2 11 10 10 qfet	LM1 18 19 2.838m
dsw1 10 2 diode	Vout1 19 123 sin(0 328.76 60 0 0 -6.17)
msw5 2 21 20 20 qfet	**Vout1 19 123 sin(0 0.0 60 0 0 -57.638)
dsw5 20 2 diode	RM2 26 28 0.59
msw3 2 31 30 30 qfet	LM2 28 29 2.838m
dsw3 30 2 diode	Vout2 29 123 sin(0 328.76 60 0
msw4 10 41 0 0 qfet	0 -126.17)
dsw4 0 10 diode	**Vout2 29 123 sin(0 0.0 60 0 0
msw2 20 51 0 0 qfet	-177.638)
dsw2 0 20 diode	RM3 36 38 0.59
msw6 30 61 0 0 qfet	LM3 38 39 2.838m
dsw6 0 30 diode	Vout3 39 123 sin(0 328.76 60 0
*waveshaping inductors	0 -248.17)
l_w1 10 15 0.1m	**Vout3 39 123 sin(0 0.0 60 0 0
l_w2 20 25 0.1m	-297.638)
l_w3 30 35 0.1m	*subcircuit for comparator
R_w1 15 16 10m	.subckt comp 1 2 9 10
R_w2 25 26 10m	rin 1 3 2.8k
R_w3 35 36 10m	r1 3 2 20meg
*current references in terms of voltages	e2 4 2 3 2 50
vref1 12 0 sin(0 30.811 60 0 0 -6.17)	r2 4 5 1k
vref2 22 0 sin(0 30.811 60 0 0 -126.17)	d1 5 6 zenerdiode1
vref3 32 0 sin(0 30.811 60 0 0 -246.17)	d2 2 6 zenerdiode2
**vref1 12 0 sin(0 201.76 60 0 0 -57.631)	e3 7 2 5 2 1

Table 12.1 (Continued)

**vref2 22 0 sin(0 201.76 60 0 0 -177.631) **vref3 32 0 sin(0 201.76 60 0 0 -297.631) eout1 13 0 15 16 100 eout2 23 0 25 26 100 eout3 33 0 35 36 100 *error signals produced as difference between *vref and eout rdiff1 12 13a 1k rdiff2 22 23a 1k rdiff3 32 33a 1k cdiff1 12 13a 1u cdiff2 22 23a 1u cdiff3 32 33a 1u rdiff4 13a 13 1k rdiff5 23a 23 1k rdiff6 33a 33 1k ecin1 14 0 12 13a 2 ecin2 24 0 22 23a 2 ecin3 34 0 32 33a 2 vtrial 5 0 pulse(-10 10 0 86.5u 86.5u 0.6u 173.6u) *gating signals for upper switches xgs1 14 5 11 10 comp xgs2 24 5 21 20 comp xgs3 34 5 31 30 comp *gating signals for lower switches egs4 41 0 poly(1) (11,10) 50 -1 egs5 51 0 poly(1) (21,20) 50 -1 egs6 61 0 poly(1) (31,30) 50 -1 *Filter is deleted because PSpice is limited to *64 nodes *lfi1 16 15b 45u	r3 7 8 10 c3 8 2 10n r4 3 8 100k e4 9 10 8 2 1 *models for zener diodes .model zenerdiode1 D (Is=1p BV=0.1) .model zenerdiode2 D (Is=1p BV=50) .ends comp *model for switch .model qfet nmos(level=3 gamma=0 kappa=0 +tox=100n rs=42.69m kp=20.87u l=2u +w=2.9 delta=0 eta=0 theta=0 vmax=0 xj=0 +uo=600 phi=0.6 vto=3.487 rd=0.19 cbd=200n +pb=0.8 mj=0.5 cgso=3.5n cgdo=100p rg=1.2 +is=10f) *model for diodes .model diode d(is=1p) *options to aid convergence .options abstol=0.01m chgtol=0.01m +reltol=50m vntol=1m itl5=0 itl4=200 *transient analysis .tran 5u 350m 300m 5u *plotting of traces .probe *Fourier analysis .four 60 12 I(l_w1) .end

Note: All comment lines must be preceded with an asterisk (*) while continued calculation statements must be preceded with a plus sign (+).

a) *Rated operation at $f_1 = 60$ Hz:* From Application Example 12.11, one obtains for the Thévenin per phase resistance $R_{TH} = 0.59\,\Omega$ and for the Thévenin per phase (linear, unsaturated) inductance $L_{TH} = X_{TH}/(2\pi \cdot 60) = 2.838$ mH. For rated slip $s_{rat} = 0.03$ pu, $f_1 = 60$ Hz the rotor per phase current is $\widetilde{I}_r' = 21.79\angle-6.17°$A, and the Thévenin (TH) per phase voltage $\widetilde{V}_{TH} = 254.75\angle1.35°$V. From Figure E12.11.1c follows the application of Kirchhoff's law for the induced phase voltage:

$$\widetilde{E} = \widetilde{V}_{TH} - \widetilde{I}_r'(R_{TH} + jX_{TH}) = 254.75\angle1.35°\text{V} - 21.79\angle$$
$$-6.17°\text{A}\cdot(0.59 + j1.07)\,\Omega = 216.02\angle-4.19°\text{V}. \tag{12.36}$$

$$\widetilde{V}_{R_r'(1-s)/s} = \frac{\widetilde{I}_r' R_r'(1-s_{rat})}{s_{rat}} = 21.79\angle-6.17°\cdot\frac{0.33(1-0.03)}{0.03} = 232.5\,\text{V}\angle-6.17°.$$

The instantaneous load phase voltages in Figure E12.11.1c are now with the amplitude:

$$\sqrt{2}\cdot232.5\,\text{V} = 328.76\,\text{V} : \text{v}_{R_r'(1-s)/s^a}(\text{t}) = 328.76\,\text{V}\cos(\omega t - 6.17°),$$
$$v_{R_r'(1-s)/s^b}(t) = 328.76\,\text{V}\cos(\omega t - 126.17°), v_{R_r'(1-s)/s^c}(t)$$
$$= 328.76\,\text{V}\cos(\omega t - 246.17°).$$

The required DC voltage of the inverter is approximately governed by the relation [5]

$$V_{DC} = 2\sqrt{2}\text{V}_1/\text{m}, \tag{12.37}$$

where $V_1 = V_{R_r'(1-s)/s} = 232.5$ V and the chosen modulation index $m = 0.7$ results in $V_{DC} = V_{bat} = 2\sqrt{2}\cdot 232.5\,\text{V}/0.7 = 939.3$ V.

The instantaneous Thévenin adjusted currents are with the amplitude $\sqrt{2}\cdot 21.79 = 30.811$ A: $i_a(t) = 30.811$ A $\cos(\omega t - 6.17°)$, $i_b(t) = 30.811$ A $\cos(\omega t - 126.17°)$, $i_c(t) = 30.811$ A $\cos(\omega t - 246.17°)$.

The output power and torque are given in Application Example 12.11. Figure 12.20b depicts the simulation results (IM phase current and phase-induced voltage).

The corresponding PSPICE program is listed in Table 12.1. Note that the PSPICE simulation confirms the phasor analysis; in addition, it provides the necessary DC input voltage to the inverter of $V_{bat} = 939.3$ V. Such high battery voltage is not feasible for electric/hybrid cars because of safety considerations, and a lower-voltage IM must be chosen for the range of $V_{bat} = (300–600)$ V. The Figure 12.20b refers to rated operation, and the (**) in Table 12.1 represents comments and changes for Figure 12.20c.

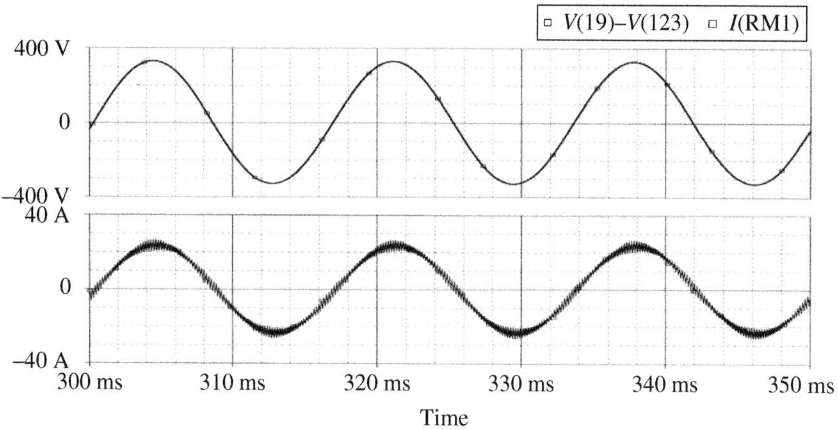

Figure 12.20b Phase input load voltage $v_{R'_r(1-s)/sa}(t) = v_{out1}(t) = V(19) - V(123)$ and phase input Thévenin adjusted current $i_a(t) = I(RM1)$ of three-phase squirrel-cage IM (see Application Example 12.11) for rated operation, unity (fundamental) displacement power factor. The current amplitude is smaller (23.6 A) than that of the analysis (30.811 A) because the inverter impedance is neglected in the analysis and the inverter operation is based on P control, causing a steady-state error.

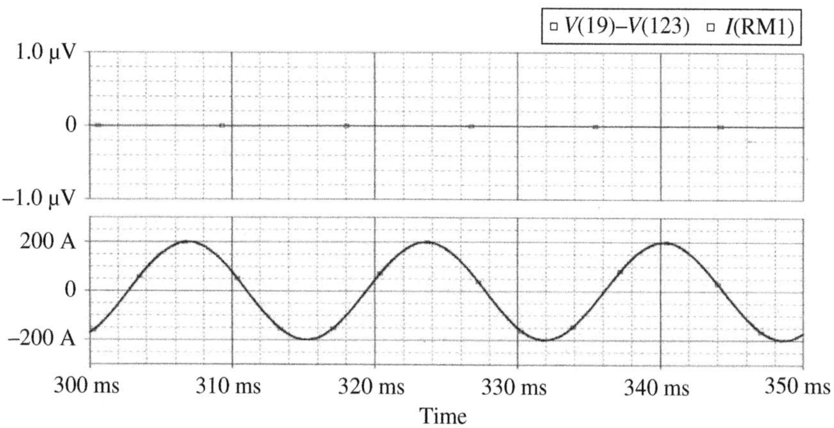

Figure 12.20c Phase input load voltage $v_{R'_r(1-s)/sa}(t) = v_{out1}(t) = V(19) - V(123)$ and phase input Thévenin adjusted current $i_a(t) = I(RM1)$ of three-phase squirrel-cage IM (see Application Example 12.11) for starting of IM at slip $s = 1$. The current amplitude is somewhat smaller (199.4 A) than that of the analysis (201.76 A) because the inverter impedance is neglected in the analysis and the inverter operation is based on P control, causing a steady-state error.

In this example, the IM is supplied by the current-controlled PWM inverter of Figure 12.20a at slightly lagging displacement (power) factor. Using consumer notation of current, the two graphs correspond to underexcited operation, delivering real power P to three-phase IM and absorbing reactive power Q at rated slip $s_{\text{rat}} = 0.03$ pu and rated frequency $f_1 = 60$ Hz.

b) *Starting at rated voltage and frequency:* At standstill (starting) the slip is $s = 1.0$ pu, $f_1 = 60$ Hz resulting in

$$\tilde{I}'_{\text{r}} = 142.69\angle -57.631°\text{A}, \tilde{V}_{\text{TH}} = 254.75\angle 1.35°\text{V},$$

$$\tilde{E} = \tilde{V}_{\text{TH}} - \tilde{I}'_{\text{r}}(R_{\text{TH}} + jX_{\text{TH}}) = 254.75\angle 1.35°\text{V}$$
$$- 142.69\angle -57.631°\text{A} \cdot (0.59 + j1.07)\ \Omega = 80.769\angle -3.283°\text{V}.$$

The voltage drop across the resistive component referred to the primary and proportional to the developed output power P_{d} is

$$\tilde{V}_{R'_{\text{r}}(1-s)/s} = \frac{\tilde{I}'_{\text{r}} R'_{\text{r}}(1-s)}{s} = 142.69\angle -57.631° \cdot \frac{0.33(1-1)}{1} = 0\,\text{V}.$$

The instantaneous load phase voltages in Figure E12.11.1c are now with the amplitude $\sqrt{2} \cdot 0\,\text{V} = 0\,\text{V}$: $v_{R'_{\text{r}}(1-s)/s^a}(t) = 0\,\text{V}\cos(\omega t - 0)$, $v_{R'_{\text{r}}(1-s)/s^b}(t) = 0\,\text{V}\cos(\omega t - 0)$, $v_{R'_{\text{r}}(1-s)/s^c}(t) = 328.76\,\text{V}\cos(\omega t - 0)$.

The required DC voltage of the inverter is approximately governed by the relation [5]

$$V_{\text{DC}} = 2\sqrt{2}\text{V}_1/\text{m}, \tag{12.37}$$

where $V_1 = V_{R'_{\text{r}}(1-s)/s} = 0\,\text{V}$ and the chosen modulation index $m = 0.7$ results in $V_{\text{DC}} = V_{\text{bat}} = 2\sqrt{2} \cdot 0\,\text{V}/0.7 = 0\,\text{V}$, as default value one chooses the battery voltage value used for rated slip $s_{\text{rat}} = 0.03$.

The instantaneous Thévenin adjusted currents are with the amplitude $\sqrt{2} \cdot 142.69 = 201.76$ A, $i_{\text{a}}(t) = 201.76\,\text{A}\cos(\omega t - 57.631°)$, $i_{\text{b}}(t) = 201.76\,\text{A}\cos(\omega t - 177.631°)$, $i_{\text{c}}(t) = 201.76\,\text{A}\cos(\omega t - 297.631°)$.

The output power and torque are given in Application Example 12.11. Note that saturation influences IM performance significantly (see Figure E12.12.1d). The change of parameters during any operating conditions (e.g. saturation during starting) exceeds the treatment of magnetics as discussed in this book.

In this case, the IM is supplied by the current-controlled PWM inverter of Figure 12.20a at lagging displacement (power) factor. Using consumer notation of current, the two graphs correspond to underexcited operation, delivering real power P to three-phase IM and absorbing reactive power Q at rated slip $s_{\text{rat}} = 1.0$ pu (start-up of IM) and rated frequency $f_1 = 60$ Hz. Note that the current of

Figure 12.20c $i_a(t) = I(\text{RM1})$ is quite sinusoidal; this is due to the high battery voltage of $V_{bat} = 939.3$ V corresponding to the modulation index $m = 0.6784$. If $m = 0.9$ were assumed, then the current would be somewhat nonsinusoidal, containing harmonic components, which are not desirable. For start-up ($s = 1$), the battery voltage could be greatly reduced, say, to $V_{bat} = 300$ V.

c) *Operation at increased frequency $f_2 = 360$ Hz with flux weakening (FW):* The resistance increase due to the skin effect is neglected. $R_s = 0.64\,\Omega$, $R'_r = 0.33\,\Omega$, $V_{L\text{-}L} = 460$ V, $f_1 = 60$ Hz, $\widetilde{V}_s = \widetilde{V}_{L-N} = \dfrac{V_{L-L}\angle 0°}{\sqrt{3}} = \dfrac{460}{\sqrt{3}}\angle 0°$V $= 265.6\angle 0°$V, $L_s = X_s/(2\pi f_1) = 1.1/377 = 2.92$ mH, $L'_r = \dfrac{X'_r}{2\pi f_1} = \dfrac{0.46}{377} = 1.22$ mH, $L_m = X_m/(2\pi f_1) = 26/377 = 68.97$ mH.

At $f_2 = 360$ Hz, the reactances are

$$X_s|_{360\text{ Hz}} = L_s 2\pi f_2 = 2.92\text{ mH} \cdot 2\pi \cdot 360 = 6.6\,\Omega,$$

$$X'_r|_{360\text{ Hz}} = L'_r 2\pi f_2 = 1.22\text{ mH} \cdot 2\pi \cdot 360 = 2.76\,\Omega,$$

$$X_m|_{360\text{ Hz}} = L_m 2\pi f_2 = 68.97\text{ mH} \cdot 2\pi \cdot 360 = 156\,\Omega,$$

$$\widetilde{Z}_{TH}\Big|_{360\text{ Hz}} = \frac{jX_m|_{360\text{ Hz}}\left(R_s + jX_s|_{360\text{ Hz}}\right)}{R_s + jX_s|_{360\text{ Hz}} + jX_m|_{360\text{ Hz}}} = R_{TH}|_{360\text{ Hz}} + jX_{TH}|_{360\text{ Hz}}$$

$$= \frac{j156(0.64 + j6.6)}{0.64 + j6.6 + j156} = (0.582 + j6.33)\,\Omega,$$

$R_{TH}|_{360\text{ Hz}} = 0.582\,\Omega, X_{TH}|_{360\text{ Hz}} = 6.33\,\Omega$ or $L_{TH}|_{360\text{Hz}} = 2.8$ mH.

$$\widetilde{V}_{TH}\Big|_{360\text{ Hz}} = jX_m|_{360\text{ Hz}} \cdot V_s/\left(R_s + jX_s|_{360\text{ Hz}} + jX_m|_{360\text{ Hz}}\right) = \frac{j156 \cdot 265.6\angle 0°}{0.64 + j6.6 + j156}$$

$$= 254.82\angle 0.226°\text{V},$$

$$\widetilde{I}_r\Big|_{360\text{ Hz}} = \widetilde{V}_{TH}\Big|_{360\text{ Hz}}\Big/\left[\left(R_{TH}|_{360\text{ Hz}} + R'_r/s_{rat}\right) + j\left(X_{TH}|_{360\text{ Hz}} + X'_r|_{360\text{ Hz}}\right)\right]$$

$$= 254.82\angle 0.226°/[(0.582 + 0.33/0.03) + j(6.33 + 2.76)] = 17.308\angle -37.30°\text{A},$$

$$\widetilde{E}\Big|_{360\text{ Hz}} = \widetilde{V}_{TH}\Big|_{360\text{ Hz}} - \widetilde{I}_r\left(R_{TH}|_{360\text{ Hz}} + jX_{TH}|_{360\text{ Hz}}\right) = 254.75\angle 0.226°$$

$$-17.308\angle -37.9°(0.582 + j6.33) = 196.23\angle -23.82°\text{V}.$$

$$\widetilde{V}_{R'_r(1-s)/s} = \frac{\widetilde{I}_r\Big|_{360\text{ Hz}} R'_r(1 - s_{rat})}{s_{rat}} = 17.308\angle -37.30° \cdot \frac{0.33(1 - 0.03)}{0.03}$$

$$= 184.68\text{ V}\angle -37.30°.$$

The instantaneous load phase voltages in Figure E12.11.1c are now with the amplitude $\sqrt{2} \cdot 184.68\,\text{V} = 261.13\,\text{V}$: $v_{R'_r(1-s)/s^a}(t) = 261.13\,\text{V}\cos(\omega t - 37.30°)$, $v_{R'_r(1-s)/s^b}(t) = 261.13\,\text{V}\cos(\omega t - 157.30°)$, $v_{R'_r(1-s)/s^c}(t) = 261.13\,\text{V}\cos(\omega t - 277.30°)$.

The required DC voltage of the inverter is approximately governed by the relation [5]

$$V_{\text{DC}} = 2\sqrt{2}V_1/m, \tag{12.37}$$

where $V_1 = V_{R'_r(1-s)/s} = 184.68\,\text{V}$ and the chosen modulation index $m = 0.7$ results in $V_{\text{DC}} = V_{\text{bat}} = 2\sqrt{2} \cdot 184.68\,\text{V}/0.7 = 746.11\,\text{V}$; see Table 12.2 and Figure 12.20d. The instantaneous Thévenin adjusted currents are with the amplitude $\sqrt{2} \cdot 17.308 = 24.474\,\text{A}$, $i_a(t) = 24.474\,\text{A}\cos(\omega t - 37.3°)$, $i_b(t) = 24.474\,\text{A}\cos(\omega t - 157.3°)$, $i_c(t) = 24.474\,\text{A}\cos(\omega t - 277.3°)$:

$$P_d\big|_{360\ \text{Hz}} = 3\frac{(17.308)^2 0.33(1-0.03)}{0.03} = 9.59\,\text{kW} \equiv 12.86\,\text{hp},$$

$$\tau_d\big|_{360\ \text{Hz}} = 3\frac{(17.308)^2 0.33}{(0.03 \cdot 1130.9)} = 8.74\,\text{Nm}.$$

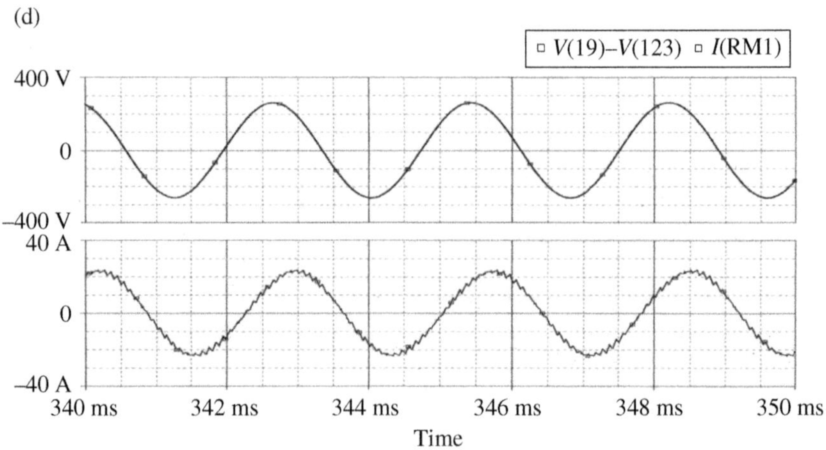

Figure 12.20d Phase input load voltage $v_{R'_r(1-s)/sa}(t) = v_{\text{out1}}(t) = V(19) - V(123)$ and phase input Thévenin adjusted current $i_a(t) = I(\text{RM1})$ of three-phase squirrel-cage IM (see Application Example 12.11) with lagging power factor at flux weakening (FW). The current amplitude is somewhat smaller (22.86 A) than that of the analysis (24.47 A) because the inverter impedance is neglected in the analysis and the inverter operation is based on P control, causing a steady-state error.

The output power and torque at 360 Hz is rather small, and therefore the same case with compensation of flux weakening (CFW) [22, 23] will be analyzed next at rated slip $s_{\text{rat}} = 0.03$ pu.

In this case, the IM is supplied by the current-controlled PWM inverter of Figure 12.20a at lagging displacement (power) factor. Using consumer notation of current, the two graphs correspond to underexcited operation, delivering real power P to three-phase IM and absorbing reactive power Q at increased frequency of $f_2 = 360$ Hz and rated slip $s_{\text{rat}} = 0.03$ pu with flux weakening.

d) *Operation at increased frequency $f_2 = 360$ Hz with compensation of flux weakening* [22, 23]: The reconfiguration of the stator three-phase winding from N_{rat} to $(N_{\text{rat}}/2)$ leads to compensation [22, 23] of flux weakening when operating at higher frequencies. The analysis is performed for $f_2 = 360$ Hz and rated slip $s_{\text{rat}} = 0.03$ pu, which requires that the phase inductances {proportional to $[(N_{\text{rat}}/2)/N_{\text{rat}})]^2$} and resistances {proportional to $[(N_{\text{rat}}/2)/N_{\text{rat}})]$} of stator and rotor are reduced – resistance increase due to the skin effect is neglected – to $R_s|_{CFW} = \frac{R_s}{2} = \frac{0.64\,\Omega}{2} = 0.32\,\Omega$, $X_s|_{CFW} = \frac{X_s|_{360\ Hz}}{4} = 6.6\,\Omega/4 = 0.165\,\Omega$,

$$R'_r|_{CFW} = \frac{R'_r}{2} = \frac{0.33\,\Omega}{2} = 0.165\,\Omega,$$

$$X'_r|_{CFW} = \frac{X'_r|_{360\ Hz}}{4} = \frac{2.76\,\Omega}{4} = 0.69\,\Omega,\ X_m|_{CFW} = \frac{X_m|_{360\ \text{Hz}}}{4} = \frac{156\,\Omega}{4}$$

$$= 39\,\Omega, \text{respectively.}$$

$$V_{L-L}|_{CFW} = V_{L-L} = 460\ \text{V}, \widetilde{V}_s|_{CFW} = \widetilde{V}_{L-N}|_{CFW} = \widetilde{V}_{L-N} = 265.6\angle 0°\text{V},$$

$$\widetilde{V}_{TH}|_{CFW} = \frac{265.6\angle 0° j39}{0.32 + j1.65 + j39} = 254.80\angle 0.451°\text{V}.$$

$$\widetilde{Z}_{TH}|_{CFW} = \frac{j39(0.32 + j1.65)}{0.32 + j1.65 + j39} = (0.29453 + j1.5854)\,\Omega,$$

$$R_{TH}|_{CFW} = 0.29453\,\Omega,$$

$$X_{TH}|_{CFW} = 1.5854\,\Omega$$

or

$$L_{TH}|_{CFW} = \frac{X_{TH}|_{CFW}}{2\pi f_2} = \frac{1.5854}{2\pi \cdot 360} = 0.7\ \text{mH}.$$

Corresponding to the above analyses, one can write:

$$\tilde{I}_r\Big|_{CFW} = \frac{254.80\angle 0.451^o}{\left[\left(0.29453 + \frac{0.165}{0.03}\right) + j(1.5854 + 0.69)\right]} = 40.93\angle - 20.983^o\ A,$$

$$\tilde{\tilde{E}}\Big|_{CFW} = \tilde{V}_{TH}\big|_{CFW} - \tilde{I}_r\Big|_{CFW}\left(R_{TH}\big|_{CFW} + jX_{TH}\big|_{CFW}\right) = 254.80\angle 0.451^0 - 40.93\angle$$

$$-20.988^o \cdot (0.29453 + j1.5854) = 226.84\angle - 13.836^o\ V.$$

$$\tilde{V}_{R_r'(1-s)/s}\big|_{CFW} = \tilde{I}_r\Big|_{CFW} R_r'\big|_{CFW}\frac{(1 - s_{rat})}{s_{rat}} = 40.93\angle - 20.988^o \cdot \frac{0.165(1 - 0.03)}{0.03}$$

$$= 218.36\ V\angle - 20.988^o.$$

The instantaneous load phase voltages in Figure E12.11.1c are now with the amplitude $\sqrt{2} \cdot 218.36$ V = 308.76 V: $v_{R_r'(1-s)/s^a}(t) = 308.76\text{ V}\cos(\omega t - 20.988°)$, $v_{R_r'(1-s)/s^b}(t) = 308.76\text{ V}\cos(\omega t - 140.988°)$, $v_{R_r'(1-s)/s^c}(t) = 308.76\text{ V}\cos(\omega t - 260.988°)$.

The required DC voltage of the inverter is approximately governed by the relation [5]

$$V_{DC} = 2\sqrt{2}V_1/m, \tag{12.37}$$

where $V_1 = V_{R_r'(1-s)/s} = 218.36$ V and the chosen modulation index $m = 0.7$ results in $V_{DC} = V_{bat} = 2\sqrt{2} \cdot 218.36\text{ V}/0.7 = 822.17$ V.

The instantaneous Thévenin adjusted currents are with the amplitude $\sqrt{2} \cdot 40.93 = 57.875$ A, $i_a = 57.875\text{ A}\cos(\omega t - 20.988°)$, $i_b = 57.875\text{ A}\cos(\omega t - 140.988°)$, $i_c = 57.875\text{ A}\cos(\omega t - 260.988°)$:

$$P_d\big|_{CFW} = 3\frac{(40.93)^2 0.165(1 - 0.03)}{0.03} = 26.813\ kW \equiv 35.942\ hp,$$

$$\tau_d\big|_{CFW} = 3\frac{(40.93)^2 0.165}{(0.03 \cdot 1130.9)} = 26.528\ Nm.$$

These values show that the CFW significantly increases by a factor of about 3.0 the power and torque at high speed compared with flux weakening resulting from the increased frequency $f_2 = 360$ Hz; see Table 12.2 and Figure 12.20e. The advantage of the PSPICE approach is that solutions for nonsinusoidal induced voltages can be obtained as well.

In this case the IM is supplied by the current-controlled PWM inverter of Figure 12.20a at lagging displacement (power) factor. Using consumer notation of current, the two graphs correspond to under-excited operation, delivering real power P to three-phase IM and absorbing reactive power Q at increased frequency of $f_2 = 360$ Hz and rated slip $s_{rat} = 0.03$ pu with CFW. Note that IMs operate at lagging power factor only (consumer reference system).

Table 12.2 PSPICE input program for PWM inverter operation at a gating signal voltage (vtria1) frequency of 5.76 kHz at lagging displacement (power) factor (consumer notation, under-excited, delivering real power to three-phase IM and absorbing reactive power) at rated slip $s_{rat} = 0.03$ pu and $f_2 = 360$ Hz operation (**), and with statements for compensation of flux weakening [22, 23].

```
Table 12.2: Inverter supplies IM
under flux
∗weakening and flux compensation
*conditions
*Current-controlled pulse-width-
*modulated
*(PWM) voltage-source inverter
*feeding an
*induction motor at srat=0.03,
*f2=360 Hz,
*with (**) flux weakening (FW) and
*with compensation of flux
*weakening (CFW)
vsuppl 2 0 882.17
**vsuppl 2 0 745.58
*switches
msw1 2 11 10 10 qfet
dsw1 10 2 diode
msw5 2 21 20 20 qfet
dsw5 20 2 diode
msw3 2 31 30 30 qfet
dsw3 30 2 diode
msw4 10 41 0 0 qfet
dsw4 0 10 diode
msw2 20 51 0 0 qfet
dsw2 0 20 diode
msw6 30 61 0 0 qfet
dsw6 0 30 diode
*waveshaping inductors
l_w1 10 15 0.1m
l_w2 20 25 0.1m
l_w3 30 35 0.1m
R_w1 15 16 10m
R_w2 25 26 10m
R_w3 35 36 10m
*current references in terms of voltages
**vref1 12 0 sin(0 24.474 360 0 0 -37.9)
**vref2 22 0 sin(0 24.474 360 0
0 -157.9)
```

```
egs6 61 0 poly(1) (31,30) 50 -1
*Filter is deleted because PSPICE is
limited to
*64 nodes
*lfi1 16 15b 45u
*lfi2 26 25b 45u
*lfi3 36 35b 45u
*rfi1 15b 15c 0.01
*rfi2 25b 25c 0.01
*rfi3 35b 35c 0.01
*cfi1 15c 26 10.3u
*cfi2 25c 36 10.3u
*cfi3 35c 16 10.3u
*parameters of power system with
line-to-
*neutral induced voltages ea, eb, ec
**RM1 16 18 0.582
**LM1 18 19 2.8m
RM1 16 18 0.29453
LM1 18 19 0.7m
**Vout1 19 123 sin(0 261.13 360 0
0 -37.30)
Vout1 19 123 sin(0 308.76 360 0
0 -20.988)
**RM2 26 28 0.582
**LM2 28 29 2.8m
RM2 26 28 0.29453
LM2 28 29 0.7m
**Vout2 29 123 sin(0 261.13 360 0
0 -157.3)
Vout2 29 123 sin(0 308.76 360 0 0
-140.988)
**RM3 36 38 0.582
**LM3 38 39 2.8m
RM3 36 38 0.29453
LM3 38 39 0.7m
**Vout3 39 123 sin(0 261.13 360 0
0 -277.30)
Vout3 39 123 sin(0 308.76 360 0 0
-260.9880)
```

(*Continued*)

Table 12.2 (Continued)

**vref3 32 0 sin(0 24.474 360 0 0 -277.9) vref1 12 0 sin(0 40.93 360 0 0 -20.988) vref2 22 0 sin(0 40.93 360 0 0 -140.988) vref3 32 0 sin(0 40.93 360 0 0 -260.988) *voltages represent load currents, measured *with shunts eout1 13 0 15 16 100 eout2 23 0 25 26 100 eout3 33 0 35 36 100 *error signals produced as difference between *vref and eout rdiff1 12 13a 1k rdiff2 22 23a 1k rdiff3 32 33a 1k cdiff1 12 13a 1u cdiff2 22 23a 1u cdiff3 32 33a 1u rdiff4 13a 13 1k rdiff5 23a 23 1k rdiff6 33a 33 1k ecin1 14 0 12 13a 2 ecin2 24 0 22 23a 2 ecin3 34 0 32 33a 2 vtrial 5 0 pulse(-10 10 0 46.5u 46.5u 0.6u +93.6u) *gating signals for upper switches xgs1 14 5 11 10 comp xgs2 24 5 21 20 comp xgs3 34 5 31 30 comp *gating signals for lower switches egs4 41 0 poly(1) (11,10) 50 -1 egs5 51 0 poly(1) (21,20) 50 -1	*subcircuit for comparator, model of *comparator: v1-v2, vgs .subckt comp 1 2 9 10 rin 1 3 2.8k r1 3 2 20meg e2 4 2 3 2 50 r2 4 5 1k d1 5 6 zenerdiode1 d2 2 6 zenerdiode2 e3 7 2 5 2 1 r3 7 8 10 c3 8 2 10n r4 3 8 100k e4 9 10 8 2 1 *models for zener diodes .model zenerdiode1 D (Is=1p BV=0.1) .model zenerdiode2 D (Is=1p BV=50) .ends comp *model for switch .model qfet nmos(level=3 gamma=0 kappa=0 +tox=100n rs=42.69m kp=20.87u l=2u + w=2.9 delta=0 eta=0 theta=0 vmax=0 xj=0 +uo=600 phi=0.6 vto=3.487 rd=0.19 cbd=200n +pb=0.8 mj=0.5 cgso=3.5n cgdo=100p rg=1.2 +is=10f) *model for diodes .model diode d(is=1p) *options to aid convergence .options abstol=0.01m chgtol=0.01m reltol=50m +vntol=1m itl5=0 itl4=200 *transient analysis .tran 5u 350m 340m 5u *plotting of traces .probe *Fourier analysis .four 360 12 I(l_w1) I(RM1) .end

Note: All comment lines must be preceded with an asterisk (*) while continued calculation statements must be preceded with a plus sign (+).

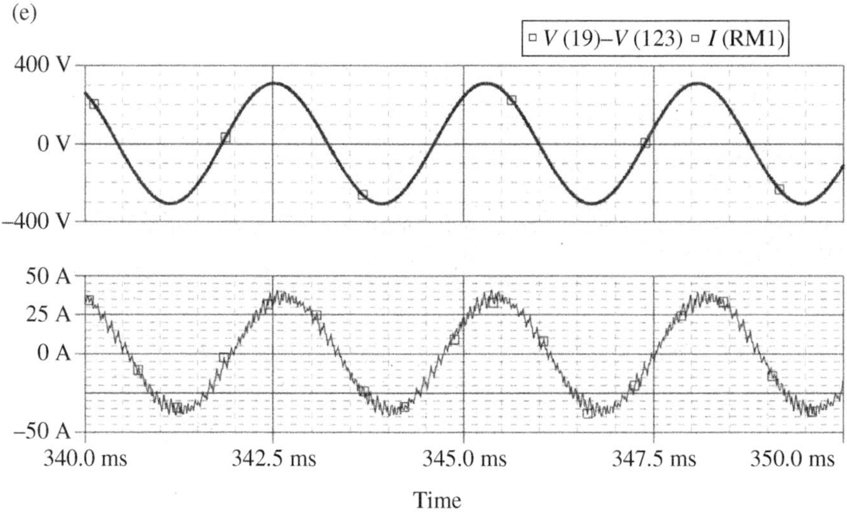

Figure 12.20e Phase input load voltage $v_{R_r'(1-s)/sa}(t) = v_{out1}(t)$ = V(19) − V(123) and phase input Thévenin adjusted current $i_a(t)$ = I(RM1) of three-phase, squirrel-cage IM (see Application Example 12.11) for compensation of flux weakening (CFW), lagging power factor operation. The current amplitude is somewhat smaller (36.08 A) than that of the analysis (40.93 A) because the inverter impedance is neglected in the analysis and the inverter operation is based on P-control, causing a steady-state error.

Conclusions

Induction machines are suitable for hybrid/electric drives provided low-voltage designs are utilized, which require battery voltages in the range of (300–600) V – similar to brushless DC machines – because the leakage inductances of such machines are small. IMs operate at lagging power factor only (consumer reference system). The compensation of flux weakening is beneficial with respect to output power and torque. German railroad traction drives employ IMs, and wind power plants use them as well. The beneficial application of CWF to PMM with respect to power and torque production is demonstrated in Application Example 12.7.

12.4.5 Optimization of Three- and Single-Phase IMs with Respect to Efficiency for Given Performance Constraints

Optimization [42–51] with respect to efficiency – for given performance constraints (e.g. power factor, efficiency, starting torque, etc.) of IMs – represents an important contribution to the reduction of energy consumption. For small ratings (with efficiencies of 70%), the efficiency improvement lies in the range of 10%, while for large ratings (with efficiencies of 90%) an efficiency gain of 2% can be reached.

12.5 Polyphase Non-salient and Salient Pole Synchronous Machines (SMs)

At steady state, SMs have a constant speed; however, transient operation is possible within the limits of the rotor position defined by the torque angle $\delta < 90°$, as can be computed from the phasor diagrams of Figures 12.21b, 12.22b, and 12.23c. The stator carries a three-phase winding (see Figures 12.8 and 12.9) similar to an IM but has a field winding installed on the rotor, which is excited by DC similar to Figure 11.7a. At steady state, that is, constant rotor speed, there is no slip ($s = 0$) between rotating stator field with the synchronous (s) angular velocity ω_s and mechanical (m) rotating rotor with the angular velocity ω_m, that is, $\omega_m = \omega_s$.

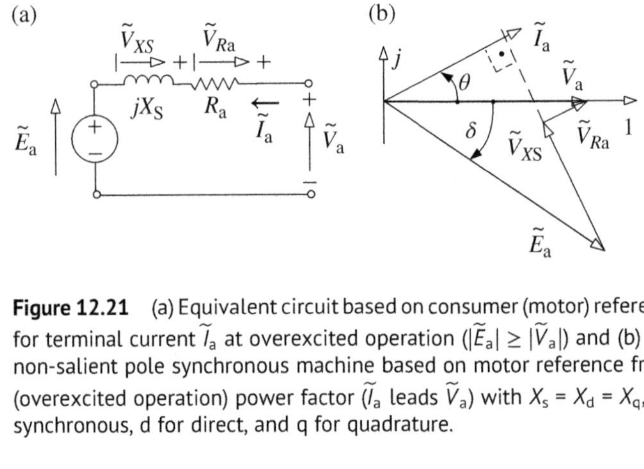

Figure 12.21 (a) Equivalent circuit based on consumer (motor) reference frame ($\tilde{V}_a$ leads $\tilde{E}_a$) for terminal current $\tilde{I}_a$ at overexcited operation ($|\tilde{E}_a| \geq |\tilde{V}_a|$) and (b) phasor diagram of a non-salient pole synchronous machine based on motor reference frame for leading (overexcited operation) power factor ($\tilde{I}_a$ leads $\tilde{V}_a$) with $X_s = X_d = X_q$, where s stands for synchronous, d for direct, and q for quadrature.

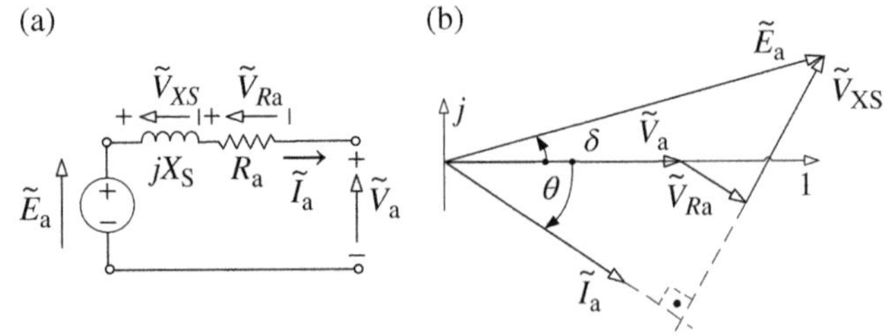

Figure 12.22 (a) Equivalent circuit based on generator reference frame ($\tilde{V}_a$ lags $\tilde{E}_a$) for terminal lagging current $\tilde{I}_a$ at overexcited operation and (b) phasor diagram of a non-salient pole synchronous machine based on generator reference frame for lagging (overexcited operation) power factor ($\tilde{I}_a$ lags $\tilde{V}_a$ and $|\tilde{E}_a| \geq |\tilde{V}_a|$).

12.5.1 Equivalent Circuits, Phasor Diagrams, and Magnetic Field Distributions Based on Polycentric Grid/Mesh Systems

All magnetic field calculations where performed with the FD method (see Chapter 11, Appendix 11.A, Figure 11.A.3a–c). Two reference frames for the definition of the armature (stator) current of SMs can be defined because in power systems SMs are used as generators and in transportation applications (e.g. TGV) they are employed as motors with regeneration of braking energy.

Consumer (Motor) and Generator Reference Frames: The equivalent circuit for a non-salient SM based on the consumer reference frame with its phasor diagram is presented in Figure 12.21a and b and in Figure 12.22a and b in the generator reference frame with the corresponding phasor diagram, respectively. The equivalent circuit and associated phasor diagram in the direct–quadrature (d–q) axis system are depicted for a salient pole SM in Figure 12.23a–c; the abovementioned

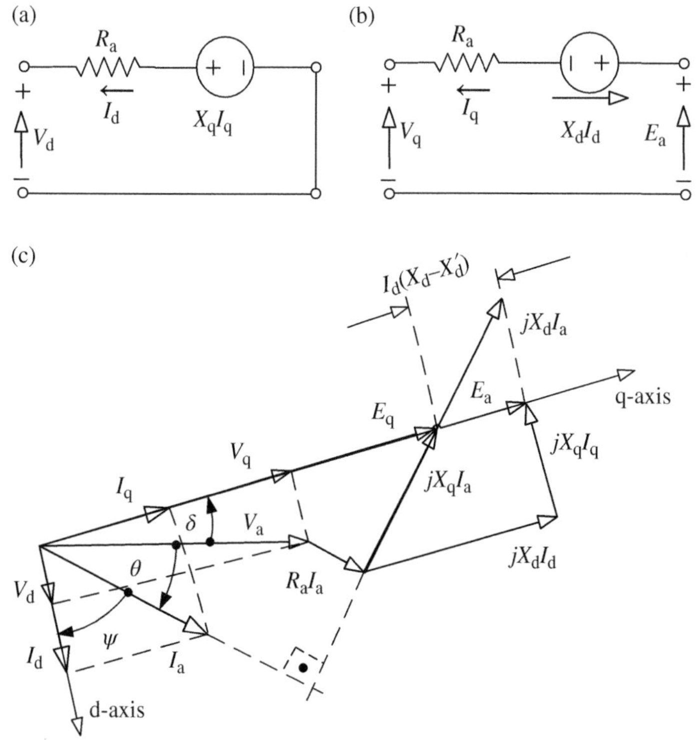

Figure 12.23 (a) d-Axis equivalent circuit, (b) q-axis equivalent circuit, and (c) phasor diagram of a salient pole synchronous machine using the generator reference frame lagging (overexcited operation) power factor with $X_d > X_q > X_d'$, and $|\tilde{E}_a| \geq |\tilde{V}_a|$.

equivalent circuit and associated phasor diagram are the same as for a PMM (see phasor diagram in Figure E12.8.2). Salient pole SMs are mostly installed as generators in hydro generation plants. The only difference between SMs and PMMs is that in the latter the $X_d < X_q$, while in the former $X_d > X_q$. Non-salient [52–58] pole SMs have a low pole number, say, $p = 2$–12, while salient pole [59–61] SMs may have a large pole number, like $p = 40$. Because of rotor slotting, the direct-axis reactance X_d for non-salient SM is somewhat larger than the quadrature-axis reactance X_q; that is, $X_s = X_d > X_q$, where s stands for synchronous. Figure 12.24 shows the rated magnetic field distribution within one period of a four-pole, non-salient pole synchronous generator. Figures 12.25 and 12.26 illustrate for different operating conditions the magnetic fields for a six-pole aircraft salient-pole synchronous machine.

Per-Unit Reactance X_s (pu): In a simplified treatment, the synchronous (s) reactance X_s is assumed to be identical with $X_d = X_q$ and mostly expressed in per unit (pu) instead of ohms (Ω). To convert the impedance – given by the sum (resistance $+j \cdot$ reactance) – from ohms to per unit, the base impedance $Z_{base}(\Omega) = V_{base}(V)/I_{base}(A) = V_{L\text{-}N}/I_{phase}$ expressed in ohms is defined, resulting in $X_s(pu) = X_s(\Omega)/Z_{base}(\Omega)$ for the per-unit reactance:

$$\widetilde{V}_a = R_a\widetilde{I}_a + jX_s\widetilde{I}_a + \widetilde{E}_a \tag{12.38}$$

$$\widetilde{E}_a = R_a\widetilde{I}_a + jX_s\widetilde{I}_a + \widetilde{V}_a \tag{12.39}$$

$$V_d = X_q I_q - R_a I_d, \tag{12.40a}$$

$$V_q = E_a - X_d I_d - R_a I_q. \tag{12.40b}$$

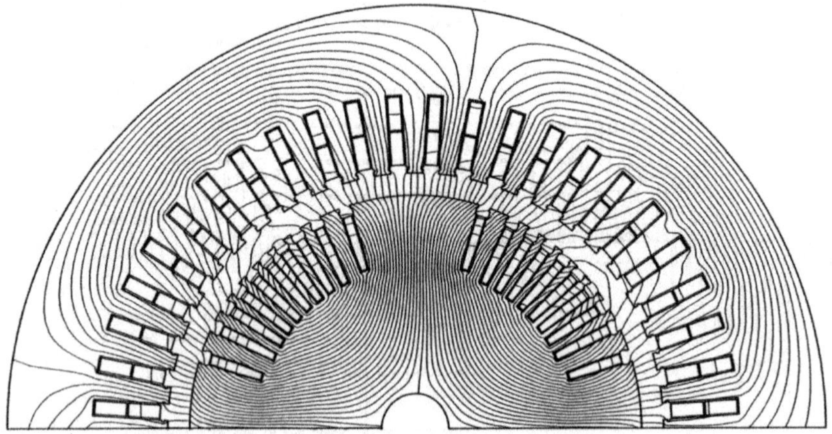

Figure 12.24 Magnetic field distribution of two pole pitches or one period of a non-salient pole synchronous generator [30, 55–57] (see Chapter 11, Appendix 11.A, Figure 11.A.3a–c).

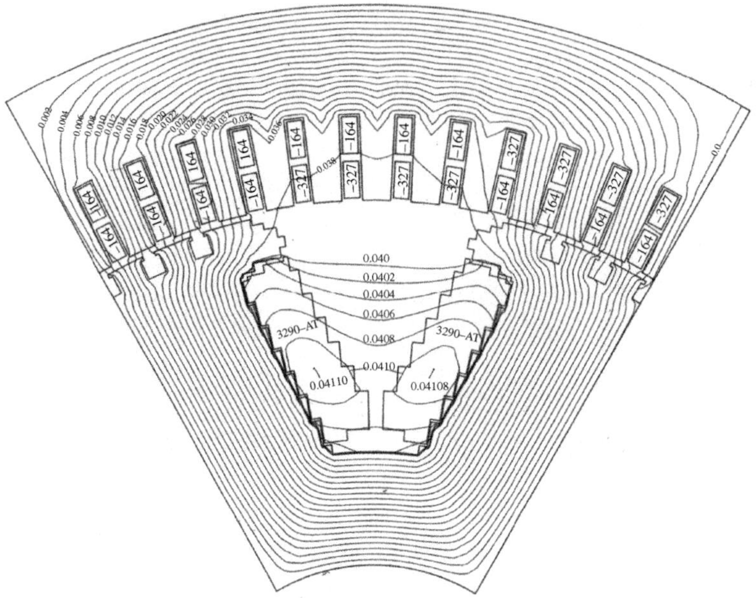

Figure 12.25 Six-pole, three-phase 250 kVA, 120/240 V, 400 Hz, and 8000 rpm aircraft [60] salient pole synchronous machine (rated operation, when stator and rotor mmf are in phase) without damper windings on the rotor member.

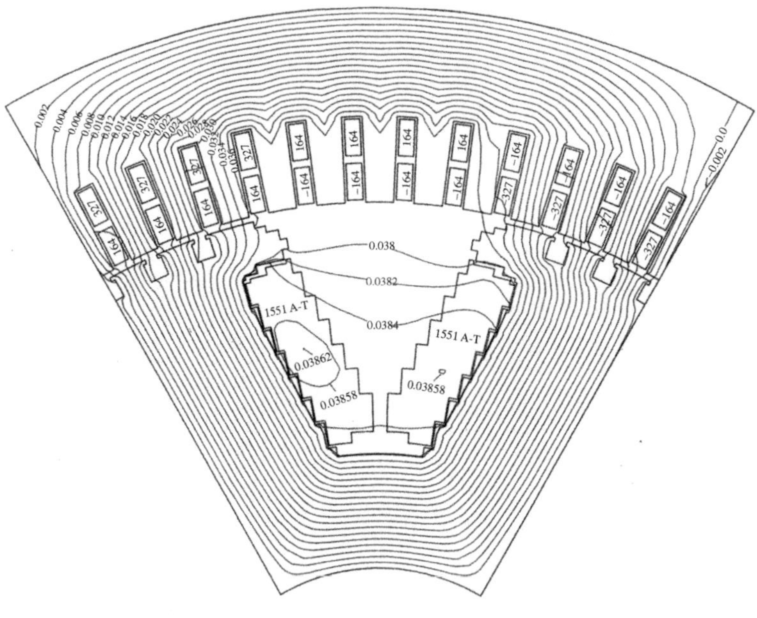

Figure 12.26 Six-pole, three-phase 250 kVA, 120/240 V, 400 Hz, and 8000 rpm aircraft [60] salient pole SM (rated operation, when stator and rotor mmf are *not* in phase) without damper windings on the rotor member.

There are two types of constructions: the non-salient (Figures 12.21a, b and 12.22a, b) and the salient pole (Figure 12.23a–c) machines. The latter develops a higher torque due to the additional saliency torque of the rotor and is used for low-speed (e.g. 180 rpm for $p = 40$ pole generators) hydro generators, while the non-salient pole type is relied on in power plants at higher speeds (e.g. 3600 rpm for $p = 2$ pole generators). For example, in France inverter-fed synchronous motors/machines are used in electric train applications, the so-called TGV.

An important design constraint for any type of electric machine is the number of slots and their dimensions. The latter aspect is given by the Carter factor [52, 62] defining the width of the slot to the width of the tooth: in short, the slot/width ratio. Figure 12.27a–c explain this design constraint based on flux distributions at no-load operation.

The Carter factor [52, 62]

$$K_C = \tau_s/(\tau_s - б) \tag{12.41a}$$

is defined by the fraction of slot variable

$$б = 1/(\pi/2)\ \tan^{-1}(b/2g) - 1.466(g/b) \cdot \log\left\{1 + (1/4)\, b^2/g^2\right\}, \tag{12.41b}$$

(a)

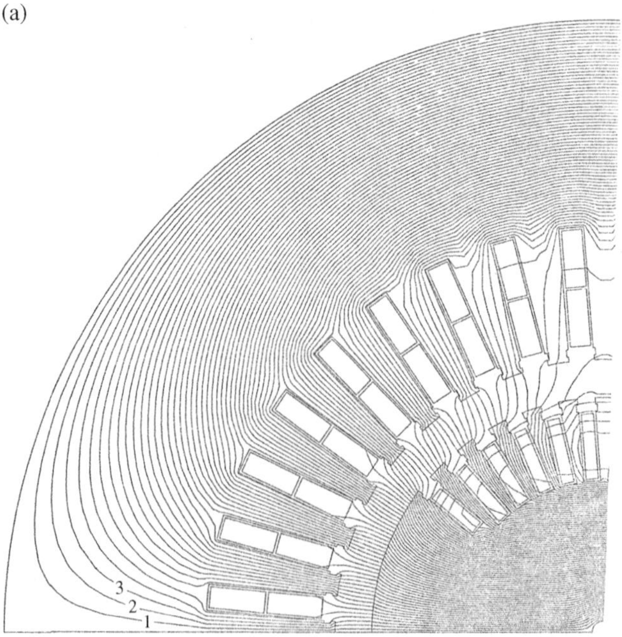

Figure 12.27a Two-dimensional no-load field of a rotating machine with a normal slot/tooth stator ratio as designed based on Carter factor [52, 62].

(b)

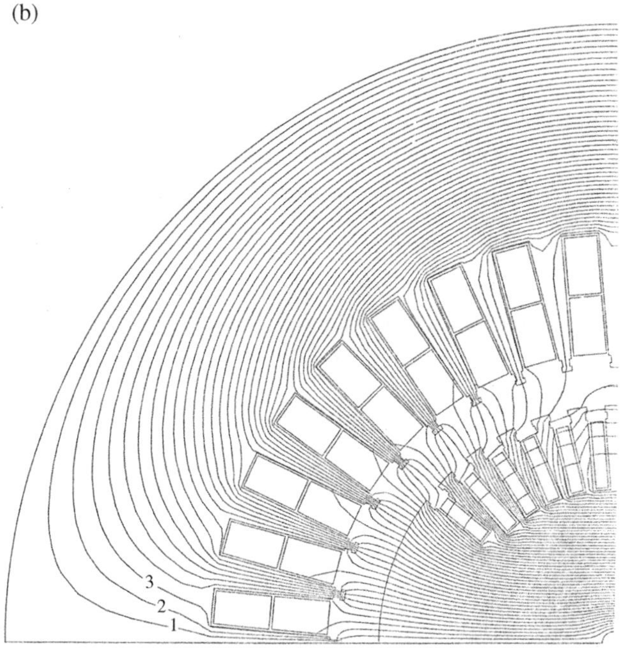

Figure 12.27b Two-dimensional no-load field of a rotating machine with a large slot/tooth stator ratio as designed based on Carter factor [52, 62].

(c)

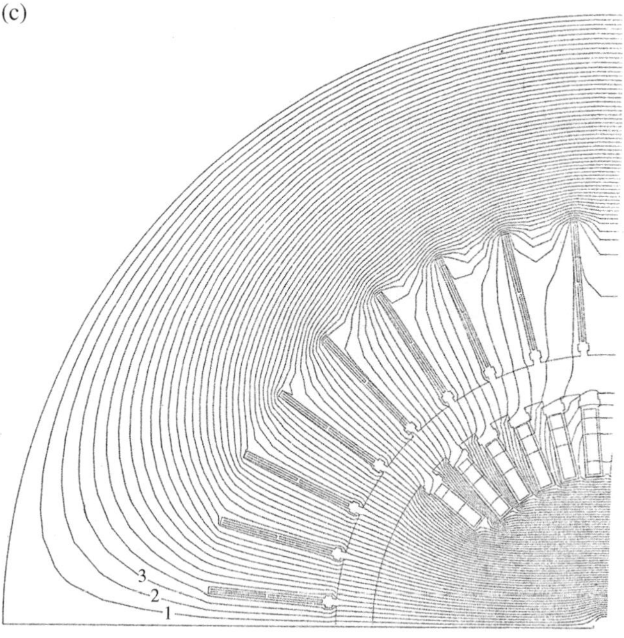

Figure 12.27c Two-dimensional no-load field of a rotating machine with a small slot/ tooth stator ratio as designed based on Carter factor [52, 62].

where b = width of slot, g = one-sided length of air gap, and τ_s is the (e.g. stator, rotor) slot pitch defined by the sum of slot and tooth widths. The geometric air gap g is increased due to stator and/or rotor slotting to become an effective air gap, as specified by Eq. (12.42):

$$g_{\text{effective}} = K_C \cdot g. \tag{12.42}$$

Figures 12.28 and 12.29a, b present an example for the two-dimensional field distribution of a non-salient pole synchronous machine at full (rated) load and the magnetic fields for the calculation of direct (X_d) and quadrature (X_q) synchronous reactances.

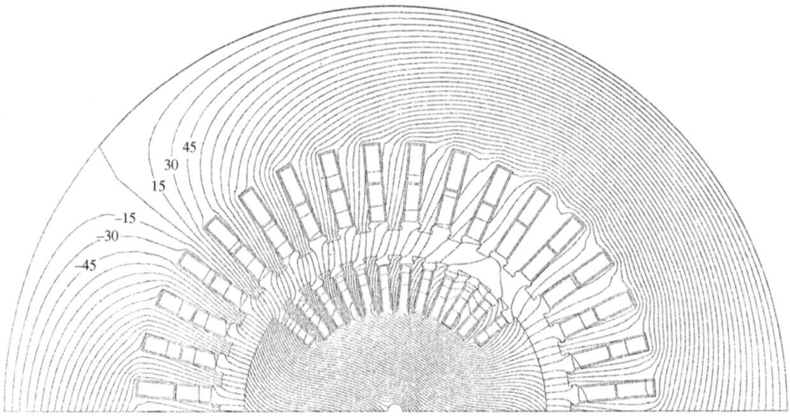

Figure 12.28 Two-dimensional flux distribution of synchronous machine at full/rated load.

(a)

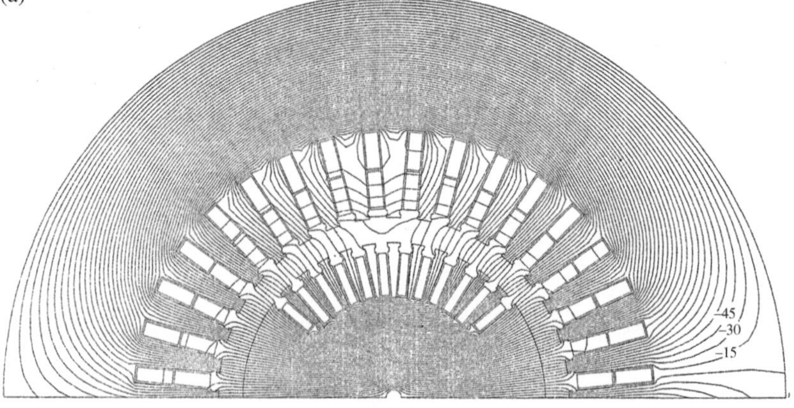

Figure 12.29a Two-dimensional flux distribution for the calculation of the direct (d) reactance X_d. The flux pattern is asymmetric due to the stator winding pitch k_w.

(b)

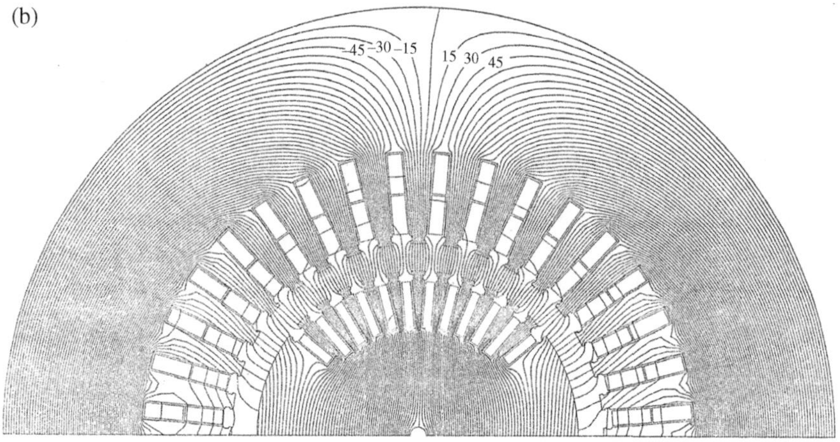

Figure 12.29b Two-dimensional flux distribution for the calculation of the quadrature (q) synchronous reactance X_q. The flux pattern is asymmetric due to the stator winding pitch k_w.

Application Example 12.13 Analysis of non-salient pole synchronous motor

Sketch and calculate for an apparent power $S = 45$ kVA, line-to-line $V_{L\text{-}L} = 360$ V, frequency $f = 60$ Hz, displacement power factor $\cos\theta = 0.8$ lagging (underexcited operation, consumer reference system), Y-connected, three-phase non-salient pole synchronous machine has a synchronous reactance $X_s = X_d = X_q = 3.0\,\Omega$ and an armature (stator) resistance $R_a = 0.2\,\Omega$ whereby the friction, windage, and iron-core losses can be neglected.

a) Sketch the phasor diagram of this motor.
b) Calculate the values of $\widetilde{V}_a, \widetilde{I}_a$, real output power P, and reactive absorbed power Q.
c) Calculate the induced voltage $\widetilde{E}_a$.
d) Determine the base impedance Z_{base} and the per unit values of R_{a_pu} and X_{s_pu}.

Solution

a) Figure E12.13.1 shows the not-to-scale phasor diagram.
b) $V_a = \left|\widetilde{V}_a\right| = \left|\widetilde{V}_{L-N}\right| = \dfrac{V_{L-L}}{\sqrt{3}} = \dfrac{360\text{ V}}{\sqrt{3}} = 208\text{ V}$. $S = 3\left|\widetilde{V}_a\right|\left|\widetilde{I}_a\right| = 45\,000\text{ VA}$, or $\left|\widetilde{I}_a\right| = \dfrac{45\,000}{3\cdot 208} = 72.12\text{ A}$ or $\widetilde{I}_a = 72.12\angle{-36.87°}\text{A}$, $P = S\cdot\cos\theta = 45\,000\cdot 0.8 = 36\text{ kW}$, $Q = S\cdot\sin\theta = S\cdot\sin(-36.87) = -27\text{ kVAr}$.
c) Accordingly in Figure 12.21a, Kirchhoff's voltage relation is

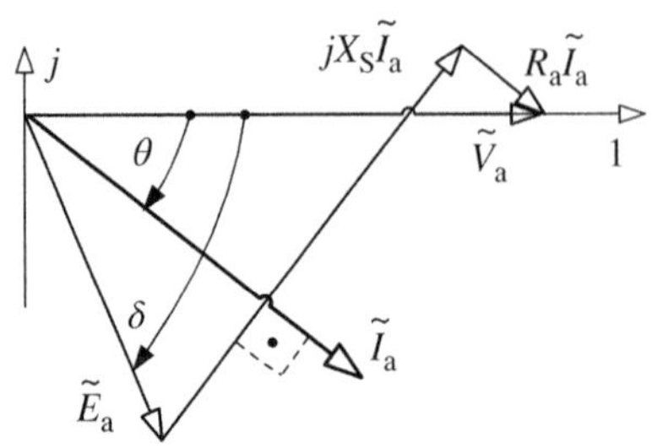

Figure E12.13.1 Phasor diagram for non-salient pole synchronous motor at lagging (underexcited operation, consumer reference system) power factor.

$$\tilde{E}_a = \tilde{V}_a - (R_a + jX_s)\tilde{I}_a = 208\angle 0° - (0.2 + j3.0)72.12\angle -36.87° = 177.43\angle -67.94°\text{V}.$$

d) $Z_{base} = \dfrac{|\tilde{V}_a|}{|\tilde{I}_a|} = \dfrac{208}{72.12} = 2.88\,\Omega$, $R_{a\_pu} = R_a/Z_{base} = 0.2/2.88 = 0.069\,\Omega$,
$X_{s\_pu} = X_s/Z_{base} = 3/2.88 = 1.04\,\Omega$.

Application Example 12.14 Inverter-fed non-salient pole synchronous motor with and without CFW [22, 23]

For the non-salient pole synchronous machine of Application Example 12.13, determine its performance at rated $f_1 = 60$ Hz operation, at $f_2 = 360$ Hz flux-weakening operation, and at $f_2 = 360$ Hz with CFW. Determine the required DC voltage V_{bat} of the current-controlled voltage source inverter as specified in Figure 12.20a.

Solution

Rated operation at $f_1 = 60$ Hz: The machine of Application Example 12.13 is supplied by a current-controlled inverter (see Figure 12.20a). A PSPICE analysis requires the voltage of the battery V_{bat}. With a modulation index $m = 0.8$, one obtains with $V_1 = V_a = 208$ V the battery voltage $V_{bat} = V_{DC} = 2\sqrt{2}V_1/m = 2\sqrt{2}\cdot 208/0.8 = 735.28$ V. By replacing the three-phase induction machine of Figure 12.20a by a three-phase non-salient pole synchronous motor with $R_a = 0.2\,\Omega$, and $L_s = L_{M1} = X_s/2\pi f_1 = 3/377 = 7.96$ mH, one obtains the PSPICE program as listed in Table E12.14.1. The time-instantaneous induced voltages are $e_a(t) = 250.89$ V $\sin(\omega t - 67.94°)$, $e_b(t) = 250.89$ V $\sin(\omega t - 187.94°)$, $e_c(t) = 250.89$ V $\sin(\omega t - 307.94°)$, and the time-instantaneous input currents are $i_a(t) = 101.98$ A $\sin(\omega t - 36.87°)$, $i_b(t) = 101.98$ A $\sin(\omega t - 156.87°)$, $i_c(t) = 101.98$ A $\sin(\omega t - 276.87°)$. See Figure E12.14.1. Neglecting losses, the output power is

Table E12.14.1 PSPICE input program for PWM inverter operation feeding a non-salient pole synchronous motor.

60Hz Current-controlled pulse-width-modulated *(PWM) voltage-source inverter *feeding a non-salient pole synchronous motor * data represent comments and operation at 60 Hz, *(**) represents operation at 360 Hz with flux weakening *(***) represents operation at 360 Hz with *compensation of flux weakening vsuppl 2 0 735.28 **vsuppl 2 0 3000 ***vsuppl 2 0 735.28 *switches msw1 2 11 10 10 qfet dsw1 10 2 diode msw5 2 21 20 20 qfet dsw5 20 2 diode msw3 2 31 30 30 qfet dsw3 30 2 diode msw4 10 41 0 0 qfet dsw4 0 10 diode msw2 20 51 0 0 qfet dsw2 0 20 diode msw6 30 61 0 0 qfet dsw6 0 30 diode *waveshaping inductors l_w1 10 15 0.1m l_w2 20 25 0.1m l_w3 30 35 0.1m R_w1 15 16 10m R_w2 25 26 10m R_w3 35 36 10m *current references in terms of voltages vref1 12 0 sin(0 101.98 60 0 0 -36.87) vref2 22 0 sin(0 101.98 60 0 0 -156.87)	*lfi1 16 15b 45u *lfi2 26 25b 45u *lfi3 36 35b 45u *rfi1 15b 15c 0.01 *rfi2 25b 25c 0.01 *rfi3 35b 35c 0.01 *cfi1 15c 26 10.3u *cfi2 25c 36 10.3u *cfi3 35c 16 10.3u *parameters of power system with line-to- *neutral induced voltages ea, eb, ec RM1 16 18 0.2 *** RM1 16 18 0.1 LM1 18 19 7.96m **LM1 18 19 7.96m ***LM1 18 19 1.989m Vout1 19 123 sin(0 250.89 60 0 0 -67.94) **Vout1 19 123 sin(0 265.56 360 0 0 -66.04) ***Vout1 19 123 sin(0 256.61 360 0 0 -67.103) RM2 26 28 0.2 *** RM2 16 18 0.1 LM2 28 29 7.96m **LM2 28 29 7.96m ***LM2 28 29 1.989m Vout2 29 123 sin(0 250.89 60 0 0 -187.94) **Vout2 29 123 sin(0 265.56 360 0 0 -186.04) ***Vout2 29 123 sin(0 256.61 360 0 0 -187.103) RM3 36 38 0.2 *** RM3 16 18 0.1 LM3 38 39 7.96m **LM3 38 39 7.96m ***LM3 38 39 1.989m Vout3 39 123 sin(0 250.89 60 0 0 -307.94)

(*Continued*)

Table E12.14.1 (Continued)

vref3 32 0 sin(0 101.98 60 0 0 -276.87) **vref1 12 0 sin(0 16.996 360 0 0 -36.87) **vref2 22 0 sin(0 16.996 360 0 0 -156.87) **vref3 32 0 sin(0 16.996 360 0 0 -276.87) ***vref1 12 0 sin(0 67.929 360 0 0 -36.87) ***vref2 22 0 sin(0 67.929 360 0 0 -156.87) ***vref3 32 0 sin(0 67.929 360 0 0 -276.87) *voltages represent load currents, measured *with shunts eout1 13 0 15 16 100 eout2 23 0 25 26 100 eout3 33 0 35 36 100 *error signals produced as difference between *vref and eout rdiff1 12 13a 1k rdiff2 22 23a 1k rdiff3 32 33a 1k cdiff1 12 13a 1u cdiff2 22 23a 1u cdiff3 32 33a 1u rdiff4 13a 13 1k rdiff5 23a 23 1k rdiff6 33a 33 1k ecin1 14 0 12 13a 2 ecin2 24 0 22 23a 2 ecin3 34 0 32 33a 2 vtrial 5 0 pulse(-10 10 0 86.5u 86.5u 0.6u 173.6u) **vtrial 5 0 pulse(-10 10 0 17.0u 17.0u 0.6u 34.6u) ***vtrial 5 0 pulse(-10 10 0 17.0u 17.0u 0.6u 34.6u) *gating signals for upper switches xgs1 14 5 11 10 comp xgs2 24 5 21 20 comp xgs3 34 5 31 30 comp	**Vout3 39 123 sin(0 265.56 360 0 0 -306.04) ***Vout3 39 123 sin(0 256.61 360 0 0 -307.103) *subcircuit for comparator .subckt comp 1 2 9 10 rin 1 3 2.8k r1 3 2 20meg e2 4 2 3 2 50 r2 4 5 1k d1 5 6 zenerdiode1 d2 2 6 zenerdiode2 e3 7 2 5 2 1 r3 7 8 10 c3 8 2 10n r4 3 8 100k e4 9 10 8 2 1 *models for zener diodes .model zenerdiode1 D (Is=1p BV=0.1) .model zenerdiode2 D (Is=1p BV=50) .ends comp *model for switch .model qfet nmos(level=3 gamma=0 kappa=0 +tox=100n rs=42.69m kp=20.87u l=2u +w=2.9 delta=0 eta=0 theta=0 vmax=0 xj=0 +uo=600 phi=0.6 vto=3.487 rd=0.19 cbd=200n +pb=0.8 mj=0.5 cgso=3.5n cgdo=100p rg=1.2 +is=10f) *model for diodes .model diode d(is=1p) *options to aid convergence .options abstol=0.01m chgtol=0.01m +reltol=50m vntol=1m itl5=0 itl4=200

Table E12.14.1 (Continued)

*gating signals for lower switches egs4 41 0 poly(1) (11,10) 50 -1 egs5 51 0 poly(1) (21,20) 50 -1 egs6 61 0 poly(1) (31,30) 50 -1 *Filter is deleted because PSPICE is limited to *64 nodes	*transient analysis .tran 5u 350m 300m 5u *plotting of traces .probe *Fourier analysis .four 60 12 I(RM1) .end

Note: All comment lines must be preceded with an asterisk (*) while continued calculation statements must be preceded with a plus sign (+).

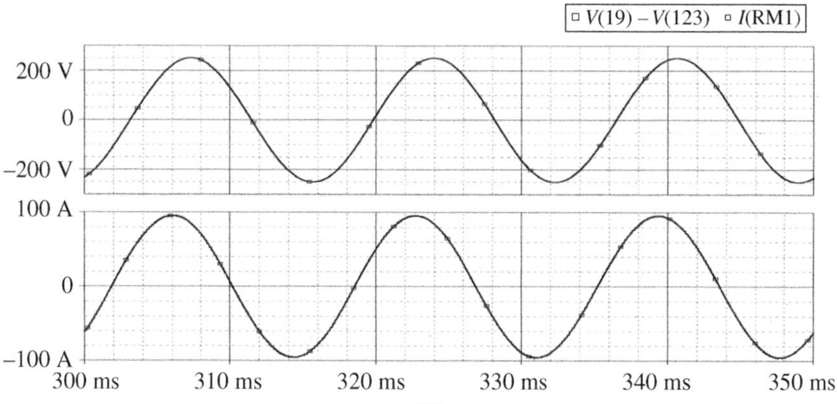

Figure E12.14.1 Input induced phase voltage $e_a(t) = V(19) - V(123)$ and input phase current $i_a(t) = I(\text{RM1})$ of three-phase, non-salient pole synchronous motor (see Application Example 12.13) at rated operation. The current amplitude is smaller (99.59 A) than that of the analysis (101.98 A) because the inverter impedance is neglected in the analysis and the inverter operation is based on P-control causing a steady-state error.

$$P = 3E_a V_a \sin \delta / X_s, \quad \text{(E12.14.1)}$$

that is, steady-state stability exists for a torque angle $\delta \leq 90°$. In this case $\delta = -67.94°$. The minus sign indicates that $\tilde{E}_a$ lags $\tilde{V}_a$, which means that this motor operates in a stable manner.

Table E12.14.1 lists the PSPICE input program for a PWM inverter feeding a non-salient pole synchronous motor. The gating signal voltage (vtrial) has the frequency of 5.76 kHz at lagging displacement (power) factor. For the power factor, consumer notation is used, which means the machine is underexcited. Real power

P is delivered to synchronous motor and absorbing reactive power Q at rated $f_1 =$ 60 Hz without (*comments) and with (**comments and changes) CFW [22, 23].

In this case, the synchronous motor is supplied by current-controlled PWM inverter at lagging displacement (power) factor. Using consumer notation of current, the two graphs correspond to underexcited operation, delivering real power P to three-phase synchronous motor and absorbing reactive power Q at rated frequency of $f_1 = 60$ Hz.

Operation at increased frequency of $f_2 = 360$ Hz with flux weakening: Determine the required DC voltage of the current-controlled voltage source inverter.

Solution

For operation in the flux-weakening motoring mode, a current $\tilde{I}_a$ is chosen so that the torque angle of $\delta = -67.94°$ is obtained as shown above and indicated below as well.

$$V_a = |\tilde{V}_a| = |\tilde{V}_{L-N}| = \frac{V_{L-L}}{\sqrt{3}} = 360\,\text{V}/\sqrt{3} = 208\,\text{V}.X_s|_{360\text{ Hz}} = 6\cdot X_s|_{60\text{ Hz}}$$
$$= 6\cdot 3\,\Omega = 18\,\Omega,\text{ neglecting skin effect } R_a = 0.2\,\Omega.$$

For reduced armature current $\tilde{I}_a|_{360\text{ Hz}} = \left(\frac{72.12}{6}\right)\angle -36.87°\text{A} = 12.02\angle -36.87°\text{A}$, one obtains the induced voltage:

$$\tilde{E}_a|_{360\text{ Hz}} = \tilde{V}_a - (R_a + jX_s|_{360\text{ Hz}})\tilde{I}_a|_{360\text{ Hz}} = 208\angle 0°$$
$$-(0.2 + j18.0)12.02\angle -36.87° = 187.81\angle -66.04°\text{V}.$$

With $V_1 = V_a = 208$ V the required battery voltage is $V_{bat} = V_{DC} = 2\sqrt{2}V_1/m = 2\sqrt{2}\cdot 208/0.8 = 735.28$ V. This value appears not to be large enough due to the large voltage drop across the synchronous inductance $L_s = L_{RM1} = 7.96$ mH at 360 Hz, and $V_{bat} = 3000$ V was chosen. The time-instantaneous induced voltages are $e_a(t) = 265.56\text{ V}\ \sin(\omega t - 66.04°)$, $e_b(t) = 265.56\text{ V}\ \sin(\omega t - 186.04°)$, $e_c(t) = 265.56\text{ V}\ \sin(\omega t - 306.04°)$, and the time-instantaneous input currents are $i_a(t) = 16.996\text{ A}\ \sin(\omega t - 36.87°)$, $i_b(t) = 16.996\text{ A}\ \sin(\omega t - 156.87°)$, $i_c(t) = 16.996\text{ A}\ \sin(\omega t - 276.87°)$.

The real input power is

$$P = 3V_aI_a\cos\theta = 3\cdot 208\cdot 12.02\cdot 0.8 = 6\,\text{kW}, \tag{E12.14.2}$$

and the reactive input power is

$$Q = 3V_aI_a\sin\theta = 3\cdot 208\cdot 12.02\cdot(-0.6) = -4.5\,\text{kVAr}. \tag{E12.14.3}$$

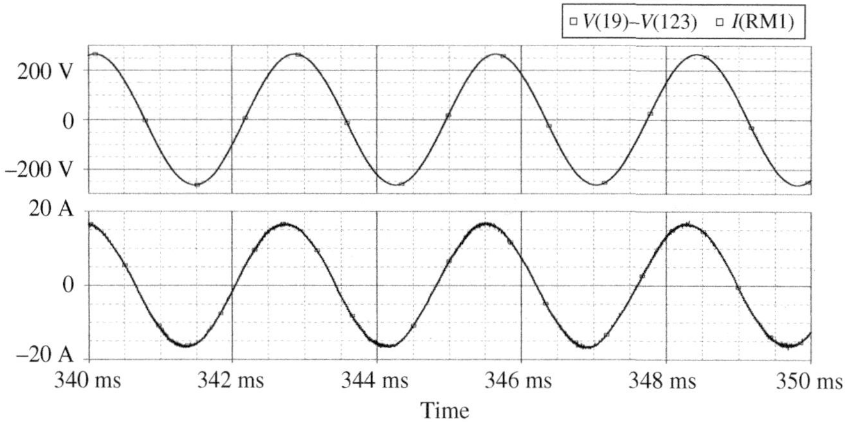

Figure E12.14.2 Input induced phase voltage $e_a(t) = V(19) - V(123)$ and input phase current $i_a(t) = I(\text{RM1})$ of three-phase non-salient pole synchronous motor (see Application Example 12.13) with flux weakening. The current amplitude is smaller (16.53 A) than that of the analysis (17.0 A) because the inverter impedance is neglected in the analysis and the inverter operation is based on P control, causing a steady-state error.

As expected, flux-weakening operation results in much reduced input powers. See Table E12.14.1 and Figure E12.14.2.

In this case, the synchronous motor is supplied by a current-controlled PWM inverter at lagging displacement (power) factor. Using consumer notation of current, the two graphs correspond to underexcited operation, delivering real power P to three-phase synchronous motor and absorbing reactive power Q at increased frequency of $f_2 = 360$ Hz with flux weakening.

Operation at increased frequency of $f_2 = 360$ Hz with CFW [22, 23]: Determine the required DC voltage of the current-controlled voltage source inverter.

Solution

The value of the current $\tilde{I}_a\big|_{\text{CFW}} = 48.04\angle -36.87°\text{A}$ is chosen so that the torque angle $\delta = -67.103°$ at rated operation and $R_a = 0.2\,\Omega/2 = 0.1\,\Omega$, $X_s\big|_{\text{CFW}} = \dfrac{X_s\big|_{360\text{ Hz}}}{4} = 4.5\,\Omega$, resulting in a synchronous inductance of $L_s = L_{M1} =$ 1.989 mH and induced voltage $\tilde{E}_a\big|_{\text{CFW}} = 208\angle 0° - (0.2 + j4.5)48.04\angle -36.87° =$ $181.48\angle -67.103°\text{V}$.

The time-instantaneous induced voltages are $e_a(t) = 256.61\text{ V}\sin(\omega t - 67.103°)$, $e_b(t) = 256.61\text{ V}\sin(\omega t - 187.103°)$, $e_c(t) = 256.61\text{ V}\sin(\omega t - 301.103°)$, and the time-instantaneous input currents are $i_a(t) = 67.929\text{ A}\sin(\omega t - 36.87°)$, $i_b(t) = 67.929\text{ A}\sin(\omega t - 156.87°)$, $i_c(t) = 67.929\text{ A}\sin(\omega t - 276.87°)$.

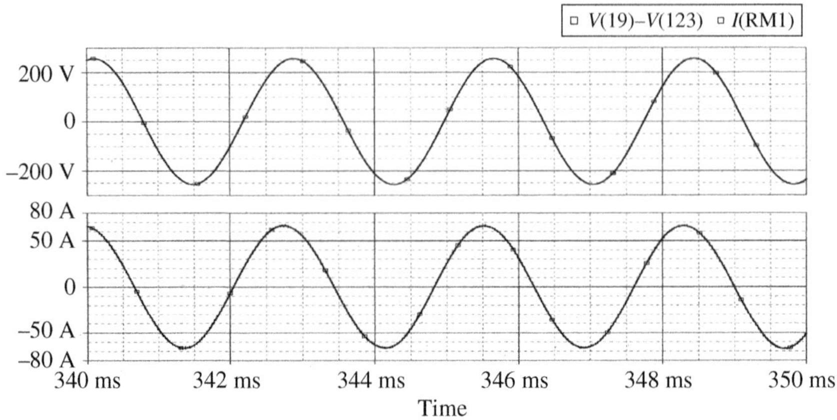

Figure E12.14.3 Phase input induced voltage $e_a(t) = V(19) - V(123)$ and phase input current $i_a(t) = I(RM1)$ of three-phase non-salient pole synchronous motor with CFW. The current amplitude is smaller (66.52 A) than that of the analysis (67.93 A) because the inverter impedance is neglected in the analysis and the inverter operation is based on P control, causing a steady-state error.

The real input power is $P = 3V_a I_a \cos\theta = 3 \cdot 208 \cdot 48.04 \cdot 0.8 = 23.982$ kW, and the reactive input power is $Q = 3V_a I_a \sin\theta = 3 \cdot 208 \cdot 48.04 \cdot (-0.6) = -17.986$ kVAr.

As expected, compensation of flux-weakening operation results in increased input power compared with flux weakening. With $V_1 = V_a = 208$ V, the required battery voltage is $V_{bat} = V_{DC} = 2\sqrt{2}V_1/m = 2\sqrt{2} \cdot 208/0.8 = 735.28$ V. The advantage of the PSPICE approach (see Table E12.14.1 and Figure E12.14.3) is that solutions for nonsinusoidal induced voltages can be obtained as well.

In this case, the synchronous motor is supplied by current-controlled PWM inverter at lagging displacement (power) factor. Using consumer notation of current, the two graphs correspond to underexcited operation, delivering real power P to three-phase synchronous motor and absorbing reactive power Q at increased frequency of $f_2 = 360$ Hz with CFW.

Conclusions

SMs are not suitable for hybrid/electric car designs because of the large synchronous inductance L_s requiring a large battery voltage in the range of 300–3000 V, in particular for field weakening operation. If CFW is employed, battery voltage in the range of 300–600 V is acceptable. SMs are used by the French railroad system TGV ("train à grande vitesse").

Application Example 12.15 Analysis of non-salient pole synchronous generator

An aircraft generator has the apparent power $S = 250\,\text{kVA}$, line-to-line voltage $V_{\text{L-L}} = 416\,\text{V}$, voltage/current frequency $f = 400\,\text{Hz}$, displacement (fundamental) power factor $\cos\theta = 0.8$ lagging (overexcited operation, generator reference system), Y-connected three-phase non-salient pole synchronous machine with a synchronous reactance $X_{\text{s\_pu}}$ ($=X_{\text{d\_pu}} = X_{\text{q\_pu}}$) $= 1.8$ pu and an armature (stator) resistance $R_{\text{a\_pu}} = 0.1$ pu. The friction, windage, and iron-core losses can be neglected.

a) Sketch the phasor diagram of this motor.
b) Calculate the quantities $\widetilde{V}_{\text{a}}$, $\widetilde{I}_{\text{a}}$, and $\widetilde{E}_{\text{a}}$.
c) Determine the real output power P and the reactive output power Q.

Solution

a) Figure E12.15.1 shows a sketch (not-to-scale) phasor diagram for lagging (current $\widetilde{I}_{\text{a}}$ lags $\widetilde{V}_{\text{a}}$ by 36.87°) generator operation. Accordingly in Figure 12.22a, Kirchhoff's voltage relation is

$$\widetilde{E}_{\text{a}} = \widetilde{V}_{\text{a}} + (R_{\text{a}} + jX_{\text{s}})\widetilde{I}_{\text{a}}. \tag{E12.15.1}$$

b) $\widetilde{V}_{\text{a}} = \widetilde{V}_{\text{L-N}} = \left(V_{\text{L-L}}/\sqrt{3}\right)\angle 0° = \left(416/\sqrt{3}\right)\angle 0°\text{V} = 240\angle 0°\text{V}$, $\quad \left|\widetilde{I}_{\text{a}}\right| = \dfrac{250\,000}{3\cdot 240}$ $= 347.22\,\text{A}$, or $\widetilde{I}_{\text{a}} = 347.22\angle -36.87°\text{A}$. Base impedance is

$$Z_{\text{base}} = \frac{\left|\widetilde{V}_{\text{a}}\right|}{\left|\widetilde{I}_{\text{a}}\right|} = \frac{240}{347.22} = 0.6912\,\text{pu}. \tag{E12.15.2}$$

$$\begin{aligned} X_{\text{s}} &= X_{\text{s\_pu}} \cdot Z_{\text{base}} = 1.8\,\Omega \cdot 0.6912\,\text{pu} = 1.244\,\Omega, \\ R_{\text{a}} &= R_{\text{a\_pu}} \cdot Z_{\text{base}} = 0.1\,\Omega \cdot 0.6912\,\text{pu} = 0.06912\,\Omega. \end{aligned} \tag{E12.15.3}$$

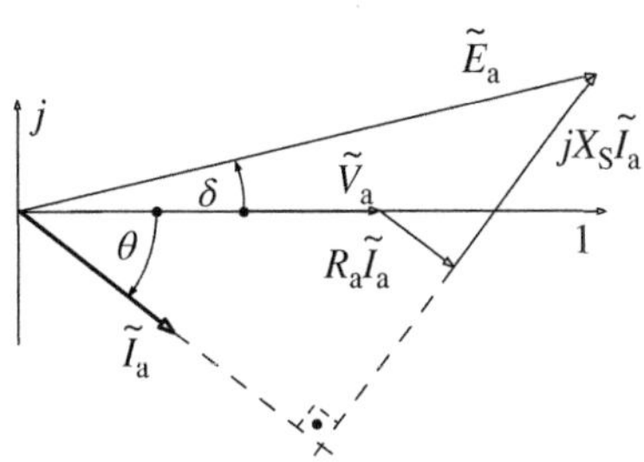

Figure E12.15.1 Phasor diagram for generator operation with lagging (overexcited, generator reference system) displacement power factor operation.

$\tilde{E}_a = 240\angle 0° + (0.0612 + 1.244)347.22\angle - 36.87° = 615.14\angle 32.573°$V. The time-dependent induced voltages are $e_a(t) = 869.81\,\text{V}\ \sin(\omega t + 32.573°)$, $e_b(t) = 869.81\,\text{V}\ \sin(\omega t - 87.427°)$, $e_c(t) = 869.81\,\text{V}\ \sin(\omega t - 207.43°)$, and the angular frequency is $\omega = 2\pi f = 2\pi \cdot 400 = 2513.2$ rad/s.

c) The real output power is $P = 3|\tilde{I}_a||\tilde{V}_a| \cos\theta = 347.22 \cdot 240 \cdot 0.8 = 66.67$ kW, and the reactive output power is $Q = 3|\tilde{I}_a||\tilde{V}_a| \sin\theta = 347.22 \cdot 240 \cdot 0.6 = 50$ kVAr.

Application Example 12.16 Non-salient pole synchronous generator feeding a rectifier

The machine of Application Example 12.15 is feeding a three-phase rectifier.

a) Perform a PSPICE analysis and determine the output DC voltage of the rectifier and its harmonic content, including the total harmonic voltage distortion (THD_v) of up to the 11th voltage harmonic.
b) Plot the DC output voltage of the rectifier as a function of the duty cycle/ratio $\delta = t_{on}/T$.

Solution

Figure E12.16.1 illustrates the circuit diagram where the non-salient pole synchronous generator feeds a three-phase rectifier consisting of a three-phase diode bridge and one self-commutated switch (e.g. IGBT). Input and output LC filters are employed to reduce the harmonic content in synchronous generator and DC load. For $f = 400$ Hz the synchronous inductance is $L_s = X_s/(2\pi f) = 1.244/(2\pi \cdot 400) = 0.495$ mH and $R_a = 0.06912\,\Omega$. Table E12.16.1 lists the PSPICE

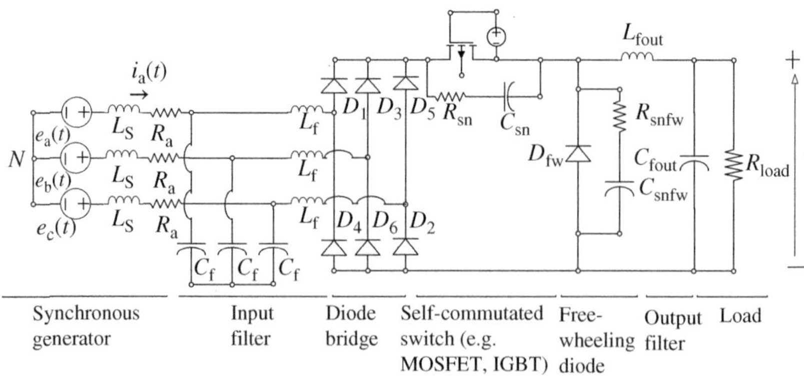

Figure E12.16.1 Non-salient pole, three-phase $f = 400$ Hz synchronous generator feeding a three-phase rectifier.

Table E12.16.1 PSPICE input program for PWM rectifier operation at a gating signal voltage (V_{mos}) frequency of 3 kHz at lagging displacement (power) factor (generator notation, overexcited), where non-salient pole synchronous generator delivers real and reactive powers to a three-phase rectifier. Duty cycle δ = 50% at a real output power P = 66.67 kW.

66.67kW controlled three-phase rectifier with self-commutated switch, *duty cycle delta=50%, *Induced voltage amplitude of generator is *869.81V, *Displacement power factor cos(theta)=-36.87 *degrees lag, generator notation *frequency of generator voltage fgen=400Hz va a st sin(0 869.81 400 0 0 32.573) vb b st sin(0 869.81 400 0 0 -87.427) vc c st sin(0 869.81 400 0 0 -207.43) * Gating signal 3 kHz Vmos gate 7 pulse(0 10 0 1u 1u 167.0u 334u) * Generator impedances Lsa a aa 0.495m Raa aa 1 0.06912 Lsb b bb 0.495m Rab bb 2 0.06912 Lsc c cc 0.495m Rac cc 3 0.06912 *Filter capacitors next to generator Cfa 1 t 50u Cfb 2 t 50u Cfc 3 t 50u Rtt t st 1G *Filter inductors between *filter next to generator and that next to *diode bridge Lfa 1 11 50u Lfb 2 22 50u Lfc 3 33 50u	D1 11 b1 dio D3 22 b1 dio D5 33 b1 dio D4 b4 11 dio D6 b4 22 dio D2 b4 33 dio * Output filter Lfout 7 0 0.1m Cfout 0 b4 100u * Load Rload b4 0 40 *MOSFET mos b1 gate 7 7 mfet * Freewheeling diode Dfw b4 7 dfree Rfw 7 fw 10 Cfw fw b4 100p * Models for diodes .model dio d(is=10p cjo=100p) .model dfree d(is=1n) *Model for MOSFET .model mfet nmos(level=3 gamma=0 +kappa=0 tox=100n rs=0 kp=20.87u l=2u +w=2.9 delta=0 eta=0 theta=0 vmax=0 xj=0 +uo=600 phi=0.6 vto=0 rd=0 cbd=200n +pb=0.8 mj=0.5 cgso=3.5n cgdo=100p rg=0 +is=10f) .tran 100u 40m 30m 65u .probe .options abstol=10m chgtol=10m reltol=0.1 +vntol=100m .end

Note: All comment lines must be preceded with an asterisk (*) while continued calculation statements must be preceded with a plus sign (+).

program for this controlled rectifier fed by a $f = 400$ Hz non-salient pole synchronous generator operating at overexcited displacement power factor (current lags induced voltage, generator reference frame). Figure E12.16.2 illustrates the DC output three-phase rectifier voltage V_{out}, the induced phase voltage $e_a(t)$, and current $i_a(t)$ of the non-salient pole, overexcited synchronous generator. Figure E12.16.3 shows the DC output voltage of the rectifier V_{out} as a function of its duty cycle $\delta = t_{on}/T$ for $0 \leq \delta \leq 1.0$ pu.

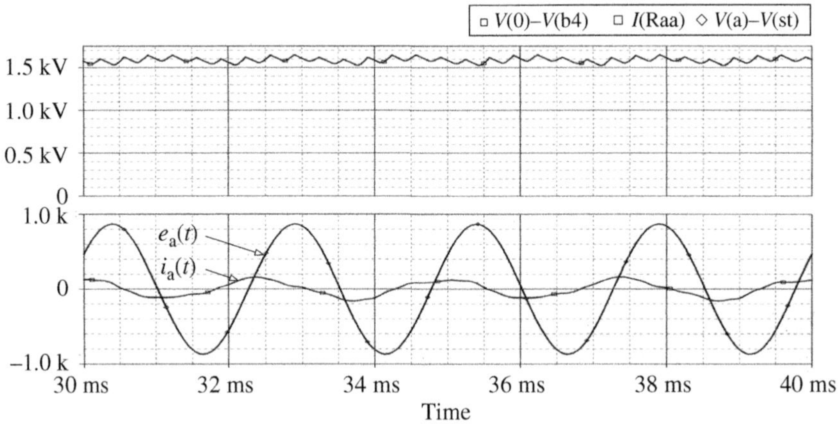

Figure E12.16.2 DC load voltage $V_{out} = V(0) - V(b4) = 1.65$ kV of three-phase rectifier, phase-induced voltage $e_a(t) = V(a) - V(st)$, and phase current $i_a(t) = I(Raa)$ supplied by $f = 400$ Hz non-salient pole synchronous generator at overexcited operation.

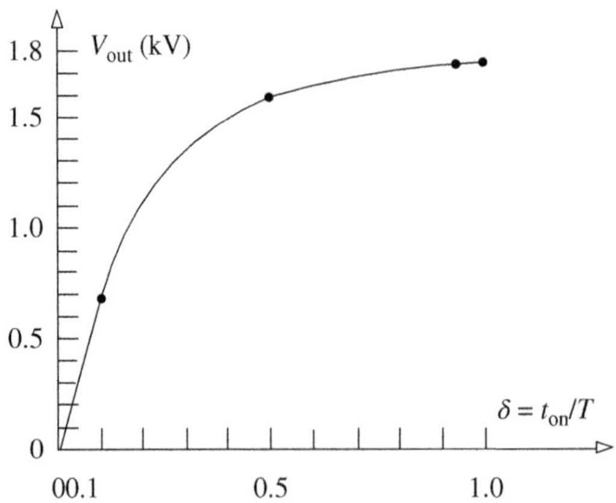

Figure E12.16.3 DC load voltage $V_{out} = V(0) - V(b4)$ of three-phase rectifier as a function of duty cycle $\delta = t_{on}/T$ for $0 < \delta < 1.0$ pu.

Conclusions

Such high-frequency generators with three-phase rectifiers are mainly employed in aircraft (see Figures 12.25 and 12.26). Low-frequency generators of the permanent-magnet and synchronous types as well as those of the induction type are used in wind power plants and railroad traction applications.

12.5.2 Applications of SMs When Independently Controlling Speed and Torque

Speed and torque of SMs can be controlled in a manner similar to that for PMMs, where an inverter is feeding the SM. For example, the TGV drive picks up via a catenary the energy from the single-phase railroad power system, rectifies the voltage and current via a rectifier, and supplies DC power to a three-phase inverter feeding a non-salient pole SM. Through changing the inverter output current and the inverter output frequency, the torque and the speed of the SM can be altered. Also, regeneration is possible if the inverter works as a rectifier and the rectifier is reconnected as an inverter, returning energy to the railroad power system [40]. Single-phase synchronous generators [63] driven by turbines generate voltages/currents at 16.667 Hz = $16^{2/3}$ Hz in addition to frequency converters (rectifiers, inverters) transforming three-phase power at 50 Hz to single-phase power supplied at 16.667 Hz within the German railroad power system. DFIGs are really SMs with nominal stator (f_s = 60 Hz) and rotor (f_r = 12 Hz) frequencies; the latter can be adjusted depending upon the required speed of the hydro turbine and the required nominal frequency f_s = 60 Hz of the power system [5, 6].

12.6 Summary

Through his invention of the revolving two-phase magnetic field, Nikola Tesla initiated the development of polyphase PMM, induction and SMs, either generators or motors. All three types are instrumental in the advance of renewable energy within the electric/hybrid car, wind power, heat pump, and refrigeration environments. This text is limited to steady-state analyses employing the combination of hand calculations and numerical solution methods such as magnetic field analyses, PSPICE, and Mathematica.

After discussing details of the revolving magnetic field theory at basic levels without resorting to complicated and complex (number) mathematical expressions, three-phase windings that generate rotating/revolving magnetic fields are discussed via examples that are used in the analysis of above-cited three types of electric machines.

Next, we highlight PMMs, which enable the design of brushless DC machines consisting of DC voltage source (e.g. battery), semiconductor converter (e.g. either rectifier or inverter), three-phase filters, and permanent-magnet materials. To date, most designs that lend themselves to renewable energy applications – in the kW to MW range – are based on the utilization of expensive (e.g. $50/[kg-force]) PMMs, because very high efficiencies (larger than 90%) can be achieved. At high speed the CFW principle [22, 23] can be applied, which yields higher powers and torques than flux weakening alone, making the response of the torque–speed control faster.

The calculations of steady-state design conditions and parameters (e.g. inductances) required for the construction of equivalent circuits and the associated phasor diagrams are based on numerical computation procedures such as the FD and FE algorithms/methods, which differ in their derivation but are identical in their analytical formulation and generate identical results if the same lattice/grid structures are employed. To support this approach, we present many two-dimensional magnetic field distributions not available in any other textbook. These will help the reader to visualize and understand magnetic fields within electromechanical energy converters. IM and SM are predominantly utilized for drives in the transportation sector and generation equipment of large ratings (MW range for railway applications and MW and GW ranges for gas and nuclear power plants, respectively). Natural gas and hydro storage plants with a fast response time (e.g. use of DFIG [5]) are still indispensable due to the intermittent operation of renewable energy power plants like PV and wind power and the nascent nature of the energy storage industry (e.g. fuel cells and flow batteries [5]). These baseload and fast response plants provide the required power supply if there is neither sunshine nor wind available as discussed in the preface so that voltage and frequency of the entire power system can be controlled within acceptable limits and the instantaneous load-generation balance – defined by the equality of instantaneous generation and consumption – is satisfied.

Problems

12.1 Rotating magnetic fields based on two and three phases.

Show that Eqs. (12.5d) and (12.5i) hold by using appropriate trigonometric relations.

12.2 Use of angular velocities ω_s, ω_m, ω_r, revolutions per second n_s^{rps}, revolutions per minute n_s^{rpm}, n_m^{rpm}, and frequencies f, f_r, f_m of rotating fields of three-phase induction and synchronous machines.

Compute the synchronous angular velocities ω_s, ω_m, ω_r, revolutions per seconds and revolutions per minute n_s^{rps}, n_s^{rpm}, n_m^{rpm}, and the various frequencies f_r, f_m of a $p = 12$ pole three-phase motor at an input voltage/current frequency of $f = 50$ Hz (see Section 12.4).

12.3 Variable-speed and torque analyses of brushless DC machine based on PSPICE.

The objective of this problem is the calculation of the battery voltage V_{bat} required for a brushless DC motor at variable speed and torque and the introduction of CFW [22, 23] resulting in high power and torque per unit of weight even at high speeds, as well as adjustable-speed (proportional to switching frequency f_1) and torque (proportional to input current) control of inverter-fed permanent-magnet machines in full-on mode analyzed with PSPICE [3]. The tendency for electric/hybrid cars is to use batteries with a voltage of $V_{bat} = 600$ V; the use of larger DC voltages in automobiles might be not desirable because of safety issues when accidents occur.

a *Rated operation:* Analyze, in a similar manner as in Application Example 12.7, a drive with $e_A(t)$, $e_B(t)$, and $e_C(t)$ amplitude of 240 V, $f_1 = 2500$ Hz, $R_s = R_1 = 0.75\ \Omega$, and $L_{s\ell} = L_{1\ell} = 75\ \mu$H. Using PSPICE [3], compute and plot the current of MOSFET Q_{AU} (e.g. i_{QAU}) and the motor current of phase A (e.g. i_{MA}) for the phase angles of the induced voltages $\theta = 0°$, $\theta = +15°$, $\theta = +45°$, $\theta = -15°$, and $\theta = -45°$. Change the phase sequence of the induced voltages so that it is different from that of the gating signals.

b *Operation with flux weakening*: Compute and plot the current of MOSFET Q_{AU} (e.g. i_{QAU}) and the motor current of phase A (e.g. i_{MA}) for the phase angle of the induced voltages $\theta = 0°$ and amplitudes of 240 V for $e_A(t)$, $e_B(t)$, $e_C(t)$, $R_s = R_1 = 0.75\ \Omega$, $L_{s\ell} = L_{1\ell} = 75\ \mu$H. The gating signal frequency of the MOSFETs corresponds to the frequency $f_2 = 5000$ Hz; that is, full-on mode operation exists.

c *Operation with CFW:* Compute and plot the current of MOSFET Q_{AU} (e.g. i_{QAU}) and the motor current of phase A (e.g. i_{MA}) for the phase angles of the induced voltages $\theta = 0°$ and amplitudes of 120 V for $e_A(t)$, $e_B(t)$, and $e_C(t)$. $R_s = R_1 = 0.38\ \Omega$, $L_{1\ell} = L_{s\ell} = 19\ \mu$H. The gating signal frequency of the MOSFETs corresponds to the frequency $f_2 = 5000$ Hz; that is, full-on mode operation exists.

12.4 Analysis of a three-phase induction motor (IM) at rated operation.

A three-phase Y-connected wound-rotor (r) IM [24] has the parameters $P_d = 8$ hp, $V_{L\text{-}L} = 230$ V, $f = 60$ Hz, six poles $p = 6$, rated (rat) slip $s_{rat} = 3\%$, and the following impedances (resistances $+ j \cdot$ reactances) per phase referred to the stator (s) circuit: $R_s = 0.5\ \Omega$, $X_s = 0.8\ \Omega$, $R_r' = 0.2\ \Omega$, $X_r' = 0.3\ \Omega$, and $X_m = 20\ \Omega$. The iron-core loss ($R_{fe} \to \infty$) and the windage/friction losses are small and can be neglected; that is, $P_{\text{iron core}} = 0$, $P_{\text{windage/friction}} = 0$, respectively. IMs always operate at lagging (consumer system) displacement or fundamental power factor due to the magnetizing (mutual) reactance X_m between stator and rotor and the stator and rotor leakage reactances X_s and X'_r, respectively.

a Draw the per-phase equivalent circuit and compute the line (L)-to-neutral (N) voltage $\tilde{V}_S = \tilde{V}_{L-N}$ and the stator current $\tilde{I}_s$ at no load, that is, $\tilde{I}'_r = 0$.

b Draw the per-phase equivalent circuit and the not-to-scale phasor diagram with all voltages and currents for any load condition.

c For rated-load and starting operations of IM with rated slip $s = s_{\text{rated}} = 0.03$ per unit (pu) and slip at starting (start) $s_{\text{start}} = 1.0$ pu, respectively, compute for the above given motor the following machine parametersand data: Thévenin (TH) voltage $\tilde{V}_{TH}$ and Thévenin impedance $\tilde{Z}_{TH} = R_{TH} + jX_{TH}$ for the stator equivalent circuit (Figure 12.18a).

d Compute the rotor current $\tilde{I}'_r$ for rated slip $s_{\text{rat}} = 0.03$ pu $= 3\%$, the induced voltage $\tilde{E}$, and the load voltage $\tilde{V}_{Rr'/s}$.

e Determine the developed (d) output power P_d in kW and horsepower hp (note 1 hp = 0.746 kW).

f Calculate the developed torque τ_{d_rat} at rated operation.

g Calculate the starting torque τ_{d_start}.

12.5 Three-phase induction motor of Problem 12.4 fed by a current-controlled voltage source inverter operating at different frequencies of $f_1 = 60$ Hz, $f_2 = 6$ Hz, and $f_3 = 360$ Hz with flux weakening at 360 Hz and CFW at 360 Hz.

Apply the current-controlled voltage source inverter of Section 12.4.4 to determine the required battery voltage V_{bat} for the inverter and the current waveshape of $i_a(t) = I(\text{RM1})$ to the three-phase induction motor of Problem 12.4 and investigate the behavior of this speed-controlled drive for:

a Rated operation at $f_1 = 60$ Hz with slip $s_{\text{rat}} = 0.03$ pu, computing $\tilde{V}_{TH}, \tilde{Z}_{TH}, R_{TH}, X_{TH}, \tilde{I}'_r, \tilde{E}, \tilde{Z}_{TH}, P_d, \tau_d, e_a(t), e_b(t), e_c(t), i_a(t), i_b(t)$, and $i_c(t)$,

b Low-speed operation at $f_2 = 6$ Hz with slip $s = 0.03$ pu, computing $X_s, X'_r, X_m, \tilde{V}_{TH}, \tilde{Z}_{TH}, R_{TH}, X_{TH}, \tilde{I}'_r, \tilde{E}, P_d, \tau_d, e_a(t), e_b(t), e_c(t), i_a(t), i_b(t)$, and $i_c(t)$,

c High-speed operation with flux weakening at $f_3 = 360$ Hz and slip $s_{\text{rat}} = 0.03$ pu, computing $X_s, X'_r, X_m, \tilde{V}_{TH}, \tilde{Z}_{TH}, R_{TH}, X_{TH}, \tilde{I}'_r, \tilde{E}, P_d, \tau_d, e_a(t), e_b(t), e_c(t), i_a(t), i_b(t)$, and $i_c(t)$,

d High-speed operation with CFW $f_3 = 360$ Hz and slip $s_{rat} = 0.03$ pu, computing X_s, X'_r, X_m, $\widetilde{V}_{TH}$, $\widetilde{Z}_{TH}$, R_{TH}, X_{TH}, $\widetilde{I}'_r$, $\widetilde{E}$, P_d, τ_d, $e_a(t)$, $e_b(t)$, $e_c(t)$, $i_a(t)$, $i_b(t)$, and $i_c(t)$,

e Starting performance with slip $s = 1$ pu at $f_2 = 6$ Hz, computing $\widetilde{V}_{TH}$, $\widetilde{Z}_{TH}$, R_{TH}, X_{TH}, $\widetilde{I}'_r$, $\widetilde{E}$, P_d, τ_d, $e_a(t)$, $e_b(t)$, $e_c(t)$, $i_a(t)$, $i_b(t)$, and $i_c(t)$.

12.6 Inverter-fed three-phase IM.

Investigate the performance of the three-phase squirrel-cage IM with the data:

Output or developed (d) power $P_d = 115$ hp, $V_{L\text{-}L} = 480$ V, Y-connected, $f_1 = 60$ Hz, $p = 4$ poles, $R_s = 0.09\ \Omega$, $X_s = 0.2\ \Omega$, $X_m = 7\ \Omega$, $R'_r = 0.07\ \Omega$, and $X'_r = 0.17\ \Omega$ when fed by a current-controlled voltage source inverter as explained in Section 12.4.4. Analyze its various performance characteristics and determine based on PSPICE the required inverter DC input voltage V_{bat}:

a Starting performance with slip $s = 1$ pu at $f_2 = 6$ Hz, computing $\widetilde{V}_{TH}$, $\widetilde{Z}_{TH}$, R_{TH}, X_{TH}, $\widetilde{I}'_r$, $\widetilde{E}$, P_d, τ_d, $e_a(t)$, $e_b(t)$, $e_c(t)$, $i_a(t)$, $i_b(t)$, and $i_c(t)$,

b Low-speed operation with slip $s_{rat} = 0.03$ pu at $f_2 = 6$ Hz, computing $\widetilde{V}_{TH}$, $\widetilde{Z}_{TH}$, R_{TH}, X_{TH}, $\widetilde{I}'_r$, $\widetilde{E}$, P_d, τ_d, $e_a(t)$, $e_b(t)$, $e_c(t)$, $i_a(t)$, $i_b(t)$, and $i_c(t)$

c Medium-speed operation with slip $s_{rat} = 0.03$ pu at $f_3 = 30$ Hz, computing $\widetilde{V}_{TH}$, $\widetilde{Z}_{TH}$, R_{TH}, X_{TH}, $\widetilde{I}'_r$, $\widetilde{E}$, P_d, τ_d, $e_a(t)$, $e_b(t)$, $e_c(t)$, $i_a(t)$, $i_b(t)$, and $i_c(t)$,

d Rated operation at $f_1 = 60$ Hz with slip $s_{rat} = 0.03$ pu, computing $\widetilde{V}_{TH}$, $\widetilde{Z}_{TH}$, R_{TH}, X_{TH}, $\widetilde{I}'_r$, $\widetilde{E}$, $\widetilde{Z}_{TH}$, P_d, τ_d, $e_a(t)$, $e_b(t)$, $e_c(t)$, $i_a(t)$, $i_b(t)$, and $i_c(t)$,

e High-speed operation with flux weakening at $f_4 = 360$ Hz and slip $s_{rat} = 0.03$ pu, computing X_s, X'_r, X_m, $\widetilde{V}_{TH}$, $\widetilde{Z}_{TH}$, R_{TH}, X_{TH}, $\widetilde{I}'_r$, $\widetilde{E}$, P_d, τ_d, $e_a(t)$, $e_b(t)$, $e_c(t)$, $i_a(t)$, $i_b(t)$, and $i_c(t)$.

f High-speed operation with CFW at $f_4 = 360$ Hz and slip $s_{rat} = 0.03$ pu, computing X_s, X'_r, X_m, $\widetilde{V}_{TH}$, $\widetilde{Z}_{TH}$, R_{TH}, X_{TH}, $\widetilde{I}'_r$, $\widetilde{E}$, P_d, τ_d, $e_a(t)$, $e_b(t)$, $e_c(t)$, $i_a(t)$, $i_b(t)$, and $i_c(t)$.

12.7 Phasor diagram and analysis of an underexcited (lagging displacement power factor, consumer reference system) three-phase non-salient pole SM as relied on for electric drives.

A $P_{3\text{-phase}} = 48$ hp, $V_{L\text{-}L} = 208$ V, $f = 60$ Hz, $\cos\theta = 0.8$ lagging power factor (consumer reference system), Y-connected non-salient pole SM has a synchronous reactance $X_s = 1\ \Omega$ and an armature resistance $R_a = 0.2\ \Omega$. The friction, windage, and iron-core losses can be neglected.

a Sketch the per-phase equivalent circuit and the (not-to-scale) phasor diagram of this motor.

b Calculate the values $\widetilde{V}_a$, $\widetilde{I}_a$, and $\widetilde{E}_a$.

c Determine the apparent power $S_{3\text{-phase}}$ and the reactive power $Q_{3\text{-phase}}$.

12.8 Phasor diagram and analysis of an underexcited (lagging displacement power factor, generator reference system) three-phase non-salient pole SM as relied on for power generation in hydro and wind power plants.

A $P_{3\text{-phase}} = 15$ hp, $V_{L\text{-}L} = 208$ V, $f = 60$ Hz, $\cos\theta = 0.8$ lagging power factor (generator reference system), Y-connected non-salient pole synchronous machine has the synchronous reactance $X_s = 1\,\Omega$ and the armature resistance $R_a = 0.2\,\Omega$. The friction, windage, and iron-core losses can be neglected.

a Sketch the per-phase equivalent circuit and the (not-to-scale) phasor diagram of this motor.

b Calculate the values $\widetilde{V}_a$, $\widetilde{I}_a$, and $\widetilde{E}_a$.

c Determine the apparent power $S_{3\text{-phase}}$ and the reactive power $Q_{3\text{-phase}}$.

12.9 Phasor diagram and analysis of an overexcited (leading displacement power factor, consumer reference system) three-phase non-salient pole SM as relied on for electric drives.

A $P_{3\text{-phase}} = 15$ hp, $V_{L\text{-}L} = 208$ V, $f = 60$ Hz, $\cos\theta = 0.8$ leading power factor (consumer reference system), Y-connected non-salient pole SM has a synchronous reactance $X_s = 1\,\Omega$ and an armature resistance $R_a = 0.2\,\Omega$. The friction, windage, and iron-core losses can be neglected.

a Sketch the per-phase equivalent circuit and the (not-to-scale) phasor diagram of this motor.

b Calculate the values of $\widetilde{V}_a$, $\widetilde{I}_a$, and $\widetilde{E}_a$.

c Determine the apparent power $S_{3\text{-phase}}$ and the reactive power $Q_{3\text{-phase}}$.

12.10 Phasor diagram and analysis of an overexcited (leading displacement power factor, generator reference system) three-phase non-salient pole SM as relied on for power generation in hydro and wind power plants.

A $P_{3\text{-phase}} = 15$ hp, $V_{L\text{-}L} = 208$ V, $f = 60$ Hz, $\cos\theta = 0.8$ leading power factor (generator reference system), Y-connected non-salient pole SM has a synchronous reactance $X_s = 1\,\Omega$ and an armature resistance $R_a = 0.2\,\Omega$. The friction, windage, and iron-core losses can be neglected.

a Sketch the per-phase equivalent circuit and the (not-to-scale) phasor diagram of this motor.

b Calculate the values of $\widetilde{V}_a$, $\widetilde{I}_a$, and $\widetilde{E}_a$.

c Determine the apparent power $S_{3\text{-phase}}$ and the reactive power $Q_{3\text{-phase}}$.

12.11 Inverter-fed three-phase non-salient pole (cylindrical rotor) synchronous motor (SM).

For the non-salient pole SM with the data: three-phases, $f_1 = 50\,\text{Hz}$, $V_{\text{L-L}} = 380\,\text{V}$, Y-connected, $X_d = X_q = X_s = 0.9\,\Omega$, armature current $I_a = 50\,\text{A}$ at $\cos\theta = 0.9$ leading (overexcited) displacement (fundamental) power factor based on consumer reference system, $R_a = 0.2\,\Omega$. Friction, windage, and iron-core losses can be neglected. Determine its performance at rated $f_1 = 50\,\text{Hz}$, at $f_2 = 300\,\text{Hz}$ with flux weakening, and at $f_2 = 300\,\text{Hz}$ with CFW operating conditions as discussed in Application Examples 12.13 and 12.14.

a Sketch the phasor diagram of this motor.

b Calculate the values of $\widetilde{V}_a$, S, $\widetilde{I}_a$, real input power P, and reactive absorbed power Q.

c Calculate the induced voltage $\widetilde{E}_a$.

d Determine the base impedance Z_{base} and the per-unit values of R_{a_pu} and X_{s_pu}.

e Apply PSPICE software program to rated operation at $f_1 = 50\,\text{Hz}$.

f Apply PSPICE software program at increased frequency of $f_2 = 300\,\text{Hz}$ to operation with flux weakening.

g Apply PSPICE software program at increased frequency of $f_2 = 300\,\text{Hz}$ to operation with CFW [22, 23].

References

1 Tesla, N. (1888). *A New System of Alternate Current Motors and Transformers.* American Institute of Electrical Engineers.

2 Schivon, M. (2016). *OrCAD Lite 17.2-2016 Capture+PSpice.* Feldkirchen, Germany: FlowCAD EDA-Software Vertriebs GmbH.

3 Roberts, G.W. and Sedra, A.S. (1997). *SPICE*, 2e. New York: Oxford University Press.

4 Rashid, M.H. (1993). *SPICE for Power Electronics and Electric Power.* Englewood Cliffs: Prentice Hall.

5 Fuchs, E.F. and Masoum, M.A.S. (2012). *Power Conversion of Renewable Energy Systems*, 1e, 2nd Printing. USA: Springer, 692 pp.

6 Masoum, M.A.S. and Fuchs, E.F. (2015). *Power Quality in Power Systems and Electrical Machines*, 2e. Elsevier/Academic Press, 1123 pp.

7 Hanna, J.W. and Fuller, J.F. (1975). *E.E 354 Laboratory Notes.* Boulder: Electrical Engineering Department, University of Colorado.

8 IBS Magnet. https://www.ibsmagnet.de/about_us/index.php (accessed 24 May 2020).

9 SHIN-Etsu Magnetics, Inc. NdFeB, SmCo. http://www.shinetsu-rare-earth-magnet.jp/e/products (accessed 24 May 2020).

10 Huang, H. and Fuchs, E.F. (1990). Successive nonlinear quadratic programming for high-power density and high-efficiency, permanent-magnet motors. *Proceedings of the International Conference on Electrical Machines*, Cambridge, MA (12–15 August 1990).

11 Huang, H., Anderson, W.M., and Fuchs, E.F. (1990). High-power density and high-efficiency motors for electric vehicle applications. *Proceedings of the International Conference on Electrical Machines*, Cambridge, MA (12–15 August 1990).

12 Fardoun, A.A., Fuchs, E.F., and Huang, H. (1994). Modeling and simulation of an electronically commutated permanent-magnet machine drive system using SPICE. *IEEE Trans. Ind. Appl.* 30 (4): 927–937.

13 Fuchs, E.F. (1992). Electronically commutated variable-speed, permanent-magnet drive train for wind power plants. *Power Eng. Rev.* 12 (12): 7–8.

14 Carlin, P.W., Fingersh, L.J., and Fuchs, E.F. (1996). Preliminary results from the variable-speed, direct-drive generating system. *Presentation at the 15th ASME ETCE Wind Energy Symposium*, Houston, TX (29 January–1 February 1996) (Abstract).

15 Yildirim, D., Fuchs, E.F., and Batan, T. (1998). Test results of a 20 kW, direct-drive, variable-speed wind power plant. *Proceedings of the International Conference on Electrical Machines*, Istanbul, Turkey (2–4 September 1998), pp. 2039–2044.

16 Fardoun, A.A. (1994). A novel wind-power train: permanent-magnet with flux weakening feeding a controlled real/reactive power converter. PhD thesis. Department of Electrical and Computer Engineering, University of Colorado, Boulder, CO.

17 Yildirim, D. (1999). Commissioning of 30 kVA variable-speed, direct–drive wind power plant. PhD thesis. Department of Electrical and Computer Engineering, University of Colorado, Boulder, CO.

18 Fingersh, L.J. (1995). The testing of a low-speed, permanent-magnet generator for wind power applications. MS thesis. Department of Electrical Engineering, University of Colorado, Boulder, CO.

19 Slomczynska, J. (1974). Nonlinear analysis of the magnetic flux distribution in the magnetized magnet stabilized in air. *IEEE Trans. Magn.* MAG-10 (4): 1109–1113.

20 Slomczynska, J. (1975). No-load magnetic-field distribution in a permanent magnet electrical machine. *IEEE Trans. Magn.* MAG-11 (6): 1720–1722.

21 Brauer, J.R., Larkin, L.A., and Overbye, V.D. (1984). Finite-element modeling of permanent magnet devices. *J. Appl. Phys.* 55 (6): 2183.

22 Fuchs, E.F. (2012). Alternating current machine with increased torque above and below rated speed for hybrid/electric propulsion systems. US Patent No. 8,183,814 B2.

23 Fuchs, E.F. (2013). Alternating current machine with increased torque above and below rated speed for hybrid/electric propulsion systems. US Patent No. 8,436,569 B2.

24 Chapman, S.J. (1999). *Electric Machinery Fundamentals*, 3e. WCB/McGraw-Hill, 716 p.

25 Fuchs, E.F. and McNaughton, G.A. (1982). Comparison of first-order finite difference and finite element algorithms for the analysis of magnetic fields, part I: theoretical analysis. *IEEE Trans. Power Appar. Syst.* PAS-101 (5): 1170–1180.

26 Fuchs, E.F. and Erdelyi, E.A. (1970). Nonlinear magnetic field analysis of DC machines, part II: application of the improved treatment. *IEEE Trans. Power Appar. Syst.* PAS-89 (7): 1555–1564.

27 Erdelyi, E.A., Fuchs, E.F., and Binkley, D.H. (1970). Nonlinear magnetic field analysis of DC machines, part III: equipotential plots drawn by computer. *IEEE Trans. Power Appar. Syst.* PAS-89 (7): 1565–1583.

28 Stratton, J.A. (1941). *Electromagnetic Theory*, 153–155. New York: McGraw-Hill Book Company Inc.

29 Reichert, K., Freundl, H., and Vogt, W. (1976). The calculation of forces and torques within numerical magnetic field calculation methods. *Proceedings of COMPUMAG Conference*, Oxford, pp. 64–74.

30 Younglove, B.L. and Johnson, S.D. (1983). *User's Manual for Program Fields*, vol. 1. Boulder, CO: Department of Electrical Engineering, University of Colorado.

31 Fuchs, E.F., Chang, L.H., Appelbaum, J., and Moghadamnia, S. (1993). *Sensitivity of Transformer and Induction Motor Operation to Power System's Harmonics*. Topical report. U.S. Department of Energy.

32 Geysen, W., DeJong, H., Fuchs, E.F., and Van Ranst, A. (1973). Flux distribution in a three-phase squirrel cage induction motor. *Rev. Electr. Belgium* 7 (8): 201–208.

33 Fuchs, E.F., Chang, L.H., and Appelbaum, J. (1984). Magnetizing current, iron losses and forces of three-phase induction machines at sinusoidal and nonsinusoidal terminal voltages, part I: analysis. *IEEE Trans. Power Appar. Syst.* PAS-103 (11): 3303–3312.

34 Fuchs, E.F., Roesler, D.J., and Chang, L.H. (1984). Magnetizing current, iron losses and forces of three-phase induction machines at sinusoidal and nonsinusoidal terminal voltages, part II: results. *IEEE Trans. Power Appar. Syst.* PAS-103 (11): 3313–3326.

35 Fuchs, E.F., Poloujadoff, M., and Neal, G.W. (1988). Starting performance of saturable three-phase induction motors. *IEEE Trans. Energy Convers.* 3 (3): 624–635.

36 Fuchs, E.F. and Hanna, W.J. (2002). Measured efficiency improvements of induction motors with thyristor/triac controllers. *IEEE Trans. Energy Convers.* 17 (4): 437–444.

37 Fuchs, E.F. and Masoum, M.A.S. (2008). Torques in induction machines due to low-frequency voltage/current harmonics. *Int. J. Power Energy Syst.* 28 (2): 212–221.

38 Fuchs, E.F., Schraud, J., and Fuchs, F.S. (2008). Analysis of critical-speed increase of induction machines via winding reconfiguration with solid-state switches. *IEEE Trans. Energy Convers.* 23 (3): 774–780.

39 Schraud, J., Fuchs, E.F., and Fuchs, H.A. (2008). Experimental verification of critical-speed increase of induction machines via winding reconfiguration with solid-state switches. *IEEE Trans. Energy Convers.* 23 (2): 460–465.

40 Mousavi, G.S.M., Tabakhpour, L.A., Fuchs, E.F., and Al-Haddad, K. (2015). Power quality issues in railway electrification: a comprehensive perspective. *IEEE Trans. Ind. Electron.* 62 (5): 3081–3090.

41 Teltsch, M. (1997). Adjustable-speed drive for an electric car with large rated torque at low rated speed. MS thesis. University of Colorado, Boulder, CO.

42 Appelbaum, J., Fuchs, E.F., and White, J.C. (1987). Optimization of three-phase induction motor design, part I: formulation of the optimization technique. *IEEE Trans. Energy Convers.* EC-2 (3): 407–414.

43 Appelbaum, J., Khan, I.A., Fuchs, E.F., and White, J.C. (1987). Optimization of three-phase induction motor design, part II: the efficiency and cost of an optimal design. *IEEE Trans. Energy Convers.* EC-2 (3): 415–422.

44 Fei, R. and Fuchs, E.F. (1989). Optimization of three-phase induction motor design by the variable metric method. *J. Tsinghua Univ. Sci. Technol.* 29 (1): 1–10.

45 Fei, R., Fuchs, E.F., and Huang, H. (1989). Comparison of two optimization techniques as applied to three-phase induction motor design. *IEEE Trans. Energy Convers.* EC-4: 651–660.

46 Fuchs, E.F., Appelbaum, J., Khan, I.A. et al. (1985). *Optimization of Induction Motor Efficiency, Volume 1: Three-Phase Induction Motors.* Report prepared by University of Colorado, Department of Electrical and Computer Engineering, Boulder, CO, for Electric Power Research Institute, Palo Alto, CA, EL-4152, Vol. 1, Research Project 1944-1.

47 Fuchs, E.F., Vandenput, A.J., Höll, J., and White, J.C. (1990). Design analysis of capacitor-start, capacitor-run single-phase induction motors. *IEEE Trans. Energy Convers.* EC-6: 327–336.

48 Huang, H., Fuchs, E.F., and Zak, Z. (1988). Optimization of single-phase induction motor design, part I: formulation of the optimization technique. *IEEE Trans. Energy Convers.* EC-3 (2): 349–356.

49 Huang, H., Fuchs, E.F., and White, J.C. (1988). Optimization of single-phase induction motor design, part II: the maximum efficiency and minimum cost of an optimal design. *IEEE Trans. Energy Convers.* EC-3 (2): 357–366.

50 Huang, H., Fuchs, E.F., and White, J.C. (1988). Optimal placement of the run capacitor in single-phase induction motor designs. *IEEE Trans. Energy Convers.* EC-3 (3): 647–652.

51 Fuchs, E.F., Huang, H., Vandenput, A.J., Höll, J., Appelbaum, J., Zak, Z., and Erlicki, M. S. (1987). *Optimization of Induction Motor Efficiency, Volume 2: Single-Phase Induction Motors.* Report prepared by University of Colorado, Department of Electrical and Computer Engineering, Boulder, CO, for Electric Power Research Institute, Palo Alto, CA, EL-4152, Vol. 2, Research Project 1944-1.

52 Fuchs, E.F. and Senske, K. (1981). Comparison of iterative solutions of the finite difference method with measurements as applied to Poisson's and the diffusion equations. *IEEE Trans. Power Appar. Syst.* PAS-100 (8): 3983–3992.

53 Fuchs, E.F. and Pohl, G. (1981). Computer generated polycentric grid design and a novel dynamic acceleration of convergence for the iterative solution of magnetic fields based on the finite difference method. *IEEE Trans. Power Appar. Syst.* PAS-100 (8): 3911–3920.

54 Fuchs, E.F. (1976). Industrial experience with the relaxation solutions of the quasi-Poissonian differential equation. *Proceedings of International Conference on Electrical Machines*, Vienna, Austria (13–15 September 1976), pp. D20-1–D20-10.

55 Fuchs, E.F. (1971). Magnetic fields in nonlinear turbogenerators at no load and balanced loads. *Third Reno Conference on Magnetic Fields*, University of Nevada, Reno, NV (2–3 September 1971) (Best Paper Prize 1972 of VDE).

56 Fuchs, E.F. and Erdelyi, E.A. (1973). Nonlinear theory of turbogenerators, part I: magnetic fields at no-load and balanced loads. *IEEE Trans. Power Appar. Syst.* PAS-92 (2): 583–591.

57 Fuchs, E.F. and Erdelyi, E.A. (1973). Nonlinear theory of turbogenerators, part II: load dependent synchronous reactances. *IEEE Trans. Power Appar. Syst.* PAS-92 (2): 592–599.

58 Fuchs, E.F. and Rosenberg, L.T. (1974). Analysis of an alternator with two displaced stator windings. *IEEE Trans. Power Appar. Syst.* PAS-93 (6): 1776–1786.

59 Fuchs, E.F. (1970). Numerical determination of synchronous, transient, and subtransient reactances of a synchronous machine. PhD thesis. University of Colorado, Boulder, CO.

60 Williams, R.W. (1974). Numerical determination of the flux distribution in salient-pole aircraft alternators. MS thesis. University of Colorado at Boulder, CO.

61 Fuchs, E.F. (1973). Lastabhängige transiente Reaktanzen von gesättigten Turbogeneratoren. *Arch. Electrotech.* 55: 263–273.

62 Carter, F.W. (1901). Airgap induction. *Electr. World* 38: 884–888.

63 Duffert, H. (1971). Large single-phase turbine-generator sets for traction power supplies. *Siemens Rev.* 45: 61–66.

Index

Introduction to Energy, Renewable Energy, and Electrical Engineering: Essentials for Engineering Science (STEM) Professionals and Students, First Edition. Ewald F. Fuchs and Heidi A. Fuchs.

Companion website: www.wiley.com/go/fuchs/renewableenergyandelectricalengineering

C

i

r

Printed and bound by CPI Group (UK) Ltd, Croydon, CR0 4YY

17/09/2026

14964316-0003